T0192532

Geotechnical Engineering for Mine Waste Storage Facilities

Geotechnical Engineering for Mine Waste Storage Facilities

Geoffrey Blight

University of the Witwatersrand, Johannesburg, South Africa

CRC Press is an imprint of the
Taylor & Francis Group, an **informa** business
A BALKEMA BOOK

Published by: CRC Press/Balkema
P.O. Box 447, 2300 AK Leiden, The Netherlands
e-mail: Pub.NL@taylorandfrancis.com
www.crcpress.com – www.taylorandfrancis.co.uk – www.balkema.nl

First issued in paperback 2020

ISBN-13: 978-0-367-57721-6 (pbk)
ISBN-13: 978-0-415-46828-2 (hbk)

Visit the Taylor & Francis Web site at
http://www.taylorandfrancis.com

and the CRC Press Web site at
http://www.crcpress.com

Library of Congress Cataloging-in-Publication Data
Blight, Geoffrey, 1934-
Geotechnical engineering for mine waste storage facilities / Geoffrey Blight.
p. cm.
Includes bibliographical references.
ISBN 978-0-415-46828-2 (hardcover : alk. paper) 1. Mineral industries–Waste disposal. 2. Waste disposal in the ground. 3. Environmental geotechnology. 4. Engineering geology, 5. Soil mechanics. I. Title.

TD428.M56B585 2010
622.028'6–dc22

2009028816

Typeset by Macmillan Publishing Solutions, Chennai, India

Table of Contents

Author Biography

Geoffrey Blight completed his Bachelor's and Master's degrees in Civil Engineering at the University of the Witwatersrand, Johannesburg and his PhD at the Imperial College of Science and Technology in London. The early years of his career were spent at the South African National Building Research Institute, Pretoria, where he was engaged in research on design, operation and safety of mine waste storage facilities, including waste rock dumps and hydraulic fill tailings storage facilities. After the disastrous failure of a coal waste dump at Aberfan, U.K. in 1966, he was sent on a country-wide inspection tour of South African mines to make sure that no similarly threatening stability situations existed locally. Eleven years later, after the disastrous failure of a hydraulic fill tailings dam at Bafokeng, north of Johannesburg, he was asked by the mining industry to write a comprehensive guide to the design, operation and closure of mine waste storage facilities. This appeared in 1979 and has been revised and updated several times since.

Geoff Blight re-joined the Department of Civil Engineering at Witwatersrand University in 1969 and the results of his continuing research and consulting work on mine waste have been widely published since. The present book has been developed from the updated mining industry guide as well as a series of post graduate courses in mine waste management that have been presented for many years.

In addition to numerous technical papers published in refereed journals and conference proceedings, he is the editor and co-author of the book "Mechanics of Residual Soil" (Balkema, 1997) and author of "Assessing Loads on Silos and Other Bulk storage Structures" (Taylor and Francis 2006).

Acknowledgements

I was first introduced to the fascinating subject of geotechnical engineering by Jere Jennings when I was a final year student at the University of the Witwatersrand in 1955. As mining was, and continues to be a major industry involving geotechnical engineering in South Africa, many of Jere's teaching examples involved the geotechnics of mine waste, a subject that has enthralled me ever since. Jere also introduced me to the subject of unsaturated soil mechanics, which, living in the water-deficient climate of South Africa, was almost inevitable and very necessary. When, a few years later, I studied at the Imperial College in London, my PhD supervisor was Alan Bishop, another of the pioneers in the fields of mine waste and unsaturated soils. It was these two engineers whose contributions to my career I now remember with gratitude.

I have also been assisted over the past fifty or so years by a number of fellow geotechnical engineers, many of whom started out as my students and became friends and professional colleagues and with whom I am grateful to have been associated. In particular, I acknowledge the friendship and assistance of Oskar Steffen, Eric Hall, John Nelson, Dick Dison, Neville Graham, Ken Lyell, Brian Wrench, Mike Smith, Derek Avalle, Dave Bentel, Gordon McPhail, Jeremy Boswell, Rod Ball, Andy Fourie, Sam Spearing, Adriaan Meintjies, Jenny Blight and Krassimir Roussev, all of whom have contributed directly or indirectly to this book.

It has also been my good fortune to have enjoyed the support of four outstanding technicians while at the University of the Witwatersrand. Bob van der Merwe, Frans Wiid, Ed Braddock and Norman Alexander always made my schemes happen both in the laboratory and in the field. In particular, most of the experimental results included in the book are the work of Norman Alexander.

Cathy Snow produced the figures and plates and patiently endured the countless changes and additions that were made.

I particularly acknowledge and am extremely grateful for the unwavering support of my wife Rhona who patiently produced the manuscript of the book.

Unless otherwise acknowledged, all photographs were taken by the author.

Geoff Blight
Johannesburg
July 2009

Acknowledgements

List of Symbols

[ ] indicates basic dimension mostly used.
(Note that the same symbol is often used for different quantities. The context will tell which is the appropriate meaning.)

1 Roman Letters

a: albedo of Earth's surface. Ratio of reflected to received radiation.
a, A: area [m^2].
A: length of arc [m].
A: Bishop-Skempton pore pressure ratio.
A_m: cross-sectional area of backfill [m^2].
A_R: cross-sectional area of steel reinforcement [mm^2].

b, B: breadth, shorter of two horizontal distances [m].
B: Bishop-Skempton pore pressure ratio.
B: bulge of geofabric under applied pressure [mm].
B_q: pore pressure ratio in cone penetrometer test $(u_d - u_e)/q_c^I$.

c, c^I: cohesion, either in total stress (c) or effective stress (c^I) terms [kPa].
c_v: coefficient of consolidation [m^2/y].
C: colour, consistency – used in soil descriptions.
C, C_d: discharge coefficient [dimensionless].
C, C_s: compressibility of soil skeleton [kPa^{-1}, m^2/kN].
C_a: area ratio of sampling tube [dimensionless].
C_{Gd}: specific heat of dry soil [kJ/kg°C].
C_h: specific heat [kJ/kg°C].
C_i: inside clearance ratio of sampling tube [dimensionless].
C_w: compressibility of water [kPa^{-1}].
C_w: specific heat of water [kJ/kg°C].
CPT: cone penetrometer test.

d: day [time].
d: depth [m].
d: drain flow [Mg/m^2].

D: decant water [Mg/m^2].
D: diameter [m].
D: number of hours of daylight.
D: particle diameter [mm]. D_{15}: size such that 15% of mass of particles are finer than D_{15}. D_{50} and D_{85} similarly.
D_e: entrance diameter [mm].
D_i: inside diameter [mm].

e: base of Naperian logarithms (2.7183 approx.).
e: void ratio [dimensionless].
E: elastic modulus [kPa].
E: evaporation [Mg/m^2, mm].
E_a: aerial evaporation [mm].
E_A, $\overline{E}_A$: A pan evaporation, mean A pan evaporation [mm].
E_B: Evaporation measured by energy balance [mm].
E_M: Menard pressuremeter elastic modulus [kPa].

f( ): denotes "function of".
fPP: flexible polypropylene.
F: factor of safety, allowance for damage by various agencies [dimensionless].
F: feed water [Mg/m^2].
F: shape factor in flow or seepage of water [dimensionless].

g: gravitational acceleration [m/s^2]. Taken as 9.81 m/s^2 or 10 m/s^2 (approximately).
G: solid particle relative density [dimensionless].
G: shear modulus [kPa].
G: soil heat [kJ/m^2].
G_s: seismic shear modulus [kPa].
GCL: geosynthetic clay liner.
GGBS: ground granulated blast furnace slag.

h: height or horizontal distance [mm, m].
h: hour [time].
h_w: height from surface to water table [m].
H: head (difference in height of free water surfaces) [m]. Subscripts, o, t, ∞ indicate times.
H: sensible heat of atmosphere [kJ/m^2].
H_s: height of solids [mm].
H_v: height of voids [mm].

i: hydraulic gradient [dimensionless, or kPa/m].
i: inclination of surface [° degrees of arc].
i_c: chord inclination [° degrees of arc].
i_o, i_L: infiltration rate [mm/h], o = initial, L = limiting.
I: interstitial water [Mg/m^2].
INT: interception [mm].
IR: irrigation water [mm].

k:	coefficient of permeability [m/s, m/y]. Subscripts h, v (k_h, k_v) etc., indicate direction of water flow.
K:	pressure coefficient.
K_A:	active pressure coefficient.
K_F:	failure pressure coefficient.
K_o:	at rest, or zero lateral strain pressure coefficient.
K_p:	passive pressure coefficient. [all above are ratios of pressures, dimensionless].
l, L:	length, longer of two horizontal distances [m].
L:	leachate flow [mm].
L_e:	latent heat of evaporation [kJ/m^2].
LL:	liquid limit [%].
LVDT:	linearly variable differential transformer.
m:	minute [time].
m_v:	vertical compressibility [kPa^{-1}, m^2/kN].
M:	mass [kg].
M:	moisture condition – used in soil descriptions.
M_A, M_B:	bending moments in pipe wall [kNm].
M_s:	mass of solid particles [kg].
M_w:	mass of water [kg].
n:	an integer: (1, 2, 3, etc.).
n:	porosity [dimensionless].
n:	number of hours of sunshine per day.
N_c:	cone factor for cone penetrometer test [dimensionless].
N_f, N_h:	number of flow tubes and number of equipotential head drops, respectively, in a flow net.
NC:	normally consolidated – description of stress history.
O:	Origin – used in soil descriptions.
OC:	overconsolidated – description of stress history.
OCR:	overconsolidation ratio.
OPC:	ordinary Portland cement.
p:	pitch of spiral [mm].
p, p^I:	average direct stress $\frac{1}{2}(\sigma_1 + \sigma_3)$, $\frac{1}{2}(\sigma_1^I + \sigma_3^I)$ [kPa].
p_I:	limit pressure [kPa].
p^{II}:	soil water suction $(u_a - u_w)$ [kPa].
P:	normal force on surface [kN].
P:	precipitation [Mg/m^2, mm].
PBFC:	Portland blast furnace cement.
PET:	polyethylene terephthalate.
PFA:	pulverized fuel ash.
PI:	plasticity index [dimensionless].
PL:	plastic limit [%].
PMT:	pressuremeter test.

q: maximum shear stress $\frac{1}{2}(\sigma_1 - \sigma_3)$, $\frac{1}{2}(\sigma_1^I - \sigma_3^I)$ [kPa].
q: rate of flow of water [m^3/s, m^3/h]. Subscripts: o = initial value, ∞ = final steady state value.
q_c: cone penetration resistance in cone penetrometer test [MPa].
q_c^I: $q_c - \sigma_v^I$: corrected cone resistance [MPa].
Q: volumetric discharge rate [m^3/s].

r_u: pore pressure ratio used by Bishop and Morgenstern (1960).
R: Radiation [W/m^2, kJ/m^2].
R: radius of circle [mm, m].
R: rate of transport of material by wind [kgh/km^3].
R: ratio of shear strengths [dimensionless].
R: recharge water [Mg/m^2, mm].
R_A: solar radiation received at outer limit of atmosphere [W/m^2].
R_d: direct solar radiation received at Earth's surface [W/m^2].
R_S: scattered sky radiation [W/m^2].
R_n: net solar radiation at Earth's surface [W/m^2].
RD: relative density [dimensionless].
RO: runoff [mm].

s: second [time].
s: shear strength [kPa].
s_u: undrained shear strength [kPa].
s_{uu}: unconsolidated undrained shear strength [kPa].
s_v, s_h: shear strength on vertical or horizontal surfaces, respectively [kPa].
S: soil or structure – used in soil descriptions.
S: supernatant water [Mg/m^2].
S: degree of saturation [dimensionless].
SD: standard deviation from the statistical mean.

t: thickness [mm].
t: time [s, h, d or y].
t_f: time to failure.
t_o: initial or starting time.
T: basic time lag (in water flow) [s, h, d].
T: dry tons of solids [Mg].
T: temperature [°C].
T: tension force [kN, kN/m].
T: time factor [dimensionless].
T: torque [kNm].

u: pore pressure [kPa]. Subscripts a, w (u_a, u_w) indicate pore air or pore water pressure. u on its own usually denotes pore water pressure.
u: wind speed [m/s, km/h].
u_{cap}: capillary pore pressure [kPa].
u_d, u_e: dynamic and equilibrium (static) pore pressures (respectively) measured in cone penetrometer test [kPa].
U: degree of consolidation [dimensionless].

U: strain energy [kNm/m^3].
U_f: strain energy stored in backfill [kNm/m^3].
U_F: fracture energy [kNm/m^3].
U_R: released strain energy [kNm/m^3].
U_s: stored strain energy [kNm/m^3].
U_z: degree of consolidation at depth z [dimensionless].
UV: ultra-violet light.

v: seepage velocity [m/s, m/y].
v: velocity [m/s, km/h].
v: vertical distance [m].
V: volume [m^3].
V, V^I: vertical normal force on surface in terms of total or effective stress.
V_a: volume of air [m^3].
V_s: volume of solid particles [m^3].
V_s: seismic shear wave velocity [m/s, km/s].
V_T: volume of thickened tailings [m^3].
V_{uT} volume of unthickened tailings [m^3].
V_v: volume of voids [m^3].
V_w: Volume of water [m^3].

w: gravimetric water content [%].
w_L: liquid limit [%].
w or W: width of space or object [m].
W: weight of defined volume of waste [kN].
W_p: water stored in pool [Mg/m^2].
W_s: weight of solid particles [g, kg].

y: year [time].

2 Greek Letters

α: planetary albedo: ratio of reflected to received solar energy at outer limit of atmosphere.
α: angle of tangency [° degrees of arc].
β: slope angle, inclination of failure surface [° degrees of arc].
γ: unit weight [kN/m^3]. Sub- or superscripts (e.g. γ_w, γ^I) indicate material (e.g. w = water) or condition of material (e.g. γ^I = submerged).
γ_d: dry density [kN/m^3].
γ_{sat}: saturated density [kN/m^3].
Γ: shear strain [dimensionless]. Subscripts (e.g. Γ_{xy}, Γ_{max}) indicate sense of rotation or magnitude.
δ, δ^I: angle of interface friction in terms of total or effective stress [°degrees of arc].
δ: declination of sun [° degrees of arc].
Δ: indicates "a small change of".
ε: direct strain. Subscripts (e.g. ε_1, ε_x) indicate direction or magnitude (e.g. ε_{max}). [dimensionless].

ε_v: volumetric strain [dimensionless].
η: viscosity of slurry [Pas].
θ: angle of direction [° degrees of arc].
λ: heat of vaporization of water [kJ/kg].
μ: coefficient of friction ($= \tan \varphi^I$ or $\tan \delta^I$) [dimensionless].
ν: Poisson's ratio [dimensionless].
π: ratio of circumference to diameter of a circle (3.1416, approx.)
ρ: density [kg/m^3].
ρ_d: dry density [kg/m^3].
ρ_w: density of water [kg/m^3].
ρ_{sub}, ρ^I: submerged density [kg/m^3].
σ, σ^I: direct stress either in total stress or effective stress terms [kPa], subscripts (e.g. σ_1, σ_2, σ_x, σ_y) indicate direction of action.
σ_{atm}: atmospheric pressure [kPa].
σ (critical): external pressure to cause pipe to collapse [kPa].
σ_y: yield stress of steel reinforcing [MPa].
Σ: summation sign, sometimes between limits, e.g. $\sum_a^b x$: limits a and b.
τ: shear stress [kPa], subscripts (e.g. τ_{xy}, τ_{yx}) indicate sense of rotation or magnitude (e.g. τ_{max}). τ_o: initial yield strength [Pa].
φ, φ^I: angle of shearing resistance either in total stress (ϕ) or effective stress (ϕ^I) terms [° degrees of arc].
φ: latitude on Earth's surface [° degrees of arc].
φ_R: angle of repose.
χ: Bishop's effectiveness parameter [dimensionless].
ψ: slope of failure (K_f) line in p, q or p^I, q^I space [° degrees of arc].
Ω: rate of rotation [radian/s]

Chapter 1

Waste engineering, characteristics of mine wastes and types of waste storage

1.1 The nature and magnitude of the mine waste storage activity

The failure of a number of mine waste storage facilities during the 1990's and early 2000's resulted in a changing attitude to the engineering management of these facilities. The reality dawned that, quite apart from potential and actual loss of life and environmental devastation, the negative publicity that resulted from these failures was seriously undermining the credibility of the mining and waste management industries, and jeopardizing their social 'licence to operate'. This has led to a greater commitment to improving the quality of mine waste management in many companies and countries and it is reasonable to say that practices have gradually improved on what they were ten or fifteen years ago. Despite this, failures of mine waste storages continue to occur wherever mine waste is produced.

Whilst the need for high quality technical design and operational plans for waste storage sites remains, this is now recognized as being insufficient for minimizing the risks of failure. To minimize the risks to which a company is exposed during the lifetime of development of a waste storage site, proper operational control is vitally necessary. Recognition of this need has led to the provision of a mandatory operations manual at many mines and to increased attention to more appropriate training and supervision of operations personnel. It is now also recognized that a mine waste storage facility is not a temporary asset, that can be disposed of at the end of its useful life. If not decommissioned and closed in an environmentally and socially acceptable manner, including a viable after use, it will constitute an ongoing liability not only to the operating company and its successors, but also to the local community, the country in which it is situated and the rest of the world.

In the last decade the activities of design and planning of operation and implementation of closure have come to be seen as a life-cycle progression. It is now generally realized that closure of waste storages does not mean abandonment, but having to ensure no increased risk to community health or to the environment for hundreds or even thousands of years. For engineers and technologists accustomed to structures or facilities with a design or operating life of 50 to 70 years, this requires unfamiliar thought processes and poses major new challenges. The engineer involved in mine waste storage is also required to work with people from many different disciplines, including botany, ecology, economics, horticulture, sociology and even local politics to achieve and implement construction, operational and closure plans that satisfy the demands of zero harm in virtual perpetuity. Providing satisfactory solutions to such open-ended problems over the decades to come will be taxing, but will provide some of the most satisfying activities engineers could wish to work in.

It will not be possible to deal in detail with all of the many components that make up a life cycle progression for a waste storage, but this book will deal with the technical issues that must be addressed. Wherever possible it will also touch on ecological, socio-political and socio-economic issues in order to alert the reader to the expectations they may have to satisfy and the factors that possibly could result in an otherwise excellent technical design or plan being regarded as unacceptable.

Many of the topics covered by this book have been developed from the guideline written by the present author for the Chamber of Mines of South Africa and published by that body (originally in 1979 and in revised form in 1983 and 1996).

The book has been written primarily for the civil engineer who is involved with mine waste storage, either on the design side or with the supervision of construction, operation and closure. However, many people become professionally involved in mine waste storage who do not have a civil engineering background or a knowledge of geotechnical engineering. For this reason, the fundamental basics of geotechnical engineering aspects have been included.

1.2 Origins and quantities of mine waste

With an expanding world market for mineral commodities such as chrome, coal, copper, diamonds, fluorspar, gold, iron, manganese and zinc, so necessary for the functioning of the modern world, mining companies are exploiting ever lower-grade ore bodies on an ever-increasing scale. Mining on a vast scale is usually necessary for profitability of a low-grade mine, and volumes of waste are commensurately large.

The actual volume of mine waste that has to be disposed of in dumps and tailings storage facilities, world-wide, is difficult to assess. In 1996 the International Commission on Large Dams (ICOLD) gave an estimate of: "almost certainly exceeds 5 thousand million tonnes per annum". Considering that some valuable commodities occur in their ores in concentrations of grams or carats per ton (1000 kg) (1 carat = 5 grams), and that many individual mines extract in excess of 50 million tons of ore per year, even ICOLD's estimate is probably much too low. For example, a single platinum tailings storage at Bafokeng, South Africa, has a storage capacity of almost 1 thousand million tons over a life of 50 years (Stuart et al., 2003), and the mine currently sends 1.5 million tons of tailings to storage per month.

Other estimated quantities of mine waste are as follows:

- van Zyl et al. (2002) reported that the world's iron, copper, gold, lead and bauxite (aluminium) mines together generated 35 thousand million tons of waste in 1995 alone.
- The South African gold mining industry produced 740 thousand tons of gold tailings in the decade from 1997 to 2006 (Chamber of Mines of South Africa, 2006).
- All gold mining waste produced in the past century in South Africa amounts to 6000 million tons, which cover a total area of 400 to 500 km^2, and contain

430 thousand tons of uranium and 30 million tons of sulphur, both of which, and especially the sulphur, have a high pollution potential (Sutton, et al., 2006).

It will be noticed that the term "mine waste storage" is used, instead of the more common "mine waste disposal". This is because advances in extractive metallurgy and increased demand and price for a commodity periodically coincide to allow a particular mine waste deposit to be re-worked and further resources to be extracted from it at a profit. As examples, some gold mine waste storages in South Africa have been re-mined and reprocessed three times in the past 100 years. Some of these deposits started out as waste rock which was unprofitable to process at the time, but was necessary to remove in order to access richer ores. Some platinum mines are now considering reprocessing their older tailings storages, and coal mines their old coal discard storages. Thus there is a realization that mine waste deposits are really storages of low grade ore. They do not consist of the waste they were formerly considered to be. Even if the grade of mineral they contain remains too low for economical extraction, there may be a present or future economic value as a construction material.

(Other terms often used synonymously are: residue deposit, or mine residue deposit (MRD), tailings storage facility (TSF), tailings deposit, tailings dam, or mine slimes dam.)

It is for this reason that there is a great reluctance in the mining industry to "dispose" of waste by placing it in locations that render the waste inaccessible for future re-processing, e.g. in "worked out" parts of a mine, except where it can be used for strata support. In time to come, not only may the waste become profitable to reprocess, but the stopes themselves may be worth re-mining to remove seams or reefs of ore previously regarded as uneconomical to mine.

Figure 1.1 illustrates one effect of this continuing process of reworking a mine when financial conditions are favourable. The Witwatersrand gold-bearing reefs outcropped at the surface over a distance of more than 160 km and were mined from the outcrop

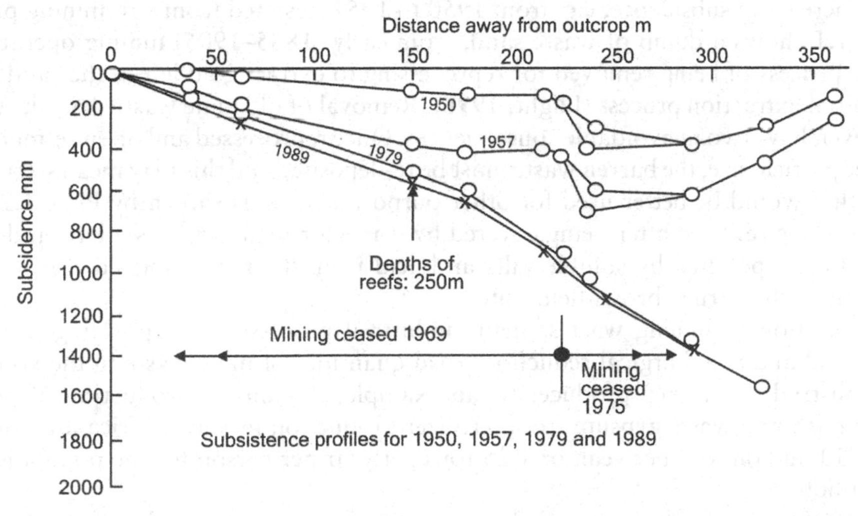

Figure 1.1 Surface subsidence profiles measured on the Witwatersrand gold mining area in South Africa that show the effect of periodic re-mining activities.

Plate 1.1 Dump of waste sand being re-mined to extract residual gold.

downwards. Figure 1.1 shows a series of subsidence profiles measured on a line at right angles to the outcrop. Subsidence ceases very soon after mining has stopped. Thus each increase in subsidence, e.g. from 1950 to 1957, resulted from a re-mining phase. Plate 1.1 shows a dump of waste sand, from early (1885–1905) mining operations, in the process of being removed for reprocessing to extract gold left in the sand after the initial extraction process (Blight, 1996). Removal of old mine waste deposits does, however, have two unavoidable consequences. Once reprocessed and often re-milled to a finer particle size, the barren waste must be re-deposited and this may mean sterilising land that would be better used for other purposes. Also, as shown by Plate 1.2, the land surface revealed after being covered by waste for many years is a barren desert, often highly polluted by soluble salts and acid from the now removed waste. It is, unfortunately, a true "brownfield" site.

In addition to mining wastes, many industrial processes, e.g. power generation from coal and metallurgical reduction, rival quantities of mine waste in the volumes of industrial waste they produce. As an example, the annual production of power station ash and waste gypsum from fertilizer production in South Africa amounts to some 50 million tons per year, or 1.25 tons per year per person for the population of 40 million.

Some mine waste can be recycled for other purposes. For example, waste rock can be used as a fill material in civil engineering works, or if the rock is sound, durable

Plate 1.2 Devastated condition of site from which mine waste has been removed after 100 years.

and unweathered, and has a satisfactory mineralogy, as aggregate for concrete and asphalt and in road layer works. Gypsum can be used to make building boards, or can be reprocessed to produce sulphuric acid and building cement. Certain slags can also be used as a partial cement replacement. Slags from iron reduction are now almost entirely consumed as cement replacements. The fine fraction of fly ash from coal-fired power stations is also widely used as a cement extender. Abandoned tailings storages have also been developed into light industrial townships. In one such development a large warehouse was constructed on the side-slope of an abandoned ring-dyke tailings storage (see Plate 1.3). The side slope was excavated to form a basement storage space. However, a few years after the building had been occupied, it was discovered that radon gas was accumulating in the underground space, and it was necessary to force-ventilate the basement to dissipate the radon as it seeped in from the surrounding tailings.

1.3 The effects of climate

As shown by Figure 1.2, 65 per cent of the earth's inhabited land surface (i.e. excluding Antarctica) is annually or seasonally water-deficient (Thornthwaite, 1948). By definition, in an annually water-deficient climate, the monthly evaporation from a free

Plate 1.3 A ware-house built into the slope of an abandoned gold tailings storage.

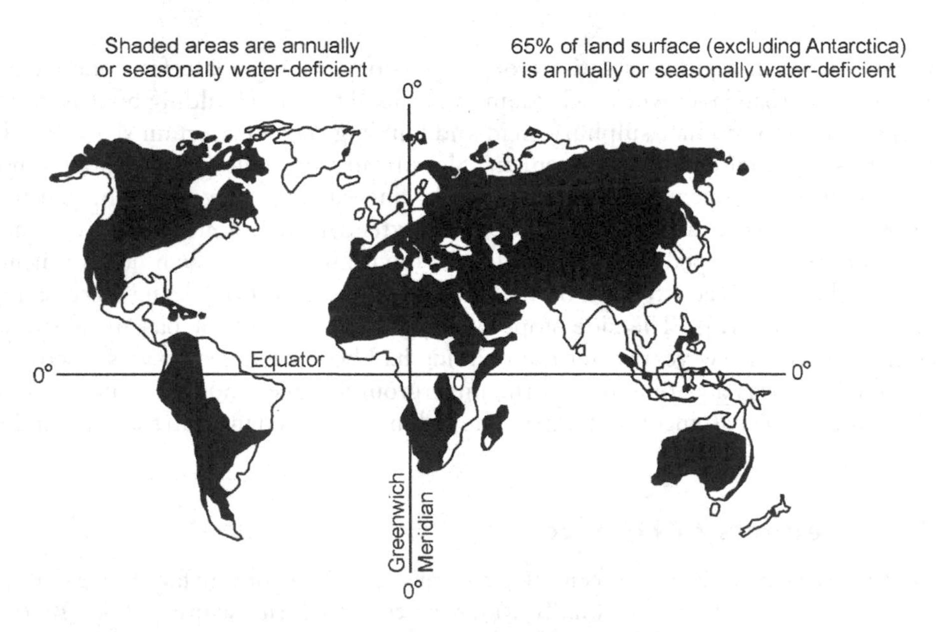

Figure 1.2 Annually and seasonally water-deficient land areas of the inhabited world.

water surface exceeds the monthly rainfall throughout the year, whereas in a seasonally water-deficient climate the annual evaporation from a free water surface exceeds the annual rainfall, although there may be seasonal excesses of rainfall over evaporation. The distribution of water-deficient areas in Figure 1.2 shows that many of the world's poorest and least developed continents and countries have water-deficient climates (e.g. Africa and western South America). Many of the water-deficient areas are highly mineralized and support major mining activity (Africa, Australia, western South America, western and central North America and Canada and Asia). Southern Europe also has a large water-deficient zone. There is thus a strong case for considering the effects of water-deficiency, and the unsaturated land surface conditions it engenders, on the management of mining waste, especially as some of the effects may, with careful design and operating practices, be used to advantage. Previous books on the subjects of mine waste management have ignored the potential effects of climatic conditions on the technology of mine waste storage, however, this book will also give attention to the environmental geotechnics of waste storage in water-deficient climates.

Engineering for the storage of mine waste has at least three aspects

- the conventional geotechnical concerns with the safety and shear stability of slopes and settlement of surfaces of waste storages or impoundments, e.g. waste rock storages, open cast mine backfill and tailings storages;
- the environmental concerns of surface erosion and transportation of air-borne dust, surface and groundwater pollution by acid mine drainage and other leachate, as well as radon gas; and
- the concerns of the local communities who may fear adverse effects on their health, their quality of life, the crops they grow and the value of their investment in property, arising from the proximity of a mine waste storage facility.

The relative importance and severity of various geotechnical, environmental and social problems varies from water surplus to water deficient climates, and from developed to developing societies.

1.4 Waste characteristics

Mine waste may arise in a number of forms: as stripped soil and coarse, broken, partly weathered rock overburden in open cast or strip-mining operations, as development waste rock in underground mining and as fine-grained tailings, the residuum of the process of comminution and mineral-extraction from ores. The various wastes are usually stored separately. The top-soil is stock-piled for eventual use in environmentally rehabilitating the dumps of coarse wastes, or the surfaces of the back-filled mining voids. The coarse broken rock is stored in dumps, either with or without compaction or is used to backfill opencast or strip-mining voids. Although generally unsaturated, because of their coarseness, any capillary water tension that may develop in these coarse wastes has little effect on their mechanical behaviour, either in terms of strength or settlement. They are also highly permeable and even in a water-deficient climate, rapid infiltration of rain renders capillary tensions ephemeral. However, capillary tension may have a considerable effect on the mechanical behaviour of fine tailings, and rain infiltration may lead to incremental collapse settlement of coarse wastes. In water surplus climates, problems relating to shear stability and water pollution or erosion may predominate. In water deficient climates, wind erosion and blowing dust may be a major environmental problem.

Figure 1.3 (Blight, 1994) shows typical particle size analyses for a range of tailings from various sources, as well as the fine-grained industrial wastes, phospho-gypsum and pulverized fuel ash (PFA) from coal-burning power stations. Some of these wastes (e.g. PFA and gypsum) can be "dry-dumped" or "stacked" either by truck or belt conveyor, although they always contain some water to prevent dust pollution arising during transport and deposition. However, most of the tailings are transported hydraulically either as a slurry, as a "thickened tailings" or a "paste" and are deposited or "beached" into hydraulic fill tailings storages where the tailings flow under their own weight, settle and consolidate to form fine-grained silty deposits. Alternative methods of deposition are to discharge a thickened tailings slurry from a single, or a series of point discharges around each of which the tailings forms a flat-sloping conical deposit; or to transport a tailings paste by conveyor belt and discharge it from a pre-constructed earthen ramp to form a viscous flow deposit. The various methods of deposition will be considered in detail later in the book.

1.5 Principles of mine waste management

This is primarily a book on geotechnical engineering as applied to mine waste storage. Nevertheless, as mentioned earlier, there is increasing public and governmental pressure to apply the highest environmental and ethical principles to all aspects of mining, including mine waste storage. The best set of general basic principles, at present, appear to be those of the International Standards Organization's Series 9 000 standards on Quality Control and Series 14 000 standards on Environmental Management Systems. Many large international mining companies are now adopting the principles of these standards.

The most important characteristic of the ISO 9 000 and ISO 14 000 Quality Control and Environmental Management Systems is that they are intended to operate on an upward (i.e. improving) time-spiral. As far as mine waste and tailings storages and associated water containment dams are concerned, this means starting with an environmental and safety policy, leading to environmental safety and operational planning, to implementation, to checking and corrective action and then to regular review by management, with the objective of spiraling upward to an improved environmental policy in future, and so on. The upward time-spiral principle is illustrated by Figure 1.4.

Environmental policy should rest securely on three cornerstones of responsibility

- The complete life-cycle of every project must be fully considered.
- The risks to safety and the environment, as well as financial risks must be realistically assessed and managed throughout the life-cycle.
- Every assessment must be broad-based, considering the benefits to the company as well as to the local population and the nation, so as to maximize the overall benefit. The local population must be consulted and involved in all assessments that may affect them in any way.

The life-cycle principle requires that the entire life-cycle of every project, from first breaking of ground to final rehabilitation for after use, be planned from the outset and

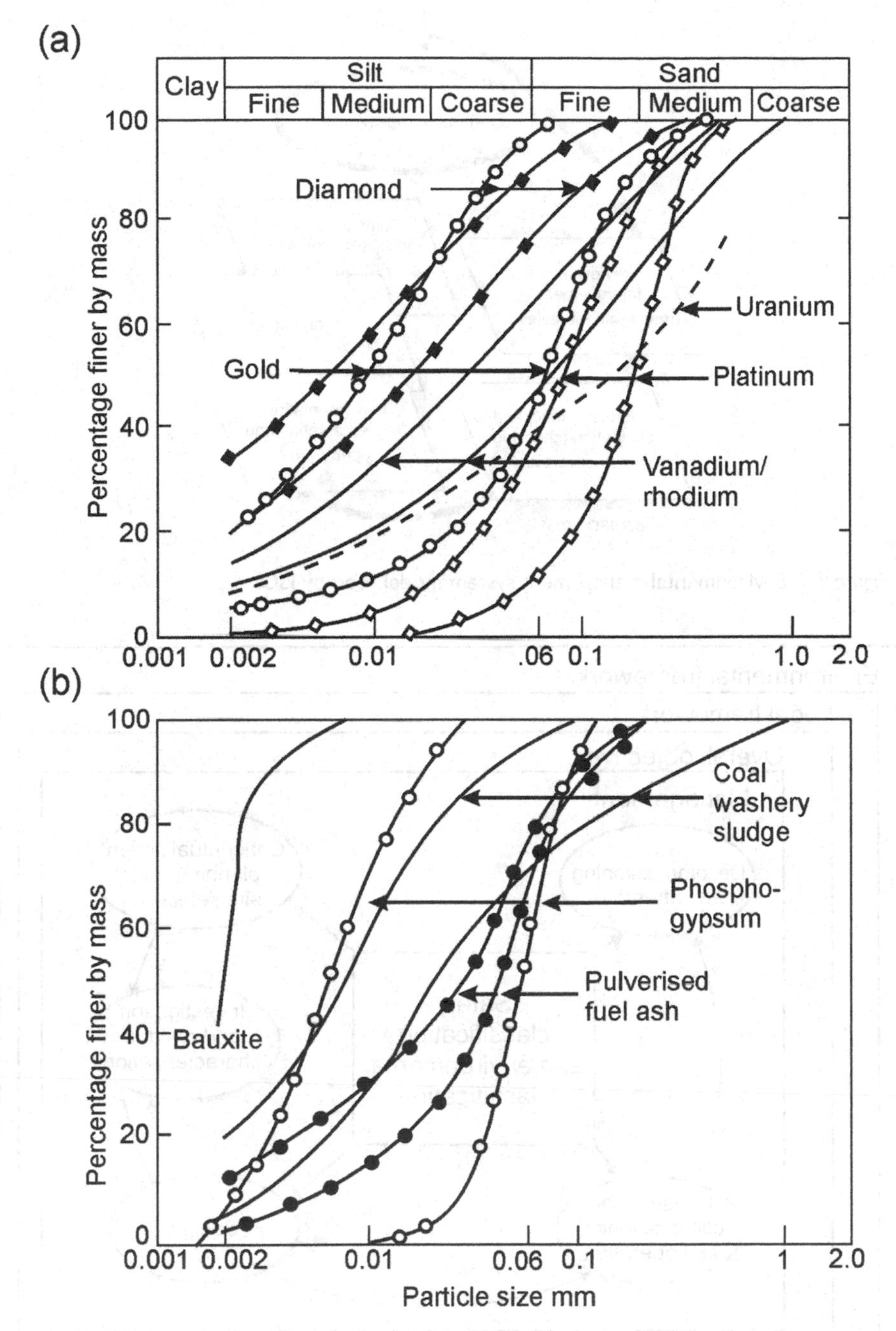

Figure 1.3 Particle size analyses for (a) typical tailings from mineral extraction of various ores, and (b) fine-grained industrial wastes.

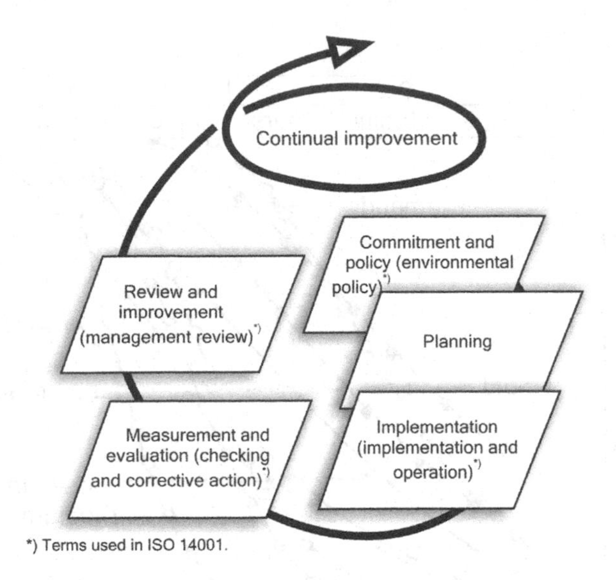

Figure 1.4 Environmental management system model used by ISO.

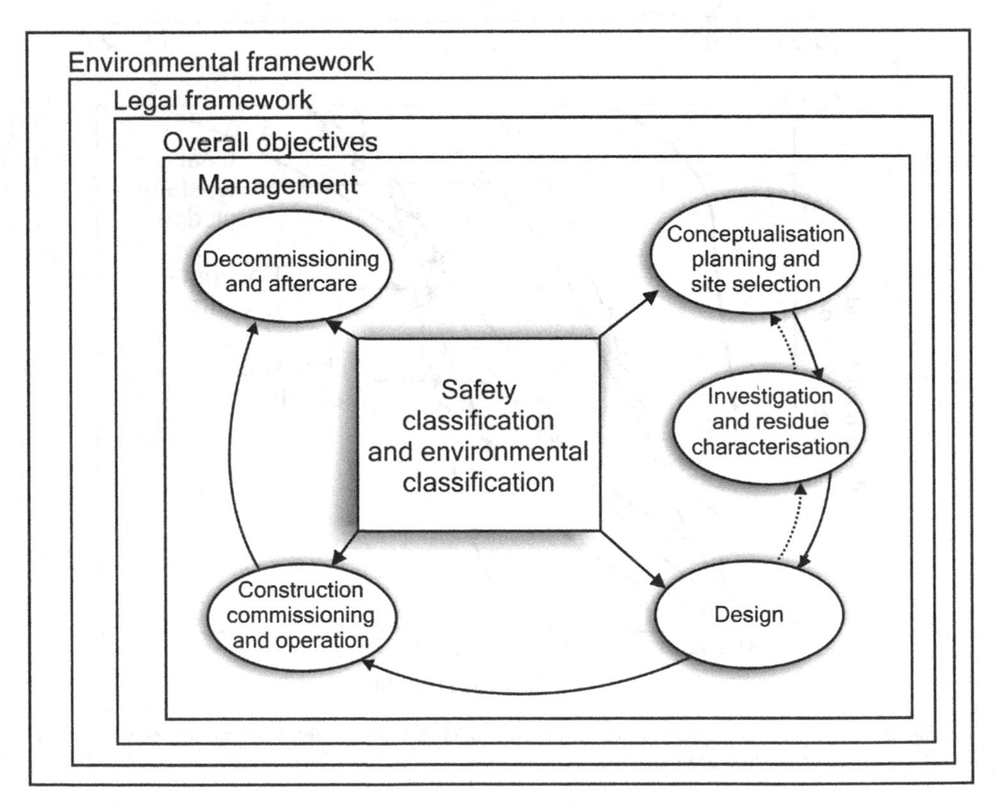

Figure 1.5 The life-cycle concept and its integration with various aspects of a mine waste management system.

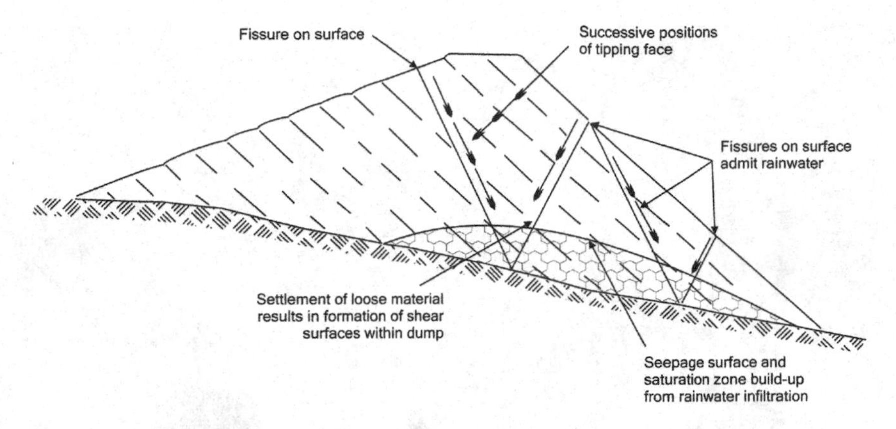

Figure 1.6 Construction of waste dump by end-tipping.

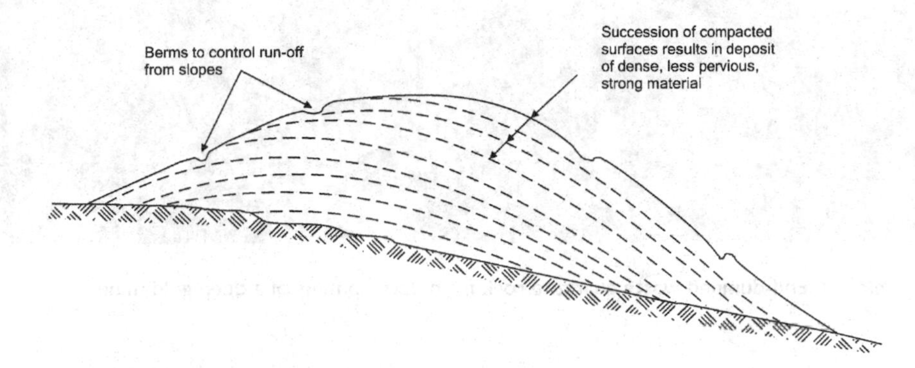

Figure 1.7 Construction of waste dump in traffic-compacted layers. (Diagram exaggerates slope angles and thickness of layers.)

that the objectives and concepts of this plan be implemented, verified and (if necessary) modified on a regular basis throughout the entire life-cycle.

Risk assessment should proceed hand-in-hand with consideration of the life-cycle. Risk cannot be assessed accurately without knowledge, therefore the sophistication of the risk assessment should evolve as the characteristics of the project evolve. Thus a simple qualitative risk assessment (or even a hazard rating) is appropriate at early stages of planning, evolving to quantified risk assessments during design, operation, closure and after-use. Figure 1.5 illustrates the life-cycle concept and its integration with continuing risk assessment for a mine waste storage.

The principle that the local population, whose lives and living conditions will be affected by a project (the interested and affected parties) should be consulted and involved at all stages of its life-cycle, is now accepted throughout the world. Every project must comply with the country's laws and regulations. Hence the national regulators must also be consulted and involved, from the outset, on an ongoing basis.

Plate 1.4 End-dumped waste quartzite rock from development of a deep gold mine.

1.6 Types of mine waste storage

Mine waste storages can broadly be divided into two main types – dumps and dams or impoundments.

1.6.1 *Mine waste dumps*

These may store barren overburden material, barren country rock, low grade mineralized material which it is not economical to process further at present or de-watered tailings from a mineral extraction operation. This type of storage is characterized by the method of deposition which may be by means of rubber-tyred or railed transport, stacking, flinging or other machinery. The defining criterion is that the material is placed mechanically in a moist or semi-dry state. It may be deposited either with or without some form of compaction. In the case of stacker, flinger or end tipping operations (see Figure 1.6), the waste is deposited without any form of compaction and an extremely large height of material may be deposited over an area in a very short space of time. Obviously, from the operational point of view, the method has considerable advantages, which is why this type of dump is used so widely. However, there are

Plate 1.5 Power station fly ash being end-dumped into the void left by an open-cast coal mine.

disadvantages that need to be borne in mind: From the point of view of stability, this method of deposition maybe unsatisfactory for the following reasons

- The material is deposited in a very loose state. It can therefore be relatively pervious and may accumulate a high moisture content as a result of rain infiltrating the dump surface.
- The imposition of a large height of material on the foundation strata in a relatively short space of time will apply high shear stresses and may lead to conditions conducive to failure of the foundation. The rapid increase in height may also generate pore pressures as a result of undrained loading of the waste. This should be investigated and taken into account in the stability considerations if it is a potential problem (see Chapter 11).
- If the surface of a dump is pervious, a significant proportion of annual precipitation may infiltrate the dump each wet season and, after percolating through the coarse residue, either flow out at the original ground level or seep into the foundation soils. If the seepage contains salts or acid dissolved from the waste, it may represent a significant source of ground and surface water pollution. However, the infiltration will usually be reduced significantly if the surface of the dump has been densified and rendered less pervious by compaction.

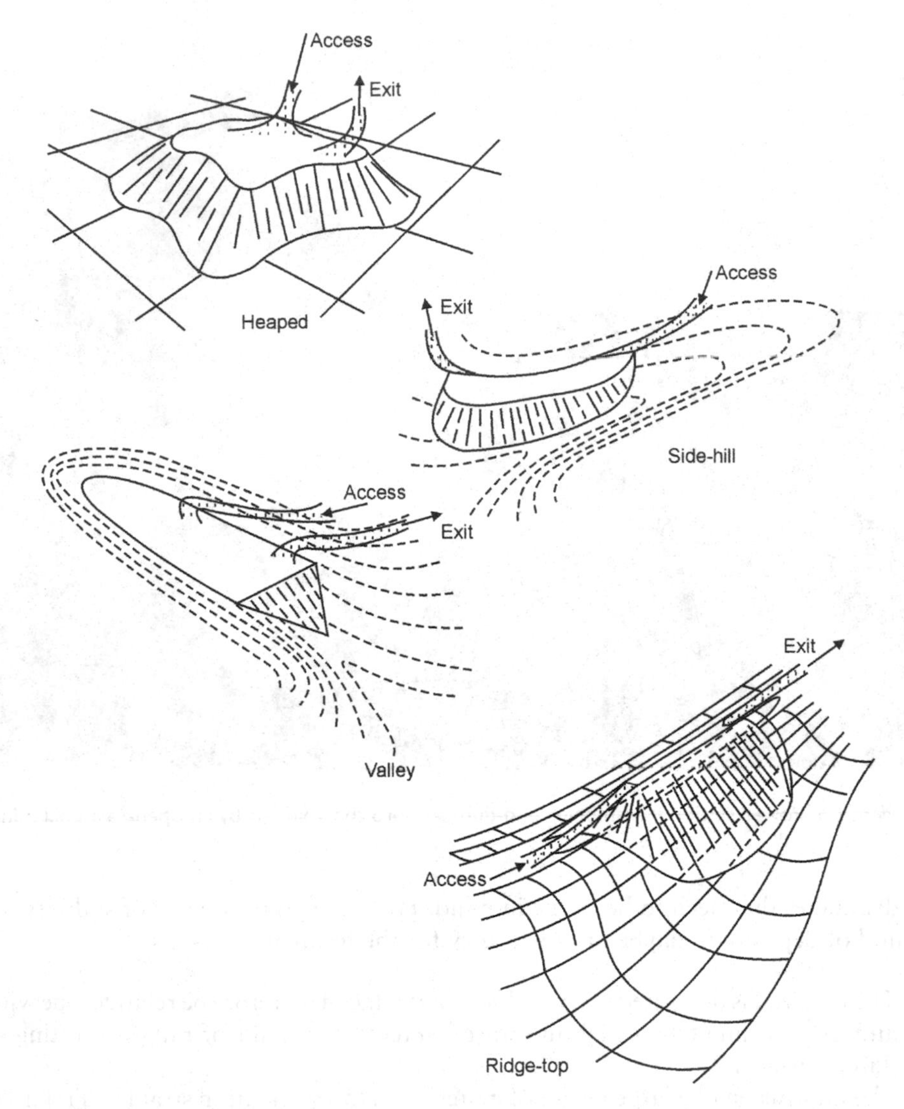

Figure 1.8 Various forms of waste dump dictated by topography. (Cartoons, not in true proportions.)

Where material is deposited by rubber-tyred vehicles, the dump can be built up in layers (see Figure 1.7) each of which receives some compaction from dumping vehicles. This means that

- The surface of such a dump is generally fairly well sealed against the ingress of moisture.
- A lot of the settlement is avoided because the material is compacted in layers as it is deposited.

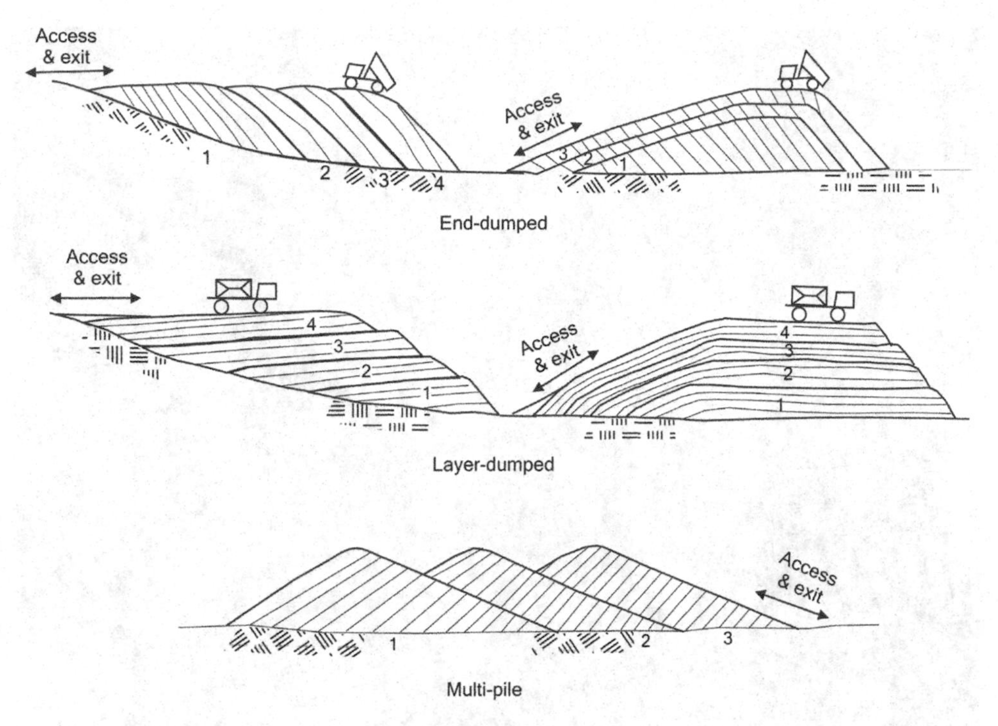

Figure 1.9 Various forms of waste dump construction (1, 2, etc. indicate order of placing). (Cartoons, not in true proportions.)

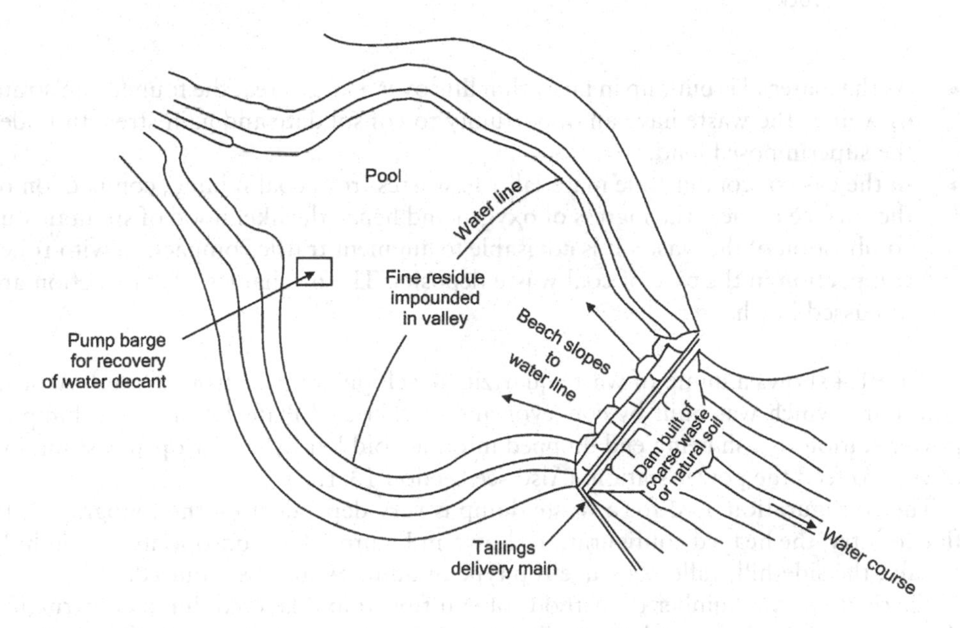

Figure 1.10 Cross-valley impoundment (note absence of spillway).

Plate 1.6 Cross-valley tailings impoundment dyke for copper tailings built of compacted waste rock.

- As the material is built up in fairly thin lifts over a large area, the foundation strata as well as the waste have an opportunity to consolidate and gain strength under the superimposed load.
- In the case of combustible material, e.g. wastes from coal mining, compaction of the surface reduces the ingress of oxygen and hence the likelihood of spontaneous combustion of the waste. It is advisable to augment traffic compaction with roller compaction in the case of coal waste deposits. The mechanics of compaction are discussed in Chapter 7).

Plate 1.4 shows a dump of waste quartzite development rock from a South African gold mine which was built by conveyor and stacker and Plate 1.5 shows a dump of power station fly ash being end dumped into the void left by earlier open cast mining of coal to feed the power station. (Also see Section 13.1.2.)

The configuration used for a waste dump is very dependent on the topography. In flat country, the heaped configuration shown in Figure 1.8 is appropriate, but in hilly terrain, the side-hill, valley or ridge-top type of dump would be required.

There are also a number of methods of end-tipped and layered dump construction that are possible. Some of these are illustrated in Figure 1.9.

Plate 1.7 Cross-valley tailings impoundment built by spigot deposition of molybdenum-copper tailings

1.6.2 *Tailings storages*

These may take a variety of forms but all are characterized by the fact that they retain or impound fine grained material which is usually deposited hydraulically, either as a slurry of low viscosity, or a more viscous thickened tailings. In certain cases the tailings are sufficiently viscous to enable them to be deposited by conveyor belt.

- 'Cross valley impoundments': An embankment or dam is built across a valley (see Figure 1.10) and the waste it impounds is used to fill the volume behind the embankment.

  Plate 1.6 shows a cross valley impoundment constructed of compacted waste rock in the Chilean Andes. Plate 1.7 is a cross valley impoundment built of tailings by upstream spigotting in the foothills of the Rocky Mountains in the U.S.A., and Plate 1.8 is a cross valley impoundment storing diamond tailings in South Africa, with the dam built of end-tipped coarse waste. (This is not a recommended method of construction, and this embankment has given major stability problems in its 60 year operational life.)
- 'Ring-dyke impoundments': An embankment is constructed to enclose a space which is then used to contain the waste (see Figure 1.11a and b). Ring-dyke impoundments are usually constructed on fairly level ground but a side hill ring-dyke may be constructed on sloping ground, in which case the closing side of the impoundment is formed by the slope of the natural ground. Partial ring-dykes

Plate 1.8 Cross-valley tailings impoundment built of dumped waste rock to contain diamond tailings.

may also be used to close off the open side of a quarry or open-pit working which has been excavated into the side of a hill.

Plate 1.9 shows a typical ring dyke storage built by paddocking in fairly flat country in South Africa. The Y-shaped causeway is the pool training wall enclosing the pool, with the penstock inlet visible near the apex of the Y.

Plate 1.10 shows a ring-dyke storage also built by paddocking on fairly flat ground, while Plate 1.11 shows a complex of ring dyke tailings storages. That a single mine has ended up with no less than eight separate tailings storages is indicative of the uncertainty of the mining industry, depending, as it does, on assessments of ore reserves, based on estimated mining costs and estimated market prices, as well as possibly unforeseen advances in both mining and extractive technology.

Plate 1.12 shows a large ring dyke storage built by spigotting. The darker shaded areas of the surface show where deposition was taking place when the photograph was taken.

- 'In-pit impoundments': These occur where waste is used to fill an existing worked-out open pit (see Figure 1.11c). Small quantities of especially noxious wastes are sometimes disposed of in impervious-lined in-pit impoundments which may be specially excavated for the purpose. (Also see Plate 1.5.)

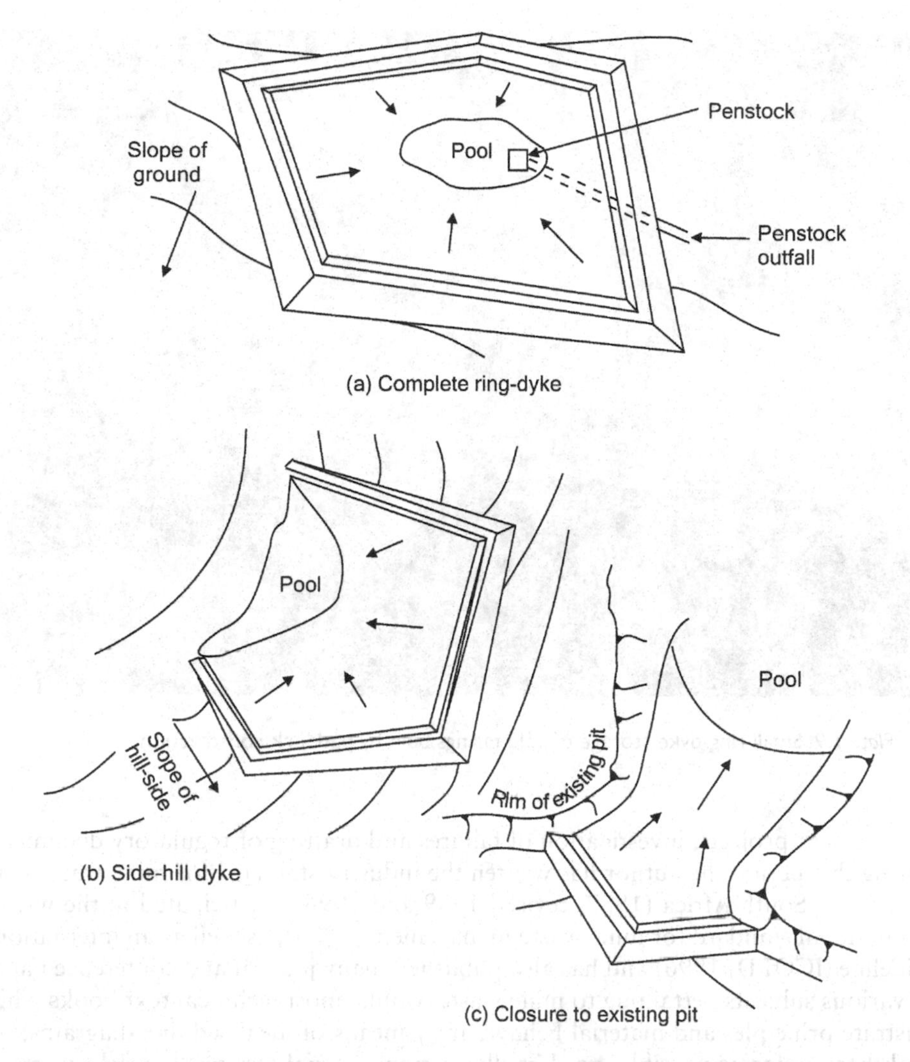

Figure 1.11 Various forms of dyke-retained storage impoundment for hydraulically placed tailings.

The types of storage impoundment listed and illustrated above are usually used for hydraulic fill deposition of low relative density slurries (see Section 4.5.4, equation 4.23d) for which particle segregation occurs as the slurry is deposited. At higher relative densities (above about 1.8) segregation does not occur and so-called single or multiple point discharge or line discharge systems are used, as illustrated in Plate 1.13. (Also, see Section 8.8.4.) Plate 1.13 shows a line discharge storage of co-disposed coarse and fine diamond tailings.

1.7 Philosophy and arrangement of this book

The book is based on 45 years of international involvement with the mine waste management industry through laboratory and field research, consulting on mine waste

Plate 1.9 Small ring-dyke storage of gold tailings built by paddock construction.

management projects, investigation of failures and drafting of regulatory documents. During this period the author has written the industry standard for mine waste management in South Africa (1987, revised 1989 and 1996), participated in the writing of a national standard for mine waste management (1973), as well as an international guideline (ICOLD) 1996) and has also published many journal and conference papers on various subjects pertaining to mine waste. Unlike most technical text books which illustrate principles and material behaviour by means of idealized line diagrams, this book has, as far as possible, used as illustrations, actual test results, field measurements and results of monitoring as well as case histories of actual behaviour or failures. This was done to illustrate the typical degree of scatter of real laboratory or field measurements which is not obvious when idealized line diagrams are used. For reasons of confidentiality, the actual mines or companies for which the work was undertaken have not been disclosed, but the descriptions of materials and situations have been given as fully as possible. Case histories have been included wherever appropriate to illustrate the practical use of the various concepts presented.

The book will be arranged generally to follow the life cycle of a waste storage. Successive chapters will deal with the following

- the process of site selection and preliminary assessment;
- geotechnical exploration of sites for development of waste storages;

Plate 1.10 Large ring-dyke storage of gold tailings built by paddock construction.

Plate 1.11 Complex of ring-dyke gold tailings storages. The storage nearest the camera on the lower left is being re-mined to extract residual gold.

Plate 1.12 Large ring dyke storage built by spigotting. Darker shaded areas of the surface are areas of current deposition.

Plate 1.13 Line discharge storage of co-disposed coarse and fine diamond tailings.

- environmental and engineering characteristics of mine waste, including stress and strain analysis and laboratory shear testing;
- measuring geotechnical properties by means of in situ tests.
- measuring the coefficient of permeability in the laboratory and in situ; the design of drainage filters, and uses and properties of geosynthetic materials in waste storages;
- principles of the compaction of soils;
- methods for constructing impounding dykes for storing hydraulically transported tailings;
- water control, safety monitoring and appraisal, carbonaceous and radioactive wastes;
- water balances for tailings storages and dry waste dumps;
- shear failure of hydraulic fill tailings storages and waste dumps;
- surface stability of waste storage slopes, erosion and erosion protection.
- back-filling with mine waste and use of mine waste as construction materials.

A detailed listing of the contents of the book is given in the Table of Contents and the Index.

References

Blight, G.E.: The Master Profile for Hydraulic Fill Tailings Beaches. *Proc., Inst. Civil Engnrs., UK*, 1994, pp. 107, 27–40.

Blight, G.E.: Minimizing Mine Waste by Making the Most of it. *Environmental Geotechnology*. Lancaster, Pennsylvania, USA.: Technomic Publishing, ISBN 1-56676-462-9, 1996, pp. 611–620.

Chamber of Mines of South Africa: (1979, revised 1983 and 1995). The Engineering Design, Operation and Closure of Metalliferous, Diamond and Coal Residue Deposits. *Guidelines for Environmental Protection*. Johannesburg, South Africa: The Chamber, 1996.

Chamber of Mines of South Africa: *Facts and Figures, 2006*. Johannesburg, South Africa: The Chamber, 2006.

ICOLD (International Commission on Large Dams): A Guide to Tailings Dams and Impoundments. Design, Construction, Use and Rehabilitation. *Commission Internationale des Grands Barrages, Bulletin 106*, Paris, France, ISBN 92-807-1590-7, 1996.

International Standards Organization. Series 9000 standards on Quality Control and Series 14000 standards on Environmental Management Systems. (Latest versions) International Standards Organization. (Many countries have adopted these standards as their national standards and they are listed as national standards of that country. e.g. Australian/New Zealand Standard AS/NZS ISO 14001: Environmental management systems – specification with guidance for use, or South African Standard Code of Practice SANS ISO 9001: Quality systems – Model for quality assurance in design, development, production, installation and servicing).

Stuart, R.J., Holmes, T.C., Jurcevici, M., Watt, I.B., & van Rensburg, G.D.J.: Influence of Tailings Dam Settlement on Decant Towers and Pipelines. In: *Tailings and Mine Waste '03*. Lisse, Netherlands: Swets and Zeitlinger, 2003, pp. 95–100.

Sutton, M.W., Weiersbye, L.M., Galpin, J.S., & Heller, D.: A G.I.S. History of Gold Mine Residue Deposits and Risk Assessment of Post Mining Land Uses in the Witwatersrand Basin, South Africa. In: *Mine Closure, 2006. 1st Int Seminar on Mine Closure*. Perth, Australia, 2006.

Thornthwaite, C.W.: An Approach Towards a Rational Classification of Climates. *Geogr. Rev. 38*, (1948), p. 55.

van Zyl, D., Sasoon, M., Digby, C., Fleurly, A.M., & Kyeyune: *Mining for the future, Report of the large volume waste component of mining and minerals sustainable development*. International Institute for Environment and Development, http://www.iied.org/mmsd/, (2002).

Chapter 2

Selection of a site for storage of mine waste

Most countries with significant mining industries have bodies of laws and regulations that govern the design, operation, closure and rehabilitation of mine and industrial waste storages to various degrees. It is obviously all-important to be aware of these laws and regulations, what they require and what they permit. Like municipal solid waste landfills, mine waste storages are dangerous structures that have been the known cause of death for 1083 innocent people over the past 80 years (see Table 11.1). Hence the risk of failure attached to a potential site should be an over-riding consideration in site selection.

The selection of a suitable, adequate site is the most important step in establishing a storage for mine waste. Once a site has been commissioned, it is nearly impossible to change it, whereas the design, operation etc. can still be changed subsequently with changing metallurgical processes, demand for the product, financial conditions, etc.

The basic objectives of site selection are as follows

- to ensure that the site is appropriately located, of adequate size and technically feasible to develop for waste storage with acceptable margins of safety;
- to ensure that achieving acceptability in terms of cost, environmental and social impacts is also feasible.

2.1 Procedure for site selection

The public (the term usually used is "Interested and Affected Parties" or AIPs) should be informed by letter, radio, television, newspaper or pamphlet that the need for a new waste storage in their area has been identified and that they will be involved and kept informed of the process and progress of the site selection. Interested and Affected Persons (or Parties), IAPs are those who will be affected in some way by the development of the proposed waste storage. They may consist of nearby residents or farmers, a residential community, central, regional and local authorities and regulators and the public at large. Local, provincial and national government forums must be regarded as IAPs as they are usually the democratically elected representatives of the public. The IAPs involved in the public participation process may change during the site selection. For example, those involved in preliminary discussions may be completely different from those involved during the feasibility study, which focuses on one or two specific sites.

The public participation process is usually initiated by a public meeting which takes place before the formal site selection process begins. The aim of the public meeting would be to invite, address, inform and involve all the IAPs. At this meeting the IAPs must be informed of the need for a waste storage. They must be informed of how the proposed site selection process will proceed and they must also be informed of the proposed site selection criteria. They must be given the opportunity to define the extent to which they wish to participate. This could include the identification or suggestion of possible candidate sites.

In order to facilitate proper liaison, a representative committee should be elected by the IAPs to maintain day to day contact with the developer of the waste storage.

A sufficient number of candidate sites should be identified to ensure an adequate consideration of possible alternatives. All the candidate waste storage sites must then be evaluated and ranked. The top ranking sites are then subjected to a more detailed, but still qualitative investigation to confirm the ranking. A feasibility study, involving preliminary estimates of storage capacity, safety classification, environmental and social impact assessments and geotechnical and geohydrological investigations, may then be carried out on the highest ranking site (or sites) to determine whether it (or they) meets (or meet) technical and cost requirements and can be developed in a socially and environmentally acceptable manner. After this has been confirmed, the IAPs and communities should be consulted for their input and acceptance of the proposed development. In all cases it is usually advantageous to consult with the AIPs and local community first, before starting any quantitative investigation of a candidate site. Should the site under consideration not prove feasible in terms of environmental or community acceptability, the rankings may be revised and the next best site is considered.

The process of site selection is only completed when a site has been judged technically feasible, as well as environmentally acceptable and acceptable to the AIPs and local community. Thereafter, detailed site investigations and all the required permitting processes can be commenced.

2.2 Preliminary assessment of required size of site

If the waste storage is required to extend the waste storage capacity of an existing mine, the annual tonnage of the waste being produced will be known, as will the density (or unit weight) of the waste, once it has been deposited in storage. From the estimated life of the mine, the total tonnage and hence volume of waste can be calculated. Hence for a given average height of the storage, (e.g. 50 m) the minimum required area will be known (e.g. 300 ha). This becomes the first criterion to be satisfied when ranking the site. The actual storage capacity of the site will depend on the shape of the site in plan and its surface contours. As the outer retaining wall of the storage is costly to build, in terms of providing under-drainage, erosion protection, supervision during the operating life of the storage and rehabilitation at the end-of-life of the storage or mine, whichever is reached first, the ratio of the volume stored to the length of the outer retaining wall should be maximized. In these terms, if the topography is suitable, a valley storage in which only one side of the waste requires containment, will usually be best. If the site is relatively flat and a ring-dyke dam is necessary, an area that approximates to a circle or a square might appear to be best. For both a square and a circle of overall plan dimension D, the ratio of area to perimeter is D/4, however for a given perimeter p, the ratio of the area of a circle to the area of a square is $4/\pi = 1.27$. For an equilateral triangle the area/perimeter ratio is D/3.46. For a rectangle of dimension D × 2D, the area to perimeter ratio is D/3, because, if an area made up of two adjacent squares has to be fenced, the fence between the two areas can be omitted. If the dimension is D × 5D, the ratio is D/2.4. Hence the larger the ratio of length to width becomes the more short cross fences are saved and the nearer the ratio of area to width approaches D/2. The most efficient containment in these terms is a long, narrow valley dam for which the ratio of area to outer wall perimeter could be very large. Obviously, a circular natural crater would require no walls at all, but these are rare and would have to be drained by siphoning or pumping. This discussion emphasizes one advantage of storing waste in an existing mine void where no retaining wall is required. The disadvantage, however, is poor accessibility for re-mining.

For a new mine, similar considerations would apply, but the rate of waste production is usually less certain. Also the life of a successful mine usually extends with time as mining and mineral extraction techniques improve. It is therefore better to be generous in choosing a site area. In fact, if area is available, it is preferable to oversize the storage than have to extend its area at a later stage. The safety margin for the proposed mine waste or tailings storage is particularly important and must receive due attention (see Section 2.7).

2.3 Possible fatal flaws in candidate sites

A "fatal flaw" is a fault or flaw peculiar to the site that will prevent it from fulfilling its intended purpose and that either cannot be remedied, or cannot be remedied with an acceptable cost.

In the initial selection of candidate sites, it is useful to bear in mind that the following may prove to be fatal flaws in a site, causing it to be rejected

- areas below the 1 in 50 year flood line; (this eliminates water courses, either perennial or seasonal, wetlands and flood plains, where damage to the planned facility or water pollution might result from encroachment of a flood on the waste storage.)
- areas in close proximity (closer than 50 m) to significant bodies of surface water, e.g., rivers, lakes, or reservoirs;
- geologically unstable areas; (These could include fault zones, seismic zones, cavernous limestone or karst areas or areas previously undermined at shallow depth where sinkholes, collapses and subsidence are likely or possible.)
- sensitive cultural, religious, ecological or historical areas; (These may include nature reserves, historic battle fields, burial grounds and other areas or features of ecological, cultural, religious or historical significance.)
- catchment areas for important water resource areas; (While, if not on a watershed, all sites are in catchments, the size, yield, importance or sensitivity of the catchment area may represent a fatal flaw.)
- areas in close proximity to land-uses which are incompatible with mine waste storage, e,g, where adequate buffer zones between the site boundaries and the nearest settled community are not possible, or areas which may attract AIP or community resistance, in other ways and for other reasons;
- any area immediately upwind, for the prevailing wind direction(s), of a residential or high value agricultural crop area, thus posing a danger of dust nuisance and pollution in dry windy weather.

2.4 Seeking and obtaining public acceptance

Public acceptance depends critically on perceptions of possible adverse impacts on quality of life, other industrial activities, local land and property values, and potential damage to the environment. Failure to attain public acceptance from interested and affected parties is always a fatal flaw for a site. Also,

- the displacement of local inhabitants will usually arouse intense public resistance;
- exposed sites with high visibility from residential areas are less desirable than secluded sites or sites naturally screened by hills or in valleys;
- the sensitivity of the environment through which the access road(s) pass(es) is also a consideration. The shorter the distance to the site through residential areas, the less potentially unacceptable the site;
- prevailing wind directions must be taken into account and new waste storages must be sited down-wind of residential areas to minimize possible dust nuisance;
- the distance to the nearest residential area or any other land-use which is incompatible with a waste storage must be considered. The greater the distance from incompatible land-uses, the lower the risk of nuisance problems and hence the lower the potential resistance to the facility.
- the assessed risk of various possible forms of failure and the possible effect on the local community and environment must also be estimated, as well as possible, and openly disclosed.

2.4.1 *Principles of public participation*

The principles of public participation are as follows

- The IAPs must be consulted and given opportunities to have a voice in the project. This requires adequate notification, and the provision of opportunities such as discussions, meetings, workshops or committees. In particular, IAPs must be allowed to define the extent to which they wish to participate in a project and to define the formal participation process they wish to see followed. This means that the IAPs must be involved during the earliest planning stages of a project. Their input must also be seen to have been taken into account in the decisions that are reached. This does not mean that all their suggestions or demands should be acceded to, but they must be given satisfactory reasons why any suggestions are considered unacceptable.
- The subject of mine waste storage is highly technical and will probably not be understood by most of the IAPs. Therefore the IAPs must be informed, or empowered, so that they can make logical decisions. In order to empower the IAPs, they have to be given access to and explanation of all relevant information, through meetings, presentations, representatives, discussions, reports and documents.
- There must be accountability for the information on which decisions are taken. This means that the information provided to the IAPs must be sufficiently detailed, accurate and understandable, so that they can sensibly participate in decision-making.

- There must be consideration of alternative options. When planning a waste storage, the advantages and disadvantages of various options including safety risks must be investigated and discussed.
- Negative impacts must be mitigated. This also involves identifying the impacts seen to be negative by the IAPs, so that they can be adequately addressed.
- The IAPs must be sure that social benefits outweigh social costs. It is the duty of the proponent of a project to demonstrate that the costs of the project will be internalized and not borne or shared by society.
- Individual rights and obligations must be upheld. Although ultimately the appropriate regulating authority must make a decision, every effort must be made during the public participation process to uphold the rights of the individual. However, the right of an individual must not be permitted to take precedence over the rights of the majority in society.

2.4.2 *Mechanisms for identifying IAPs*

The objective of informing the public will usually be to identify as many IAPs as possible. In particular, the objective is to identify a reliably representative group of IAPs. Typical IAPs would include central government departments, provincial and local municipal government departments, other local authorities, residents in the near vicinity, Non Governmental Organizations (NGOs) having an appropriate function, and others.

There are many mechanisms for identifying IAPs. Established lists such as commercial lists, memberships or ratepayers lists can be used. Networking, or chain referral, is another option for identifying IAPs. This uses key players such as local authorities and adjacent property and business owners to identify other possible IAPs. In identifying IAPs and obtaining input from a disadvantaged or indigenous community, such issues as literacy levels, language barriers, level of community structures, and social and cultural biases must be taken into account. IAPs from disadvantaged or indigenous communities can be notified and identified by using traditional methods of community participation, by appointing locally based organizations to hold meetings, workshops and interviews; by means of illustrated posters or loudhailers; and by identifying key players within the communities. IAPs can also be notified and asked to come forward using announcements in the press, or by means of publications, television, radio, pamphlets, exhibitions, newsletters, direct mail and public notices. Public meetings, workshops, 'open houses', telephone canvassing, newspaper advertisement, surveys and questionnaires, and advisory groups are all mechanisms to ensure IAP notification and involvement.

2.4.3 *Stages at which IAPs must be involved*

The key stages at which IAPs must be consulted are as follows

- at the announcement of the intention to develop a waste storage, once possible sites have been identified; (At this stage consultation with IAPs may disclose fatal flaws in some possible sites.)
- once a provisional ranking of possible sites has been made; (The provisional ranking must be finalized in consultation with the IAPs.)

Table 2.1 Candidate Site Ranking Matrix

	Siting Criteria								
Candidate Sites	*Storage area*	*Length of outer wall*	*Pumping head*	*Distance to plant*	*Social acceptability*	*Safety risk*	*Environmental impact*	*Etc*	*Total Score*
Site 1									
Site 2									
Site 3									
Site 4									
etc									

- during the feasibility study for the preferred sites; (Those IAPs who would be most affected by the proposed waste storage must be included in the feasibility consultations.
- once the feasibility study has been completed, the IAPs must participate in the final choice of a site.

2.5 Preliminary ranking of candidate sites

It is useful to rank candidate sites by means of a simple numerical ranking matrix such as that illustrated by Table 2.1. Nominal ranking values (e.g. 0 to 5) are assigned under each heading of the matrix and the total score assists in deciding on one or two preferred sites for more detailed investigation. Table 2.1 is merely an example of a siting matrix. A real siting matrix may include many more criteria.

The evaluation and scoring process must be carried out by a multi-disciplinary committee (including representatives of AIPs and the community affected by the project, where appropriate), and is the outcome of consensus reached through discussion (a Delphi process).

2.6 Site feasibility study

The feasibility study examines the feasibility of building the required waste storage on the preferred site or sites. At this stage, the necessary characteristics for the storage site (minimum total capacity required, acceptable safety risk, necessary access, maximum pumping head, flood hydrology, etc) are matched with the characteristics of the site. The objectives of a feasibility study are to help in a final choice of site, also

- to identify any critical design or other important factors and address them satisfactorily;
- to confirm that the site has no fatal flaw associated with it;
- to satisfy the statutory regulators and other IAPs that the site is indeed suitable for development as a mine waste storage, based on available information.

A site must be acceptable in terms of safety and environmental impact and acceptable to the IAPs (i.e. socially acceptable) for it to be feasible for development.

Table 2.2 "Zone of Influence" Risk Rating Scheme for a Waste Storage Site

1	*2*	*3*	*4*	*5*
Number of residents in zone of influence	*Usual number of workers in zone of influence*[1]	*Value of property in zone of influence*[2]	*Depth to underground mine working*[3]	*Risk Rating*
0	<10	0–2.5 million	>200 m	Low
1–10	10–100	2.5–25 million	50–200 m	Medium
>10	>100	>25 million	<50 m	High

1 Not including workers employed solely for the purposes of operating the waste storage.
2 The value of property shall be the replacement value in terms of all property, other than that associated with the facility (in local currency units, figures in Table 2.2 are in 2007 US$, for illustration).
3 The potential for collapse of the waste storage into the underground workings effectively extends the "zone of influence" to below ground level.

The extent of the feasibility study will depend on the physical complexity of the actual site and the sensitivity of the surrounding environment. Components of the study are described below. All activities need to be explored to a preliminary degree in order to establish the site feasibility. If feasibility is shown, this assists with the final choice of site. For the design of the waste storage on the chosen site, the feasibility stage investigations need to be extended to provide more detail.

2.7 Risk analysis

A number of simple methods for qualitative risk analysis are in use. Table 2.2 illustrates a simple, qualitative "zone of influence" method which may be suitable for a very preliminary ranking of sites. The "zone of influence" referred to in Table 2.2 is the zone likely to be affected by a slope failure or over-topping of the waste storage on the site concerned. The boundary of the zone of influence for waste storages placed by hydraulic filling can be determined as follows

- upstream or uphill of any point on the perimeter, the lesser of a distance of 5 h from the toe (where h = planned final height of the waste at the point under consideration) or the distance to the point where the ground level exceeds h/2 above the elevation of the toe, at that point on the perimeter;
- on sides parallel to the ground slope – a distance of 10 h from the toe;
- downstream of the lowest point on the perimeter, a distance determined from the greater of 100 h; or
- 2 × steepest ground slope in percent measured over a distance of 200 m from the lowest point on the perimeter, multiplied by h, with a minimum of 0.5 km and a maximum of 6.0 km, where h is the maximum design height. The 6 km distance may have to be increased, depending on the topography and density of inhabitation down-stream. (When the tailings dam at Bafokeng mine failed in 1974, significant quantities of tailings flowed for 45 km downstream of the dam (Blight, 2000)). (Also see Section 11.3.1.)

Table 2.3 "Consequences of Failure" Risk Rating Scheme

Potential Incremental Consequences of Failure		
Loss of Life	*Economic, Social, Environmental*	*Risk Rating*
Large increase expected*	Excessive increase in social economic and/or environmental losses.	Very high
Some increase expected*	Substantial increase in social, economic and/or environmental losses.	High
No increase expected	Minimal social, economic and/or environmental losses. Losses generally limited to the owner's property, damages to other property are, by agreement, payable by the owner and therefore socially acceptable.	Low

* The loss of life criterion which separates the High and Very High categories may be based on risks which are acceptable or tolerable to society, often taken to be a probability of an accidental death of 1 in 100 000 in any one year. Wong (2005), however, gives a much higher probability of 1 in 10 000 per year as the limit of acceptability for fatalities. Consistent with this tolerable risk, the minimum criterion for a Very High hazard rating would result in an annual probability of failure of less than 1 in 100 000 (US. National Research Council, 1989, Canadian Dam Safety Association, 1995). A higher probability than 1 in 100 000 would constitute a fatal flaw.

Alternatively, risks may be classified qualitatively in accordance with the foreseeable consequences of failure. The consequences of failure of a waste storage on a site are evaluated in terms of potential for

- loss of life;
- economic values of other losses and/or damage to property and facilities, (including those owned by the mine and by others), other utilities e.g. dams, power stations or waste storages downstream as well as loss of power generation or water supply;
- other less quantifiable consequences related to social, cultural and environmental damages. (Where appropriate, provisional monetary costs can be assigned to social, cultural and environmental impacts.)

The consequences are those assessed as being incremental to (or in addition to) the impacts that would occur under the same natural conditions (flood, earthquake or other event) but without the presence, or presence and failure of the waste storage. The type of consequence (e.g. loss of life, or economic loss) with the highest rating determines which risk category is assigned to the storage if constructed on that site.

Table 2.3 shows an alternative qualitative risk classification scheme

Evaluation of potential economic, social and environmental losses, both with and without failure of the waste storage, should be based on studies that consider existing and anticipated future downstream population and industrial development and land uses.

The extent and detail of the investigations described in what follows (environmental impact report, waste or tailings characterization and site investigation) depend on the hazard rating of the site.

In Hong Kong, with its steep residual granite soil slopes, tall buildings and exceptionally high population density, acceptable risks of slope failures and their consequences

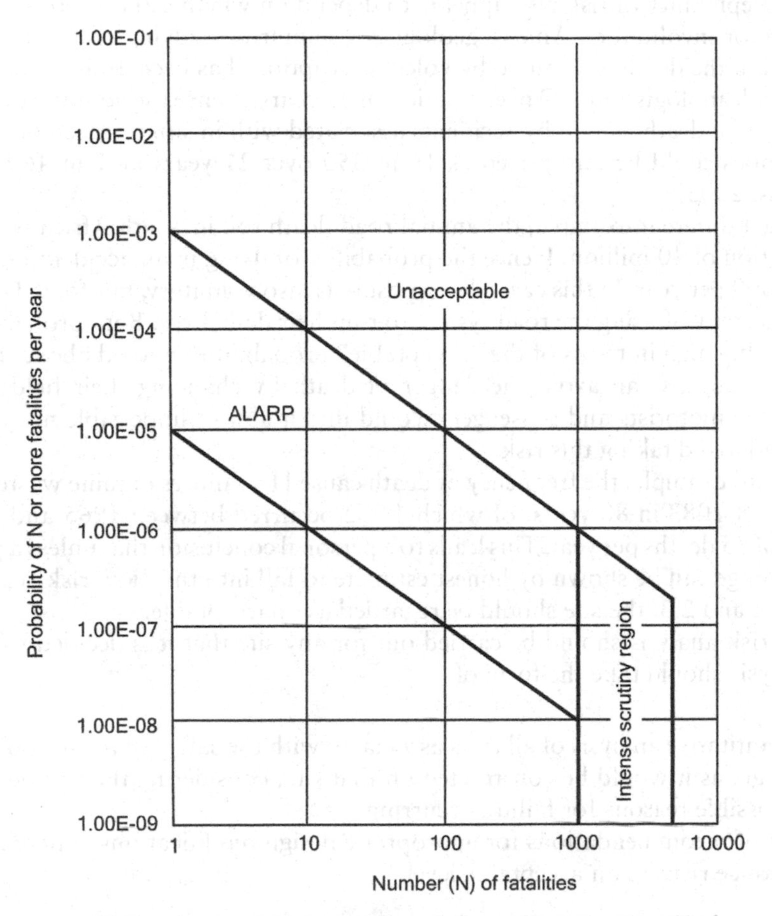

(1) The above societal risk criteria are to be used in conjunction with a reference toe length of the natural hillside of 500m.
(2) If a development is affected by more than 500m toe length of natural terrain, an appropriate linear scaling factor should be used to scale up the risk criteria. For example, in the case of a large development affected by natural terrain with a toe length of 5km, then the above societal risk criteria should be increased by one order of magnitude.
(3) If the development is affected by less than 500m toe length of natural terrain, then the same criteria as proposed above are taken to apply (i.e. the criteria will not be scaled down).
(4) The societal risk criteria are intended to aid decision-making and not intended to be mandatory.

Figure 2.1 Acceptable risks of fatalities and corresponding frequencies of occurrence for failures of natural slopes in Hong Kong (Ho, et al., 2000).

of likely fatalities have been formalized as shown in Figure 2.1. (Ho, et al., 2000). In this figure, the letters ALARP stand for “As Low As Reasonably Practicable”. Note that the highest acceptable probability of a single death is 1 in 100 000, while for more than one death the acceptable probability is less than this.

The acceptability of risk also appears to depend on whether the exposure to risk is voluntary or involuntary. Among geologists voluntarily studying eruptions of volcanoes in situ, the death toll caused by volcanic eruptions has been 16 in a participating body of volcanologists of 350 over a period of 21 years. Hence the voluntarily accepted probability of death caused by accidents associated with in situ research on eruptions of volcanoes could be interpreted as 16 in 350 over 21 years or 1 in 460 per year. (Williams, 2003).

Taking a different example, the annual road death toll in South Africa is 10 000 in a population of 40 million. Hence the probability of dying in an accident on the road is 1 in 4 000 per year. In this case, the exposure is also voluntary, but forced on people by the necessity of using the road system to run their daily lives. Both probabilities are unacceptably high in terms of the "acceptable" probabilities quoted above. However, the volcanologists can avoid the danger of death by changing their field of study, whereas the motorists and passengers would find it almost impossible not to use the roads, and avoid taking this risk.

As a third example, the frequency of death caused by failures of mine waste storages was at least 1083 in 80 years, of which 1 072 occurred between 1965 and 1996, an average of 35 deaths per year. This leads to a personal conclusion that unless a proposed waste storage can be shown by honest estimate to fall into the "low risk" category in Tables 2.2 and 2.3, the site should be regarded as unacceptable.

A full risk analysis should be carried out for any site that it is decided to use. The risk analysis should take the form of

- a quantitative analysis of all risks associated with the safety of the proposed waste storage, as it would be constructed on that site, considering the consequences of all possible reasons for failure occurring;
- a list of recommendations for appropriate design modifications, where necessary to reduce risks to an acceptable level.

Figure 2.2 shows the number of tailings storage failures ascribed to various causes over a period of about 80 years. This could be of assistance in identifying possible risks. (US Committee on Large Dams, 1994). It is unfortunate that Figure 2.2 does not give more specific information, and particularly information concerning the chains of events that lead to the final breaching of the impoundments. It is very seldom that a failure results from a single cause. Examples of chains of events could be:

- Design freeboard compromised while making essential repairs to decant tower → unseasonal rainstorm occurs → impoundment overtops → gulley erosion occurs → slope failure follows → flow failure occurs.
- Commodity price increases → mine increases production → rate of rise of tailings storage is increased → degree of consolidation of tailings reduces → slope failure occurs.

Very often, the first link in the chain of events could have been avoided with proper planning, design, maintenance or at the expense of share-holders' profits. As examples, the decant tower could temporarily have been replaced by a pump barge pumping

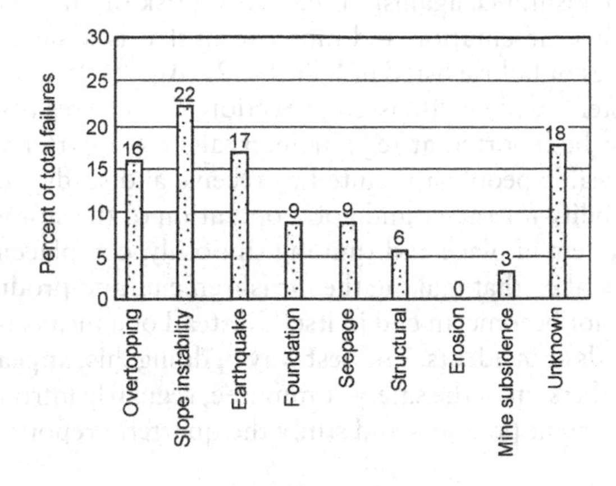

Figure 2.2 Analysis of causes of tailings dam failures (USCOLD, 1994).

through a temporary decant system, or a supplementary tailings storage could rapidly have been commissioned to accommodate the increased tailings production.

It should be noted particularly that tailings dams differ fundamentally from dams built for water supply or flood control, in two important ways:

1. Dams, other than tailings dams, are usually built completely, to store a predetermined full supply level, in a continuous process that takes place over a period of a few years. Most tailings dams, however, are continually under construction from the time the first tailings are placed until the dam or storage is decommissioned, possible 30 to 60 years later.
2. In water storage dams there is a sharp distinction between the structural material that forms the dam (concrete, rockfill or earth) and the material stored (water). However, most tailings dams are constructed of the material they store and there is hardly any distinction or gradation between the material that forms the structure and the material that is stored. These two distinctions make tailings dams inherently more risky to build than conventional dams, because:

- The building of the tailings storage may take three or more working lifetimes to complete which means that a large turnover of staff occurs over the life-cycle of the storage. Unless excellent staff training schemes are instituted and maintained, the level of skills, knowledge and interest of the staff responsible for a tailings storage tend to decline with time, and hence, through complacency and ignorance of the staff the tailings storage building process becomes more risky.
- Quality control levels are relatively easy to maintain at a high level in an intensive building operation that lasts only a few years. In the long drawn out process of building a tailings storage, however, understanding of the aims of the process and hence quality control itself is very difficult to maintain at a high level.

The only effective insurance against an increasing risk of failure is to implement an ongoing system of instrumentation and monitoring that is designed to guard against the most likely causes of failure listed in Figure 2.2. (Also see Table 11.2.) Monitoring and monitoring systems will be discussed in Section 9.3. The results of the monitoring measurements must be reported at regular intervals (e.g. quarterly) to a safety committee of knowledgeable people appointed to receive and study the reports and who accept the responsibility for safety and good operation of the storage.

Even with this system in place and running smoothly, complacency can erode vigilance. Care must be taken that making the measurements and producing the quarterly safety reports does not become an end in itself, instead of a means of guarding against safety-reducing trends or incidents. The best way of doing this, appears to be to periodically rotate the membership of the safety committee, regularly introducing new-comers who will question current practices and study the quarterly reports with unjaded eyes and minds.

2.8 Environmental impact report

The Environmental Impact Report will include the Environmental Impact Assessment, and the Consequences of Failure.

In order to make the environmental impact assessment and to assess the environmental consequences of failure, it will be necessary to draw on information obtained in the preliminary residue characterization and the preliminary site investigation (Sections 2.9 and 2.10 which follow this section). Hence this section must be read in conjunction with the later sections. The process used for the environmental impact assessment and the consequences of failure are as follows

2.8.1 *Environmental impact assessment*

This makes use of accepted methodology to assess the potential impacts of a project on the environment. The objectives of the Environmental Impact Assessment (EIA) are

- to identify the various ways in which the waste storage is likely to affect its receiving environment (i.e. the extent of the environment likely to be affected by the storage);
- to assist the designer to address the mitigation or elimination of any identified impacts by means of design.

EIAs should be undertaken according to the principles briefly listed below
There must be

- informed decision-making;
- accountability for information on which decisions are taken;
- accountability for decisions taken;
- a broad meaning given to the term environment (i.e. one that includes physical, biological, social, economic, cultural, historical and political components);
- an open, participatory approach in the planning of proposals;
- consultation with interested and affected parties (IAPs);
- due consideration of alternative options;
- an attempt to mitigate negative impacts and enhance positive aspects of proposals;

- an attempt to ensure that the 'social costs' of development proposals (those borne by society, rather than the developers) are outweighed by the 'social benefits' (benefits to society as a result of the actions of the developers);
- democratic regard for individual rights and obligations;
- compliance with these principles during all stages of the planning, implementation and decommissioning of the proposed development (i.e. from 'cradle to grave'), and
- the opportunity for public and specialist input in the decision-making process.

The EIA must be regarded and undertaken as an aid to site selection and later, as an aid to design of the waste storage on the finally selected site, and not as an end in itself, or as a means of satisfying bureaucratic requirements.

Initial steps leading towards an EIA are as follows

- establish purpose (or need) for the proposal;
- establish policy requirements;
- establish legal requirements;
- establish administrative requirements;
- consult with relevant authorities;
- identify and notify IAPs;
- consult with IAPs;
- identify and consider alternatives;
- identify and consider issues, opportunities and constraints for alternatives;
- consider mitigatory options;
- consider management plan options.

2.8.2 *Steps in the EIA*

- Scoping. This determines the extent of and approach to the investigation. The developer of the proposed waste storage together with the relevant authorities and the IAPs determine which alternatives and issues should be investigated; the procedure that should be followed, and report requirements.

As a result of the scoping, it may be agreed with the IAPs and the regulatory authorities that no further investigation is needed. When engaged in the initial ranking of sites, a scoping for each site is usually all that is required for ranking.

- Investigation. This is guided by the scoping decisions and is intended to provide the authorities, IAPs and designers with enough information on the positive and negative aspects of the proposal, and feasible alternatives, with which to make a decision.

Investigations more extensive than the scoping would only be made on top-ranked sites, as part of the feasibility investigation. The investigation may need to be extended for the purposes of assisting the design, once the final choice of site has been made.

Table 2.4 Environmental Impact Identification Matrix

Action or results of residue deposition →	*Blowing dust*	*Danger of flow slides*	*Noise and additional traffic on roads during construction*	*Pollution of surface and ground-water*	*Radon gas exposure*	*Spills of tailings slurry*	*Etc etc.*
Possible impact on: ↓							
Agriculture							
Aquifer Recharge							
Archaeological Site							
Indigenous Forest							
Industrial Development							
Recreation							
Residential Areas							
Surface Water							
Etc., etc.							

Scoring system Uncoloured = Zero Interaction
+ or Green = Positive Impact, i.e., desirable
− or Red = Negative Impact, i.e., undesirable
0 or Orange = Impact requiring further investigation

2.8.3 *Methodology for EIA*

There are many methods of Environmental Impact Assessment available, but most of them either depend on, or have as starting point, a checklist of considerations that should form part of the design process. The checklist may simply be used to ensure that all aspects of the design and all possible impacts on the environment are taken into consideration, or it may be used more effectively to identify possible interactions between operations and processes at the residue deposit and the surrounding environment.

In order to identify interactions, use may be made of a two-dimensional environmental impact identification matrix (see Table 2.4). These matrices usually list the project actions along a horizontal axis and the possible impacts on various aspects of the environment on a vertical axis. In order to be effective, the matrices normally have to be large and complex. Table 2.4 is a simple example.

Actions that would form part of the matrix would include those linked to the following phases of the life-cycle of the proposed project

- site preparation and construction;
- operation;
- closure and rehabilitation;
- after-use.

There are numerous ways of scoring an environmental impact identification matrix. One typical and very simple scoring method is illustrated in Table 2.4. The rectangle forming each intersection between a vertical and a horizontal column corresponds to a possible interaction, relevant to the project under consideration. The intersection is then assigned a score, e.g. green or a plus sign to indicate a positive or favourable impact, red or a minus sign to indicate a negative or adverse impact, and orange (amber) or a zero sign to indicate no impact, or the need for further investigation.

The matrix should be used to identify possible impacts at scoping and at final EIA stages.

The selection of the actions and impacts that make up the axes of the matrix must be undertaken by a team having multi-disciplinary representation and including, representatives of the local authority, and the IAPs. The matrix must also be scored by this team, each rating being the result of rational discussion and consensus.

Once the matrix has been prepared and scored, a report must be provided that interprets the matrix for the reader and describes how each major adverse impact will be ameliorated or, preferably, eliminated by the design and operation of the waste storage. This report is referred to as "The Environmental Impact and Risk Assessment Report". It must be emphasized that the object of the EIA is to ensure that the design and operation of the waste storage is optimized, to ensure the least adverse impact on the surrounding environment and affected communities, while taking economic considerations and feasibility into account. It is also emphasized that the EIA forms part of the Risk Assessment (see, e.g. Table 2.3) and vice versa.

2.8.4 *Consequences of failure or failure to function safely*

In addition to assessing the environmental impact of the waste storage on its receiving environment and adjusting the design so as to eliminate or reduce adverse impacts, it is also necessary to consider the consequences of the failure or failure to function safely of any of the components of the waste storage, or its ancillaries, such as a penstock or return water dam. These considerations should also be reported in the Environmental Impact Assessment Report. Consequences of failure or failure to function safely may be trivial for one site, but disastrous for another, and need to be considered for every site, or possible site.

The following is a list of some failure modes or functional failures that might or might not prove disastrous, depending on site and storage characteristics, and for which the consequences, on a particular site, should be considered. The list is not intended to be exhaustive, and each site and proposed storage design will generate its own particular list.

- bursting of slurry delivery pipe;
- collapse of penstock shaft or outfall;
- drowning of worker or trespasser in return or polluted water dam or solution trench, or by falling down penstock shaft;
- dust blowing from deposit;
- erosion of berm on outer slope;
- failure of slope under-drainage;
- flow failure of outer slopes;

- liquefaction of outer slopes by seismic action or static liquefaction;
- overtopping of tailings storage dam;
- overtopping of return or polluted water dam;
- piping failure of outer slopes;
- piping failure of return or polluted water dam;
- piping along penstock outfall;
- piping into penstock tower;
- poisoning of worker or trespasser by drinking process or seepage water;
- pollution of groundwater by seepage from deposit;
- pollution of stream/river/wetland by seepage;
- shear failure of outer slopes;
- spontaneous combustion of waste.

2.9 Preliminary geotechnical characterization of waste

The properties of the waste together with the topography of the site, the required storage capacity and the geohydrological and geotechnical characteristics of the site determine the design requirements to a large extent. Hence the characteristics of the waste must be known or estimated before preliminary designs can be made to check the adequacy of a site. More often than not, if the waste storage is required for a new mine, no tailings will have been produced at the time this information is required. Alternatively, tailings will have only been produced at pilot plant scale and may not be representative of the tailings that will be produced by the production plant. Hence it is often necessary to rely on information on tailings produced at other mines from a similar ore. Obviously, the applicability of such data must be verified as soon as possible and the design and operation of the storage modified as necessary. If the waste storage is required to extend the waste storage capacity of an existing mine, the properties of the waste will either be known or it will be possible to determine them by sampling and testing current waste production or waste from an existing and current storage.

Properties of the waste that are generally required are

- bulk density of a coarse waste or settled density of a tailings slurry or thickened tailings;
- chemical analysis of the process water (including information on the presence of toxic substances, e.g. acids, cyanide or heavy metals);
- compaction characteristics (density-water content) for a coarse or semi-dry waste
- consolidation characteristics (C_c and c_v);
- mineralogy of the waste, particularly the presence and identification of clay minerals;
- particle size analysis (or grading);
- permeability or hydraulic conductivity;
- shear strength parameters (c' and Φ');
- viscosity-water content characteristic for a tailings slurry or thickened tailings.

The actual parameters that will be required depend on the type of waste storage (i.e. coarse waste, fine tailings or co-disposed coarse waste and fine tailings). Details of the testing required to establish the geotechnical properties and examples of typical properties and parameters will be given in Chapter 4.

2.10 Preliminary site investigation

This investigation is designed to enable site feasibility and suitability to be assessed. It also provides information for the EIA scoping (see Section 2.8.2), for the EIA and Environmental consequences of failure or failure to function safely.

2.10.1 *Extent and detail of investigation*

The ideal situation is that enough data should be gathered and analysed to ensure that additional exploratory work is unlikely to add significantly to the level of understanding considered necessary for the site under investigation. The extent and detail will also depend on the risk rating of the site and the findings of the environmental impact and risk assessment report. Because these two documents will be based on the preliminary site investigation, the ideal situation cannot occur in a single stage, and it is necessary to investigate in stages. A preliminary stage will suffice for the site selection and feasibility studies, but a further, more detailed and extensive stage will be needed for design, should the site be selected for use.

Four aspects should be covered by a site investigation report. These are the

- geology as revealed by the largest scale geological map available;
- physical geography, i.e. the observable surface and meteorological features associated with the site;
- near-surface features which have to be exposed by means of excavation or drilling before they can be assessed;
- miscellaneous issues, such as the geological setting, i.e. present or past surface or underground mining, seismicity or karst (cavernous limestone) and other relevant features associated with the site.

2.10.2 *Geology*

Geological maps show the rock formations underlying the site, below the surface soils. These are important as they often influence the type of soil developed on the site. Geological maps will also indicate the presence of features such as faults and dykes.

2.10.3 *Physical geography*

The investigation should cover not only the immediate area to be occupied by the proposed waste storage, but also the surrounding area to the extent that it is likely to be affected, or is likely to affect the storage. e.g., If blowing dust is a potential problem (a function of tailings characteristics, climate and prevailing wind intensity and direction) then features of the extent of land likely to be affected by the dust should be investigated. If there is an existing or planned water reservoir upstream or downstream of the site, its potential effect on, e.g., ground water levels around the waste storage and flooding in the event of its failure or the possibility of passing extreme floods without the flood affecting the storage, should be investigated, as well as the potential reciprocal effect of the waste storage on a downstream water reservoir.

Information on the area, proximity and percentage of surrounding area of each of

- built-up land or isolated dwellings or other buildings or structures;
- arable land;
- grazing land;
- wetlands, and;
- wilderness.

in the area likely to be affected by the waste storage is also required, as is information on

- historical or previous land productivity (in terms of crop tonnages per hectare or grazing capacity in head of livestock per hectare));
- evidence of land misuse or deterioration caused by natural processes, and
- existing structures, roads and other facilities.

2.10.4 *Topography and surface drainage*

- Appropriate topocadastral data must be provided in the form of a topographic map or maps. These must include all significant topographic features including surface contours. Most important are the drainage patterns, including seasonal and perennial streams and the distances to the nearest important water courses, wetlands and rivers. Rock outcrops and surface soil must also be recorded here, as well as the 1 in 50 year floodline, if there is one. Any evidence of surface instability should also be noted, as this could constitute a fatal flaw. For example Plate 2.1 shows surface instability on the surface of a slope. The series of parallel sub-horizontal scarps shows that the entire slope surface is unstable and moving periodically.
- Information on the amount of surface water associated with the site is also necessary. This can be provided as a map of the main catchment area in which the site is located which shows the boundaries of the catchment, the boundaries of the sub-catchments for the site and the route of drainage from the site. Also the mean annual run-offs for the main catchment and the sub-catchments for the site, as well as normal flows in affected streams and water-courses during dry weather. Also, the peak flood levels and volumes for return periods of 1 in 20, 1 in 50 and 1 in 100 years for the main and sub-catchments.
- Surface water quality – background water quality sampling will be required. Surface water quality must be determined by sampling both upstream and downstream of the proposed site. Analysis of the samples must be performed to the satisfaction of the regulatory authorities. These data will provide background information on surface water quality prior to any deposition of waste.
- Surface water usage – a survey must be conducted to assess the purpose for which the surface water is used as well as the quantities used and to assess the strategic or community value of the water body.

Plate 2.1 View of unstable surface of a natural slope.

2.10.5 *Infrastructure and man-made features*

Existing and planned regional infrastructure such as roads, railways or airports must be indicated. Of particular importance in the consideration of sites for waste storages are earthworks or other structures that affect the natural drainage system or whose removal could result in spoil suitable as a construction material. Elements such as landfills, sewage works, cemeteries or existing waste storages, may have specific exploration requirements.

2.10.6 *Climate*

Relevant climatic data must be provided to assist with water balance studies for the planned waste storage, flood studies, dust dispersion studies, etc. Records of monthly maximum and minimum temperatures, rainfall and pan evaporation, wind speed and direction (in the form of a wind rose), must be obtained from the nearest meteorological stations to the site.

More specifically, the following are required

- mean, maximum and minimum monthly and yearly precipitation for the site and the number of days in each month with measurable precipitation;
- maximum rainfall intensities per month for storms of 1 hour, 24 hours, 24 hours in 50 years and 24 hours in 100 years;

- mean monthly wind directions and speeds, as well as hourly wind directions and wind speeds with the maximum 1 minute gust per hour (usually in the form of monthly wind roses).
- mean, maximum and minimum monthly evaporation pan measurements.

2.10.7 *Flora and Fauna*

All existing vegetation on the site must be described and mapped, whether it be original indigenous vegetation or exotic vegetation, such as plantations, crops or fallow agricultural land.

For flora, information is usually required on

- dominant species;
- endangered or rare species;
- invader or exotic species.

For a virgin site, a vegetation map is required.

For fauna, information is required on the populations of

- dominant species, and
- endangered or rare species.

Most of the above information can be obtained from published or easily obtainable works, including topographic and other maps, orthophotos, reports and books, climatic statistics, existing air photos, the internet e.g. Google Earth, etc. Published information must, however be verified, updated and elaborated upon by on-site observation and examination. The latest available information must always be used. (e.g. the satellite and aerial photographs given by Google Earth are not always up to date).

2.10.8 *Stratigraphy and lithology*

Sub-surface features – Information on sub-surface features such as soil and rock profiles or ground water can usually be gained only by means of profiling boreholes or test pits.

Geophysical and remote sensing techniques can be used to guide the siting of test-pits and boreholes. These give initial insight into the geological and geohydrological characteristics of a site. Geophysical techniques e.g. resistivity and gravity surveys are particularly useful in the location of water-bearing or cut-off features such as dykes, faults and geological contacts. Thermal scanning can be used to locate near-surface subterranean caverns and potential sinkholes in karst areas, shallow undermining, etc.

Additional geophysical surveys, pump and recharge tests and even radioactive or isotope tracer studies might be required, particularly where sites are proposed in or close to karst areas or near strategic water resources.

The information required in this section is often available from published or existing geological maps and reports. It must, however, be supplemented in all cases with field data, comprising borehole logs or profiles. The site must first be described in terms of the regional geology. This indicates where it fits into the regional stratigraphy.

Thereafter, the stratigraphic and lithological features adjacent to and immediately beneath the site must be examined and described.

2.10.9 *Tectonics, lineaments and structures*

The presence and disposition of any geological faults, predominant joint and fracture dips and strikes and other linear features, resulting from the intrusion of dykes or from steeply dipping strata, must also be described and indicated on the maps and cross-sections referred to above. Appropriate photo interpretation, using the best available aerial photography should be undertaken and reported on where considered relevant.

2.10.10 *Soils*

Agricultural

A description is required of the surface soil types on the site as well as their fertility, erodability and depth following a locally or internationally recognized agricultural soil classification. The dry-land production as well as the irrigation potential of the surface soils must be evaluated and reported. If the site is used for waste storage, this soil can be stripped, stockpiled and used in the eventual rehabilitation of the storage surface.

Geotechnical

- Access to the first 2 m of the soil profile can usually be gained by test pits or trenches (properly shored if necessary). If greater depths are required, augering, wash-boring or sounding techniques, e.g. cone penetration or vane shear tests, may be used.
- Soil profiles must be carefully examined, described and recorded by a locally or internationally accepted method.
- In situ permeability and other geotechnical tests – any natural soil layer that is suitable for use as an aquitard to prevent water pollution should be tested for permeability. Examples of suitable test methods are borehole infiltration tests of various types, double ring infiltrometer tests and tests such as the Guelph in situ permeameter.
- Other tests may include soil indicator tests for the purpose of soil identification (particle size analysis, Atterberg limits and clay mineralogy), as well as compaction tests (standard Proctor) for future earth works.

It may be necessary to measure the shear strength of soils to establish or design for the stability of the outer slopes of the waste storage.

2.10.11 *Geohydrology*

All available geohydrological data and any restrictions affecting ground water use in the area must be identified and must form part of the site investigation report.

- Ground water morphology and flow – The depth of the ground water phreatic surface and its seasonal fluctuations, particularly the position of the wet season high elevation, including seasonal and perennial springs and seepages, as well as the presence of any perched water tables, must be determined. The gradient and general flow direction(s) of the ground water and other relevant data must be determined and possibly illustrated by appropriate maps and cross-sections. In addition, all significant geological features and inferred structures must be explored to determine the possible presence and importance of preferred ground water flow paths.

- Ground water quality – The background quality of the ground water, both up gradient and down gradient of the proposed site, must be determined prior to any waste deposition. A comparison of pre-deposition and post-deposition ground water quality will provide an indication of the impacts of the waste storage on ground water quality.
- Ground water usage – A census of existing boreholes and wells (a hydrocensus) must be conducted. Abstraction rates, yield, depth, age (by tritium test) and the purpose for which the water is used must also be obtained, with a view to assessing the strategic or community value of the water resource. A clear indication must be given of the perceived reliability of such census data and a definite distinction made between guesswork and factual information. Cognizance must also be taken of the source of the information, and this must be stated in the report.
- Sensitive sites – Where waste storages are considered or proposed in areas that are characterized by potentially strategic aquifers, or where ground water is, or may be used in the future, special caution must be exercised.
- Undermined areas – Existing underground mines must be identified, delineated and examined to establish the effect of their presence on ground water flows and potential subsidence.
- Seismicity –The risks and implications of natural or mining-induced seismicity must be addressed. If the waste storage is to be sited in an area where natural earthquakes are known to occur, the effect of the maximum credible earthquake must be taken into account in design of the storage and its appurtenant works.
- Rehabilitated open-cast mines – Open-cast mines associated with the site, whether rehabilitated or otherwise, must be identified, delineated and properly described.
- Potential for future mining – The possibility of future mining activities should be assessed.
- Sinkholes and surface subsidences – Areas where sinkholes or surface subsidences occur should have been avoided during the site selection process as these constitute fatal flaws.

2.10.12 *Air quality*

A survey of existing air quality at the site should be made, including existing sources of wind-blown dust and particulates from other sources, flue and other gases, existing odours, etc. In the case of mine waste storages, wind-blown dust (and possibly radon gas) will be the most likely form of air pollution, and the extent of dust generation and dispersal will depend on the waste characteristics, the mode of operation of the waste storage and the extent to which the waste surfaces are protected against wind erosion.

In every case, the impact of wind-blown dust should be considered to a distance of at least 10 km from the down-wind site boundary.

2.10.13 *Noise*

Present noise levels and noise sources around the site should be noted and if necessary measured as a record of the "pre-development" noise background.

2.11 Final site selection

The site is finally selected considering the information gained from the

- public consultation;
- a final site ranking matrix;
- the site feasibility study.

together with additional considerations not touched on here such as the pumping head for hydraulic fill tailings deposits, prevailing road gradients for truck haulage by road, etc. Numerical ranking schemes (e.g. Table 2.1) are also often useful in making the final choice of one of two feasible sites.

2.12 Examples of disastrous selection of sites

Many of the waste storage failures that have been listed in Table 11.1 would not have caused disasters if the storages had not been sited so as to become "disasters waiting to happen". Referring to the numbers listed in Table 11.1:

(2) The fly ash dump that failed at Jupille, Belgium, was sited immediately up-hill of the village that was devastated by fly ash flowing down into the village and killing 11 villagers.

(3) The El Cobre copper tailings storage in Chile which failed in an earthquake, killed 300 people when the tailings flowed downhill and buried a village.

(4) At Aberfan, in the U.K., a similar disaster happened when coal waste, stored on a hillside sloping at 12° and overlooking Aberfan village, failed and flowed into the village, demolishing the primary school and killing 144 people, including 116 children.

(6) At Buffalo Creek, U.S.A., coal waste had been dumped across a stream. When the stream came down in flood, overtopping the waste, 118 people were killed by the ensuing flood of coal waste.

(9) The Stava disaster in Italy, occurred when two tailings dams, sited on very steep ground with a slope of 12°, failed by overtopping and the resulting tailings flow killed 268 people in houses and villages downstream.

(12) The village of Merriespruit in South Africa had originally been sited uphill of the proposed tailings storage. To save money on the access road to the village, the sites of the tailings storage and the village were interchanged, and when the tailings dam failed, 17 deaths were caused in the village.

In every case, if thought had been given to the consequences of a failure, the waste deposit could have been sited so that if it failed, no lives or property out of the mine area would have been endangered.

References

Blight, G.E.: Management and Operational Background to Three Tailings Dam Failures in South Africa. In: W.A. Hustrulid, M.J. McCarter, D.J.A. van Zyl (eds): *Slope Stability in Mining*. Littleton, Colorado, U.S.A.: Society for Mining, Metallurgy and Exploration, Inc., ISBN 0-87335-194-0, 2000, pp. 383–390.

Canadian Dam Safety Association: *Dam Safety Guidelines*. Edmonton, Canada: Dam Safety Association, 1995.

Ho, K., Leroi, E., & Roberds, W.: Quantitative Risk Assessment: Application, Myths and Future Directions. Geoeng 2000. In: *Int. Conf. Geotech. Geol. Eng.* Lancaster, Pennsylvania, U.S.A.: Technomic Publishing, ISBN 1-58716-067-6, Vol.1, 2000, pp. 269–312.

U.S. National Research Council: *Improving Risk*. Communication Committee on Risk Perception and Communication. Washington D.C., U.S.A.: National Academy Press, 1989.

US National Committee on Tailings Dams 1994: Tailings Dam Incidents. In: R.C. Lo, E.J. Klohn (eds): *Design Against Tailings Dam Failure, Proc. Int. Symp. Seismic Environ. Aspects Dams Design*, Santiago, Chile, 1996, pp. 35–50.

Williams, S.: *Surviving the Volcano*. London, U.K.: Abacus, ISBN 0 349 11367X, 2001.

Wong, W.: How did that happen? In: *Engineering Safety and Reliability*. London, U.K.: Professional Engineering Publishing, 2005.

Chapter 3

Geotechnical exploration of sites for development of mine waste storages

Potential sites for waste storages should be explored in detail by qualified and experienced professional geotechnical engineers or by qualified professional engineering geologists who have specialized in work of this nature. The final decision on the suitability and feasibility of a site cannot be taken until subsurface conditions at the site have been fully explored and established.

The following information should be collected in respect of every potential site for a waste storage

3.1 Soil engineering survey

A detailed soil engineering survey should be made of the site in order to establish the different types of soil present and the areas and boundaries within which they occur.

- Typical soil profiles, depths to bed rock, the position of the ground water table, preliminary estimates of the permeability or impermeability of the soils and bed rock and their in situ shear strength should all be established.
- Any indications of the presence of dykes, faults, sills and other geological features which were not revealed on available geological maps should be carefully recorded.
- All potential problems with soils in the area should be noted (e.g. the presence of collapsing sands, karst, soft clays, dispersive soils).
- Because of the large surface area typically covered by waste storages (up to 1 000 ha or more), geophysical methods are attractive for site investigation and in particular as a means of extending the basic information provided by test pits or boreholes: In karst or undermined areas gravimetric surveys and thermal imaging are useful in locating possible sinkhole or subsidence areas which correspond to gravimetric lows or thermal contrasts. Microseismic methods are useful for locating the depth to bed rock. Resistivity surveys may be used to obtain information on water table depths.

3.2 Soil engineering data

The soil engineering information is most readily obtained by means of large diameter (about 1 metre) boreholes which can be drilled rapidly and economically by means of a pile hole auger. Holes dug by a back actor shovel may also be used.

However, safety considerations limit the depth of an unsupported hole to 1.5 m.

Whenever a person enters a test hole for the purpose of recording the soil profile, all recognized safety precautions must be taken, including supporting the sides of the hole to prevent possible collapse.

Particular care should be taken when entering a hole that

- has penetrated the water table and into which water is seeping;
- has stood open for more than eight hours since excavation;
- is possibly filled with poisonous or inert gas.

With regard to the latter point, holes dug in the vicinity of an existing waste storage or sanitary landfill should be regarded with particular suspicion, as they may fill with methane and carbon dioxide in the case of a landfill, or with carbon dioxide or other gases in the case of a waste storage. A grass or bush fire passing over the top of a hole may also cause it to fill with carbon dioxide and carbon monoxide, both of which are heavier than air.

Safety demands that no-one ever descends into a hole if he is alone, that the observer wear a hard hat and that he be linked to the surface by a rope fastened around his body by means of a harness designed to keep his head uppermost.

If large size exploration holes are not feasible, it may be necessary to explore the site by means of small diameter (75 mm, 100 mm or 150 mm) boreholes. The actual siting, distribution and number of holes will depend on the characteristics of a particular site and should be decided by the geotechnical engineer or engineering geologist on the basis of available knowledge of the local geology and from features visible at the surface. Because the outer slopes of waste storages are generally critical as far as the design is concerned, exploration boreholes will tend to be concentrated around the perimeter of the proposed storage area.

In addition to the geological information which is collected during the course of the site survey, the following information should also be obtained

- any evidence of local seismicity and the magnitude of possible seismic events in the area (available from the local or national geological or seismic survey department);
- mean and extreme rainfall distributions for the area as well as mean and extreme monthly pan evaporation (available from the local or national weather office);
- maximum rainfall intensities for each month for 1 hour, 24 hours, 24 hours in 50 years and 24 hours in 100 years storm events;
- mean monthly wind direction and speed and hourly wind direction and speed with the maximum one minute gust speed in each hour (a series of monthly wind roses is preferable, if available).

3.3 Detailed information for design of slopes & seepage control

The following detailed information is required to enable the slopes around the perimeter of the waste storage to be designed as well as to make provision for possible seepage control measures

- The permeability of the foundation soil or of particular strata in the foundation soil: The permeability is best measured by means of the in situ tests which are described in Section 6.4. Supplementary measurements of permeability may be made in the laboratory. A sufficient number of permeability or infiltrometer tests should be performed around the perimeter and covering the site of the residue deposit to ensure that representative permeability values are available for design purposes.

- The shear strength of the foundation soil: In order to design the stability of the slopes of a waste storage it is necessary to have information on the shear strength of the foundation material, and in particular the shear strength of any unusually weak strata. The shear strength may be measured using a number of in situ methods of testing of which the most suitable appear to be the vane shear test and the pressuremeter which are described in Sections 5.1 and 5.2. Alternatively, consolidated undrained or consolidated drained triaxial tests or shear box tests on undisturbed specimens taken from the field may be performed in order to establish the shear strength parameters of the foundation soil in terms of effective stresses. These tests are described in Section 4.5. Whichever methods are used, sufficient tests should be made to establish the shear strength properties of all the soil types that occur under the perimeter slopes of the waste storage. (Note: In situ test methods are most suitable in conditions of high water table where the imposition of a tailings deposit will not alter the degree of saturation of the soil to any great extent. Where water levels are deep (greater than 5 metres) and in dry climates where the building of a residue deposit is likely to have a pronounced local effect on moisture conditions in the soil, (a wetting of the soil) laboratory methods of measuring shear strength of undisturbed specimens taken from the foundation soil may be more suitable as the effect of changing the ground water regime can be simulated in these tests).
- In many cases it may be necessary to consider both short term and long term stability conditions for a waste storage, depending on the method by which the deposit is built, the speed of building, the characteristics of the foundation soil and other factors. If both long term and short term stability are to be considered it may be necessary to have information not only on the in situ strength of the foundation soil in terms of total stresses but also to measure the strength parameters in terms of effective stresses by means of triaxial shear or shear box tests in the laboratory. (See Section 4.5).
- The compressibility of the foundation strata is not usually relevant but with certain soils, e.g. loose, potentially collapsing sands, it may be necessary to estimate the settlement that will be undergone by the foundation strata in order to check that this will not cause distress to the slopes or drainage filters at the perimeter of the waste storage. If compressibility is thought to be a problem, the compressibility of the foundation strata or any particular stratum may be measured in the laboratory by means of standard oedometer tests. (See Sections 4.5.5 and 4.6).

3.4 Profile description

Because all soil sampling involves some degree of disturbance, it is necessary to distinguish facts related to the soil, from facts related to, or affected by, the method of inspection or sampling. The engineering properties of the material in its natural field state may be difficult to assess, except by including observations of the behaviour of excavation equipment, drills, or probes during the investigation (e.g. by the rate of advance of a drill, or rate of penetration of a driven probe).

Often, for legal purposes or because clients believe that they are getting more cost-effective information, factual reporting only is required. In practice, however, some degree of interpretation is essential. Whatever the case, reporting should always clearly distinguish between factual information and interpreted information. For example, classification information is always based on interpretation. It is good practice to indicate interpretation clearly on logging sheets, either by use of parentheses or by clearly marked sections for "interpretation".

3.4.1 *Profile description procedures*

Procedures for profile description and soil sampling have been developed by a wide variety of organizations, and are set out in various procedure manuals or codes of practice. Most of these procedures were developed many years ago and still represent acceptable recording procedures (e.g. Cook and Newill, 1988, Hvorslev, 1948, Jennings, Brink and Williams, 1973, Brink, 1979). Particular procedures may be adapted or extended for other purposes. There is a temptation to include detail for its own sake rather than for specific purposes. This detail may, however, become valuable later on, for reasons not earlier anticipated.

The following procedure list is not exhaustive, but can serve as a basis for most purposes. It may be supplemented or reduced according to requirements. The appropriate sections of locally applied codes of practice may also be substituted. The final choice for use of any such lists lies with the person responsible for the fieldwork.

3.4.2 *Site records*

The following should be recorded

- general description of site, general location, vegetation types and distribution across site, access routes (if remote from existing towns or villages);
- dates of site investigation;
- weather;
- precise location details (co-ordinates, marks,beacons, reference features);
- all field activities (diary, logging forms, equipment used).

Descriptions should be recorded for the following aspects of the soil:

- Moisture condition (M)
- Colour (C); (Best described based on standard colour charts e.g., the well-known Munsell soil colour charts. Multicolours, e.g. mottling, spotting, striping may also occur.)
- Consistency (C);
- Soil (e.g. clay, silt, sand, gravel, sandy silt, silty clay, etc.) (S);
- Structure or fabric (zoning, fissuring, slickensiding, cementing, quartz veins, nodules etc.) (S);
- Origin (transported or residual). If transported, likely transport agent, e.g. wind-blown, delta deposit, fluvial. If residual, parent rock, e.g. quartzite, granite. (O).

The mnemonic MCCSSO is useful to remember when recording data for a soil profile. Field notes must include entries under each of M, C, C, S, S and O. Field logs will include information from fieldwork alone. Therefore, descriptions of plasticity, moisture condition, etc. will be qualitative field descriptions. Subsequent laboratory testing may cause modification of the plasticity description, and allow inclusion of numerical moisture content values.

The soil profile must be observed in a freshly dug trial hole or inspection pit and will consist of layers which can be discerned by means of changes in moisture condition, colour, and consistency, the presence or absence of joints or fissures, or by changes in the grain size distribution.

A convenient procedure for observation is to fasten the zero end of a measuring tape at the surface on the northern side of the hole so that the observer can orientate himself as he observes the depth, thickness, direction of dip, etc. of the various layers, even when his head is below the soil surface.

As the observer descends the hole by means of a chain ladder or bosun's chair the moisture condition, colour, consistency, structure, soil type and origin of each layer of the soil profile (MCCSSO) should be observed and noted as well as the depth and thickness of each.

The observer can either note his observations in a field note book or a small voice recorder, or (a better, safer practice) can call up his observations to an assistant on the surface who records the observations, and prompts the observer to provide information under each of the categories MCCSSO. The assistant will also quickly become aware and act if the observer encounters any difficulties, e.g. a partial collapse of the hole, or a loss of consciousness or confusion as a result of encountering poisonous or asphyxiating gas.

(MCCSSO) Moisture condition

This is recorded as dry, slightly moist, moist, very moist or wet.

The moisture condition provides a useful indication of water requirements for compaction, should the soil be used in construction as fill. Dry and slightly moist materials will require additional water to attain the optimum moisture content for compaction. Moist materials are near the optimum moisture content while very moist soils will require drying. Wet soils are generally only found below the water table.

(MCCSSO) Colour

Colour is used for describing the soil and for identifying the same layer in different holes. Colours are difficult to describe so that the reader will "visualise" the same colour in his mind's eye as the observer describing the colour. It is useful to use a standardised soil colour chart such as the Burland (1958) chart. This is conveniently small in size and suitable for carrying in an overall pocket. Plate 3.1 shows the Burland chart which has the segments of colour on one side and the colour names on the reverse. The actual size is 80 mm × 80 mm.

The natural colour as seen in the freshly exposed soil, i.e. the colour 'in profile', should be noted, e.g. 'light grey mottled yellow', the predominant colour being noted first.

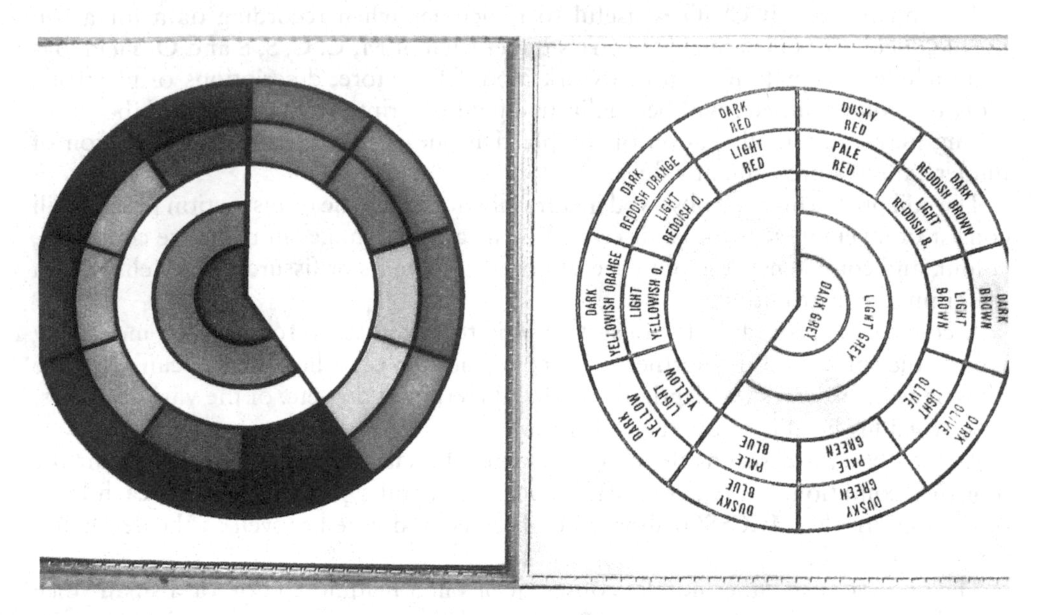

Plate 3.1 Burland's pocket colour chart.

(MCCSSO) Consistency

Consistency is a measure of the strength of the soil and is based on the effort required to dig into the soil with a geological pick, or to mould it in the fingers. Different sets of terms are used to describe the consistence of granular non-cohesive soils and clayey cohesive soils, as in Table 3.1.

(MCCSSO) Structure

'Intact' indicates an absence of fissures or joints.

'Fissured' indicates the presence of closed joints. The joint surfaces are frequently stained with iron and manganese oxides.

'Slickensided' indicates the presence of closed fissures which are polished or glossy and sometimes striated.

'Shattered' indicates the presence of open fissures. Shattered soils are usually in a dry state. Fissures will usually narrow or close when the moisture content of the soil increases.

'Stratified, laminated, foliated': Many residual soils show the structure of their parent rock. Observation of this structure may provide identification of the parent rock.

(MCCSSO) Soil type

The soil type is based on the predominant grain size.

'Boulders' are particles of rock larger than 100 mm size. The rock types and range of sizes should be recorded as should the presence of a matrix in the voids, if present.

Table 3.1 Categories of soil consistency

Consistency of			
Granular Soils		*Cohesive Soils*	
Very loose	Very easily penetrated by geological pick.	Very soft	Easily moulded by fingers. Pick head can easily be pushed in.
Loose	Small resistance to penetration by geological pick	Soft	Moulded by fingers with some effort. Sharp end of pick can be pushed in 30–40 mm
Medium dense	Considerable resistance to penetration by geological pick.	Firm	Very difficult to mould with fingers. Sharp end of pick can be pushed in 10 mm.
Dense	Very high resistance to penetration by geological pick.	Stiff	Cannot be moulded by fingers. Slight indentation by pick.
Very dense	High resistance to repeated blows of geological pick.	Very stiff	Slight indentation produced by blow of pick point.

The matrix should also be described. The shape of the boulders should be described as this often aids the interpretation of the origin. Terms are

'well-rounded' (nearly spherical)
'rounded' (tending to spherical)
'sub-rounded' (corners rounded off)
'sub-angular' (corners removed)
'angular' (corners sharp or irregular)

'Gravel' consists of particles of rock between 100 mm and 2.5 mm. The method of description follows that for boulders.

'Sand' consists of particles between 2.5 mm and 0.074 mm (100 mesh sieve) size. Most of the particles sizes are visible to the naked eye.

'Silt' consists of particles which are between 0.074 mm and 0.002 mm (2 μm) size. Individual particles cannot be distinguished by the naked eye, but the material feels gritty when rubbed against the teeth with the tongue.

'Clays' consists of particles smaller than 0.002 mm (2 μm) size. Clays feel greasy or soapy when wet.

Most natural soils have a combination of one or more particle size ranges. When describing such a soil, the adjective refers to the lesser size range, and the noun to the predominant size, e.g. a silty clay is a clay containing some silt. A silt-clay has approximately equal proportions of silt and clay and a clayey silt is a silt containing some clay.

(MCCSSO) Origin

An attempt should be made to determine the origin of the soil in each layer of the soil profile. On many sites a pebble marker, or layer of gravel indicates the boundary

Table 3.2 Possible origins of transported soils

Transported soil type	*Transport by*	*Usual soil type*
talus	gravity (from cliff face or upper steep slope)	unsorted angular gravel and boulders
hill wash	sheet wash (erosion from hillside)	boulders gravel sand clayey sand silt clayey silt clay
alluvium gulley wash	river stream gulley flow	gravel sand silt clay
lacustrine deposit	stream depositing in pan pond lake	sand silt clay
estuarine deposit	rivers and tides	sand silt clay
aeolian deposit	wind	fine sand*
littoral deposit	waves	medium and coarse sand

*Clayey or silty sand in ancient deposits that have partially decomposed in situ.

between transported and residual soils. This generally represents a stratum of free drainage which, if drainage is required, may be retained and usefully employed for providing a free flow of water. It also indicates a level below which soil behaviour may be predicted from experience with similar decomposed rock types.

Transported soils

Possible origins of transported soils are as shown in Table 3.2, generally ranked from highest to lowest topographic elevation of occurrence.

Residual soils

A knowledge of the local geology and reference to geological maps, provides a guide to the origin of residual soils on any site. Residual soils may be recognized by the preservation of the primary rock structures inherited or relict from the parent rock, e.g. bedding planes, amygdales or characteristic rock jointing structures.

It is sometimes possible to identify primary and secondary minerals characterizing the mineralogical composition of their parent rock, e.g. residual soils derived from granite will contain quartz grains, mica flakes and kaolinite derived from the decomposition of feldspars. Those formed from rocks such as dolerite or diabase, will contain no quartz (other than vein quartz) and will consist entirely of clay and silt. Amygdales in residual soil will identify its derivation from a volcanic lava such as andesite or basalt. Residual soils developed from sedimentary parent rocks are usually easy to identify from their inherited or relict structure, e.g. bedding, particle size distribution across bedding layers, etc.

3.4.3 *Recording of soil profiles*

The left hand side of the profile record should consist of a section, with the depth drawn to scale, which shows the various strata of the profile. Soil types and structure are indicated by the standard symbols shown in Figure 3.1. The section should also record the presence of such features as water tables and the pebble marker as well as the full depth of penetration of the hole.

To the right of each stratum in the section, a full description (MCCSSO) should be recorded, as shown in Figure 3.2.

Water table

It is important to establish and record the depth of the water table or phreatic surface on the virgin site. If there are any indications of the presence of a perched water table, this should be noted and an attempt made to establish its depth as well. The depth of a water table is indicated by the level at which water trickles into the hole. However this level may be that of a perched water table, not the main water table. Also, in soils of low permeability, the flow of water towards the hole may be too slow to show as a trickle. It is therefore always best to install a stand-pipe piezometer in the profile inspection hole before back-filling it. The stand-pipe could be as simple as a length of plastic electrical conduit with the lower end perforated with 3 mm diameter holes over the bottom half metre and the perforated length wrapped in needle-punched felt geofabric. If possible, the hole around the perforated end should be backfilled with clean sand or gravel before back-filling and compacting the rest of the hole with the spoil that came out when digging or drilling it.

It may take several days or even a week for the water level in the piezometer to stabilize. The water level in the stand-pipe can easily be established by pushing a length of small-bore plastic tubing down the standpipe to the bottom. If there is water in the tube, the sound of bubbling will be heard when the tube is blown down by mouth. By blowing and slowly withdrawing the tube, the level of the water can be established as the point at which the bubbling sound stops. The water table depth is then the length of small-bore tube remaining in the standpipe, measured to ground level.

3.5 Simple in situ tests and soil sampling

It is usually convenient to augment the visual and qualitative soil assessment described in section 3.4 by simple in situ tests that can be carried out in the test hole at the same

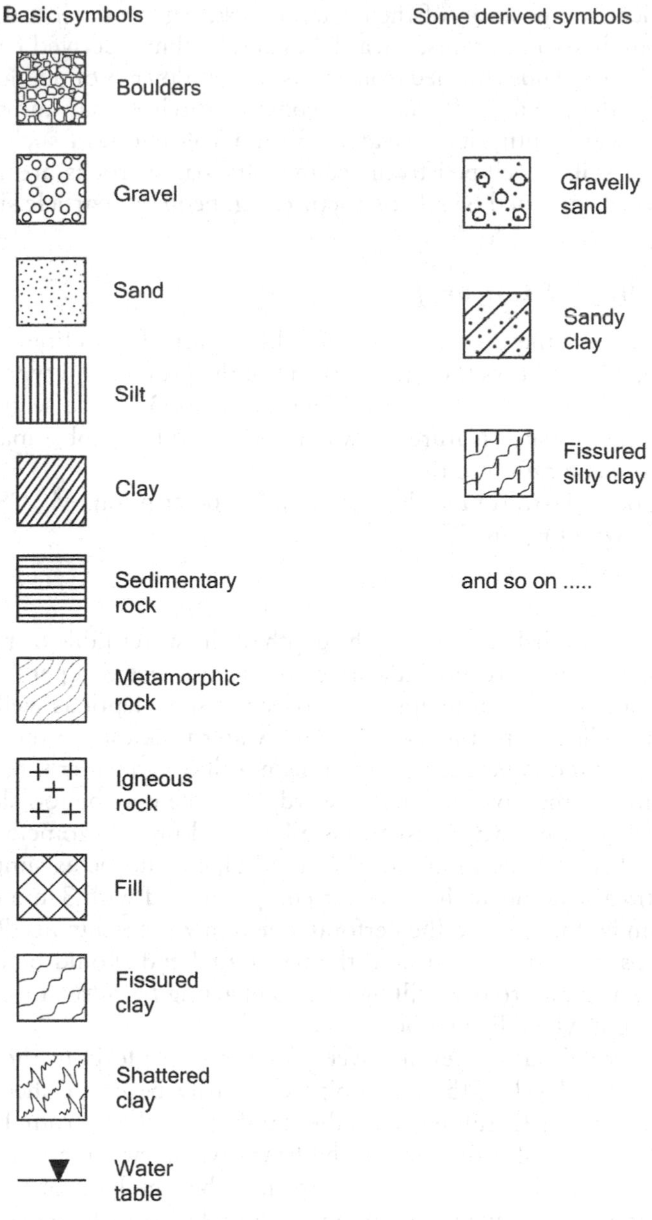

Basic symbols

Boulders

Gravel

Sand

Silt

Clay

Sedimentary rock

Metamorphic rock

Igneous rock

Fill

Fissured clay

Shattered clay

Water table

Some derived symbols

Gravelly sand

Sandy clay

Fissured silty clay

and so on .....

Figure 3.1 Symbols used for recording soil profiles.

time as the visual assessment. The hand held penetrometer and hand vane instruments are particularly useful for this purpose.

Sampling can also be carried out to obtain material for index and compaction testing. These are usually disturbed samples, taken by digging out sufficient soil for the purpose of the tests from the sides of the test holes at appropriate depths. The quantity required

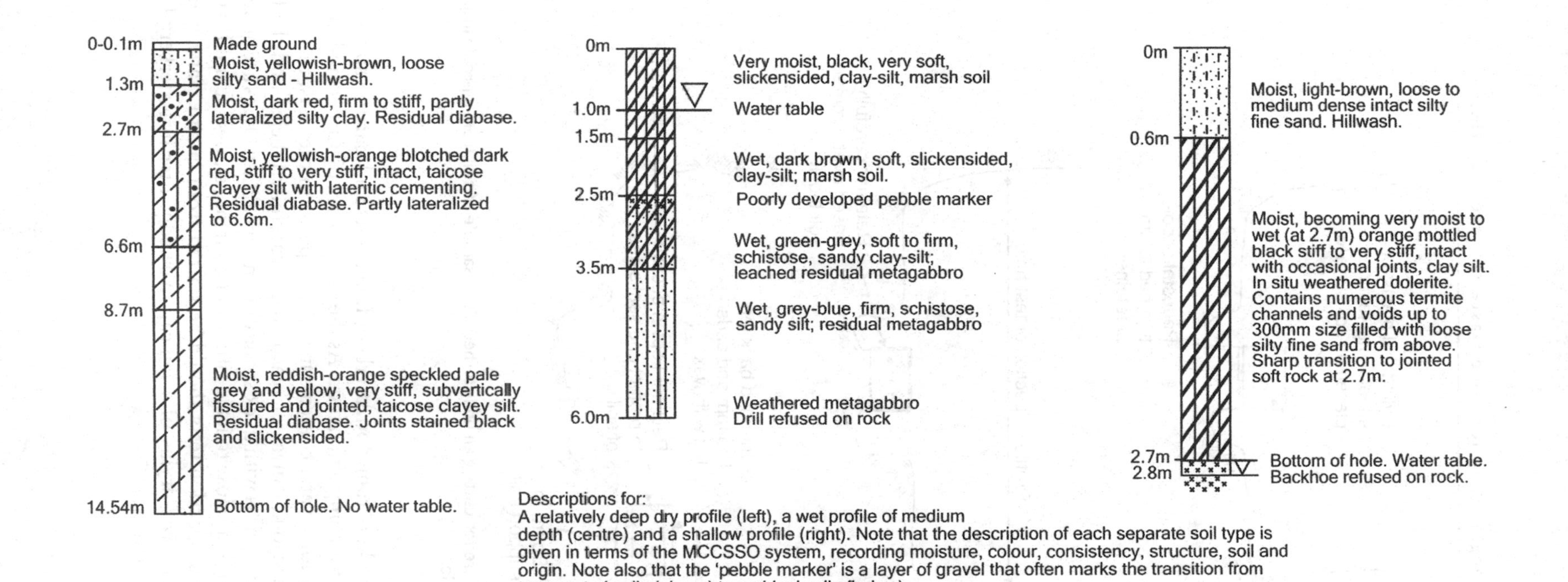

Figure 3.2 Examples of three soil profile records.

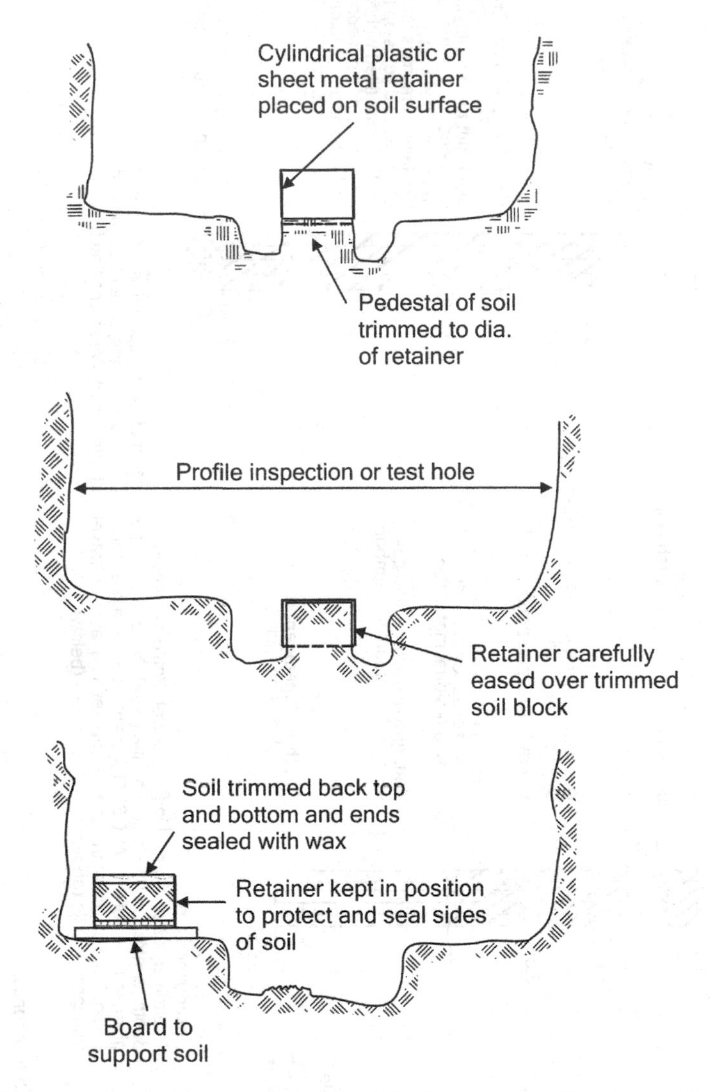

Figure 3.3 Procedure for cutting an undisturbed block sample of soil from a profile inspection or test hole.

will vary from a few kilograms for Atterberg Limit tests and particle size analyses to 20 to 30 kilograms for compaction tests. As the samples are disturbed and the soil will be remoulded at various water contents in order to perform the tests, the samples are usually collected in strong canvas or plastic bags, carefully labeled and the bags are closed by tying their mouths with thick string or thin rope. Plastic bags are preferable as the soil should be kept as close to its in situ water content as possible until it is tested. All Atterberg and compaction tests should be started from in situ water content, as air-drying or (worse) oven-drying of a soil can markedly alter its Atterberg limit and compaction characteristics.

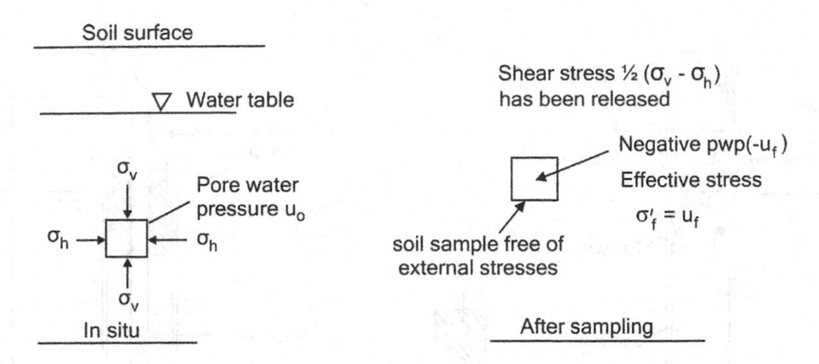

Figure 3.4 Changes in stress on a soil below the water table caused by sampling.

It may also be convenient to take undisturbed block samples for laboratory testing to establish shear strength and compressibility parameters. Block samples can only, of course, be taken above the water table. To do this, it is convenient to trim a pedestal of soil out of the side or bottom of the test pit (depending on available space) to fit an open-ended cylinder (e.g. a 200 mm or 300 mm diameter by 200 or 300 mm long piece of thin-walled plastic or thin steel pipe). The cylinder is worked down over the soil pedestal by trimming away excess soil. Once the cylinder fits over the soil pedestal, the pedestal is cut off at its base and excess soil trimmed away. The ends of the cylinder can then be sealed with purpose-made caps, by means of melted paraffin wax, or even by wrapping in several layers of "cling-wrap" plastic wrapping.

3.6 Taking undisturbed soil samples for laboratory testing

The purpose of taking "undisturbed" samples is to subject the soil in the laboratory to the stress changes and drainage conditions it is envisaged will be applied in the field by the prototype structure, i.e. the soil sample is a model of the soil in the field. Therefore it must represent the in situ soil as closely as possible in structure, void ratio and water content. Sampling procedures must be suited to the material being sampled.

3.6.1 *Sampling of saturated clay soils below the water table*

The relationship between void ratio e, water content w and particle relative density G for a saturated soil is:

$$e = wG$$

Hence a saturated soil is incompressible (e is constant) provided the water content does not change. In a clayey soil, removal of a sample from below the water table to the surface should not result in a change of water content as the soil will retain the water by capillarity. The soil will, however, undergo a distortion as a result of the release of the in situ shear stress, as illustrated by Figure 3.4.

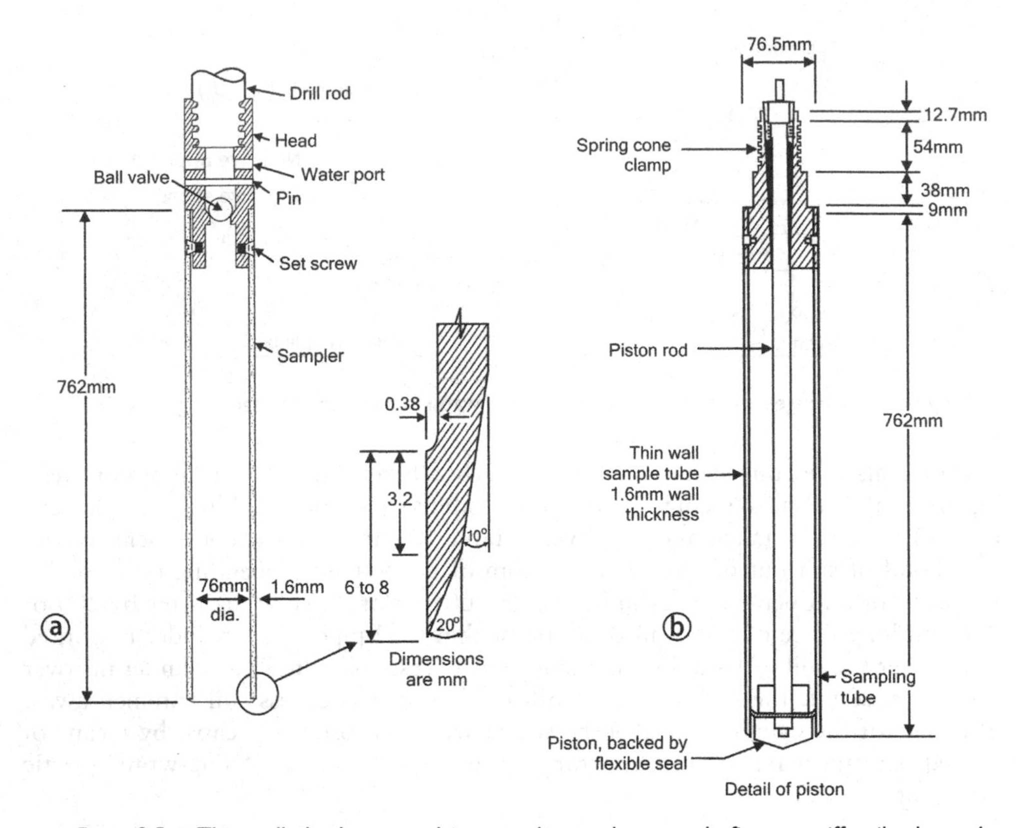

Figure 3.5 a: Thin-walled tube open-drive sampler used to sample firm to stiff soils above the water table. b: Thin-walled stationary piston sampler, suitable for soft soils and soils below the water table.

3.6.2 *Sampling firm to stiff saturated clays above the water table*

The usual method is to use a 76 mm internal diameter thin-walled open-drive sampler like that illustrated in Figure 3.5a. The sample is taken by hydraulically pushing the sampler into the soil at the bottom of an augered borehole. Care must be taken not to "overdrive" the sampler, i.e. not to compress the soil by pushing in the sampler further than the length of empty sampler tube. The drill rods are then twisted to shear off the sample at its base. The sampler is withdrawn, and the sample extruded carefully into a plastic film tube, placed in an appropriately marked and identified cardboard tube and waxed to seal and support it during transport to the laboratory and storage before testing. The soil must be extruded from the sample tube in the same direction as it moved into the tube, i.e. first in, first out.

3.6.3 *Sampling soft saturated clays and silts*

When an open-drive sampler is pushed into the soil, the sample is released from the side restriction provided by the surrounding soil, and tends to expand during the initial stages of penetration. However, as the tube is pushed in further, side friction builds up

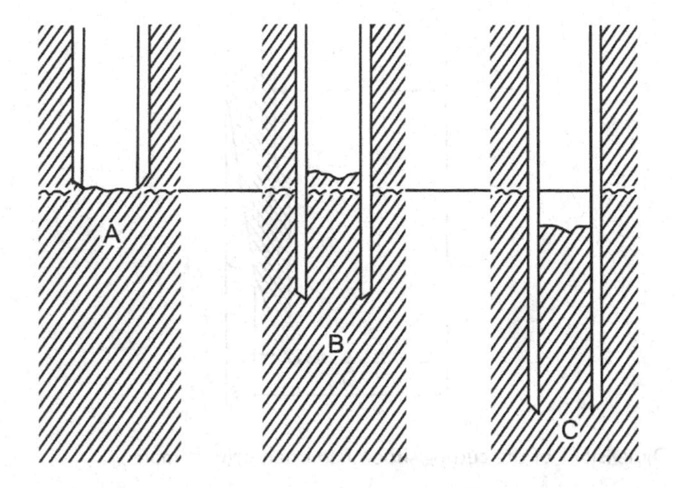

Figure 3.6 Effects of pushing an open-drive sampling tube into an undisturbed soil.

inside the tube and the sample is then slightly compressed, as illustrated in Figure 3.6 by the progression A, B, C. To prevent the disturbance caused by these changes in lengths, piston samplers, as shown diagrammatically in Figure 3.5(b) are used. The piston is initially flush with the cutting edge. Rods connect the piston to the surface: When the sampler is ready for sampling the piston rests on the surface of the soil to be sampled. The piston is then clamped in position and the sampling tube is pushed past it. Thus, while the sampling tube penetrates into the soil the piston remains stationary at the original level of the top of the sample, ensuring that the length of the sample cannot change during sampling. An automatic locking device or spring cone clamp is incorporated in the sampler head so that the piston will not be forced into the empty sampler tube by hydrostatic pressure before driving and cannot slip down while the sampler is being pulled up after the clamp at the surface has been released. This procedure is known as fixed piston sampling.

Another system sometimes used is that of a free piston. In this case the piston is pushed forward but the rods are not connected to the surface. This means that, when the sampling tube is pushed down, the piston "floats" on the top of the sample. When the sampler is withdrawn the piston automatically locks and the sample is retained in the tube.

Although the piston sampler was developed for sampling soft clays, it has also been used successfully to sample soft saturated silts, in particular mine tailings below the phreatic surface.

3.6.4 *Important soil sampler dimensions and their function*

Open-drive and piston sample tubes have a number of characteristics that can be quantified and are of importance in obtaining good undisturbed samples. These are:

The amount of disturbance caused to the soil in the sampling operation depends to a large degree on the "area ratio" between the cross-sectional area of the metal annulus

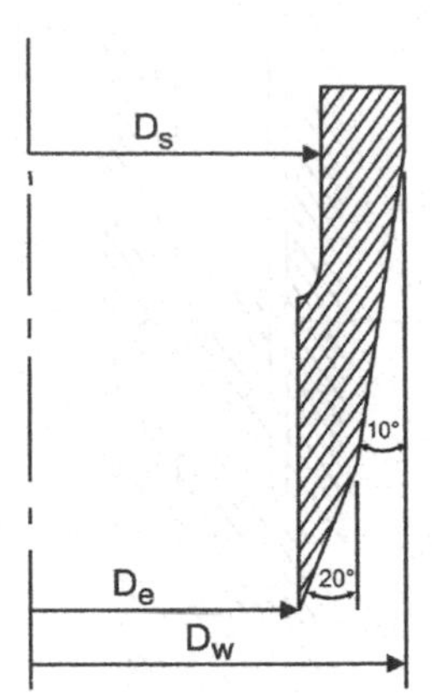

Figure 3.7 Dimensions of the cutting shoe of an open-drive sampler.

of the sampling tube and the cross-sectional area of the sample. For the entrance to the sampling tube shown in Figure 3.7 the area ratio C_a is

$$C_a = (D_w^2 - D_e^2)/D_e^2 \leq 10\%$$

It has been shown that, for large samplers, the disturbance due to displacement of soil by the tube itself is very small with an area ratio of 10% (Hvorslev, 1948). To achieve such a low area ratio the sampling tubes are made from thin-walled, seamless, high strength steel tubing.

Another cause of disturbance is friction on the inside of the tube, which also limits the length of sample which can be taken with a minimum of disturbance. This internal friction can be reduced by providing the sampling tube with a smooth interior surface having a low coefficient of friction and by making the diameter of the cutting edge D_e slightly smaller than the inside diameter of the sampling tube D_s. This clearance is expressed by the inside clearance ratio,

$$C_i = (D_s - D_e)/D_e = 0.5\% \text{ to } 1.5\%$$

The Recovery Ratio RR is defined by

$$RR = \text{Length of sample/Distance sampler was driven}$$

The Recovery Ratio should lie within 97% to 103% for satisfactory samples. Samples with RR's outside the limits of 95% to 105% should be rejected for strength or compressibility tests.

References

Brink, A.B.A.: *Engineering Geology of Southern Africa, 5 Vols.* Pretoria, South Africa: Building Publications, ISBN 0 908423 04 7, 1979.

Burland, J.B.: *A Simple Soil Colour Chart for Soil Profiling Purposes.* Final year project, Department of Civil Engineering. Johannesburg, South Africa: University of the Witwatersrand, 1958.

Cook, J.R., & Newill, D: The Field Description and Identification of Tropical Residual Soils. 2nd *Int. Conf. Geomech. Tropical Soils, Singapore*, Vol. 1, 1988, pp. 3–10.

Hvorslev, M.J.: *Subsurface Exploration and Sampling of Soils for Civil Engineering Purposes.* Vicksburg, Miss., U.S.A.: Waterways Experiment Station, 1948.

Jennings, J.E., Brink, A.B.A., & Williams, A.A.B.: Revised Guide to Soil Profiling for Civil Engineering Purposes in Southern Africa. *The Civil Engineer in South Africa Jan* (1973), pp. 3–12.

Chapter 4

Environmental and engineering characteristics of mine waste, including stress and strain analysis and laboratory shear testing

4.1 Characteristics having environmental impact

The following information should be ascertained concerning those characteristics of the waste that may have environmental effects:

- toxicity of the waste, e.g. its content of fluorine, arsenic, lead, nickel or similar toxic or heavy metal salts;
- possible radioactivity of the waste and other potential health hazards, e.g. free asbestos content;
- the soluble solids content of the waste and its acidity or alkalinity as it leaves the mill;
- the effect of weathering and oxidation on the above. (For example, insoluble sulphides may be oxidized to produce soluble sulphates and sulphuric acid.)
- The possibility of spontaneous combustion of the waste.
- The possibility of the soluble chemical content of the waste precipitating out in the underdrain system of the storage, e.g. by the precipitation of insoluble Fe_2O_3 gels in the drainage filters or adversely affecting the properties of pond linings. (e.g. Clay may become rigid and brittle when in contact with calcium hydroxide or phosphoric acid, or a sodium bentonite may be converted to a less impermeable calcium bentonite.)
- Decomposition of certain contents of the waste may produce radioactive combustible or toxic gases, e.g. radon, or methane which will emanate from the storage.
- If the waste has adverse chemical properties, it may be possible by a simple pretreatment to improve or eliminate these characteristics. For example, an initially acid waste could be neutralized by adding lime prior to deposition.

Testing to establish chemical characteristics does not fall within the scope of this book, but falls within the field of geochemistry. However, all of the above characteristics may influence the physical design of the waste storage and the designer should be aware of any potential problems before starting the engineering testing necessary for the design.

4.2 Engineering characteristics

The following information on the engineering characteristics of the waste should be obtained:

4.2.1 *The particle size analysis of the waste*

In this regard the possibility of separating the waste into coarse and fine fractions should be considered. If it is possible to do this relatively cheaply (for example by cycloning) it may be possible to use the coarse fraction of the waste to build the outer slopes of the deposit while depositing the finer less free-draining fraction of the waste within this outer retaining embankment. The particle size distributions of wastes may vary widely from boulder-sizes in the case of waste rock, down to clay-sizes in the case of kimberlite tailings or other tailings derived from weathered rock. Another possibility is to consider combining a coarse waste with a fine waste and co-disposing of the mixture in a single operation. An example could be the co-disposal of coarse coal discards with fine coal washery sludge, both produced by the same colliery.

Figure 1.3 summarized the particle size analyses for a number of tailings from the extraction of various mineral products. These analyses are only intended to illustrate the range of types of grading found in tailings. The actual particle size distribution for a waste will depend on factors such as the fineness to which the ore is milled, the mineralogy and degree of weathering of the ore, the type of milling process, the separation or extraction process, etc. etc. Gradings may be very variable both at a single mine and from mine to mine, as well as being variable on a single tailings storage. For example, Figure 4.1 compares the variability of particle size analyses for gold tailings from mines in the Witwatersrand of South Africa, within the height of a single tailings storage and from storage to storage for 12 different mines. This illustrates the possible variability of tailings from one generic type of ore.

4.2.2 *The permeability of the whole material and of its coarse and fine fractions*

If an existing storage of the waste is available, the permeability is best measured by in situ means (Chapter 6). If, however, the waste storage is a new one, laboratory permeability measurements should be carried out on waste obtained from pilot plant operations. If it is planned to deposit the waste hydraulically and there is a possibility of stratification of the deposit occurring as a result of the hydraulic deposition as well as a variation of grading down the length of the hydraulic beach, laboratory measurements of the permeability parallel and normal to the stratification should be performed, as the difference between these two coefficients will influence not only the shape of the seepage or phreatic surface at the perimeter of the waste storage, but also possible seepage losses through the floor (or footprint) of the storage.

4.2.3 *The shear strength characteristics of the waste*

These may either be established in situ by means of the vane shear test, the pressuremeter test, the cone penetrometer or piezocone or other forms of in situ test, if an existing deposit of the waste is available for this purpose. Otherwise, they can be

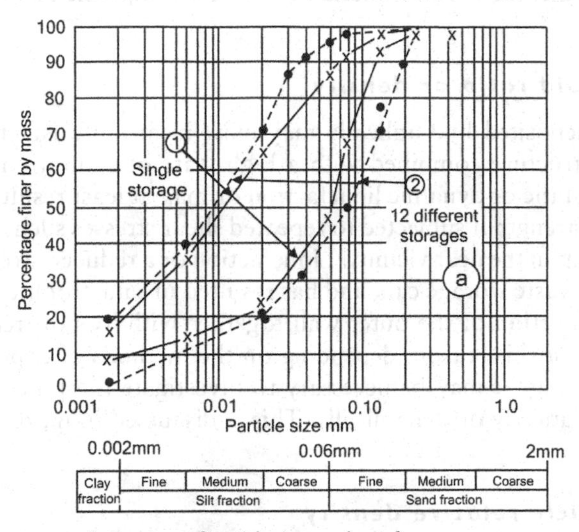

a: Comparison of particle size envelopes for
1. variation within the height of a single tailings storage
2. variation from mine to mine for 12 different tailings storages.

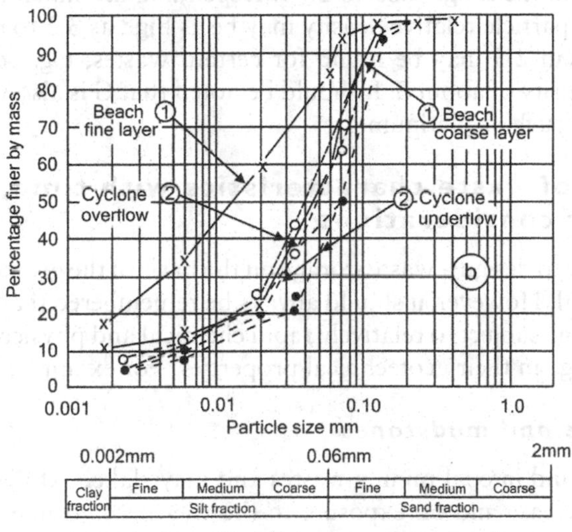

b: Comparison of particle size separation produced by
1. gravitational sorting on tailings beach
2. cyclone separation.

Figure 4.1 Variability of particle sizes within a milled quartzite gold tailings.

established by means of consolidated drained or undrained laboratory triaxial shear tests or shear box tests. The coefficient of consolidation of the waste affects the rate at which the storage may safely be built as this parameter controls the rate of drainage of the waste and hence its rate of gain of strength. The coefficient may be measured in the laboratory or (semi-empirically) in situ by means of the piezocone test. The theory of consolidation and methods of consolidation testing are explained later in Section 4.6.2 of this chapter.

4.2.4 *In situ void ratio or density*

Wastes that are deposited hydraulically may exist in an unsatisfactorily loose state in situ. A loose structure combined with a high state of saturation may render the material liable to static or dynamic liquefaction or may at least result in a very significant reduction in strength if subjected to repeated shear stresses such as those resulting from a seismic event in the near vicinity. Liquefaction or a reduced strength of the outer embankment of a waste storage can, and has resulted in catastrophic failure, or severe settlement and distortion of the outer wall together with loss of freeboard followed by over-topping. For this reason, depending on the methods of deposition and management of the storage it may be necessary to investigate the potential of the waste to liquefy either statically or dynamically. This is discussed in more detail in sections 4.5.8 and 4.5.9.

4.2.5 *The particle relative density*

The particle relative density G has an influence on the density or unit weight of the waste and hence on volume/mass calculations. For many wastes the particle relative density has values in the range 2.65 to 2.75. Depending on mineralogy (e.g. for platinum tailings), the particle relative density may be as high as 3.5 to 4. Particle relative densities of less than 2.5 may be found for certain wastes, e.g. coal wastes have a particle relative density of about 2. It should be noted that G is also variable in a single generic type of ore such as platinum ore.

4.3 Changes of waste characteristics with time, and other considerations

It is often necessary to design a waste storage on the basis of the geotechnical properties of the fresh material. However, it should always be remembered that, after deposition, many wastes become subject to relatively rapid chemical and physical weathering with a consequent change in their geotechnical properties. For example:

4.3.1 *Siltstones and mudstones*

These are often found in coal-mining wastes and may slake, exfoliate and expand as a result of stress release and on exposure to the weather and may relatively rapidly revert to their original unconsolidated state as silty clays and clays. Hence an initially coarse, porous and free-draining material may, within a few years, deteriorate to a fine-grained impervious material which is not free-draining and has a reduced shear strength. Severe settlement and piping erosion may occur as the lumps of rock slake and disintegrate, opening up passages for water flow.

4.3.2 *Partly weathered igneous rocks*

Rocks such as dolerite and kimberlite may also undergo rapid deterioration on stress release and exposure to the weather and change from pervious granular materials to impervious clayey materials within a few years.

Plate 4.1 Burning coal discard dump.

4.3.3 *Coal wastes*

These are also subject to weathering. Particles of siltstone and mudstone will deteriorate as described above. Pyrite, often contained by the waste, will weather. Unless precautions are taken, the waste may become subject to spontaneous combustion. Burning may completely alter the waste's geotechnical properties. In general, clay minerals will be altered to stable fused or semi-fused materials and the geotechnical properties will be considerably improved. This, however, may be at the expense of unacceptable air and water pollution, and possibly uneven settlement and the formation of depressions and dangerous sink-holes in the storage slopes and top surface. Plate 4.1 is a view of a burning coal discard dump.

4.3.4 *Wastes containing metallic sulphides (e.g. pyrite or chalcopyrite)*

These will change with time, as a result of weathering and oxidation of the sulphide content and disintegration of shaly components. The frictional characteristics of waste rock will probably deteriorate with time as the shale content disintegrates. Gold mine waste, usually alkaline when deposited, may become acidic as its sulphide content weathers. If shaly material has been milled, the shale may deteriorate to clay which will adversely affect the geotechnical properties of the waste over time.

4.3.5 *Waste samples from initial stages of mining*

The initial design of a waste storage is often based on the properties of material obtained from a metallurgical pilot plant. This may not resemble the waste eventually produced from the proto-type plant. Also, initial samples may consist of weathered, near surface ore. As mining goes deeper, unweathered ore with very different geotechnical properties, either more or less favourable, may be produced. There is always the possibility that the metallurgical process will change with improving extraction technology, and in so doing, change the properties of the tailings.

4.3.6 *Other considerations*

Although the following considerations would probably not affect the design details of a waste storage they should be taken into account as they may have an overall influence on the siting, shaping and dimensions of the storage.

- Most wastes must be regarded not as waste, but as low-grade ore. There is therefore always a possibility that the storage will be reclaimed or re-mined for the extraction of different minerals or residual minerals. For example, after copper extraction, tailings of the copper ore mined at Phalaborwa in South Africa, which is also rich in phosphates, is used to produce phosphoric acid for agricultural fertilizer manufacture. Waste gypsum arising from phosphoric acid manufacture may be used to produce building products or even to reclaim sulfur.
- It may be possible that the storage will be reclaimed for use as underground fill or to refill opencast surface workings. This is happening at present in South Africa.
- There may be a possible use for certain wastes as fill for civil engineering applications, e.g. in road or airfield construction or as concrete or road aggregate. Almost all of the waste quartzite rock produced in over a century of gold mining in South Africa's Witwatersrand area, has been used as concrete and road aggregate. It is now very scarce and has become a highly sought after construction material.
- It may be possible to rehabilitate the waste storage for agricultural or recreational after-use.
- There may be a possibility that the area could be used in the future as a site for residential housing or industrial development. For example, Plate 1.3 shows a commercial building constructed on the side slope of an abandoned gold tailings storage.

All of the above have occurred in various parts of the world in the past, and are even more likely to occur in the future. If any of these possibilities exist it is worthwhile recognizing them at an early stage and incorporating into the design of the waste storage such features as will facilitate this possible after-use, especially if this can be done without additional cost and with negligible effect on the normal operation of the waste storage.

However, it must be realized that certain wastes contain components that are toxic or may become toxic in time, or are radioactive, and thus cannot be used in the ways suggested. After-use of certain waste storages may not be feasible except at a cost considered excessive. As an example, measurements taken in the basement of the

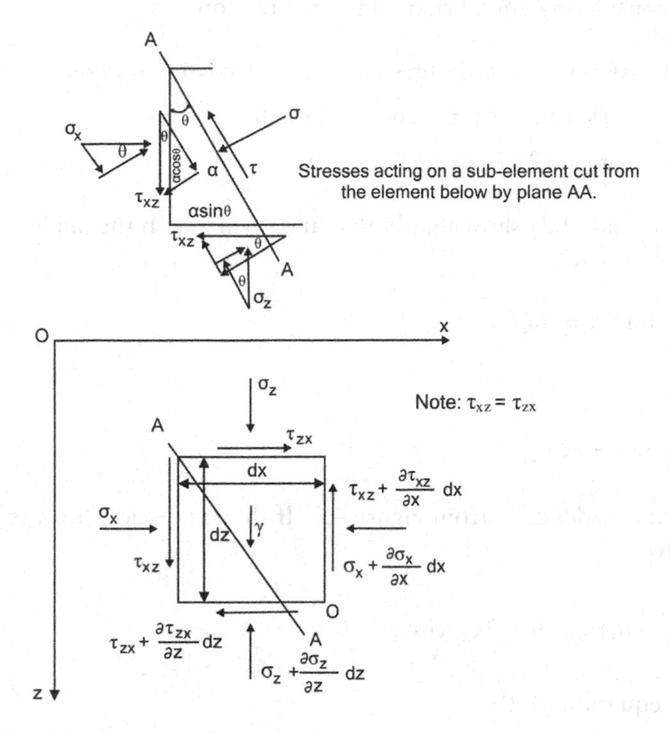

Figure 4.2 Stresses acting on an element of particulate material in a waste storage.

building illustrated in Plate 1.3 showed that it was accumulating radon gas, and a forced ventilation system had to be installed to make the space safe.

4.4 Analysis of stresses and strains and the principle of effective stress

4.4.1 *Principal stresses*

Figure 4.2 shows (above) a sub-element cut from the larger element below by a plane AA inclined at angle θ to the vertical. The vertical and horizontal shear stresses are $τ_{xz} = τ_{zx}$ and the element is held in equilibrium by the inclined direct stress σ and the corresponding shear stress τ acting on the inclined face. For force equilibrium in the σ-direction:

$$σ.α = σ_x α \cos θ \cos θ + σ_z α \sin θ \sin θ - τ_{xz} α \sin θ \cos θ - τ_{xz} α \cos θ \sin θ$$

$$σ = σ_x \cos^2 θ + σ_z \sin^2 θ - 2τ_{xz} \sin θ \cos θ$$

Following some manipulation, this can be written:

$$σ = ½(σ_x + σ_z) + ½(σ_x - σ_z) \cos 2θ - τ_{xz} \sin 2θ \qquad (4.1)$$

Similarly, by considering equilibrium in the τ-direction:

$$\tau\alpha = \sigma_x\alpha\cos\theta\sin\theta - \sigma_z\alpha\sin\theta\cos\theta + \tau_{xz}\alpha\cos\theta\cos\theta - \tau_{xz}\alpha\sin\theta\sin\theta$$

$$\tau = (\sigma_x - \sigma_z)\sin\theta\cos\theta + \tau_{xz}(\cos^2\theta - \sin^2\theta)$$

$$\tau = ½(\sigma_x - \sigma_z)\sin 2\theta + \tau_{xz}\cos 2\theta \quad (4.2)$$

Equations (4.1) and (4.2) show that both σ and τ vary with the angle θ.
For example, if $\theta = 0$,

$$\sigma = \sigma_x \quad \text{and} \quad \tau = \tau_{xz}$$

If $\theta = 90°$,

$$\sigma = \sigma_z \quad \text{and} \quad \tau = \tau_{xz}$$

Both results can be deduced from Figure 4.2. If the expression for σ is differentiated with respect to θ,

$$\frac{d\sigma}{d\theta} = (\sigma_x - \sigma_z)\sin 2\theta + 2\tau_{xy}\cos 2\theta$$

That is, from equation (4.2):

$$\frac{d\sigma}{d\theta} = 2\tau$$

Maximum and minimum values for σ will occur when $\tau = 0$, i.e. when

$$\tan 2\theta = 2\tau_{xz}/(\sigma_z - \sigma_x) \quad (4.3)$$

The maximum and minimum values for σ are known as the major and minor principal stresses, respectively, and are denoted σ_1 and σ_3. (In a three-dimensional stress system, the third stress is called the intermediate principal stress, denoted σ_2.)

It follows from equation (4.2) that the maximum value for τ will occur when $\theta = 45°$ ($2\theta = 90°$) and that the maximum value of τ will be

$$\tau_{max} = ½(\sigma_1 - \sigma_3) \quad (4.4)$$

Finally, it also follows that if the orientation of the x and z axes in Figure 4.2 is chosen so that $\sigma_z = \sigma_1$ and $\sigma_x = \sigma_3$, i.e. so that x and z become the principal axes 1 and 3:

$$\sigma = ½(\sigma_1 + \sigma_3) + ½(\sigma_1 - \sigma_3)\cos 2\theta \quad (4.5a)$$

and

$$\tau = ½(\sigma_1 - \sigma_3)\sin 2\theta \quad (4.5b)$$

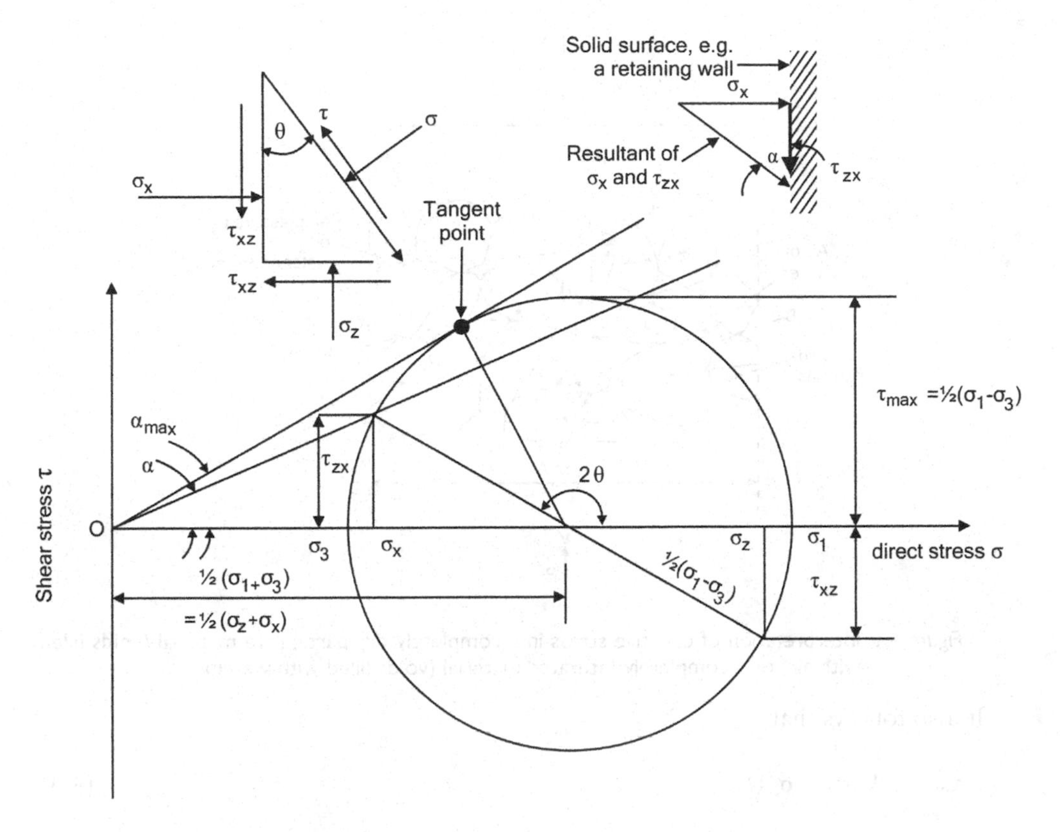

Figure 4.3 The Möhr stress circle.

4.4.2 *The Mohr stress diagram*

The stress equations set out above can be represented graphically by a very useful construction known as the Mohr stress circle or stress diagram, which is illustrated in Figure 4.3. In this diagram, direct stresses (σ, σ_z, etc) are plotted horizontally, and shear stresses (τ, τ_{zx}, etc) are plotted vertically. The Mohr circle for a given system of two-dimensional stress is centred on the horizontal axis at the point representing the mean principal stress $½(\sigma_1 + \sigma_3)$ and has a radius $½(\sigma_1 - \sigma_3)$. It is then simple to derive the various stress relationships of section 4.4.1 directly from the geometry of the circle diagram.

For example:

$$\sigma_x = ½(\sigma_1 + \sigma_3) - ½(\sigma_1 - \sigma_3)\cos 2\theta \tag{4.6a}$$

$$\tau_{zx} = \tau_{xz} = \pm ½(\sigma_1 - \sigma_3)\sin 2\theta \tag{4.6b}$$

Angle 2θ is set off anti-clockwise from the direction of the direct stress or σ axis

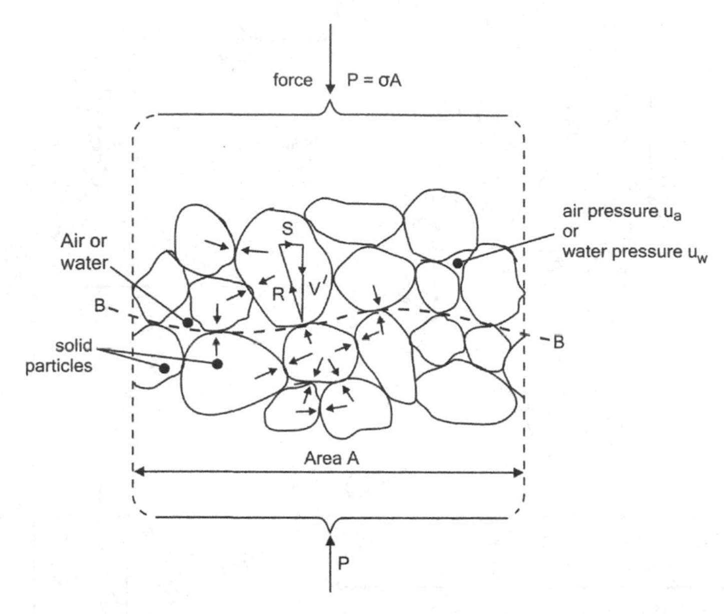

Figure 4.4 Interpretation of effective stress in a completely dry particulate material (voids filled with air) or a completely saturated material (voids filled with water).

It also follows that

$$\tau_{max} = \tfrac{1}{2}(\sigma_1 - \sigma_3) \tag{4.4}$$

Referring to the stressed element to the right of Figure 4.3, α is the angle that the resultant of σ_x and τ_{zx} makes with the vertical side of the stressed element and is known as the angle of obliquity, or simply as the obliquity. The Mohr diagram shows that a maximum value of α corresponds with tangency to the Mohr circle of a line inclined at α to the σ-axis and maximum obliquity is given by

$$\sin \alpha_{max} = (\sigma_1 - \sigma_3)/(\sigma_1 + \sigma_3) \tag{4.7}$$

4.4.3 *The principle of effective stress*

Particulate materials like soil and mine waste, are two or three-phase materials consisting of solid particles with interstitial voids that can be completely filled with water, or air or partly with air and partly with water. The solids consist of mineral grains of, e.g., coal, waste rock or tailings. Stresses or loads carried by multi-phase materials are supported by the sum of

- forces between particles, acting through the points of contact, and
- the pressures of the interstitial fluids.

The sharing of load in a completely saturated or completely dry material is illustrated by Figure 4.4 in which P represents a vertical external load and wavy line BB represents a surface that passes through all the inter-particle points of contact and cuts through the fluid-filled voids in which the fluid pressure is u. At each contact the forces acting are a shear force S, a vertical force V^l, and their resultant force R. The areas of inter-particle contact are very small and to a first approximation, vertical equilibrium across area A requires that

$$P = \Sigma(\text{over } A)V^l + Au$$

In stress terms, dividing by A gives

$$\sigma_v = \Sigma V^l/A + u$$
$$\text{or } \sigma_v = \sigma_v^l + u, \quad \text{and} \quad \sigma_v^l = \sigma_v - u \qquad (4.8)$$

σ_v is the total or applied vertical stress,
u is the pore pressure (either pore air or pore water) because it acts within the pores or interstices, and
σ_v^l is the intergranular vertical stress, the sum of the vertical components of the intergranular forces, divided by A.

In a material saturated with water, which is the more usual condition in a fine-grained mine waste such as a tailings, the pore pressure is a pore water pressure u_w, whereas in a coarse-grained waste such as a waste rock, the pores are more usually filled with air at pressure u_a. This may also apply to fine grained dry industrial wastes such as pulverized fuel ash (pfa) produced by a power station.

It was first observed experimentally by Terzaghi (1943) that the strength and volume change of a particulate material are controlled by σ^l, rather than σ. σ^l is therefore called the "effective" stress. Terzaghi's principle of effective stress can be stated as:

"The change in volume or the maximum resistance to shear stress of an element of particulate material depends directly on the effective stress σ^l, and not on its components, the total stress σ and the pore pressure u."

In a water-saturated material, equation (4.8) applies with $u = u_w$. In dry materials u_w is replaced by u_a. In many cases, the pore spaces are relatively large (e.g. in waste rock) and the pore air pressure remains at zero (relative to atmospheric air pressure). In these cases, if the rock is dry, the total stress numerically equals the effective stress. However, in dry fine powders such as pulverized fuel ash, air can be trapped within the pores and pore air pressures significantly greater than atmospheric pressure can develop. Here, the full effective stress equation (4.8) becomes operative with $u = u_a$.

Some materials are commonly stored wet or damp, (e.g. coarse coal discards) and the pore space is partly filled with air and partly with water. Here the effects of both air and water pressure may become important. Free water (i.e. water not absorbed by waste particles) or air trapped in the pores of the waste may both exert significant pore pressure. More commonly, however, water is held by capillary tension in damp, i.e. wet but not saturated, material. The consequences of the presence of these capillary water tensions, will now be considered in more detail.

In an unsaturated soil, the pore space contains both water and air. Because of the great affinity of water for mineral surfaces, the water is always contained around the points of contact of the solid grains and the water surfaces form air-water menisci

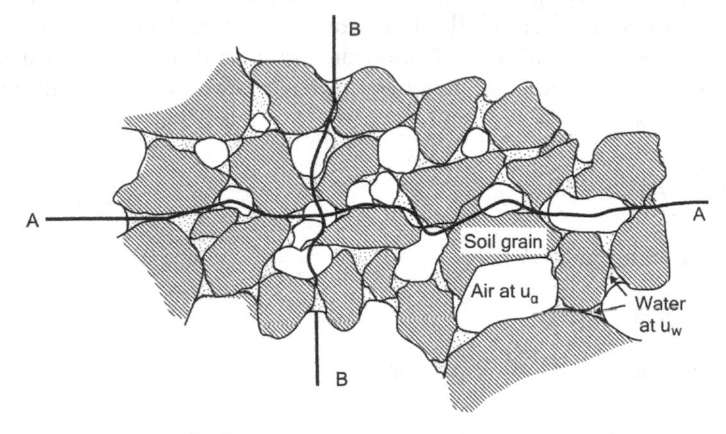

a: Idealised concept of the structure of an unsaturated soil.

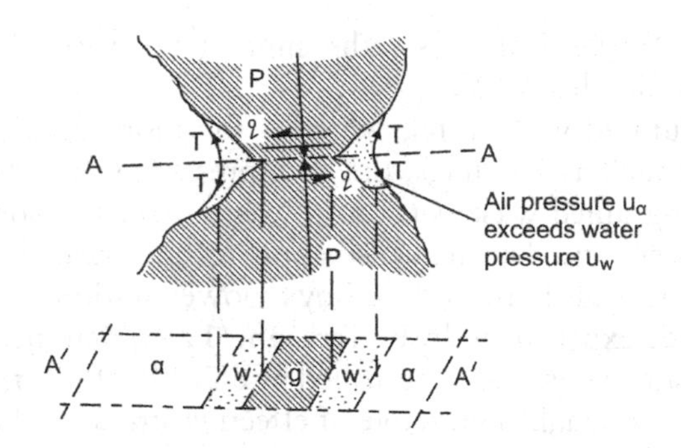

b: Forces at an intergranular contact.
α, w, g are the areas or air, water and grain contact intersected by plane AA and projected onto plane A'A'.

Figure 4.5 The structure of an unsaturated soil.

curved in two principal directions (parallel and normal to the plane of the contact area). Diagrammatically, a section through an unsaturated soil would be as shown in Figure 4.5. There is now the complication, as shown in the enlarged section through a grain-to-grain contact, that the pore air and pore water pressures u_a and u_w differ as a result of the meniscus curvature. Also, the intergranular stresses are augmented by the surface tensions (T) that act at the liquid-to-air interfaces. The net result is that the effective stress equation becomes

$$\sigma^1 = (\sigma - u_a) + f(u_a - u_w) \tag{4.9}$$

where $f(u_a - u_w)$ is a function of the difference in pressure of the pore air and the pore water, including the effect of the T-forces.

$(u_a - u_w)$ is termed the soil water suction. Equation 4.9 is often written

$$\sigma^l = (\sigma - u_a) + \chi(u_a - u_w) \quad (4.9a)$$

which is known as Bishop's effective stress equation,

$$\text{or } \sigma^l = \sigma - \chi u_w - (1 - \chi)u_a$$

which separates out the effects of u_a and u_w. χ is called the Bishop effectiveness parameter.

Note that if the soil is water-saturated, there is no u_a and $\chi = 1$, i.e. $\sigma^l = \sigma - u_w$. If the soil is completely dry, there is no u_w and $\chi = 0$, i.e. $\sigma^l = \sigma - u_a$. Further reference may be found in Bishop and Blight (1963), Fredlund and Rahardjo (1993) and Khalili, et al. (2004).

4.4.4 *Direct strains, shear strains and principal strains*

When an element of material is subjected to stresses, as illustrated in Figure 4.2, it undergoes changes of dimension (movements or strains). These are of two kinds. The direct stresses σ cause direct compressions (or direct compressive strains) while the shear stresses τ cause shear distortions (or shear strains). Direct strains are defined as

$$\text{Direct strain} = \text{shortening/original length, or } \varepsilon = \Delta L/L_o \quad (4.10a)$$

Shear strains are defined by the change in angle (i.e. the distortion) caused by the stress, e.g. if what was originally a right angle changes by an angle Γ, then

$$\text{Shear strain} = \Gamma$$

Shear strain is usually symbolized by lower case Greek gamma (γ). However, in geotechnical engineering γ usually symbolizes unit weight (kN/m^3). To avoid confusion, shear strain has been denoted by capital gamma (Γ).

Figure 4.6 illustrates the two types of strain. Figure 4.6a illustrates the compressions caused by applying principal direct stresses σ_1 and σ_3. No distortion occurs in the directions of σ_1 and σ_3, but the original lengths of the sides of the element change from unity to $(1 - \varepsilon_1)$ and $(1 - \varepsilon_3)$ where ε_1 and ε_3 are the corresponding principal strains. The diagonal AA shortens and rotates to aa. The rotation $(\varepsilon_1 - \varepsilon_3)$ (expressed in radians) equals the maximum shear strain Γ, i.e.

$$\Gamma_{max} = (\varepsilon_1 - \varepsilon_3) \quad (4.10b)$$

If the applied direct stresses are not principal stresses, see Figure 4.6b, distortions also occur in the directions of the direct stresses, as shown.

The relationship between direct strains and shear strains can also be shown by means of a Mohr circle diagram, the Mohr strain circle, illustrated in Figure 4.7. Note that the strain circle is very similar to the stress circle, except that direct strain ε replaces

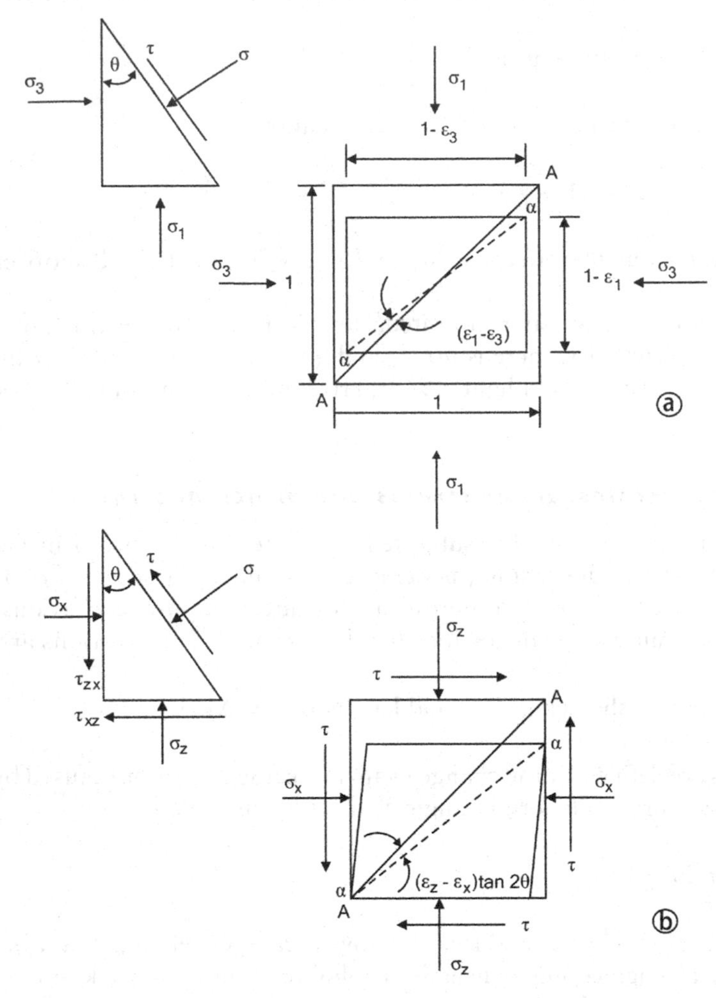

Figure 4.6 **Direct strains (compressions) and shear strains (distortions) under the actions of stress.**

direct stress σ, and the vertical axis represents ½ (shear strain). Thus the radius of the strain circle is

$$\Gamma_{max}/2 = ½(\varepsilon_1 - \varepsilon_3)$$

or $$\Gamma_{max} = \varepsilon_1 - \varepsilon_3$$

Whereas with the stress circle

$$\tau_{max} = ½(\sigma_1 - \sigma_3)$$

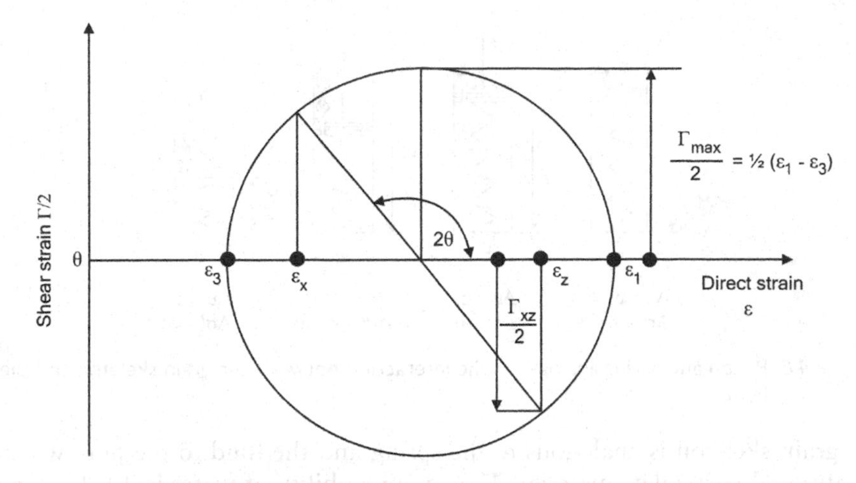

Figure 4.7 Möhr strain circle.

As was the case with the stress circle, relationships between principal and other direct and shear strains can be found from the geometry of the strain circle. For example:

$$\varepsilon_1 = ½(\varepsilon_z + \varepsilon_x) - ½(\varepsilon_z - \varepsilon_x) \sec 2\theta \text{ (if } 2\theta > 90°)$$

4.4.5 *Interaction between the grain skeleton and the pore fluid*

The simplest analogy to the interaction between the pore fluids and the grain skeleton is given by the cylinder and spring model illustrated in Figure 4.8. Figure 4.8a represents a fluid-filled rigid walled cylinder closed by a piston which in turn is supported by a coil spring. There is a small orifice in the piston which can be closed by a valve. In Figure 4.8b a stress σ has been applied to the piston with the orifice closed. If the compressibility of the fluid is C_f and that of the spring C_s then, neglecting friction between the piston and cylinder, the strain ε_o is given by

$$\varepsilon_o = \Delta u C_f = \Delta\sigma^l C_s$$

$$\text{or} \quad \Delta u/\Delta\sigma^l = C_s/C_f \tag{4.11}$$

where Δu and $\Delta\sigma^l$ are the respective changes in stress in the pore fluid and the spring. If the compressibility of the fluid is very much less than that of the spring, Δu would be far greater than $\Delta\sigma^l$ and

$$\Delta u = \sigma \text{ (approximately)}$$

If the orifice is now opened, fluid will escape from the cylinder and the spring will compress (Figure 4.8c) until (Figure 4.8d) the pressure in the fluid is zero (gauge). At this stage

$$\varepsilon_f = \Delta\sigma^l C_s \text{ and, since } u = 0 \ \Delta\sigma^l = \sigma \tag{4.12}$$

and the spring will carry all of the load.

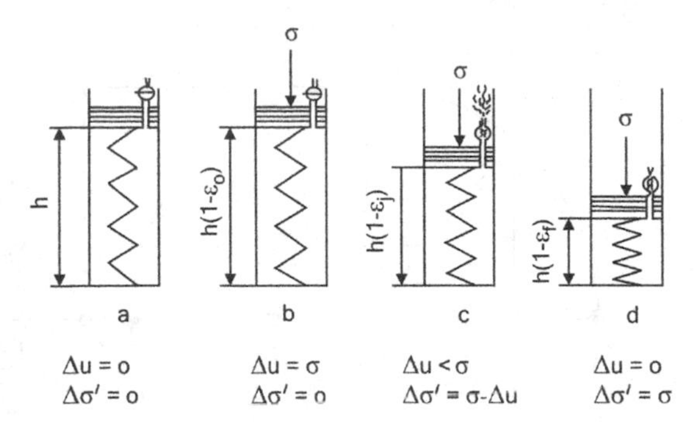

Figure 4.8 Piston and spring analogy for the interaction between the grain skeleton and the pore fluid.

The grain skeleton is analogous to the spring and the fluid to the pore water in a fully saturated particulate material. The compressibility of water is $4.8.10^{-4}$ mm^2/N while a typical grain skeleton compressibility is 10^{-1} mm^2/N. Hence, for a saturated particulate material loaded under undrained conditions,

$$\Delta\sigma^l/\Delta u = 1/208 = 0.5\%$$

In other words in Figure 4.8b the pore water carries 99.5% of the applied stress. If water is allowed to drain from the pores in the skeleton, i.e. if the material consolidates, stress will be progressively transferred from the pore water to the grain skeleton until all the stress is carried by the skeleton, i.e. the material will be consolidated under the applied stress.

The stress in the grain skeleton can be regarded as the effective stress which controls the shear strength and changes in volume of the soil.

Volumetric strains (ε_v) in a particulate material are governed by the equation:

$$-\varepsilon_v = C\Delta(\sigma - u) = C\Delta\sigma^l \tag{4.13}$$

in which C is the grain skeleton compressibility, and the minus sign shows that the volume decreases as σ^l increases. The shear strength (τ_f) is usually taken to be governed by the Coulomb equation

$$\tau_f = (\sigma - u)\tan\phi^I + c^I = \sigma^l \tan\phi^l + c^l \tag{4.14}$$

in which φ^l is the angle of shearing resistance of the soil and c^l is a cohesion representing the shear strength when $\sigma^l = 0$.

According to equations (4.13) and (4.14) neither the shear strength nor the volume of a saturated particulate material can change unless the effective stress changes. Also, if σ^l increases, the strength of the material will increase while the volume decreases and vice-versa. Note from equation (4.8) that σ^l can be altered by altering σ, keeping u constant or by varying u, keeping σ constant. The liquefaction of mine tailings is an extreme example of changing the effective stress by varying the pore water pressure.

In this case the pore pressure increases under constant total stress causing the effective stress to reduce. As a consequence the tailings loses strength and can fail and flow like a heavy liquid.

Figure 4.9 illustrates the applicability of the cylinder and spring analogy to a saturated tailings. Figure 4.9a shows changes in pore water pressure and volumetric strain corresponding to increases of isotropic total stress under undrained conditions. After a small initial compression, probably of gas in the voids, the volume remained sensibly constant while the increases in pore pressure were almost exactly equal to the increases in total stress, i.e. $\Delta u = \Delta\sigma$, both results indicating that the effective stress in the tailings remained virtually constant.

Figure 4.9b shows the results of undrained triaxial compression tests carried out on four specimens of saturated tailings at different total stresses. The very small increase in shear strength indicates that the effective stress in the tailings was almost unaffected by increasing total stresses. If the test results are plotted in terms of effective stresses, the four test results reduce to a closely spaced group, showing that the effective stress at failure and the shear strength were virtually the same for all four specimens of tailings.

4.4.6 *Relationships between triaxial total stress and pore pressure in undrained conditions*

While this chapter will address the subject of laboratory shear and consolidation testing of mine waste materials, Chapter 5 will deal with in situ (or field) testing.

The object of field and laboratory testing is to simulate, as closely as possible, stress conditions in the prototype waste structure or natural soil foundation stratum, and hence to measure and record the behavior of the material concerned under expected service conditions. As the behavior of soil and mine wastes is governed by variations in total stress and pore pressure, it is useful to have an analytical link between the two so that changes in pore pressure resulting from a given change in total stress can be predicted. The most convenient way of obtaining such a relationship is to consider the behavior, under undrained conditions, of an idealized perfectly elastic particulate material. The grain skeleton is assumed to be perfectly elastic and isotropic and the pore space to be water saturated. Under undrained conditions, the material is subjected to changes in major, minor and intermediate total stresses of $\Delta\sigma_1$, $\Delta\sigma_3$ and $\Delta\sigma_2$ (see Figure 4.10). If the resulting change in pore pressure is Δu, the changes in principal effective stresses are

$$\begin{aligned} \Delta\sigma_1^{\mathrm{l}} &= \Delta\sigma_1 - \Delta u \\ \Delta\sigma_2^{\mathrm{l}} &= \Delta\sigma_2 - \Delta u \\ \Delta\sigma_3^{\mathrm{l}} &= \Delta\sigma_3 - \Delta u \end{aligned} \tag{4.15}$$

If $\varepsilon_1, \varepsilon_2, \varepsilon_3$ are the strains in the three principal directions, and E and ν are Young's modulus and Poisson's ratio for the grain skeleton, applying the generalized Hook's law:

$$E\varepsilon_1 = [\Delta\sigma_1^{\mathrm{l}} - \nu(\Delta\sigma_2^{\mathrm{l}} + \Delta\sigma_3^{\mathrm{l}})] \tag{4.16}$$

and similarly for ε_2 and ε_3

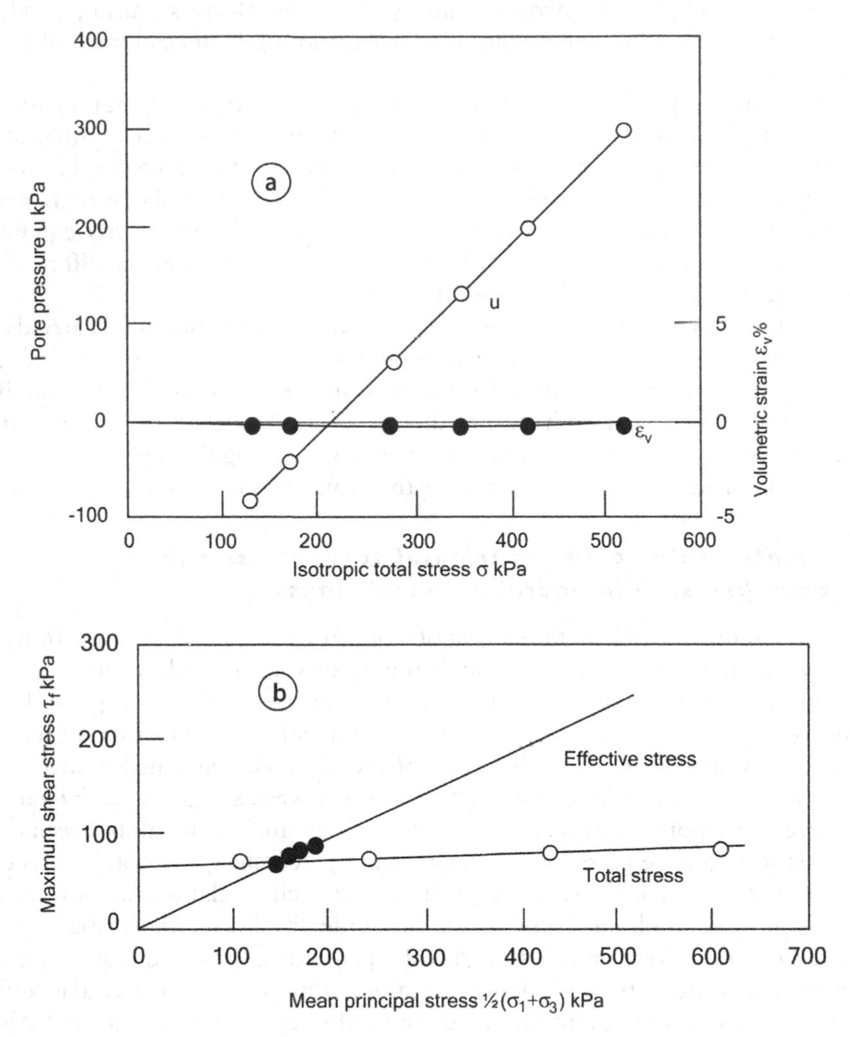

Figure 4.9 The applicability of the principle of effective stress to a saturated particulate material: a: effects of undrained compression under isotropic total stress; b: effects of plotting shear test measurements in terms of total and effective stresses.

For small strains,

$$-\varepsilon_v = \varepsilon_1 + \varepsilon_2 + \varepsilon_3 = \text{volumetric strain of grain skeleton}$$

i.e. from equation (4.16)

$$-E\varepsilon_v = (1 - 2\nu)(\Delta\sigma_1^l + \Delta\sigma_2^l + \Delta\sigma_3^l) \qquad (4.17)$$

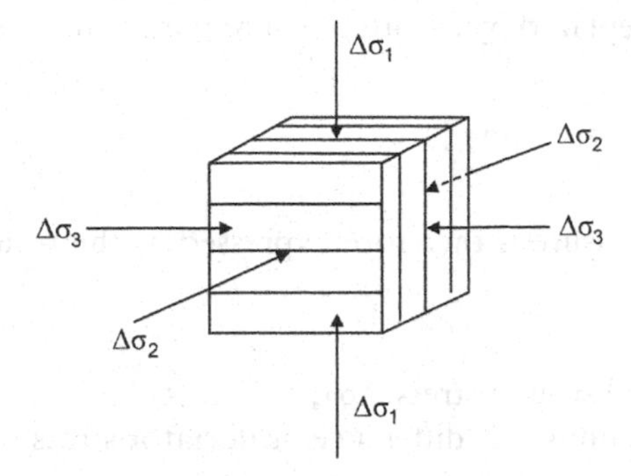

Figure 4.10 A system of triaxial stress applied to an undrained saturated particulate material.

This can be rewritten as:

$$-3\varepsilon_v = C_s[\Delta\sigma_1 + \Delta\sigma_2 + \Delta\sigma_3 - 3\Delta u] \quad (4.17a)$$

$C_s(1 - 2v) = 3E\Delta\sigma_1^I = 3E(\Delta\sigma_1 - u)$, etc.

i.e. $C_s(1 - 2v) = 3E$ where C_s is the compressibility of the grain skeleton.

Due to the increase in pore pressure Δu, the pore water will have compressed by

$$[-\varepsilon_v]\text{water} = C_w\, n\, \Delta u \quad (4.18)$$

Where C_w is the compressibility of water, and n is the volume of water per unit overall soil volume = the porosity of the grain skeleton.

As no drainage has taken place,

$$\varepsilon_v = [\varepsilon_v]\ \text{water}$$

and $3C_w n\Delta u = C_s[\Delta\sigma_1 + \Delta\sigma_2 + \Delta\sigma_3 - 3\Delta u]$

$$\text{or } \Delta u = C_s/(C_s + nC_w)[\Delta\sigma_1 + \Delta\sigma_2 + \Delta\sigma_3]/3 \quad (4.19)$$

$$\text{if we write } B = 1/(1 + nC_w/C_s) \quad (4.20)$$

$$\text{Then } \Delta u = B[\Delta\sigma_1 + \Delta\sigma_2 + \Delta\sigma_3]/3 \quad (4.19a)$$

In a saturated material, C_w is very small compared with C_s/n and B tends to 1. i.e. A saturated material is virtually incompressible under undrained conditions and $B = 1$ very closely.

In a standard triaxial compression test $\Delta\sigma_2 = \Delta\sigma_3$
and $\Delta u = B[\Delta\sigma_1 + 3\Delta\sigma_3 - \Delta\sigma_3]/3$

$$\text{or } \Delta u = B[\Delta\sigma_3 + (\Delta\sigma_1 - \Delta\sigma_3)/3] \quad (4.19b)$$

This applies to an idealized elastic grain skeleton. For actual particulate materials, the elastic constant 1/3 is replaced by a pore pressure parameter A so that

$$\Delta u = B[\Delta\sigma_3 + A(\Delta\sigma_1 - \Delta\sigma_3)] \tag{4.19c}$$

The change in pore pressure is therefore expressed as the result of two independent stress impositions:

- a change in equal all-round stress $\Delta\sigma_3$;
- a change in maximum stress difference or deviator stress of $(\Delta\sigma_1 - \Delta\sigma_3)$.

The pore pressure parameters A and B provide a useful dimensionless method of relating applied stress changes to changes in pore pressure.

4.5 The behaviour of mine waste materials subjected to shear

4.5.1 *The effects of stress history on the behaviour of soils during shear*

Most mine waste materials are deposited in a loose condition and then become densified in the course of time, e.g.

- Hydraulically deposited tailings settle out from a slurry in a very loose state and then become denser as subsequent layers are deposited over them or as they drain and dry at the surface in the sun.
- Coarse wastes such as waste rock or soil overburden are deposited in a loose state and then become densified by the weight of subsequent deposition or by mechanical compaction.

The strength and pore pressure behaviour of particulate materials is largely controlled by the relationship between the present average effective stress in the material and the greatest average effective stress to which the material has ever been subjected since deposition. This relationship is referred to as the "stress history" of the soil. The ratio of maximum effective stress to present effective stress is known as the over-consolidation ratio (OCR).

Figure 4.11 shows a typical relationship between void ratio or water content and average effective stress as a saturated particulate material is first compressed from point O to an effective stress σ^{1}_{m}, allowed to swell back to point A and then recompressed to σ^{1}_{m} and beyond.

At points 1 and 4, $OCR = 1$, and the material is normally consolidated (NC).
At points 2 and 3, $OCR = \sigma^{1}_{m}/\sigma^{1}$, and the material is overconsolidated (OC).

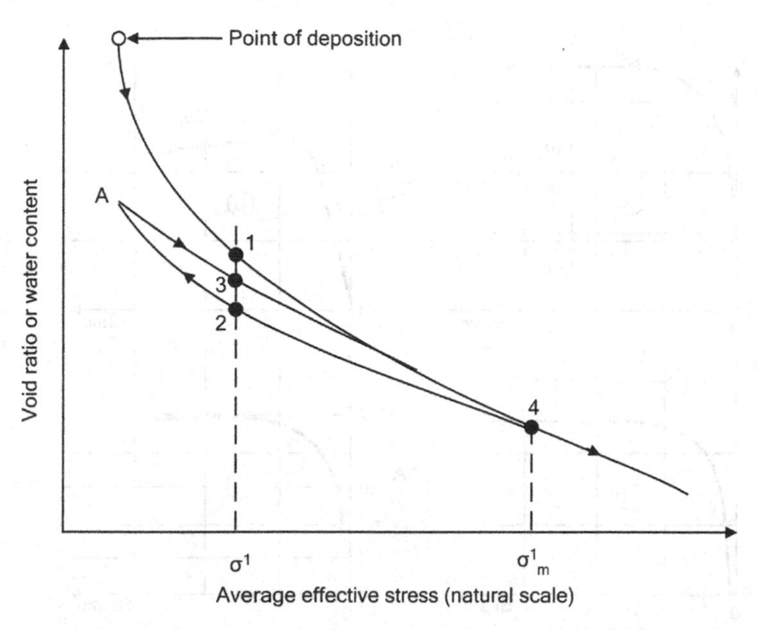

Figure 4.11 Illustration of the terms "normally" and "overconsolidated". Material is normally consolidated (NC) at points 1 and 4 and overconsolidated (OC) at points 2 and 3.

4.5.2 *Relationships between stress and strain, and parameter A and strain*

The parameter A (equation 4.19c) varies with strain and, at a particular strain, depends on the stress history of the particulate material. Figure 4.12a shows typical stress-strain and pore pressure-strain curves for undrained shear of a normally consolidated saturated fine tailings. Diagram A shows a typical variation of σ_1, during shear, while B shows the variation of the principal stress ratio σ_1^1/σ_3^1. It will be noted that maximum σ_1^1/σ_3^1, corresponding to maximum obliquity angle α, usually occurs before maximum $(\sigma_1 - \sigma_3)$. Diagrams A and C are drawn to the same scale and it will be seen that the increase in pore pressure Δu is almost equal to the rise in $(\sigma_1 - \sigma_3)$. This is characteristic of fine grained normally consolidated materials. Diagram D shows the variation of parameter A with strain. At low strains, the curve starts off close to the theoretical value for an elastic material of 0.33 (equation 4.19b). Parameter A then rapidly increases to a value close to unity at failure. In materials having a high void ratio, A can rise as high as 1.3 to 1.4.

Figure 4.13a shows corresponding diagrams for undrained shear of a heavily over-consolidated saturated tailings. Diagrams A and B are generally similar to those of Figure 4.12a, but there is a substantial difference in Diagrams C and D. In C, the pore pressure at first rises and then starts to decrease just before maximum σ_1^1/σ_3^1 is reached. The pore pressure continues to decrease until large strains have been reached. The value of A once more starts off at about +0.33 and then decreases continuously, becoming negative. In some materials a maximum value of $(\sigma_1 - \sigma_3)$ is never reached. As long as the pore pressure continues to drop, the effective stress increases and the

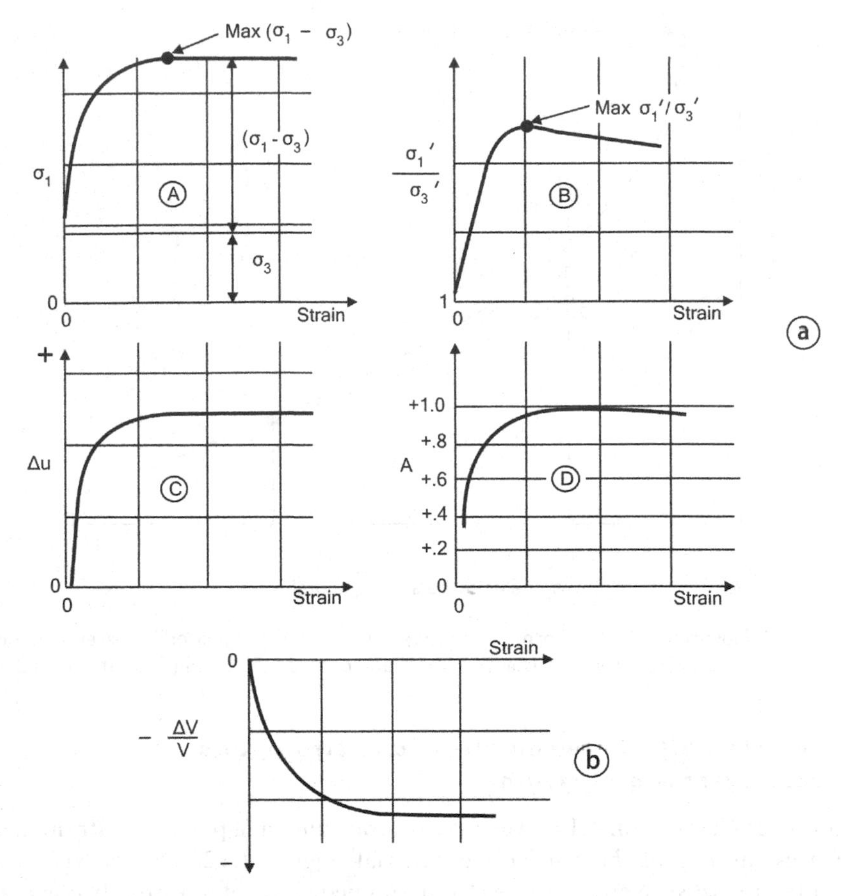

Figure 4.12 a: Typical stress and pore pressure changes during undrained shear of a normally consolidated (NC) fine tailings. b: Volumetric strain during drained shear of a normally consolidated tailings.

material gains in strength. The effect of stress history on the parameter A is shown in Figure 4.14a. As the OCR increases, A at failure decreases, becoming negative at OCR values exceeding about 4.

4.5.3 *The effects of drainage during shearing*

So far, only the shear behaviour of particulate materials under undrained conditions has been considered. Under fully drained conditions no pore pressures develop during shear but the material changes in volume. Generally, the direction of volume change during drained shear is the converse of the direction of pore pressure change during undrained shear. This is illustrated in Figures 4.12b and 4.13b and in Figure 4.14b.

In many practical problems, mine waste may be loaded so rapidly that if failure occurs, it occurs with negligible drainage. Examples of such rapid loading are:

- the loading on the foundation strata caused by the rapid advance of the tip face of a waste rock dump;
- the loading caused by a rapid rate of rise of a hydraulic fill tailings dam.

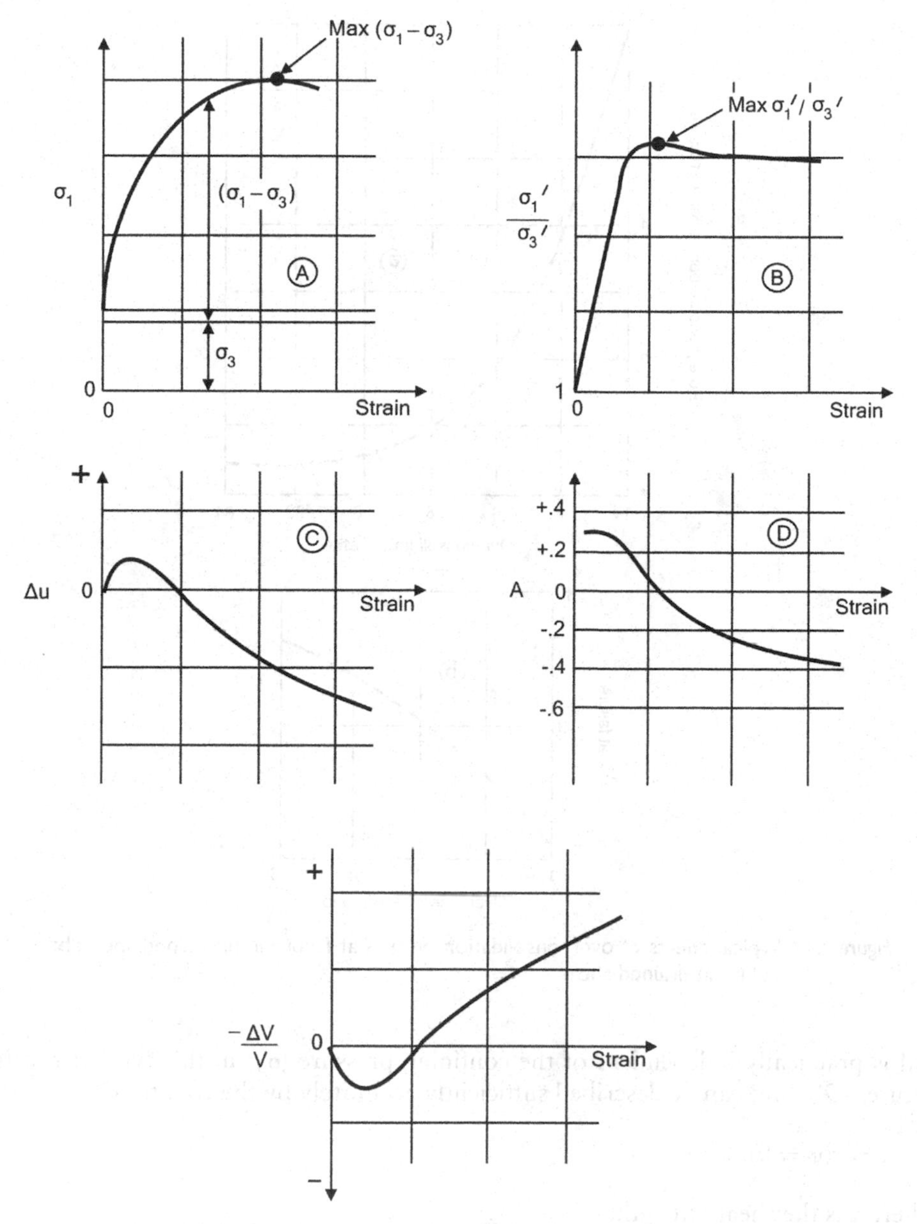

Figure 4.13 a: Typical stress and pore pressure changes during undrained shear of a heavily overconsolidated fine tailings (OCR = 30); b: Volumetric strain during drained shear of a heavily overconsolidated tailings.

These conditions can be simulated in laboratory or in situ tests by shearing the material undrained and without consolidating it any further than its original effective overburden stress before the rapid loading was applied.

If a series of unconsolidated undrained shear tests are carried out on samples of a saturated tailings, all taken from the same depth, it is found that in terms of total stresses an almost horizontal strength envelope is obtained. The strength of the

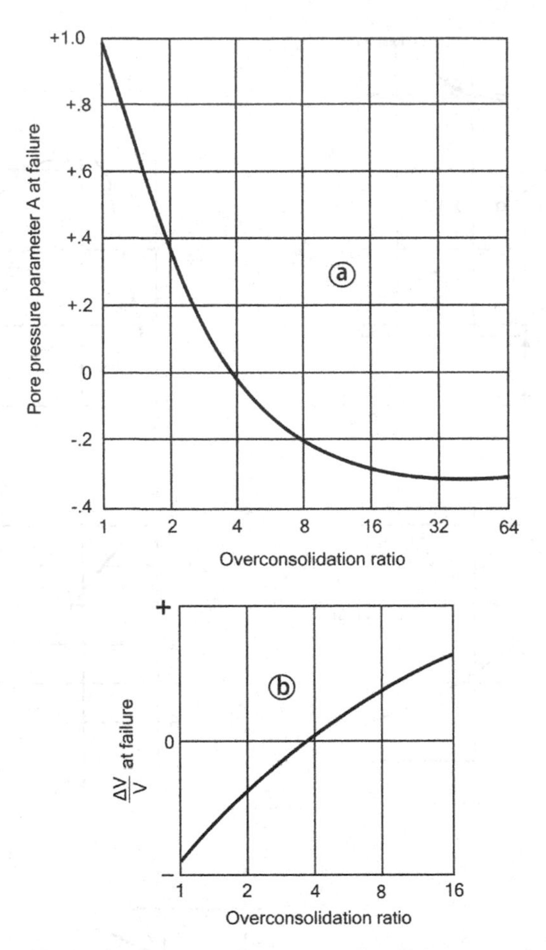

Figure 4.14 Typical effects of overconsolidation on: a: A at failure in undrained shear; b: ΔV/V at failure in drained shear.

soil is practically independent of the confining pressure (σ_3) in this type of test (see Figure 4.9b) and can be described sufficiently accurately by the equation

$$\tau = c(\varphi = 0) \tag{4.20}$$

where τ is the shear strength,
c is the cohesion, and
φ is the angle of shearing resistance, all in terms of total stresses.

4.5.4 *Strain, volume, mass and weight relationships between water, solids and voids*

To consider the strains that may occur in a particulate material such as a mine waste, it is necessary to establish a terminology that describes the material's internal geometry.

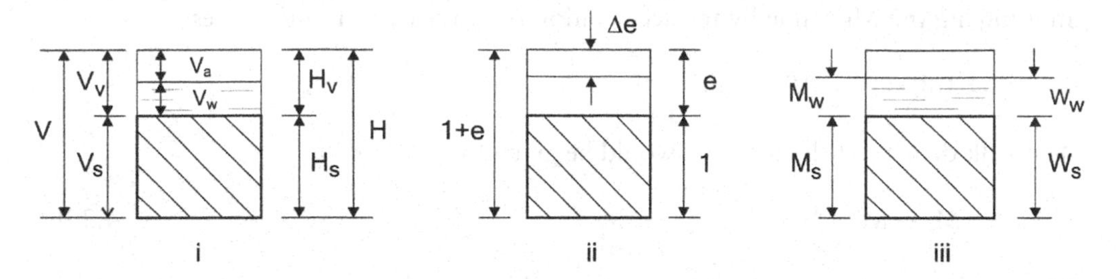

Figure 4.15 Describing internal geometry of a particulate material.

Referring to Figures 4.4 and 4.5 the solid material and void space can conceptually be lumped together as shown in Figure 4.15. Diagram (i) in Figure 4.15 shows the total volume of solids lumped together as V_s, which for a given cross-sectional area A would have height H_s. Similarly the voids would have volume V_v or height H_v. The void ratio e is defined as (volume of voids) ÷ (volume of solids), or

$$e = V_v/V_s = H_v/H_s \tag{4.21}$$

The total volume is $V = V_v + V_s = e\ V_s + V_s = V_s(1 + e)$ and hence for a unit volume of solids (e.g. $V_s = 1\,m^3$).

$$V = 1 + e$$

If the material were to be compressed (diagram (ii)), all volume change would occur in the voids as the solids are usually relatively incompressible, and if the voids or void height were to change by Δe, the volumetric strain would be

$$\varepsilon_V = \Delta e/(1 + e) \tag{4.22}$$

Diagram (iii) shows the mass and weight of the solids lumped together. (The pore air has both mass and weight, but this is negligible in comparison with that of the solids and water.) The dry bulk density of the material is defined as the dry mass (mass of solids) per unit volume

$$\rho_d = M_s/V\,(\text{units kg/m}^3) = M_s/(1 + e) \tag{4.23a}$$

and the unit weight as

$$\gamma_d = W_s/V = \rho_d g(\text{units kN/m}^3) = W_s/(1 + e) \tag{4.24a}$$

where g is the acceleration of gravity (g can be taken as $10\,m/s^2$ for practical purposes.)

As an example, if $\rho_d = 1800\,kg/m^3$ ($=1.8\,Mg/m^3$ or $1.8\,Tonne/m^3$ or T/m^3) then multiplying the Mg value by the acceleration of gravity $g = 10\,m/s^2$ gives

$$\gamma_d = 1.8 \times 10 = 18\,kN/m^3$$

The bulk density of the material would be given by

$$\rho = (M_s + M_w)/V \qquad (4.23b)$$

The gravimetric water content of the material is given by

$$w = M_w/M_s \qquad (4.25)$$

hence

$$\rho = M_s(1 + w)/V$$

and for a unit volume of solids

$$\rho = M_s(1 + w)/(1 + e)$$

or $\rho = \rho_d(1 + w)$ and $\rho_d = \rho/(1 + w)$

Similarly,

$$\gamma = \gamma_d(1 + w) \quad \text{and} \quad \gamma_d = \gamma/(1 + w)$$

The particle relative density, G = mass of solids ÷ mass of an equal volume of water.

In terms of Figure 4.15(iii), $G = M_s/\rho_w$. Hence

$$\rho = G(1 + w)/(1 + e) \cdot \rho_w \text{ and } \rho_d = G/(1 + e) \cdot \rho_w \qquad (4.23c)$$

Similarly,

$$\gamma = G(1 + w)/(1 + e) \cdot \gamma_w \quad \text{and} \quad \gamma_d = G/(1 + e) \cdot \gamma_w \qquad (4.24b)$$

The extent to which the pores of the particulate material are filled with water is described by the degree of saturation S = volume of water ÷ volume of voids $= V_w/V_v = V_w/(V_a + V_w)$

Because $V_w = M_w/\rho_w$ and from equation (4.25), $M_w = wM_s = wG\rho_w$, and

$$S = wG/e, \text{ or } eS = wG \qquad (4.26)$$

Hence γ can be expressed as

$$\gamma = G(1 + eS/G)/(1 + e) \cdot \gamma_w = (G + eS)/(1 + e) \cdot \gamma_w \qquad (4.27)$$

And similarly for ρ

The expression for γ_d (equation 4.24)b) follows from equation (4.27) by setting S = 0 and the density of the saturated soil follows by setting S = 1, i.e.

$$\gamma_{sat} = (G + e)/(1 + e) \cdot \gamma_w \quad \text{and} \quad \gamma_d = G/(1 + e) \cdot \gamma_w \tag{4.24c}$$

To summarize:

$$\begin{aligned} w &= M_w/M_s \\ e &= V_v/V_s \\ G &= M_s/\rho_w \\ S &= V_w/(V_a + V_w) = V_w/V_v \\ eS &= wG \text{ (in saturated soil, } e = wG) \\ \gamma &= (G + eS)/(1 + e) \cdot \gamma_w \\ \gamma_d &= G/(1 + e) \cdot \gamma_w = \gamma/(1 + w) \\ \gamma_{sat} &= (G + e)/(1 + e) \cdot \gamma_w = G(1 + w)/(1 + wG) \cdot \gamma_w \end{aligned}$$

The submerged unit weight (or density) would be the apparent density of the soil submerged in water, which would be

$$\begin{aligned} \gamma_{sub} \text{ (or } \gamma') &= \gamma_{sat} - \gamma_w = [(G + e) - (1 + e)]/(1 + e) \cdot \gamma_w \\ &= (G - 1)/(1 + e) \cdot \gamma_w \end{aligned} \tag{4.28}$$

For ρ, ρ_d, ρ_{sat}, ρ_{sub}, replace γ_w with ρ_w

Another term that is often used is the porosity, n

$$n = \text{Volume of voids} \div \text{total volume}$$

$$n = e/(1 + e)$$

Another concept that is used quite frequently is that of relative density defined by

$$RD = \rho/\rho_w = \gamma/\gamma_w$$

RD is a useful term for describing the consistency of slurries of tailings. From equation (4.23)c

$$RD = G(1 + w)/(1 + e)$$

and because e = wG in a saturated slurry,

$$RD = G(1 + w)/(1 + wG) = (G + e)/(1 + e) \tag{4.23d}$$

EXAMPLE As an example, if the solids density of quartz is 2650 kg/m^3, then the volume of 1000 kg (= 1 Mg) of quartz solids in a quartz sand would be

$$V_s = 1000\,kg/2650\,kg/m^3 = 0.377\,m^3$$

If the overall volume of the dry quartz sand is 0.650 m^3,

$$\text{then the volume of voids, } V_v = 0.650 - 0.377 = 0.273\,m^3$$

$$\text{and the void ratio} \quad e = V_v/V_s = 0.273/0.377 = 0.723$$

$$\text{The porosity is} \quad n = e/(1+e), \text{ and } n = 0.723/1.723 = 0.420$$

If water is mixed with the 1000 kg of sand to give a water content of w = 8% by dry mass, the mass of the water will be $M_W = 0.08 \times 1000 = 80$ kg and the volume of water will be $V_w = 80\,kg/1000\,kg/m^3 = 0.08\,m^3$.

The dry density of the sand will be $\rho_d = 1000\,kg/0.650\,m^3 = 1538\,kg/m^3$

If, after adding the 8% of water, the volume of the sand increases to 0.715 m^3,

$$\text{the dry density will be} \quad \rho_d = 1000/0.715 = 1398\,kg/m^3$$

$$\text{and the bulk density will be} \quad \rho = 1080/0.715 = 1510\,kg/m^3$$

$$\text{The void ratio will now be} \quad e = (V - V_s)/V_s = (0.715 - 0.377)/0.377 = 0.897$$

and the degree of saturation $S = V_w/V_v = 0.08/(0.715 - 0.377) = 23.7\%$

$$\text{Note that} \quad G = 2650/1000 = 2.65$$

$$\rho_d = 1510/(1+0.08) = 1398\,kg/m^3$$

$$\rho_{sat} = (2.65 + 0.897)/(1 + 0.897) \times 1000 = 1869\,kg/m^3$$

$$\text{or} \quad \rho_{sat} = 1000(1 + 0.715 - 0.377)/0.715 = 1871\,kg/m^3$$

The two values should be identical, but differ slightly because of rounding errors.

For a saturated soil $w = e/G = 0.897/2.65 = 0.338$ (33.8%)

$$\text{Hence} \quad \rho_{sat} = 2.65(1 + 0.338)/(1 + 0.338 \times 2.65) \times 1000 = 1869\,kg/m^3$$

$$\text{Finally,} \quad \rho_{sub} = 1869 - 1000 = 869\,kg/m^3$$

4.5.5 *The "at rest" pressure coefficient K_o*

Figure 4.16 represents an element of particulate material that has been poured centrally into a circular cylinder. Ignoring the effects of friction on the walls of the cylinder, the stresses acting on the element are vertical total and effective stresses σ_v and σ^1_v and radial horizontal total and effective stresses σ_h and σ^1_h. The vertical stress is often approximated by the overburden stress or the weight per unit area of the column of

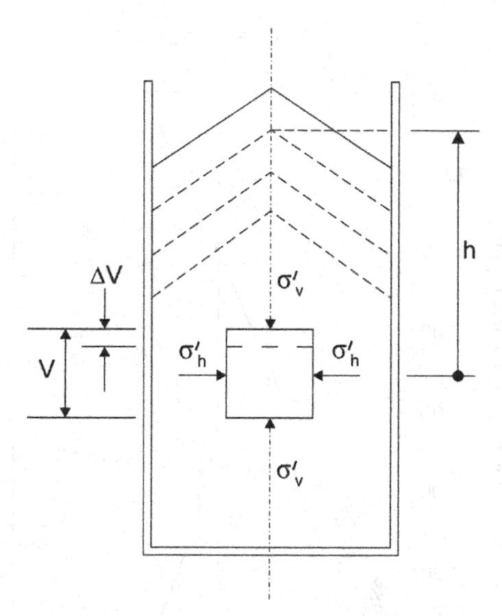

Figure 4.16 Compression of material poured into a cylindrical container.

material above the element. For the situation sketched, σ_v will be approximated by the weight per unit area of a column of material of height h, i.e.

$$\sigma_v = \gamma h\,(kN/m^3 \times m = kN/m^2 = kPa) \qquad (4.29)$$

The horizontal stresses σ_h and σ^l_h in Figure 4.16 will depend on the extent to which the material in the cylinder is able to move or strain horizontally. If, due to the rigidity of the walls relative to the contained material, no lateral strain is possible, any vertical compression Δv that occurs, resulting in a vertical strain $\Delta v/v$, will also be the volumetric strain, i.e.

$$\varepsilon_v = \Delta v/v = \Delta V/V$$

For the situation where lateral strain is prevented, the ratio of the effective stresses σ^l_h to σ^l_v is defined as the zero lateral strain pressure coefficient or the "at rest" pressure coefficient:

$$K_o = \sigma^l_h/\sigma^l_v \text{ with zero lateral strain} \qquad (4.30)$$

Figure 4.17 shows the measured relationship between horizontal stress σ^l_h and σ^l_v for compression with zero lateral strain of a fine grained tailings. The ratio between σ^l_h and σ^l_v, that is K_o, decreases from 0.65 at low stresses to just less than 0.5 at $\sigma^l_v = 600$ kPa and then remains constant.

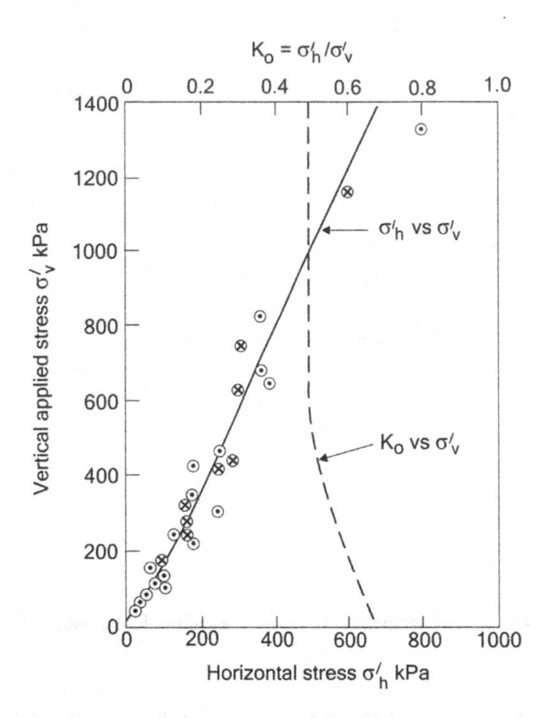

Figure 4.17 Stresses σ_v^l and σ_h^l during K_o compression of a fine-grained tailings.

Figure 4.18 shows data for the compression and unloading of a coal waste under K_o conditions. The loading and unloading stages were quite distinct with an almost constant K_o (= 0.48) for loading and a different constant K_o (= 1.0) for unloading.

Figure 4.18 can also be used to illustrate the concepts of "normal consolidation" and "over-consolidation". During first loading (AB in Figure 4.18) the specimen is normally consolidated. During unloading (BCA in Figure 4.18) it is over-consolidated. If the specimen were to be re-loaded, it would remain over-consolidated until σ_v^l exceeded its largest previous value, i.e. in this case 190 kPa. At values of σ_v^l greater than 190 kPa, the coal waste would become normally consolidated again.

4.5.6 *Apparatus for measuring compression and compressive strains and the stress ratio K_o for particulate materials. Typical results of measurements*

Two types of apparatus can be used for measuring the compression properties of particulate materials such as mine waste. The first is the oedometer, the principles of which are illustrated in Figure 4.19a. The specimen is contained within a rigid metal ring and sandwiched between porous plates that allow pore water, compressed by the load, to escape. The compressive load is applied via a rigid loading platen and the compression movement is measured by means of a dial gauge or a displacement transducer (e.g. a linearly variable differential transformer, or LVDT).

Figure 4.20 shows typical results of oedometer compression tests on a fine-grained tailings. Figure 4.20a shows the progression of compression or settlement with time

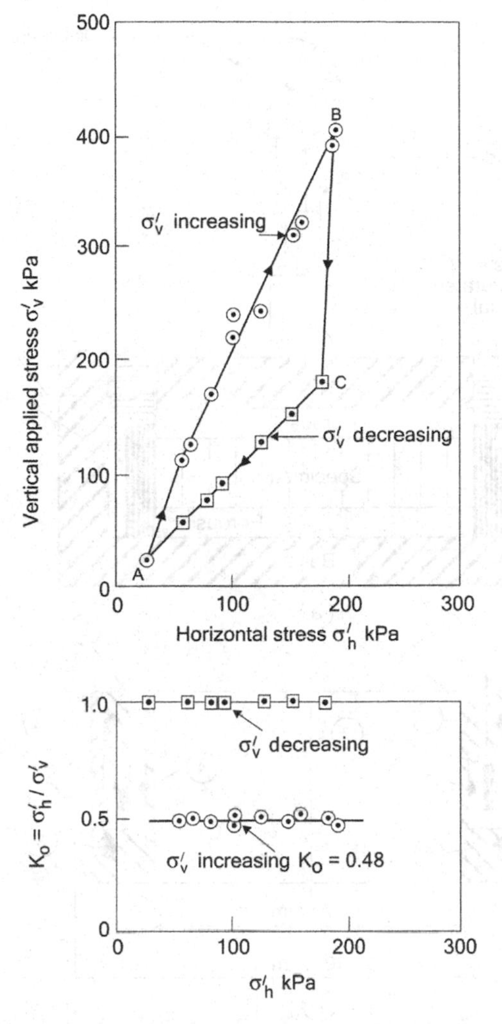

Figure 4.18 Stresses and stress ratios during K_o compression of coal waste with a gravelly sand particle size distribution.

after the application of a load increment. Compression can only take place by expulsion of water from the voids of the tailings (see analogy in Figure 4.8), and because of the small dimensions of the interconnections between pore spaces, this process takes a considerable time. In Figure 4.20a, time has been plotted to a square root scale to compress the length of the diagram. The vertical axis is plotted in terms of the compression or settlement of the specimen measured in mm. Referring to Figure 4.15, the compression represents changes in void height H_v. The very rapid initial compression corresponds to compression of air or gas in the void space, and is referred to as "instantaneous" compression. Each load increment is usually left in place for a 24 hour period (1440 minutes) for the drainage process to be completed. However, in

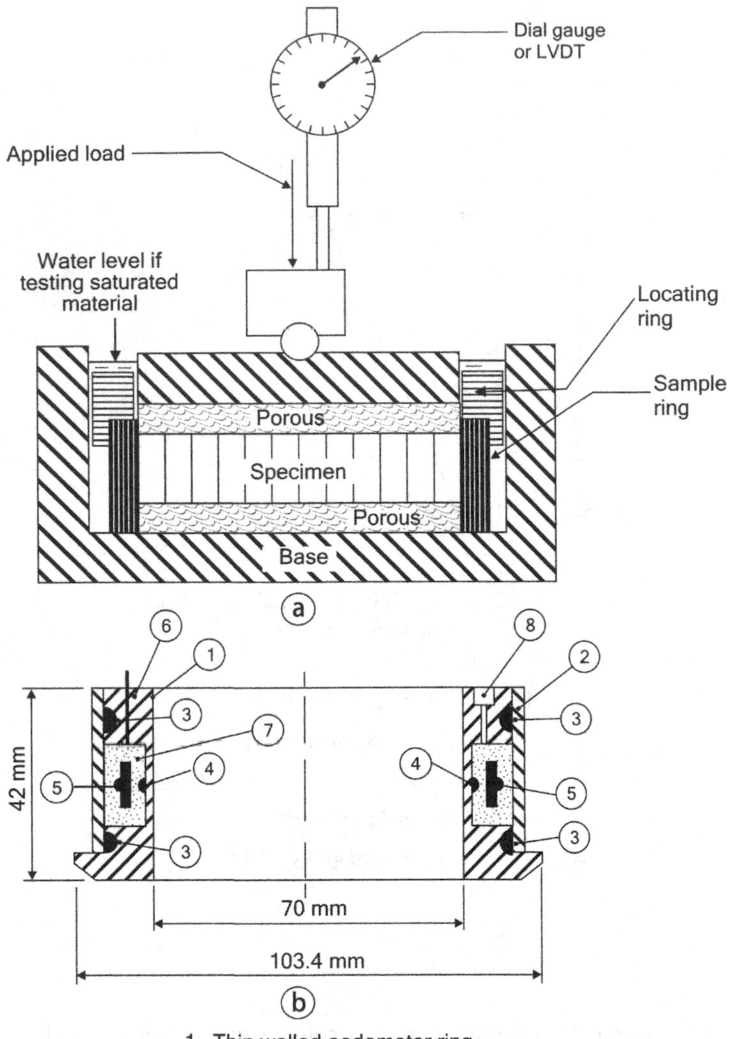

Figure 4.19 Section showing principal components of: a: oedometer apparatus; b: K_o measuring ring.

this case, it is clear that further settlement would have occurred even after the elapse of 24 hours.

Figure 4.20b shows the effect of increasing applied vertical stress on void ratio, as well as the effect of a slightly differing initial void ratio. In this figure, the applied

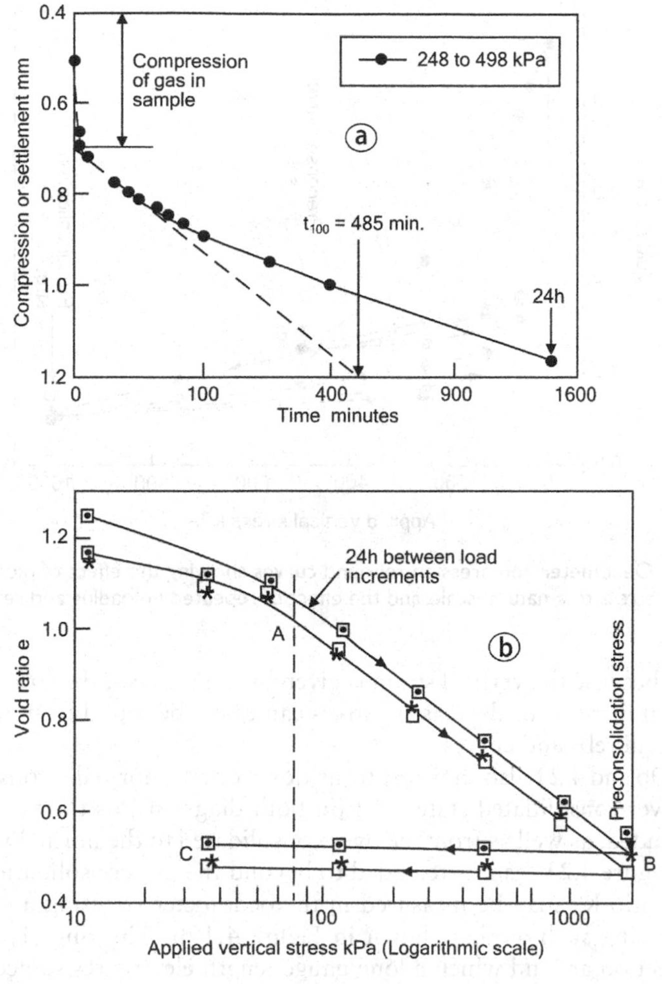

Figure 4.20 Typical oedometer results for tests on tailings: a: $\sqrt{}$time-settlement curve; b: void ratio-applied stress curves. (3 separate samples and tests.)

stress has been plotted to a logarithmic scale. It should be noted that once the stress exceeded 80 kPa, the relationship between the log of stress and void ratio was linear. When the load was reduced, very little of the compression was recovered. Note that this diagram represents a near-equilibrium condition in which the pore pressure is close to zero (atmospheric pressure), i.e. the stresses are both effective and total stresses.

In Figure 4.21 the applied stress (for a different material to that for Figure 4.20) has been plotted to an un-compressed natural scale. This diagram shows that the compressibility of the tailings actually continually reduces as the vertical stress is increased (i.e. the tailings becomes stiffer with increasing effective stress). The diagram also shows that if the tailings are unloaded and reloaded between two stresses (in this case load 500 kPa, or load 1000 kPa) very little change of void ratio occurs. It should also

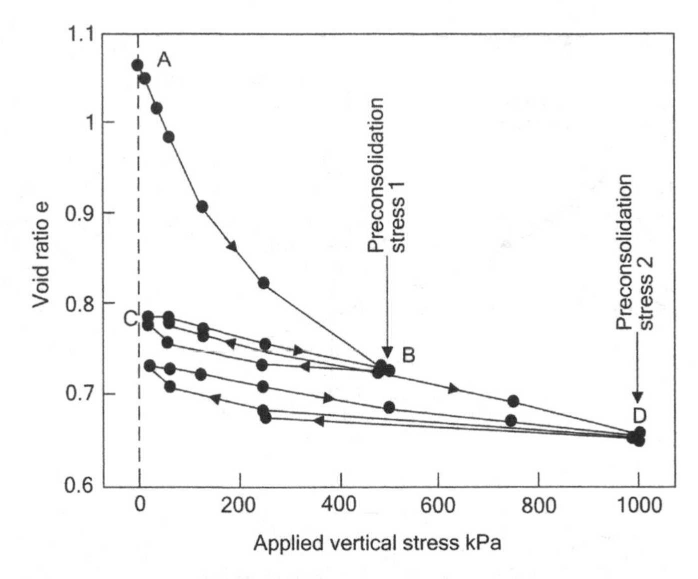

Figure 4.21 **Oedometer compression-rebound curves showing the effect of plotting the applied stress to a natural scale, and the effect of repeated unloading and reloading.**

be noted that because the vertical strain is given by $\Delta e/(1 + e_o)$, the strain corresponding to either an increase or decrease in stress can easily be found from the oedometer compression and rebound curves.

Figures 4.20b and 4.21 also show the transition from the normally consolidated state (AB) to the over-consolidated state (BC) (in both diagrams) as the tailings is loaded and then unloaded, as well as from the over-consolidated to the normally consolidated state (BD in Figure 4.21) as it is reloaded to beyond the pre-consolidation stress.

The stress ratio K_o may be measured in the oedometer by using an instrumented K_o measuring ring such as that shown in Figure 4.19b. The ring (1) has a central thin-walled section around which a long gauge-length electric resistance strain gauge is wrapped and glued to the metal ring (4). As vertical stress is applied to the specimen, horizontal stress (and therefore strain) is generated within the ring. By applying an air pressure to the sealed toroidal cavity (7), the strain in the wall (indicated by the temperature-compensated strain gauge (4, 5)) can be maintained at zero. Hence the lateral stress on the thin-walled section of the ring equals the pressure applied within the cavity. Knowing both $\sigma^{\prime}_{v}$ and $\sigma^{\prime}_{h}$, K_o $(=\sigma^{\prime}_{h}/\sigma^{\prime}_{v})$ can be determined. The K_o measurements shown in Figures 4.17 and 4.18 were made by means of a K_o ring like that illustrated in Figure 4.19b.

The second type of apparatus is the triaxial cell, the principle of which is illustrated in Figure 4.22. The specimen is cylindrical, usually with a height to diameter ratio of 2. The specimen is mounted on a pedestal and supports a loading cap. It is isolated from its surroundings by a thin latex rubber sleeve, pressure-sealed to the pedestal and loading cap by means of tight-fitting rubber O-rings. The triaxial cell is a cylindrical pressure vessel that fits over the latex rubber-encapsulated specimen and is sealed to the base of the cell by means of tie-bars, wing-nuts and a rubber O-ring seal. The

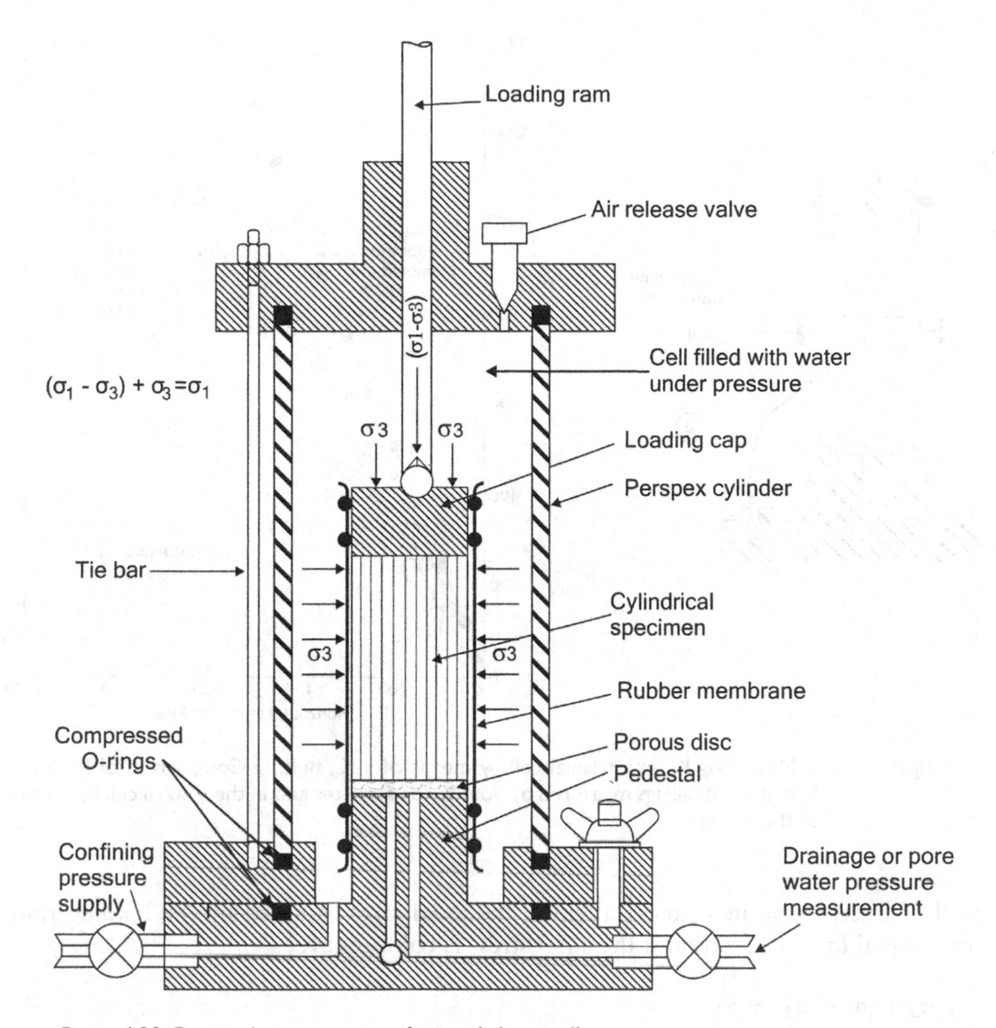

Figure 4.22 **Principal components of triaxial shear cell.**

wall of the cell is a transparent Perspex cylinder so that the behaviour of the specimen can be observed during the test. (The classic reference to triaxial testing is Bishop and Henkel, 1962).

The cell is filled with water that is then pressurized to apply the minor principal stress σ_3. The pore water in the specimen communicates with the outside of the cell via a porous ceramic disc on the pedestal of the cell and a hole drilled through the pedestal. As stress is applied to the specimen, the pore water can either be maintained at atmospheric pressure (zero gauge), in which case $\sigma_3 = \sigma_3^{\prime}$, or the pore water pressure can be measured, or a controlled pore water pressure can be applied from outside the cell.

A loading ram passes through a gland in the top of the cell and can be used to apply an axial stress to the top of the specimen. This can be either an additional compressive stress, or a tensile stress. Because σ_3 acts downwards on the top of the specimen, as

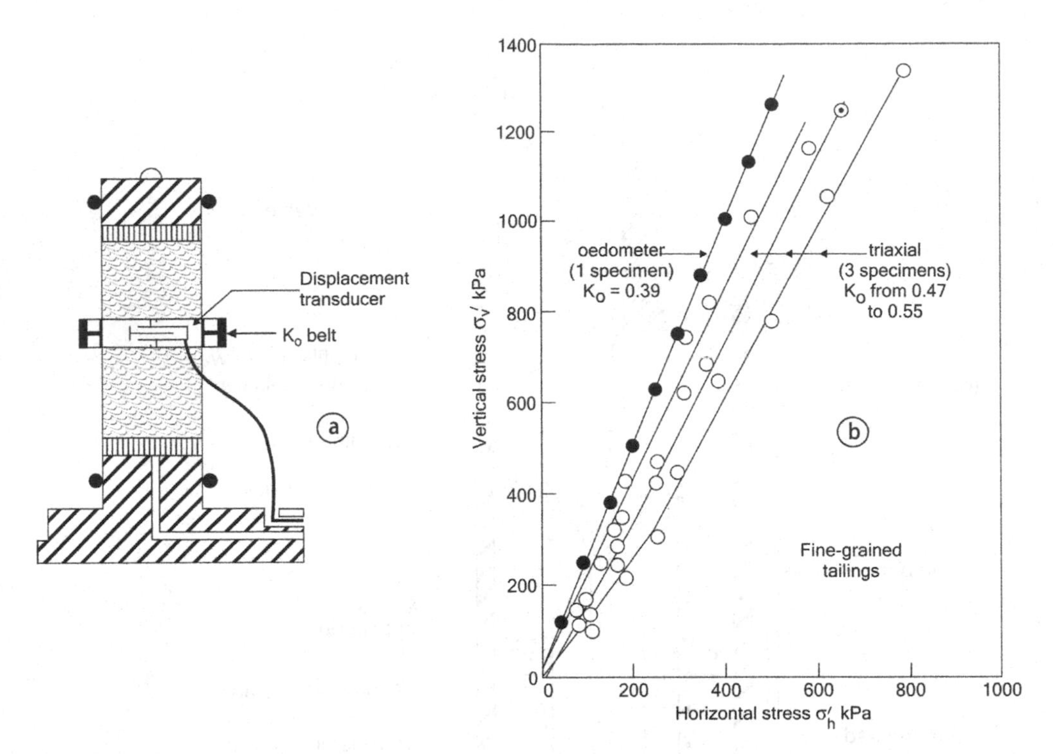

Figure 4.23 a: Measuring K_o in a triaxial cell by means of a K_o belt. b: Comparison of K_o for a fine-grained tailings measured by both K_o oedometer and in the triaxial cell by means of the K_o belt.

well as laterally on its cylindrical sides, the axial stress applied by the loading ram must equal $(\sigma_1 - \sigma_3)$ so that if the axial stress is compressive,

$$\sigma_3 + (\sigma_1 - \sigma_3) = \sigma_1$$

If the axial stress is tensile, the cell pressure becomes the major principal stress σ_1 and the axial stress is

$$\sigma_1 - (\sigma_1 - \sigma_3) = \sigma_3$$

The axial strain of the specimen can be assessed by measuring the movement of the loading ram into or out of the cell, and the volume change can be assessed by measuring the volume of water (at pressure σ_3) flowing into or out of the cell (which is calibrated for volume changes resulting from changes in pressure). Otherwise the apparatus can be fitted with an inner stress-free cell (σ_3 acting inside and out) which is used to measure volume changes.

The stress ratio K_o can be measured in the triaxial cell using a so-called K_o belt, a light-weight hinged or flexible band of metal that fits around the triaxial specimen at mid-height, as shown in Figure 4.23a. The vertical stress σ_1 and the horizontal stress σ_3 are simultaneously increased in such a way that the diameter of the specimen remains constant, as indicated by the displacement transducer on the K_o belt. This maintains the zero lateral strain condition. K_o for unloading can also be measured by simultaneously decreasing both σ_1 and σ_3. The drainage port at the base of the triaxial specimen (Figure 4.22) is left open to atmosphere while measuring K_o to ensure that the pore water pressure remains at zero (gauge). Figure 4.23b shows a comparison of K_o values measured in a triaxial cell with a K_o belt and by means of an oedometer fitted with a K_o ring. As the diagram shows there is a difference between the two types of measurement, as well from one set of triaxial measurements to the next which amounts to 40% of the lowest value. Like all measurements on particulate materials, K_o is a parameter that is affected by many variables and an absolute value cannot be determined. Also, as shown previously in Figure 4.17, at low stresses, the value of K_o may be stress dependent. For the triaxial measurements shown in Figure 4.17, K_o varies from 0.67 at low stresses to 0.5 at higher stresses. Thus it is important to carry out K_o measurements in the stress range for which the value of K_o will be required. This is also evident from the measurements in Figure 4.23.

It is generally a simpler procedure to use the K_o oedometer ring for measuring K_o, rather than the K_o belt in a triaxial compression test. However, most oedometers test a smallish specimen with dimensions of 70 to 80 mm diameter by 25 to 35 mm thick. A general rule of thumb is that the size of the largest particle in the material tested should not exceed one tenth of the minimum dimension of the test specimen. This limits the largest particle tested in an oedometer to 2.5 to 3.5 mm and rules out its use for mine wastes of gravel or larger particle size. It is only just acceptable to test coarse sand grain sizes in an 80 mm diameter oedometer. Standard triaxial cells are available for specimens of up to 150 mm in diameter by 300 mm high, however, setting a limit on maximum particle size of up to 15 mm. For coarser materials, a larger diameter oedometer can be improvised from a short section of steel pipe. Lateral and vertical strains in the pipe wall can be measured by means of electric resistance strain gauges and hence σ^{l}_{h} and σ^{l}_{v} be deduced. (σ^{l}_{v} cannot be taken as (applied load) ÷ (cross-sectional area) as the friction on the walls of the pipe can significantly reduce the vertical stress.)

4.5.7 *The shear strength and shear strain behaviour of particulate materials*

Suppose that a triaxial specimen of mine waste (e.g. see Figure 4.22) is subjected to a constant value of σ_3 with the specimen drainage valve open so that $\sigma_3 = \sigma^{l}_3$ at all times. If $(\sigma_1 - \sigma_3)$, and hence σ_1, is then increased (see Figure 4.24a), the specimen will strain axially as shown in Figure 4.24b.

Figure 4.3 shows that $(\sigma_1 - \sigma_3)$ is twice the maximum shear stress in the specimen and dividing by the constant σ^{l}_3 simply normalizes this shear stress. When $(\sigma_1 - \sigma_3)$ reaches a maximum, the specimen will fail in shear and the maximum obliquity angle α_{max} (equation 4.7) will also be maximized. It has been found experimentally that the shear stress at failure (τ_f) of a particulate material is usually linearly related to the normal stress (σ_f) on the plane on which failure occurs by the Coulomb equation (equation 4.14) which is illustrated in Figure 4.24c: (also see Figure 4.9.)

In the special case where $c^{l} = 0$, which is always the case for NC materials

$$\phi^{l} = \alpha_{max} \quad \text{(see Figure 4.3)}$$

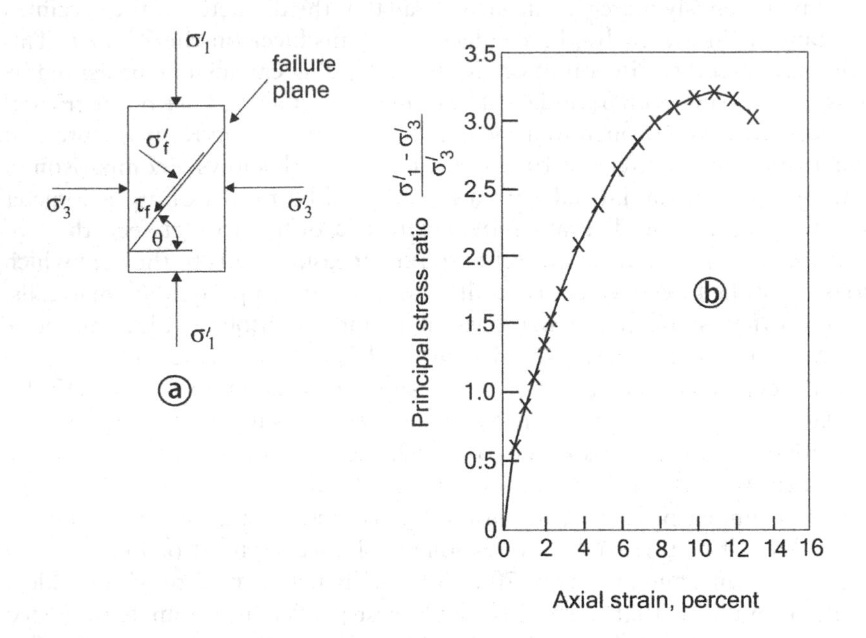

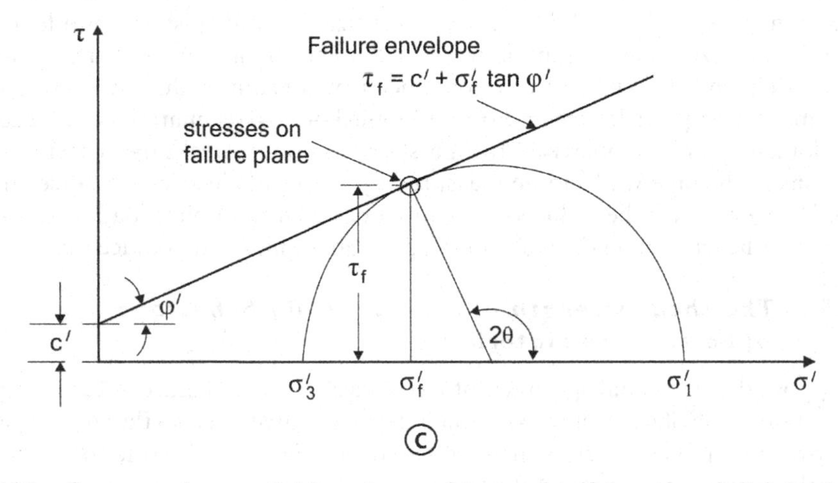

Figure 4.24 a: Stresses acting on triaxial specimen at failure and on failure plane. b: Relationship between stress ratio $(\sigma_1 - \sigma_3)/\sigma_3^1$ and axial strain. c: Coulomb failure envelope.

Note that the failure envelope is not always a straight line, but may be curved. The shear and direct stresses on the failure plane (for $c^I = 0$) are given by

$$\tau_f = \frac{1}{2}(\sigma_1^1 - \sigma_3^1)\sin 2\theta, \text{ and} \tag{4.28}$$

$$\sigma_f = \frac{1}{2}(\sigma_1^1 + \sigma_3^1) + \frac{1}{2}(\sigma_1^1 - \sigma_3^1)\cos 2\theta \tag{4.29}$$

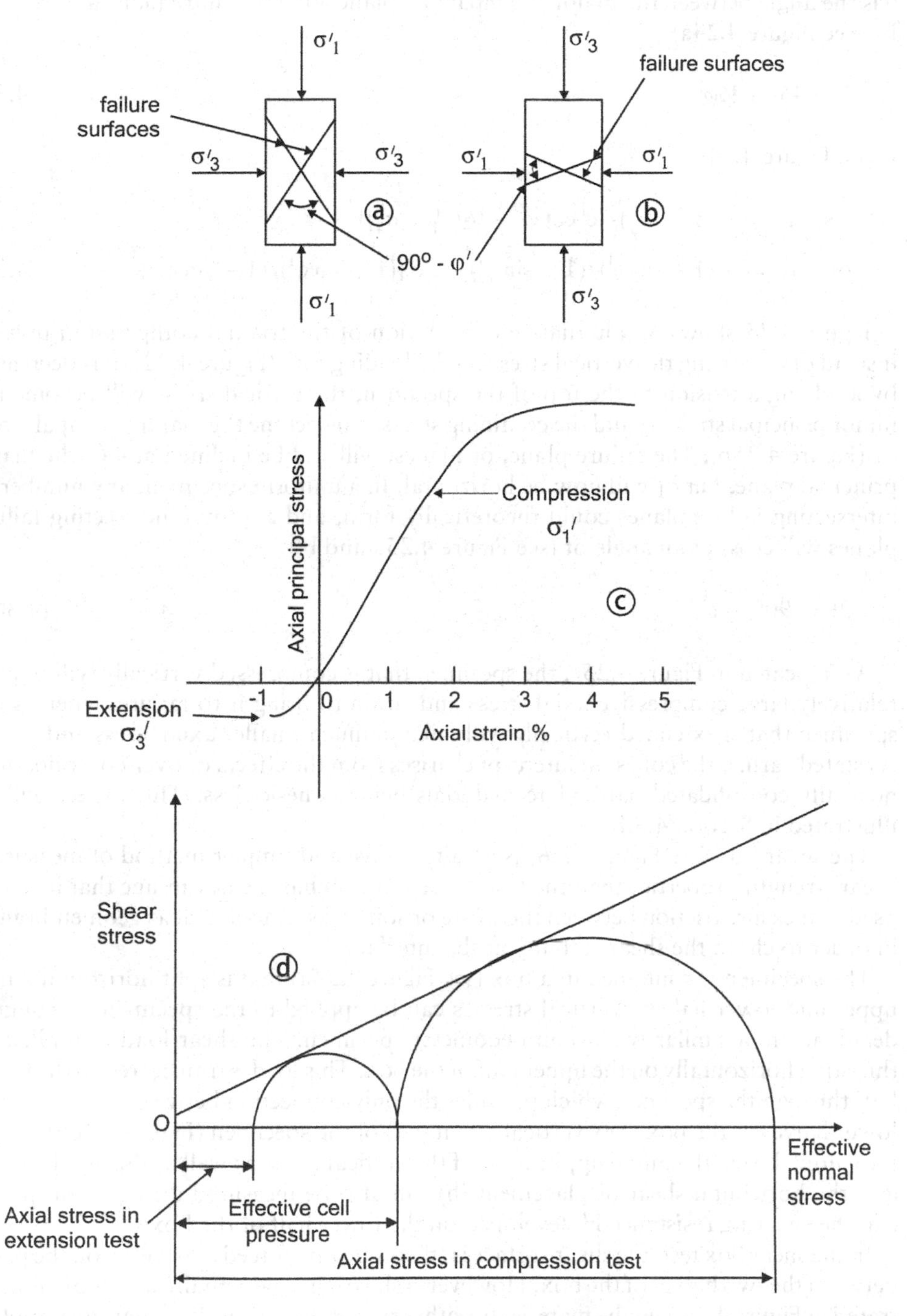

Figure 4.25 **Interpretation of triaxial shear test results: a: triaxial compression; b: triaxial extension; c: triaxial stress-strain relationships; d: Möhr circles at failure.**

θ is the angle between the major principal (σ'_1) plane and the failure plane and is given by (see Figure 4.24a):

$$\theta = 45^\circ + \tfrac{1}{2}\varphi^l \tag{4.30}$$

From Figure 4.24:

$$\sin\varphi' = \tfrac{1}{2}(\sigma_1^l - \sigma_3^l)/\{c^l \cot\varphi^l + \tfrac{1}{2}(\sigma_1^l + \sigma_3^l)\}$$

$$\text{or} \quad \sigma_1^l = \sigma_3^l(1+\sin\varphi^l)/(1-\sin\varphi^l) + 2c^l\{(1+\sin\varphi^l)/(1-\sin\varphi^l)\}^{1/2} \tag{4.31}$$

Figure 4.25 shows an alternative explanation of the triaxial compression test. If, instead of increasing the vertical stress on the loading ram (Figure 4.22), it is decreased by applying a tension to the top of the specimen, the vertical stress will become the minor principal stress σ_3 and the confining stress will become the major principal stress σ_1 (Figure 4.25b). The failure plane, or planes, will still be inclined at θ to the major principal plane, but σ_1 will now be horizontal. In a uniform specimen, any number of intersecting failure planes could theoretically form, and any two intersecting failure planes will cross at an angle of (see Figure 4.25a and b):

$$2\theta = 90^\circ - \varphi^I \tag{4.30a}$$

As indicated in Figure 4.25c, the specimen that is compressed vertically will require relatively large compressive axial stress and strain to bring it to failure, whereas the specimen that is extended vertically will fail at a much smaller axial stress and strain. As stated earlier, the cohesion intercept c^l arises from the effects of over-consolidation, normally consolidated particulate materials being cohesionless. This aspect will be illustrated in Section 4.5.8.

The shear box (see Figure 4.26) is an alternative and simpler method of measuring shear strength properties than the triaxial cell. It also has the advantage that it can be used to measure friction between the waste or soil and surfaces such as geomembranes, in order to check the shear stability at the interface.

The specimen is contained in a box (see Figure 4.26a) that is split horizontally into upper and lower halves. Vertical stresses can be applied to the specimen by hanging deadloads in a similar way to an oedometer specimen. The shear load is applied by thrusting horizontally on the upper half of the box. This load is transferred to the lower half through the specimen which provides the only connection between the upper and lower halves of the box. The vertical strain (v/t) of the specimen (Figure 4.26b) can be measured during the initial application of the vertical stress as well as during shearing, and the horizontal shear displacement (h) can also be measured during shearing, as can the shearing resistance H developed on the lower half of the box.

In the shear box test, the shear surface is (theoretically) forced to develop on the plane between the two halves of the box. However, a shearing zone actually develops, as illustrated in Figure 4.26b, in the material on either side of this plane. The quantities applied or measured are the normal (vertical) stress σ_N^l and the shear stress τ on the potential failure plane. (All tests are done in a drained condition, and the pore pressure is maintained at zero.) At failure, these become σ_f^l and τ_f as defined in Figure 4.24c. In other

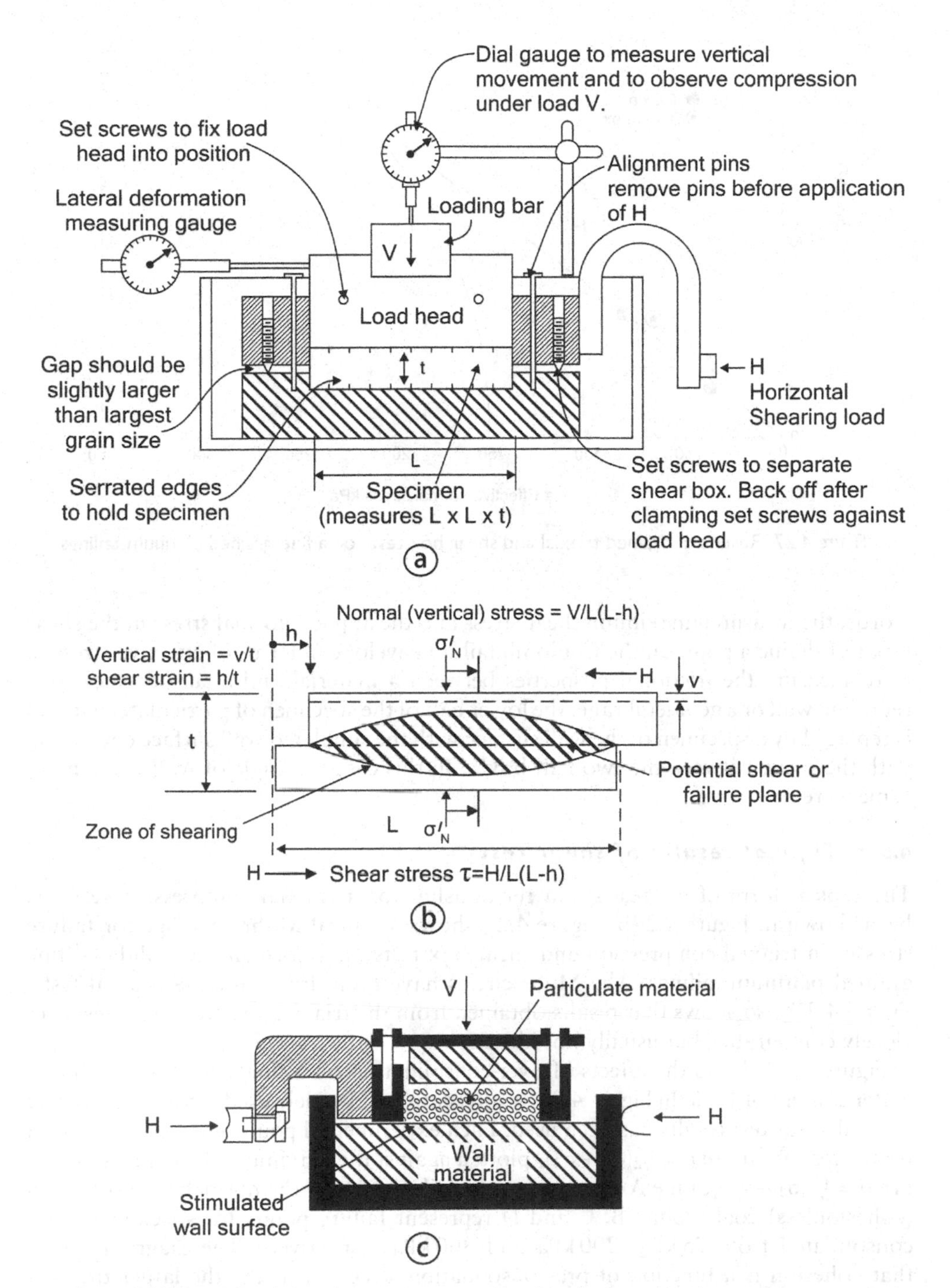

Figure 4.26 Principle of shear box apparatus: a: shear box, b: shear zone in specimen; c: measuring angle of wall friction or interfacial friction.

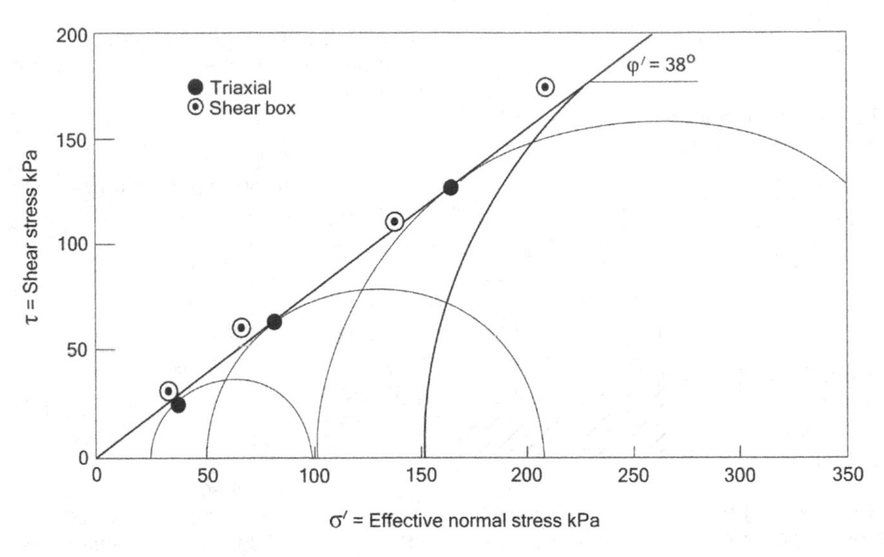

Figure 4.27 Results of drained triaxial and shear box tests on a fine-grained platinum tailings.

words, the measured maximum shear stress and the applied normal stress in the shear box test define a point on the Coulomb failure envelope in terms of effective stresses.

To measure the frictional properties between a material and a surface such as a retaining wall or a geomembrane, the lower half of the specimen of particulate material is replaced by a specimen of the wall surface with the simulated wall surface coinciding with the plane between the two half boxes. In this case the angle of wall friction (δ^{I}) is measured.

4.5.8 *Typical results of shear tests*

The typical form of a stress-strain relationship for a triaxial compression test has been shown in Figure 4.24b. Figure 4.27 shows a typical Mohr envelope for failure stresses in triaxial compression and shear box tests on a normally consolidated fine grained platinum tailings. The Mohr circles have been drawn for the triaxial tests. Figure 4.27 also shows that results obtained from the triaxial and shear box tests are closely comparable, but usually not identical.

Figure 4.28 shows the effects of over-consolidation on a fine coal slurry having a water content of 14%. In Figure 4.28 an alternative and widely used method of plotting triaxial shear test results has been used. Each experimental point represents the mean total stress at failure $p = \frac{1}{2}(\sigma_1 + \sigma_3)_f$ plotted against the maximum shear stress at failure $q = \frac{1}{2}(\sigma_1 - \sigma_3)_f$. Line AA represents the failure line for the normally consolidated (cohesionless) coal. Points B, C and D represent failure points for specimens over-consolidated from 75 kPa, 200 kPa and 300 kPa respectively. The diagram shows that cohesion is a function of pre-consolidation stress – usually, the larger the pre-consolidation the larger the cohesion. Because the coal was moist, but not saturated, capillary stresses existed between the coal particles, and it was these capillary stresses that (in this case) gave rise to the cohesion. Because the capillary stresses were not

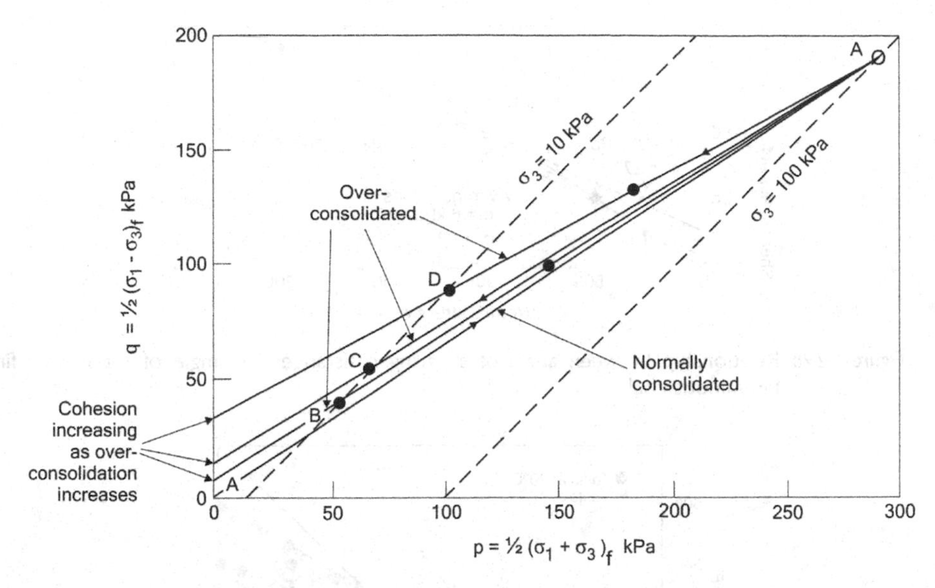

Figure 4.28 p–q plot showing effect of overconsolidation on total stress cohesion in an unsaturated fine coal slurry (14% water content).

measured, Figure 4.28 is plotted in terms of total stress, even though the pore air pressure was zero throughout all tests.

The Coulomb equation (4.14) implies that the shear strength is a linear function of the effective stress and that φ^l is always constant. This is approximately true for most particulate materials. However, if the individual grains are friable or relatively weak, they will tend to crush progressively as the average effective stress p^l is increased. In this case Φ^l may change (usually reduce) as p^l increases. Figure 4.29a, for example, shows the (shear box) Mohr envelope for a bituminous coal. The angle of repose represents the value of Φ^l for a particulate material in its loosest state, and is usually the minimum value Φ^l assumes. In the case of coal, coal wastes and materials with weak grains, e.g. waste rock consisting of a partly weathered shale, φ^I may be variable. As shown here for coal, φ^I reduces from 39° at low values of σ^I_N to 27° at higher values. The angle of repose has an intermediate value of 36°.

Figure 4.29b shows a further example of the curved Mohr envelope that is usually associated with coal and coal wastes. In this case the material was a fine coal duff which at the time was unsaleable and therefore regarded as a waste. The results of the shear box tests showed a considerable scatter and it was for this reason that so many tests (40) were carried out. At low stresses the value of φ^1 was 55°, but the envelope flattened to between 15° and 30° at higher stresses. Because the tests were made to investigate the stability of a large (70 m high) dump of the waste duff that threatened housing established at its toe if it failed, (see Section 5.3.4) and because the in situ duff coal had heated to over 70°C as a result of spontaneous heating, the first series of 40 tests at a testing temperature of 20°C was supplemented by a further limited series of 5 tests at 100°C. As shown in Figure 4.29b, the heated tests (H) were weaker than those done at 20°C, but also indicated a value of φ^1 of 15°, plus an equivalent

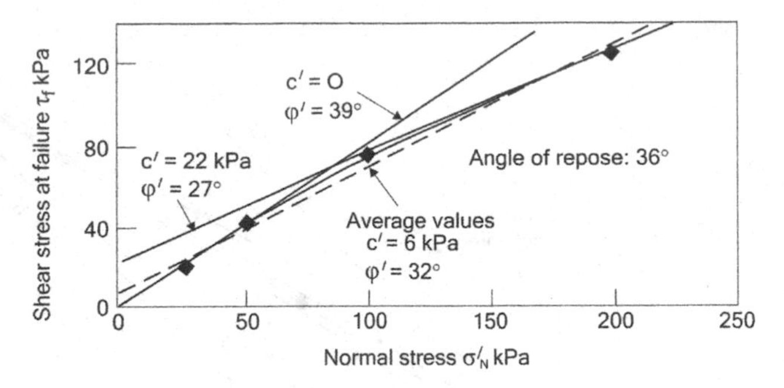

Figure 4.29a Relationship between angle of shearing resistance and angle of repose for fine bituminous coal.

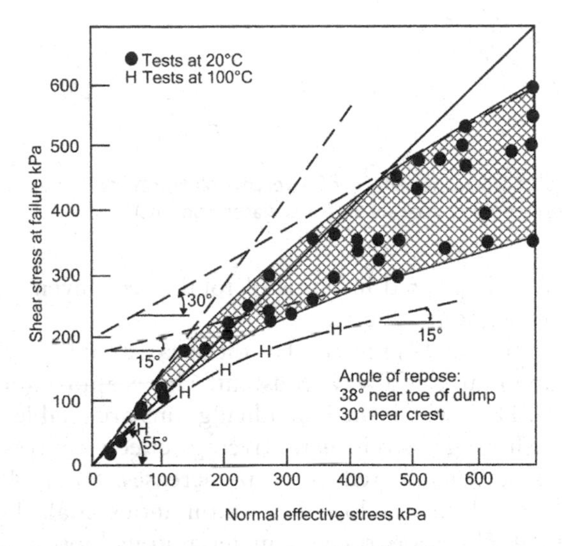

Figure 4.29b Möhr failure envelope for a waste coal duff with strengths measured at both 20°C and 100°C.

"cohesion" of 100 kPa as a result of the curvature of the envelope. The angle of repose of the tipping face of the dump varied from 38° near the toe to 30° near the crest with an average of 34°. Stability analyses indicated that the average shear strength required for stability (factor of safety exceeding 1.0) was 84 kPa which at low normal stresses falls within that part of the strength envelope for which φ^1 was at least 45°. At higher normal stresses, 84 kPa lay well below the envelope of failure shear stresses.

4.5.9 *Contractant and dilatant behavior during shear*

Figure 4.12 showed typical pore pressure behaviour for a normally consolidated tailings, in which the pore pressure increases continually when sheared undrained. If this material is subjected to drained shear, its volume reduces or contracts. The opposite type of behaviour is exhibited by a heavily over-consolidated tailings. When sheared

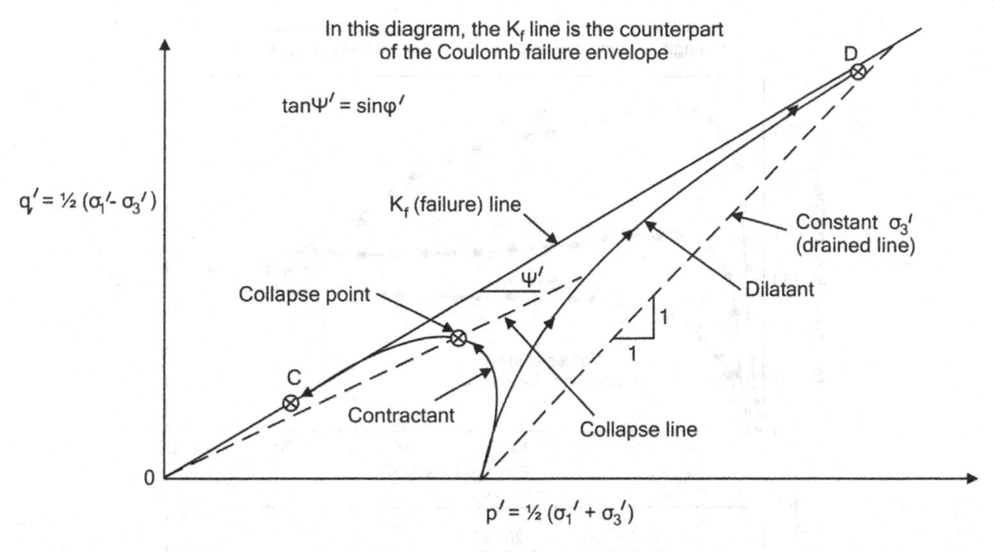

Figure 4.30 Stress paths on a p^l-q^l plot showing contractant and dilatant undrained shear behaviour of a saturated tailings.

undrained, the pore pressure initially increases, but then starts to decrease, and decreases continually as failure is approached and passed. In drained shear, the OC material increases in volume, or dilates. Because of its tendency to contract, the behaviour of the NC tailings is termed "contractant", while that of the OC tailings is "dilatant".

Stress paths for the two types of behaviour are shown schematically in Figure 4.30. In the case of dilatant behaviour, the continually decreasing pore pressure causes p^l to increase continually, and hence the shear strength q^l also continues to increase. In the case of contractant behaviour, the increasing pore pressure causes p^l to decrease, and after a maximum value of q^l has been reached, the specimen starts to lose strength. This is a particularly dangerous phenomenon in the case of hydraulic fill tailings storages, because, if the overall factor of safety against shear failure is close to unity, the shear strength reduction over even a limited part of a potential failure surface could cause catastrophic overall failure to occur. (In Figure 4.30, note that the slope of the K_f-line, which is the counterpart of the Coulomb failure envelope in the Mohr circle plot, is denoted ψ^l and that $\tan \psi^l = \sin \Phi^l$.)

Although the pore pressure and strength behaviour of the contractant and dilatant specimens is so completely different, both stress paths end on the same K_f-line (at C for contractant and D for dilatant). This can lead to confusion and error if the difference between contractant and dilatant behaviour is not recognized and only the shear strength parameters for the K_f-line are quoted. It must be emphasized that contractant materials are strain-softening and dilatant materials are strain-hardening, and a clear distinction must be made between the two. If the K_f-line is used to estimate a drained shear strength, for instance, point D would be predicted, not realizing that if the material is contractant and undrained shearing occurs, starting from the same value of p^l the strength will be given by point C.

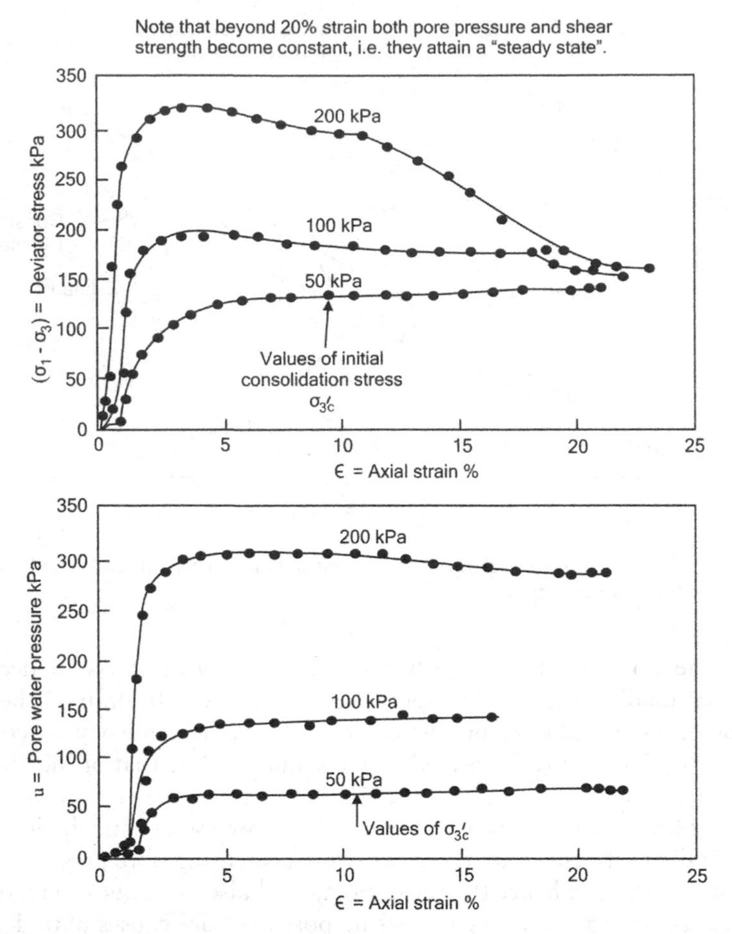

Figure 4.31 Deviator stress versus strain and pore pressure change versus strain for a normally consolidated tailings that is contractant during shear.

Figure 4.30 can also be used to introduce the concept of the "collapse point", the point on the stress path for a contractant material at which maximum shear stress is reached, before the material softens and loses strength (Sladen, et al., 1985). A line joining a series of collapse points forms the "collapse line".

Figure 4.31 shows a set of data for three consolidated undrained triaxial shear tests on a saturated platinum tailings. The specimens were all trimmed from a single piston tube sample for which the in situ effective stress was about 50 kPa. The upper diagram shows the deviator stress $(\sigma_1 - \sigma_3)$ versus axial strain, while the lower diagram shows the corresponding pore water pressure versus axial strain. Figure 4.32 shows the corresponding stress paths in $p^l - q^l$ space. It will be noted that losses of strength, after maximum $(\sigma_1 - \sigma_3)$, occurred for the specimens consolidated to 100 and 200 kPa, but the one consolidated to 50 kPa reached an approximately constant position on the K_f-line after an axial strain of more than 20%. The figure also shows the collapse line which, in this case, lies very close to the K_f-line.

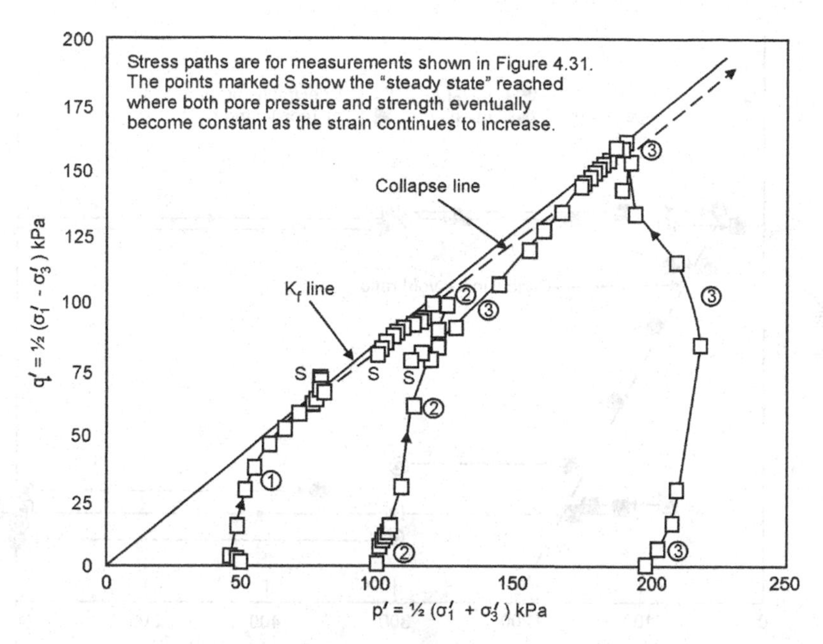

Figure 4.32 Stress paths for a normally consolidated tailings that is contractive during shear and loses strength after reaching failure.

The two major factors that decide on whether shear behaviour will be dilatant or contractant are the initial void ratio and the initial effective stress. This statement tends to be supported by Figure 4.32 which could be interpreted as showing that a material consolidated to 50 kPa was neither dilatant or contractant, but neutral, whereas the same material consolidated to a higher initial effective stress became contractant. However, it must be remembered that void ratios in hydraulically placed tailings can be highly variable from layer to layer of deposition and the specimens tested at higher consolidation stresses may simply have had higher initial void ratios. The fact of the matter is that tailings deposited hydraulically can be very variable in properties.

Figure 4.33 shows the relationship between changes of $p^l = \frac{1}{2}(\sigma_1^l + \sigma_3^l)$ and void ratio during consolidated undrained shear of nine undisturbed specimens of platinum tailings (open and closed circles). The results indicate the variability of properties found from tests on undisturbed specimens, and also show a tentative critical void ratio line that divides dilatant behaviour (to the left) from contractant behaviour (to the right). Figure 4.34 shows similar data for tests on undisturbed block samples cut from the failure surface of the Merriespruit gold tailings storage that suffered a catastrophic flow failure in 1994 (see Section 11.3.2 and e.g. Wagener, et al., 1997, Fourie et al., 2001). In terms of variability, the characteristics of the platinum and gold tailings appear very similar, and only one test result (in Figure 4.34) indicates a very large loss of strength once peak q′ had been passed.

Figure 4.34 shows a "change of state" line (Ishihara, et al., 1991) which joins the points on the stress paths of the three slightly dilatant specimens at which the initial contractant behaviour, with increasing pore pressure, goes over to dilatant behaviour,

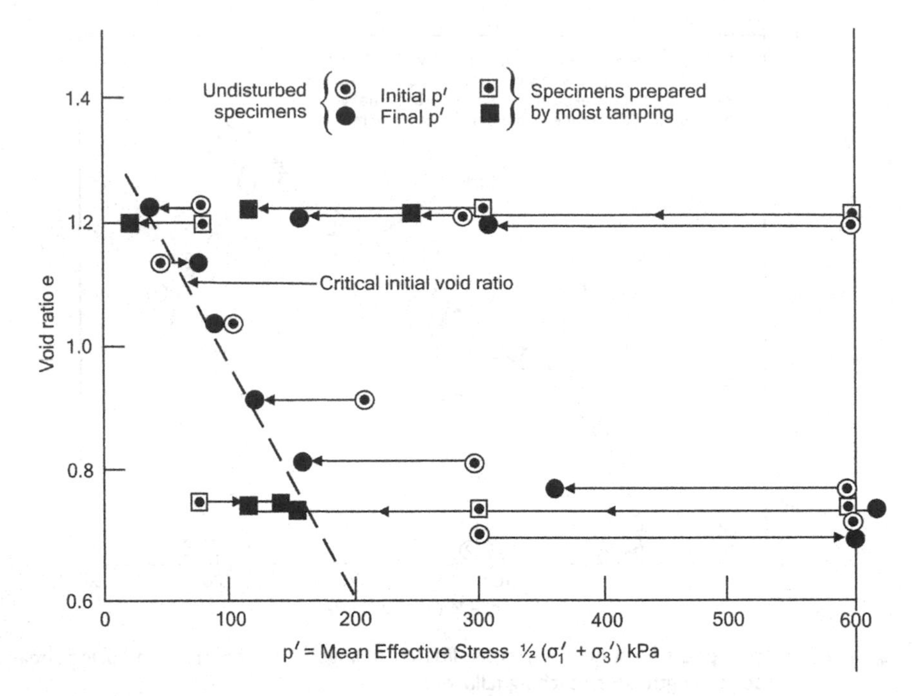

Figure 4.33 Effective stress changes in undisturbed and laboratory-prepared specimens of platinum tailings subjected to consolidated undrained triaxial shear.

with the pore pressure decreasing as the shear stress increases (See Figures 4.13a and 4.43a, also.) In other words, the initial contractant state changes to a dilatant state.

It is very difficult to prepare high void ratio, contractant specimens even in the laboratory, and the results may also be scattered and contradictory, as shown in Figures 4.33 and 4.35. The specimens prepared in the laboratory at $e = 1.20$ gave results that were consistent in that all specimens showed a loss of strength after passing maximum q′. Of results for specimens prepared at $e = 0.75$, however, one showed dilatancy, rather than contractancy from an initial $p^1 = 80$ kPa and the largest reduction of strength occurred with the specimen that started with $p^1 = 600$ kPa.

Figure 4.35 has also been used to show that the K_f-line (in terms of effective stresses) alone does not distinguish contractant from dilatant behaviour if the stress paths are not shown. The K_f-line in terms of total stress, however, shows the effect of the collapse to low strengths, as the specimens pass the collapse line and approach the steady state at which pore pressure and shear strength remain constant. (See Figures 4.31 and 4.32.)

Results of this type of testing must be viewed with caution, as the contractivity may be exaggerated in comparison with what is likely to happen in the prototype tailings storage.

4.5.10 *The effects of repeated loading on shear behaviour*

There are several situations in which repeated loads may be applied to mine waste. The most obvious is earthquake loading in areas that are seismically active, the second is

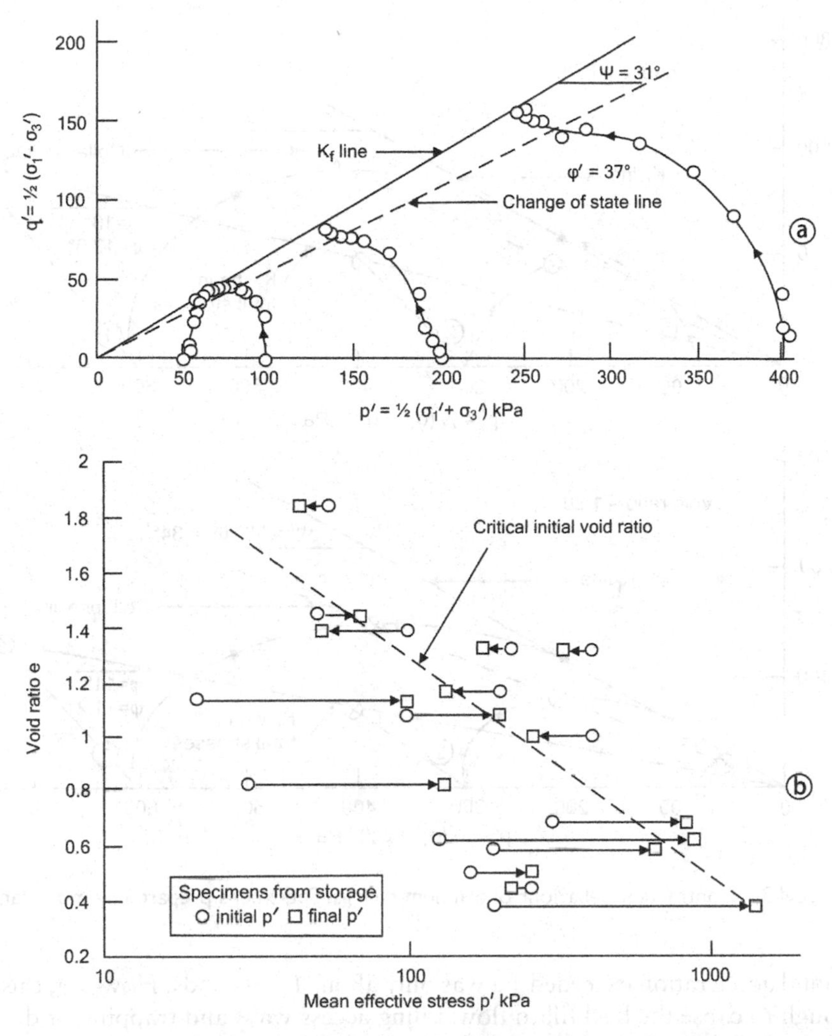

Figure 4.34 **a: Stress paths for consolidated undrained triaxial shear of undisturbed tailings specimens from the Merriespruit tailings storage. b: Effective stress changes in undisturbed tailings specimens from Merriespruit during unconsolidated undrained triaxial shear.**

loading due to mining-induced seismicity. Typical accelerations caused by earth-quake loading are shown in Figure 4.36a, an accelerogram for the Olympia, Washington, USA earthquake of April 13, 1949. Note that the peak acceleration was 0.16 g. The frequency of shaking at maximum acceleration was slightly more than 2 Hz. The second accelerogram, shown in Figure 4.36b is for a mining-induced seismic event, and represents the sort of acceleration that could be applied to an underground backfill installed for strata support. The characteristic of Figure 4.36b is completely different, with a peak horizontal acceleration of 7.7 g at a frequency of 300 Hz. Whereas the duration of shaking for the natural earthquake for which the acceleration exceeded 0.1 g was about 20 minutes, for the mining-induced shake the duration for which the

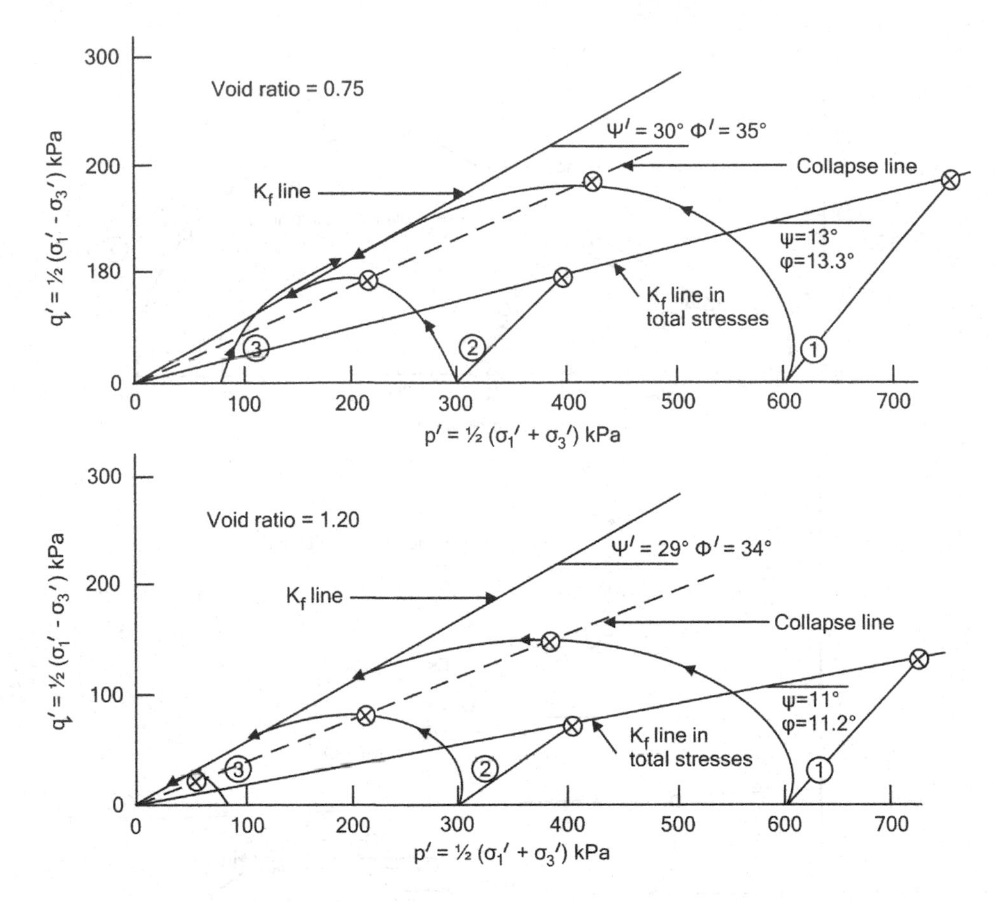

Figure 4.35 Contractant behaviour of platinum tailings. Specimens prepared by moist tamping.

horizontal acceleration exceeded 1 g was only about 1.5 seconds. However, this might be enough to cause the backfill to flow, filling access ways and trapping or drowning workers underground.

When a fine grained saturated high void ratio tailings is loaded in triaxial shear, the pore pressure and strain will increase as shown (Δu, loading) in Figure 4.37a. If the load is removed, only a small part of the strain will be recovered and the pore pressure increment will also only be partly recovered (Δu, unloading). If the load is re-applied so that the initial loading strain increment is doubled, the pore pressure will increase and decrease again on unloading. If the process of loading, straining and unloading is repeated, a pseudo steady-state condition will be approached in which the increment of pore pressure in loading approaches that on unloading (i.e. Δu loading $=$ Δu unloading).

Figure 4.37b shows the results of dynamic undrained shear tests on five nominally identical specimens of high void ratio saturated tailings. The numbers 1, 2, 3, etc represent data points for the cumulative pore pressure in the loading cycles as strain accumulated under a load cycle consisting of increasing q^1 from zero to q^1 to zero to

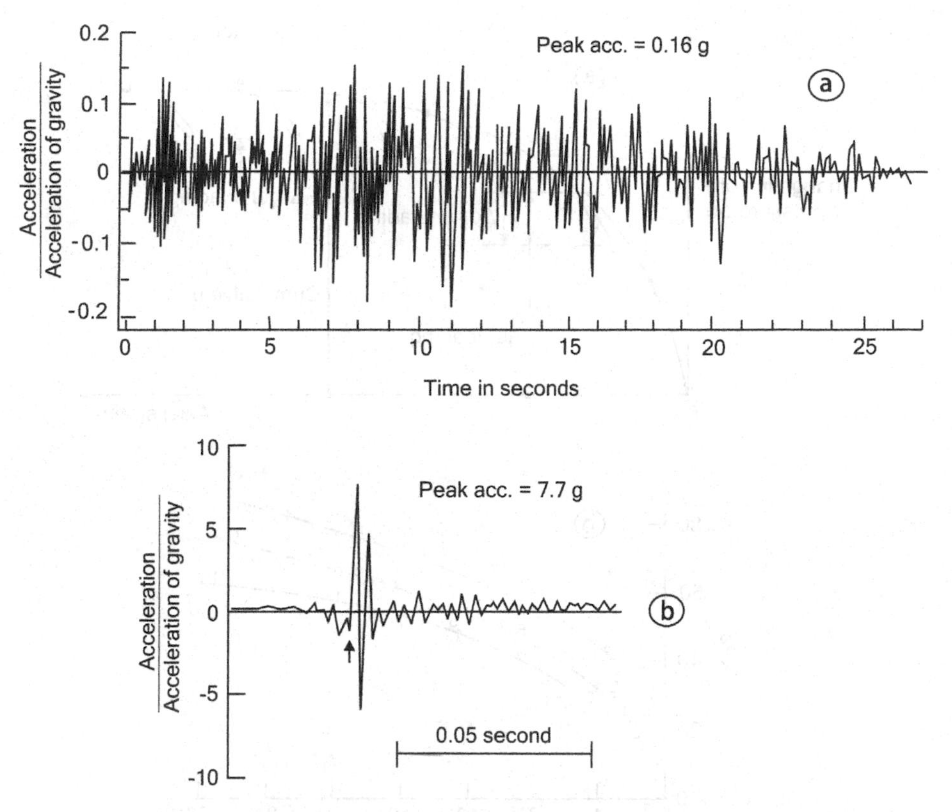

Figure 4.36 a: Accelerogram for Olympia, Washington earthquake of April 13, 1949. b: Accelerogram for horizontal component of underground seismic event (rock burst).

$q^l + \Delta q^l$ to zero, etc. Specimens all had nominally the same void ratio. Note that all specimens approach a steady state pore pressure as the axial strain approaches 30%.

Figure 4.38 shows the stress path in $p^l - q^l$ space for a loose saturated sandy silt in which a constant increment of q^l was repeatedly applied. The diagram also shows the equivalent (schematic) stress path for monotonic loading. In this test, the specimen was able to support the applied shear stress until, on the thirteenth application, the stress path reached the K_f-line and the specimen began to fail. On the fifteenth cycle the specimen collapsed.

Figure 4.39 compares the effects of dynamic shear loading with monotonically increasing shear stress on a saturated cycloned gold tailings which is in use as an underground fill to provide strata support in ultra-deep (3000 m and deeper) gold mines. The backfill is designed to be dense and have a low void ratio and it is therefore intended to behave dilatively. Figure 4.39a shows the results of tests on two samples subjected to monotonically increasing undrained shear loading. Sample 1 was initially consolidated to 80 kPa and (surprisingly) contracted as it approached the K_f-line. The specimen was then consolidated to 150 kPa and after showing some initial contraction, began to dilate and followed the K_f-line upwards to a value of $q^l = 340$ kPa.

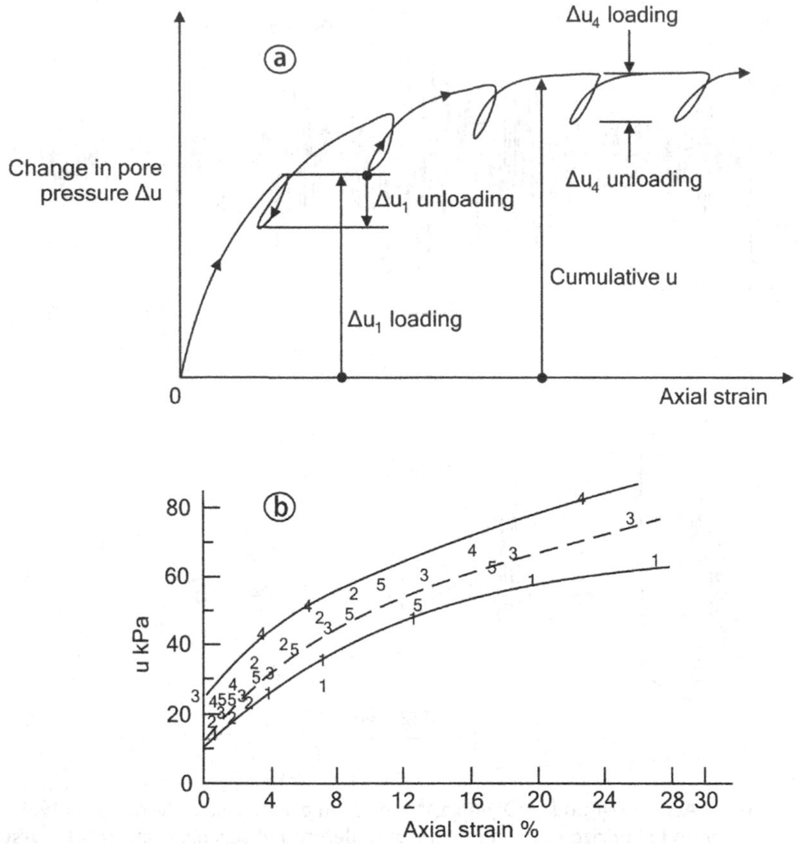

Figure 4.37 a: Schematic of increases of pore pressure when high void ratio tailings is subjected to repeated undrained loading. b: Dynamic undrained shear tests on 5 nominally identical specimens of tailings to illustrate degree of reproducibility of phenomenon of pore pressure increase.

Sample 2 (consolidated to only 25 kPa) dilated immediately it was loaded and followed the K_f-line upwards to values of q^l beyond 400 kPa.

Figure 4.39b shows the results of dynamic undrained shear loading of two specimens of the same material that had been tested under monotonically increasing loading (Figure 4.39a). The loading cycle increased q^l from zero to q^l, to zero to $q^l + \Delta q^l$, to zero. In some tests, the return to zero load was omitted. After consolidating each specimen isotropically, dynamic shear stresses were applied to the undrained specimens. After applying 25 cycles of a particular shear stress q^l, at a frequency of 0.5 Hz, the shear stress was increased and another 25 cycles applied. In Figure 4.39b each point represents the end condition of a 25 cycle loading episode.

It will be seen that the stress paths for dynamic tests look very similar to those for monotonically increasing shear. In each case, the stress path moved through a change of state and up the K_f-line for the material. However, the K_f-line for dynamic shear was somewhat steeper than that for static shear. It did not appear to make a difference

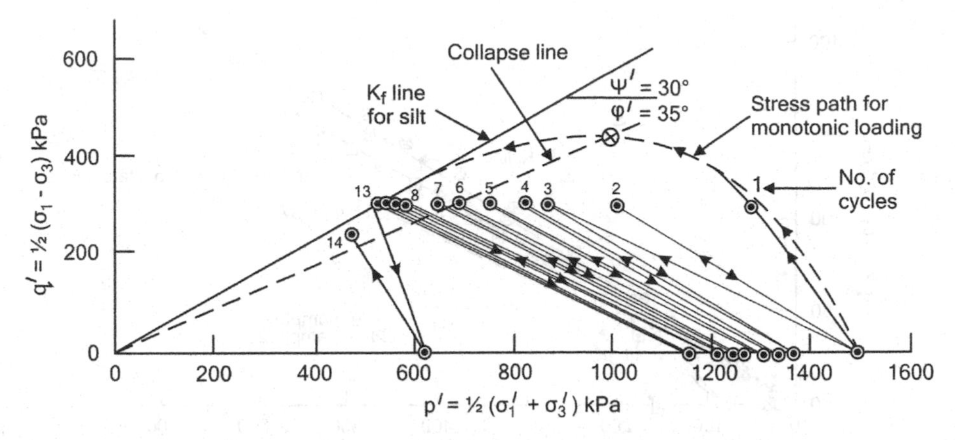

Figure 4.38 Stress path for dynamic shear test on loose saturated natural sandy silt subjected to a repeated constant loading of $(\sigma_1 - \sigma_3)$ to zero to $(\sigma_1 - \sigma_3)$.

if the specimens were cycled under a cyclically increased shear load or under cyclically increased and decreased shear load, as indicated in Figure 4.39b.

All specimens showed similar behaviour whether they were at first cyclically loaded well below the K_f-line, or whether the very first loading brought them onto the K_f-line. It should, however, be noted that the axial strains that corresponded to the stress paths in Figure 4.39 were large, being more than 20% when the tests were terminated.

It may be concluded that whether a particulate material is tested undrained under monotonically increasing shear stress or cyclically applied shear stress, there is no fundamental difference in the mechanics of shearing. The loose specimen tested under a repeated constant shear stress behaved very similarly to the denser specimens tested under monotonically increasing or fluctuating applied shear stress.

4.5.11 A generalization of the effects of applying monotonically increasing or fluctuating shear stress under undrained conditions

Static or dynamic liquefaction has been identified as the key cause of catastrophic flow failures in a number of tailings dams (e.g. Blight, 2000, Fourie et al., 2001, Jefferies and Bean, 2006). Figure 4.40 (Olson and Stark, 2003) generalizes the mechanisms of static and dynamic liquefaction in a saturated cohesionless tailings that has a sufficiently large void ratio to be contractant when subjected to shear stress. Diagram a represents the strain response when the tailings are subjected to a static, cyclically applied or seismic shear stress under undrained conditions. In diagram a, point A represents the result of applying a static in situ shear stress that can be matched by the available shear resistance. If the shear stress is increased sufficiently to reach the maximum available undrained shear strength, at B, the yield shear stress under undrained conditions, (e.g. by depositing another lift of material), any attempt to increase the shear stress further can only result in a decrease of available shear stress resistance, and the tailings will spontaneously strain from B to C. On diagram b, the stress path will move from the initial point A, to B on the yield strength envelope, to C on the failure

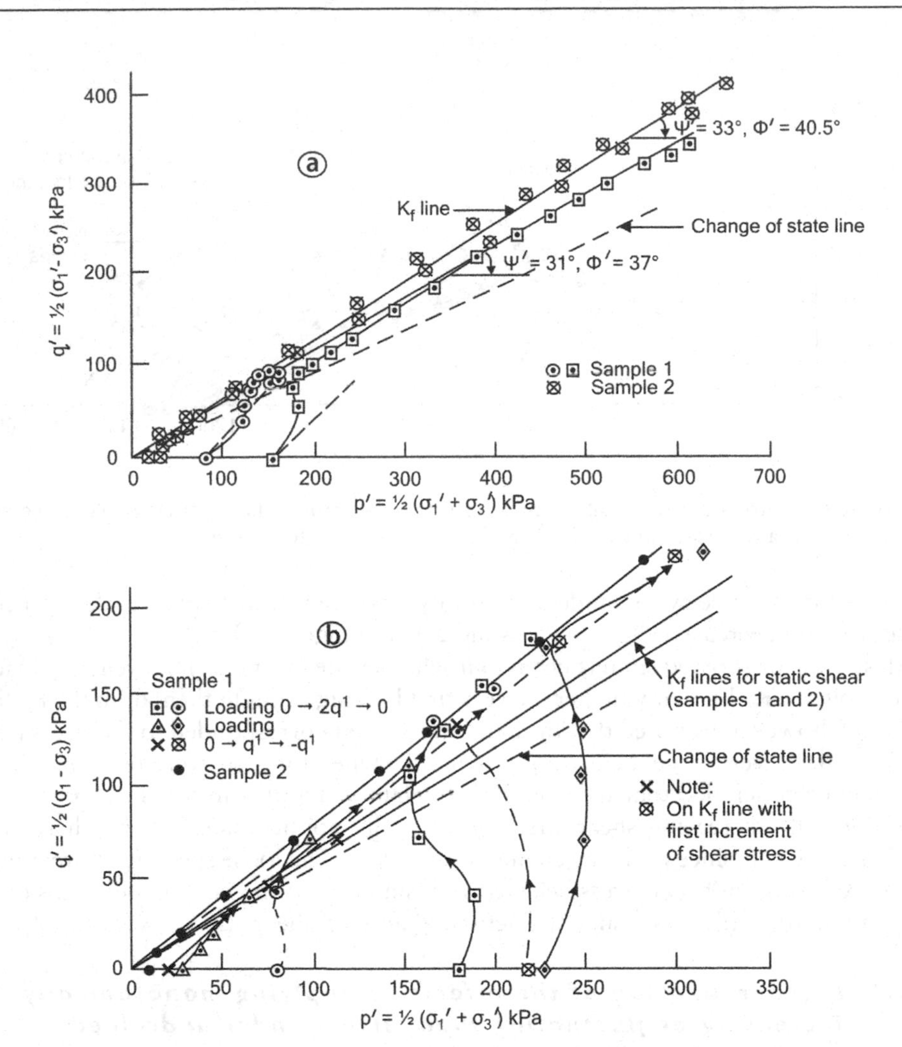

Figure 4.39 a: Stress paths for monotonically increasing shear loading of a typical underground tailings fill. b: Stress paths for dynamic undrained shearing of same fill as in a.

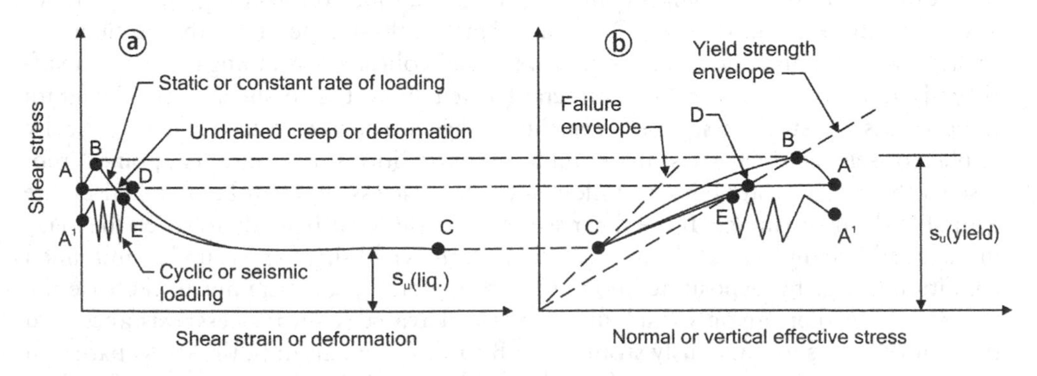

Figure 4.40 The mechanism of liquefaction in a saturated contractant tailings.

envelope and beyond. The available shear strength may decrease substantially, and the strain or deformation increase to a large degree. Alternatively, if (in both diagrams a and b) the tailings creeps under static shear stress A from A to D, then when the creep strain path reaches point D, the strain will spontaneously increase to C. A third possibility is that cyclic or seismic stress applications, together with their accompanying irreversible strain, will move a point such as A^1 to point E. Once again, the tailings will spontaneously lose strength and the strain (diagram a) or stress (diagram b) will move towards C. Once the strain or stress path passes point B, progression to point C may be very rapid, resulting in sudden and possibly catastrophic failure.

4.6 The process of consolidation and pore pressure re-distribution in laboratory shear tests

4.6.1 *Laboratory testing for compressibility*

The compressibility of natural clays, silts and clayey sands as well as tailings is usually measured by means of the oedometer test:

A cylindrical disc of undisturbed or compacted natural soil or tailings (usually 75 mm to 100 mm in diameter and 18 mm to 25 mm thick) is carefully trimmed from an undisturbed block or tube sample (for 75 mm tube samples, the sample ring should be 70 mm dia) so as to fit snugly into a polished brass or stainless steel sample ring. (The rings are sometimes lined with Teflon to reduce side friction.)

The specimen is compressed between rigid porous plates, the whole, for saturated samples, being immersed in water (see Figure 4.19a.) As the specimen is immersed in water at atmospheric pressure, the pore pressure at the end of each loading period is zero and hence the vertical effective stress $\sigma^1_v = p =$ applied vertical stress when the load is at equilibrium, i.e. when settlement under a particular load increment is complete.

Loads are usually applied in a series of increments increasing logarithmically in magnitude, e.g. 20, 50, 100, 200 etc., kPa. Each increment is left in place until the specimen has ceased to compress. Conventionally, this usually means 24 hours, but some clays may take longer to reach equilibrium (e.g. Figure 4.20a). Conversely equilibrium may be reached more quickly with more pervious soils. If the interest is in the rate at which settlement will take place, time-settlement curves are observed for a number of load increments within the stress range that is of concern. It is convenient to plot the time axis of the observations to a square root scale. This compresses the axis and also transforms the initial portion of the curve into a straight line. Figure 4.20a shows a typical compression-square root of time curve, for an oedometer test which often show an instantaneous compression. Compression then proceeds linearly (on a square root of time plot) for a time and then slows down and levels off. In some very plastic clays the curve does not level off, but the clay continues to compress even after all excess pwp has dissipated (this is known as secondary consolidation). The theory of consolidation is dealt with in Section 6.11.

The relationship between applied vertical stress p or σ^1_v and the vertical strain $\Delta h/h$ or the void ratio e is used to estimate the magnitude of the settlement under a given foundation loading. Typical test results are shown in Figures 4.20b and 4.21. If the stress axis is plotted to a logarithmic scale, the settlement curve is linearized (Figure 4.20b) and a log σ^1_v scale is usually adopted for this reason.

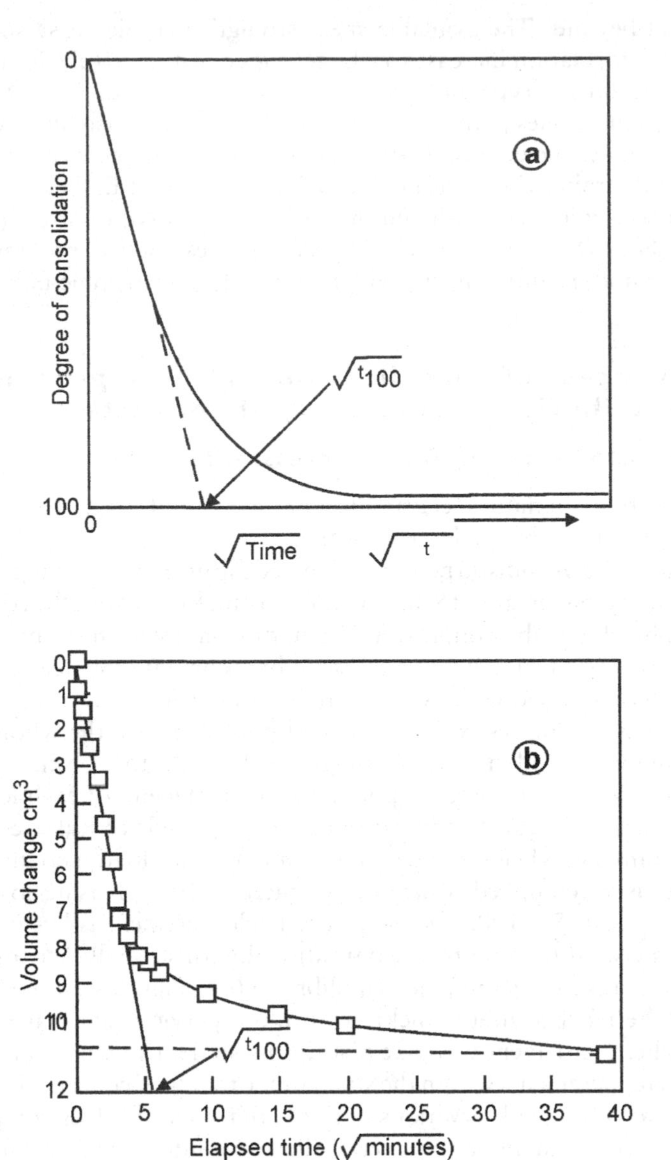

Figure 4.41 Evaluation of coefficient of consolidation from consolidation curve for triaxial test. a: Form of curve (degree of consolidation versus √time), and b: example of measured curve (volume change versus √time).

4.6.2 *Calculating the coefficient of consolidation, c_v*

(See Section 6.11, equation (6.14a)

During the consolidation stage of the triaxial or shear box test the variation of volume with time is recorded. A typical time-volume change curve for triaxial consolidation is shown diagrammatically in Figure 4.41a As already explained, a square root scale is used on the time axis as this results in a straight line plot for the initial stages

Table 4.1 Calculation of coefficient of consolidation c_v from t_{100} measured in triaxial test, shear box or oedometer

Type of Test	*Drainage Surfaces at*	$t_{100} =$
Triaxial	One end only (Single end drainage)	$\frac{\pi h^2}{c_v}$
Shear box, oedometer or triaxial	Both ends (Double end drainage)	$\frac{\pi h^2}{4c_v}$
Triaxial	Both ends and radial boundary (All round drainage)	$\frac{\pi h^2}{4c_v} \cdot \left(1 + \frac{2h}{R}\right)^2 = \frac{\pi h^2}{100c_v}$ if h = 2R

of consolidation. The volume change is expressed as a percentage of the total volume change occurring. Figure 4.41b shows an example of an actual triaxial consolidation curve for a tailings.

The straight line portion of the curve is extended to cut the 100 per cent consolidation ordinate at an abscissa of $\sqrt{t_{100}}$. From the value of t_{100} and Table 4.1 below, a value of c_v can be calculated. This factor will have the units $[L^2T^{-1}]$, e.g. cm^2/s. In Table 4.1:

h = height of specimen
2R = diameter of cylindrical specimen

Table 4.1 gives relationships between t_{100} and c_v that allow c_v to be calculated for drainage from one end of a triaxial specimen only (the drainage boundary would normally be at the base of the triaxial specimen, see Figure 4.22), from both ends of a triaxial specimen (a porous disc would be located at the top of the specimen with an external drainage tube to the base of the triaxial cell), or from both ends and the radial boundary (porous discs would be provided top and bottom and a layer of longitudinally slotted filter paper would be wrapped around the soil specimen to provide the radial drainage).

In the case of a shear box or oedometer, porous discs are provided at the top and bottom of the soil specimen to give drainage from both ends.

As an example, c_v values will be calculated for the test results shown in Figures 4.20 and 4.41b.

For the oedometer test results in Figure 4.20, $t_{100} = 485$ minutes, sample thickness h = 2.5 cm and therefore, with drainage at both ends:

$$c_v = \frac{\pi h^2}{4t_{100}} = \frac{\pi(2.5)^2}{4 \times 485 \times 60} \text{cm}^2/\text{s} = 0.0169\,\text{cm}^2/\text{s}$$

$$= 0.0169 \times 10^{-4} \times 3600 \times 24 \times 365 = 53\,\text{m}^2/\text{y}$$

For the single end drainage curve of Figure 4.41b, the specimen height was 38 cm = h, $t_{100} = 36$ minutes

$$c_v = \frac{\pi h^2}{t_{100}} = \frac{\pi(38)^2}{36 \times 60} = 2.1\,\text{cm}^2/\text{s}$$

$$= 2.1 \times 10^{-4} \times 3600 \times 24 \times 365 = 6600\,\text{m}^2/\text{y}$$

4.6.3 *Variability of measured coefficients of consolidation*

The coefficient of consolidation (see equations 6.16 and 6.17) is defined by

$$c_v = k(1 + e)/a_v\gamma_w$$

it is therefore affected by the variability of the coefficient of permeability k, the void ratio e and the compressibility of the particulate material a_v, all of which vary with and are affected by particle size distribution.

Figures 4.42a and b show collected measurements of c_v for platinum tailings and for gold tailings from the Gauteng region in South Africa. For the platinum tailings, c_v varies from about 20 000 m^2/y close to the surface and near the edge of the storage to 100 m^2/y at 30 m depth. As the inset diagram shows, this is not really a variation with depth, but a variation with distance from the point of deposition, as a depth of 30 m down a hole drilled vertically from the current point of deposition corresponds to a point 120 m towards the pool at 30m depth. The variation in c_v results from the gravitational sorting of particle sizes that occurs as the tailings slurry is beached. For the gold tailings, similar trends of c_v with depth are apparent. Because the gold tailings are ground finer, the reduction of c_v with depth (or distance away from the point of deposition) is less, being by a factor of about 20 rather than the 200 for platinum tailings. The mean value for gold tailings is 200 m^2/y, but values vary from just over 400 m^2/y to 20 m^2/y. A second contributing cause of the variability is probably variability in particle size analyses because of differences in the fineness of grinding (see Figure 4.1) from mine to mine, and also the inclusion of varying amounts of shaly rock in the ore.

It is therefore dangerous to assume for design purposes that a generic value of c_v applies to tailings of a particular product or geological origin, e.g. a particular gold tailings is quartzitic, and therefore all gold tailings are quartzitic and have similar values of c_v.

4.6.4 *Concerning rates of testing and times to failure in drained and undrained triaxial testing*

The rigid loading platens used in conventional triaxial cells cause frictional constraint at the ends of the test specimen that results in the pore water pressure at the ends of the specimen being larger than it is in the central region in which the shearing action takes place. (A similar effect occurs in a shear box specimen, see Figure 4.26b.) The magnitude of the pore pressure difference that can occur is illustrated in Figure 4.43a for a triaxial test on an unsaturated compacted clay in which the pore water pressure u_w was measured simultaneously at the base and by means of a probe at mid-height of the 200 mm high specimen. In this example, the difference in pore pressure between

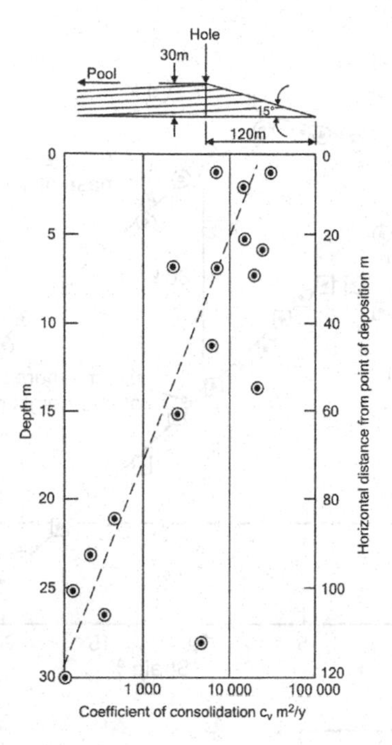

Figure 4.42a Coefficient of consolidation of platinum tailings and its variation with depth and distance from point of deposition.

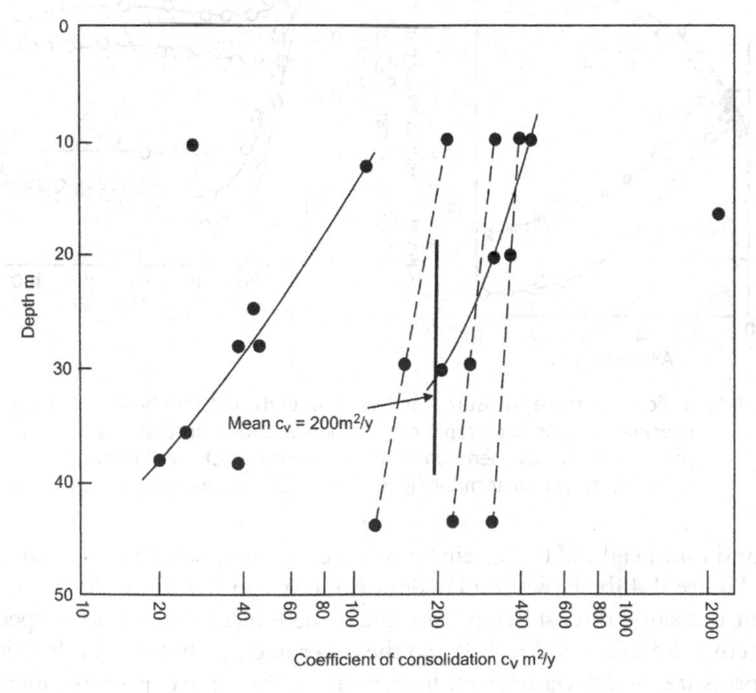

Figure 4.42b Collected measurements of the coefficient of consolidation of gold tailings and its variation with depth.

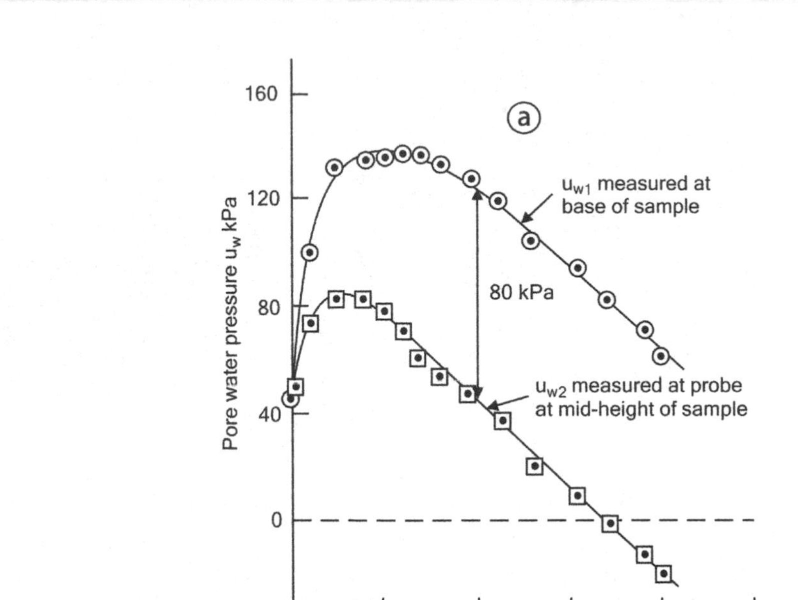

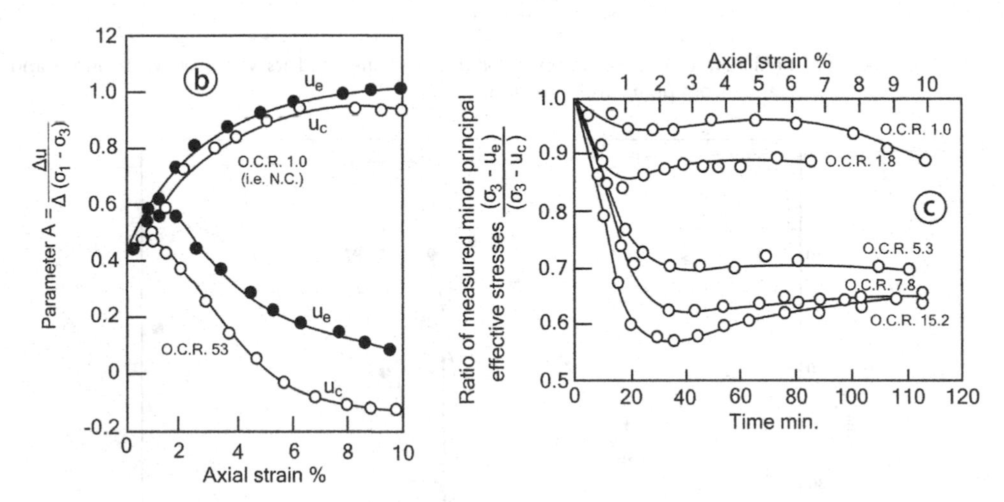

Figure 4.43 a: Pore pressure measured at base (or end) and mid-height of a compacted clay specimen strained at a rapid rate. b: Similar data (expressed as A versus strain) for a pair of triaxial specimens, one NC and the other OC. c: Ratio of $\sigma_3^{\prime}$ measured at end (e) to $\sigma_3^{\prime}$ measured at mid-height (c) as OCR was increased from 1.0 to 15.2.

the base and mid-height of the specimen was a consistent 80 kPa over a range of strain of 15%. Figure 4.43b shows similar data for tests on the same material, that were strained at the same rate, starting from the same effective stress. One specimen was normally consolidated (OCR = 1.0) and the other had an OCR of 5.3. In this diagram, the pore pressures and deviator stress have been combined to express the measurements as the parameter A versus strain (see equation (4.19)c).

Figure 4.43c shows the effect of an increasing OCR on the ratio of the difference between σ_3^1 measured in terms of the pore pressure at the end (or base) of a series of triaxial specimens to σ_3^1 measured in terms of the pore pressure at the centre (or mid-height). The diagram shows that not only does the difference between centre and ends persist for all OCRs, but it increases with increasing OCR.

The problem of pore pressure differences between the ends and mid-height of a triaxial specimen also applies to drained testing, where the assumption is that the rate of straining is sufficiently slow to ensure that pore pressures caused by shearing (see equation (4.19c) dissipate as straining or loading occurs. In both undrained and drained testing, the error can be eliminated by testing sufficiently slowly to allow the pore pressure between ends and centre of the specimen to equalize, by movement of water from the zone of higher pore pressure.

The problem of deciding on a suitable time to failure can be approached theoretically by means of consolidation theory (e.g. Gibson and Henkel, 1954, Bishop and Henkel, 1962), but the theoretical solution assumes that the drained surfaces are 100% efficient, which they are not (e.g. Blight, 1963) and for this reason, an experimental solution was developed. This appears in Figure 4.44. The input data consists of the c_v, the specimen dimensions and the drainage conditions.

Figure 4.44a shows (upper diagram) experimentally established equalization curves A and B for drained triaxial shear tests equipped either with all-round filter paper drains or with double end drainage. These have been drawn in terms of the degree of drainage versus the time factor $T = c_v t_f / H^2$, where t_f is the time to reach failure and **H is half the sample height.** The lower diagram shows a similar experimental equalisation curve for undrained tests with all-round drains. It will be noted that the equalisation curve for drained tests with all-round drains is almost identical with that for undrained tests with all-round drains. Theoretically, if the time factor T_{90} for 90% equalisation in tests with all-round drains is 0.04, T_{90} for tests with double end drainage, in which the minimum drainage path length is twice as long should be 4 times larger, i.e. T_{90} should be 0.16. In fact, the experimental value is 20 times larger, i.e. 0.8. This is because the theory assumes that the peripheral all-round filter paper drains are 100% efficient, whereas their efficiency is only 20%.

Figure 4.44b shows the information from Figure 4.44a redrawn in the form of a simple chart for which the degree of equalisation is 95%.

As far as pore pressure equalization is concerned drained tests with double end drainage and undrained tests without circumferential filter paper drains behave alike. Shear box tests are drained tests with double end drainage. The time to failure for 95% degree of drainage can be calculated from:

$$t_f = 1.6\,H^2/c_v$$

Drained and undrained tests with double end drainage and circumferential drains (i.e. all-round drains) behave alike and the time to failure can be calculated from:

$$t_f = 0.07\,H^2/c_v$$

To convert a time to failure to a rate of strain it is usually advisable to do a trial test to establish the strain at failure for the particular material being tested.

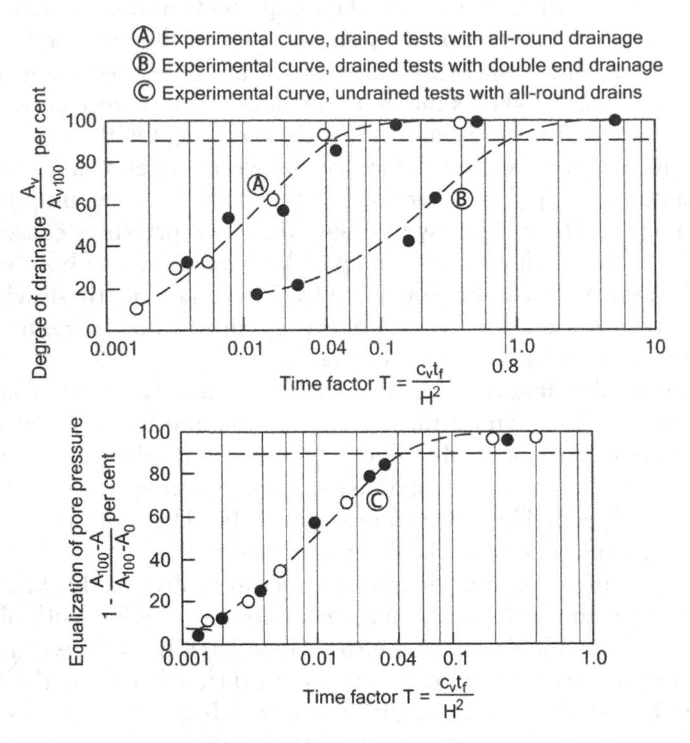

Figure 4.44a The relation between degree of drainage and time factor in drained tests (top) and the relation between equalization of pore pressure and time factor in undrained tests (bottom).

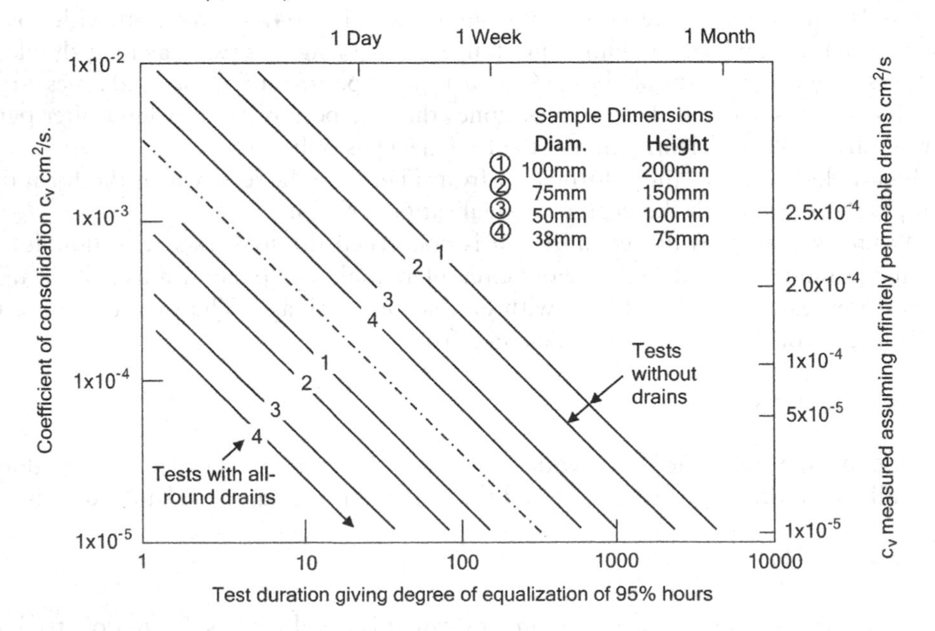

Figure 4.44b Chart for finding test duration giving 95% pore pressure equalization.

4.7 The strength and viscosity of tailings at large water contents

The water content of a tailings can vary from several hundred per cent when it is delivered from the mineral extraction plant to the tailings storage, to less than 10 per cent in desiccated tailings in place in the storage. So far, the strength at water contents of a few tens of per cent and strength values of tens to hundreds of kiloPascals has been considered. In this section, the strength properties of tailings slurries at large water contents and low strengths (ones to tens of Pascals) will be considered.

4.7.1 *Relationships between viscosity, shear strength, and water content*

A knowledge of the relationship between the viscosity, shear strength and the water content of tailings slurries is important in the assessment of

- head losses in slurry pumping lines,
- the slope assumed by thickened slurries when deposited by the thickened-discharge or paste method, and
- the flow pattern of a slurry after it has escaped through a failure breach in the outer wall of a tailings dam.

A wide range of water contents is of interest. For deposition in a conventional tailings storage, the water content of the tailings during transportation from the plant to the storage is typically in the range of 150 to 100 per cent by mass of dry solids, a relative density (RD) for the slurry of 1.3 to 1.5. For deposition by the thickened-discharge or paste method, the water content is typically about 50 per cent (RD 1.7 to 1.8). For the slurry filling of underground mining excavations, the water content ranges from 50 to 30 per cent (RD 1.7 to 1.9). Actual water contents depend very much on the particle-size distribution of the slurry and the nature of clay minerals present. The relative densities are, of course, also influenced by that of the solid particles (G).

A number of methods are in common use for the measurement of viscosity, in absolute terms. The variable-speed coaxial-cylinder type of viscometer has been found to be suitable for the measurement of the absolute viscosity of tailings slurries, and all the measurements reported here were made with instruments of this type (which is shown diagrammatically in Figure 4.45).

The viscosity of Newtonian fluids is defined by the relationship

$$\tau = \frac{d\Gamma}{dt}\eta \qquad (4.32)$$

in which τ is the shear strength of the fluid, η is its viscosity, and

$\frac{d\Gamma}{dt}$ is the rate of shear strain.

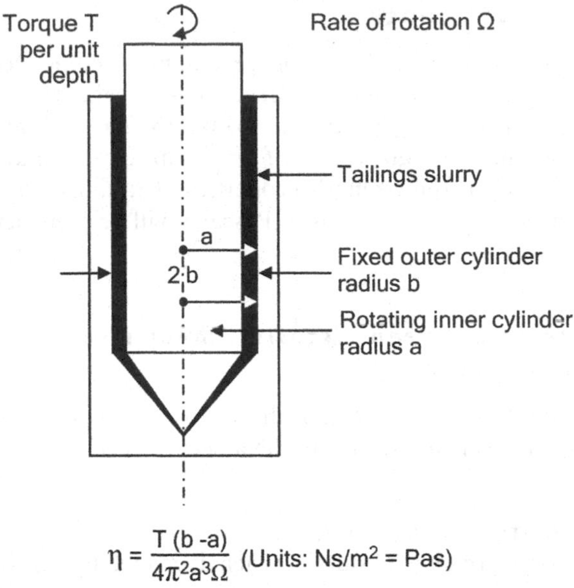

Figure 4.45 Principle of co-axial cylinder viscometer.

However, tailings slurries are non-Newtonian and behave more as Bingham plastics, for which viscosity is related to shear strength by

$$\tau = \tau_0 + \frac{d\Gamma}{dt}\eta \tag{4.32a}$$

In equation (4.32)a τ_0 is a yield shear strength that applies when $d\Gamma/dt = 0$.

The coaxial-cylinder viscometer relies on the torque generated by shear resistance of the slurry, at a given rate of rotation, to indicate shear strength and viscosity. If the conventional interpretation of the viscometer readings is followed (i.e. according to equation (4.32)), what is actually measured is an apparent viscosity η_a.

$$\eta_a = \eta + \frac{\tau_0}{d\Gamma/dt} \tag{4.33a}$$

If the true viscosity is required, the value of τ_0 has first to be established from the relationship between τ and $d\Gamma/dt$. The true viscosity can then be determined from

$$\eta = \frac{\tau - \tau_0}{d\Gamma/dt} \tag{4.33b}$$

As most flow relationships have been derived for viscous fluids and not Bingham plastics, the apparent viscosity, η_a, is in fact a useful engineering property for approximating the behaviour of a slurry to that of a Newtonian fluid.

Figure 4.46a, which shows the properties of a tailings slurry from a diamond (kimberlite) operation at a water content of 125 per cent, illustrates the relationship between shear strength and rate of shear strain, and indicates a yield shear strength, τ_0, of 3.5 Pa. The fact that the shear strength does not increase linearly with the rate

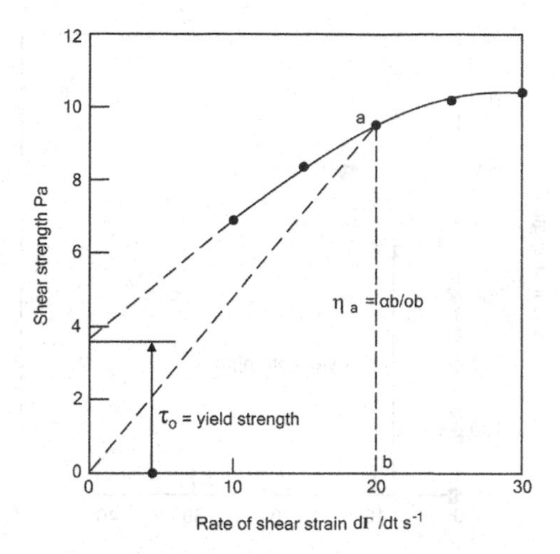

Figure 4.46a **Relationship between shear strength and rate of shear strain for diamond tailings at a water content of 125%.**

of shear strain shows that, although the behaviour of the slurry approximates that of a Bingham plastic, it is actually more complex than equation (4.32a) would indicate.

Figure 4.46b is a second example of a measured yield strength, this time for a water purification sludge (suspended clay removed from a river water by flocculating it with lime). This diagram shows the initial extremely small shear strain that occurs before the slurry, at a water content of 280%, yields at a shear stress of 380 Pa (10 times larger than the yield stress of the diamond tailings at a water content of twice as much!).

Figure 4.47 shows relationships between apparent viscosity, η_a, and rate of shear strain (calculated from equation (4.32)) and between 'true' viscosity, η, and rate of shear strain (from equation (4.32)a). It is again clear that equation (4.32a) is only an approximation to the real behaviour of the slurry.

Figures 4.48 and 4.49 show the results of a series of shear strength and viscosity measurements on a slurry of platinum tailings. The measurements were made as part of an investigation into the flow of tailings through a breach in a tailings dam, and the water contents correspond to settled and partly consolidated tailings that have become liquefied by shear disturbance. Figure 4.48 shows relationships between shear strength and rate of shear strain for a range of water contents. Figure 4.49 shows relationships between apparent viscosity, η_a, and rate of shear strain and between 'true' viscosity, η, and rate of shear strain. The pattern is similar to that shown by Figure 4.47.

It appears from Figures 4.47 and 4.49 that, while the water content has a major influence on the yield shear strength, τ_0, and the apparent viscosity, η_a, it has much less influence on the 'true' viscosity, η.

As an alternative, the apparent viscosity of tailings slurries can be characterized by an empirical equation of the form

$$\eta_a = nK\left(\frac{d\Gamma}{dt}\right)^{n-1} \tag{4.34}$$

where n and K are empirical parameters.

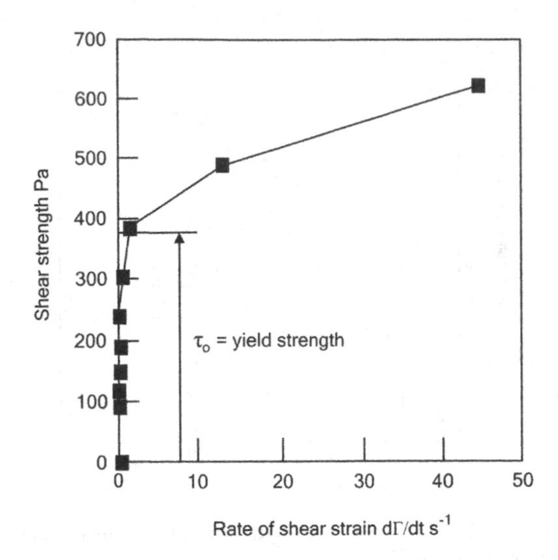

Figure 4.46b Relationship between shear strength and rate of shear strain for water purification sludge at a water content of 280%.

Integration of this equation with respect to (dΓ/dt) gives the result for shear strength:

$$\tau = \tau_0 + K(d\Gamma/dt)^n \tag{4.34a}$$

The shear strengths shown in Figures 4.46a and b and 4.48 are considerably less than those measureable by means of the shear box or the triaxial shear test (less than 5 Pa in Figure 4.48, as compared with the lowest strength of 50 kPa in Figure 4.34). However, there is a continuous spectrum of strengths with varying water content between these two ranges. This is illustrated by Figure 4.50 which shows the results of shear strength measurements on a gold tailings over a range of water contents from 17 to 69%.

The measurements cover a range of shear strengths from nearly 30 kPa down to less than 0.01 kPa. Because of the large range of strengths covered in the figure, the shear strengths have been plotted to a natural scale in the upper diagram and to a logarithmic scale in the lower one in order to display the data more clearly. Strengths above 1 kPa were measured by means of a laboratory hand vane, while those below 0.1 kPa were measured by means of a viscometer. Unfortunately, the measurement ranges of the two instruments do not overlap, but the data appear continuous. The strength of the tailings declines very rapidly with increasing water content. For example, an increase in water content of 5% in these tailings decreased the shear strength from 8 kPa to 1 kPa, while a further increase of 5% decreased it to 0.3 kPa. In the slurry range, an increase of water content from 45 to 50% was enough to reduce the strength from 0.05 kPa (50 Pa) to 0.01 kPa (10 Pa).

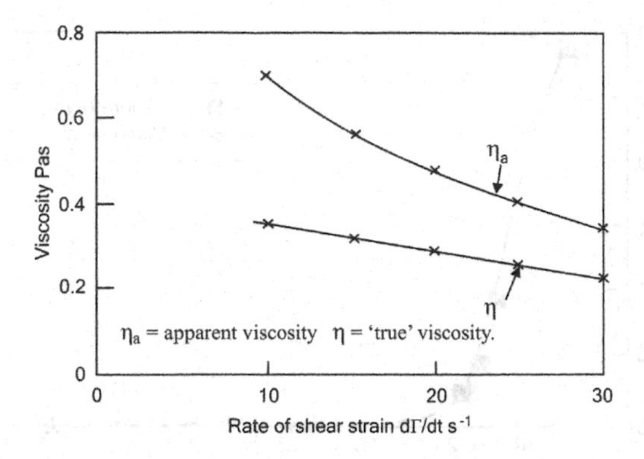

Figure 4.47 Relationship between viscosity and rate of shear strain for diamond tailings at a water content of 125%.

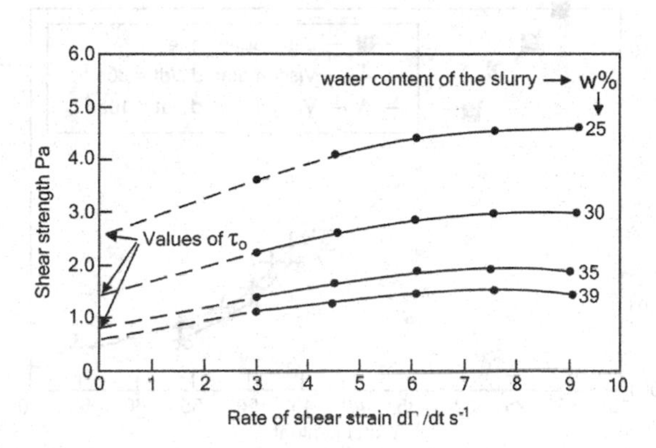

Figure 4.48 Measurements of shear strength of platinum tailings showing the influence of varying water content.

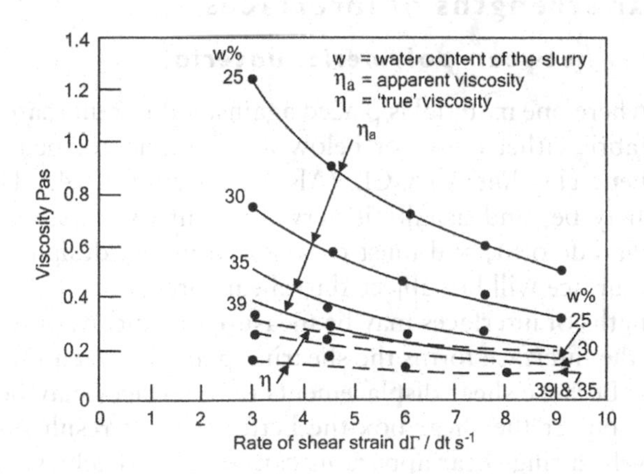

Figure 4.49 Measurements of viscosity of platinum tailings showing the influence of water content.

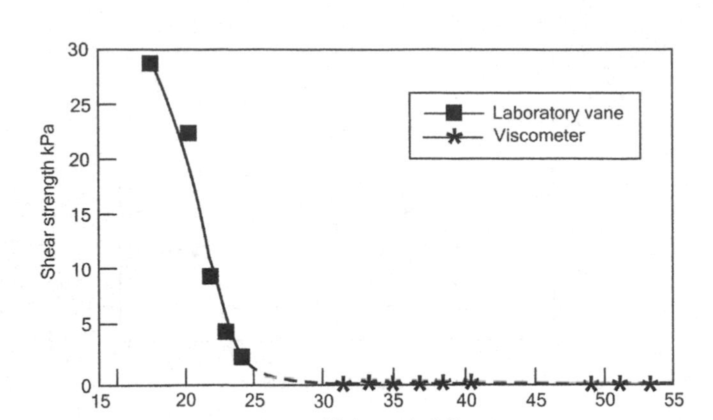

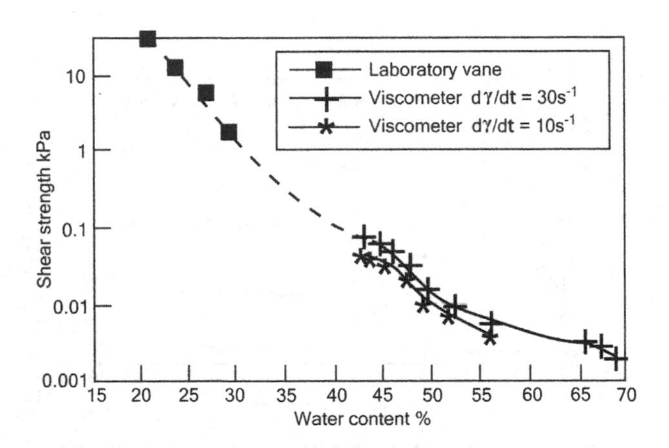

Figure 4.50 Variation of strength with water content for gold tailings.

4.8 The shear strengths of interfaces

4.8.1 *Interfaces between polymeric materials*

Interfaces occur where one material is placed against a different material, for example, a protective geofabric either above or below a polymeric geomembrane, or tailings above a geosynthetic clay liner or GCL. (Also see Section 6.14.) The shear strength of the interface may be, and usually is very different to the shear strengths of the materials on either side of it, and must be assessed in any design situation in which such a two-sided surface will be subjected to shear forces.

The shear strengths of interfaces may be measured by means of tests in a shear box modified so that the interface forms the shearing plane between the upper and lower halves of the box. Because shear displacements on interfaces may become quite large (tens of mm), the bigger the shear box the better the test result (other things being equal). Alternatively, a ring shear apparatus can be used to achieve the required large shear displacements.

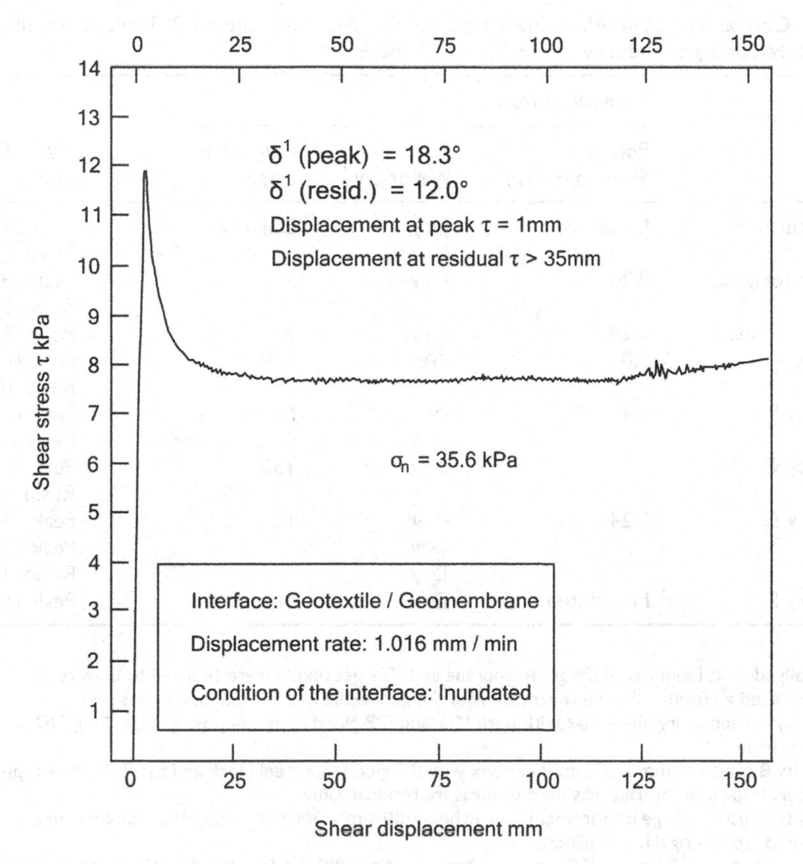

Figure 4.51 Typical result for large displacement shear box test of fPP geomembrane to PET geotextile interface.

Figure 4.26 shows a shear box, as it is used to measure an interface shear strength between two flexible sheets of material. The test specimen is replaced with two blocks either of metal or wood with specimens of the two materials (e.g. geofabric and geomembrane) glued to their surfaces, and with upper and lower thicknesses adjusted so that the plane of the interface accurately coincides with the split between the upper and lower halves of the shear box. If the interface between a flexible sheet and a particulate material (e.g. tailings) is to be measured, the block-mounted sheet of material occupies the lower half of the box and the upper half is filled with tailings. Tests are usually carried out with the interface inundated with water, preferably leachate seepage water similar in soluble solid content and chemical analysis to that of the waste being investigated.

Figure 4.51 shows a typical shear stress versus shear displacement relationship for an interface between an fPP (flexible polypropylene) geomembrane and a PET (polyethylene terephthalate) non-woven geofabric. The test was performed in a 300 mm × 300 mm shear box, in this case with the PET fabric supported on a sand bed

Table 4.2 Comparison of interface shear tests for fPP geomembrane on PET felt geotextile. Tests by manufacturer, designer and universities A, B, C, D and E.

	Condition of tests			
Tested by	*Rate of Shear (mm/min)*	*Wet or dry*	*Maximum Displacement mm*	*Angle of inter-face friction* δ^I
Manufacturer	Unknown	Dry	Unknown	Peak 29° Resid.22°
Designer (original)	0.24	Dry	7	Peak 28° Resid. 21°
Designer (repeat)	0.24	Dry	7	Peak 16°
University A	1.0	Wet	180	Peak 16°–17° Resid 10°–13°
University B	0.4	Wet	Not stated reversal used	Peak 14° Resid. 8°
University C	1.0	Wet	150	Peak 17° Resid. 12°
University D	0.24	Wet Dry Dry	10	Peak 14° Peak 16° Resid. 10°
University E	Not stated	Dry	Not stated	Peak 16°

Notes:
1. Nominally identical samples of fPP geomembrane and PET geotextile were supplied to universities A to E.
2. Designer used a 60 mm × 60 mm shear box with the geotextile and fPP clamped in position.
3. University A used a ring shear box with both PET and fPP glued to the opposing rings. Ring OD was 180 mm. Ring 25 mm wide.
4. University B used a 60 mm × 60 mm shear box with fPP glued to a steel block and the PET was wrapped around a steel block. Repeated shearing was used to measure residual value.
5. University C used a "large displacement" shear box (300 mm × 300 mm specimen), with PET on a sand bed, fPP was "attached" to the rigid upper platen.
6. University D used a 100 mm × 100 mm shear box with both fPP and PET glued to 10 mm thick plywood plates.
7. University E used a 300 mm × 300 mm shear box. Shear displacement was not stated. fPP was "held in place" by a plywood sheet (no glueing or clamping mentioned.) PET "rested" on coarse dry sand bed.
8. Summary of tests on samples of nominally identical fPP to PET interfaces: Range: Peak (dry) δ^I : 16°–17° Range: Peak (wet) δ^I : 14°–17°
Range: Residual (dry) δ^I : 10°–12° Range: Residual (wet) δ^I : 8°–13°

and the geomembrane attached to a rigid upper block (see Table 4.2). Note the sharp peak in the shear stress at a displacement of about 5mm. Three criticisms of this test are:

- The interface was dry and not inundated as it probably would be in the field application.
- The rate of shear displacement of 1mm/minute was probably too rapid. A lower rate of displacement would almost certainly have reduced the magnitude of the peak shear stress and at rates of displacement occurring in the field, the peak would probably have been entirely absent.
- A test in a 300 mm long shear box is probably only valid to a displacement of about 50 mm. Thereafter, corrections for the reduced area of contact across the shear plane would become far too inaccurate to be valid. (At a shear displacement

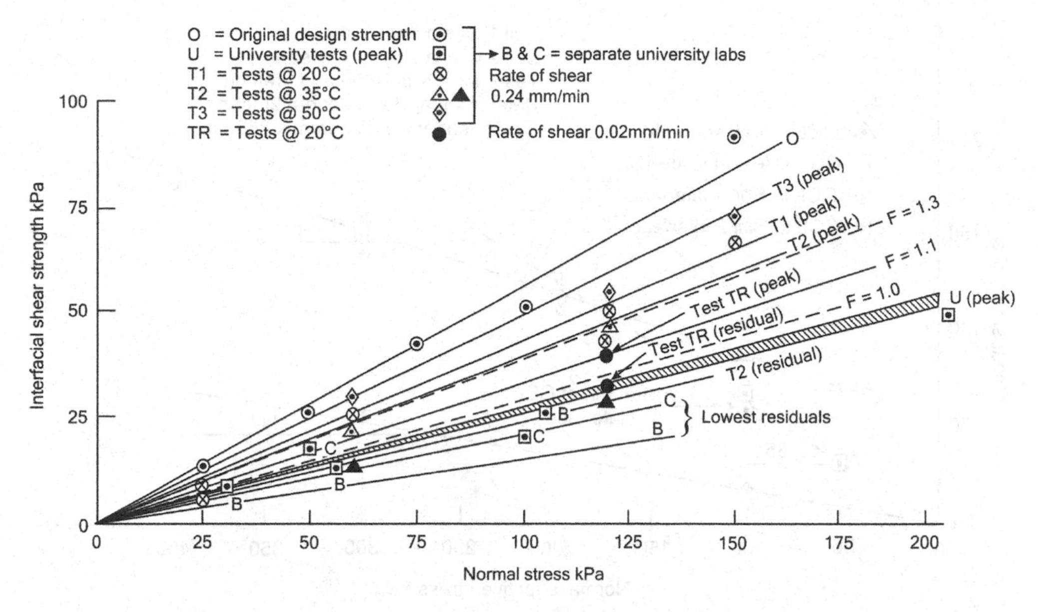

Figure 4.52 Summary of shear test results for fPP-to-PET geomembrane to geotextile interfaces.

of 50 mm, the contact area is only $250 \times 300 = 7500\,mm^2$, 75% of the original area. At a displacement of 150 mm it is only 50% of original area.)

Because interface shear testing has not been standardized as to specimen size and shape and rate of shear displacement, measurements of interface shear strength can be very variable, as illustrated by Table 4.2. In an attempt to resolve a dispute between a designer and his client as to the correct value for the angle of interfacial friction δ^I between an fPP geomembrane and a PET geofabric, nominally identical specimens of the two materials were sent to five universities (on three different continents) with requests that they determine δ^I. Their results, together with the manufacturer's value and those determined by the designer, are collected in Table 4.2 and Figure 4.52. As these show, residual values ranged from 22° to 8° ($\tan \delta^I$ from 0.4 to 0.14). The average slope of the steepest section of the ground on which this interface was laid was 16°. As there was no evidence of excessive movement on the interface, the actual value of residual δ^I must have been at least 16°.

As shown in Figure 4.52, two additional variables that were considered by University D were temperature and rate of displacement (not recorded in Table 4.2). It was found that as the temperature rose above 20°C, interfacial friction initially decreased, but at 50°C the fPP had softened sufficiently to cause adherence between the PET and fPP surfaces, causing δ^I to increase to above its value at 20°C. Reducing the rate of shear displacement from 0.24 mm/min to 0.02 mm/min (1.2 mm/h) caused the absolute value of δ^I to decrease and the difference between δ^I (peak) and δ^I (residual) also decreased. This is obviously an area of knowledge that requires further research, and careful consideration before measured values of δ^I are accepted as realistic and

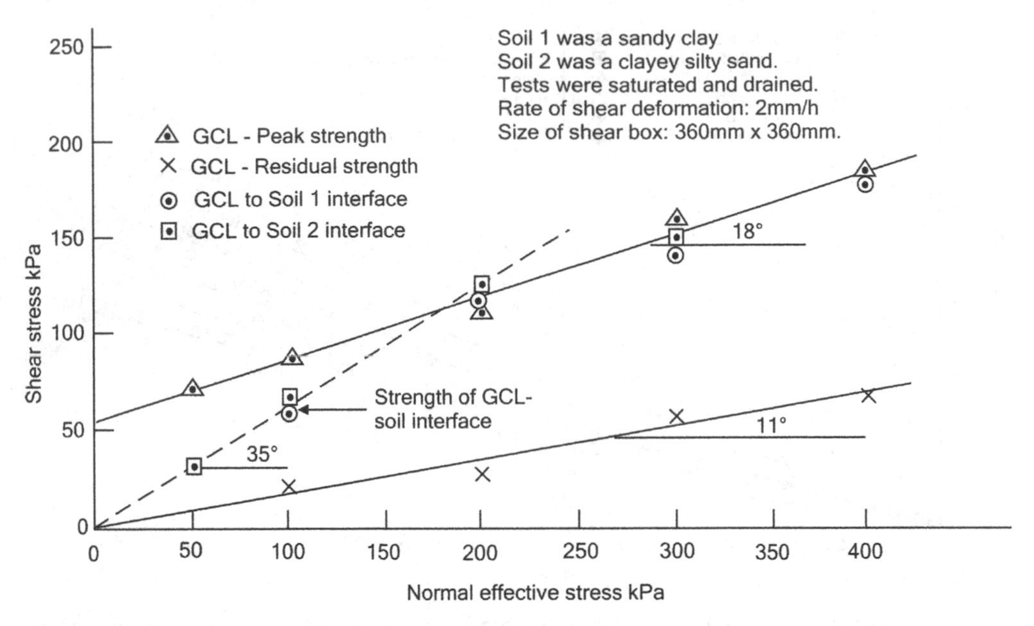

Figure 4.53 Shear strength envelopes for stitched GCL bentonite core and interfaces between GCL and two soils.

representative properties. It appears that residual values, although conservative, may be over-conservative if they are still to be reduced by a factor of safety.

4.8.2 *Interfaces between GCLs and soils and shear within the thickness of a GCL*

Figure 4.53 summarises the results of saturated 360 mm × 360 mm shear box tests on interfaces between two soils (a clay and a clayey silty sand) and a stitched GCL, as well as tests of the shear resistance of the stitched bentonite layer forming the core of the GCL. In the GCL-soil tests the GCL was supported by a layer of compacted soil and loaded via a roughened sheet of steel. To test the internal shear resistance of the core the GCL was sandwiched between two sheets of steel (with their surfaces roughened to the extent of a wood-working rasp) and aligned so that shearing took place through the core. For all tests the specimen was placed under load, inundated with water and allowed to consolidate, and the bentonite to hydrate for 24 hours before testing at a displacement rate of 2 mm/h.

The results in Figure 4.53 show that there was very little difference between the three series of tests, except at normal stresses of less than 175 kPa. The cohesion intercept of 50 kPa was provided by the stitching linking the outer geofabric layers of the GCL together. Once these had broken after a displacement of 30 to 40 mm the strength of the interface was provided by the residual strength of the bentonite core ($\varphi^{I} = 11°$).

Figure 4.54 shows a shear stress versus shear displacement relationship for the bentonite layer in the GCL on its own. The apparent yield at a shear stress of 155 kPa

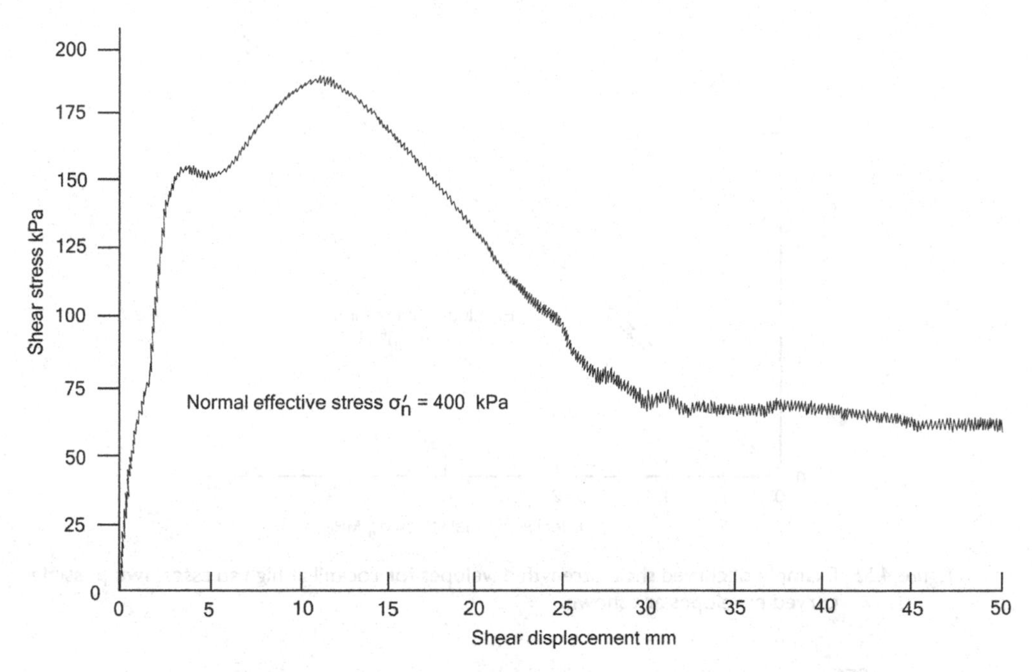

Figure 4.54 Typical shear stress versus shear displacement curve for hydrated GCL sheared through thickness of saturated bentonite core in 360 × 360 mm shear box. Rate of shear 2 mm/h. Areas of yield along displacement axis were caused by snapping of stitching.

and the spikes in the shear stress record beyond this were caused by the clearly audible progressive snapping of the stitching.

Figure 4.53 shows that at normal stresses of less than 175 kPa the stitching gave the GCL a strength greater than the shear strength of the GCL-soil interface. Once 175 kPa was exceeded, the GCL-soil interface was stronger than the core of the GCL and failure occurred through the GCL core, which then became the weakest part of the interface.

4.9 The shear strength of waste rock

Information on the shear strength of waste rock is almost non-existent in the literature. However, a lot of research has been done into the properties of rockfill for use in rockfill dams and this can be used as a guide to the properties of waste rock. In the late 1960s and 1970s a large number of rockfill dams were constructed in various parts of the world and large triaxial cells, testing specimens of up to 1.15 m in diameter by 2.3 m high were built for the purpose of testing these coarse granular materials. Marsal in particular, carried out exhaustive investigations of the shear strength and compressibility of rockfill. His investigations were reported in detail in 1973.

Rockfill has a curved shear failure envelope that is caused by the progressive breakage of particles as confining and normal stresses are increased. Figure 4.55 shows a typical shear failure envelope (de Mello, 1977) together with the type of equation

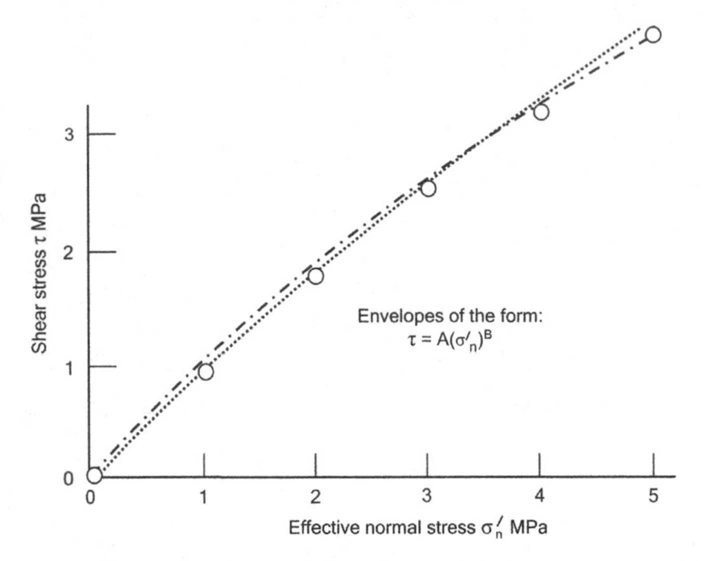

Figure 4.55 Example of curved shear strength envelopes for rockfill at high stresses. Two possible curved envelopes are shown.

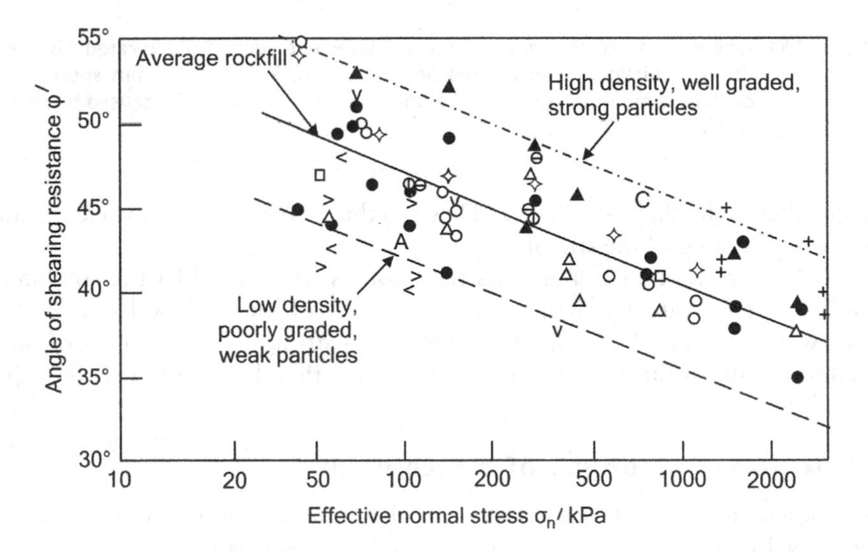

Figure 4.56 Collection of values of $\varphi^{\prime}$ for shear resistance of rockfill at various values of effective normal stress $\sigma_n^{\prime}$.

used to describe the relationship between shear stress τ and normal effective stress $\sigma_n^{\prime}$, Figure 4.56 (Leps, 1970) collects together a large number of the published results of testing rockfill in shear. Because of its variability, the usual presence of weathered particles, etc., the lower limit to this envelope (i.e. $\varphi^{\prime} = 35°$ to $45°$) should be used when investigating the shear stability of waste rock dumps. In the absence of better information, the angle of repose of an end-tipped face of the waste rock in question (usually between 35 and 40°) can be used as an estimate of strength. However, for soft rocks, for example coal and coal shales, the angle of shearing resistance may be considerably different to the angle of repose (see Figures 4.29a and b). It is therefore sound practice to apply a factor of safety directly to the tangent of the angle of repose, i.e. to take $\tan\varphi^{i} = \tan\varphi_{R}/F$ where F could be 1.2 to 1.3.

4.10 Strain softening of "dry" coarse mine wastes

Bishop (1973) drew attention to the phenomenon of 'bulking' of unsaturated sands and gravels when deposited without compaction, a phenomenon long known in concrete technology with relation to volume batching of aggregate. In general terms, if a given mass of dry cohesionless sand or gravel is deposited loosely, it will assume a certain volume and void ratio. If the water content is gradually increased prior to depositing the material, the volume of the mass (and hence its void ratio) will increase up to a critical water content after which the volume will decrease again. When the material is saturated, it will have approximately the same volume and void ratio as when it is dry and this void ratio is also reached if an initially damp material is subsequently saturated. Bulking is well-illustrated by the results shown in Figure 4.57a and b for mixtures of the coarse gravel and sand-fractions of a diamond mining waste. The two sets of curves in Figure 4.57a were prepared with different compactive efforts, and hence had different initial void ratios, but regardless of initial void ratio, showed much the same maximum change in void ratio as bulking proceeded. At water contents approaching saturation, the void ratios were much the same as the initial values. Note that the sand content of the material had little effect on the bulking, but the addition of sand did affect initial void ratios for the same compactive effort. Subsequent tests on these materials (Figure 4.57b) showed that specimens prepared at void ratios of 1.0 or above were contractant in consolidated undrained triaxial shear. Below a void ratio of 1.0, the materials were neutral to dilatant.

Figure 4.57c shows bulking results presented by Bishop (1973) for waste from the Aberfan tip (see Section 11.10.2), which show the percent decrease in volume on saturation. Figure 4.58 shows results for a triaxial shear test on bulked colliery waste (Dawson et al., 1998) which was set up at a void ratio of 0.51, consolidated isotropically to 0.40 under an effective stress of 200 kPa and sheared undrained (although it is not clear at what stage the specimen was saturated). The strain-softening behaviour was very similar to that shown in Figure 4.35.

'Dry' mine wastes are often deposited in a damp, bulked condition without compaction. A subsequent increase in water content caused by heavy or continuous rain or some other source of water can cause a tendency for a sudden decrease in void ratio with its consequent strain-softening and loss of shear strength.

There is a current trend for mines to compact their dry wastes. In the case of colliery wastes, this is done to reduce the air permeability of the waste and thus prevent spontaneous combustion, sustained by the entry of oxygen. Some mines sluice their coarse wastes with waste mine water as a means of disposing of waste water. The sluicing causes the rock to compact, reducing its tendency to contract, but may unfortunately increase acid seepage and the escape of nitrates, produced by explosives, from the base of the dumps, leading to undesirable, but preventable surface and ground water pollution.

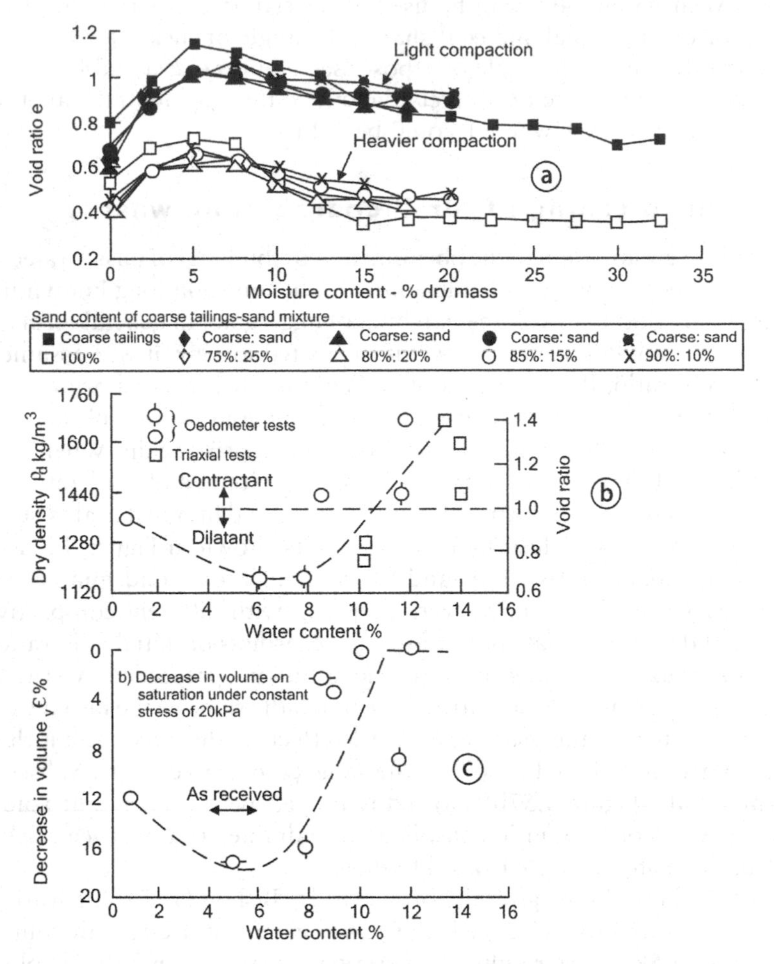

Figure 4.57 a&b: Bulking curves for diamond tailings. c: Bulking effect in coal waste from Aberfan. (Decrease in volume when saturated under $\sigma_v = 20\,kPa$.)

4.11 The mechanics of unsaturated waste materials

So far, Chapter 4 has considered only saturated waste materials, apart from introducing the equation for effective stress in unsaturated materials (equation 4.9). As mentioned in Section 1.3, over much of the world's surface, potential annual evaporation exceeds annual rainfall and many of these water-deficient regions (e.g. Angola, Arizona, Australia, Chile, China, Namibia, South Africa, Spain, Zambia and Zimbabwe) support considerable mining and industrial activity. Hence fine grained mining wastes, although usually saturated when first deposited in storage, become unsaturated as a result of their exposure to desiccating weather. The subject of desiccation by solar energy will also be touched on in Sections 10.7 and onwards in Chapter 10.

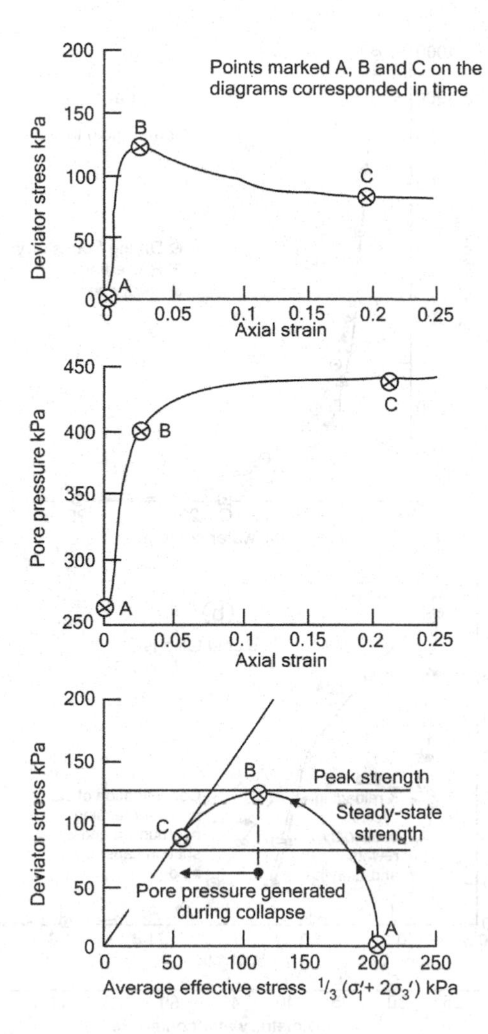

Figure 4.58 Typical isotropically consolidated undrained triaxial shear test on coarse mine waste.

4.11.1 *Retention of water by fine wastes*

This subject has importance because it directly affects the water storage capacity of every waste storage. Figure 4.59a shows the suction-water-content or water retention relationship for drying and re-wetting of a coarse gold tailings (see coarse side of grading envelope in Figure 1.3a.) The suction-water-content relationship can be measured by a number of techniques, but in the most basic method a specimen of saturated tailings or other waste of interest is set up in a triaxial cell (Figure 4.22) in which the basal porous disc has a very fine pore size, is sealed to the base (usually using epoxy resin) and its pore space is fully saturated with water. The rubber membrane is omitted and an air pressure is applied within the cell. The pore water pressure is maintained at zero (atmospheric pressure) (under the sealed-in fine porous disc) and

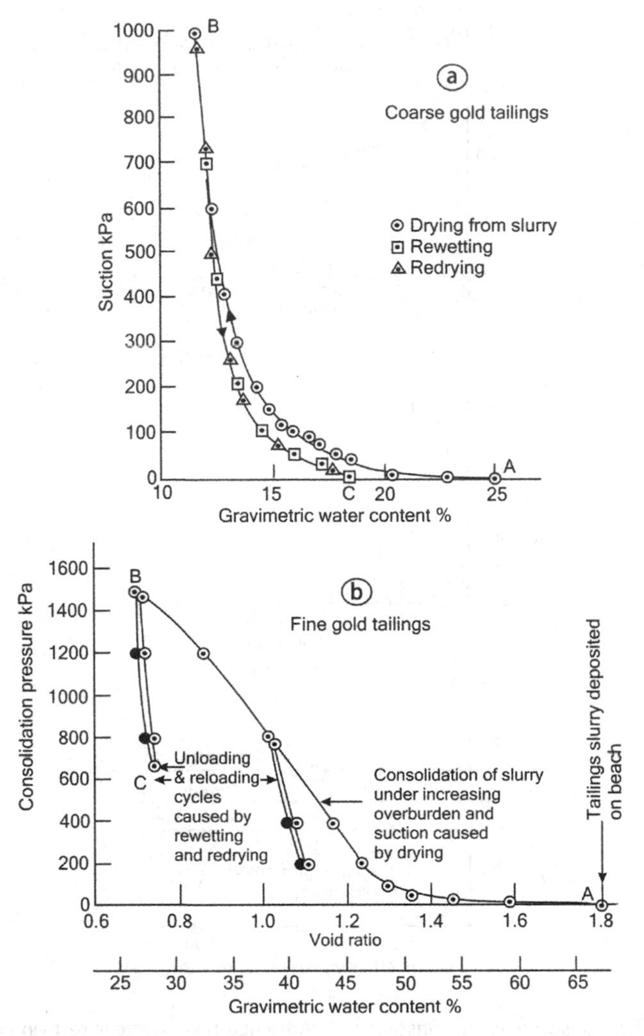

Figure 4.59 **a: Suction-water content relationship for a coarse gold tailings. b: Void ratio-consolidation pressure relationship for a fine gold tailings.**

the lead from the pore water pressure outlet is connected to a volume measuring gauge (see Figure 6.2a). Referring to equation (4.9)a, in the test $(\sigma - u_a) = 0$ and therefore $\sigma^I = \chi(u_a - u_w)$. But u_w is also zero and hence, when the specimen has come to stress and pore pressure equilibrium by draining into the volume measuring gauge, or drawing water from the gauge, $\sigma^I = \chi u_a$ and the suction $= u_a$. Thus the water content can be related to the suction, and the suction-water-content curve can be constructed.

Referring to Figure 4.59a, as the tailings dried from a slurry, they shrank at very low suctions until they reached a water content of about 15%. Further drying developed suctions in the pore water that increased substantially with decreasing water content. Re-wetting caused the suction-water content curve to retrace its path with a small

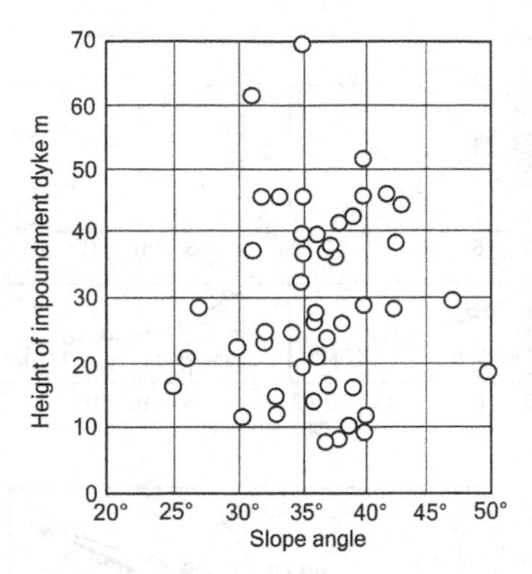

Figure 4.60 Relationship between average outer slope angle and height for 46 stable gold tailings storages in the water deficient climatic zone of South Africa.

amount of hysteresis, and thereafter, re-wetting and re-drying followed substantially the same suction-water content path. This is typical for most tailings.

Figure 4.59b shows an example of the consolidation behaviour of a fine gold tailings (on the fine side of the envelope shown in Figure 1.3a). The process is closely related to that shown in Figure 4.59a and occurs simultaneously with the processes of drying and re-wetting as successive layers are deposited in a tailings impoundment.

After deposition on the beach of a tailings impoundment, a tailings slurry becomes subject to two processes, either sequentially or simultaneously: The water content and hence void ratio is decreased by evaporation engendered by solar radiation, and also by compression under the superimposed load of additional layers of slurry. The net effect is that the tailings consolidate under the combined effect of drying suction and increasing overburden, with each re-wetting cycle causing some over-consolidation. In Figures 4.59a and b, lines AB are closely related and so are lines BC. The main difference is that the total stress in Figure 4.59a remains at zero while the suction varies, whereas in Figure 4.57b both suction and total stress contribute to the consolidation pressure.

4.11.2 *The shear strength of unsaturated tailings and stability of slopes of unsaturated tailings*

In order to consider the geotechnical properties of shear strength and volume change for an unsaturated soil in terms of applied total stresses and suction, it is necessary to have a relationship linking these variables, i.e. an effective stress relationship is required, of which equation (4.9) is the most widely recognized form.

A striking demonstration of the strength that can be developed by slopes of unsaturated tailings in a semi-arid climate is given in Figure 4.60 where the average slope angles of 46 stable ring dykes in South Africa, built of gold tailings and retaining gold tailings to their full height, have been plotted against the heights of the slopes. All of these impoundments were operated in such a way that the bulk of the deposits

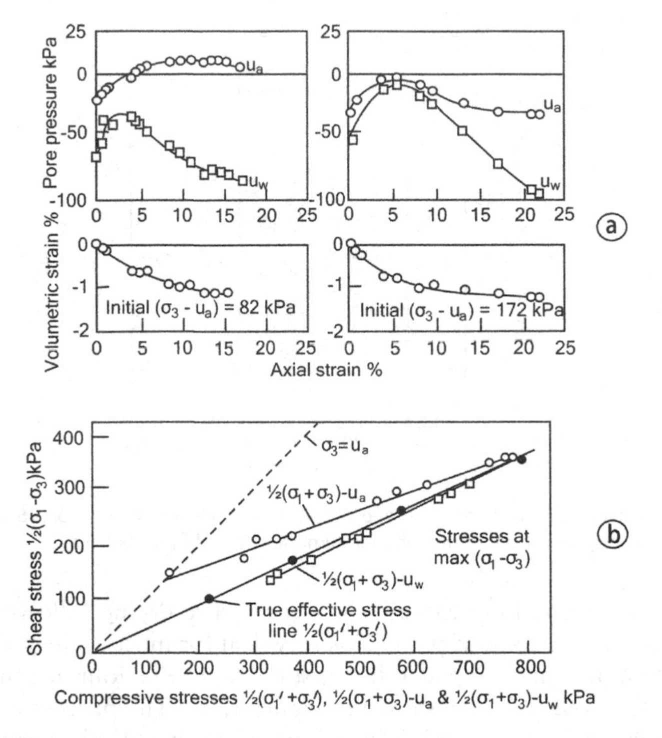

Figure 4.61 **Results of measurements of shear strength for unsaturated soil. a: Pore pressure changes during undrained shear. b: Triaxial shear strength at constant water content.**

(and especially the all-important retaining embankments) were maintained in a partly saturated state at all times. This so-called semi-dry operation was achieved by providing adequate under-drainage to the retaining embankments to depress the phreatic surface, by keeping the pool area of the hydraulic fill remote from the embankments and by adopting a slow rate of rise (about 2 m per year) so that successive layers of deposition could consolidate and dry thoroughly before being covered by newly deposited material. (See Section 8.8.2 for further discussion.) Plate 12.1 illustrates one of these excessively steep slopes.

This practice, which is possible only in a water-deficient climate, enables stable embankments to be constructed that may have almost incredibly steep slopes, especially considering that gold tailings is a cohesionless silt with an angle of shearing resistance of 35°. However, the practice of building tailings storages at the maximum possible outer slope angle has come to an end, as these slopes are almost impossible to rehabilitate in such a way that continuing maintenance is not required. The current tendency is to move towards slope angles of 12° to 18°, which are considerably more erosion resistant (See Chapter 12.)

In measuring the shear strength of an unsaturated soil in terms of the effective stress components $(\sigma - u_a)$ and $(u_a - u_w)$, both u_a and u_w must be measured, or alternatively, u_a can be controlled to a constant value and u_w can be measured. Figure 4.61 shows (a) typical measurements of u_a, u_w and volumetric strain during undrained tests on

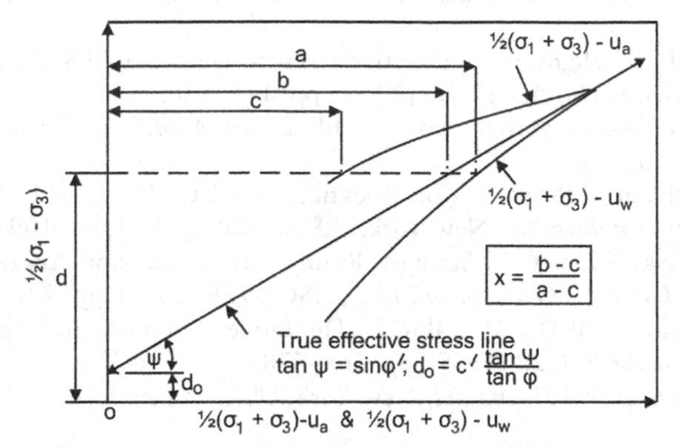

Figure 4.62 Calculating values for the effectiveness parameter X.

an unsaturated soil at two values of $(\sigma_3 - u_a)$, and (b) the failure stresses, recorded as $1/2(\sigma_1 - \sigma_3), 1/2(\sigma_1 + \sigma_3) - u_a$ and $1/2(\sigma_1 + \sigma_3) - u_w$. The "true effective stress line" can be obtained from tests on the same soil in a saturated state so that true effective stress conditions apply to it. The effectiveness parameter χ can be calculated as shown in Figure 4.62.

References

Bishop, A.W., & Henkel, D.J.: *The Measurement of Soil Properties in the Triaxial Test*, 2nd edition. London, U.K.: Edward Arnold, 1962.

Bishop, A.W., & Blight, G.E.: Some Aspects of Effective Stress in Saturated and Partly Saturated Soils. *Geotechnique 13* (1963), pp. 177–197.

Bishop, A.W.: The Stability of Tips and Spoil Heaps. *Q. J. Eng. Geol. 6(3 & 4)* (1973), pp. 335–376.

Blight, G.E.: The Effect of Non-Uniform Pore Pressures on Laboratory Measurements of the Shear Strength of Soils. Laboratory Shear Testing of Soils. *ASTM Special Technical Publication 361* (1963), pp. 173–184.

Dawson, R.F., Morgenstern, N.R., & Stokes, A.W.: Liquefaction Flow Slides in Rocky Mountain Coal Mine Waste Dumps. *Can. Geotech. J. 35* (1998), pp. 328–343.

de Mello, V.F.B.: Reflections on Design Decisions of Practical Significance to Embankment dams. *Geotechnique 27(3)* (1977), pp. 279–355.

Fourie, A.B., Blight, G.E., & Papageorgiou, G.: Static Liquefaction as a Possible Explanation for the Merriespruit Tailings Dam Failure. *Can. Geotech. J. 38* (2001), pp. 707–719.

Fredlund, D.G., & Rahardjo, H.: *Soil Mechanics for Unsaturated Soils*. New York, USA: Wiley, 1993.

Gibson, R.E., & Henkel, D.J.: Influence of Duration of Tests at Constant rate of Strain on Measured "Drained" Strength. *Geotechnique* (4) (1954), pp. 6–15.

Ishihara, K., Verdugo, R., & Acacio, A.A.: Characteristics of cyclic behaviour of sand and post-seismic stability analyses. In: 9th *Asian Regional Conf. Soil Mech. Found. Eng., Bangkok, Thailand, Vol.12*, 1991, pp. 17–40.

Jefferies, M., & Been, K.: *Soil Liquefaction, A Critical State Approach*. Abingdon, UK: Taylor and Francis, 2006.
Khalili, N., Geiser, F., & Blight, G.E.: Effective Stress in Unsaturated Soils: Review with New Evidence. *Int. J. Geomech., ASCE 4(2)* (2004), pp. 115–126.
Leps, T.M.: Review of Shearing Strength of Rockfill. *J. Soil Mech. Found. Div. – ASCE 96(SM5)* (1970), pp. 351–365.
Marsal, R.J.: Mechanical Properties of Rockfill. In: R.C. Hirschfeld, S.J. Poulos (eds): *Embankment-Dam Engineering*. New York, U.S.A.: Wiley, 1973, pp. 109–200.
Olson, S.M., & Stark, T.D.: Yield Strength Ratio and Liquefaction Analysis of Slopes and Embankments. *J. Geotech. Geoenviron. Eng., ASCE 129* (2003), pp. 727–737.
Sladen, J.A., D'Hollander, R.D., & Krahn, J.: The Liquefaction of Sands, a Collapse Surface Approach. *Can. Geotech. J.* 22 (1985), pp. 564–578.
Terzaghi, K.: *Theoretical Soil Mechanics*. New York, USA.: Wiley, 1943.

Chapter 5

In situ shear strength testing of tailings and other waste materials and its interpretation

Although laboratory shear testing is essential and usually the only option at the stage of design, it has the following disadvantages:

- If design is based on the properties of material produced by a pilot plant, these may not be representative what will eventually exist in the prototype storage structure.
- The stress conditions in situ are not known with any accuracy and therefore may not be correctly simulated in the laboratory tests.
- If so-called "undisturbed" specimens from an existing storage are being tested, they will inevitably have suffered some disturbance in being sampled, i.e. in releasing the stresses they carried in situ, being trimmed and packed, being subjected to shock and vibration during transportation to the laboratory, lying (usually horizontally) in storage awaiting testing, being trimmed for testing and mounted in the shear box or triaxial. The list of unknown effects should probably be even longer than this.

  A study of the undisturbed sampling of tailings carried out recently in Canada (Wride et al., 2000) concluded that only samples that had been frozen in situ, sampled and transported to the laboratory, then trimmed and mounted in the triaxial and placed under stress, before carefully thawing the sample, could be regarded as reasonably "undisturbed". Large diameter fixed piston samples came a poor second in quality.
- Sampling and testing is expensive and inevitably, what are usually judged to be a "sufficient number of samples" are really the absolute minimum required to get sensible-looking results. The Wride, et al. study also observed that deposits of tailings that appeared to be homogeneous, were actually very variable in density, and therefore also variable in other relevant properties. To quote Wride, et al., "This variation must be recognized in the design of sampling and laboratory testing programs to properly characterize the deposits, and the interpretation of in situ testing data".
- In the sampling process, there is an inevitable tendency to lose or discard the "worst" or "poor" parts of the sample which are really the parts that should be tested. The material that is eventually tested is therefore also not representative of what exists in situ, but an optimistic selection from the material that will really control the shear stability of the storage.

The alternative, in situ testing, is also not the perfect process that is needed, but has several advantages over laboratory testing.

- Far more test measurements can be carried out at much closer depth intervals than can be achieved with soil sampling followed by laboratory testing.
- If tests are carried out at fixed depth intervals, the strength sampling is unbiased to the extent that the measurement is made regardless of the quality of the material that happens to exist at that depth.
- With some apparatus, e.g. the cone penetrometer and the pressuremeter, pore pressures can be measured during the test and hydrostatic or "at rest" pore pressures can also be measured. Total stress lateral pressures can be estimated from the results of pressuremeter tests. This allows in situ effective stresses to be estimated as well as providing a check on pore pressures measured by means of fixed position piezometers.
- A disadvantage is that the material which is being tested at a particular depth cannot be identified except by recovering corresponding profiles of samples, preferably "undisturbed".

The ideal arrangements are therefore to carry out in situ tests backed up with laboratory tests on undisturbed samples, or (vice versa) laboratory tests backed up with in situ tests.

Several types of apparatus can be used to test mine tailings and other reasonably fine-grained wastes in situ. These include various types of shear vane, cone penetrometers and pressuremeters. Plate 5.1 shows a track-mounted drillrig suitable for use on the surface of a tailings storage. Cruder tests such as the standard penetration test (SPT) can also be used, but interpretation of the results is entirely empirical, and the difficulty lies in knowing if the available correlations are valid for the material being tested. The following is a description of some in situ tests that have been found suitable for testing mine wastes, together with typical results of measurements.

5.1 The shear vane test

5.1.1 *Principle of the vane test*

The vane usually consists of four thin rectangular spring steel plates attached at right angles to a torque rod in a cruciform. (See Figure 5.1.) The vane is pushed vertically into the soil. A torque is applied to the rod by means of a torque head or torque wrench, causing the blades to rotate and thus producing a shear failure along a cylindrical surface. The shear strength is calculated from the measured torque required to shear the tailings or other material along the cylindrical surface.

The vane shape commonly regarded as 'standard' has blades with a height to overall width ratio, $H/D = 2$. The actual values of H and D depend on the strength of the material to be tested. For example, a vane with $H = 100$ mm and $D = 50$ mm can be used for strengths between about 50 and 70 kPa.

Vane tests may be carried out at the bottom of a pre-bored hole or by pushing the vane into the ground from the surface to the required depth. The latter procedure is rarely possible in tailings as their strength is usually far too high to push in the vane.

Plate 5.1 Track-mounted drill rig being used for in situ tests on a tailings storage.

Figure 5.1 is a diagram of the usual vane shear arrangement for use in mine and industrial wastes.

5.1.2 *Interpretation of measured torque*

It is normally assumed that failure in the vane shear test takes place on a cylindrical surface which coincides with that generated by the vane blades.

The blades of the vane shear apparatus are usually rectangular in shape and the axis of rotation of the vane is usually vertical. The torque applied to the vane shear apparatus measures the sum of the shear torques developed on the vertical cylindrical surface and the horizontal circular surfaces traced out by the rotating vane. The relationship between applied torque T and the strengths giving rise to the component shear torques is

$$\mathrm{T} = \frac{\pi D^2}{2}\left[Hs_v + \tfrac{1}{3}Ds_h\right] \tag{5.1}$$

where D is the diameter and H the height of the cylinder of rotation of the vane, and s_v and s_h are, respectively, the shear strengths developed on the vertical and

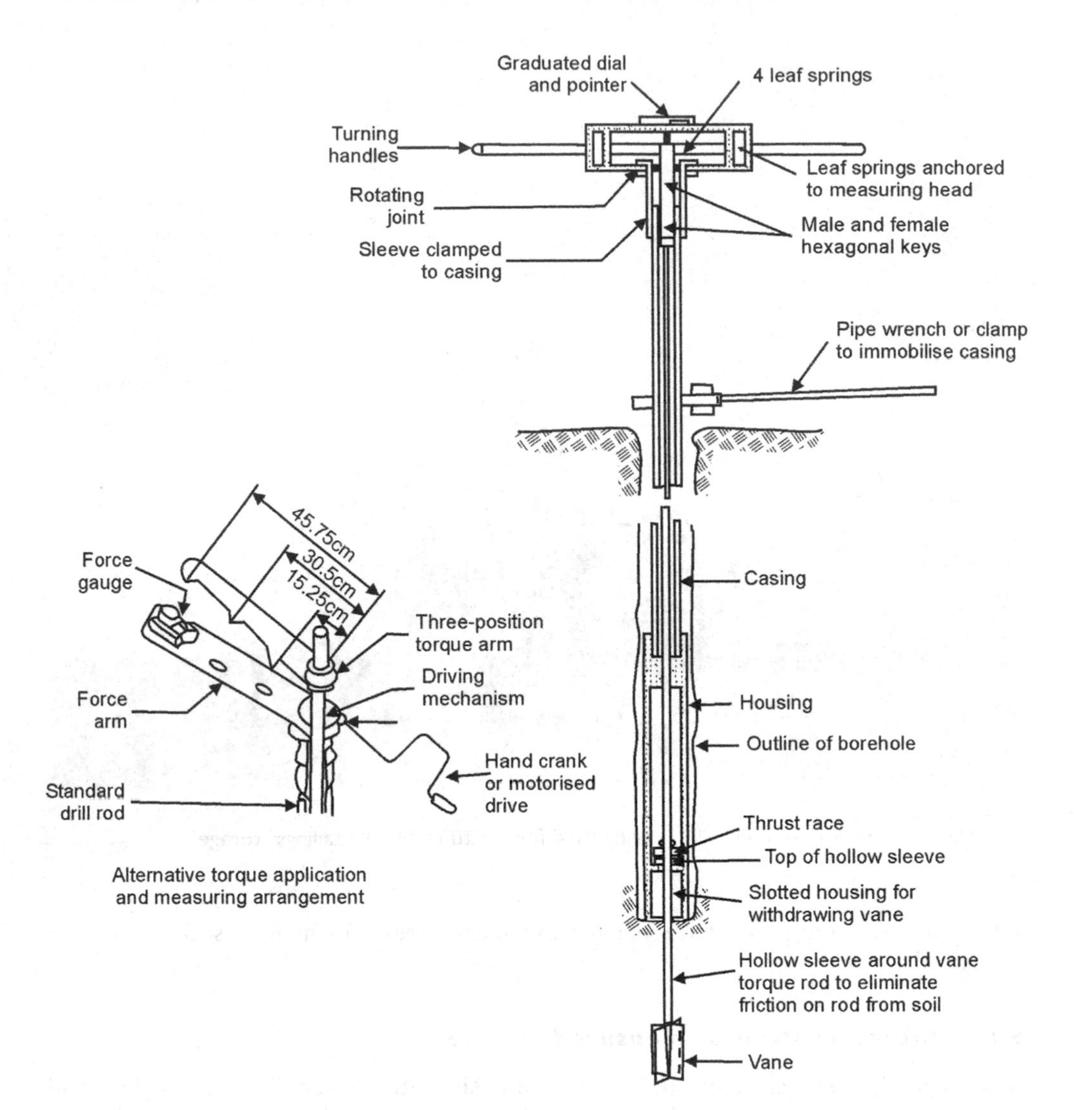

Figure 5.1 The shear vane apparatus.

horizontal surfaces. If s_v and s_h are both taken as equal to s, which is the conventional assumption:

$$T = \frac{\pi D^2}{2} s[H + \tfrac{1}{3}D]$$

and if H = 2D

$$T = \frac{7}{6}\pi D^2 s \ or \ s = \frac{6T}{7\pi D^2} \tag{5.2}$$

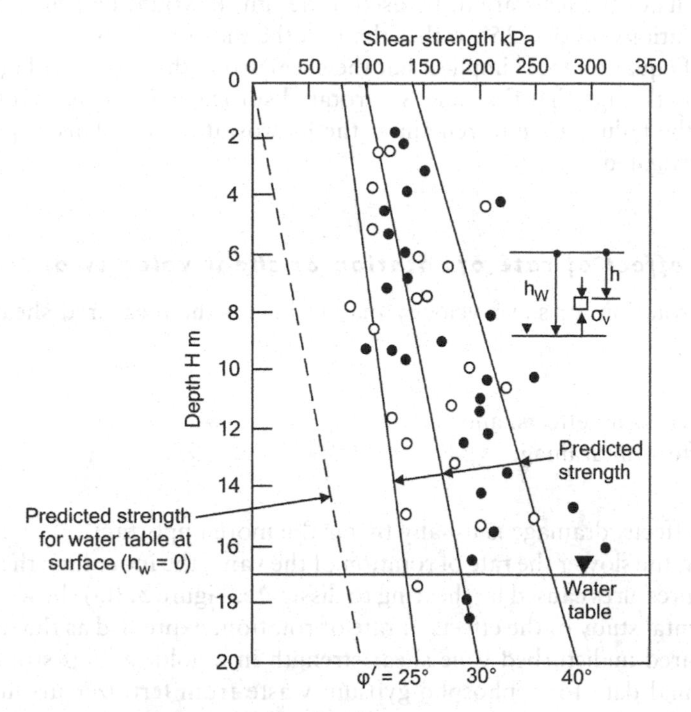

Figure 5.2 Profile of vane shear strength in a storage of gold tailings where the water table was 17 m deep. The data represent measurements in two adjacent holes 2 m apart by open and closed circles.

Strength profiles measured by means of the vane shear test agree quite well with strengths predicted on the basis of shear strength parameters measured in the laboratory and estimated in terms of in situ vertical effective stresses.

Figure 5.2 shows a profile of vane shear strength measured in a gold tailings storage where the phreatic surface (or water table) was 17 m deep. The results were obtained from two holes 2 m apart, and were calculated on the basis of equation (5.2) which assumes that the strength of the deposit is isotropic. The predicted strength lines for $\varphi^l = 25°$, 30° and 40° were calculated from the equation

$$\tau = [\gamma_{sat}h + \gamma_w(h_w - h)] \tan \varphi^l = \sigma^l \tan \varphi^l \quad (5.3)$$

where $\sigma^I = h(\gamma_{sat} - \gamma_w) + h_w\gamma_w$ in which γ_{sat} is the saturated unit weight of the tailings (18.5 kN/m^3) and γ_w is the unit weight of water. The equation assumes that the tailings remain fully saturated from the water table up to the surface, where the pore pressure is $\gamma_w h_w$ and the shear strength is $\tau = \gamma_w h_w \tan \varphi^l$.

Figure 5.2 illustrates one of the big advantages of a group of in situ tests – they clearly show the variability of the tailings with both depth and lateral extent which may not be appreciated or apparent from the limited number of results usually available for laboratory tests on undisturbed samples.

If the results of the tests are to be used in design, it would be prudent to base the design calculations on $\varphi^1 = 25°$, rather than on the mean $\varphi^1 = 30°$.

The use of equation (5.3) implies that the shear strengths shown in Figure 5.2 were drained strengths, i.e. that the vane was rotated sufficiently slowly to allow the pore pressure in the failure zone to remain at the hydrostatic level. This requirement will now be investigated.

5.1.3 *The effect of rate of rotation or shear velocity of the vane*

The rate of rotation, or shear velocity may influence the measured shear strength in two ways:

- via soil viscosity effects, and
- by its effect on drainage.

Of the two effects, drainage is usually by far the more important.

Obviously, the slower the rate of rotation of the vane, the more time there will be for excess pore pressures caused by shearing to dissipate. Figure 5.3(a) shows the results of an experimental study of the effects of rate of rotation, expressed as the time to failure t_f, on measured undisturbed vane shear strength in a gold tailings storage, together with additional data for a phospho-gypsum waste from fertilizer manufacture. The time to failure t_f, the coefficient of consolidation of the waste c_v and the vane diameter D have been combined as a dimensionless time factor $T = c_v t_f / D^2$ (see section 4.6.4). The mean line through the data represents an experimental relationship between vane strength and time to failure that can be used to decide on a suitable time to failure which will result in measuring a drained shear strength. For example, for a 90% degree of drainage $T = 0.7$ and if $c_v = 200\,m^2/y (=200 \times 10^4/365 \times 1440) = 3.8\,cm^2/min$ and $D = 6\,cm$

$$t_f = \frac{0.7 \times 36}{3.8} = 6.6\,min$$

If, on the other hand, the tests are performed in a clay with $c_v = 2\,m^2/y$ ($0.038\,cm^2/min$), $t_f = 660\,min$ or 11 h, which is very difficult to do without using a slow, adjustable speed motor drive to turn the vane at a very slow rate. If $t_f = 6\,min$ for the clay, $T = 0.038 \times 6/36 = 0.0063$ and the corresponding degree of drainage would be about 15%. In other words, the measured shear strength would essentially be an undrained strength.

A possible soil viscosity effect is illustrated by the measurements shown in Figure 5.3b. Here, the estimated value of c_v was $1\,m^2/y$ ($0.019\,cm^2/min$) and the time factor for the slowest test result shown in Figure 5.3b was $T = 0.019 \times 60/36 = 0.003$. Hence even the slowest test was effectively undrained. The decrease in strength corresponding to an increase in t_f from 0.1 to 60 minutes was therefore probably due to soil viscosity effects, but amounted to only 10% of the strength for $t_f = 0.1\,min$. This justifies the statement that drainage causes the more important time-effect.

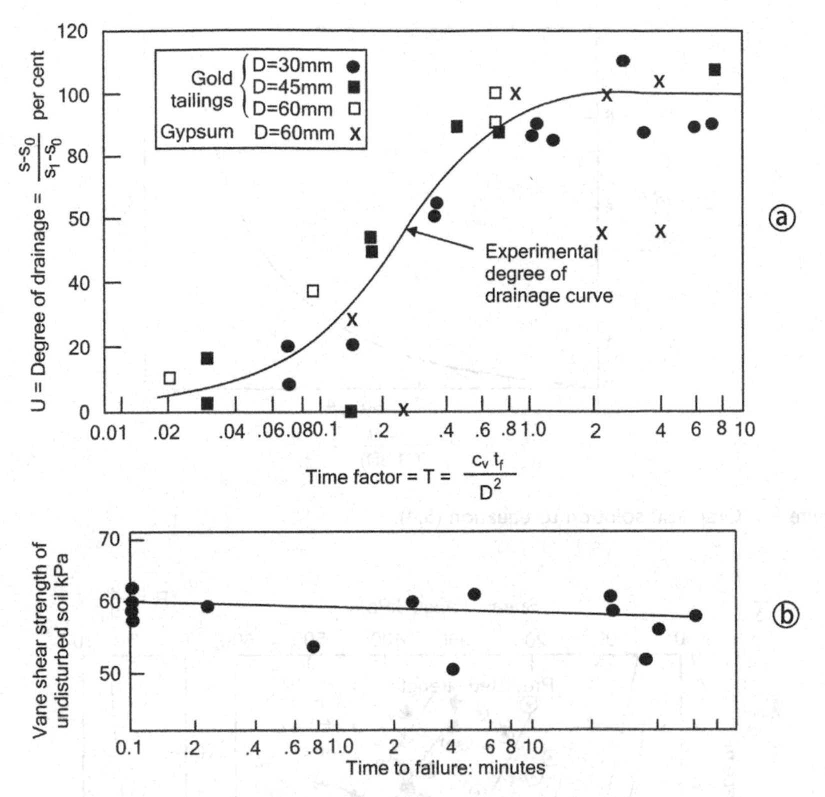

Figure 5.3 a: Experimental relationship between degree of drainage U and time factor T for vane shear tests in a gold tailings and a waste gypsum storage. (D = vane diameter, 2D = vane height), c_v = coefficient of consolidation, t_f = time to failure, s = vane strength at t_f, s_o = vane strength at t_f = 5 seconds, s_l = vane strength at t_f for test with longest duration. b: Effect of time to failure on vane shear strength of a stiff clay.

5.1.4 *Vane shear strength when $s_v \neq s_h$*

When the shear strength of a soil differs with the direction of the shear surface, the components s_v and s_h can be resolved by carrying out two parallel series of vane shear tests using vanes having different proportions of blade height H to blade width D. The biggest practical difference between H and D is to have the first vane with H = 2D and the second vane with H = D/3. In this case, the ratio of torques measured in the same soil with the two vanes will be

$$T(2D)/T(D/3) = (1 + 6R)/(1 + R) \tag{5.4}$$

where $R = s_v/s_h$

Figure 5.4 shows a graphical solution to equation 5.4. It should be noted that if T(2D)/T(D/3) is 2 or less, R is effectively zero, which means that s_v. the shear strength on vertical surfaces, is very low.

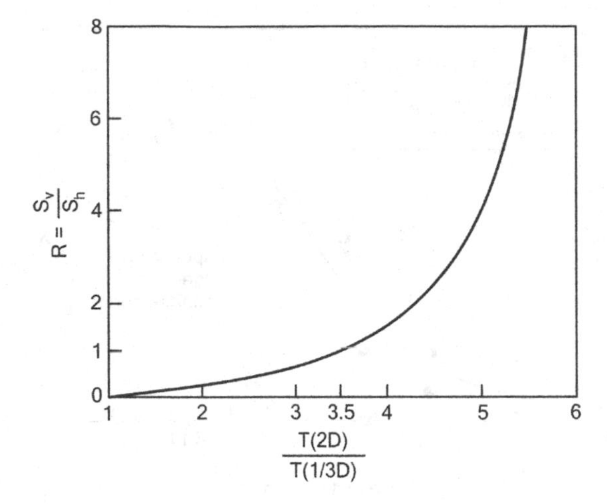

Figure 5.4 Graphical solution to equation (5.4).

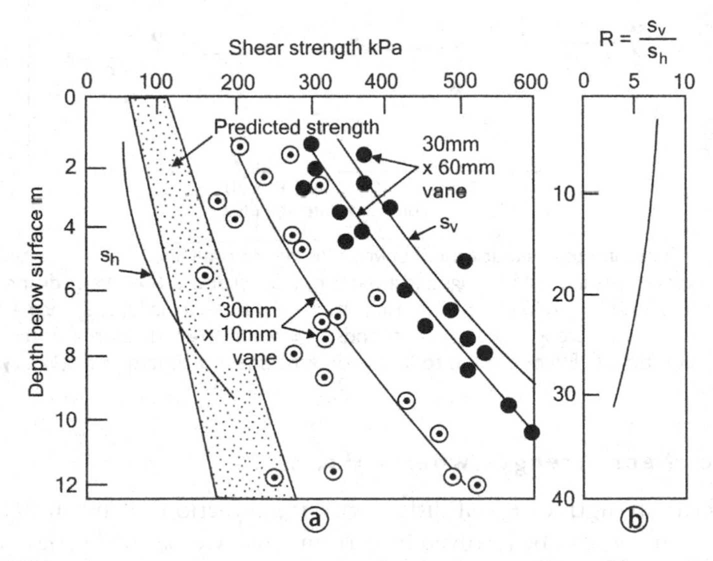

Figure 5.5 Vane shear tests in a compacted soil embankment using a 30 mm dia × 60 mm high, and a 30 mm dia × 10 mm high vane. Shear strengths s_h and s_v were calculated from Figure 5.4.

Figure 5.5 shows the results of a set of vane shear measurements in a compacted earth dam embankment. Diagram (a) shows the vane measurements made by means of a 30 mm diameter by 60 mm high vane and a vane measuring 30 mm diameter by 10 mm high, as well as the mean trends for s_h and s_v In this embankment the strength on vertical surfaces far exceeded that on horizontal surfaces, because of the high horizontal stresses residual from compaction.

Figure 5.6 shows a set of vane measurements in a storage of mine tailings using only one vane with dimensions of 45 mm diameter by 90 mm high. The predicted strength

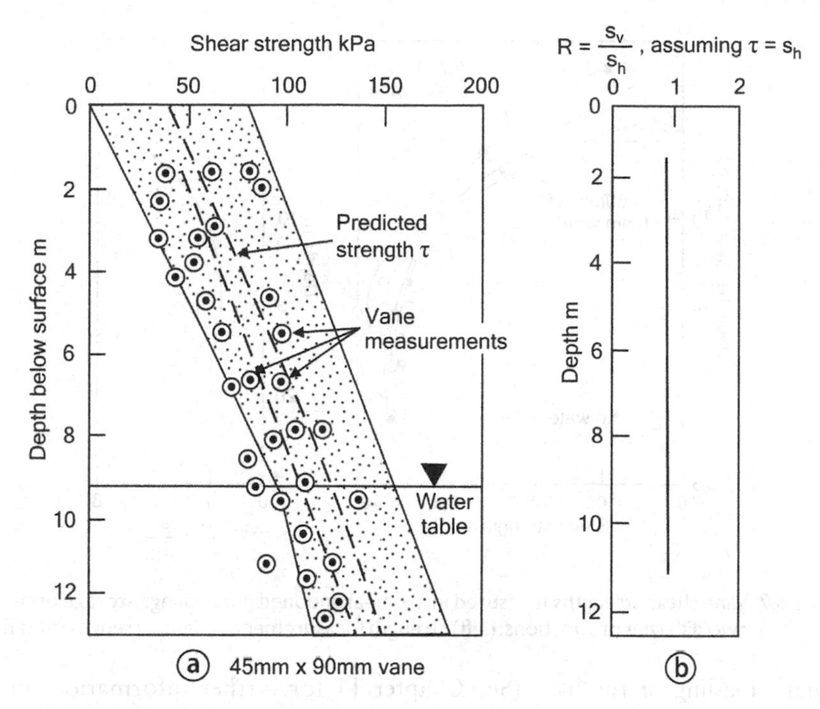

Figure 5.6 Vane shear tests in a storage of gold tailings. The ratio of strengths, R, was calculated on the assumption that the predicted strength, τ, equalled s_h and R was then calculated.

was calculated by means of equation (5.3). Because the term $[\gamma_{sat}h - \gamma_w(h_w - h)]$ represents a vertical effective stress, the predicted shear strength τ can be assumed to represent the strength on a horizontal plane, s_h. If τ is then used to calculate R, the result is that s_v is almost equal to, but slightly less than s_h. That is, the strength is almost isotropic.

On the other hand, Figure 5.7 shows vane shear measurements made in a gold tailings storage that had been abandoned 40 years previously and in which there was no phreatic surface, or water table. In this case, by using vanes of three different proportions (30 mm × 10 mm, 30 mm × 30 mm and 30 mm × 60 mm) the ratio (R) of s_v to s_h was found to vary from 3 at 5 m below surface to 1 at about 17 m and below.

5.1.5 *The effect of shear disturbance or remoulding*

It was shown in Figures 4.35 and 4.40 that tailings with a high void ratio tend to contract on shearing, and the same applies to similar materials subjected to the vane test. Figure 5.8a shows typical vane shear results for a residual gold tailings with both undisturbed or peak shear stress s_u and remoulded or residual shear stress s_R shown. The decrease of strength on remoulding is expressed as the "sensitivity ratio" $= s_u/s_R$. In this case the sensitivity ratio averaged about 2.5, enough to reduce a factor of safety of, say, 1.5 against shear failure to below 1.0 and to induce catastrophic failure if a progressive liquefaction of the tailings were to be induced by static liquefaction,

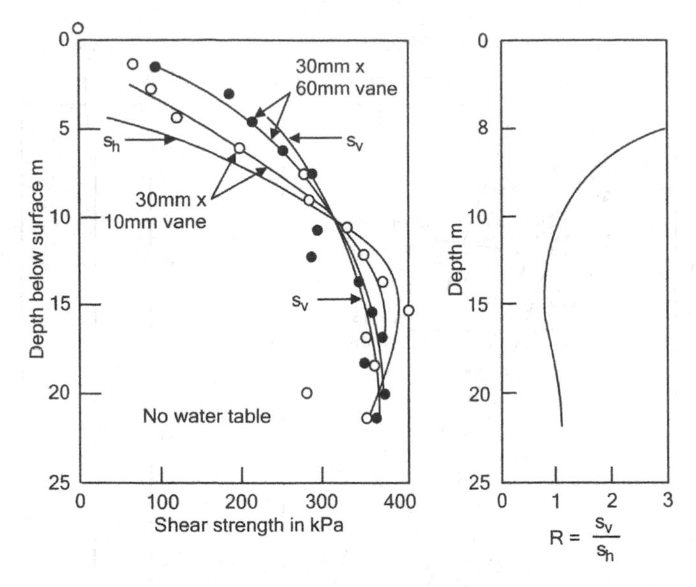

Figure 5.7 **Vane shear strengths measured in a long-abandoned gold tailings storage using vanes of two different proportions: (left) strength measurements; (right) strength ratio $R = s_v/s_h$.**

earthquake loading or the like. (See Chapter 11 for further information on failure of tailings storages.) Figure 5.8b shows similar data on undisturbed and remoulded vane shear tests on an industrial sludge. For this material, the sensitivity ratio was 2 down to a depth of 8m. Below this, the sensitivity rapidly increased to 10 as a result of an increase in the undisturbed shear strength, and an accompanying decrease in remoulded strength. This could result in a very dangerous situation if not recognized and if the effects of the fall of shear strength as a result of shear straining were not designed against.

5.1.6 *Vane shear strength tests in natural soils*

Vane shear tests are also often used to assess the shear strength of natural soil strata when considering the overall shear stability of a tailings or waste rock storage. Figure 5.9 illustrates a set of in situ vane shear tests made in a stiff fissured clay soil that formed the foundation stratum for a waste rock dump. The dump had failed in shear with the failure surface passing through the foundation and the measurements were made as part of the failure investigation. (See Section 11.10.) It will be seen that the vane strengths for the undisturbed soil are much larger than the remoulded strengths. Also, the strengths measured in laboratory triaxial and shear tests agreed reasonably well with the remoulded vane strengths. The explanation is that the strength along fissures in the clay is very much less than that of the intact clay between fissures. When the vane is rotated in the undisturbed soil the failure surface is forced to cut through intact material. When the vane has been turned a number of times to remould the failure surface, it creates an artificial fissure. Similar considerations apply to the small-scale laboratory specimens. Each one usually contains at least one natural fissure, and the failure occurs preferentially along the fissure. This is confirmed by the computed

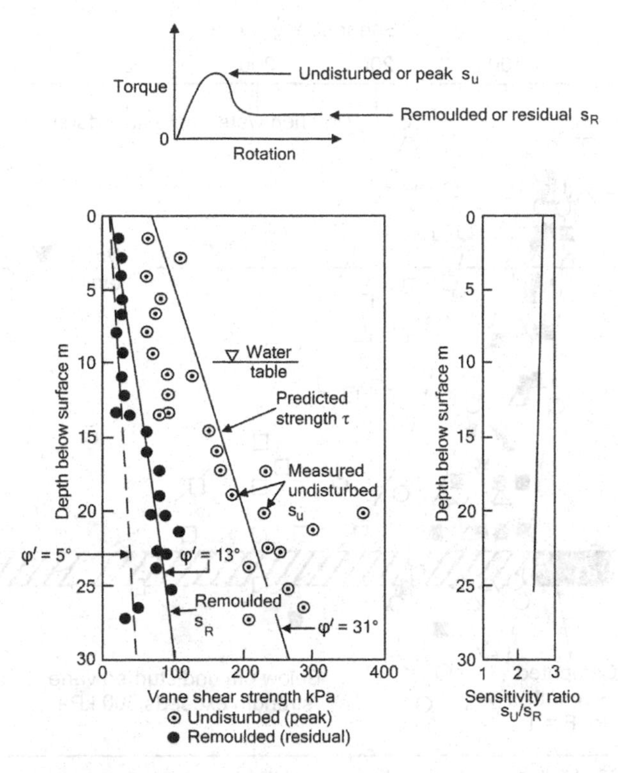

Figure 5.8a **The effect of shear disturbance or remoulding on vane shear strength of gold tailings. Sensitivity ratio s_u/s_R indicates the fall in shear strength from peak (undisturbed) value to residual (remoulded) value.**

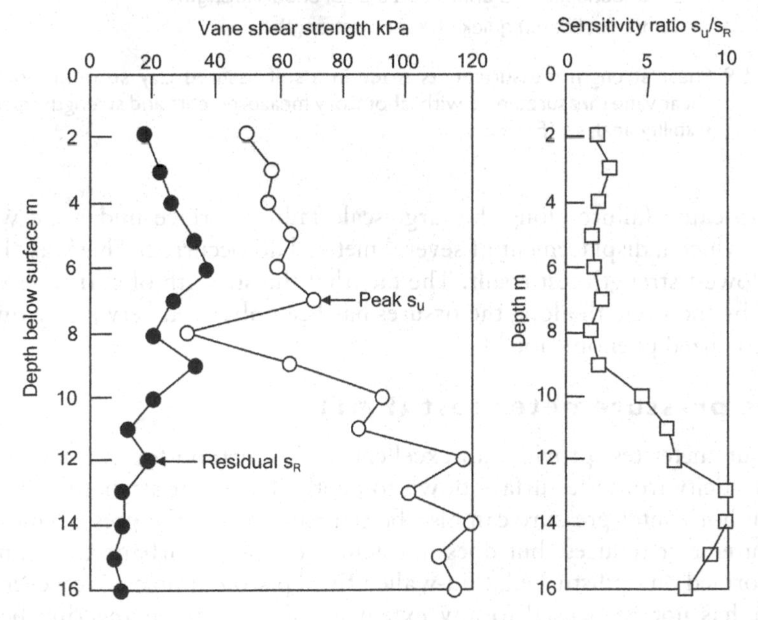

Figure 5.8b **The effect of remoulding on the vane shear strength of an industrial sludge showing the sensitivity ratio s_u/s_R. This material is much more variable with depth than the gold tailings and sensitivity ratios are as high as 10.**

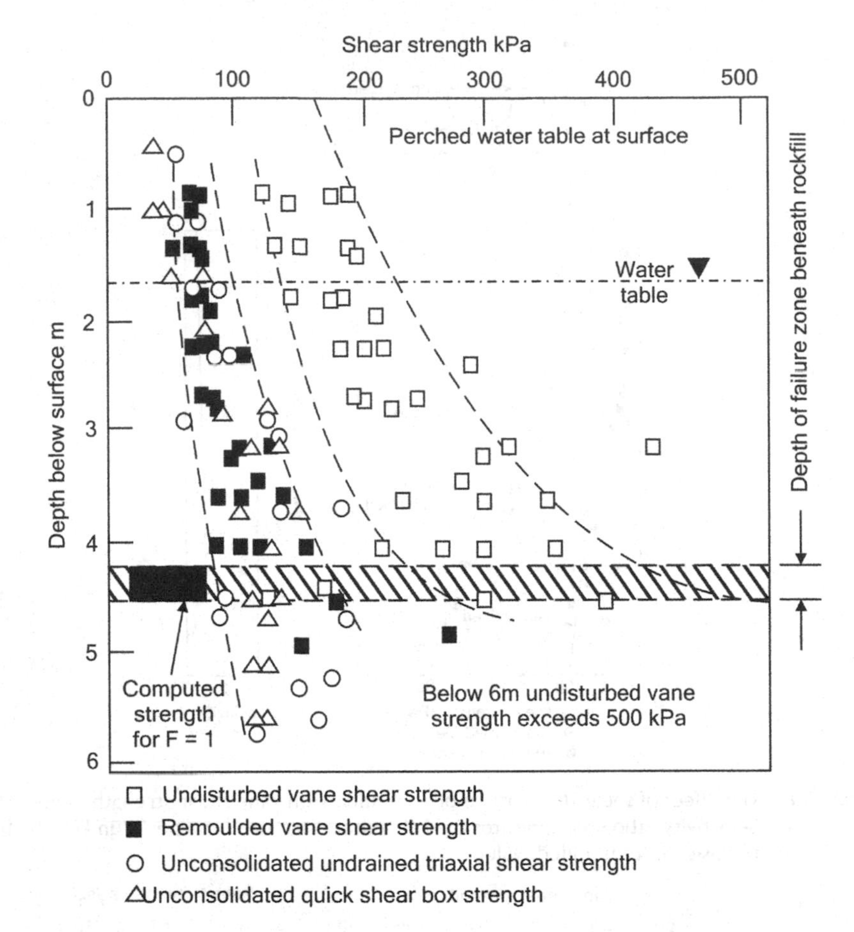

Figure 5.9 Shear strength measurements made in a stiff fissured clay stratum which compare shear vane measurements with laboratory measurements and strength calculated from stability analysis (F = 1).

strength to cause failure along the large-scale failure surface under the waste rock dump, on which a displacement of several metres had occurred. This was slightly less than the lowest strength test result. The fact that the strength of stiff fissured clays is governed by the strength along the fissures has been observed very many times and is a well recognized phenomenon.

5.2 The pressuremeter test (PMT)

The pressuremeter test provides an excellent in situ method for evaluating the modulus of elasticity from the surface down to depth. The shear strength of tailings and the in situ horizontal pressure can also be estimated using the pressuremeter. Use of the pressuremeter reduces, but does not eliminate soil disturbance, as compared to tests performed on undisturbed, thin-walled fixed piston samples. However, the pressuremeter has not been used to any extent in mine waste engineering because the

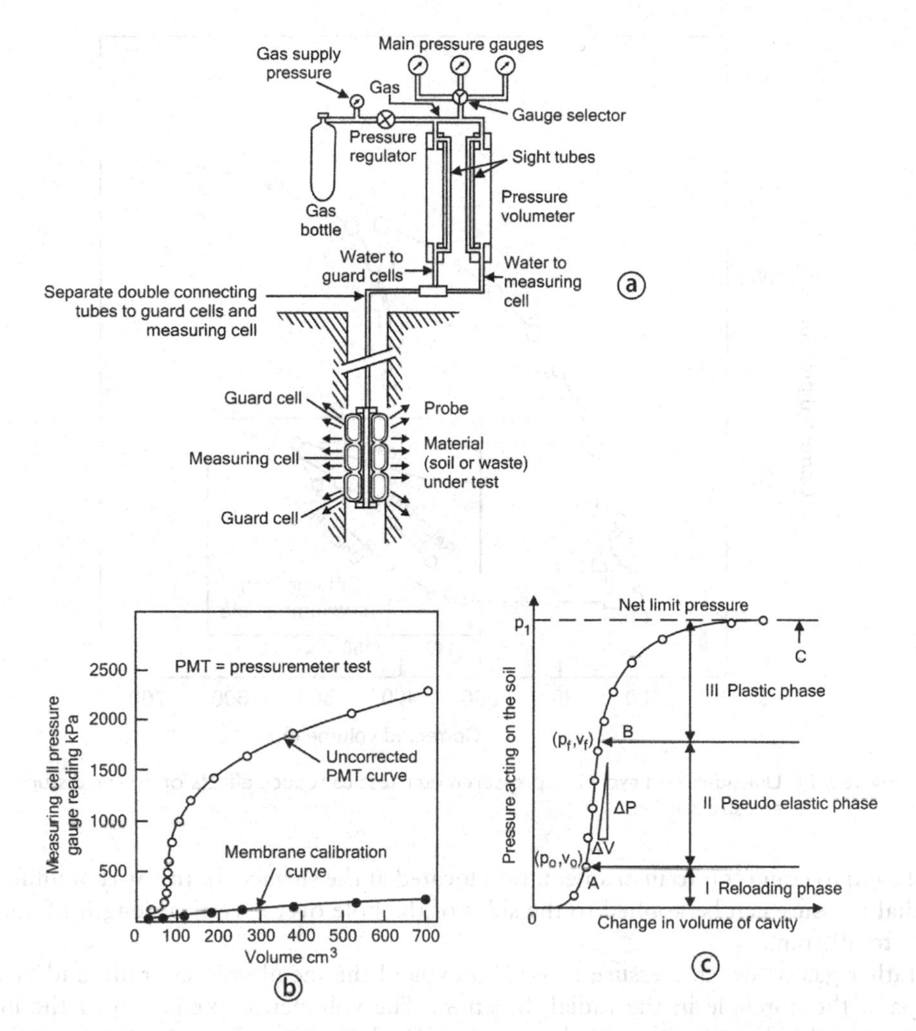

Figure 5.10 **Arrangement of the Menard-type pressuremeter and typical measurements. a: Principal components of the apparatus. b: Basic measurements. c: Interpretation of measurements.**

simpler forms of the instrument require to be inserted into and withdrawn from a hole that will not collapse during testing, and this is often difficult to arrange in tests below the phreatic surface. However, a stable hole can be achieved by means of drilling muds.

Figure 5.10 shows the arrangement of the simplest form of pressuremeter, the Menard pressuremeter. There are self-boring versions of the pressuremeter available and pore pressures can also be measured in situ in more elaborate versions.

The pressuremeter test can be regarded as a generalized form of lateral bearing load test. The pressuremeter, consisting of a metal cylinder covered by a cylindrical rubber membrane, is inserted into an uncased borehole having a slightly larger diameter than the pressuremeter. The membrane is expanded radially by increasing the fluid pressure inside the device. Cables and/or pipes are used to lower the presssuremeter down the

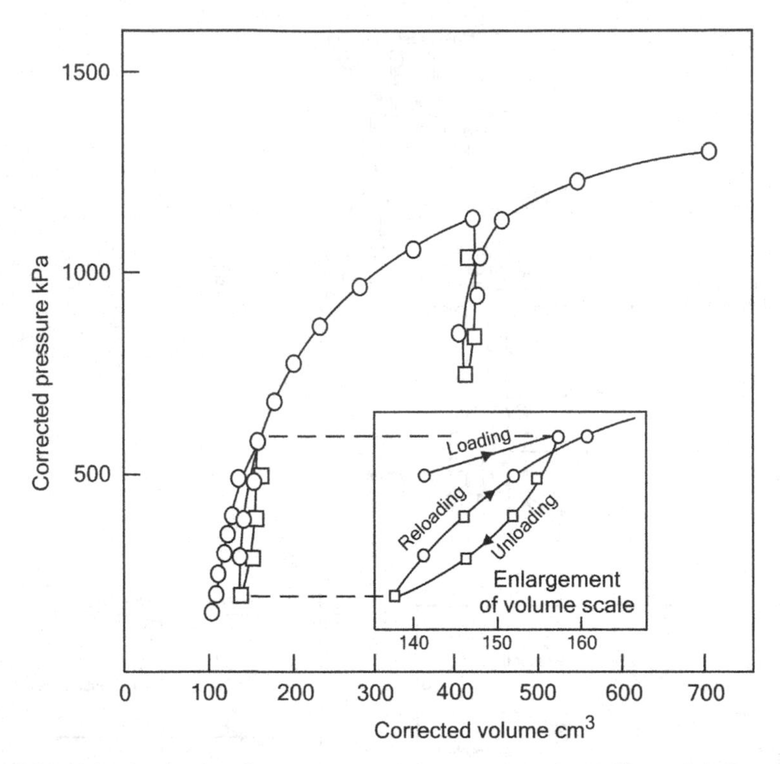

Figure 5.11 Unload-reload cycle in a pressuremeter test to reduce effects of drilling disturbance and stress release.

hole and to connect it to instrumentation located at the surface. In this way, a uniform radial pressure can be applied to the sides of the hole over a vertical length of about 300 to 800 mm.

Either gas or water pressure is used to expand the membrane laterally and hence expand the borehole in the radial direction. The volumetric expansion of the hole is measured indirectly either by determining the change in volume of the pressurized fluid by means of a calibrated volume gauge or else by deformation gauges consisting of strain-gauged spring steel fingers that follow the membrane surface radially as it expands. For a given radial expansion pressure, the change in diameter of the hole can be related, through theoretical and empirical analysis, to the horizontal modulus of elasticity of the soil in the vicinity of the pressuremeter. Referring to Figure 5.10a, the two guard cells are used to ensure that the central section over which the expansion is measured, expands uniformly over almost its whole height. Figure 5.10b shows a typical measured relationship between expansion pressure and volume (note the large values of pressure required.) The membrane has a finite resistance to expansion and this has to be subtracted from the measured expansion pressure. Figure 5.10c shows the corrected expansion curve and its interpretation:

- In phase I, the soil is being re-compressed to its original in situ radial lateral pressure.

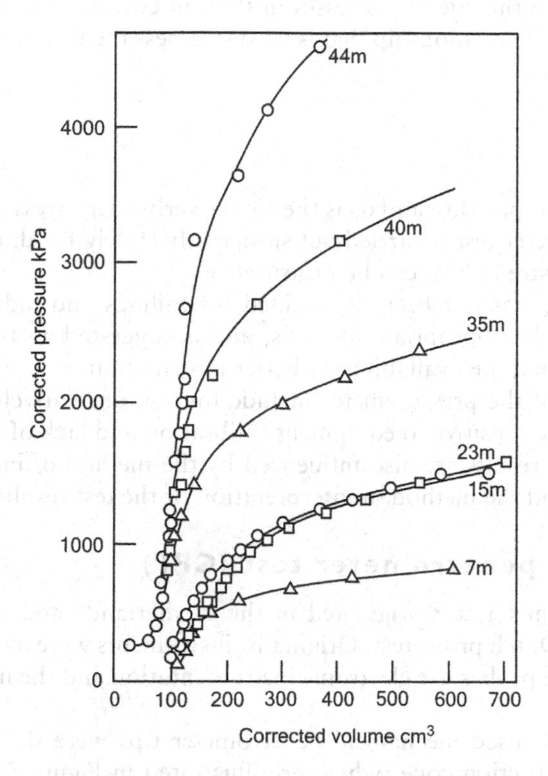

Figure 5.12 Typical set of pressuremeter tests at increasing depths in a soil profile.

- In phase II the soil compresses linearly with increasing pressure, and
- in phase III, the soil begins to fail and deforms plastically until it reaches the limit of volume change that can be accommodated by expansion of the membrane.

Because the Menard test is carried out in a pre-drilled and uncased hole, the soil forming the walls of the hole has been disturbed both by the forming of the hole and by stress release. To reduce this effect, it is usual to cycle the pressure, with the typical result shown in Figure 5.11. The steeper slope of the recycled curve can then be used to calculate the pressuremeter elastic modulus E_M for the soil from the equation:

$$E_M = (1 + v) \cdot 2V\Delta p/\Delta V \qquad (5.5)$$

where v = Poisson's ratio, is usually taken as $\frac{1}{3}$, $\Delta p/\Delta V$ = the slope of the (reloading) pressuremeter curve, and V = the cavity or expansion volume for the pressure range over which E_M is being measured.

The recycled modulus is typically 2 to 2.5 times larger than the modulus for first loading.

Figure 5.12 shows a typical set of pressuremeter curves measured at various depths in a profile of a natural soil. The diagram shows how the shear strength (related to the net limit pressure in Figure 5.10c) and the stiffness of the soil increase as the depth increases and the effective stresses in the soil correspondingly increase. In the Menard-type test, the relationship that is used to assess the unconsolidated undrained shear strength s_{uu} is:

$$s_{uu} = \frac{p_l - \sigma_v}{N_c} \tag{5.6}$$

where p_l is the limit pressure and σ_v is the total overburden stress.

If the pressuremeter test is carried out sufficiently slowly for drained conditions to prevail, a drained strength s_D can be measured.

The value of N_c has not been determined for tailings and industrial wastes, but $N_c = 9$ appears to be appropriate for soils, and is suggested as an interim value for mine wastes, pending the availability of better information.

Disadvantages of the pressuremeter include the use of relatively expensive equipment which is quite sensitive to equipment calibration and lack of operator expertise. Pressuremeter test results are also influenced by the method of installing the device, test procedures, and the method of interpretation of the test results.

5.3 The cone penetrometer test (CPT)

The cone penetrometer test originated in the Netherlands and is still occasionally referred to as the Dutch probe test. Originally, instruments were mechanical in nature, but presently, cone probes use electronic instrumentation and the mechanical probe is rarely used.

The most widely used mechanical penetrometer tips were the Delft mantle cone and the Begeman friction cone (which are illustrated in Figure 5.13). It is useful to consider these early types of cone, as they explain the principles used in more modern instruments. Both have a mantle of slightly reduced diameter attached above the cone. A sliding mechanism allows the downward movement of the cone relative to the outer casing. The force on the cone (i.e. the penetration resistance) is measured as the cone is pushed downward by applying a thrust on the inner push rods. If the cone is equipped with a friction sleeve, a second measurement is taken when the flange engages in the friction sleeve and cone and friction sleeve are pushed down a further increment together.

In the electrical penetrometer tip the cone is fixed and the cone resistance and the force on the friction sleeve are measured by means of electronic force transducers attached or built into the instrument. (See Figure 5.14.) Electric cables threaded through the outer casing or other suitable means (e.g. solid state memory) transmit the transducer signal to a data recording system. The electronic cone penetrometer tip permits continuous recording of the quantities measured over each push rod-length interval.

Electronic piezometers are vastly superior to the mechanical type. They permit of measurements with higher precision and with optimum repeatability. They also allow the simultaneous recording of cone resistance, pore pressure and sleeve friction. There is the option to install an inclinometer into the cone to check the verticality of the sounding. The string of relatively flexible rods may undergo considerable bending resulting in a drift of the penetrometer tip when used at depths exceeding 30 m, or when penetrating alternate hard and soft layers such as often occur in tailings deposits.

Note that the friction sleeve cone extends in two increments to reach the extended position shown in b. The cone is advanced by (69-33.5)=34.5mm and the friction sleeve is then dragged down by 45mm to measure cone + sleeve resistance. System is only effective if cone resistance does not decrease by penetrating into a layer of different strength as the sleeve is extending.

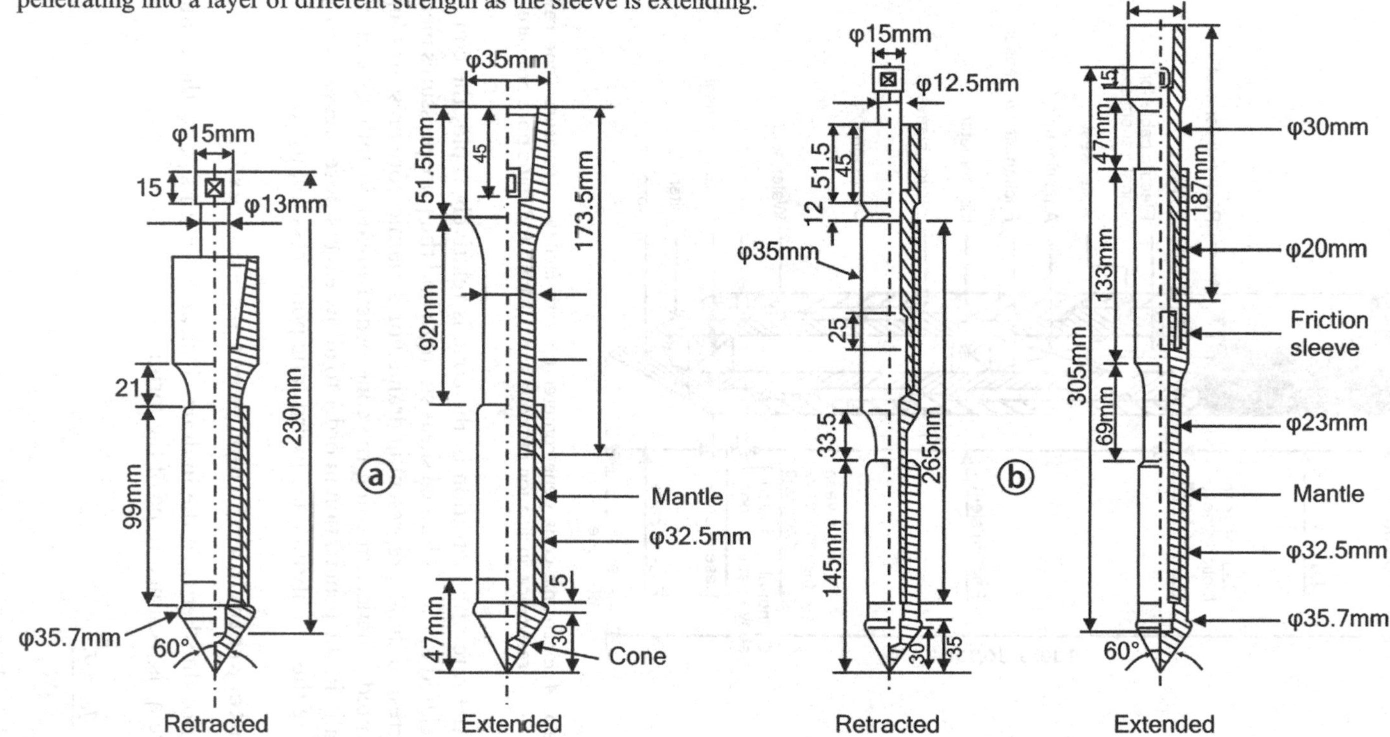

Figure 5.13 Two common types of mechanical cone: a: Delft mantle cone, b: Begeman friction sleeve cone.

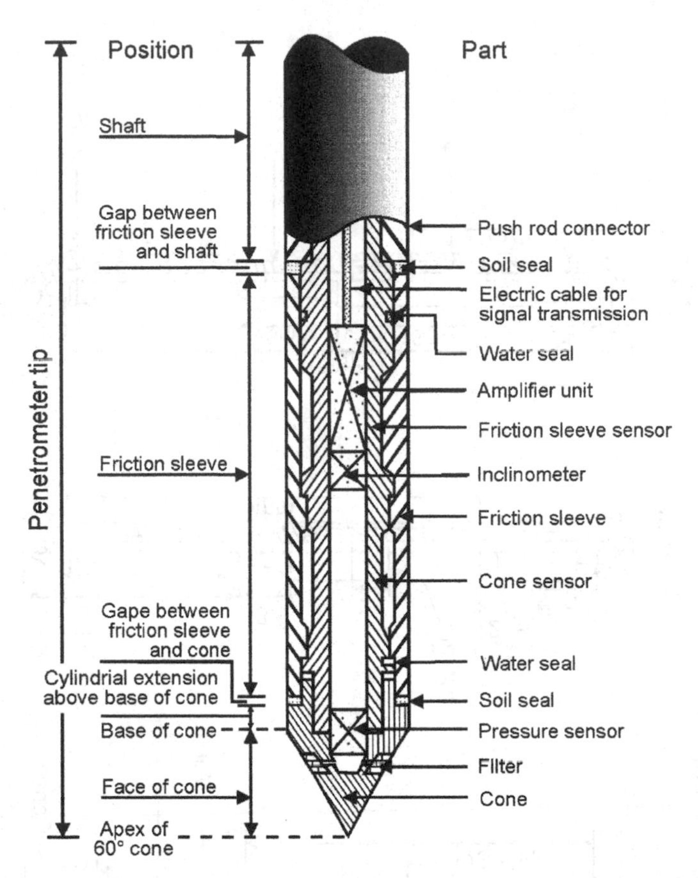

Figure 5.14 Electronic cone penetrometer tip with facility to measure cone resistance, sleeve resistance, inclination of probe and dynamic and static pore pressure.

Because the rate of penetration of the cone is relatively rapid, the cone resistance is usually related to the undrained shear strength and the pore pressures measured by the cone, correspondingly, represent undrained or dynamic pore pressures (u_d). However, it is accepted practice, in using the cone penetrometer in tailings storages, to pause periodically in the penetration in order to allow excess pore pressures to dissipate and to measure the equilibrium, i.e. hydrostatic pore pressure, u_e.

5.3.1 *Interpretation of cone penetrometer results*

The relationship between the undrained shear strength s_u and the cone penetration resistance q_c is assumed to be of the form:

$$s_u = \frac{q_c - \sigma_v^I}{N_c} \tag{5.7}$$

Where σ_v^I is the effective overburden pressure at the depth of the cone and N_c is a bearing capacity or cone factor.

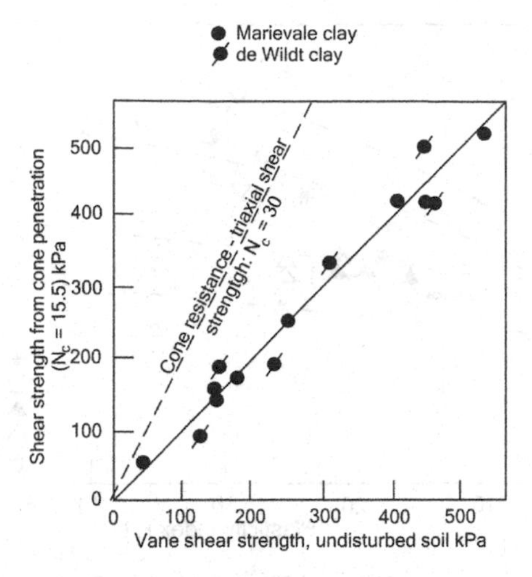

Figure 5.15a **Correlation between cone penetrometer resistance and undisturbed vane shear strength for two indurated clays.**

σ_v^I, is the effective overburden stress given by $(\sigma_v - u_e)$, i.e.

$\sigma_v^I = h(\gamma - \gamma_w) + h_w\gamma_w$

in which h is the depth of the cone below the soil surface and h_w is the depth of the water table below the surface. (see Figure 5.17).

$(q_c - \sigma_v^I)$, is referred to as the "corrected (or effective) cone penetration resistance", q_c^I.

The values of N_c have mainly been established by correlating shear strengths measured in the laboratory with cone penetrometer measurements. Most, if not all of the correlations relate to measurements in saturated clays (see Lunne et al., 1997) with a few measurements in heavily overconsolidated clays (Thomas, 1965) and hard, lime indurated clays (Blight, 1967). Even for clays there is great uncertainty, as values for N_c derived from a correlation between triaxial shear strength and cone resistance may be twice those for a correlation between in situ vane shear strength and cone resistance. This happens because the triaxial test measures the shear strength along fissures in the heavily overconsolidated and desiccated clays, whereas the vane blades and cone are constrained to shear through the intact clay between fissures. The strength along fissures is usually only half or less of the strength of the intact clay. Figure 5.15a shows the correlation between undisturbed (peak) vane shear strength and strength from cone penetration tests which indicated $N_c = 15.5$, established for two stiff desiccated clays by Blight (1967). However the correlation between q_c and triaxial shear strength indicated $N_c = 30$. This shows the uncertainty involved in using q_c to estimate in situ shear strength, and is the reason the author prefers not to use cone penetrometer measurements for estimating in situ strength. The results obtained by Thomas (1965) and Marsland and Quarterman (1982) were very similar in variability and uncertainty. (See Figure 5.15b). Marsland and Quarterman's values for N_c were correlated with the Plasticity Index (PI) of the various clays. The results are scattered but extrapolate to N_c = roughly 15 at PI = 0.

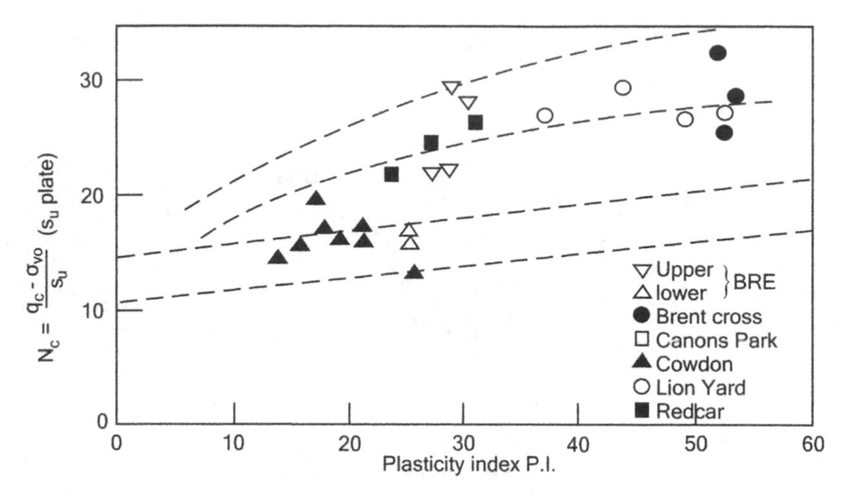

Figure 5.15b Correlation between N_c and Plasticity Index (PI) for stiff clays and soft rock with shear strength s_u measured by means of plate loading tests.

N_c for normally and overconsolidated clays has also been correlated with triaxial shear strengths via a pore pressure ratio, B_q, given by (Lunne et al., 1997):

$$B_q = (u_d - u_e)/q_c^I \tag{5.8}$$

in which u_d = dynamic pore pressure and u_e = equilibrium pore pressure. The relationship between B_q and N_c is summarized by Figure 5.16. The original experimental points are widely scattered between and outside of the two limiting lines shown in Figure 5.16, but the N_c values cover a range from the theoretical Prandtl bearing capacity value (Prandtl, 1921) of

$$N_c = \pi + 2 = 5.14$$

to a maximum of about 30.

Because of the uncertainty in choosing a suitable value of N_c, the cone penetration test should not be used to estimate the in situ shear strength of tailings. Nevertheless, the test is widely used in tailings storages because its measurements reveal the variability of the tailings, and show up hard and soft layers with amazing detail (see Figures 5.18, 5.19 and 5.20). The probe can also be used very easily to establish profiles of equilibrium pore pressure u_e in areas of a storage that are not equipped with piezometers.

5.3.2 *Interpretation of cone resistance in cohesionless sands and silts*

The cone penetrometer was originally developed for tests in saturated sands and silts and the greatest body of experience relates to its use in these materials. The following analysis relates to the system of assumed shear surfaces illustrated at top left in Figure 5.17.

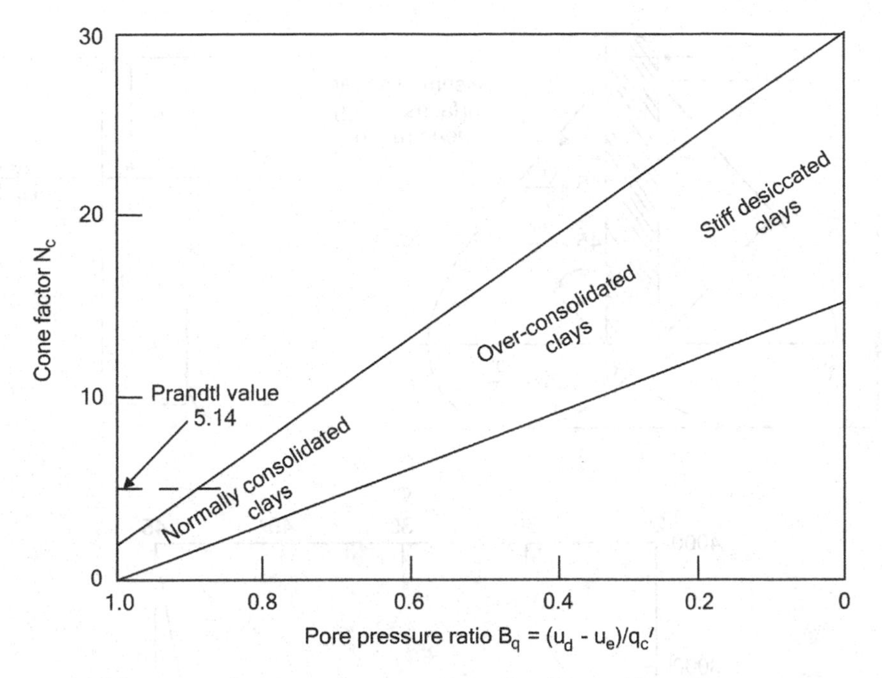

Figure 5.16 Summarised relationship between cone factor N_c and pore pressure ratio $B_q = (u_d - u_e)/q_c^I$.

The expression for the bearing capacity of the cone tip is:

$$q_C^I = \tfrac{1}{3}\sigma_V^I \tan^2(45° + \varphi^I/2)\exp(2\pi \tan\varphi^I)$$

or

$$\frac{q_C^I}{\sigma_V^I} = \tfrac{1}{3}\tan^2(45° + \varphi^I/2)\exp(2\pi \tan\varphi^I) \tag{5.9}$$

Values of q_c/σ_v^I corresponding to a range of values of φ^I are given in Table 5.1 and have been graphed in Figure 5.17.

Robertson and Campanella (1983) proposed the following empirical relationship between peak values of φ^I and q_c^I/σ_v^I, which they established by means of comparisons between laboratory shear tests and CPT tests carried out in a laboratory calibration tank:

$$\varphi^I = \tan^{-1}[0.1 + 0.38\log_{10}(q_c^I/\sigma_v^I)] \tag{5.10}$$

As shown in Table 5.1, except at the start of the comparison, the empirical values are considerably bigger than values based on equation (5.9). The comparison stops at $\varphi^I = 47°$ because that is the upper limit of the measured values of φ^I. The difference may arise because the theoretical values do not include an allowance for the effects of dilation of the sand. The lowest measured φ^I by Robertson and Campanella was 30° and the value could also have included a component caused by dilation.

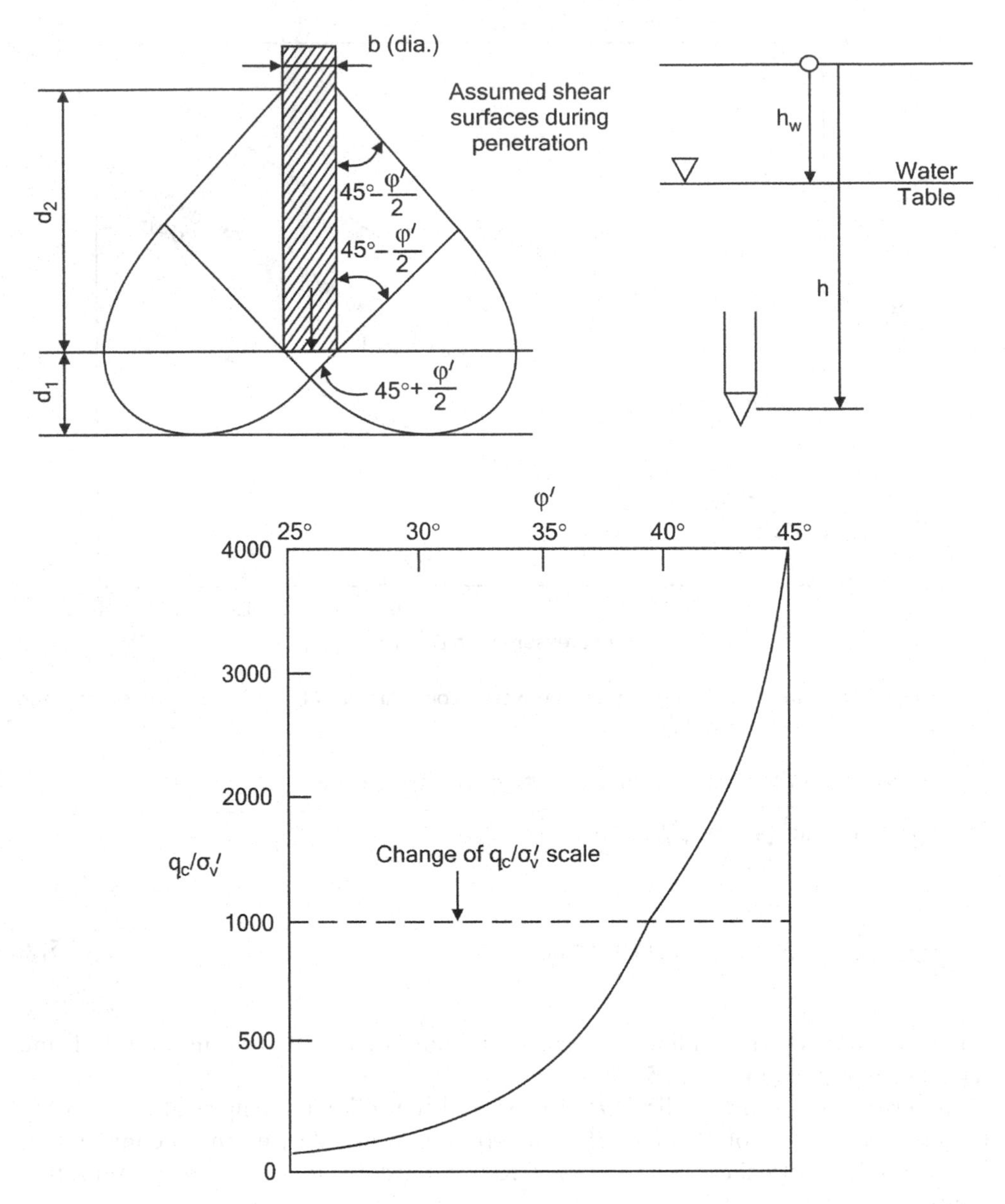

Figure 5.17 Interpretation of cone resistance measurements in a cohesionless sand or other cohesionless particulate material.

A second empirical equation is due to Kulhawy and Mayne (1990):

$$\varphi^{I} = 17.6° + 11 \log_{10}(q_{c1}) \tag{5.11}$$

in which

$q_{c1} = q_c/(\sigma_v^I/\sigma_{atm})^{0.5}$ and σ_{atm} = atmospheric pressure in the same units as σ_v^I.

Table 5.1 Tabulated Values of φ^I and q_c^I/σ_v^I for equations (5.9) and 5(10)

q_c^I/σ_v^I	φ^I (Eqn. 5.9)	φ^I (Eqn. 5.10)
3	15°	16°
26	20°	32°
60	25°	38°
93	27.5°	40°
147	30°	43°
236	32.5°	45°
391	35°	47°
663	37.5°	
1165	40°	
2125	42.5°	
4057	45°	

(Note that $q_c^I = q_c - \sigma_v^I$)

However equation (5.11) cannot be compared with equations (5.9) and (5.10) in general, because it is not possible to assign unique values of q_c and σ_v^I to the ratio q_c^I/σ_v^I in order to obtain the corresponding $(\sigma_v^I/\sigma_{atm})^{0.5}$.

But, following Table 5.1,

$$\begin{aligned} \text{if} \quad q_c &= 1000\,\text{kPa and } q_c = 3\sigma_v^I, \text{then } \sigma_v^I = 333\,\text{kPa}, \\ q_{c1} &= 1000/\surd 3.33 = 548\,\text{kPa}, \text{and } \varphi^I = 48^\circ \\ \text{if} \quad q_c &= 1000\,\text{kPa and } q_c = 391\sigma_v^I, \text{ then } \sigma_v^I = 2.5\,\text{kPa}, \\ q_{c1} &= 1000/\surd 0.025 = 6329\,\text{kPa, and } \varphi^I = 59^\circ \end{aligned}$$

Hence equation (5.11) does not seem to agree with either equation (5.9) or (5.10), nor does it seem to give realistic values for φ^I.

The comparison of theoretical with empirical values is not very encouraging and equation (5.9) seems the more realistic. Also, as shown in Section 5.3.4, good agreement can be obtained between strengths measured by means of the vane shear test and corresponding strengths predicted from measured values of q_c^I via equation (5.9).

5.3.3 *Cone penetrometer tests in tailings and other wastes*

Figure 5.18 shows the results of a typical cone penetrometer test on a deposit of platinum tailings. The diagram shows the variation of cone resistance (q_c in MPa), dynamic pore pressure (u_d in kPa) and equilibrium pore pressure (u_e) with depth. The tailings storage has been established on a clay foundation which in turn rests on a more permeable weathered rock. To measure the equilibrium pore pressures, the downward progression of the cone is halted and the dynamic pore pressure is allowed to dissipate until a steady (i.e. equilibrium) value u_e is attained. This approximates to the static pore pressure in the surrounding tailings (or clay, in this example), and is the value that should be used to estimate σ_v^I.

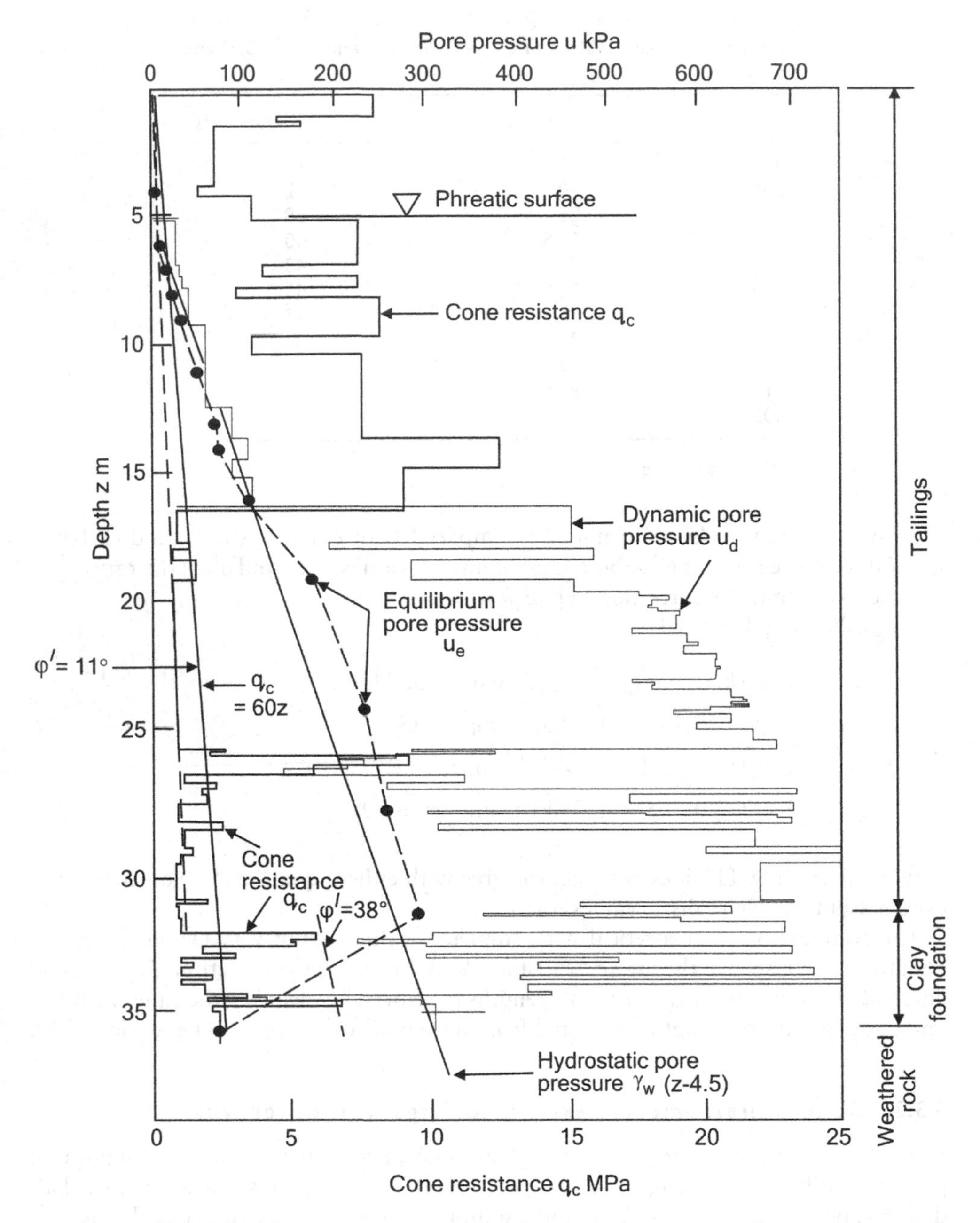

Figure 5.18 Results of cone penetration test in platinum tailings storage. Cone resistance q_c in MPa, pore pressure u in kPa. Dynamic pore pressure u_d as continuous line, equilibrium pore pressure u_e as solid circles.

Re-arranging equation (5.7) and noting that a predicted value for $s_u = \sigma_v^I \tan \varphi^I$:

$$q_c = s_u N_c + \sigma_v^I = \sigma_v^I (N_c \tan \varphi^I + 1), \text{ and } \tan \varphi^i = (q_c/\sigma_v^I - 1)/N_c \quad (5.7a)$$

The most striking features of Figure 5.18 are:

- the extremely variable nature of both dynamic pore pressure and cone resistance, and
- the extremely low values of the cone resistance that occur below a depth of 16 m

It is also apparent from Figure 5.8 that there is a small excess pore pressure over hydrostatic pore pressure in the tailings between 16 and 31 m depth, and that the excess pore pressure is probably slowly draining through the basal clay layer into the more permeable weathered rock below the clay. It should also be noted that high dynamic pore pressures are associated with low cone resistances, indicating a contractant tendency (at depths of 17 m, 19–25 m, 28–32 m, etc.). Conversely, high cone resistances are associated with low dynamic pore pressures, indicating a dilatant tendency (at depths of 0–16 m, 26 m, 32 m). If the tailings were to suffer liquefaction, it would be most likely to occur in the zones of low cone resistance, contractant tailings.

If we take γ_{sat} for the tailings as 20 kN/m^3, and the relationship between u_e and depth z as $u_e = \gamma_w(z - 4.5)$, we could take the average value of q_c in the lower part of the profile as $q_c = 60$ zkPa. Applying equation (5.7)a:

at $z = 32$m, $\sigma_v^I = 32 \times 20 - 10 \times 27.5 = 365$ kPa. With $N_c = 20$,

$\tan \varphi^I = (60 \times 32/365 - 1)/20 = 0.21$, and $\varphi^I = 12°$.

For the high value of $q_c = 6000$ kPa at $z = 32$ m

$\tan \varphi^I = (6000/365 - 1)/20 = 0.77$, and $\varphi^I = 38°$

Both of which seem reasonable for contractant and dilatant materials, respectively.

This does not address the remarkably high q_c values down to 16 m. It is known that these arise because in recent years, the mine has disposed of quenched slag along with the tailings. This acts as a hydraulic cement and produces a high cohesion in the tailings. This thick cemented layer might not prevent an overall failure of the dam if the deeper layers liquefy, as the top layers might simply ride on the heavy liquid below them and fail as a huge block.

Figure 5.19 shows a second example of a cone penetrometer test record, this time for gold tailings. This test record was observed during the investigation of the failure of the Merriespruit tailings storage. The test was located near to the decant tower of the storage and therefore represents much finer material than the platinum tailings of Figure 5.18 (see Figure 1.3a for a comparison of the particle size analyses of platinum and gold tailings. The grading of the material represented in Figure 5.19 lies on the fine side of the envelope for gold tailings.).

Figure 5.19 also shows the great variability of cone resistance and dynamic pore pressure that is typically recorded in a tailings storage. Over the top 3 m the cone resistance shows that there is a crust formed by desiccation of the surface. This was formed while the storage was being "rested, without deposition, prior to the failure, to

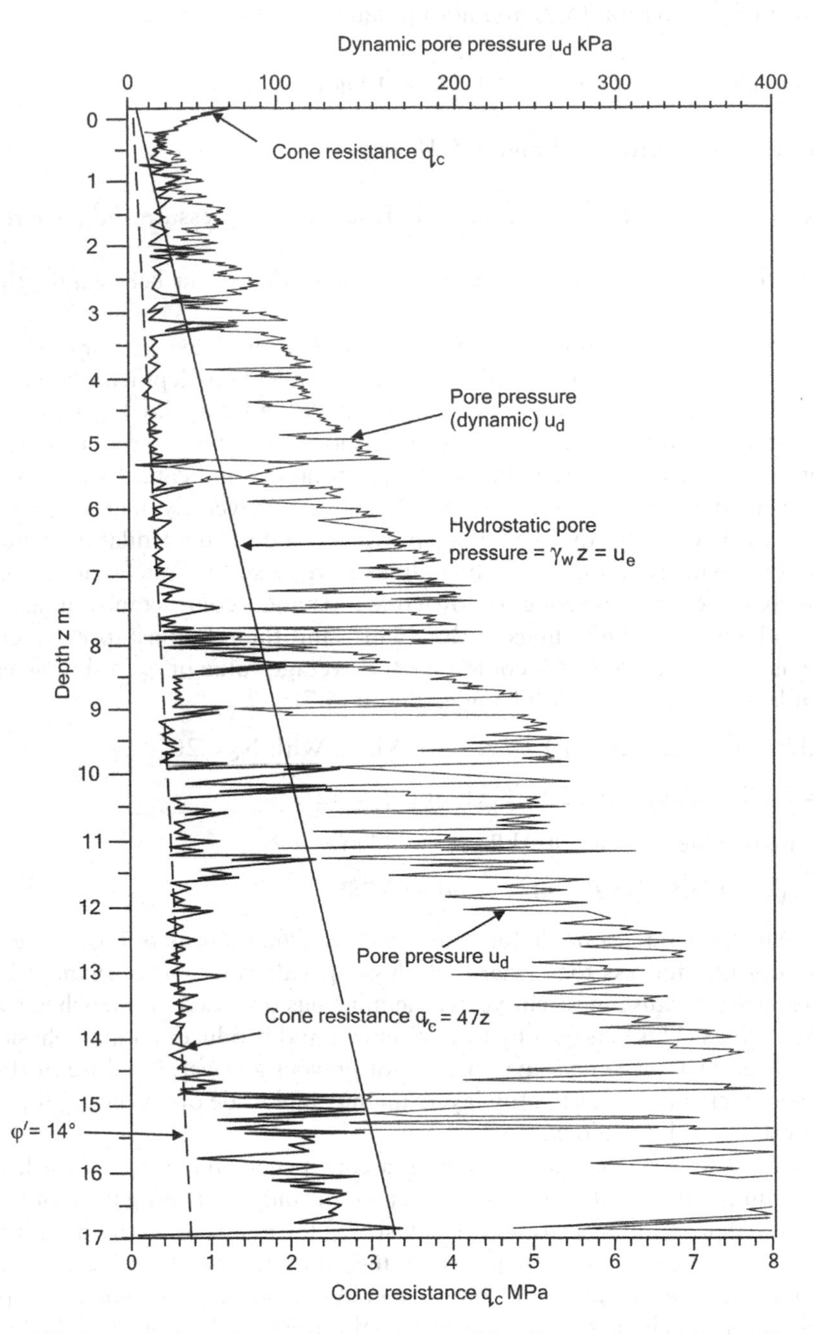

Figure 5.19 Record of cone penetration test in gold tailings storage. No equilibrium pore pressures were measured. Hydrostatic pore pressure was calculated.

allow the phreatic surface to subside. The record shows the presence of older desiccated horizons that existed at depth intervals of 1 to 2 m (e.g. 5.5, 8, 9, 10, 11.5 and 13 m). Below 15 m the cone had probably entered the foundation strata. The drop in cone resistance at 17 m shows where withdrawal of the cone started. Also note that in the gold tailings, a sharp increase in cone resistance coincides with a sharp decrease in dynamic pore pressure, and vice versa.

Applying equation (5.7)a to Figure 5.19, the mean q_c/z was 47 kPa/m and $\gamma_{sat} = 18\ kN/m^3$. At $z = 15$ m: $\sigma^I_v = (18 - 10)15 = 120$ kPa, and with $N_c = 20$:

$$\tan \varphi^I = (47 \times 15/120 - 1)/20 = 0.24, \text{ and } \varphi^I = 14°.$$

This value would also be appropriate for a liquefiable material.

The cone penetrometer used to generate the data in Figures 5.18 and 5.19 was not equipped with a friction sleeve (see Figure 5.14). The friction sleeve measures the large strain, i.e. residual or remoulded strength of the tailings, which is analogous and comparable with the remoulded strength measured by the vane shear apparatus (e.g. Figures 5.8a and b). Figure 5.20 shows profiles of cone resistance, sleeve resistance and dynamic pore pressure measured in the same storage of industrial sludge as that referred to in Figure 5.8b. For reasons of accessibility, the probe was made near the crest of the outer slope of the waste storage. The storage had been constructed by upstream methods with the outer retaining shell constructed of coarser wastes. The outer shell sloped at 40° to horizontal. As shown by the diagram inserted in Figure 5.20, this means that the first 8m of probing passed through the coarse, higher strength shell material and then passed into the sludge below this. The change in material characteristics is shown very clearly. Below 8m, the probe encountered a number of thin coarser zones that are shown by local increases of both q_c and f_s. At the lower end of the probing, the cone was about 30m horizontally from the crest of the outer slope. No dynamic pore pressure was encountered in the (presumably) pervious shell. Once the cone passed into the high water content, lower permeability sludge dynamic pore pressures were recorded. Cone resistance was about 1MPa and the minimum sleeve resistance about 25 MPa. The strength of the sludge did not increase with depth below 8 m. The diagram shows two examples of applying equations 5.7 and 5.8 together with Figure 5.16, in order to establish values of N_c and s_u. It must be emphasized that this is purely as an example of the procedure. Because the correlations between B_q and N_c were established for natural clays, it is not necessarily valid to apply them to a material that is not a natural clay. Nevertheless, the example shows how uncertain the estimates of both N_c and s_u are. It also shows that s_u and f_s come out at the same order of magnitude, but cannot really be compared with each other except statistically, if a large amount of data are available.

5.3.4 *Comparison of cone penetration and vane shear tests in a waste coal duff*

Most tailings fall into the silt or fine sand particle size range and hence the analysis in Section 5.3.2 should be applicable, but is seldom used, The results should perhaps be treated with caution as φ^I values found by this analysis are reportedly different (smaller or bigger) than corresponding vane or triaxial measurements would indicate. However,

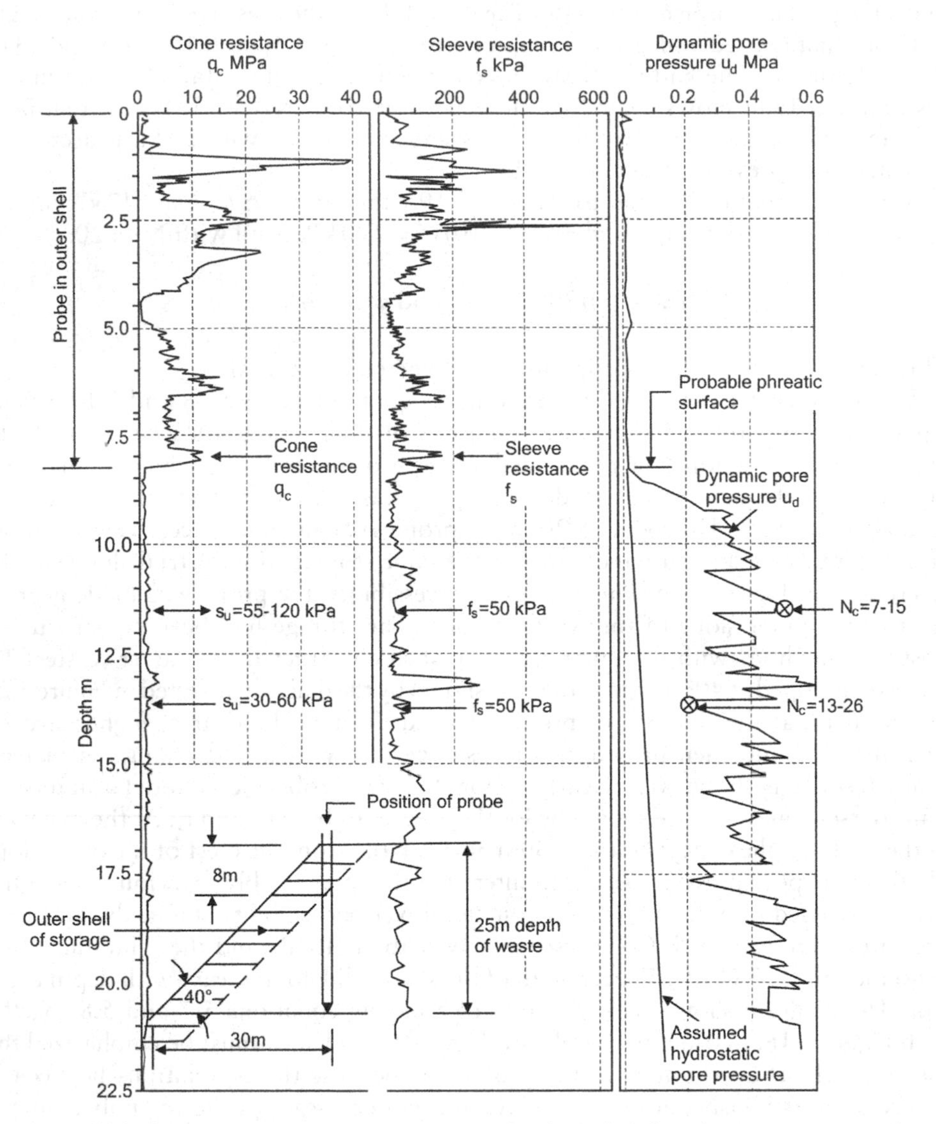

Figure 5.20 Cone penetration test in a storage of industrial sludge, showing variations of cone resistance, q_c, sleeve resistance, f_s, and dynamic pore pressure u_d.

in the one comparison made by the author, strengths calculated from cone resistance values agreed very well with values measured by means of the vane apparatus. A description of this comparison between tests on a waste coal duff follows:

The waste anthracite coal duff consisted of the following particle sizes

12 mm–2 mm	40% by mass
2 mm – 0.06 mm	50%
finer than 0.06 mm	10%

The duff had a loose dry density of 770 kg/m^3 (dry unit weight = 7.7 kN/m^3). The material was coarser than would usually be considered suitable for testing by either the shear vane or the cone penetrometer, but because of the 60% sand and silt sizes it contained, it was likely that the overall behaviour would be governed by the fine particles and the tests were undertaken on the basis of this expectation. The dump had also become heated by spontaneous combustion and was thought to be burning internally. On closer examination, what had been thought to be smoke turned out to be steam issuing from the dump surface at a temperature of 77°C. Because of the possibility that higher temperatures would be encountered at depth, the cone apparatus at the end of its outer casing, but without any push-rods, was equipped with a maximum thermometer with its bulb located just above the cone and the improvised temperature probe was pushed into the body of the dump, pausing for 20 minutes at each depth to allow the temperature at the thermometer to equilibrate with that in the surrounding coal, before withdrawing the thermometer, at the end of a string, to read the temperature. Temperatures down to 46 m turned out to be about 75°C. The probe and its thermometer were left in place overnight at 46m and substantially the same temperature was recorded the next day. The temperature measurements are shown in Figure 5.21. A second, less detailed temperature profile at a second location confirmed the first. The application of Charles' law showed that the pore air pressure at 46 m was unlikely to exceed $(273° + 75°)/(273° + 20°) = 1.19$ atmospheres (absolute) or 0.19 atmospheres (=19 kPa) of excess pore pressure at 46 m depth (an overburden stress of $46 \times 7.7 = 354$ kPa). Hence the effect of pore air pressures on shear strength could be disregarded.

Vane shear tests proceeded with measurements of peak or maximum torque being recorded at progressively larger depths, followed by remoulded torque. This was repeated at another 3 locations. The limit of strength of the vane blades was reached at 31 m. Figure 5.22 records the results of the 4 vane shear strength profiles.

The cone penetrometer measurements (3 separate profiles) were then carried out, down to a depth of 39 m. Figure 5.23 compares the results of the vane and cone testing. The cone results have been presented as the predicted shear-strength in terms of $\gamma h \tan \varphi^1$ with the unit weight $\gamma = 8$ kN/m^3 and φ^1 calculated from equation (5.9) together with Figure 5.17. As shown by Figure 5.23, the cone penetrometer results agreed very well with those for the mean remoulded vane shear strength.

5.3.5 *Seismic shear wave velocity measurements*

The capability of the cone penetrometer can be very usefully extended by including an accelerometer mounted internally behind the cone (Robertson, et al., 1986, Vidic, et al., 1996). An instrumented hammer is used to strike a rigid plate, weighed down by one of the supporting jacks of the drill rig, to provide the seismic source. The travel time of the seismic shear wave between the instant of the hammer blow at the surface and the CPT cone at depth is then measured and recorded to allow the low-strain shear modulus of the intervening material, natural soil or tailings, to be calculated from:

$$G_s = \gamma V_s^2 / g \tag{5.12}$$

where G_s is the seismic shear modulus, γ is the bulk unit weight of the material, V_s the shear wave velocity and g the gravitational acceleration. Dimensionally:

$$[N/m^2] = [N/m^3] \times [(m/s)^2]/[m/s^2]$$

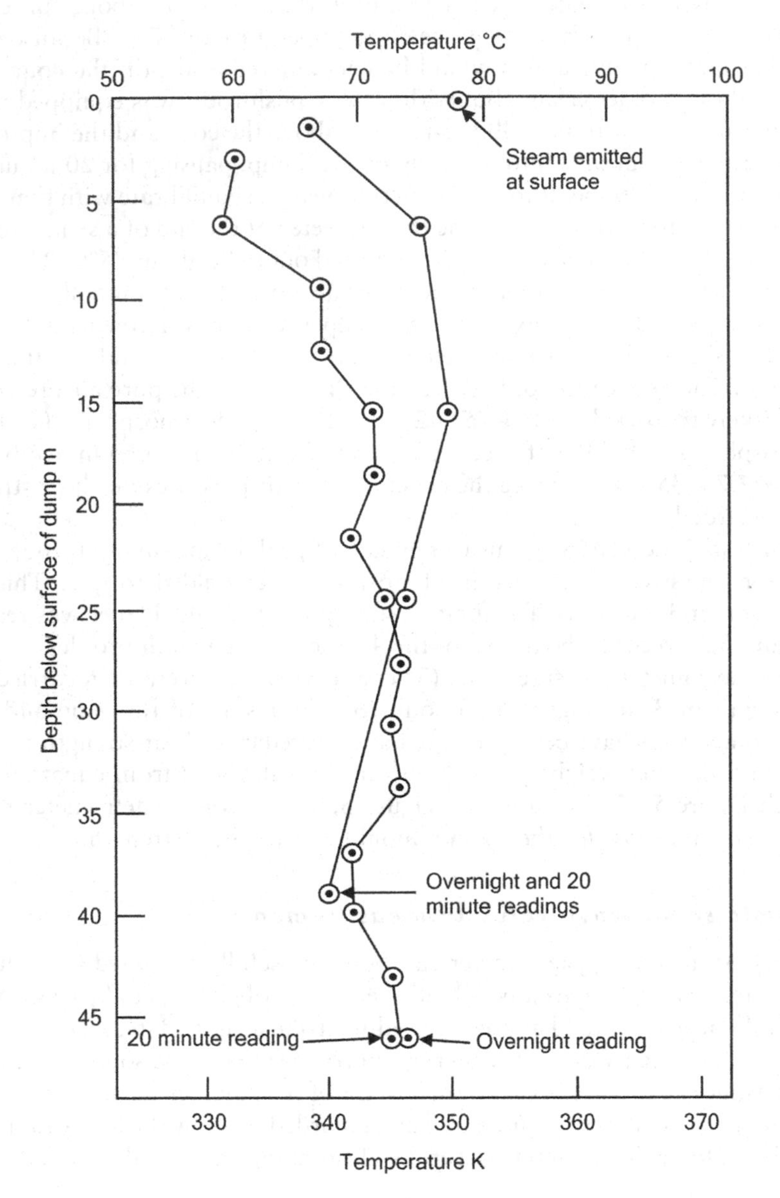

Figure 5.21 Temperatures measured in situ in an anthracite duff dump.

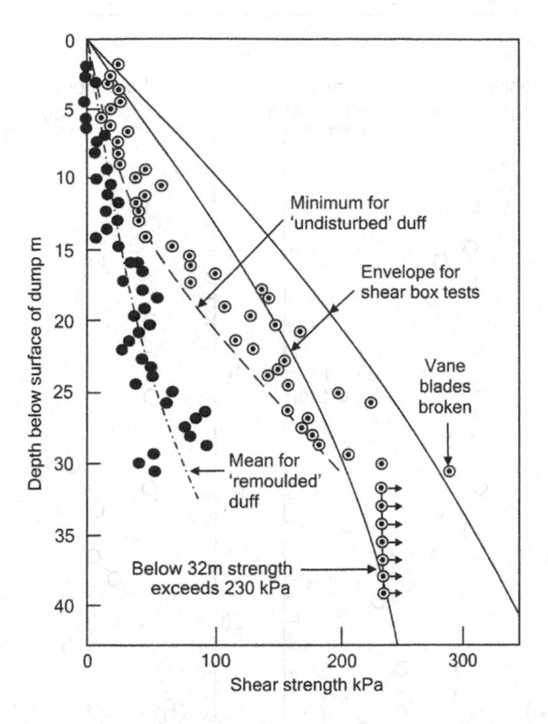

Figure 5.22 Vane shear strength measurements in a dump of waste coal duff.

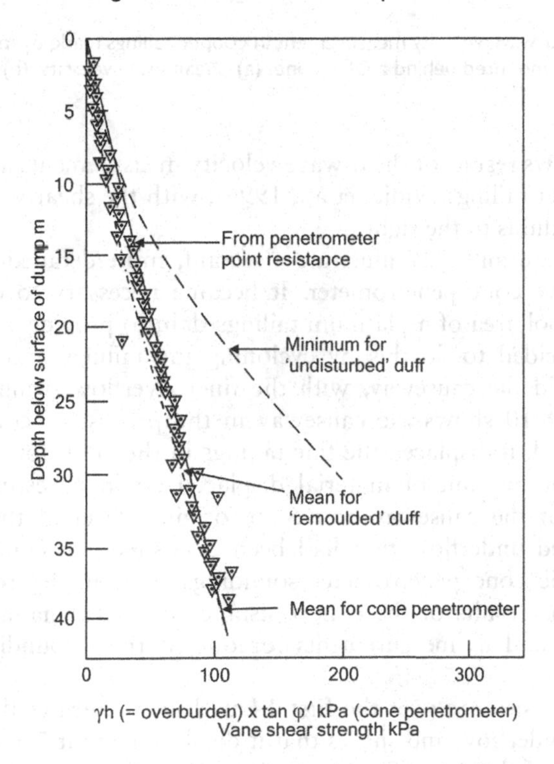

Figure 5.23 Comparison of mean in situ strength of waste coal duff measured by means of vane apparatus (broken lines) with predicted strength from cone resistance via equations (5.7) and (5.7)a.

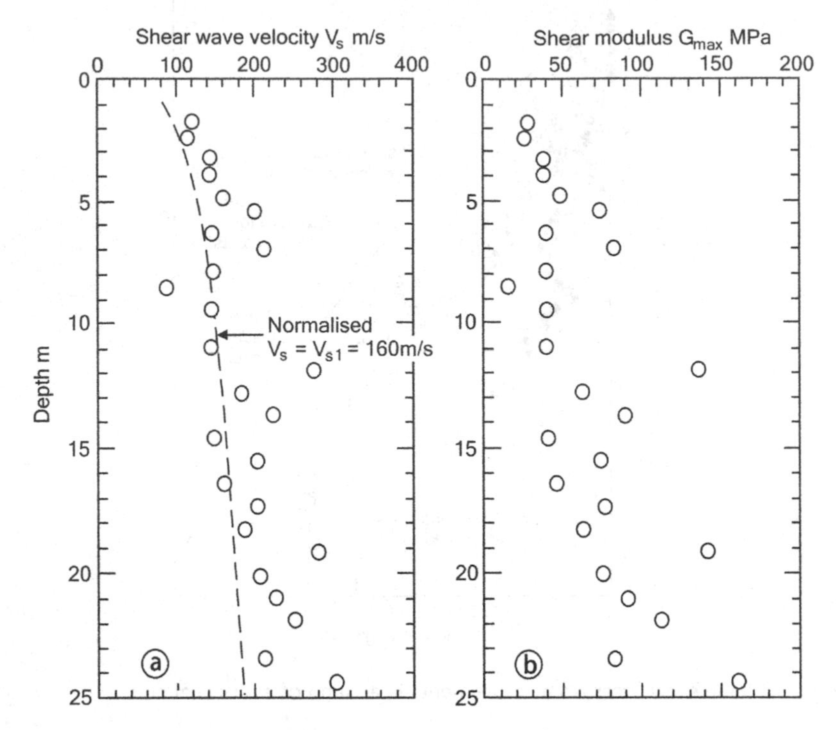

Figure 5.24 Shear wave velocity measurement in copper tailings made by means of an accelerometer mounted behind a CPT cone. (a) Shear wave velocity. (b) Shear modulus.

Figure 5.24 shows results of shear wave velocity measurements at various depths in a storage of copper tailings (Vidic, et al., 1996), with the shear wave velocity on the left and shear modulus to the right.

Figures 5.25, 5.26 and 5.27 illustrate a second, more detailed application of the seismic shear wave cone penetrometer. It became necessary to construct a causeway across the pool area of a platinum tailings dam to provide access to the decant tower. It was decided to do this by cycloning the tailings and using the coarser underflow to build the causeway, with the finer overflow being directed into the pool area. Plate 8.10 shows the causeway in the process of construction. As the causeway advanced, it displaced the fine tailings in the pool, forming a pronounced "mud wave". The amount of material displaced could be estimated by comparing the volume of the causeway above the original level of the pool floor with the total estimated underflow that had been deposited. To confirm this estimate, a series of seismic cone penetrometer soundings were made from the surface of the causeway. The results of the cone resistance q_c and dynamic and equilibrium pore pressure u_d and u_e measurements for one of these soundings are shown in Figure 5.25.

The high cone resistance over the first 14 m clearly indicates the thickness of the coarser cyclone underflow and shows that it displaced about 7 m of the fine tailings forming the floor of the pool. The water table in the sand was quite clearly located

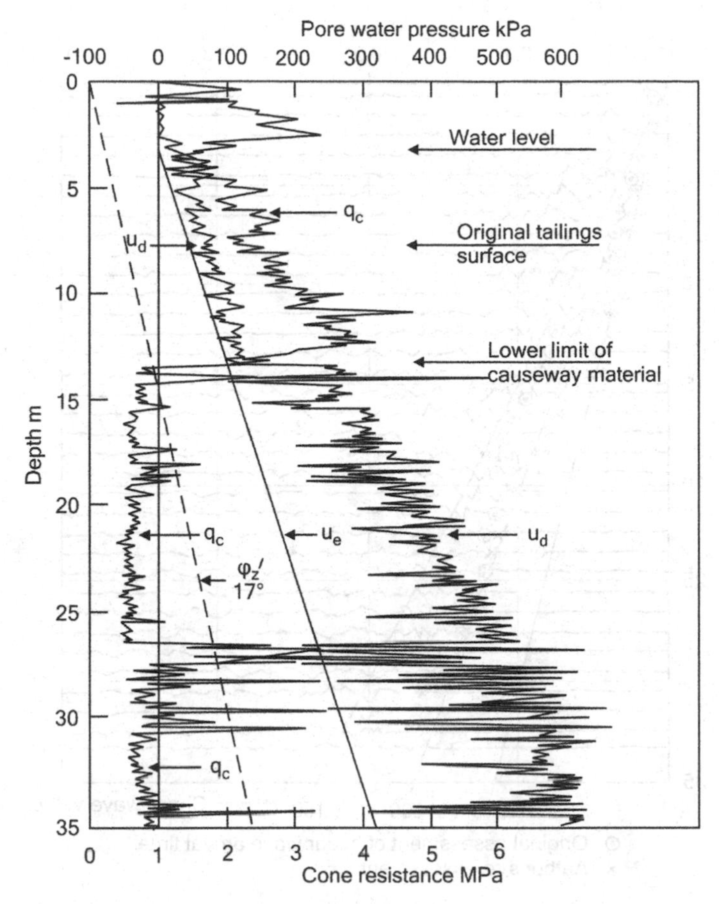

Figure 5.25 Cone penetration probing through cyclone underflow causeway into softer tailings below.

at 2 m below the causeway surface, and dynamic pore pressures were small. Once the cone entered the fine tailings underlying the sand, q_c reduced dramatically and u_d increased equally dramatically. A harder layer was penetrated at a depth of 27 m with weaker tailings beneath it.

Figure 5.26 shows the accelerometer traces recorded as the cone was advanced downwards. The arrival time of the shear wave as it travelled from surface to cone level is indicated by the first disturbance to occur in the trace at each depth. As shown in the diagram, inclined lines can be drawn to indicate various constant shear wave velocities, 300 m/s, 200 m/s, etc.

Figure 5.27 shows the variation of shear wave velocity with depth interpreted from the traces in Figure 5.26. The circles represent the original interpretation of shear wave velocity from Figure 5.26. When the author drew the lines of constant shear wave velocity on Figure 5.26, it was noticed that not all of the data in Figure 5.27 agreed with these, and a re-assessment was made, which is shown by diagonal crosses.

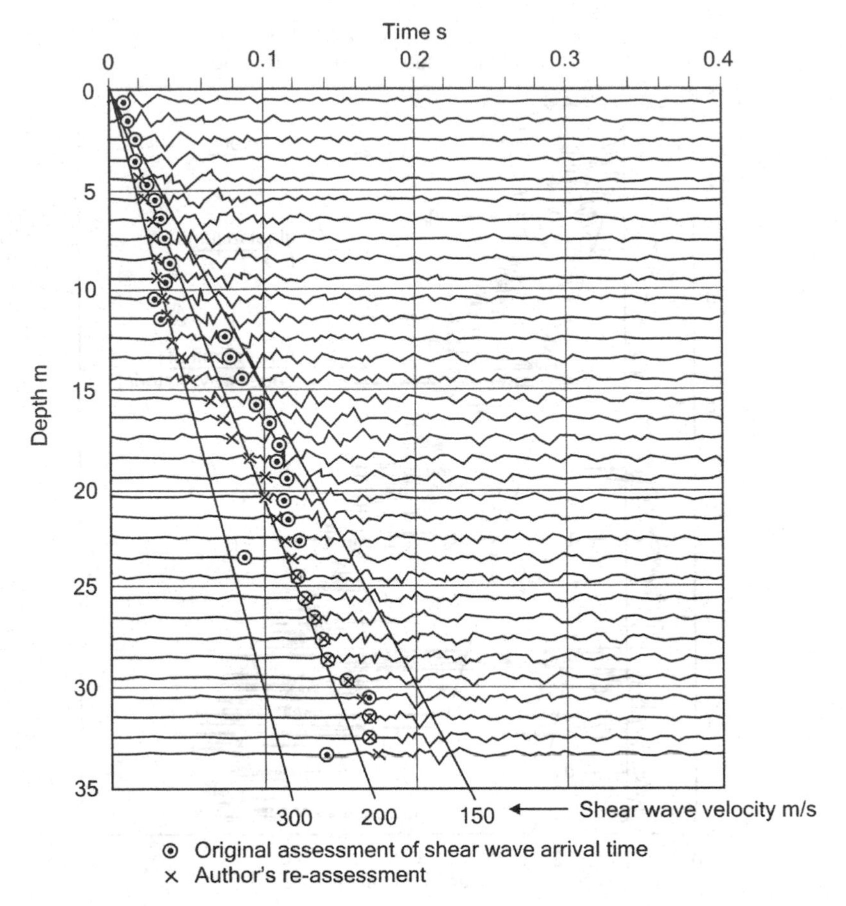

Figure 5.26 Accelerometer traces corresponding to the cone penetrometer sounding shown in Figure 5.25.

This is quite a lot different from the original assessment. The high shear wave velocities of 250 to 300 m/s extend to 15 m depth which agrees quite well with the depth of high cone resistance shown in Figure 5.25. Examining Figures 5.26 and 5.27 closely, it will be seen that the depths of the various traces do not coincide exactly with the depths at which the corresponding shear wave velocities have been plotted. In particular, the velocity that was plotted at 15 m in Figure 5.27 is shown at 14.5 m in Figure 5.26. This improves the fit between Figures 5.27 and 5.25. Below 15 m, the revised shear wave velocity was fairly constant with depth and the harder layer at 27 m depth does not show on Figure 5.27.

(This section has also illustrated an important principle: Test results should never be accepted at face value. To be credible, they must be supplied with sufficient details to allow the result to be checked, i.e. to carry out checks to ensure that the numerical results are correct and that the tests have been performed by means of the correct methods, correctly applied.)

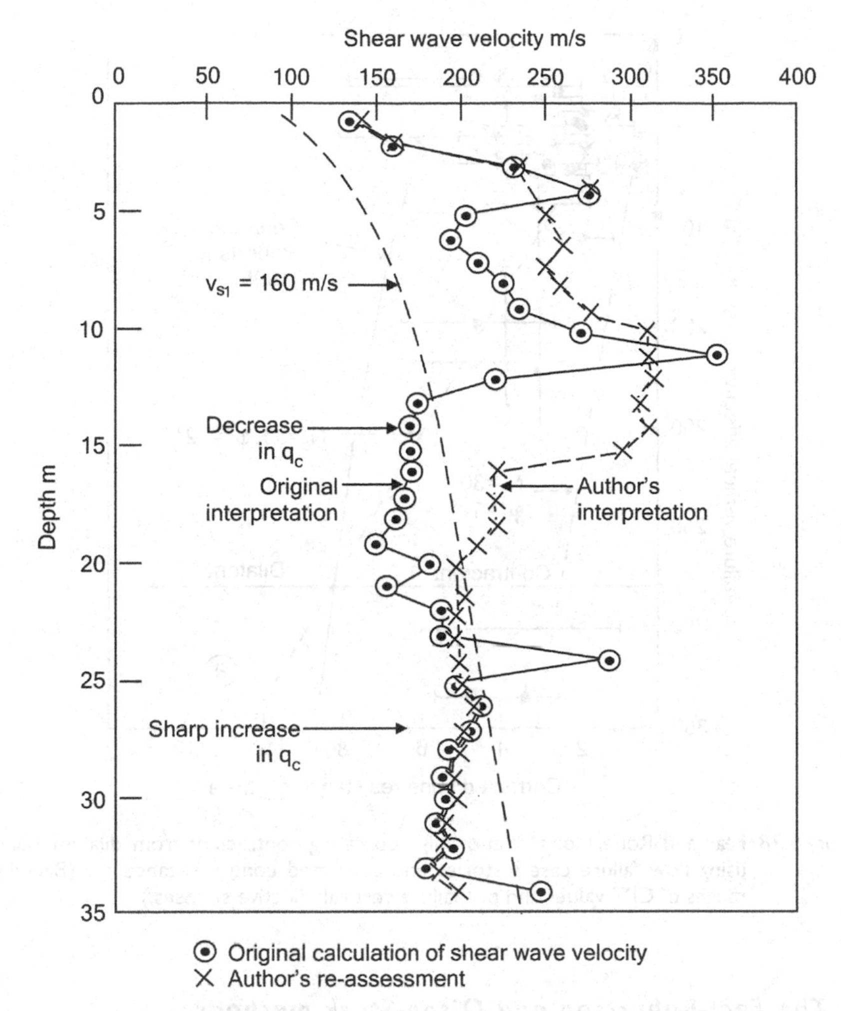

Figure 5.27 Variation of shear wave velocity with depth from interpretation of shear wave arrival time in Figure 5.26.

5.4 Estimation of potential for liquefaction from cone penetration test results

It is suspected that liquefaction failures of hydraulic fill tailings storages usually initiate in relatively thin, high void ratio layers that have formed during periods when the relative density of the tailings slurry was unusually low or the rate of rise was unusually high, or both. This can and does occur when problems are experienced in operating the mineral extraction plant. The fine detail of the variation of cone resistance with depth that it is possible to see with a continuous record of cone resistance (see Figures 5.18, 5.19 and 5.20) shows up these weak, high void ratio layers very clearly. The cone penetrometer is thus an ideal instrument to investigate the potential for liquefaction of an existing tailings storage.

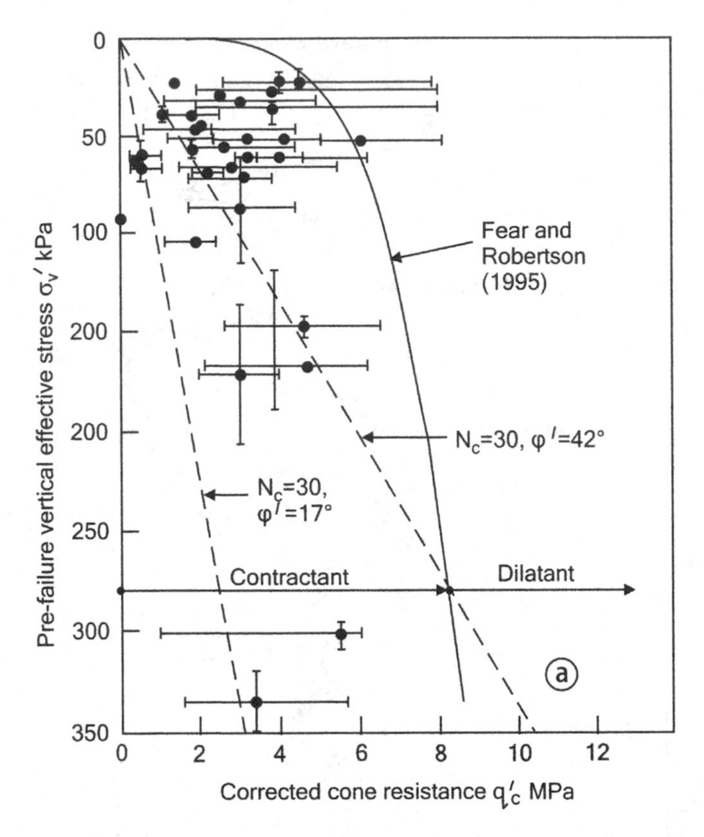

Figure 5.28 Fear and Robertson's relationship separating contractant from dilatant conditions using flow failure case histories and corrected cone resistance q_c^I. (Bars indicate ranges of CPT values and pre-failure vertical effective stresses.)

5.4.1 *The Fear-Robertson and Olson-Stark methods*

Two well-documented methods are available for assessing potential liquefiability from CPT measurements. The older is due to Fear and Robertson (1995) and the more recent due to Olson and Stark (2003). The two methods are summarized in Figures 5.28 and 5.29.

Figure 5.28 shows the Fear and Robertson method in which the pre-failure vertical effective stress, i.e. the effective overburden stress σ_v^I, is plotted against the corrected cone resistance (see equation (5.7)). A boundary line divides the area of the plot into two zones, one of which represents contractant, and therefore liquefiable materials, the other dilatant and therefore non-liquefiable materials. The boundary, in fact, is the equivalent of the critical state line, plotted in a different space. the Fear-Robertson diagram (originally derived via laboratory and standard penetration test (SPT) results and subsequently converted for use with cone penetrometer test (CPT) results) has been tested by Olson and Stark (2003) by applying it to a number of case histories of liquefaction failures, the data for which are shown in Figure 5.28. The bars in the

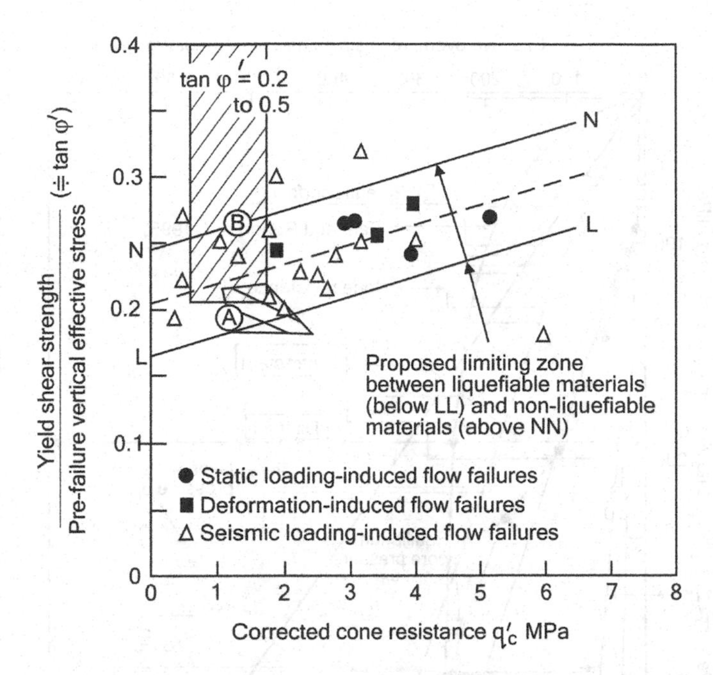

Figure 5.29 Olson and Stark's comparison of yield and mobilized strength ratios and corrected CPT cone resistance for liquefaction flow failures.

diagram show the uncertainty inherent in making an assessment from a case history, but also support the applicability of the diagram.

The Fear-Robertson boundary has been superimposed on the original Figures 5.18 and 5.19 in Figures 5.30 and 5.31, in which the depth (z) axes of the original plots have been converted to effective overburden stress σ_v^I, and the cone resistance q_c has been corrected to q_c^I. (Note that in Figure 5.31, the relationship between σ_v^I and q_c^I is linear, with the phreatic surface being at $z = 0$. γ, has been taken as $15\,kN/m^3$, making $\gamma^l = 5\,kN/m^3$ and $\sigma_v^I = 75\,kPa$ at $z = 15\,m$. In Figure 5.30, γ has been taken as $20\,kN/m^3$, and $\gamma^l = 10\,kN/m^3$. Because the phreatic surface was at a depth z of about 4.5 m, with zero pore pressure above this level, $\sigma_v^I = 90\,kPa$ at $z = 4.5\,m$. Below this level σ_v^I increased by 10 kPa per m to reach 395 kPa at 35 m.

It will be seen from Figure 5.30 that only the highest cone resistances that occur where σ_v^I is less than 200 kPa, fall into the dilatant zone. All of the tailings for which σ_v^I exceeds 200 kPa are predicted to be contractant and therefore potentially liquefiable. In Figure 5.31 the whole depth of material is predicted to be potentially liquefiable.

For the lower limit of cone resistance in Figure 5.30, at $\sigma_v^I = 400\,kPa$, φ^I was previously estimated to have been 11°, whereas the spikes of higher cone resistance were estimated to represent $\varphi^I = 38°$.

Similarly, for Figure 5.31, the value of φ^I was previously estimated at 14°.

On the face of it, the low values of 11° and 14° for φ^l in Figures 5.30 and 5.31 seem to indicate a liquefiable material, but the high value of 38° in Figure 5.30 seems unlikely to show liquefiability.

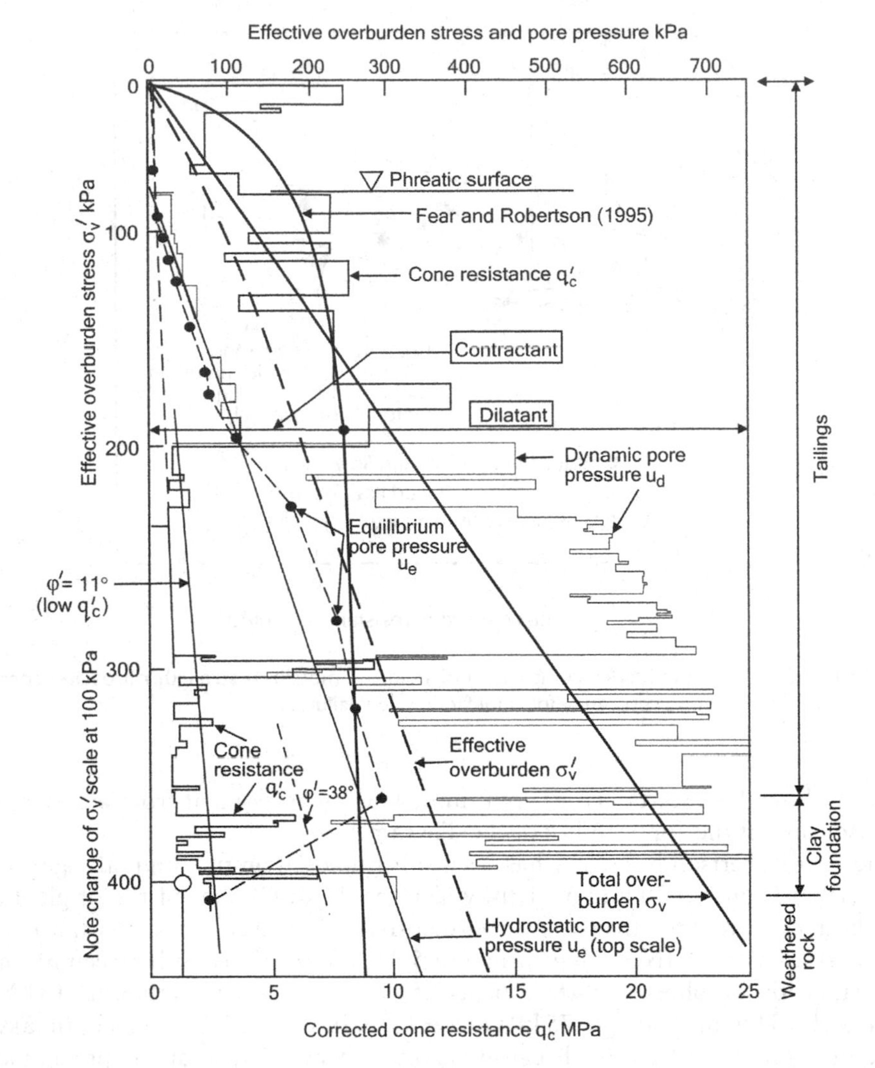

Figure 5.30 Fear and Robertson's line superimposed on Figure 5.18. (Vertical scale is now σ^I_v and q^I_c has been corrected for σ^I_v.)

The Olson and Stark plot depends on the ratio:

$$\frac{\textit{Yield shear strength}}{\textit{Pre}-\textit{failure vertical effective stress}} = q^I_c/\sigma^I_v = \tan\varphi^I$$

The Olson-Stark relationship was derived by back-analysing a large number of liquefaction failures, resulting in the band of observations shown in Figure 5.29. The relationship is more complex than it appears, because both axes involve the shear strength of the liquefiable material.

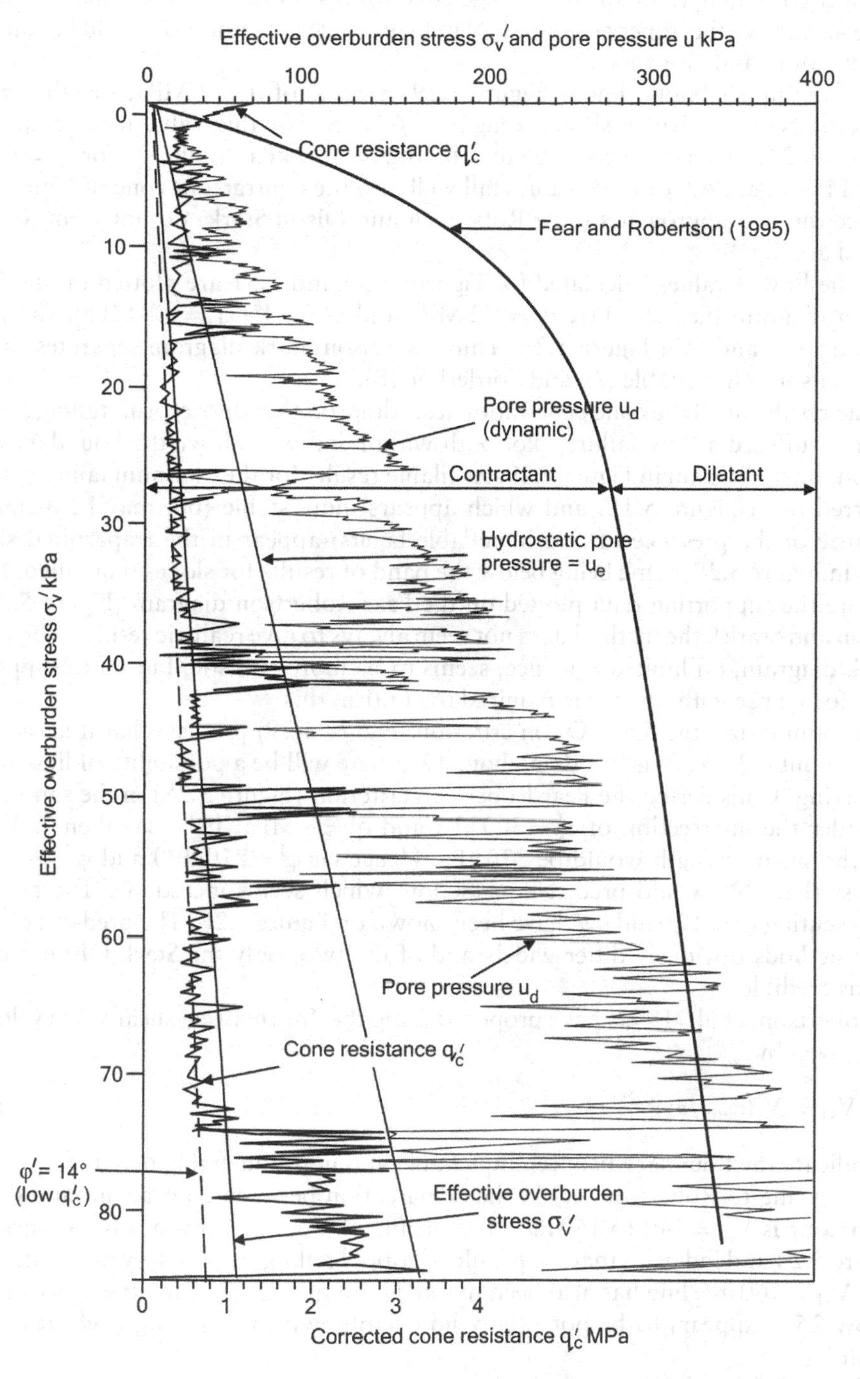

Figure 5.31 Fear and Robertson's line superimposed on Figure 5.19. (Vertical scale is now σ_v^I and q_c^I has been corrected for σ_v^I.)

For a particular value of the average cone tip resistance, if the calculated value of tan φ^l lies above the upper envelope NN in Figure 5.29, the tailings should be safe from liquefaction, and vice versa.

For the upper bound line in Figure 5.29, a value of $q_c^I = 1$ MPa, together with a value of $N_c = 15$ gives a shear strength of 67 kPa. For this value to correspond to tan $\varphi^l = 0.24$, σ_v^I must be given by $\sigma_v^I = 67/0.24 = 279$ kPa. Similarly, for $q_c^I = 6$ MPa, $\sigma_v^I = 1176$ kPa. Both of these values fall well into the contractant zone of Figure 5.28. Hence the two approaches (Fear-Robertson and Olson-Stark) do not seem to correspond at all.

If the low φ^l values calculated for Figures 5.30 and 5.31 are plotted on the Stark-Olson diagram (i.e. $\varphi^l = 11°$, $q_c^I = 1.2$ MPa and $\varphi^l = 14°$, $q_c^I = 0.8$ MPa), the points appear as A and B in Figure 5.29. Thus the Olson-Stark diagram separates the two materials into liquefiable (A) and borderline (B).

The results of all the cone penetration tests done on the Merriespruit tailings storage (which suffered a flow failure) plot both within and well above the boundary of the Olson-Stark diagram in Figure 5.29. Available results for the platinum tailings storage referred to in Figure 5.18, and which appears quite stable (but may be metastable because of the presence of thin liquefiable layers) appear in the trapezoidal shaded area in Figure 5.29, some being below the band of results for slopes that failed. Hence despite the supporting data plotted on the Fear-Robertson diagram (Figure 5.28) by Olson and Stark), the method does not seem always to give realistic results. The Olson-Stark diagram, on limited evidence, seems to be more realistic, but more supporting data for its use with tailings is required to confirm this.

To summarize, the Stark-Olson criterion (Figure 5.29) predicts that if tan φ^I is less than about 0.3, i.e. φ^I is less than about 17°, there will be a possibility of liquefaction occurring. Considering the Fear-Robertson criterion (Figure 5.28) in the same terms: consider the intersection of $\sigma_v^I = 300$ kPa and $q_c^I = 8$ MPa. If N_c is taken as high as 30, the shear strength would be 270 kPa. Hence tan $\varphi^I = 270/300$ and $\varphi^I = 42°$. Any φ^I less than 42° would predict liquefaction, which seems incredible. The two lines representing $\varphi^I = 17°$ and 42° have been shown on Figure 5.28. The predictions of the two methods obviously differ widely and of the two, only the Stark-Olson criterion seems credible.

Robertson, et al. (1992) have proposed using the "normalised shear wave velocity", V_{s1}, given by

$$V_{s1} = V_s(\sigma_{atm}/\sigma_v^I)^{0.25} \tag{5.13}$$

to indicate the boundary between liquefiable and non-liquefiable materials.

According to Robertson, et al., the demarcation between contractant and dilatant behaviour is $V_{s1} = 140$ to 160 m/s. The profile of $V_{s1} = 160$ m/s has been marked on Figure 5.24 and indicates that the profile of copper tailings is probably not contractant. The $V_{s1} = 160$ m/s line has also been marked on Figure 5.27. Only the weak material below 25 m appears to be potentially liquefiable, which is a completely reasonable result.

Figures 5.25 and 5.27 represent data from the same probing. The Stark-Olson limiting line ($\varphi^I = 17°$) for $N_c = 20$ and $\gamma = 20$ kN/m^3 has been marked on Figure 5.25. In this case Robertson et al.'s predicted limit between liquefiable and non-liquefiable material almost coincides with the Stark-Olson prediction, for $N_c = 20$. Hence, on available evidence, the Stark-Olson criterion and Robertson et al. shear wave velocity criterion seem to give the best available indication of liquefiability.

5.4.2 Summary and conclusions

Although laboratory tests have the advantage of having better control over applied stress and drainage conditions, there is always considerable uncertainty concerning the effects of sample disturbance on the results. This relates particularly to the void ratio of the sample vis-à-vis the void ratio existing in the field. This is one of the biggest difficulties arising in testing samples taken from tailings storages.

In situ testing, on the other hand suffers from the disadvantages of lack of knowledge of true stress conditions, and to a lesser extent drainage conditions. The effects of drainage can be handled by either testing sufficiently quickly to ensure an undrained strength result, or sufficiently slowly to ensure a drained result. The effects of disturbance from drilling to get the in situ device into place are not negligible, but are believed to be less than those attendant on the field sampling, transportation and storage, and laboratory trimming operations.

The big advantage of field testing is that more data can be collected from more points in the strata being investigated. As a tailings or other waste storage may cover hundreds of hectares in area and possibly have a history of deposition going over 100 years, the only way to gain a useful knowledge of the storage and the properties of its contents is to carry out as many tests as possible both with respect to financial constraints and with regard to available time.

References

Blight, G.E.: Observations on the Shear Testing of Indurated Fissured Clays. In: *Geotech. Conf. Oslo, Norway*, 1967, pp. 97–102.

Fear, C.E., & Robertson, P.K.: Estimating the Undrained Strength of Sand: A Theoretical Framework. *Can. Geotech. J.* 32(4) (1995), pp. 859–870.

Kulhawy, F.H., & Mayne, P.W.: *Manual on Estimating Soil Properties for Foundation Design*. Report EL-6800. Palo Alto, U.S.A.: Electric Power Research Institute, 1990.

Lunne, T., Robertson, P.K., & Powell, J.J.M.: *Cone Penetration Testing in Geotechnical Practice*. London, U.K.: Chapman and Hall, 1997.

Marsland, A., & Quarterman, R.S.T.: Factors Affecting the Measurement and Interpretation of Quasi-Static Penetration Testing in Clays. In: *2nd Eur. Symp. Penetration Testing, Amsterdam, Netherlands*, 1982, pp. 697–702.

Olson, S.M. & Stark, T.D.: Yield Strength Ratio and Liquefaction Analysis of Slopes and Embankments. *J. Geotech. Geoenviron. Eng. – ASCE 129(8)* (2003), pp. 727–737.

Prandtl, L.: Uber die Eindringungsfestigkeit (Harte) plastischer Baustoffe und die Festigkeit von Schneiden. *Zeitschrift von angewese Mathematik en Mechanik* (1921).

Robertson, P.K., Campanella, R.G., Gillespie, D., & Rice, A.: Seismic CPT to Measure In-situ Shear Wave Velocity. *J. Geotech. Eng., ASCE 112(8)* (1986), pp. 791–903.

Thomas, D.: Static Penetration Tests in London Clay. *Geotechnique XV*(2) (1965), pp. 177–185.

Vidic, S.D., Beckwith, G.H., Mayne, P.W., & Burns, S.E.: Seismic CPT Profiling of Mine Tailings Dams. *Int. Symp. Seismic Environ. Aspects Dams Design, Santiago, Chile*, Vol. 1, 1996, pp. 109–118.

Wride, C.E., Hofmann, B.A., Sego, D.C., Plewes, H.D., Konrad, J.M., Biggar, K.W., Robertson, P.K., & Monahon, P.A.: Ground Sampling Program at the CANLEX Test Sites. *Can. Geotech. J. 37* (2000), pp. 530–542.

Chapter 6

Measuring the coefficient of permeability in the laboratory and in situ, Seepage flow nets, drains and linings, geosynthetics, geomembranes and GCL's

6.1 Measuring permeability

6.1.1 *The coefficient of permeability*

In the middle years of the nineteenth century, Darcy showed, by means of experiment, that the rate of flow of fluid, either water or air (gas) is proportional to the gradient of pressure head in the direction of flow, i.e. the loss of pressure head per unit of distance in the direction of flow. Darcy's law can be expressed as

$$v = ki, \quad \text{or} \quad k = v/i \tag{6.1}$$

where v is the velocity of flow with dimensions [m/s]

k is called the coefficient of permeability

i is the gradient of the pressure head [m/m = dimensionless] the pressure head is the pressure divided by the unit weight of water ($N/m^2.m^3/N = m$)

Thus the units of k are [m/s].

In the engineering of waste storages, the permeabilities of both the construction materials used to contain or drain the waste, as well as the waste itself need to be established. Construction materials could include the compacted soil used for preliminary earthworks – toe walls, erosion control paddocks, etc., as well as features such as compacted soil retaining dykes and embankments, compacted clay linings and filter drains. The permeability of the tailings also needs to be known in order to design drainage measures for the storage, construct flow nets etc. In order to construct a flow net, both the permeabilities for vertical (cross layer) flow and horizontal (along layer) flow need to be known.

It was pointed out by the analogy shown in Figure 4.8, sub-sections 4.4.5 and 4.4.6 that the effective stress in a mine waste or a natural soil, cannot change unless pore fluid flows into or out of the pore spaces, or unless the grain skeleton becomes distorted by the action of a shear stress under undrained conditions. Section 4.5 considered the influence of changes of effective stress on the shear strength of mine wastes, but did not consider the time-dependent process of expulsion or imbibition of pore fluid which is known as consolidation or swelling. The effects of these processes in producing the states of normal or overconsolidation and their effects on shear strength were considered (i.e. Figures 4.11 to 4.14 and 4.28), but the process of flow of pore fluid was not considered. The processes of consolidation and swelling affect both the strength and volume change of mine wastes as well as the natural foundation soils or fill embankments used to store and contain them. In triaxial or shear box testing, redistribution of pore pressure occurs. It is essential to understand this process and to allow for its effects if accurate strengths and pore pressures are to be measured.

The coefficient of permeability is very much affected by the structure and macro-structural features of a natural soil (shrinkage cracks, fissures, layering, alternating coarse and fine layers, occasional coarse layers, etc); and equally those of a waste (principally shrinkage cracks and fissures, alternating coarse and fine layers and horizontal gradations from coarse to fine). As a result of this sensitivity, small scale measurements of permeability (usually laboratory measurements) tend to be unreliable because a small scale specimen is unlikely to contain a macro-feature, and if it does, the true overall effect of a series of such macro-features will not be correctly simulated. Hence, in general, larger scale field measurements are preferred. However, at the design stage for a new storage, there may be no alternative to measuring permeabilities in the laboratory.

6.1.2 *Basic types of permeameter*

There are two basic types of laboratory apparatus – rigid walled and flexible walled permeameters. Originally, permeameters consisted of a simple cylinder of metal or plastic material which contained the particulate material to be tested. This is the rigid walled permeameter. However it was realized that leakage between the wall and the sample could badly affect measurements, resulting in apparent permeabilities that may be considerably greater than true values. The solution adopted was to measure the permeability on a specimen contained in a triaxial cell where the flexible triaxial membrane, forced against the specimen by the cell pressure effectively prevents leakage down the side wall. This is the flexible walled permeameter. For specimens containing sharp particles, however, such as crushed gravels, or for very large particles such as waste rock, it is difficult to avoid punctures in the flexible membrane and to overcome this, a hybrid rigid-flexible wall consisting of a rigid cylinder, lined with resilient rubber can be used. In testing highly pervious large particled materials a steel cylinder lined with a resilient material, such as a relatively impervious closed cell foam sheet (e.g. flexible polyurethane foam) can be used to prevent side wall leakage.

6.2 Observed differences between small scale and large scale permeability measurements

Day and Daniel (1985) conducted comparative field and laboratory measurements of permeability on two clays. Test ponds were constructed in the field, and samples were later retrieved from the test liners for laboratory measurements. Measurements of seepage rate were made for the pond as a whole, and by means of single and double ring infiltrometers. Tests using both rigid and flexible walled permeameters were made on block and tube samples of the clay compacted in situ, and also on samples compacted in the laboratory. Effective confining stresses in the laboratory were about 100 kPa and seepage gradients ranged from 20 to 200. Day and Daniel found that values of permeability deduced from seepage losses from the ponds were 900 to 2000 times larger than permeabilities measured in the laboratory, but only 1.2 to 1.9 times larger than field infiltrometer measurements.

Chen and Yamamoto (1987) also carried out a comparison of field and laboratory permeability measurements, using infiltrometers and porous probes in situ, and flexible-walled permeameters in the laboratory. For the laboratory tests, effective stresses were about 200 kPa and the seepage gradient was 180. They found field permeabilities were 10 times larger than laboratory values. Elsbury et al. (1990) made a comparison of field and laboratory permeability measurements on a highly plastic clay. They found that double ring infiltrometer tests gave slightly lower permeabilities than did seepage rates from a test pond. However, compaction in the field with a vibratory roller resulted in a clay with a permeability ten times larger than one compacted using the same roller without vibration. Permeabilities measured in the laboratory used seepage gradients of 20 to 100 and effective stresses of 15 to 70 kPa. Permeabilities measured in the field proved to be between 10 000 and 100 000 times greater than values measured in the laboratory.

Pregl (1987) has stated that a permeability measured in the laboratory serves as an index of material quality but is not directly related to the permeability of a lining in the field. The permeability in the field will always be less than that measured in the laboratory (according to Pregl) because the seepage gradient used in laboratory tests is usually of the order of 30 whereas that in the field approximates to unity. Also, the Darcy coefficient of permeability is not constant with seepage gradient.

It is apparent from these studies that there are several possible reasons why a permeability measured in the field may differ from one measured in the laboratory:

- A large area exposed to seepage is more likely to contain defects in the form of cracks and more permeable zones than is a small area.
- If the Darcy coefficient of permeability is not constant with flow gradient, the use of different seepage gradients in the field and laboratory will result in different field and laboratory values.
- A similar remark applies to effective stresses. A specimen subjected to a high effective stress can be expected to show a lower permeability than a similar one with a low effective stress.

According to the classical form of the Darcy equation (6.1), one is led to believe that k is a constant for all values of the flow gradient i. Pregl (1987) has pointed out that this is not always so, but rather that k may increase with increasing i. The set of measurements shown in Figure 6.1 confirm this observation. The soil was a clayey sand residual from weathered granite and the measurements show that the seepage flow rate increases at a greater rate than the hydraulic gradient. For this set of data, the value of k at a hydraulic gradient of 1 was 1×10^{-4} cm/s, while at a hydraulic gradient of 20, k was 3×10^{-4} cm/s. Hence when measuring permeability in the laboratory, a flow gradient that is as close to that which will prevail in the field should always be used.

Nevertheless, it is possible to obtain reasonable agreement between field and laboratory permeability tests, as shown by Table 6.1 which compares the results of the ponding tests (shown later in Figure 6.5b) with a number of laboratory test results.

All laboratory permeability tests were of the constant head flexible wall triaxial type. The average effective stress was kept at 3 kPa for all tests and the seepage

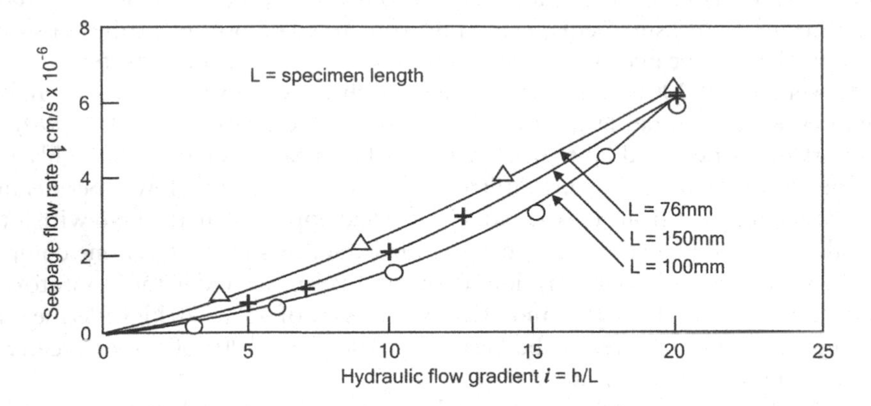

Figure 6.1 Observed relationships between flow rate q and flow gradient i.

Table 6.1

Range of Mean Permeability from Pond Test (cm/s × 10^{-6})	*Range of Mean Permeability from Laboratory Tests (cm/s × 10^{-6})*
59 to 81	37 to 93

gradient at unity. This stress was the lowest value that could be controlled reliably in the laboratory and was similar to the effective overburden stress in the pond tests of 0.5 m (only 4 to 5 kPa).

The table compares the permeability values measured in the laboratory on specimens compacted to the same dry density as the upper 150 mm of soil in the ponds. This set of measurements shows that it is possible, with correct interpretation and correct testing, to estimate in situ permeabilities reasonably closely from the results of laboratory tests.

However, it must be noted that the permeability of this granite soil is higher than the permeabilities that were considered by, e.g. Day and Daniel. It is noted from the literature that discrepancies between field and laboratory permeabilities generally appear to increase as the soil becomes less permeable, and presumably therefore, discontinuities and defects may play a greater role in modifying the permeability of large volumes of the soil.

6.3 Laboratory tests for permeability

The two main types of permeability test are the constant head or constant gradient test and the falling head or reducing gradient test. As mentioned earlier, permeability values determined in the laboratory do not necessarily represent the in-situ behaviour of residual soils. This is particularly true for undisturbed soils where the relatively small size of laboratory samples is inadequate to incorporate the various geological discontinuities, e.g. permeable and impermeable veins and fissures and other relict structures, present in a weathering profile, permeable layers and shrinkage cracks in

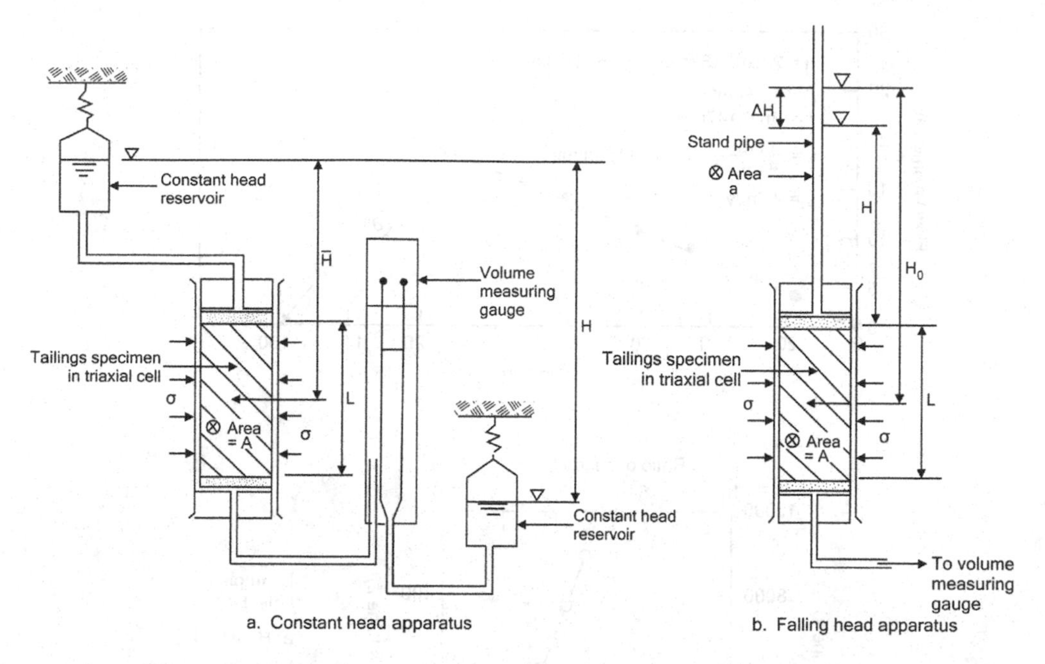

Figure 6.2 **Layout of laboratory constant and falling head permeability tests.**

a transported profile or a compacted layer. It should be noted that Darcy's law is only partially valid and that the coefficient of permeability of a soil can vary considerably with the flow gradient (Pregl, 1987), It is therefore important to use a similar flow gradient in the laboratory to that likely to occur in the field.

Conventional permeability tests in the laboratory may however be applicable to compacted soils and more uniformly structured natural soils in situ, especially where permeability is determined on both horizontally and vertically trimmed samples. It is then possible to estimate the overall, or effective permeability for uniformly textured soils. Laboratory tests, unlike field tests, also have the advantage of providing an indication of the variation in the coefficient of permeability with changes in effective stress. This data is often important for the design of earthworks and is generally not available from field tests. Constant head permeability tests coupled with pore pressure dissipation tests to measure the coefficient of consolidation c_v, carried out in a triaxial apparatus are particularly useful in this regard. Such tests permit the determination of permeability at various effective stresses as the sample undergoes consolidation (or swelling) and its relation to compression and void ratio (Tan, 1968, Garga, 1988).

In constant head tests the permeability is measured directly by maintaining a small pore pressure (10–20 kPa or 1 to 2 m of water head) differential across the sample and by applying Darcy's law when a steady state flow rate is achieved. The coefficient of permeability can be determined by the following expression which is simply the defining equation (4.32) for Darcy's law expressed in directly measurable quantities:

$$k = \frac{q_{\infty} L}{A \Delta H} = v/i \qquad (6.2a)$$

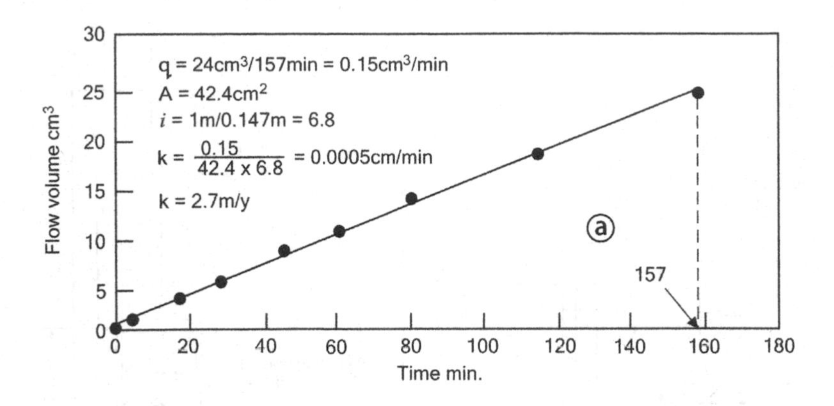

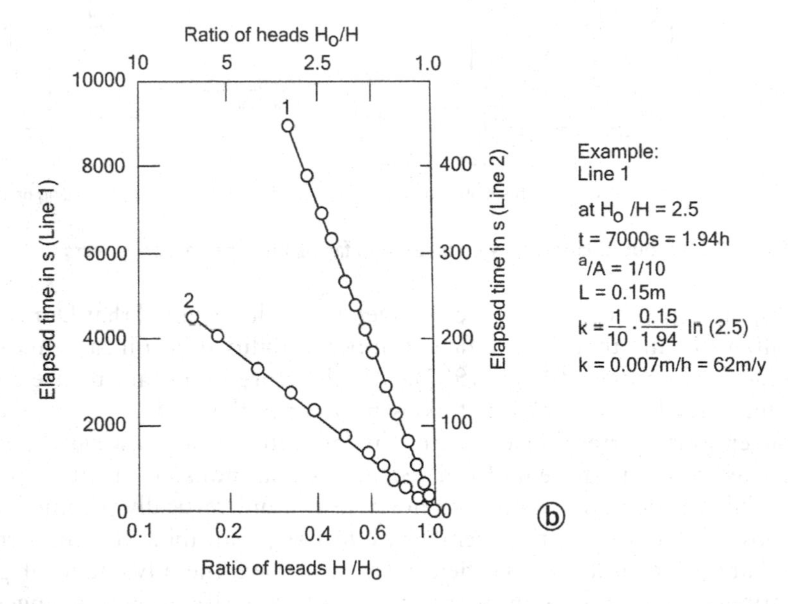

Figure 6.3 Typical results for permeability tests on tailings: a: constant head test, b: falling head test.

Table 6.2 Typical values of permeability for various soils

Coefficient of Permeability k cm/s				
	10^2 10	1.0 10^{-1} 10^{-2}	10^{-3} 10^{-4} 10^{-5} 10^{-6}	10^{-7} 10^{-8} 10^{-9}
Types of soil	Clean gravel	Clean sands, clean sand and gravel mixtures	Very fine sands, silts, mixtures of sand, silt & clay	Homogeneous clays

where q_∞ = steady state rate of flow (volume per unit time),
A = area of cross-section of the sample,
q_∞/A = v,
L = length of sample,
ΔH = constant differential head across the sample,
ΔH/L = i.

The layout of the apparatus for the constant head test is shown in Figure 6.2a, in which (H–$\overline{H}$ = ΔH, and that for the falling head test is shown in Figure 6.2b.

For the falling head test, the water flowing through the sample is supplied from the open standpipe of cross-sectional area a. If the head falls by ΔH in time ΔT the rate of flow is v = aΔH/Δt which equals AkH/L. Written differentially,

$$-a\frac{dH}{H} = \frac{Ak.}{L} \cdot dt$$

Integrating between times t = 0 and t = t, and heads H_o and H gives

$$k = \frac{aL}{At} \cdot \ln(H_o/H) = 2.3\frac{aL}{At} \cdot \log_{10}(H_o/H) \qquad (6.2b)$$

where H_o = H at t = 0   H = H at t = t

In less pervious soils or waste materials, it may take a very long time for q to attain the steady state value q_∞. It may also take a long time for an a accurately measureable quantity of flow to accumulate. For this reason, falling head tests are usually preferred for less pervious soils or wastes. Not only is it not necessary to wait for equilibrium flow, but the falling head can be magnified by increasing the ratio A/a of sample area to standpipe area.

Figure 6.3a shows typical results for a constant head permeability test on a tailings specimen, in which the cumulative flow has been plotted against elapsed time. The flow gradient was 10 kPa over a length of 0.147 m, i.e. i = 6.8 and k = 2.7 m/y.

Figure 6.3b shows results for falling head permeameter tests on two tailings specimens. Here, the ratio of the heads (H/H_o) is plotted to a logarithmic scale against the elapsed time t. k can then be calculated from equation (6.2)b.

Table 6.2 shows the range of coefficients of permeability to be expected for various types of soil. The values are given in cm/s. To convert these values to m/y, multiply by 315 360. For example, 1×10^{-6} cm/s = 0.32 m/y.

6.4 Methods for measuring permeability in situ

The most common techniques to measure permeability in the field involve some form of either constant head or falling head testing in unlined or lined (i.e. cased) boreholes. It is very common to obtain an estimate of k values by performing simple falling head tests in the drill stem at various depths as the drilling proceeds. In special cases, permeability testing can also be carried out using inflatable packers to isolate specific zones for testing, or by installing sealed hydraulic piezometers at different depths.

Many near surface soils have sufficient cohesion to permit a testing hole to be opened up with a manual or machine operated auger, without the need of a casing and to remain stable during permeability testing. A continuous-flight auger drill can be used successfully in unsaturated soils where speedy drilling may be necessary.

Other methods for drilling in cohesive soils e.g. undisturbed residual soils, include wash boring, percussion-hammer drilling and rotary drilling. Whatever the method used, it is important that the inside surface of the hole used for permeability testing be free of loose or remoulded (smeared) material. Smear must be removed by brushing the sides of the hole or surging it with water.

The most frequently used direct measurements of in-situ permeability can be divided into two major groups: those which feed water into the ground and those which extract water. The feed-in tests may be used above or below the water table, while the extraction tests can only be conducted below the water table. Because, in mine waste engineering, the interest is usually to measure the permeability of near-surface soils, above the water table, extraction tests are seldom used to evaluate permeability for the design of waste storages. However, extraction or feed-in tests can be used in situ in tailings, but usually, it is necessary to case the hole with a pervious casing, e.g. a perforated plastic pipe wrapped in geofabric, to keep the hole open.

6.4.1 *Permeability from surface ponding or infiltration tests*

Ponding tests are suitable for measuring the coefficient of permeability for vertical flow and are thus appropriate for obtaining permeability values for the design of waste deposits including checking the permeability of compacted clay liners. The layout and geometrical requirements for a ponding test are shown in Figure 6.4a.

The main pond, on which the seepage measurements is made is surrounded by edge ponds (or moats) so that the effects of essentially vertical flow are measured. It is essential that the position of the regional water table be known and that the dimensions of the ponds be related to the depth of the water table. The ponding test referred to in Figure 6.5a consisted of four ponds surrounded by a moat. One of the four was lined and served as an evaporation measuring pond. Each pond had four water level observation points. The pond complex is shown, during preparation, in Plate 6.1. The scatter evident in Figure 6.5a resulted from the difficulty of measuring small changes of water level accurately and also difficulty in compensating the measured seepages for evaporation losses which also depend on measuring small changes of water level. Temperature changes also affected the accuracy of the measurements by causing the plastic pipes (used as small stilling ponds against wind effects) to change in length. Thus an increase in temperature caused an apparent increase in seepage rate, and vice versa.

A subsidiary pond with an impervious liner is used to allow for the effects of precipitation and evaporation during the course of the test. The geomembrane liner must be covered with soil or waste similar to the floor of the prototype reservoir or pond to give a similar reflective albedo. This is absolutely essential, as the rate of evaporation is often of the same order as the seepage rate. As an alternative to separately measuring the evaporation rate, a layer of lubricating oil can be used to cover the surface of the water to eliminate evaporation. Seepage rates may be of the order of mm/day and it is extremely difficult to measure seepage accurately. The measuring system must be protected from the effects of temperature and wind if accurate results are to be achieved. Figure 6.4b shows a suggested system to do this. After filling the ponds the net infiltration v = (change in water level of main pond – change in water level of sealed pond) should be observed and plotted against time t (see Figure 6.4c). Obviously, any replenishment of water in the ponds either by rain precipitation or other means must

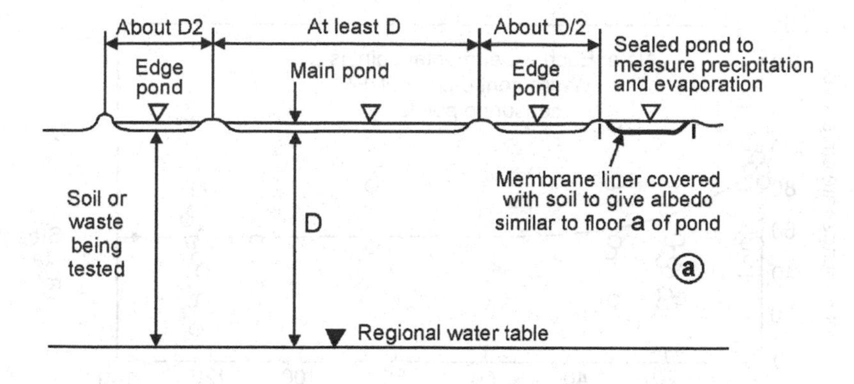

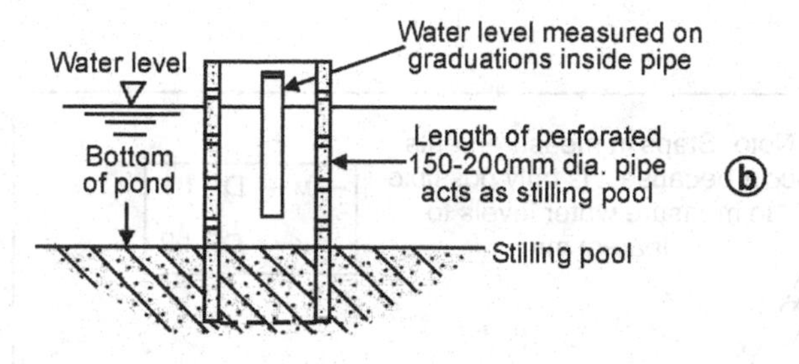

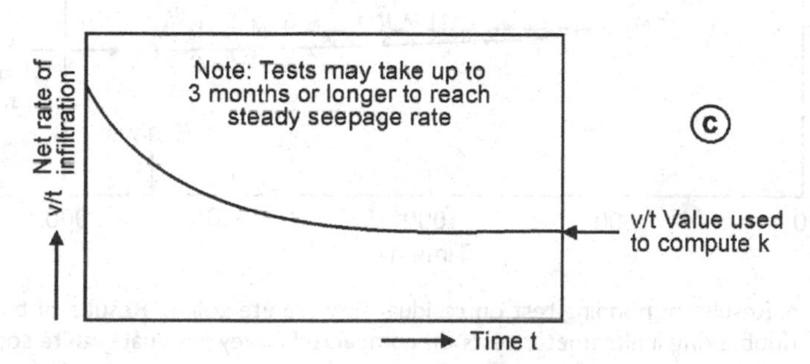

Figure 6.4 a: Layout of ponding test to measure permeability for vertical flow, in situ. b: Method of shielding water level measurements from effects of wind. c: Variation of infiltration rate with time after filling pond.

also be taken into account. The calculated coefficient of permeability is based on the net rate of infiltration once this has reached a steady value.

The single or double-ring infiltrometer test is a smaller and cheaper variant of a full-scale ponding test (See Figure 6.6). For the double ring test a pair of concentric sheet metal rings are set and sealed into the ground surface and filled with water. The outer ring serves the same purpose as the edge pond and reduces edge effects. To give reasonable values of permeability, the diameter of the outer ring should not be less than 1200 mm and that of the inner ring, 600 mm.

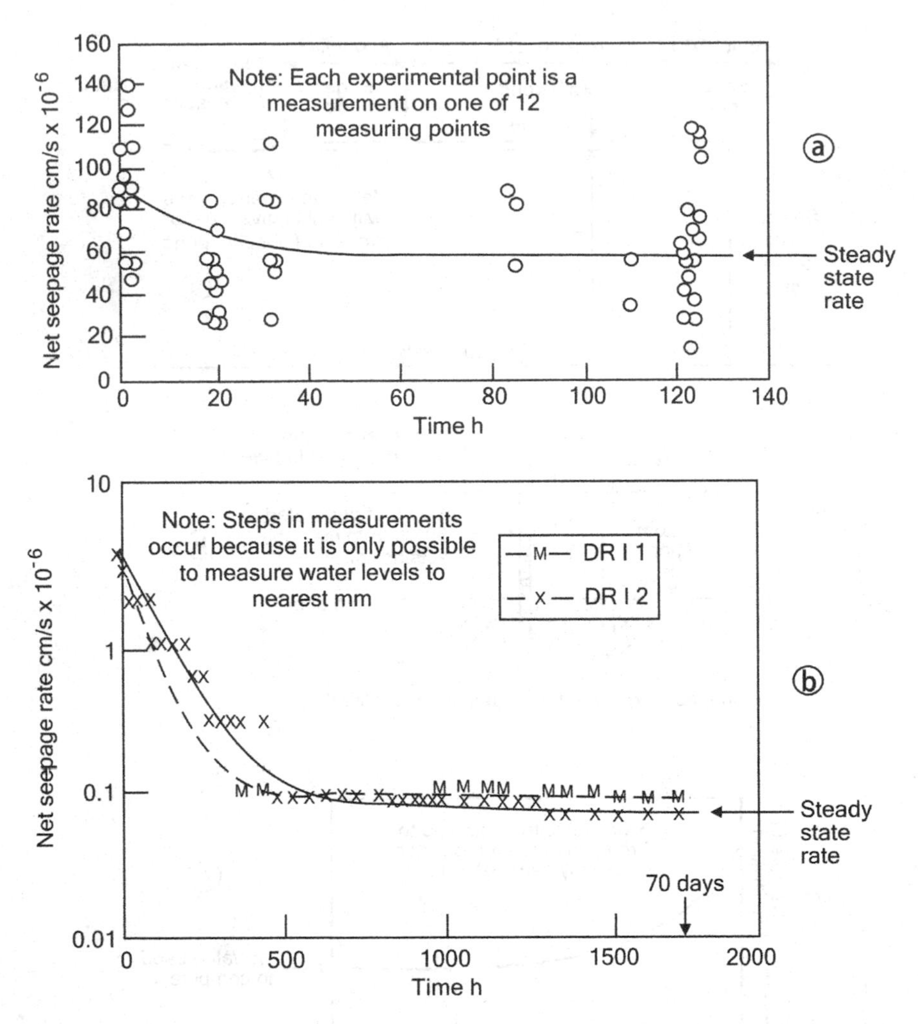

Figure 6.5 a: Results of ponding test on residual silty granite soil. b: Results of two side-by-side double ring infiltrometer tests on compacted clayey residual granite soil.

The two forms of ring infiltrometers popularly used are open and sealed. Four variations are illustrated in Figure 6.6. Open rings are less desirable because with a low conductivity soil, it is difficult to separate the drop in water level of the pond caused by seepage from changes caused by evaporation and rainfall.

With sealed rings, however, very low rates of flow can be measured. Single-ring infiltrometers allow flow to spread laterally beneath the ring, complicating the interpretation of test results. Single rings are also susceptible to the effects of temperature variation: as the system heats up, the whole of it expands and as it cools down, the whole system contracts. This can lead to erroneous measurements when the rate of flow is small.

The sealed double-ring infiltrometer has proved the most successful. The outer ring forces the infiltration from the inner ring to be closer to one dimensional. Covering

Plate 6.1 Seepage ponds constructed for large scale infiltration test.

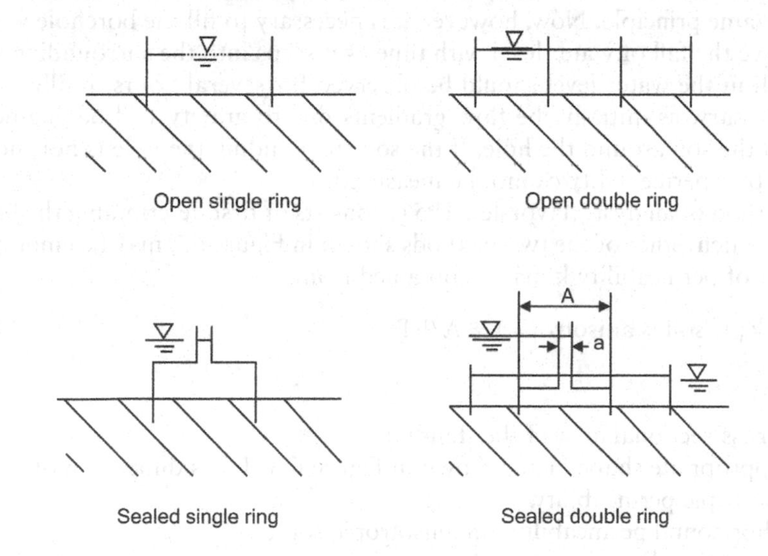

Figure 6.6 Open and sealed single and double ring infiltrometers.

the surface virtually eliminates the effects of evaporation and rainfall, and the use of small diameter standpipes make more accurate flow measurements possible. With a standpipe, the flow rate can be measured more accurately. If the plan area of the inner ring is A, and that of the standpipe is a, the drop in level is magnified by the ratio A/a.

The average coefficient of permeability of the stratum between the soil surface and the regional water table at depth D is given by

$$k = \frac{v}{t} \cdot \frac{D}{D+d} \tag{6.3}$$

where v is the net infiltration in time t and d is the average depth of water in the pond during the test.

Note that d only becomes significant if D is not much greater than d. For example, if d = D, k = 1/2 (v/t), if d = D/10, k = 10/11(v/t).

Figure 6.5b shows the results of a pair of sealed double ring infiltrometer tests carried out on a compacted clay liner. Note that the results are repeatable from test to test with much less scatter in the measurements than in Figure 6.5a. Also note the length of time taken to achieve a steady flow rate.

6.4.2 Permeability from borehole inflow or outflow

Variable head tests

If the water table is close to the surface, the permeability may be estimated by observing the rate of rise of water in a borehole which penetrates the water table. This method suffers from the disadvantage that the permeability measured is primarily that for horizontal flow. Thus the presence of a highly permeable horizon such as the pebble marker or a sandy layer may have an overriding effect on the results.

If the borehole does not intercept the water table, measurements can be carried out using the same principle. Now, however, it is necessary to fill the borehole with water and observe the fall of water level with time as it seeps into the surrounding soil. The rate of fall in the water level should be observed for several hours, refilling the hole when necessary, as initially the flow gradients due to gravity will be augmented by suction in the soil around the hole. If the soil surrounding the hole is not thoroughly wetted, a true permeability cannot be measured.

The method of analysis (Hvorslev, 1951) consists of first determining the basic time lag T, for which either of the two methods shown in Figure 6.7 may be employed. The coefficient of permeability k may be obtained from:

k (or k_h if soil is anisotropic) = A/FT

where

A = cross-sectional area of the standpipe,
F = appropriate shape factor shown in Figure 6.8 (F has dimension of length),
k = isotropic permeability,
k_h = horizontal permeability (in anisotropic soil),
T = basic time lag.

Where the soil is anisotropic, the ratio of $k_h/k_v = m^2$ must be estimated*, or obtained from laboratory tests. However, it should be noted that the error in evaluating the permeability of the soil due to error in selection of m is less than the inherent error in a falling head test. (*Note that in this section, m denotes a dimensionless ratio of horizontal to vertical permeability, not the unit of metres.)

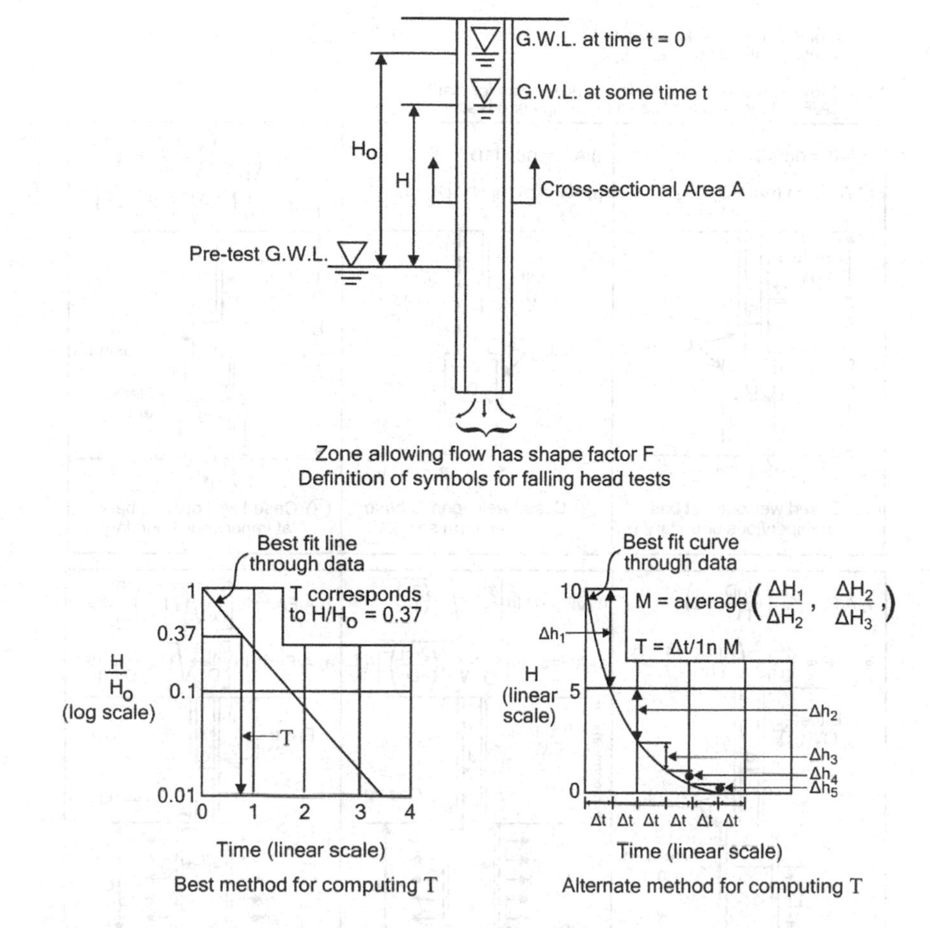

Figure 6.7 Calculation of basic time lag, T, for variable head tests.

Constant head tests

This analysis (Hvorslev, 1951) can be used for any feed-in test where the inflow during a test under a constant head becomes constant over time, i.e. when steady state flow conditions are achieved or approached. The analysis consists of the following steps:

Determination of the steady state conditions:

An approximate value of the steady flow rate can be found from observed changes in flow rate with time, as follows: If H denotes the constant height of the water in the test hole above the base of the test zone, q denotes the flow rate, and t denotes the time, then q at steady state can be obtained from a plot of q versus log $(1/\sqrt{t}$, as t becomes large. A typical plot of this kind (Garga, 1988) is shown in Figure 6.9.

Determination of the effective head at test zone, H_c:

For all cases except the packer test, this is the constant head of water above the test zone. In the case of packer tests, the height of the column of water above the test zone

i = isotropic conditions: $k_h = k_v = k$

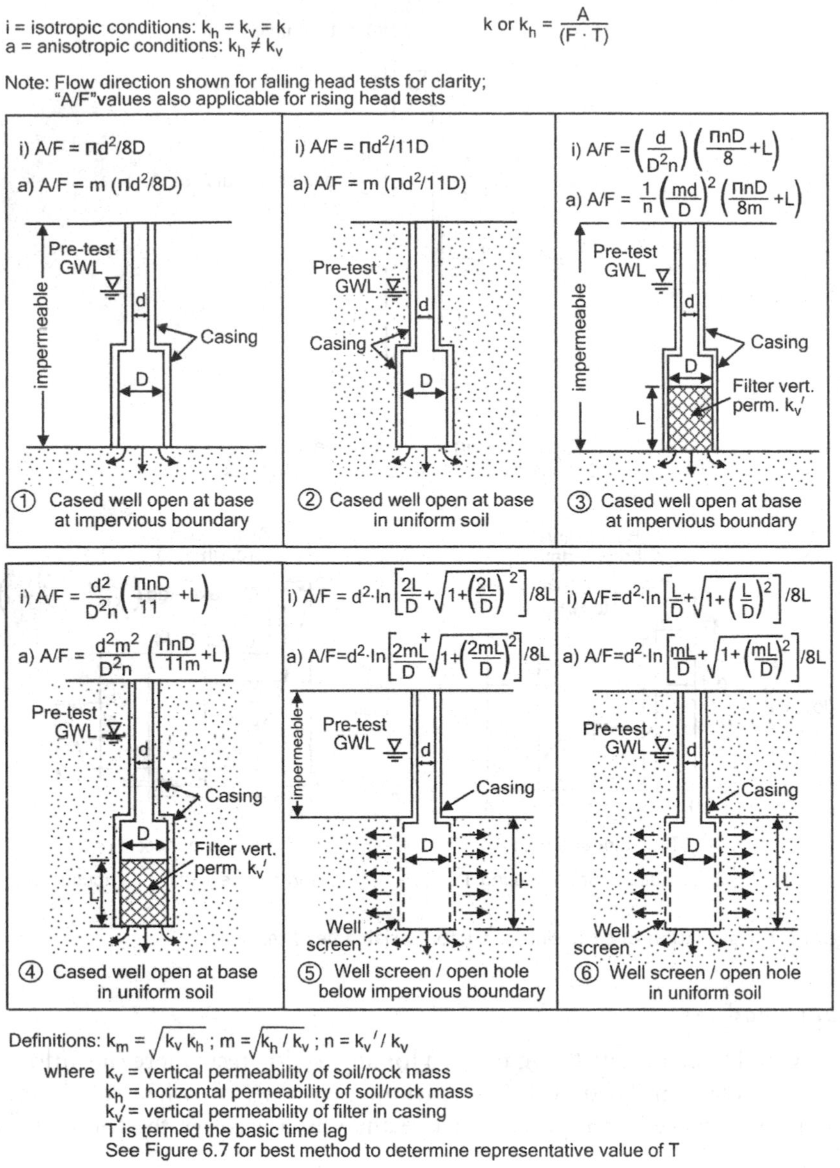

a = anisotropic conditions: $k_h \neq k_v$

$k \text{ or } k_h = \frac{A}{(F \cdot T)}$

Note: Flow direction shown for falling head tests for clarity; "A/F" values also applicable for rising head tests

① Cased well open at base at impervious boundary

i) $A/F = \pi d^2/8D$

a) $A/F = m(\pi d^2/8D)$

② Cased well open at base in uniform soil

i) $A/F = \pi d^2/11D$

a) $A/F = m(\pi d^2/11D)$

③ Cased well open at base at impervious boundary

i) $A/F = \left(\frac{d}{D^2 n}\right)\left(\frac{\pi n D}{8} + L\right)$

a) $A/F = \frac{1}{n}\left(\frac{md}{D}\right)^2\left(\frac{\pi n D}{8m} + L\right)$

④ Cased well open at base in uniform soil

i) $A/F = \frac{d^2}{D^2 n}\left(\frac{\pi n D}{11} + L\right)$

a) $A/F = \frac{d^2 m^2}{D^2 n}\left(\frac{\pi n D}{11m} + L\right)$

⑤ Well screen / open hole below impervious boundary

i) $A/F = d^2 \cdot \ln\left[\frac{2L}{D} + \sqrt{1 + \left(\frac{2L}{D}\right)^2}\right] / 8L$

a) $A/F = d^2 \cdot \ln\left[\frac{2mL}{D} + \sqrt{1 + \left(\frac{2mL}{D}\right)^2}\right] / 8L$

⑥ Well screen / open hole in uniform soil

i) $A/F = d^2 \cdot \ln\left[\frac{L}{D} + \sqrt{1 + \left(\frac{L}{D}\right)^2}\right] / 8L$

a) $A/F = d^2 \cdot \ln\left[\frac{mL}{D} + \sqrt{1 + \left(\frac{mL}{D}\right)^2}\right] / 8L$

Labels: Pre-test GWL; d; D; L; Casing; impermeable; Filter vert. perm. k_v'; Well screen

Definitions: $k_m = \sqrt{k_v k_h}$; $m = \sqrt{k_h / k_v}$; $n = k_v' / k_v$

where k_v = vertical permeability of soil/rock mass

k_h = horizontal permeability of soil/rock mass

k_v' = vertical permeability of filter in casing

T is termed the basic time lag

See Figure 6.7 for best method to determine representative value of T

Figure 6.8 Shape factors for variable head tests.

is adjusted for head losses in the water hose and couplings as well as for any additional pressure head supplied by the pump.

Determine the shape factor F for a given test configuration from Figure 6.10, then:

$$k = \frac{qc}{FH_c}$$

where q_c is the constant flow under steady state conditions:

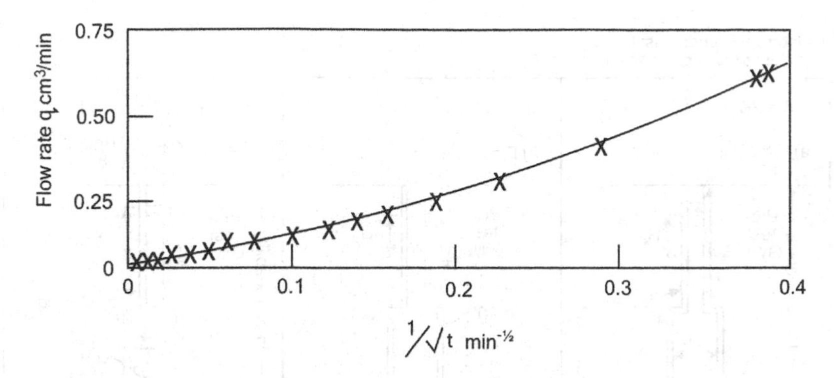

Figure 6.9 Typical plot of q versus $1/\sqrt{t}$ from in situ constant head permeability tests.

The analysis of in-situ constant head permeability tests conducted in sealed hydraulic piezometers is also common practice. This analysis takes into account the compressibility of the soil and the resultant volume change as an excess (or deficit) pressure head is applied. A graph of the rate of flow q versus $1/\sqrt{t}$ is plotted as the test progresses (see Figure 6.9). Because $1/\sqrt{t}$ reduces as t increases, the steady state value of q is approached at $1/\sqrt{t}$ $t=0$. It should be noted that it is not necessary to obtain a continuous record of the flow for the entire duration of the test. It is sufficient to monitor the flow rate over small time intervals periodically as the test proceeds. If t_1 and t_2 are the times (from commencement of the test) over which the flow rate is measured, then q can be plotted against $2/(\sqrt{t_1}+\sqrt{t_2})$. At large times, the expression for the coefficient of permeability reduces to:

$$k = \frac{q_\infty}{F\Delta h} \tag{6.4}$$

where

q_∞ = flow rate as t becomes large (or $1/\sqrt{t}$ approaches zero)
Δh = constant head applied during the test
F = shape factor

The shape factors for cylindrical tips (Olson and Daniel, 1981), with length L and diameter D, are shown in Figure 6.11.

6.5 Estimation of permeability from field tests

Often only a rough preliminary estimate of the soil permeability is required, in which case useful results can be obtained from a simple soak-away test in a test-pit. The test-pit is filled with water, and the fall in water level is recorded over a period of several days. The hole must be covered over with a metal or plastic sheet to prevent evaporation losses and surrounded by a mound to prevent surface water from running in. The measurements can then be analysed approximately to assess the order of magnitude of the permeability. Figure 6.12 shows a set of data and a calculation for a soak-away test performed on a clayey silt residual from a mud-rock. The object of the

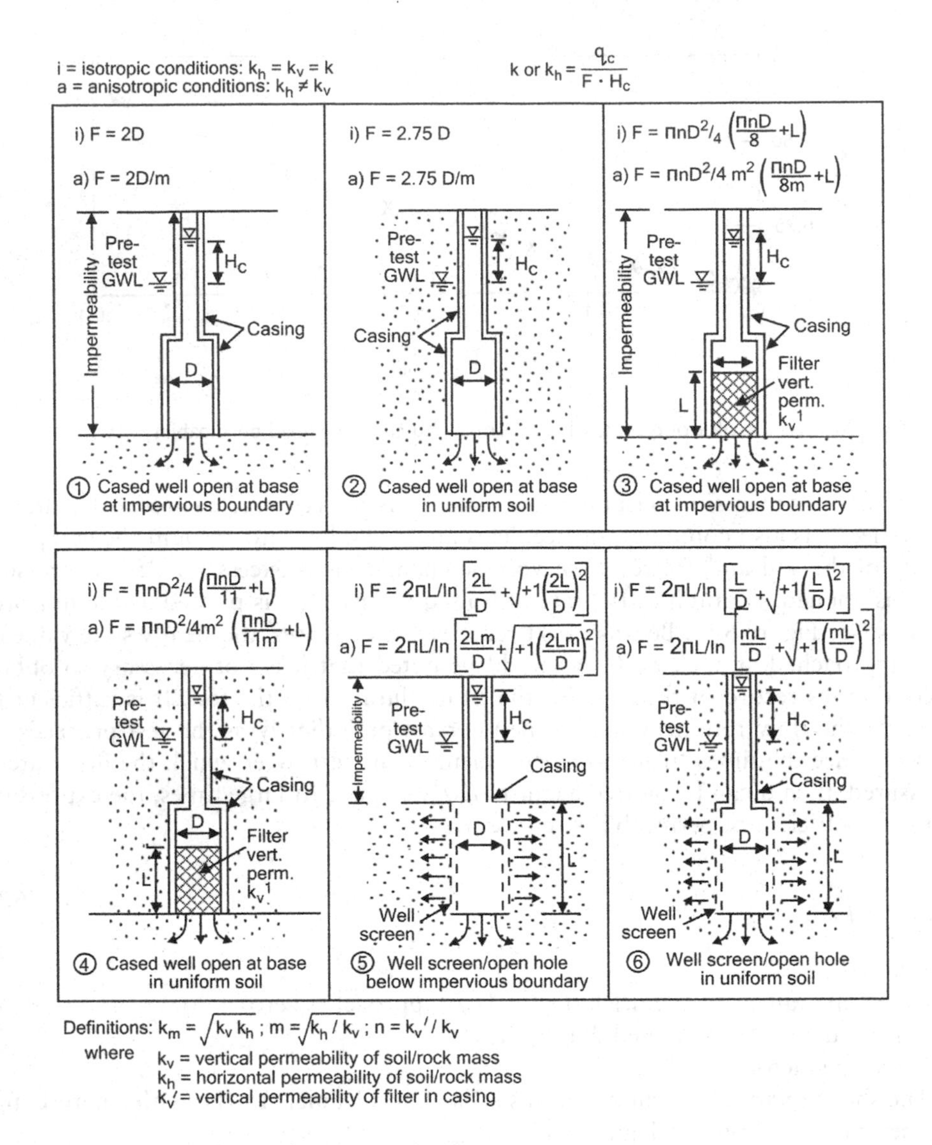

Figure 6.10 Shape factors for constant head tests.

test was to see if a low-permeability clay liner would be required at the site (required permeability 0.1 m/y). The result of 30 m/y showed that a liner would be required.

The determination of field permeability by Matsuo et al.'s (1953) method has been widely applied to compacted soils in earth embankment construction. The test appears to work well in soils with permeability in the range of 10^{-4} to 10^{-6} cm/s (30 to 0.3 m/y). This simple method consists of excavating of a large rectangular test pool of width B and length L. The sides of the excavation may be sloped if required. The flow rate q necessary to maintain a constant level H in the pool is monitored. The seepage flow in this case is a three dimensional flow. In order to obtain a two dimensional estimate

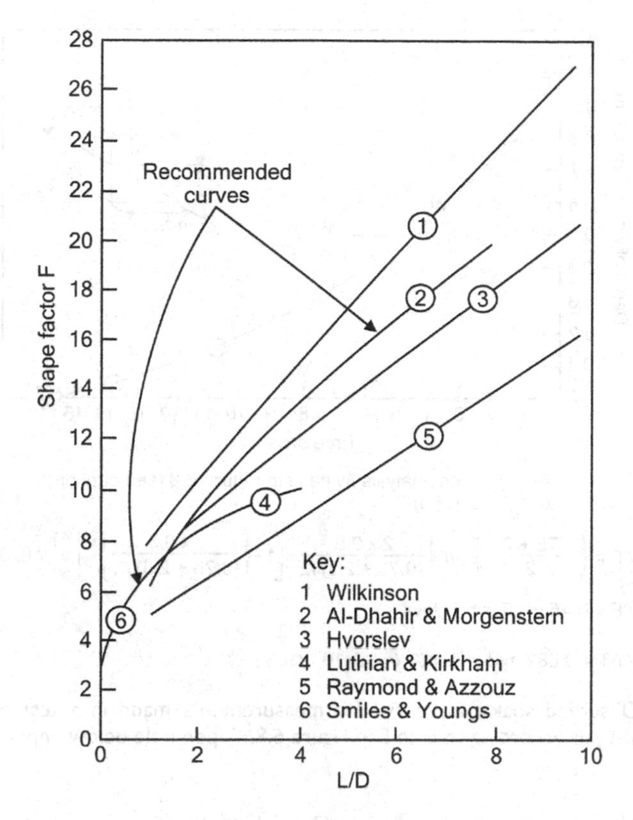

Figure 6.11 Shape factors used by various investigators.

of the flow, the pool is next enlarged to twice the initial length ($L_1 = 2L$). The new flow rate, q_1 required to maintain the same constant level H is noted. By subtracting the two flow rates, the effect of flow near both ends of the lateral cross-section can be eliminated, and the average discharge per unit length may be calculated as follows:

$$q_{ave} = \frac{q_1 - q}{(L_1 - L)} \tag{6.5}$$

The range of in-situ permeability coefficient, depending on whether the flow is perpendicularly downwards or horizontal, may be obtained from the following simple expressions:

$$k_v = \frac{q_{ave}}{B - 2H} \quad \text{and} \quad k_h = \frac{q_{ave}}{B + 2H} \tag{6.6}$$

Figure 6.13 shows the results of a series of tests using Matsuo et al.'s method to explore possible anisotropy in permeability. As shown by the diagram, the calculated permeability proved to be constant regardless of the ratio of wetted side area A_s to

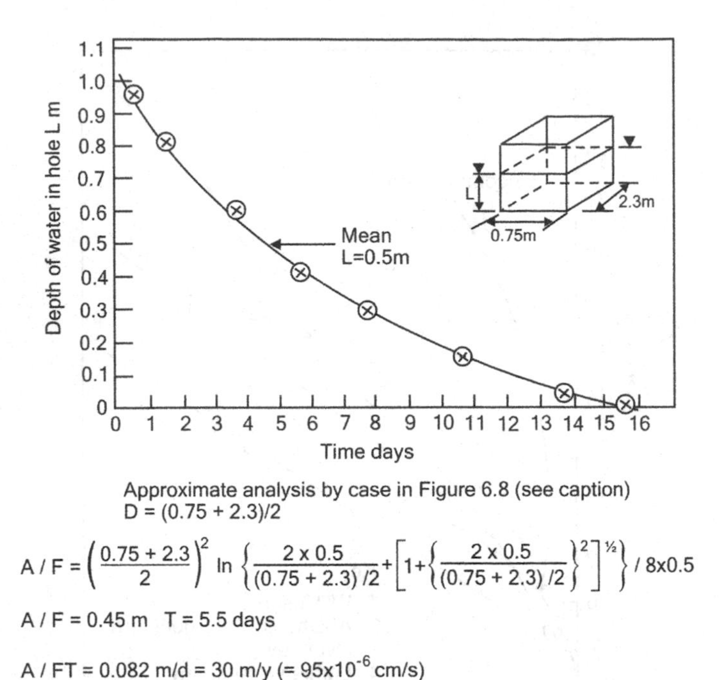

Figure 6.12 Observed soak-away curve for measurements made in a test pit. (Approximate analysis according to case 5 in Figure 6.8, "open hole below impervious boundary".)

base area A of the pit. It could thus be concluded that in this case the permeability was isotropic for all intents and purposes.

6.6 Large-scale permeability tests using a test pad

The most realistic way of testing the permeability of a compacted clay layer is to incorporate a test pad, or a number of such pads into the prototype installation. The pad can be constructed in accordance with the design and specifications for the prototype. If tests on the pad (or pads) are successful, it (or they) can be incorporated into the prototype and monitored on a long term basis as a means of checking on the long-term performance of the clay liner. Figure 6.14 shows a schematic layout for a typical test pad (USEPA, 1989). The dimensions of such a test pad should be generous, and at least 25 m × 25 m or 31.6 m × 31.6 m (to give an area of 1000 m^2).

6.7 The permeability of tailings

So far, the discussion has been aimed mainly at measuring the permeability of natural or compacted soils used to construct the containing walls or floor of a mine waste storage. There have been relatively few systematic sets of permeabilities for tailings published in the literature. Figure 6.15a shows a set of measurements of the coefficient of permeability for platinum tailings. It will be seen that the permeability decreases from about 15 m/y near the surface to 0.15 m/y at a depth of 30 m, a reduction by

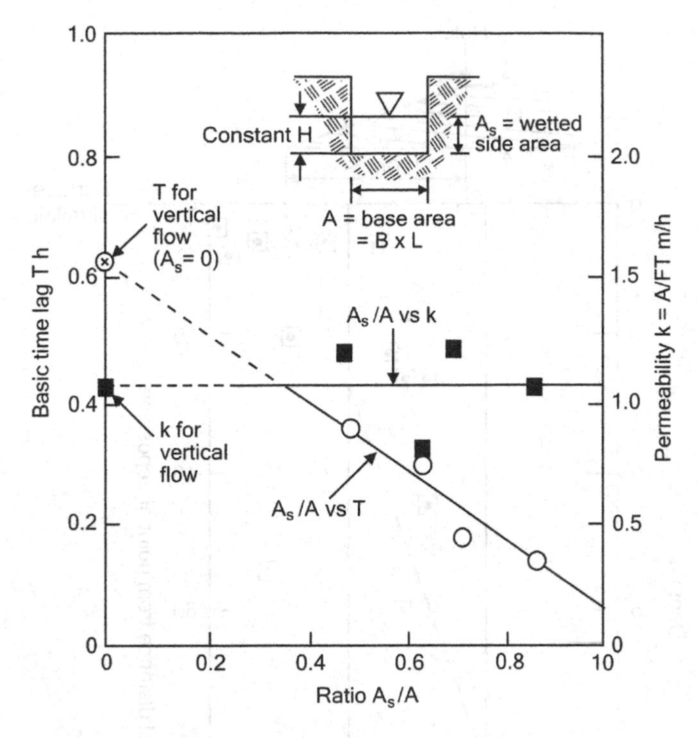

Figure 6.13 Variation of basic time lag with A_s/A in test based on Matsuo, et al.'s method.

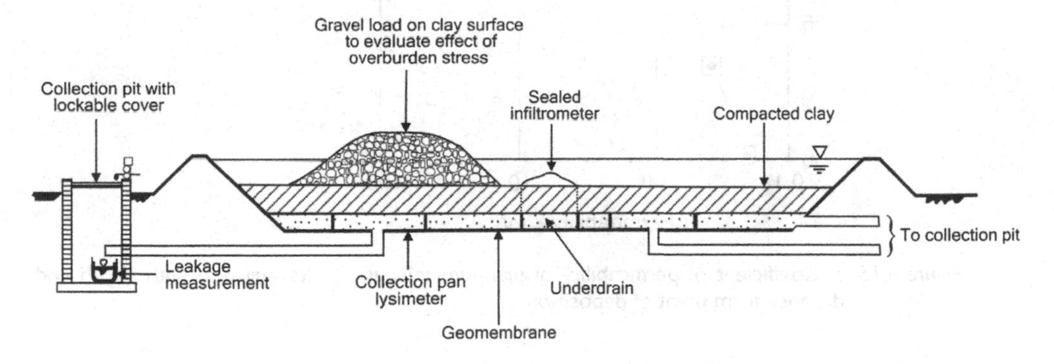

Figure 6.14 Large-scale permeability test by means of a test pad.

a factor of 1 in 100. The main reason for this is that the hole that was sampled was located at the crest of the outer slope. As the slope angle averaged 15°, a depth of 30 m corresponds to a horizontal distance of $30\cot 15° = 112$ m in from the crest. (See sketch in Figure 6.15a). Hence gravitational particle size sorting on the beaches has caused the tailings to become progressively finer and less permeable as the vertical depth increases.

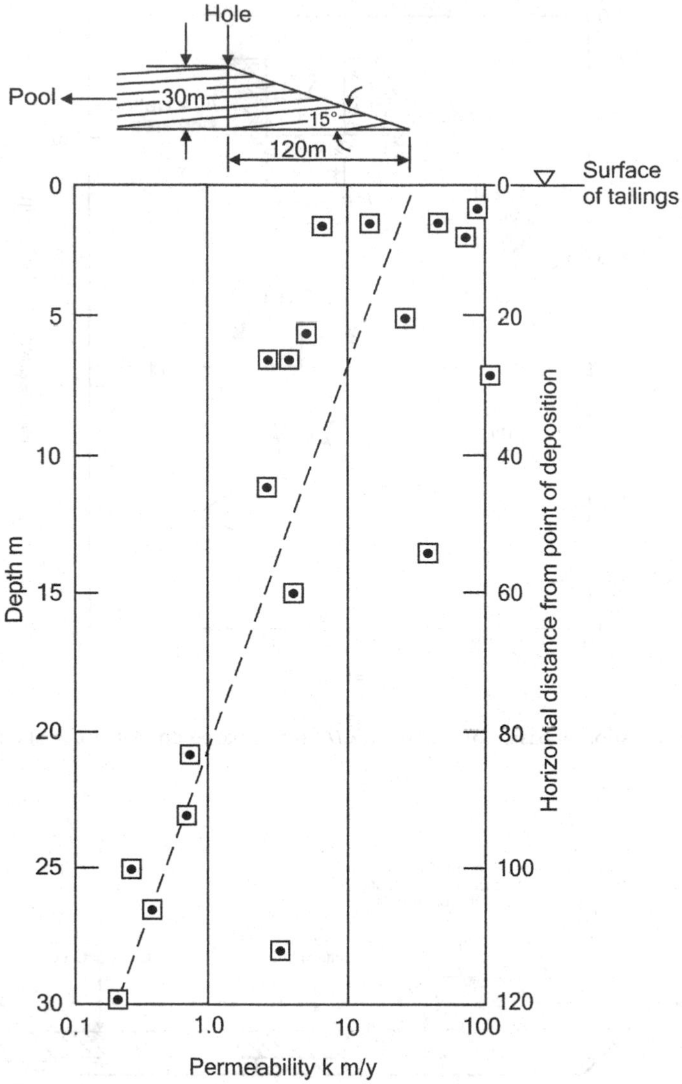

Figure 6.15 a: Coefficient of permeability of platinum tailings and its variation with depth and distance from point of deposition.

Figure 6.15b shows a similar set of data for gold mine tailings. The data were collected for five gold tailings impoundments in the Witwatersrand region of South Africa (labeled dam 1 to 5) and show a decrease of permeability with depth similar to that shown in Figure 6.15a All of the measurements were made in the laboratory (as were those for c_v shown in Figure 4.45) and hence are subject to the reservations expressed in section 6.2. However, they have the advantage of including values of both horizontal and vertical permeability, and of showing potential differences from one source of tailings to another, as well as demonstrating the likely variation with increasing depth.

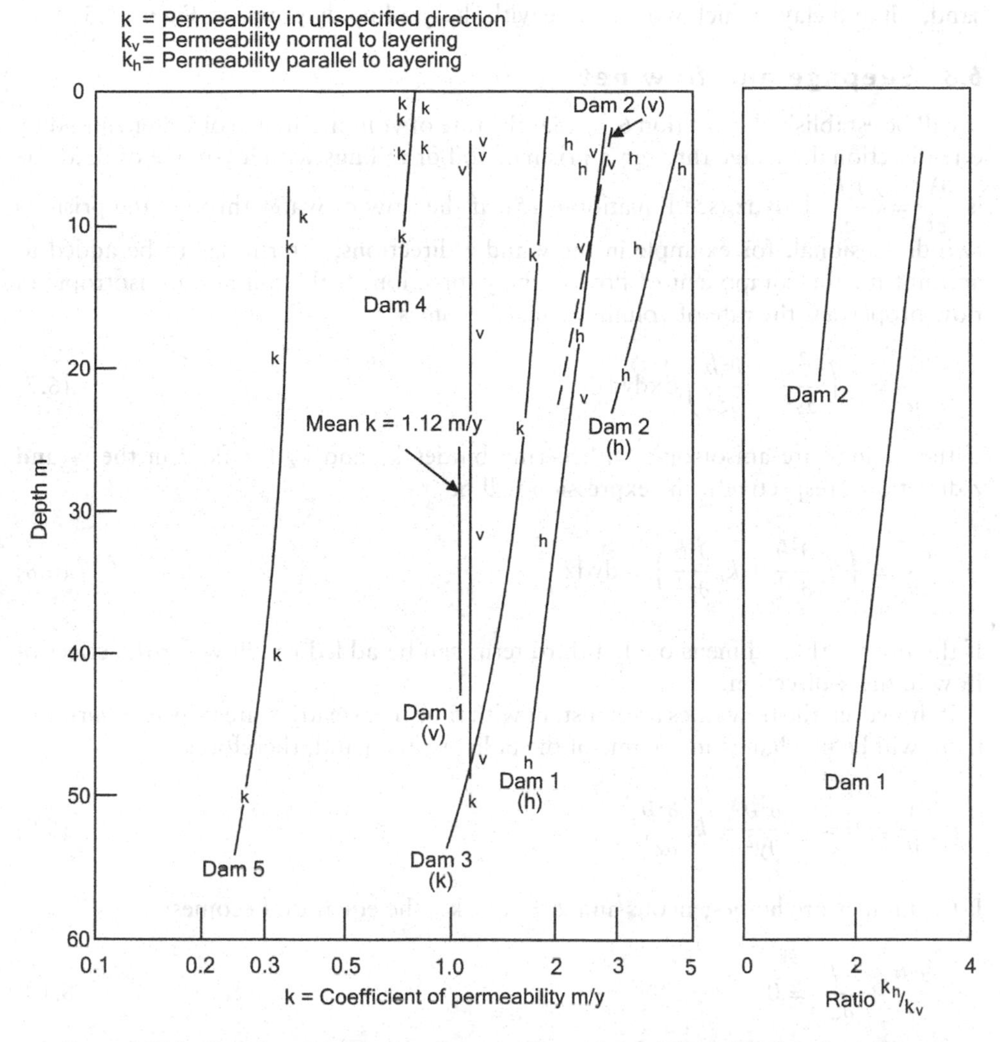

Figure 6.15 b: Variation of coefficient of permeability with depth for gold tailings from five storages and the ratio of permeabilities for horizontal and vertical flow.

The right hand diagram in Figure 6.15b shows the ratio of permeabilities for horizontal (along layer) flow to vertical (cross layer) flow. This appears to be between 1.5 and 3 for hydraulically deposited gold tailings. The left hand diagram shows that increasing overburden stress has a relatively minor effect on k and also that k may vary by a factor of 10 or more from one storage to another. This degree of variability is somewhat less than that shown for c_v in Figure 4.42 where the data cover a range of 20 times.

Taking 1 m/y as representative of permeabilities for gold tailings, and referring to Table 6.2:

$$1\ \text{m/y} = 1/315360 = 3.2 \times 10^{-6}\ \text{cm/s}$$

Table 6.2 shows that this would be typical for "very fine sands, silts and mixtures of sand, silt and clay" which would agree with the grading diagrams in Figure 1.3.

6.8 Seepage and flow nets

It will be established in Section 6.11 that the rate of volume change of water caused by one-directional seepage through a prism of soil or tailings with a volume of dx.dy.dz is $\frac{\partial V}{\partial t} = k\frac{\partial^2 h}{\partial z^2}$ dxdydz (see Equation 6.15). If the flow of water through the prism is two-dimensional, for example in the y and z directions, a term has to be added to account for the component of flow in the y-direction. If the tailings are isotropic in flow properties, the rate of volume change becomes

$$\frac{\partial V}{\partial t} = \mathrm{k}\left(\frac{\partial^2 h}{\partial y^2} + \frac{\partial^2 h}{\partial z^2}\right) \mathrm{dxdydz} \tag{6.7}$$

If the tailings are anisotropic with permeabilities k_y and k_z for flow in the y- and z-directions, respectively, the expression will be

$$\frac{\partial V}{\partial t} = \left(k_y\frac{\partial^2 h}{\partial y^2} + k_z\frac{\partial^2 h}{\partial z^2}\right) \mathrm{dxdydz} \tag{6.8}$$

If the flow is three dimensional, a third term can be added to allow for the effect of flow in the x-direction.

If, however, the flow rates are constant with time (i.e. steady-state flow is occurring), there will be no change in volume of the tailings prism and, therefore:

$$\frac{\partial V}{\partial t} = 0 = k_y\frac{\partial^2 h}{\partial y^2} + k_z\frac{\partial^2 h}{\partial z^2} \tag{6.9}$$

If the tailings are homogeneous and $k_y = k_z = k_x$, the equation becomes:

$$\frac{\partial^2 h}{\partial y^2} + \frac{\partial h}{\partial z^2} = 0 \tag{6.10}$$

Equation (6.10) is a Laplacian equation, the solution to which can be represented by a network of orthogonal lines (i.e. lines intersecting at right angles) that represent "flow lines" and "equipotential lines."

6.8.1 *Flow nets for homogeneous tailings*

Equation (6.10) can be solved for a given set of boundary conditions by means of a very simple and instructive method in which a net-work of flow lines and equipotential lines is constructed that satisfy the required boundary conditions. Figure 6.16 illustrates the simplest possible form of flow net. The two-dimensional constant head permeability test specimen represented in the figure has unit depth into the page and has been divided into 5 flow tubes, of equal area, by the dashed vertical flow lines. This indicates that the flow velocity is uniform over the width of the specimen. The total length L of

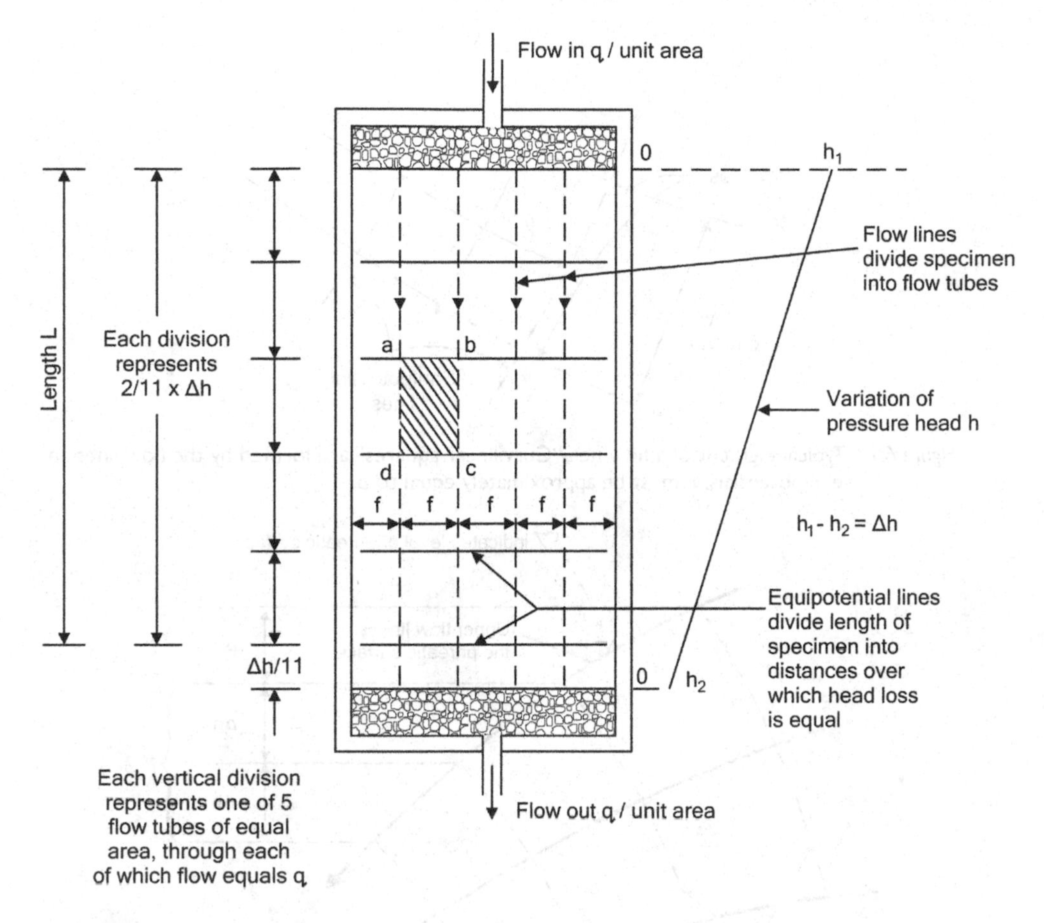

Figure 6.16 Simple representation of a flow net.

the specimen, over which the head lost is Δh, is divided into equal distances (except for the last step) over which the head loss is equal, i.e. the gradient of head is constant, and the horizontal lines represent "equipotentials". Each volume (such as abcd) between a pair of flow lines and a pair of equipotentials is thus a scale model of the total specimen. The outer boundaries of the permeability specimen also represent flow lines. It should be noted that the equation

$$\frac{\partial^2 h}{\partial y^2} = 0 \tag{6.10a}$$

simply means that $\frac{\partial h}{\partial y} = \text{constant}$, which is what Figure 6.16, with its constant head gradient represents.

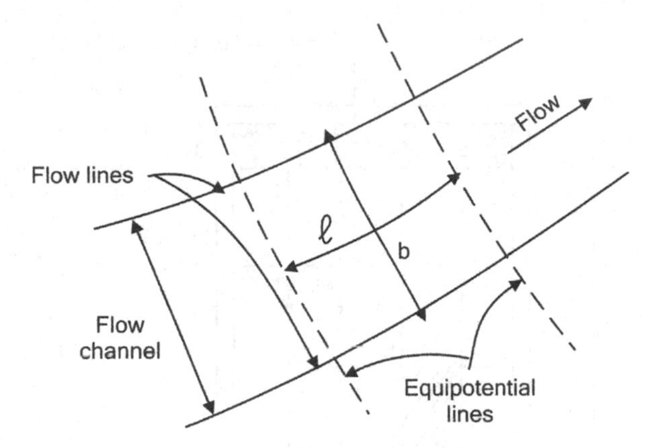

Figure 6.17 Typical element of a flow net. "Curvilinear squares" are formed by the flow lines and equipotentials. ℓ must be approximately equal to b.

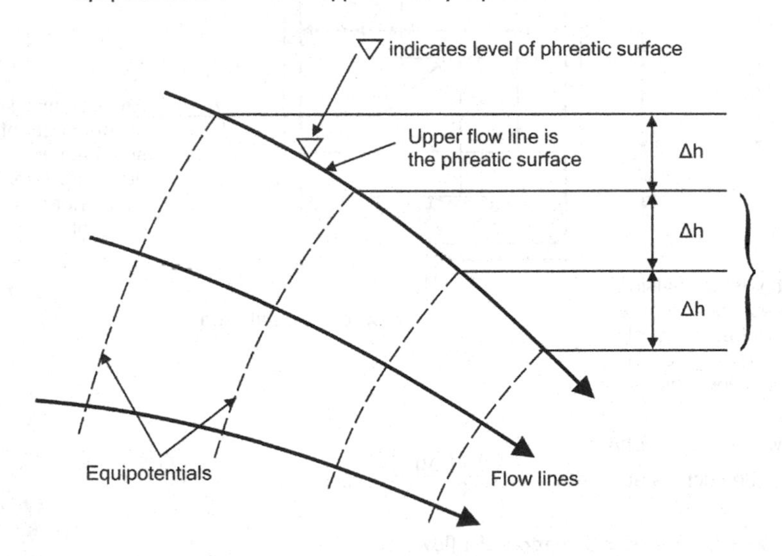

Figure 6.18 Upper part of the flow net in the outer slope of a hydraulic fill tailings storage.

More generally, flow lines and equipotentials are non-parallel and curved, as shown in Figure 6.17. As in the case of the flow net shown in Figure 6.16, the flow through every tube is the same, and the head lost over a distance in the direction of flow such as ℓ is the same for all squares in the flow net. Hence the total flow quantity through the flow net per unit depth normal to the plane of the flow net will be:

Q = soil permeability (k) × total head lost (h) × number of flow tubes (N_f) divided by number of head drops (N_h), or

$$Q = khN_f/N_h \tag{6.11}$$

If k is in m/y and h is in m, Q will be in m^3/y per m depth of specimen into the page.

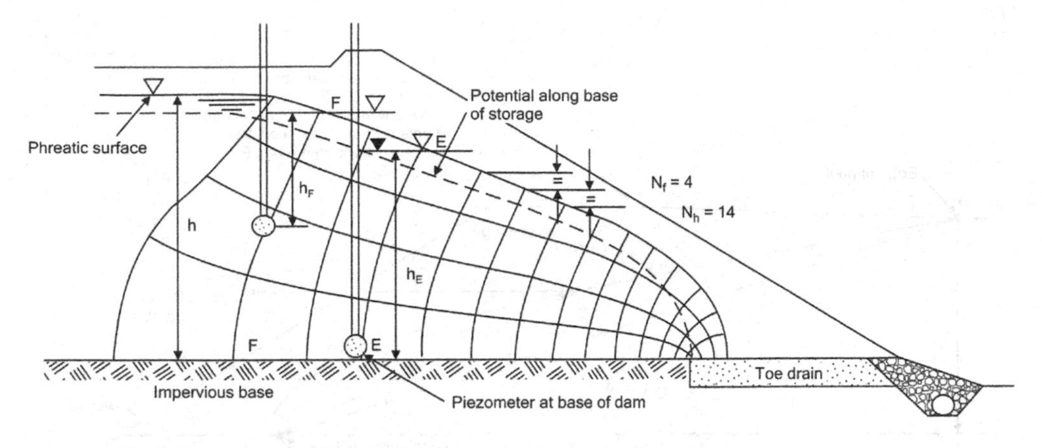

Figure 6.19 Flow net for the outer wall of a hydraulic fill tailings storage.

Other rules applying to flow nets are as follows:

- a horizontal phreatic line (i.e. a level water surface on which the water pressure is atmospheric) is an equipotential, and therefore no lateral flow of water can occur;
- a sloping phreatic line such as occurs within the outer slope of a hydraulic fill tailings storage, is a flow line. As shown in Figure 6.18, equipotentials intersect the top flow line at equal vertical intervals, i.e. equal decrements of potential Δh. Because the phreatic line is at atmospheric pressure, any loss of head must equal the corresponding loss of elevation.

Figure 6.19 represents part of a flow net within the outer wall of a hydraulic fill tailings storage. Some of the characteristics of the flow net are as follows:

1. The base is impervious, therefore it has to represent a flow line and all equipotentials will intersect it at right angles.
2. Potential losses between equipotentials are equal along the phreatic surface.
3. If $k = 0.03$ m/y, $h = 30$ m, and because $N_f = 4$ and $N_h = 14$, $Q = 0.03 \times 30 \times 4/14 = 0.26\ \text{m}^3$/y per m length of wall.
4. Piezometers located at the base of the wall will register the potential at the bottom of each equipotential. Because of the curvature of the equipotentials, a line showing the variation of the potential at the base of the dam will be located below the phreatic surface. For example:

The water level in a piezometer with its porous tip located at the base of the storage on a particular equipotential EE, would stand at the level at which EE intersects the phreatic surface. The pore pressure at the base of EE would be $\gamma_w h_E$. Similarly, the pore pressure at the tip of the piezometer located on equipotential FF would be $\gamma_w h_F$. Note that in both cases, the pore pressure is less than that corresponding to the depth of the piezometer below the phreatic surface.

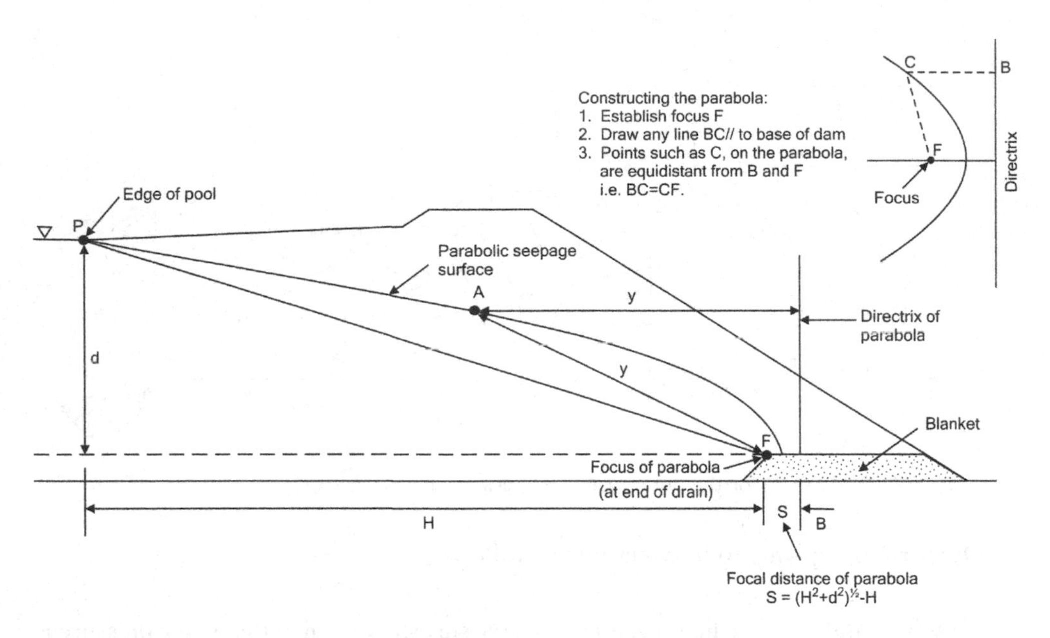

Figure 6.20 Approximate phreatic surface for a storage with a toe drain.

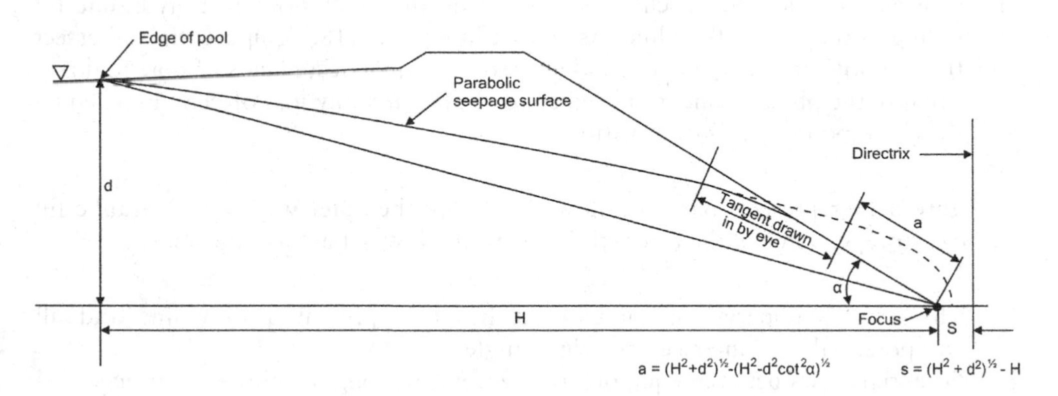

Figure 6.21 Approximate phreatic surface for a storage with no toe drain.

The phreatic or seepage surface can be located by constructions due to Casagrande (1937) which are illustrated in Figures 6.20 for a tailings storage with a toe drain and Figure 6.21 for a storage without a toe drain. In both cases the permeability of the tailings is assumed to be isotropic ($k_y = k_z = k_x$), and the base of the storage is considered impermeable. (Figures 6.17, 6.18, 6.20 and 6.21, which appear in many texts, have all been taken from Casagrande's remarkable 1937 paper.)

The sketching procedure is as follows (referring to Figure 6.19):

1. Draw out the section of the storage to scale.
2. Locate the phreatic surface using the construction shown in Figure 6.20.

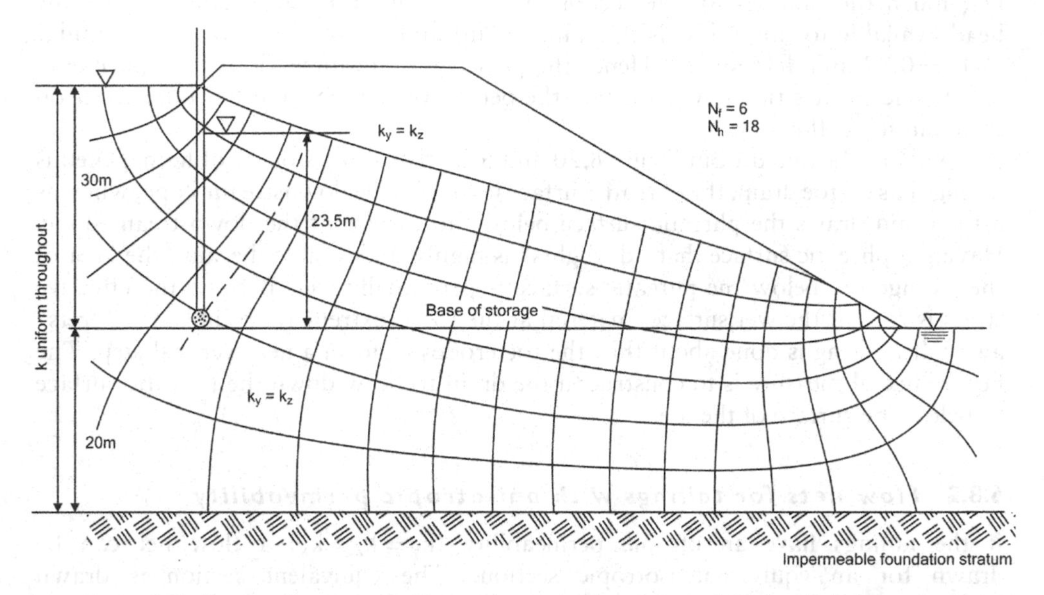

Figure 6.22 Flow net for storage of homogeneous tailings (permeability $k_y = k_z$) resting on a layer of identical material which, in turn, rests on an impervious foundation stratum.

3. Using a soft pencil, sketch in a series of flow lines to divide the space between the base and the phreatic surface into a whole number of flow tubes (4 in Figure 6.19).
4. Sketch in a set of equipotentials, bearing in mind that they must start at right angles to the base and end at right angles to the phreatic surface. They should also form a series of "curvi-linear squares" with the flow lines, as illustrated by Figure 6.17.
5. Progressively improve the geometry of the first sketch of the flow net by erasing and re-drawing lines until the sketch meets the requirements of a flow net as closely as possible.

Many computer packages exist that can be used to produce flow nets, but it is usually quicker to sketch a flow net, as usually only an approximate solution to the flow net is required. In most cases, at the flow net stage, only a rough estimate is required for pore pressure and flow quantity. The actual quantities can be measured by means of piezometers and drain flows, once the storage is in operation. The effect of variation in the dimensions, boundary conditions, etc. can be assessed quickly and easily.

Figure 6.22 shows an example of a flow net for a storage of homogeneous tailings resting on a layer of homogeneous material (which could be similar tailings or natural soil) having an identical permeability. (S.A. National Building Research Institute, 1959). If the tailings have a permeability $k = 0.03$ m/y, the height of the top of the storage is 30 m, the thickness of the lower layer is 20 m, the number of flow tubes is $N_f = 6$ and the number of equipotential steps is $N_h = 18$, then the quantity of flow will be:

$$Q = 0.03 \times 30 \times 6/18 = 0.3\,\text{m}^3/\text{y per m length of the storage.}$$

i.e., much the same as for the section shown in Figure 6.19. This is because the head available to cause flow is the same (30 m) and the ratio N_f/N_h is very similar (6/18 = 0.33 and 4/14 = 0.29) Hence the presence of an impervious base (in Figure 6.19) concentrates the flow paths into the permeable tailings, but has little effect on the quantity of flow.

It will also be noted from Figure 6.20 and 6.21 that if the storage of homogeneous tailings has no toe drain, the phreatic surface intersects the downstream slope, whereas a toe drain draws the phreatic surface below the surface of the downstream slope. Having a phreatic surface that "daylights" is highly undesirable, because the area of the storage toe below the phreatic surface is perpetually wet. Because the effective stress is zero at the wet surface, material at the toe is extremely erodible and "frets" away. If nothing is done about this, the toe erodes to form a near-vertical step. The best remedial measure is to construct a toe drain to draw down the phreatic surface to below the surface of the toe.

6.8.2 *Flow nets for tailings with anisotropic permeability*

If the tailings have anisotropic permeability ($k_y \neq k_z \neq k_x$) a flow net can be drawn for an equivalent isotropic section. The equivalent section is drawn by keeping the vertical (y) scale unchanged, but transforming the horizontal (z) scale so that

$$z(\text{transformed}) = z\sqrt{(k_y/k_z)} \tag{6.12}$$

An orthogonal flow net can be drawn to suit the transformed section, and the flow quantity can be calculated from equation (6.10) as before. Figure 6.23 shows the graphical transformation and also the flow net when it is re-drawn with the z-scale re-transformed to its natural value. If $k_y = 0.03$ m/y and $k_z = 0.3$ m/y, i.e. $k_z = 10\,k_y$ the flow quantity would be

$$Q = k^I \times 30 \times 5\tfrac{1}{2}/10 = 16.5k^I$$

The equivalent isotropic permeability would be

$$k^I = \sqrt{(k_y.k_z)} = \sqrt{(0.03 \times 0.3)} = 0.095\,\text{m/y}$$

Hence $Q = 0.095 \times 16.5 = 1.6\,\text{m}^3/\text{y}$ per m length

As a result of the 10:1 ratio of k_y to k_z, the flow increases by about 1.6/0.3 or a factor of 5:1. (The sections in Figures 6.22 and 6.23 are not identical, but close enough to make the comparison.

The construction on Figure 6.22 shows that the potential at the level of the base of the storage just to the left of the crest would be 23.5 m if the permeability were isotropic. A similar construction on Figure 6.23 shows that if $k_z = 10k_y$, the potential in a similar position would have been 19 m, a 19% decrease. Hence the higher horizontal permeability appreciably reduces potentials and hence pore pressures in the

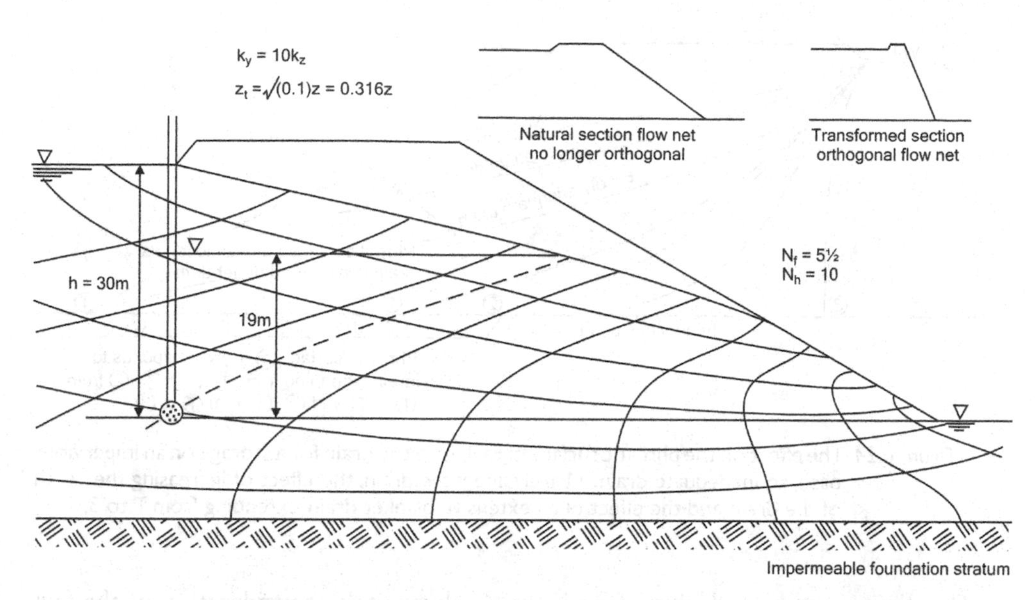

Figure 6.23 Flow net for storage of tailings having anisotropic permeability, ($k_y = 10k_z$). Storage rests on deposit of Identical anisotropic tailings which, in turn, rests on impervious foundation.

slope. Reference to Figure 6.15 will show that the assumption of $k_z = 10k_y$ is probably extreme. The overall conclusion is that anisotropy of tailings permeability will usually not have a particularly important effect on a tailings slope. There are always exceptions, and if the properties of the tailings are such that k_z may exceed k_y by a factor of 100 to 1000, a careful examination of the effects should be made. (For example, cases have been reported where k_y is effectively zero, in which case the flow lines become almost horizontal and seepage exits on the slope over almost its entire height.)

6.8.3 The effect on the position of the phreatic surface of drainage beneath the toe of a tailings storage on an impervious base

As illustrated by Figures 6.22 and 6.23, if the permeability of the base of a tailings storage is equal to or less than that of the tailings, it is likely that the phreatic surface will emerge on the downstream slope. Figure 6.24 shows the effects of providing a toe drain that extends under the toe to various distances. Providing a small toe drain may not be completely effective in depressing the phreatic surface to below the slope, but a drain becomes progressively more effective as its width increases. If a full-width blanket drain is provided, the whole of the outer wall will be drained and therefore its stability will be much enhanced. However, full blanket drains are extremely costly to construct, and for that reason, various other forms of drain are often installed. As illustrated by Figure 6.25, partial blanket drains can take the form of grid or ladder, or herringbone drains.

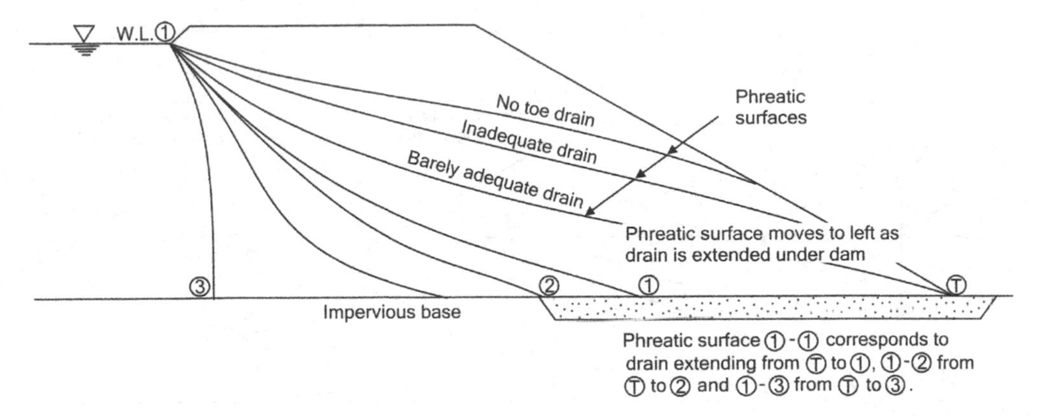

Figure 6.24 The effect on the phreatic surface of having no toe drain for a storage on an impervious base, an inadequate drain, a barely adequate drain, the effect of increasing the width of the drain and the effect of an extensive blanket drain extending from T to 3.

The dimensions and placing of drains must always take consideration of the fact that although drains are usually designed to be effective when the storage is high, they must also be effective when it is low. Situations such as that illustrated by Figure 6.26 must be avoided.

6.9 The design of filter drains

The drains referred to in section 6.8.3 have to be designed as filter drains consisting of a series of layers having increasing permeabilities, with the final layer surrounding a perforated drainage or seepage collection pipe.

Filter drains are used to significantly reduce seepage pressures in soils over short distances. This implies that filter layers may be subjected to large seepage gradients. Filter materials must therefore be capable of preventing fine particles of tailings adjacent to the filter and upstream of it from moving into or through the voids of the filter. A filter layer must also be significantly more permeable than the material it is filtering or retaining in order to reduce seepage gradients significantly over a short distance.

Design rules for filters have been established empirically and are based on relationships between the particle sizes and particle size distributions of successive layers. The requirements (illustrated in Figure 6.27) are as follows:

1. D_{15} of the filter must be less than 5 times D_{85} of the material to be retained.
2. D_{15} of the filter must be between 4 and 20 times D_{15} of the material to be retained.

These two requirements ensure that particles will not move from the retained material into the filter, and also that the filter will be significantly more permeable than the material retained.

3. D_{85} of the filter must be less than 5 times D_{15} of the filter.

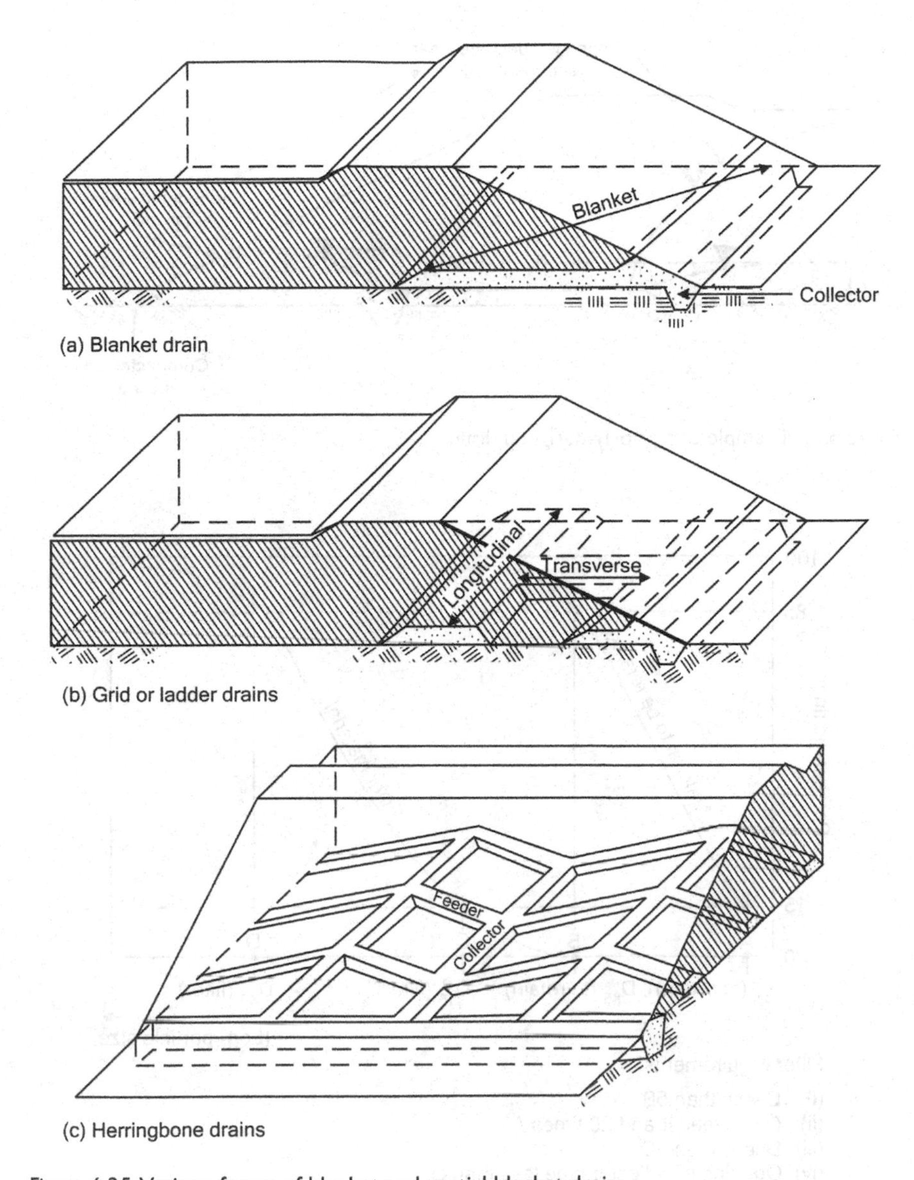

Figure 6.25 Various forms of blanket and partial blanket drain.

This ensures that the grading of each filter layer will be stable, and that fine particles in the filter will not migrate in the direction of seepage and as a result either move into the next (coarser) layer, or else block the pores in the layer from which they originate.

Where a filter material is drained by a perforated or slotted pipe, the criterion to prevent material from the filter from being washed through the openings in the pipe is:

4. D_{85} of the filter must be greater than the size of the opening (width of slot or diameter of hole) in the pipe. A similar criterion to 4 is used to select a geofabric

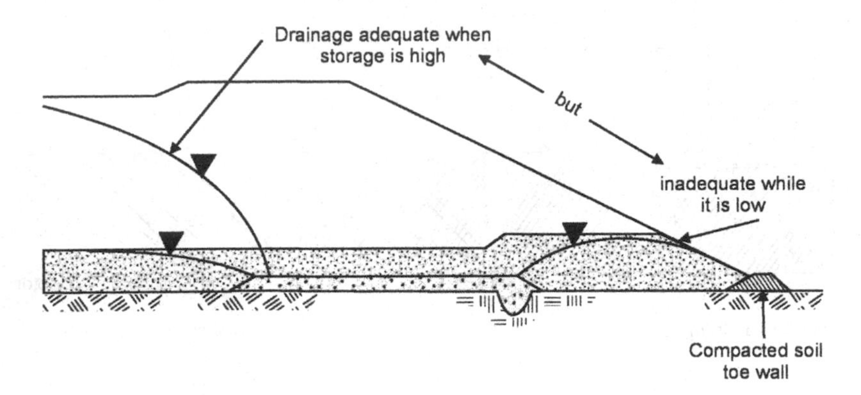

Figure 6.26 Example of a poorly-designed drain.

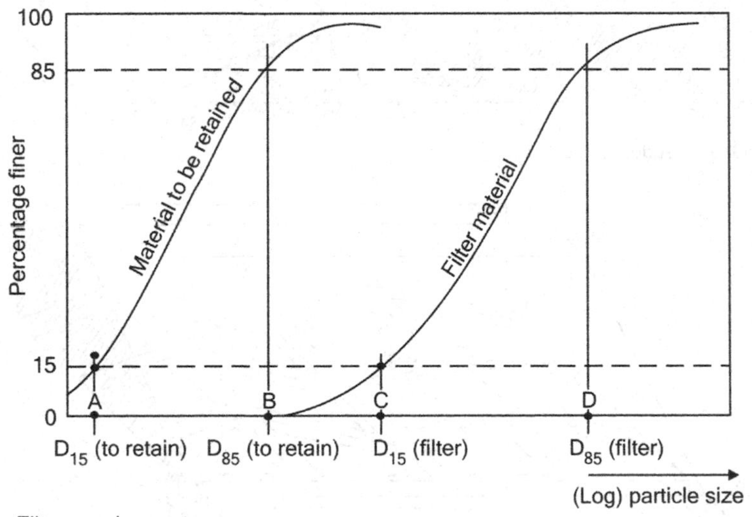

Figure 6.27 Summary of filter design requirements.

for a particular purpose. Reference should, however, be made to the relevant supplier's design information. (Also see Section 6.14.3.)

(In the "filter rules" above, D_{15} is the particle size of a material such that 15% by dry mass of material is finer than D_{15}, and similarly for D_{85}.)

For example, the first layer of a filter retaining a gold tailings (for which $k = 10^{-5}$ cm/s) might be a medium sand (with $k = 10^{-2}$ cm/s). The second layer may be a fine gravel (with $k = 10$ cm/s). The gravel would surround a perforated drainage pipe having openings of an appropriate maximum size.

In terms of particle sizes, suppose for the gold tailings:

$$D_{15} = 0.005\,\text{mm and } D_{85} = 0.05\,\text{mm}$$

For the **first filter layer** to comply with rule 1:

$$D_{15}(\text{filt1}) < 5 \times 0.05 = 0.25\,\text{mm, say } 0.2\,\text{mm.}$$

For this layer to comply with rule 2:

$$D_{15}(\text{filt 1})/D_{15}(\text{tail}) = 0.2/0.005$$
$$= 40, \text{ this violates rule 2, therefore } D_{15}(\text{filt 1}) \text{ must be reduced.}$$

$$\text{Make } D_{15}(\text{filt 1}) = 0.1\,\text{mm, with } D_{85}(\text{filt 1}) = 0.4\,\text{mm.}$$

For the **second filter layer**

Rule 1 is: $D_{15}(\text{filt 2}) < 5 \times 0.4 = 2$ mm

Rule 2 is: $D_{15}(\text{filt 2})/D_{15}(\text{filt 1}) = 2/0.1 = 20$, which complies with rule 2.

Make $D_{15}(\text{filt 2}) = 2$ mm with $D_{85}(\text{filt 2}) = 8$ mm

Rule 3 is: $D_{85}(\text{filt 2})/D_{15}(\text{filt 2}) = 4$, which complies.

To comply with rule 4, perforations in the drainage pipe must be ≤ 8 mm.

Figure 6.28 shows typical details of double filter layer drains, with a toe drain above and a segment of a ladder or grid drain below. Only the main collecting segments of the drain would be equipped with collection pipes. Where the collected flow is expected to be small, filter layer 2 can be used as the collector.

The surface of the in situ soil should be compacted to provide a firm base for building the drains and toe wall. Filter layers must have a substantial thickness (200 mm is suggested) and side slopes are made at the angle of repose of the filter material.

With hydraulic deposition, the first tailings to reach and contact the side or surface of a filter drain is usually the finest fraction which may enter and block or "blind" the surface of filter layer 1 (which is designed for the average grading of the tailings). To prevent this from happening, and also to protect the filter layers from accidental disruption before they are covered by deposited tailings, the drains should be covered with a substantial protective layer of compacted tailings. (Drains may lie exposed for several months, or even years after their construction but before being covered by tailings deposition.)

To protect the drain from possible upward flow of seepage from the foundation soil into the drain, filter layer 1 is extended over the area of contact of the drain with the soil. Plate 6.2 shows a filter drain being prepared at the toe of an extension to a tailings storage. In this case, the first filter layer is a geofabric, and the second layer, of fine

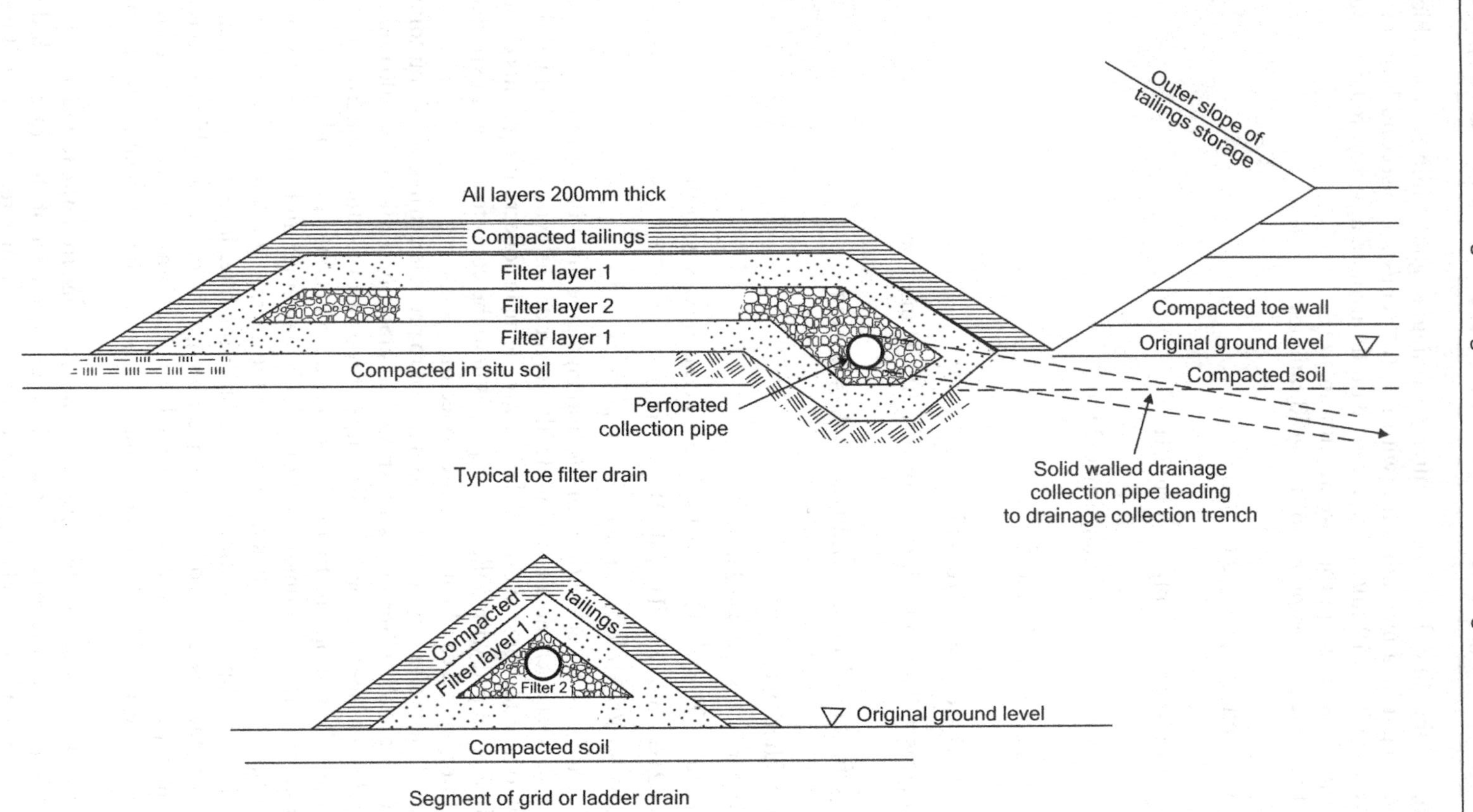

Figure 6.28 Typical construction of filter drains.

Plate 6.2 Constructing a ladder filter drain at the toe of a tailings storage.

gravel is being placed. The third layer will be a coarser gravel and will contain the collector pipe.

6.10 Calculation of seepage rates through tailings storages

Figure 6.29 represents a section through a hypothetical tailings storage in which there is shallow water ponded at the surface over the tailings. The storage has a semi-pervious liner to limit seepage into the underlying soil which is highly pervious, and the regional ground water table is at a shallow depth. The diagram to the right shows the total potentials p at various depths in the storage and its foundation stratum. The potential at any depth is the sum of the elevation above an arbitrary datum and the water pressure or pore water pressure u, both expressed in metres of water head. Thus the potential p_0 between the water surface and the regional water table is the same as that at the tailings surface:

$$p_0 = d_0 + d_1 + d_2 + d_3$$

at depth $(d_0 + d_1)$ the pore pressure is

$$u = p_1 - (d_2 + d_3)$$

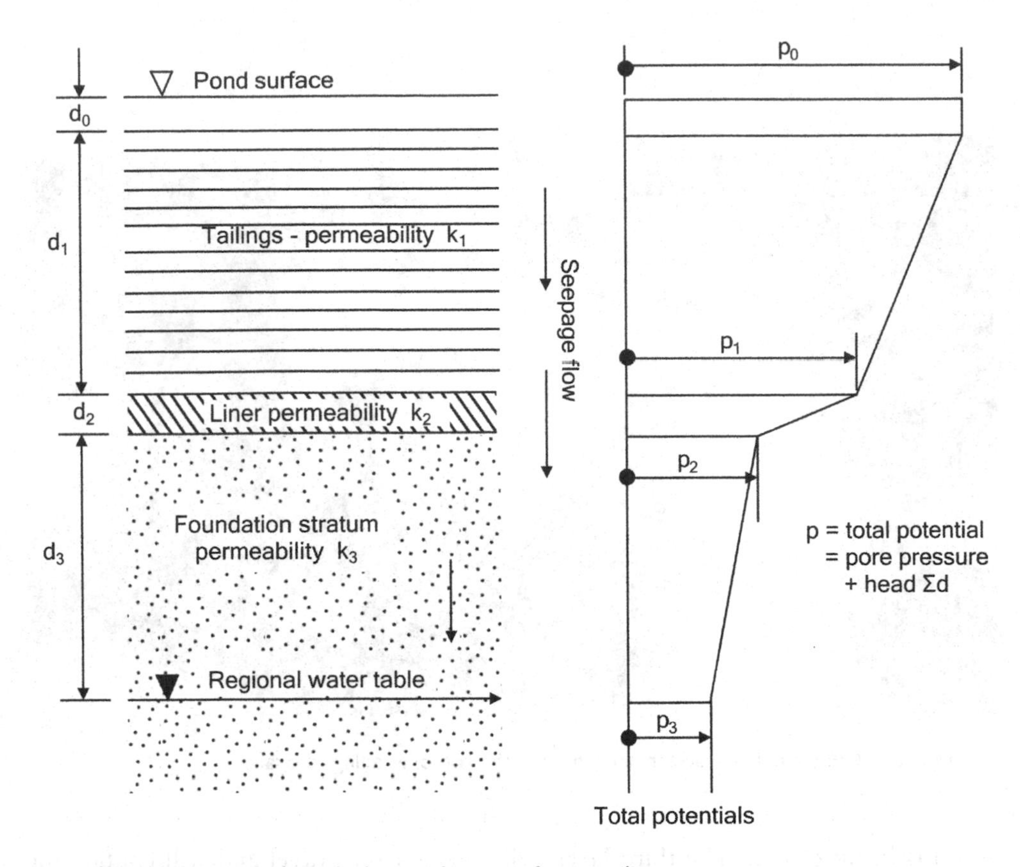

Figure 6.29 Example for calculation of seepage from a tailings storage.

To simplify the analysis, the dimensions used in Figure 10.9 have been replaced by potentials p_0 to p_3, where the potential p = (pressure head + elevation). The analysis is as follows:

For continuity of flow, the velocity of flow through each of the layers must be the same. Applying Darcy's law:

$$v = \text{flow velocity} = k_1 \cdot \frac{p_0 - p_1}{d_1} = k_2 \cdot \frac{p_1 - p_2}{d_2} = k_3 \cdot \frac{p_2 - p_3}{d_2}$$

Dimensionally: v = [m/s · m/m = m/s]

The potential at the regional water table (p_3) can be regarded as a datum value and set to zero. Also $p_0 = (d_1 + d_2 + d_3)$, expressing potential in terms of head, i.e.

$$\frac{k_1}{d_1}[(d_0 + d_1 + d_2 + d_3) - p_1] = \frac{k_2}{d_2} \cdot (p_1 - p_2) = \frac{k_3}{d_3} p_2 = v \qquad (6.13)$$

where v is the seepage rate per unit area of the residue deposit.

Solving for p_1:

$$p_1 = \frac{k_1(d_0 + d_1 + d_2 + d_3)}{d_1\left[\dfrac{k_2 k_3}{k_3 d_2 - k_2 d_3} + \dfrac{k_1}{d_1}\right]} \tag{6.14}$$

whence v may be found.

EXAMPLE 1 For a shallow pool with a large thickness of tailings:

$d_0 = 1\,m$
$d_1 = 20\,m$ $k_1 = 3\,m/y$
$d_2 = 0.5\,m$ $k_2 = 0.03\,m/y$
$d_3 = 5\,m$ $k_3 = 300\,m/y$

From equations (6.13) and (6.14)

$p_1 = 18.9\,m$ $v = 1.14\,m/y$

The pore pressure distributions corresponding to examples 1, 2 and 3 are found by subtracting the physical height above the water table from the corresponding potential. For example 1, $p_1 = 18.9\,m$ and the corresponding pore pressure

$$u_1 = 18.9 - (d_2 + d_3), \text{ i.e. } u_1 = 18.9 - (5 + 0.5) = 13.4\,\text{m of water.}$$

$$p_2 = vd_3/k_3 = 0.02\,\text{m and } u_2 = 0.2 - 5 = -5\,\text{m of water.}$$

That is, the pore pressure is negative from near the bottom of the liner down to the water table. Figure 6.30 shows the pore pressure diagrams for examples 1, 2 and 3 plotted to scale.

EXAMPLE 2 Contrast this result with that of the following for a deep pool, with water replacing most of the tailings:

$d_0 = 20\,m$
$d_1 = 1\,m$ $k_1 = 3\,m/y$
$d_2 = 0.5\,m$ $k_2 = 0.03\,m/y$
$d_3 = 5\,m$ $k_3 = 300\,m/y$
$p_1 = 25\,m$ and $u_1 = 25 - 5.5 = 19.5\,m$ $v = 1.5\,m/y$
$p_2 = v \cdot d_3/k_3 = 0.025\,m$ and $u_2 = 0 - 5 = -5\,m$

That is, the seepage loss is only slightly changed, because it is controlled by flow through the liner.

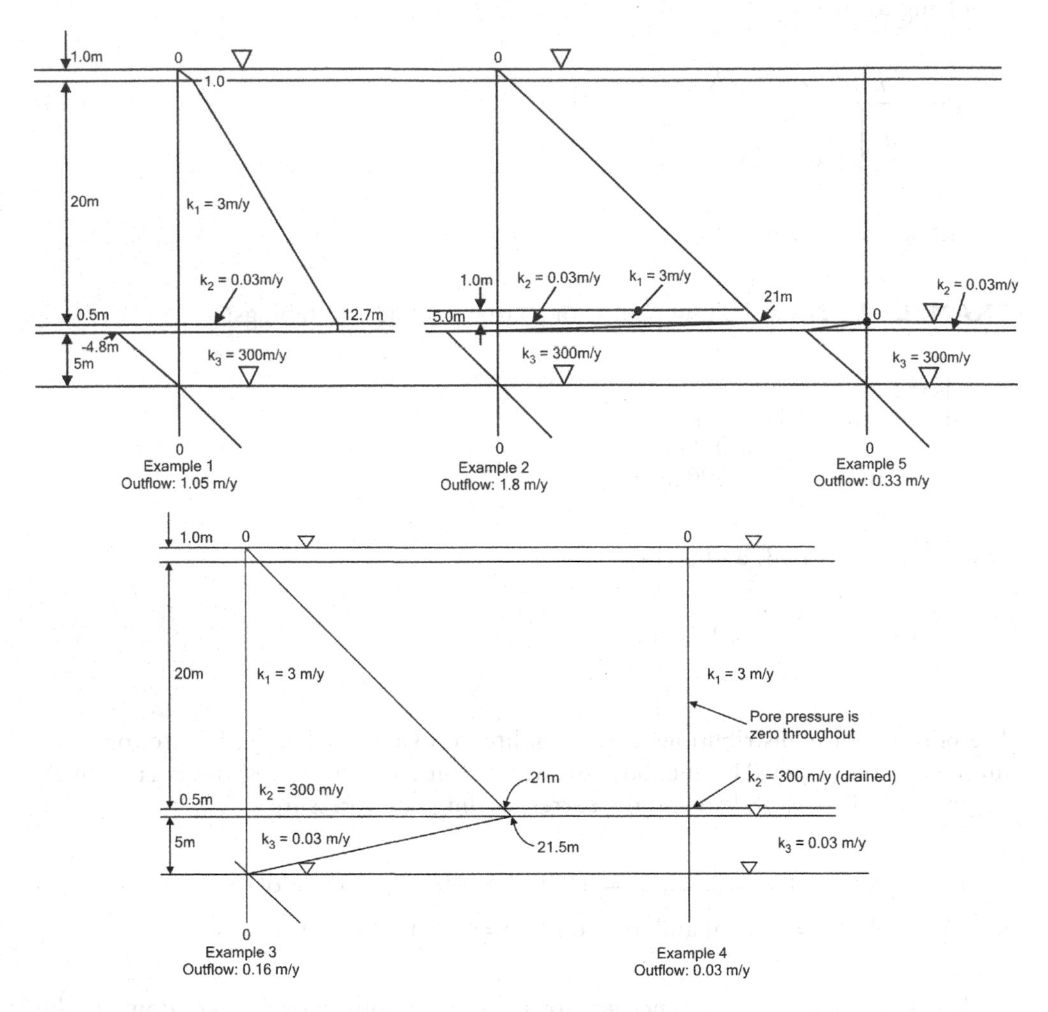

Figure 6.30 Pore pressure profiles for examples 1, 2, 3, 4, 5 in Section 6.10.

EXAMPLE 3 Now consider the case where the liner is changed from a semi-pervious to a highly permeable layer, and the foundation is made impermeable:

$d_0 = 1\,m$

$d_1 = 20\,m$ $\quad k_1 = 3\,m/y$

$d_2 = 0.5\,m$ $\quad k_2 = 300\,m/y$ The rate of

$d_3 = 5\,m$ $\quad k_3 = 0.03\,m/y$

Then $p_1 = 27.6\,m$ and $u_1 = 27.6 - 5.5 = 22.1\,m$ $\quad v = 0.16\,m/y$

seepage is now controlled by flow through the thick impervious foundation stratum.

EXAMPLE 4 This is similar to example 3, but the pervious zone with $k_2 = 300\,m/y$ is now drained so that the pore pressure is zero within this layer. With this modification, the potentials in every layer become equal to the height above the water table, and the pore pressures become zero from between the water tables to the pond surface. Downward seepage takes place at a seepage gradient of unity, and the rate of outflow through the tailings would be $k_1 = 3$ m/y, but the rate of flow into the foundation would be only 0.03 m/y compared with 0.16 m/y in Example 3. This shows that having a drain above an impervious layer can be very effective in reducing the escape of polluted water from a tailings storage.

EXAMPLE 5 In the last example, the effect of a drain immediately above a relatively thin semi-pervious liner is analysed by placing a water table, i.e. $u = 0$ immediately above the liner shown in Figure 6.29, i.e. d_0 and d_1 no longer exist,

$d_2 = 0.5$ m $\quad k_2 = 0.03$ m/y
$d_3 = 5$ m $\quad k_3 = 300$ m/y

Then $p_1 = 5.5$ m and $v = 0.33$ m/y, whereas in Example 1, $v = 1.14$ m/y

Hence the installation of a drain immediately above a thin semi-pervious liner is also very effective in reducing the escape of water from a tailings storage, into the foundation strata, because it reduces the seepage potential above the liner.

Figure 6.30 shows all the pore pressure distributions, drawn to scale, for examples 1 to 5.

6.11 The process of consolidation and pore pressure re-distribution

6.11.1 *The Terzaghi theory of consolidation*

So far, only steady-state flow has been considered. However, there are many problems that require the consideration of non-steady state or transient flow. These problems are usually handled by the application of Terzaghi's consolidation theory.

The theory of consolidation is usually used to predict time-settlement relationships for buildings on compressible soil, settlement and pore pressure dissipation in earth embankments, etc. In mine waste engineering, its primary use is for calculating safe rates of rise for hydraulic fill tailings dams. The theory can also be used to estimate suitable rates of testing in triaxial and shear box tests. (see Sections 4.6.2, 4.6.3 and 4.6.4).

The relatively simple Terzaghi consolidation theory was first published in 1925 and is still considered sufficiently accurate for much of geotechnical work. The assumptions in the theory are as follows:

- The soil is saturated and homogeneous.
- Darcy's law applies.
- The coefficient of permeability is unaffected by changes in void ratio.
- The flow of water is one dimensional and so is the compression or swell of the consolidating material.
- Volume changes are caused only by changes in void ratio which occur in accordance with the principle of effective stress.

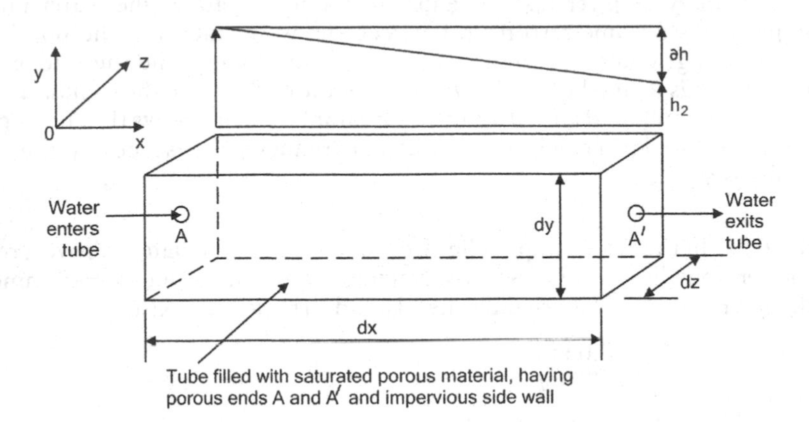

Figure 6.31 Prism of saturated soil with permeable ends and impermeable sides subject to axial flow in the x-direction.

Referring to Figure 6.31, the flow gradient will vary along the length of the tube and its rate of variation with distance will be:

$$\frac{\partial i}{\partial x}$$

If the flow gradient at the centre of length of the tube is:

$$i = -\frac{\partial h}{\partial x} \text{(The minus sign indicates that h decreases as x increases.)}$$

The flow gradient at entry to the tube is:

$$\text{i (entry)} = -\frac{\partial h}{\partial x} + \frac{\partial i}{dx} \cdot \frac{dx}{2} = -\frac{\partial h}{\partial x} + \frac{\partial^2 h}{\partial x^2} \cdot \frac{dx}{2}$$

Therefore, flow into the tube is $k\left(-\frac{\partial h}{\partial x} + \frac{\partial^2 h}{\partial x^2} \cdot dx/2\right) dydz$

Similarly, flow out of the tube is $k\left(-\frac{\partial h}{\partial x} - \frac{\partial^2 h}{\partial x^2} \cdot dx/2\right) dydz$

Subtracting flow out from flow in, the time rate of change of the volume of water in the tube is

$$\frac{\partial V}{\partial t} = k\frac{\partial^2 h}{\partial x^2} dxdydz \tag{6.15}$$

Simultaneously, the time rate of change of void ratio of the material in the tube will be

$$\frac{\partial}{dt}\left(\frac{e}{1+e} \cdot dxdydz\right)$$

The two time rates of change of volume must be equal, and

$$k\frac{\partial^2 h}{\partial z^2} = \frac{1}{1+e} \cdot \frac{\partial e}{\partial t}$$

The head causing flow is equivalent to the excess pore pressure causing flow, i.e. the excess pressure over and above the equilibrium static pore pressure, divided by the unit weight of water, so

$$\frac{k}{\gamma_w} \cdot \frac{\partial^2 u}{\partial z^2} = \frac{1}{1+e} \cdot \frac{\partial e}{\partial t} \qquad (6.16)$$

Defining a_v = soil compressibility $= -\frac{\partial e}{\partial \sigma^I} = \frac{\partial e}{\partial u}$ (following the principle of effective stress: $\sigma^I = \sigma - u$ and if σ is constant, $\partial \sigma^I = -\partial u$)

$$\frac{k(1+e)}{a_v \gamma_w} \cdot \frac{\partial^2 u}{\partial z^2} = \frac{\partial u}{\partial t}$$

Defining $\frac{k(1+e)}{a_v \gamma_w}$ as the coefficient of consolidation, c_v

$$\frac{c_v \partial^2 u}{\partial z^2} = \frac{\partial u}{\partial t} \qquad (6.17)$$

This is the one-dimensional Terzaghi consolidation equation which strictly applies only if the total stress σ is constant.

The solution to equation (6.17) depends on the boundary conditions and the pore pressure u_o initially set up in the consolidating material. For a commonly occurring situation where u_o is constant with depth z at time $t = 0$, the solution is given by

$$u = \Sigma(m = 0 \rightarrow \infty)(2u_o/M.\sin(Mz/H)\exp(-M^2\,T)) \qquad (6.18)$$

In this equation $M = 1/2\pi(2m+1)$ and m is any integer from 1 to infinity.

$$T = \frac{c_v t}{H^2}$$

where c_v is the coefficient of consolidation and H is the drainage path length, the shortest distance water would have to move to escape from the consolidating layer into a drainage or pervious layer. T is the time factor.

The tatio

$$U_z = 1 - u_z/u_0 \qquad (6.19)$$

is called the consolidation ratio at any depth z for a given time factor T. When $u_z = u_o$ at time $t = 0$, $U_z = 0$ and when $u_z = 0$, $U_z = 100\%$.

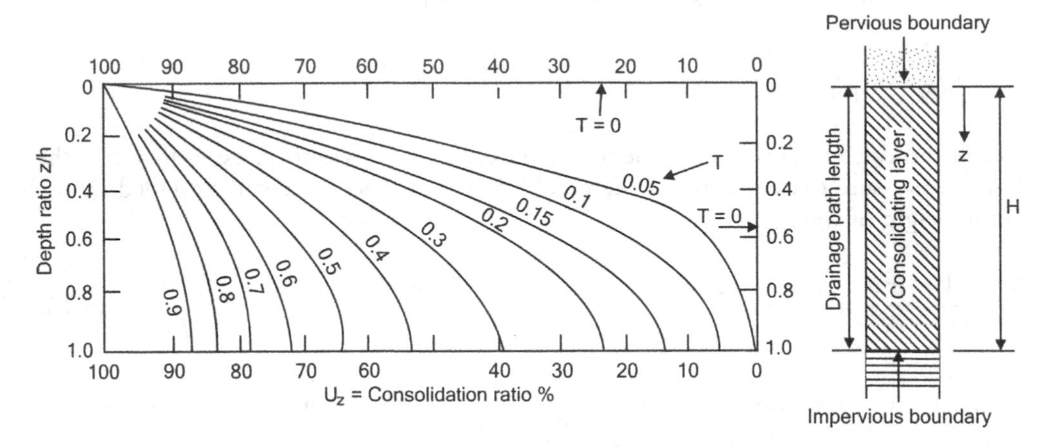

Figure 6.32 Solution to one-dimensional Terzaghi consolidation equations (6.18) and (6.19).

Equation 6.19 is shown graphically in Figure 6.32. Each z/H versus U_z contour corresponds to a constant value of T. The top horizontal axis and the right hand vertical axis constitute the $T = 0$ and $U_z = 0$ contour, and the left hand vertical axis represents the $T = \infty$ contour representing $U_z = 100\%$. Note that to obtain the solution for a layer of thickness H draining through top and bottom pervious boundaries, the drainage path length becomes H/2 and the time factor is calculated as $T = 4c_v t/H^2$. The values of U_z on the $z/H = 1.0$ axis then become those for mid-depth of the layer.

The average degree of consolidation U is the average value of U_z over the depth H. For the case of $u = u_0$ at $t = 0$, U is given by

$$U = 1 - \Sigma(m = 0 \rightarrow \infty)(2/m^2 \cdot \exp(-M^2\ T)) \tag{6.20}$$

and Figure 6.33 shows equation (6.20) graphically for a range of values for the ratio R of u_o (top of layer) to u_o (bottom of layer) where the top boundary is pervious and the bottom boundary is impervious. (Janbu, Bjerrum and Kjaernsli, 1956). The solution that derives specifically from equation (6.20) is that for $R = 1.0$.

It is important to remember that u, u_o u_z, etc. represent excess pore pressures, i.e. the amount by which the actual pore pressure exceeds the hydrostatic pore pressure. If the actual pore pressure at depth z below a level phreatic surface is u_w, then

$$u_w = \gamma_w z + u \tag{6.21}$$

This is particularly important in considering the mechanics of hydraulic fill tailings because the material is deposited in a saturated state and consolidates under the overburden of material deposited over it. When consolidation is complete, the pore pressure u_w does not equal zero, but $\gamma_w z$, and it is only after the phreatic surface has subsided by the depth z that u_w will have declined to $u_w = 0$.

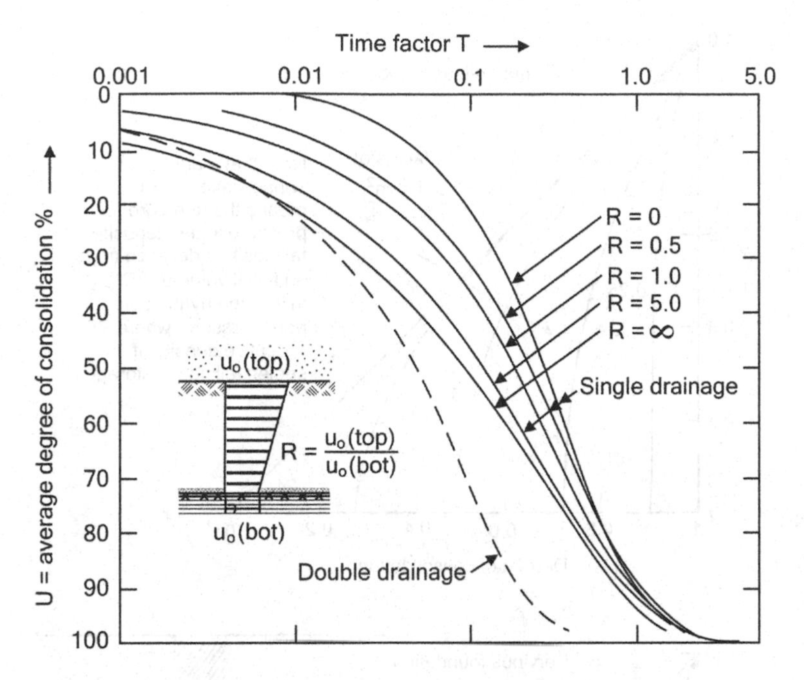

Figure 6.33 Relationship between degree of consolidation and time factor.

6.11.2 *Gibson's and Schiffman's extensions to Terzaghi's theory of consolidation*

The consolidation theory considered in Section 6.11.1 deals with bodies of material that have dimensions fixed in time, although they may be undergoing small time-dependent strains. However, in a hydraulically deposited tailings storage, and even in a waste dump that is deposited in sub-horizontal layers, the thickness or depth of waste is continually increasing. Hence the stresses at a particular point, as well as the drainage path length also increase with time. Gibson (1958) published an extension to Terzaghi's theory that takes account of the effects of continual deposition. His extension includes an additional term to account for the increase in load and drainage path length, the extended governing equation being, in terms of excess pore pressure u,

$$c_v \frac{\partial^2 u}{\partial z^2} = \frac{\partial u}{\partial t} - \gamma^I \frac{dh}{ht} \tag{6.22}$$

where γ^I is the submerged density of the consolidating material and dh/dt is the rate of increase of material depth h with time, i.e. the rate of rise. The equation is derived assuming that strains are small and that c_v is constant with increasing stress. In terms of the pore pressure u_w, the equation becomes:

$$c_v \frac{\partial^2 u_w}{\partial z^2} = \frac{\partial u_w}{dt} - \left[\gamma_w \frac{dH}{dt} + \gamma' \frac{dh}{dt}\right] \tag{6.22a}$$

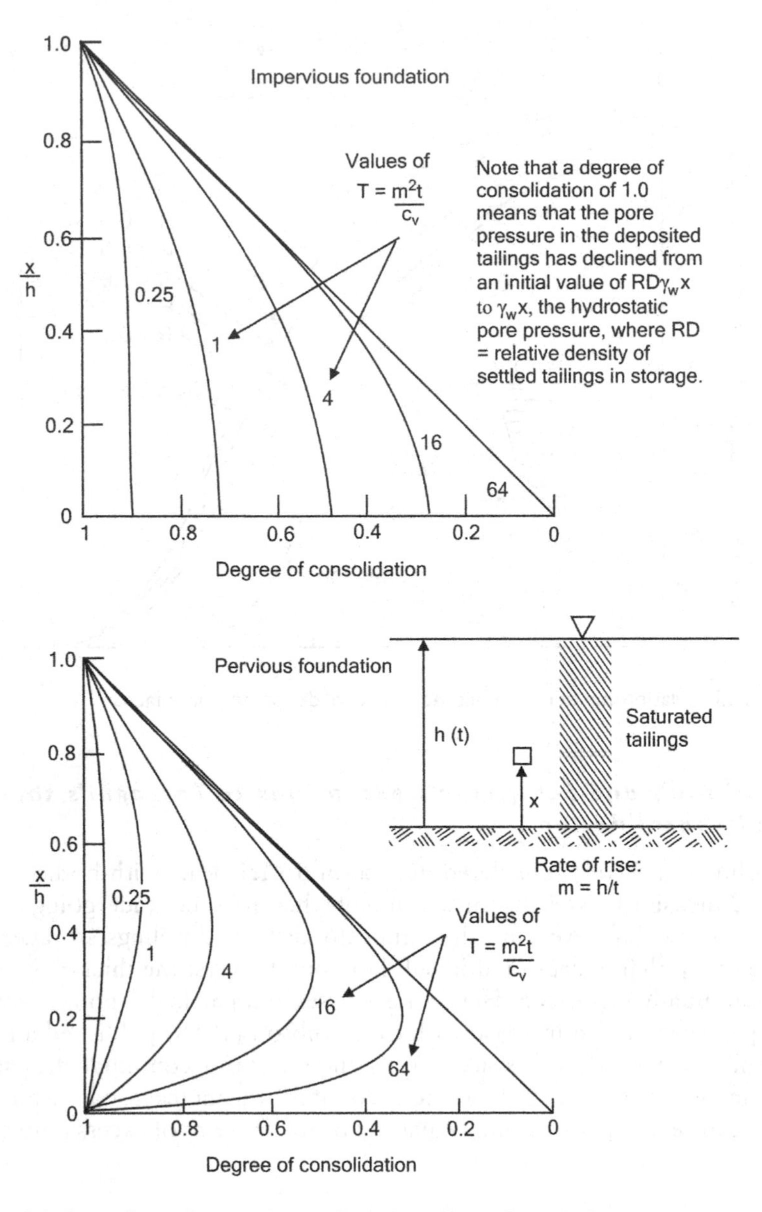

Figure 6.34 Gibson's small strain theory for tailings being deposited at a constant rate of rise.

where γ_w is the density of water and H is the depth of water above the base. If $h = H$, the phreatic surface coincides with the surface of the deposited material.

The solution to equations (6.22) and (6.22)a is shown in graphical form in Figure 6.34. for both impervious and pervious foundations.

As examples, suppose that $c_v = 100\,m^2/y$ and $m = 5\,m/y$,

1. Find the degree of consolidation at x/h = 0.4 for both impervious and pervious foundations after 10 years

$$T = \frac{m^2 t}{c_v} = \frac{5 \times 5 \times 10}{100} = 2.5$$

For the impervious foundation, from Figure 6.34, the degree of consolidation U is about 0.6 and for the pervious foundation, it is about 0.8.

2. What will the corresponding effective overburden stresses be if $\gamma_{sat} = 18\,kN/m^3$? For $h = 5\,m/y \times 10y = 50\,m$, $x/h = 0.4$ corresponds to a depth below surface of 30 m.

   For the impervious foundation, $\sigma_v = 30 \times 18 = 540\,kPa$ and we have to assume that with zero consolidation $u_o = \sigma_v$ and hydrostatic pore pressure $= 30 \times 10 = 300\,kPa$. Hence for $U = 0.6$, u will be $(1 - 0.6) \times (540 - 300) + 300 = 396\,kPa$. Therefore $\sigma_v^I = 540 - 396 = 144\,kPa$.

   For the pervious foundation, $U = 0.8$, $u = 0.2\ (540 - 300) + 300 = 348$ and $\sigma_v^I = 540 - 348 = 192\,kPa$.

3. If the angle of shearing resistance for the normally consolidated tailings is 28°, what will be the shear strength τ on a horizontal surface for the two cases?
   Impervious foundation: $\tau = 144 \tan 28° = 76\,kPa$.
   Pervious foundation: $\tau = 192 \tan 28° = 102\,kPa$.

In this way the entire shear strength profile could be predicted and used, for example, to predict the factor of safety against failure of the slopes at a rate of rise of 5 m/y for 10 years.

In consolidating, tailings can undergo reasonably large strains. Schiffman, et al., (1988) have also extended the theory to cases of finite strain, where c_v varies with void ratio. They applied the extended theory to the analysis of a large copper tailings storage. The results of their calculations by both small strain and finite strain versions are shown in Figure 6.35.The void ratio at deposition was believed to have been 1.45 which, assuming $G = 3.0$, corresponds with $\gamma = 18\,kN/m^3$. On this basis, the total overburden stress line would have intersected the base of the storage at $u_w = 900\,kPa$. It therefore appears from Figure 6.35 that:

1. consolidation was negligible down to a depth of about 30 m, and
2. there was relatively little difference between the small strain and large strain theories except below 30 m, where the finite strain theory was the more conservative.

6.12 Basal impervious liners and surface cover layers or caps

Because of the large areas usually covered by tailings and industrial waste storages, they are presently not always equipped with impervious underliners. In fact, the requirement for underliners depends on both local regulations, which vary widely from country to country, as well as the characteristics of the material being stored. For

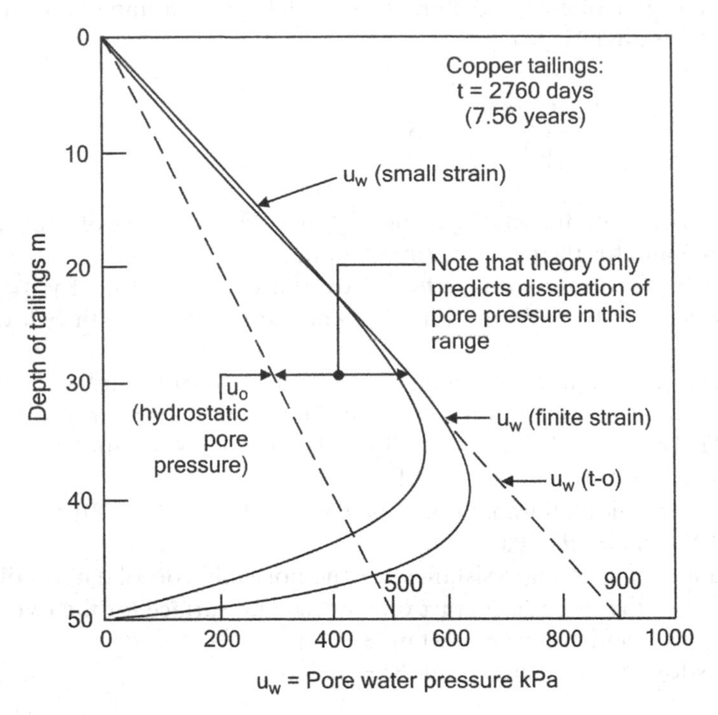

Figure 6.35 Schiffman-Gibson consolidation theory applied to consolidation of copper tailings storage on a pervious foundation.

example, waste phospho-gypsum which contains toxic fluorides and traces of fluorsilicic acid would always be required to be stored over an underliner, whereas diamond tailings which usually contain no polluting substances, would not. Seepage from waste storages may also be unacceptably saline or highly acid (in the case of sulfidic ores). Environmental regulations protecting groundwater are becoming much more stringent in most countries, and it is inevitable that impervious linings will become more and more widely required in the future.

The requirement to provide an impervious underlining over the whole area of a waste storage immediately implies that the lining should be covered by a system of underdrainage to extract and collect seepage reaching the base of the storage. This is necessary for several reasons:

- There is a need to extract and collect the potentially polluting or toxic leachate before any can escape through the lining, both as a result of the lining's finite permeability and through the leaks that will inevitably occur as a result of imperfections and damage during installation. Collection allows the extracted seepage to be treated and discharged to stream or to be recirculated to the mineral extraction plant, where the water recovery can be an important benefit in arid climates. In any case, potential pollution can be both prevented and treated at its source.
- Drainage will reduce the seepage gradient across the lining. If no drainage system exists, the seepage gradient across a liner of, for example 1 m thickness, could reach

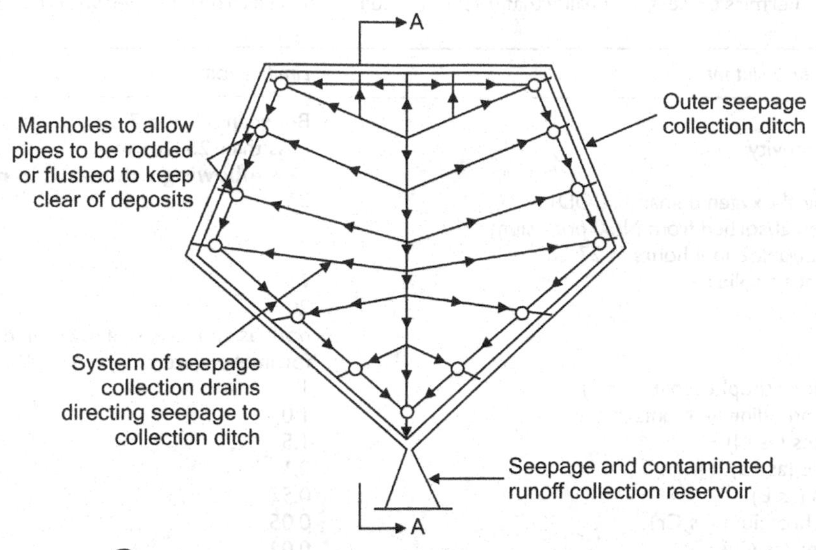

ⓐ Diagrammatic plan showing seepage collection system

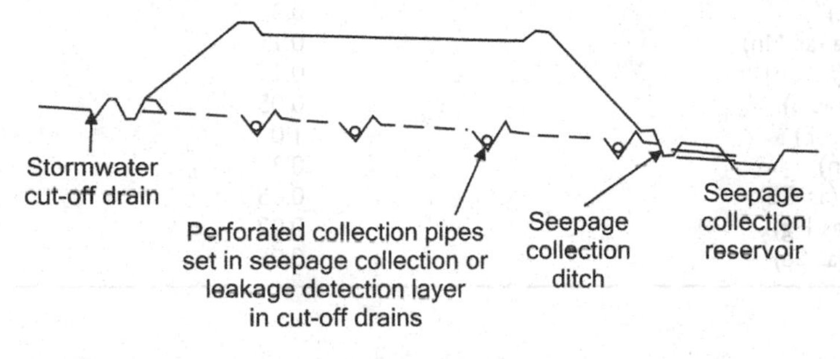

ⓑ Section AA through waste storage

Figure 6.36 Diagrammatic layout of seepage collection system.

30 to 50 or more. (See example 1 in Figure 6.30 where the gradient across the liner is $(12.7 + 5)/0.5 = 35.4$.) Reducing the gradient to closer to 1.0 by installing an efficient drainage system (example 4 in Figure 6.30) will proportionately reduce the potential seepage loss and allow savings to be made by installing a less elaborate underlining.

- The deposited waste will consolidate more rapidly and become more shear stable and less liable to potential liquefaction if deposited on a pervious, drained surface. This may allow a higher rate of rise to be used in building the storage. The period during which the phreatic surface subsides when the storage is retired from service will also be much reduced. (See Figure 6.34).

As mentioned earlier, where waste storages are required to be lined, the linings are often specified in terms of a succession of materials, layer thicknesses, etc. The example

Table 6.3 Permissible seepage quality that does not require a storage to be equipped with an impervious liner

Property or pollutant	*Limiting maximum values*
• pH	Between 5.5 and 7.5
• Conductivity	Less than: 250 mS/m
	The following values are in mg/litre
• Chemical oxygen demand (COD)	30
• Oxygen absorbed from N/80 potassium permanganate in 4 hours at 27°C	5
• Suspended solids	10
• Sodium	20 both as an increase above value for intake water
• Soluble orthophosphate (as P)	1.0
• Free and saline ammonia	1.0
• Nitrates (as N)	1.5
• Arsenic (as As)	0.1
• Boron (as B)	0.5
• Total chromium (as Cr)	0.05
• Copper (as Cu)	0.02
• Phenols	0.01
• Lead (as Pb)	0.1
• Iron (as Fe)	0.3
• Manganese (as Mn)	0.1
• Cyanides (as Cn)	0.1
• Sulphides (as S)	0.05
• Fluoride (as F)	1.0
• Zinc (as Zn)	0.3
• Cadmium (as Cd)	0.05
• Mercury (as Hg)	0.02
• Selenium (as Se)	0.05

shown in Figure 6.37 is typical of current liner requirements, although the requirements vary from country to country and even locally from state to state. The information that appears below is typical of supplementary information required by a regulator before allowing a liner to be dispensed with:

Waste storages and return water reservoirs or lagoons that produce or store seepage that does not meet the requirements listed in Table 6.3 are required to be equipped with an impervious liner.

When a waste storage produces seepage that does not meet the requirements listed in Table 6.3 a liner must be provided that limits the measured outflow rate from the storage to less than 0.03 m/y (30 mm/y or 1×10^{-7} cm/s).

The following is typical of supplementary requirements for the construction of an acceptable liner:

The liner must be a composite liner incorporating compacted clay, impervious geomembrane and drainage elements, with the geomembrane above, and in intimate contact with the clay liner.

A lining layer, constructed of compacted soil of low permeability, must be so constructed that it permits no more than the above specified maximum rate of seepage outflow to pass through its layers. Clay liner components must be compacted to a minimum dry density of 95% Standard Proctor maximum dry density, at a water content of Proctor optimum to Proctor optimum +2%. Any soil used for a compacted soil liner must have a minimum Plasticity Index (PI) of 10 and a maximum that will not result in excessive desiccation cracking. The maximum particle size must not exceed 25 mm. (Also see Sections 7.4 and 7.8.)

In addition, the following information is required by the regulator:

- Full particle size analysis (sieve and hydrometer tests).
- Double hydrometer test, using the seepage water, to ensure that the soil will not become dispersive when exposed to the seepage.
- Atterberg limits.
- Shear strength tests in terms of effective stresses on soil compacted at Proctor optimum water content to Proctor maximum dry density. Shear tests are to be either drained, or undrained with measured pore pressures. The soil is to be saturated with the seepage in question before the start of shearing.
- Permeability measurements in triaxial cells are also required on saturated soil, compacted as above, and using seepage as the permeant.

Because the liner will usually have to be designed at a time when only laboratory test data are available, the expected outflow rate will usually have to be based on permeability coefficients measured in the laboratory on specimens constituted in the laboratory. The seepage water used in the above tests must either be actual seepage from a similar storage, or a simulated seepage containing maximum permissible values of the expected contaminants.

These estimates must, however, be validated by field tests once the liner has been constructed. It must also be remembered that small-scale laboratory measurements could underestimate the permeability of a liner by as much as two orders of magnitude (see Section 6.2).

To validate the design, in situ permeability tests using double ring infiltrometers must be carried out on every compacted soil layer that forms part of a liner. The diameter of the inner ring of such an infiltrometer must be at least 600 mm, while the diameter of the outer ring must be twice that of the inner ring. The infiltrometer must be covered and sealed with plastic sheeting to prevent loss of water by evaporation (see Section 6.4.1).

The clay components of liners must fully meet the above maximum permissible outflow rate of 0.03 m/y and must be shown by test to comply with this prior to installing the overlying geomembrane.

Every liner system is made up of a series of elements that can be assembled in various ways to provide the necessary degree of protection to the ground water system. For all underlined waste storages, the base must be shaped and sloped so that all seepage entering a seepage drain above the liner is directed to a seepage collection reservoir from which it can be pumped back to the mineral extraction plant (see Figure 6.36 for a diagrammatic layout of such a system).

The seepage collection system is a system of drains, bunds or ditches covered by the seepage collection layer. It is equipped with suitable filter drains and collection pipes that direct the gravity flow of seepage or leakage to defined collection points or sumps, from which it can be collected for re-circulation to the mineral extraction plant.

Any drain, whether open or covered, that is used to transfer seepage from the seepage collection system to the seepage reservoir must be properly lined. This should be by means of a properly laid 2 mm thick geomembrane liner with joints welded to the same specification as for a hazardous waste liner, or equivalent. The lining of open drains should be covered by a protective layer (e.g. of concrete interlocking blocks) to protect against physical damage and deterioration by UV (ultraviolet) attack.

The liner design for a seepage collection reservoir must be to the same specification as the waste storage liner, but without the leachate collection layer, and must include a 2 mm thick geomembrane liner, laid directly on the surface of the uppermost clay layer. The liner should be protected from physical damage (by, e.g., wind) and UV attack by providing a protection layer (e.g. of concrete interlocking blocks).

The leakage detection and collection system is designed to intercept, collect and remove any seepage that passes the barrier of the upper liner. This leakage is then directed to a separate leakage collection sump, where the quantity and quality can be monitored and from which accumulated leakage can be removed. The leakage detection system must be divided into area zones so that the source of any collected leakage can be identified.

A typical design for an impervious underliner is shown in Figure 6.37a. The liner consists of the following layers, working downwards from the underside of the tailings or other waste:

A-layer: A filter layer which, depending on the particle size distribution of the tailings or other waste, would either be a single 150 mm layer, or a double layer. Where possible the A-layer should be covered with 150 mm of compacted tailings to prevent it from being blinded by the extremely fine fraction of hydraulically placed tailings. The A-layer can be a continuous blanket or can be placed as a grid drain with the grid segments designed according to the filter design rules summarized in Section 6.9 and Figure 6.27.

B-layer: A seepage collection layer comprising a 150 mm thick layer of single-sized gravel or crushed stone having as large a size as possible that complies with the filter criteria (given in Section 6.9 and Figure 6.27) as regards its interaction with the A-layer above it and any collection pipes it may incorporate.

C-layer: A 2.0 mm thick impervious flexible membrane or geomembrane laid with welded seams that are tested for leaks.

D-layer: A 150 mm thick compacted clay liner layer. This must be compacted to a minimum density of 95% Standard Proctor* maximum dry density at a water content of Proctor optimum to optimum +2%. Permeabilities must be such that the specified outflow rate is not exceeded. Interfaces between B-layers must be lightly scarified to assist in bonding the layers together.

* 0.945 litre cylindrical compaction mould, 2.5 kg compaction hammer dropped 300 mm on each layer in mould. Compaction in 3 layers each compacted with 25 blows (compactive effort = 595 kNm/m^3).

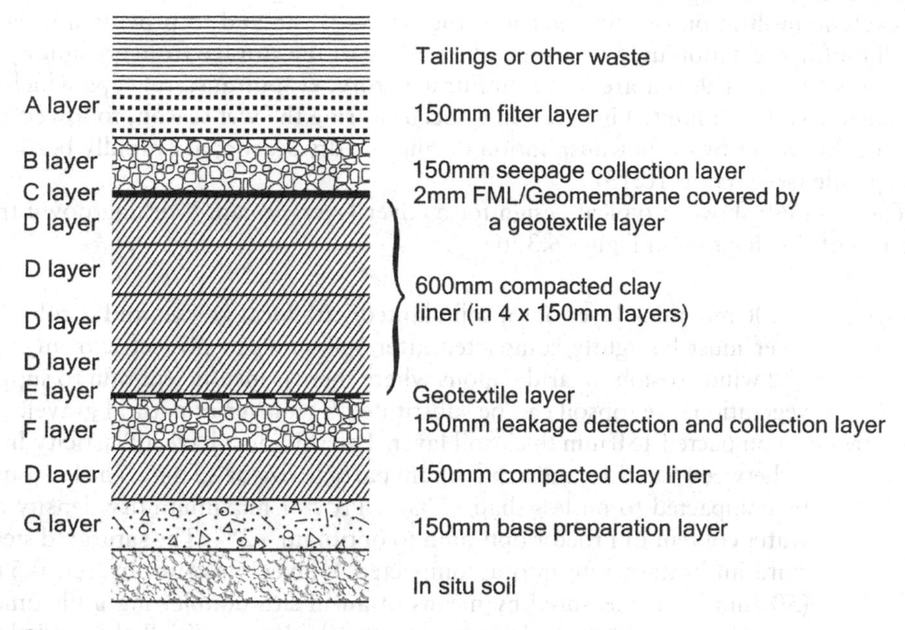

ⓐ Typical design for a drained impervious under-liner

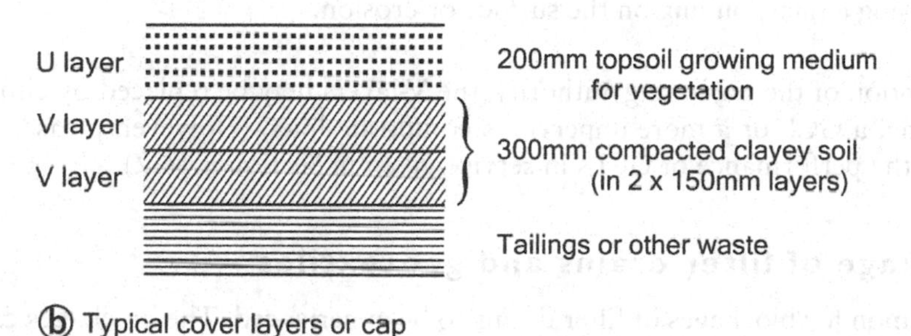

ⓑ Typical cover layers or cap

Figure 6.37 Typical designs for: a: drained impervious underliner, b: cover or capping layers.

The surface of every clay liner layer must be graded towards the seepage collection drains at a minimum gradient of 5%. At the discretion of the regulating authority B-layers may be replaced by a geomembrane, a GCL**, or a composite liner.

E-layer: This is a layer of heavy duty geotextile laid on top of any F-layer to protect it from contamination and blocking by fine material from above.

F-layer: A leakage detection and collection layer. This is always located below an E-layer and above a D-layer. It has a minimum thickness of 150 mm and will consist of single-sized gravel or crushed stone.

** Note that at the time of writing, GCLs (geocomposite or geosynthetic clay liners) incorporating bentonite were suspected of being subject to severe deterioration by desiccation and interaction with seepage. (See Section 6.14.2.)

Cover layers or caps for waste storages are usually designed to have two functions: to exclude infiltration of rainwater into the stored waste and to provide a growing medium for vegetation used to protect the surface of the storage from erosion by the elements. In water deficit areas, ISE (infiltrate, store, evapotranspire) caps which are designed to allow a limited infiltration to occur during the wet season, followed by a loss of this water by evapotranspiration during the dry season, are rapidly becoming acceptable (see section 10.13).

Figure 6.36b shows a typical design for a cover layer, or cap. Working down from the top of the diagram in Figure 6.37b:

U-layer: A 200 mm thick layer of topsoil planted with local grasses and shrubs. The layer must be lightly compacted after spreading to protect it from water and wind erosion. In arid regions where there is insufficient rain to support vegetation, the topsoil can be substituted by a layer of natural gravel.

V-layer: A compacted 150 mm thick soil layer. The soil must have a Plasticity Index of between 5 and 15 and a maximum particle size of 25 mm. The layer must be compacted to no less than 85% of Proctor maximum dry density at a water content of Proctor optimum to optimum +2%. The saturated steady state infiltration rate into a compacted V-layer must not exceed 0.5 m/y (500 mm/y) as measured by means of an in situ double ring infiltrometer test. The surface on which every V-layer is laid must initially be graded at a minimum of 3% so as to shed precipitation to a series of surface drains that must be provided to remove runoff from the surface of the storage without causing either ponding on the surface or erosion.

At the discretion of the regulating authority, the V-layers may be replaced by either a geomembrane, a GCL or a more impervious composite liner. (However, please see comments on the performance of GCLs in service given in Section 6.14.2).

6.13 Blockage of filter drains and geotextiles

It is quite common for blockages of filter drains to be experienced. The blockages can develop as a result of a physical break-down of the filter system that may occur because the filter layers have not been correctly designed or because incorrect materials were used in the layers during its construction. A more common cause is the deposition of chemical or bio-chemical precipitates within the voids of the filter layers, openings in geotextile separator layers, or within the perforations in the walls of the seepage collection piping. An example of the effect of such void or pore blockage is shown in Figure 6.38 (Fourie, et al, 1994). The diagram shows the decline with time of the flow rate through a 200 mm layer of sand filtered by single layers of three different geotextiles in laboratory tests:

NW = nonwoven needle-punched polyester fiber, 1.5 mm thick,
W1 = woven polypropylene 0.3 mm thick, and
W2 = woven polypropylene 1.1 mm thick (also see Section 6.14). The permeating liquids were clean water and a landfill leachate.

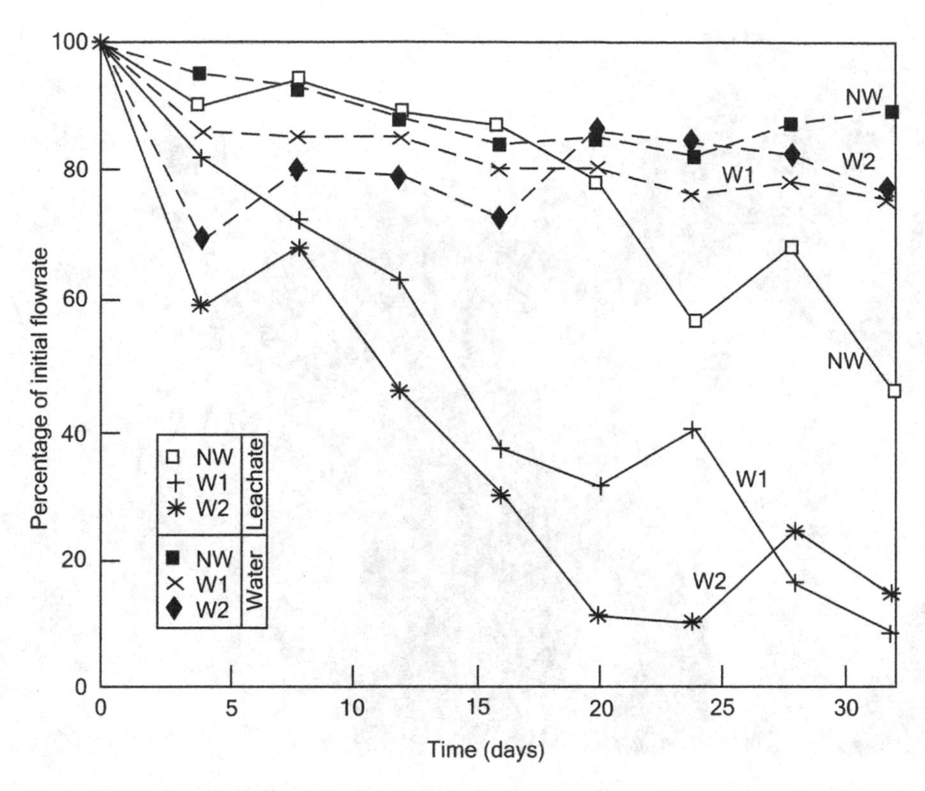

Figure 6.38 Example of effect of progressive blockage of woven and non-woven geotextiles during permeation by clear water and landfill leachate.

The 20% decline in through-flow rate that occurred with the permeation of water, probably resulted from blockage of pores caused by the migration of fine particles out of the sand layer and into the geotextiles. The increased decline of flow rate that occurred with the leachate was the result of precipitation of iron oxide out of the leachate, as well as the growth of bacterial bio-slime within the pores or openings of the geotextiles. Blockage was sufficient to reduce the flow rate by 90% in the case of the woven materials and over 50% for the non-woven material within only a month.

Other data by Metcalfe et al. (1995) on the reduction in through-flow rate of geotextiles exhumed after various periods of service as separators in filter drains (3–16 years of service) are as follows:

needle punched non-woven: 50–98% reduction,
heat-bonded non-woven: 89–93% reduction and
woven slit film: 27–98%

There appears to be very little literature on this problem in the field of hydraulic fill tailings storages, but the author has witnessed complete or near-complete blockages caused by deposits of ochre (yellow to red-coloured iron-rich slimes or gels). In one case a layer of graded stone chips in a filter blanket had been completely blocked

Plate 6.3 Layer of gravel in a filter blanket drain blocked by ochre deposited in pores of gravel.

with a deposit of ochre. (see Plate 6.3). The filter blanket had been exposed by gulley erosion that had cut through the filter, otherwise it may not have been suspected that the drain had been blocked.

However, a great deal of research on the problem of blocking has been done in the fields of agricultural drainage and the collection and drainage of leachate from sanitary landfills. An extensive review of the subject has been given by Rowe and VanGulck (2004). Conclusions, based in part on this review, are as follows:

- Field and laboratory studies show that clogging is related to the type and content of dissolved substances in the seepage, as well as the grain size and grain size distribution, initial pore volume and the presence of a separator between the tailings or other waste and the seepage collector. The separator may be the perforated wall of a collector pipe or a geofabric wrapping between the tailings and the first filter layer.
- An accumulation of precipitated insoluble material within filter gravel layers or separator layers decreases the overall permeability of the filter drain. When the drain permeability decreases to the point where the phreatic level starts to rise, the drain has functionally failed. (The functioning of a drain can be checked by regular measurements of the rate of discharge and also by placing a piezometer

either within the drain itself or closely adjacent to the drain. A continuing rise in piezometric level will indicate a decreasing functionality of the drain.)

- It follows from the above that it is advantageous to use drainage materials with large individual voids, thus maximizing the volume of clog material per void that can be accommodated without seriously affecting the drain's functionality. This will also reduce the surface area per void available for chemical and biochemical slimes to attach to. (Note that two gravels having the same particle size distribution and particle shape, but with different particle sizes will have the same porosity [void volume per unit volume of gravel] but the individual voids will be larger in the coarser material. The coarser gravel will have a smaller particle surface area per unit volume of gravel than the finer gravel.)
- Restoring a blocked or clogged filter drain may prove extremely difficult. If the blockage is within the collector pipes, it may be possible to clear then by rodding (see Figure 6.36) or flushing by means of a probe consisting of small diameter pipes with a high pressure radial disc spray at the leading end. The probe is pushed along the main collector pipes which are flushed by the passing radial jet. If the seat of the blockage is the filter layers themselves, the only remedy is either to excavate an accessible drain, such as a toe drain (although safety considerations will usually rule this out) or to replace the drain by perforated drainage pipes pushed or jetted into the slope. (Plate 9.10 shows two drain outlets, one of which is flowing strongly, the other of which is blocked and needs maintenance attention.)

6.14 Geosynthetic materials

There are many specialized publications that deal comprehensively with geosynthetic materials (e.g. Daniel and Koerner 2007) and what follows is a very brief summary of what geosynthetic materials are available and some fundamentals of their composition, characteristics and uses in mine waste engineering.

6.14.1 *Geomembranes*

A geomembrane is a sheet or membrane of material of uniform thickness that is used as an impervious layer or seal to replace or augment the function of a compacted clay layer. Geomembranes are made from a range of polymeric materials ranging from what are commonly known as "plastic" or polymeric materials to bitumen. Table 6.4 gives a brief summary of some of the polymeric materials and their properties, that are in common use for manufacturing geomembranes.

Bituminous membranes have the advantage of being produced in the field, where they can be made seamless to fit any shape of surface. Disadvantages are that unless field supervision is very strict, quality may be poor. The permeability is intrinsically higher than for factory-formed membranes and bituminous membranes are difficult to lay on surfaces with more than a moderate slope (15–20°).

Factory-made membranes of polymeric material can have consistently high quality if factory supervision is good, and are intrinsically completely impervious. Disadvantages lie mainly with the necessity to lay the pre-formed material in sheets that have to be seamed together and also with their low puncture resistance. It is the inevitable presence of defects, such as holes, splits and leaking seams, that results in preformed geomembranes having low but finite permeabilities as laid in the field (see Table 6.4).

Table 6.4 Composition and properties of commonly used types of geomembrane

Mnemonic	*Name in full*	*Main components*	*Physical properties*			*Advantages*	*Disadvantages*
			Density	*Coefficient of thermal expansion*	*Permeability to water in situ*		
HSA HMA	Hot-sprayed or hot-mixed asphalt	Bitumen binder: 7–12% fine mineral filler: 2–4% balance: mineral aggregate	Bitumen: 1 025 kg/m^3 at 20–25°C Asphalt: 2500–2600 kg/m^3	Asphalt: 0.04–0.09 mm/m°C	30–50 mm/y	Can be laid as seamless hot-mix asphalt or seamless spray and chip layer. Less technical skill required for good result	Higher permeability than preformed membranes. Requires excellent supervision of installation for good result
PVC	Polyvinyl chloride	PVC resin: 60% plasticizer: 30% carbon black: 2–5% balance: filler	950 kg/m^3	0.05–0.07 mm/m°C	Up to 15 mm/y (Koerner, 2001)	Factory-made product, consistency of quality can be well-controlled	Formulation, raw materials and manufacture critical for excellent quality.
HDPE	high density polyethylene	PE resin: 96% carbon black: 3% balance: additives, e.g. anti-oxidants	940 kg/m^3	0.14 mm/m°C			Weak points are seams and poor puncture resistance as well as high coefficient of thermal expansion, resulting in wrinkling during hot times and developing tension during cold weather
LLDPE	linear low density polyethylene	PE resin: 95% carbon black: 2% balance: additives, e.g. anti-oxidants	920 kg/m^3	0.20 mm/m°C		More flexible, greater elongation and less prone to stress cracking than HDPE	
fPP	flexible polypropylene	PPresin: 85–95% carbon black: 3% fillers: 0–10% balance: additives	910 kg/m^3	0.11 mm/m°C		More flexible than HDPE & LLDPE. Can be reinforced by sandwiching fabric layer between two fPP layers	
GCL	Geosynthetic clay liner Geocomposite liner	Layer of powdered clay or (clay+adhesive) sandwiched between upper and lower layers of geo-textile			15 mm/y or more (Benson, et al., 2007)		

Plate 6.4 Wrinkles in HDPE geomembrane caused by expansion of HDPE in heat of sun.

It also means that pre-formed geomembranes must always be backed up with a compacted impervious soil layer, as shown in Figure 6.37a, and also by a leakage collection layer to warn of, identify and locate areas of serious leakages, should they occur as a result of tearing or puncturing of the membrane.

Polymeric geomembranes can be seamed by welding or glueing (by means of a suitable and compatible adhesive.) Seams can be tested destructively by cutting out sample sections of seam (randomly selected) and testing them in the laboratory for shear and peeling strength. Nondestructive field testing can also be used (see Daniel and Koerner, 2007, for details of available methods).

The high coefficient of thermal expansion of all polymeric materials (about 5 times that of soil) is a major problem with laying the membranes, as a change of only 10°C can cause a strain of as much as 0.2% or 20 mm per 10m (see Table 6.4). As a result, geomembranes tend to wrinkle when the temperature rises and pull taut when it falls. Referring to Figure 6.37a, this means that if a B layer is placed over a C layer at a cold time of day, wrinkles will be formed when the geomembrane heats up. These may be trapped to form pervious, interlinked passages between the geomembrane and the underlying compacted clay D layers, if the overlying B-layer is placed while the geomembrane is hot. Hence a leak through a puncture in a wrinkle surface can spread over a large area of the top D layer.

(Plate 6.4 shows typical wrinkles caused by expansion of an unprotected geomembrane lining of a return water reservoir in the heat of the sun).

An additional problem that relates to membranes on slopes is that polymeric materials have a relatively low coefficient of friction. This can be overcome to a certain extent by texturing the surface, but the thermal problem will persist (in muted form) even when the geomembrane has been covered by a protective layer, such as the B layer in Figure 6.37a. Diurnal temperature changes at the surface will affect materials to a depth of about 200 mm. As the surface temperature may vary diurnally by up to 25°C, this would correspond to close to a variation of $50/200 \times 25 = 6°C$ in the buried geomembrane. This is enough to cause repeated daily movement between the geomembrane surfaces and the material within which it is sandwiched, resulting in decreased frictional resistance. (Section 4.8.1 discusses the shearing resistance of a fPP-to-geotextile interface, fPP = flexible polypropylene.)

Geomembranes must obviously be laid on a smoothed clay surface without any protruding stones that might cause puncturing and must be covered immediately with a protective layer after installation and seaming has been completed and satisfactorily tested. The protection layer may be the overlying geotextile and seepage collection layer (layer B in Figure 6.37a), or a purpose-made cement-stabilized sand layer if the geomembrane will not be covered by tailings or other waste for a period of a year or more. The presence of the geomembrane and its protective layer also acts to protect the underlying compacted clay C layers from desiccation and physical damage.

6.14.2 *Geosynthetic clay liners*

Geosynthetic clay liners or geocomposite liners (GCLs) consist of a sandwich of dry powdered clay between two layers of geotextile. The clay and the geotextile covering are held together by stitching, needle punching or adhesive. The clay is usually bentonite as this has a very low permeability and a high potential to swell when hydrated. The exchangeable cation of the bentonite is usually sodium as sodium bentonite has a lower permeability and a larger swell potential than calcium bentonite, the other commonly occurring form of bentonite.

As the GCL is produced in a roll, it has to be joined or jointed when installed. The joints are made by overlapping adjacent sheets by a minimum distance of 200 mm and covering the lap area of the under-sheet with a thin layer of dry bentonite powder. This layer must be continuous, but sufficiently thin so that the clay, once hydrated will not be extruded from the joint by pressure from overlying deposition.

Because of the potential for hydration of the bentonite, installation of GCL's can only take place in stable, dry weather when there is no danger of the GCL being prematurely wetted by rain or by absorbing water from an earlier rainfall. Once installed, the GCL must immediately be covered with its protective layer, so that when it next rains, hydration can take place under an overburden stress, so that swell of the clay is inhibited. Unchecked swelling may cause disruption or disintegration of the GCL sandwich.

Two additional problems that may affect GCLs are that the shear strength of hydrated bentonite is low, and hence situations in which shearing parallel to the GCL will occur should be avoided or carefully checked for shear stability. (See section 4.8.2 for the results of typical shear tests parallel to the plane of a stitch-bonded GCL.) Also, if the seepage contains a high concentration of calcium ions, these may exchange with the sodium ions in the bentonite, converting it to the more pervious calcium bentonite.

Plate 6.5 Shrinkage and cracking in bentonite layer of GCL after hydration under 100 kPa and shearing under 400 kPa normal stress, followed by desiccation. Scale to left is in mm.

This may cause the permeability of the GCL to increase beyond the figure of 15 mm/y given in Table 6.4.

Recently published research (Benson, et al, 2007) has shown that gross increases in the permeability of GCLs can arise by the replacement of the original sodium cations in the bentonite by calcium and magnesium combined with the effects of desiccation of the clay during dry seasons. In one study, the measured permeability of a sodium bentonite GCL when installed, ranged from 1 to 2.5 mm/y. After 12–18 months service, this had increased to 440 to nearly 30 000 mm/y. In other words the GCL would pass all the infiltration that penetrated to it. In terms of percolation rate measured in field lysimeters, the initial percolation rate was below 13 mm/y and this had increased to 200 to 450 mm/y after 12–18 months service. The conclusion from this study was that GCLs should be used only where means exists to ensure that cation exchange and/or dehydration cannot occur. Periodic desiccation is almost impossible to avoid when a GCL is used as an impervious layer in a cap or cover.

An alternative to a conventional GCL suggested in this study is to use a GCL with a "geofilm" or thin geomembrane bonded to one face of the sandwich.

Work by Han et al. (2008) has also shown that landfill leachates can increase the permeability of GCLs by factors of more than 200. These workers, whose tests were

Table 6.5 Typical physical properties of geosynthetic filtration, drainage or reinforcement elements

Type of element	*Purpose*	*Mass per m² (thickness under 0.5 kPa tension) g/m² (mm)*	*Throughflow rate under 100 mm water head l/m²s (mm/s)*	*Tensile strength* kN/m*		*Elongation at failure %*		*CBR penetration load and elongation*		*Pore size μm*
				Warp	*Weft*	*Warp*	*Weft*	*Load kN*	*Elongation %*	
Non-woven polyester fibre	Filtration, separation of materials, protection	110 (0.7)	220	8	10	22	30	1.1	30	250
		210 (1.1)	150	8	15	70	55	1.8	25	240
		400 (2.1)	130	15	12	50	50	2.5	20	190
		750 (4.4)	110	50	40	50	50	6.5	20	100
Woven polypropylene twirled tapes	Filtration, separation of materials, protection	110 (0.65)	40	35	15	20	15	2.3	15	700
		205 (1.1)	65	40	35	20	10	5.6	18	500
		270 (1.1)	80	60	50	20	10	6.2	18	400
Woven polyester yarn geogrid	Reinforcement of elements subject to tension, e.g. geo-membranes on a sloping base	450	N/A**	110	30	18	80	N/A	N/A	N/A
Netting pipe	Collection pipes in filter drains	500(5)	see Section 6.14.3 for information on crushing resistance							10 mm
				Compressive stress without flattening						
Drainage grids	Pervious collection layers in filter drains	500 (7.5)	70	200 kPa		N/A	N/A	N/A	N/A	750

* Note that strengths are "index values" measured in short term (10–20s duration) tests
** Not applicable

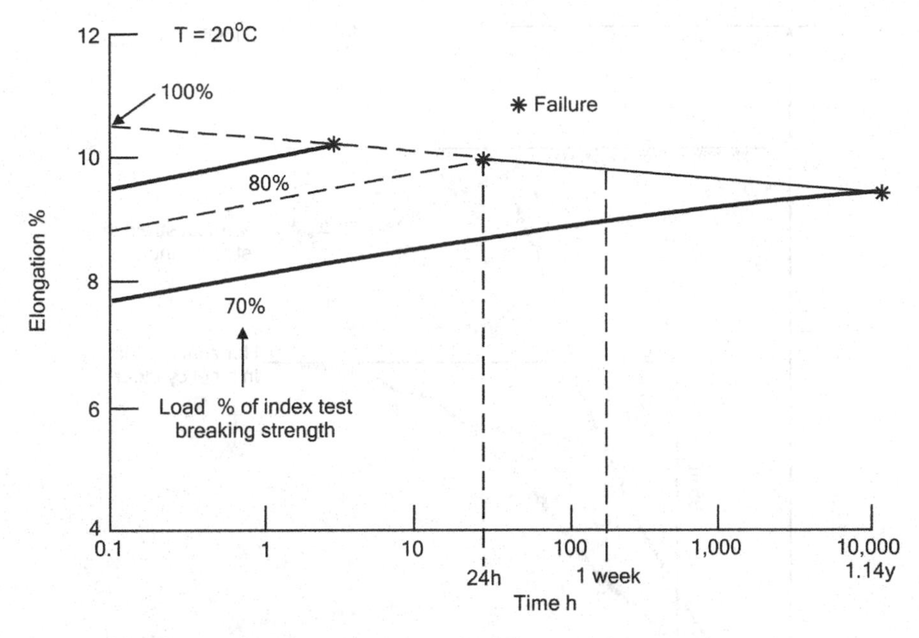

Figure 6.39a: Sustained-load creep tests causing failure in polyester yarn.

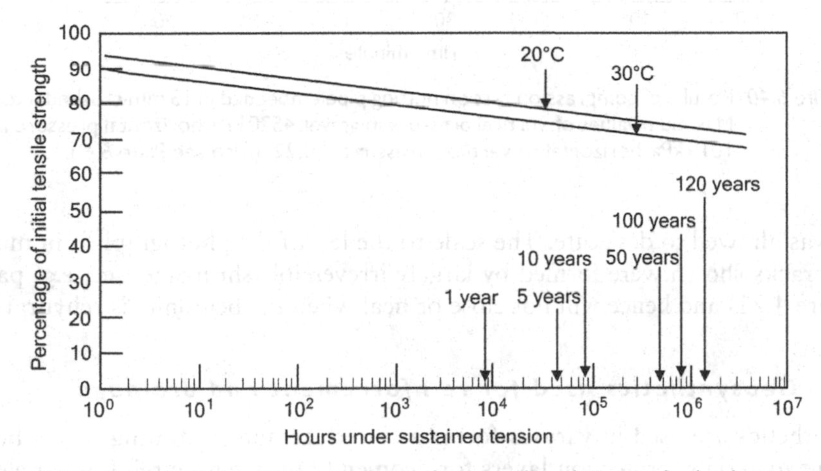

Figure 6.39b: Typical relationship between tensile strength and time under load at two temperatures for woven polyester yarn geogrids.

of a short term nature, also warn that the adverse effect of permeation by leachate could increase over the design life of a GCL.

As an indication of the effect of dehydration on a GCL that has been hydrated under pressure, Plate 6.5 shows the bentonite layer from a specimen of GCL that was hydrated under a pressure of 100kPa, then consolidated to 400 kPa and sheared drained. After the test, the upper geofabric layer was peeled off, and the bentonite

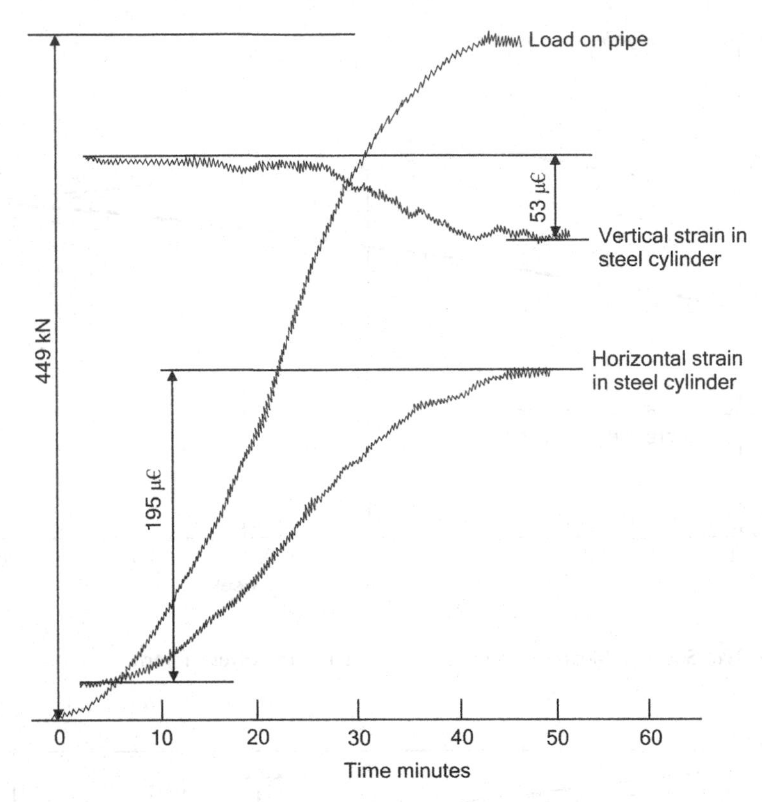

Figure 6.40 Results of compression test on netting pipe embedded in 13 mm crushed filter gravel. Maximum values of: vertical pressure in gravel: 4520 kPa, horizontal pressure in gravel: 1010 kPa, horizontal to vertical stress ratio: 0.22. (Also see Plate 6.6.).

layer was allowed to desiccate. The scale to the left of the photograph is in mm. The 1 mm cracks shown were formed by largely irreversible shrinkage (see, e.g. path BC in Figure 4.21) and hence will not close or heal when the bentonite is rehydrated.

6.14.3 *Geosynthetics used for reinforcement and drainage*

Geosynthetics are used in various forms as interface and separating layers between different materials, protection layers for geomembranes, tensile reinforcing elements where geomembranes are laid on slopes and seepage collection elements in filter drains. Table 6.5 gives typical physical properties for some of these materials and elements.

It must be emphasized that the strength properties listed in Table 6.5 were measured in short term standardized "index" tests where the time to failure is typically 10 to 20 seconds. In practical use, these elements and materials are required to sustain loads indefinitely. Because all polymeric materials creep under sustained load, and because it is probable that the material will sustain some damage during installation in the field, actual design loads must be substantially less than the "index" strengths listed in most suppliers' catalogues. Figure 6.39a (Jewell, 1996) shows the effect of

Plate 6.6 Compression test on HDPE netting pipe embedded in 13 mm gravel.

time of loading on a polyester yarn similar to that commonly used in the weaving of geogrids for reinforcing against sustained tensile forces. The figure shows that the 24 hour strength of the yarn was only 77% of its 0.1 hour strength. The strength had declined to 70% of its short term strength after a year (a reduction factor of 1.4). Jewell recommended that a further reduction factor of 1.5 be applied to allow for accidental damage during installation and other strength-reducing factors arising from burial in soil. This would reduce the design strength to 47% of the nominal strength, and after applying an additional load factor of 1.5, to 31% of the nominal value. Hence the long term, load factored strength of a nominal 110 kN geogrid would be only 34 kN in this case. Figure 6.39b shows a more generalized relationship between time of loading and strength reduction for polyester geogrids which includes the effect of temperature in service.

The consensus among authorities at present is that the design tensile strength T (design) for any polymeric material should be calculated as:

$$T\,(\text{design}) = T(\text{index})/F_c \times F_i \times F_e \times F_{dl} \qquad (6.23a)$$

where

T (index) is the short term index strength
F_c is the allowance for creep (1.5–2)
F_i is the allowance for damage during installation (1.5–2)
F_e is the allowance for environmental factors (e.g. UV or chemical attack (1.5–2))
F_{dl} is the allowance for design life (1.1 for 60 years – 1.2 for 120 years)
The minimum reduction factor is thus 3.7 and the maximum is 9.6.

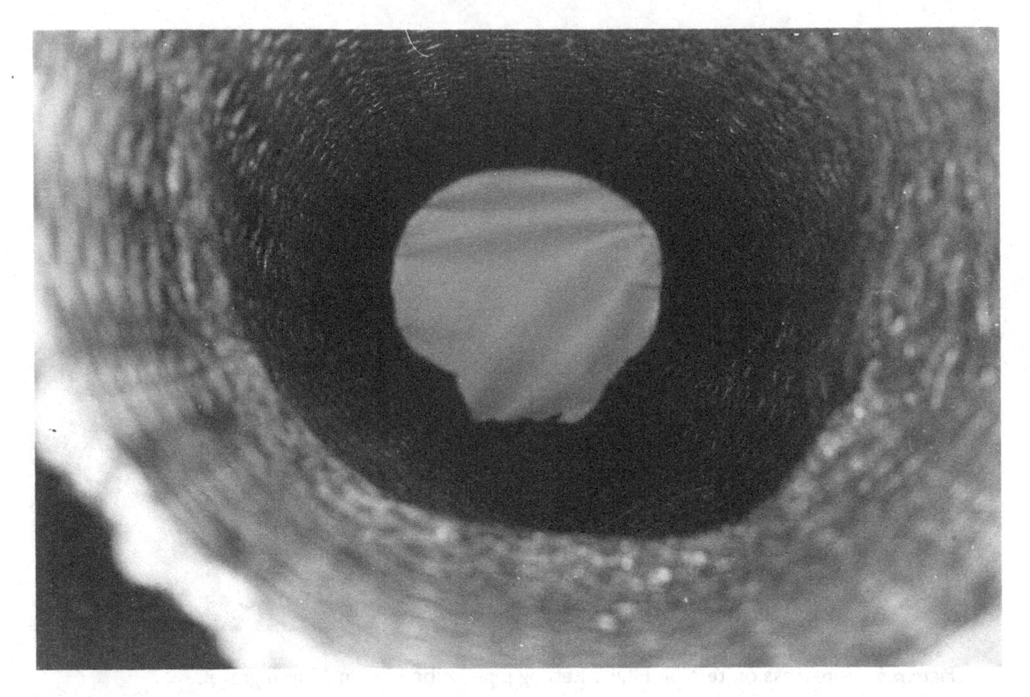

Plate 6.7 Shape of pipe when stresses in gravel were: $\sigma_v = 4250$ kPa, $\sigma_h = 1010$ kPa.

The netting pipes also require some elaboration. The pipes are extruded in various diameters and consist of HDPE mesh, usually with a thickness of 4 to 5 mm. The pipes are not rigid in themselves and can easily be flattened with the hand. However, when buried in a gravel-filled seepage collection trench, they allow a compression cylinder to form in the granular material surrounding them. This cylinder, operating as a granular compression ring has a surprisingly high strength, and is virtually uncrushable, as its resistance arises from the stressed gravel particles that form a compression-resistant double voussoir arch.

Plate 6.6 shows a section of netting pipe set up in a 350 mm diameter steel cylinder. The pipe (projecting from the right side of the upright steel cylinder) was embedded either in 13 mm crushed stone or in a fine sand. Strain gauges were mounted on the steel cylinder at the level of the pipe's centre-line so that the vertical and horizontal stresses in the stone surrounding the pipe could be measured. Deformations of the pipe under load could be observed through the open end of the pipe.

The granular fill was loaded by means of a plunger covering the full area of the retaining cylinder at its surface and simultaneous recordings were made of the applied load and the vertical and horizontal strains in the pipe. These could be used to estimate the vertical and horizontal stresses in the crushed stone, and hence on the granular compression ring around the pipe. Figure 6.40 shows the recordings of load and strain increases with time made in a test on a 75 mm diameter netting pipe embedded in crushed stone. Whether the pipes were embedded in stone or sand made very little difference, none of the pipes collapsed, even though the loads were taken to excessive levels. Plate 6.7 shows the shape of the pipe for which the test results appear in Figure 6.40, photographed as it was under the maximum load, of nearly 450 kN, corresponding to stresses in the gravel of $\sigma_v = 4520$ kPa, $\sigma_h = 1010$ kPa.

References

Al-Dhahir, Z.A.R., & Tan, S.B.: A note on the One-Dimensional Constant Head Permeability Test. *Geotechnique 8(4)* (1968).

Benson, C.H., Thorstad, P.A., Jo, H.-Y., & Rock, S.A.: Hydraulic Performance of Geosynthetic Clay Liners in a Landfill Final Cover. *J. Geotech. Geoviron. Eng, ASCE 133(7)* (2007), pp. 814–827.

Bishop, A.W. & Henkel, D.J.: *The Measurement of Soil Properties in the Triaxial Test*. London: Edward Arnold Publishers, 1962.

Casagrande, A.: Seepage Through Dams. *Contributions to Soil Mechanics, 1925–1940*. Boston, U.S.A.: Boston Society of Civil Engineers, 1937, pp. 295–336.

Chen, H.W., & Yamamoto, L.D.: Permeability Tests for Hazardous Waste Management with Clay Liners. *Geotechnical and Geohydrological Aspects of Waste Management*. USA.: Lewis Publishers, 1987, pp. 229–243.

Costa Filho, L.M., & Vargas Jr. E.: Topic 2.3. Hydraulic Properties. *Peculiarities of Geotechnical Behaviour of Tropical Lateritic and Saprolitic Soils*, Progress Report (1982–1985). Brazilian Society of Soil Mechanics, 1985, pp. 67–84.

Daniel, D.E.: Earthen Liners for Land Disposal Facilities. In: *Proc. ASCE Specialty Conf. Geotech. Practice Waste Disposal*. New York: ASCE, 1987, pp. 21–39.

Daniel, D.E., & Koerner, R.M.: *Waste Containment Facilities*, 2nd edition. Virginia, U.S.A.: ASCE Press, (ISBN 13: 978-0-7844-0859-9, ISBN 10: 0-7844-0859-9), 2007.

Day, S.R., & Daniel, D.E.: Hydraulic Conductivity of Two Prototype Clay Liners. *ASCE J. Geotech. Eng 111(8)* (1985), pp. 957–970.

Elsbury, B.R., Daniel, D.E., Straders, G.A., & Anderson, D.C.: Lessons Learned from Compacted Earth Liners. *J. Geotech. Eng.ASCE 16(11)* (1990), pp. 1641–1659.

Fourie, A.B., Kuchena, S.M., & Blight, G.E.: Effect of Biological Clogging on the Filtration Capacity of Geotextiles. In: *5th Int. Conf. Geotextiles, Geomembr. Relat. Products*, Singapore. ISBN 961-00-5822-Vol. 5, 1994, pp. 721–724.

Garga, V.K.: Effect of Sample Size on Consolidation of Fissured Clay. *Can. Geotech. J. 25(1)* (1988), pp. 76–84.

Gibson, R.E.: The Progress of Consolidation in a Clay Layer Increasing in Thickness with Time. *Geotechnique 8* (1958), pp. 171–182.

Han, Y.-S., Lee, J.-Y., Miller, C.J., & Franklin, L.: Characterization of Humic Substances in Landfill Leachate and Impact on the Hydraulic Conductivity of Geosynthetic Clay liners. *Waste Manag. Res.*, (In press), (2009).

Hvorslev, M.J.: *Time-lag and Soil Permeability in Groundwater Observations*. Bulletin 36. Vicksburg, USA: USWES, 1951.

Janbu, N., Bjerrum, L., & Kjaernsli, B.: *Soil Mechanics Applied to Some Engineering Problems*. Publication 16. Olso, Norway: Norwegian Geotechnical Institute, 1956.

Jewell, R.A.: *Soil Reinforcement with Geotextiles*, CIRIA. Special Publication 123. London, U.K.: CIRIA, 1996.

Matsuo, S., Hanmachi, S., & Akai, K.: A Field Determination of Permeability. In: *3rd Int. Conf. Soil Mech. Found. Eng, Zurich, Switzerland*. Vol. 1, 1953, pp. 268–271.

Metcalfe, R.C., Holtz, R.D., & Allen, T.M.: Field Investigations to Evaluate the Long-term Separation and Drainage Performance of Geotextile Separators. In: *Geosynthetics "95*, Nashville, Tennessee, U.S.A., 1995, pp. 561–567.

Olson, R.E. & Daniel, D.E.: Measurement of the Hydraulic Conductivity of Fine-Grained Soils. *Permeability and Groundwater Contaminant Transport, STP 746, ASTM* (1981), pp. 18–64.

Rowe, R.K. & VanGulck, J.F.: Filtering and Drainage of Contaminated Water. In: A. Fourie (ed): *Filters and Drainage in Geotechnical and Environmental Engineering*. Johannesburg, South Africa: University of Witwatersrand Press, 2004, pp. 1–64.

Schiffman, R.L., Vick, S.G., & Gibson, R.E.: Behaviour and Properties of Hydraulic Fills. In: D.J.A. van Zyl, S.G. Vick, (eds): *Hydraulic Fill Structures A.S.C.E. Geotechnical Special Publication No. 21*, New York, U.S.A., 1988, pp. 166–202.

South African National Building Research Institute: *An Investigation into the Stability of Slimes Dams with Particular Reference to the Nature of the Material of Their Construction and the Nature of their Foundation*. Pretoria, South Africa: The Institute, 1959.

Tan, S.B.: *Consolidation of Soft Clays with Special Reference to Sand Drains*. PhD Thesis, University of London, 1968.

Terzaghi, K.: *Erdbaumechanik auf bodenphysikalischer grundlage*. Vienna, Austria: Deuticke, 1925.

United States Environmental Protection Agency: *Requirements for Hazardous Waste Landfill Design, Construction, and Closure*. Seminar Publication, USEPA 625 4-89/022, 1989, pp. 96–98.

Chapter 7

The mechanics of compaction

7.1 The compaction process

Compaction is a process whereby a granular or particulate material is densified by expending energy on it. *Traffic compaction* results from the work done as vehicles travel over the surface, not with the specific purpose of compacting it, but in order to discharge some other function, e.g. to dump more material. *Roller compaction* arises from the deliberate trafficking of a surface by a vehicle specifically designed to expend energy in compaction, i.e. a compactor or roller. The energy input may arise simply from the weight of the roller moving down as it compresses the fill, or there may be an additional energy input, e.g. resulting from the falling of the roller as with an impact roller or the gyration of an eccentric weight as with a vibrating roller. Figure 7.1 illustrates these three ways of expending compactive energy. In addition, rollers may have a variety of surfaces that contact the soil in different ways. Smooth wheeled, footed or pneumatic tyred rollers as well as grid-rollers are all used.

Compaction occurs because the solid particles of fill are forced closer together, thus expelling air and reducing the void volume. It is not common for compaction to result in the expulsion of water from a fill. Some breakage of particles may occur during compaction and the resulting fines then partly fill the reduced void space.

For a given energy input, the density achieved depends on the water content of the fill. In general, an optimum water content will exist at which a given energy input will result in a maximum dry density. If the energy input is changed, both the optimum water content and the maximum dry density will change. As shown by Figure 7.2a, as the energy input is increased, the maximum dry density increases and the optimum water content at which it is achieved decreases.

The largest effect on dry density occurs during the first few passes of a roller. Thereafter as indicated by Figure 7.2b, the effect reduces exponentially. It is seldom worthwhile applying more than five or six passes of a roller. If the required effect is not obtained with this number of passes, either the water content of the fill is too high or too low or the energy input per roller pass is too low. (Also see Section 7.10).

Compaction results from the imposition by the roller of compressive and shear stresses on the fill material. These stresses are largest immediately under the roller and reduce or disperse with increasing depth below the surface. The type of stress dispersion that occurs is illustrated by Figure 7.2c. Rollers will usually not produce any appreciable effect at a depth greater than about 1 m below the surface. Because of this, if a reasonably uniform state of compaction is required throughout a fill,

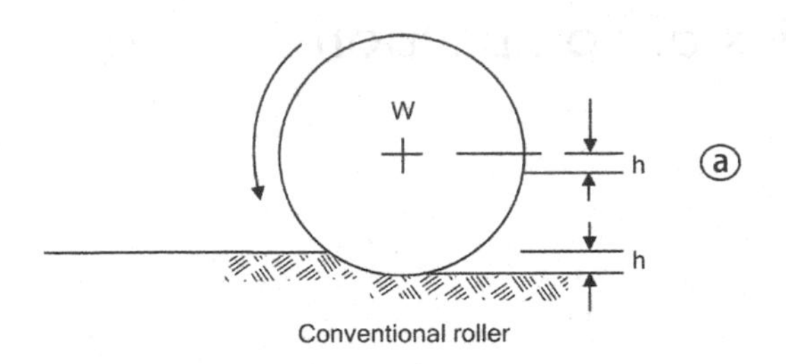

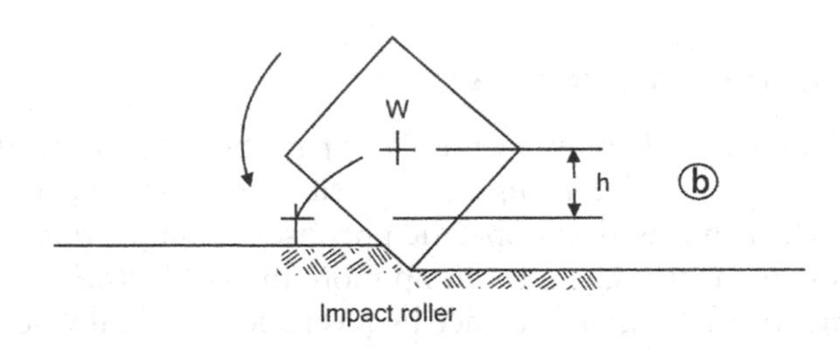

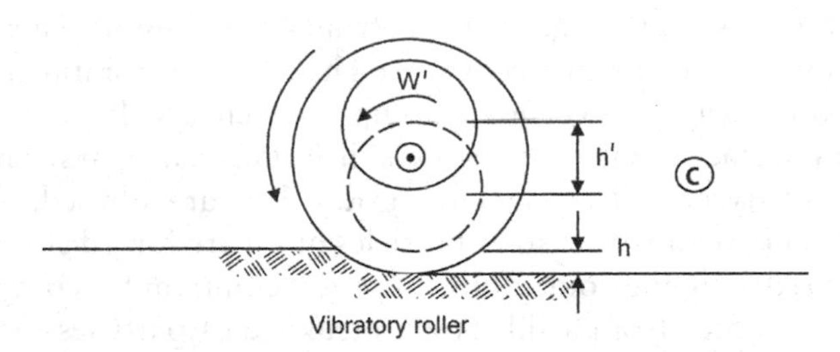

Figure 7.1 Three ways of expending compactive energy: a&b: Weight W of roller moves down distance h expending energy Wh on soil. c: Additional energy $W^{l}h^{l}$ is expended by eccentric weight W^{l}.

the material must be spread in layers not exceeding 300 mm in loose thickness, which will compact to a thickness of about 200 mm.

Smooth-wheeled and pneumatic-tyres are not used to any extent for the earthworks of tailings storages, being more suited to the extra-heavy compaction required by road layer works. Impact rollers are usually used to densify deep loose sandy soils, and might be required to densify the foundation layer for a tailings storage embankment, if this consists of a loose sandy or silty stratum. Grid rollers are useful if the soil contains aggregations of particles such as often occur in a partly-weathered residual soil and in coal waste, and which require to be broken down by compaction. Footed rollers with feet long enough to completely penetrate the loose layer being compacted are favoured for compacting relatively thin impervious clay layers for the reasons shown in Figure 7.3 (USEPA, 1989).

Because smooth wheeled rollers give a very smooth surface finish, they are usually used to finish off the top surface of a clay liner that will be covered by a geomembrane.

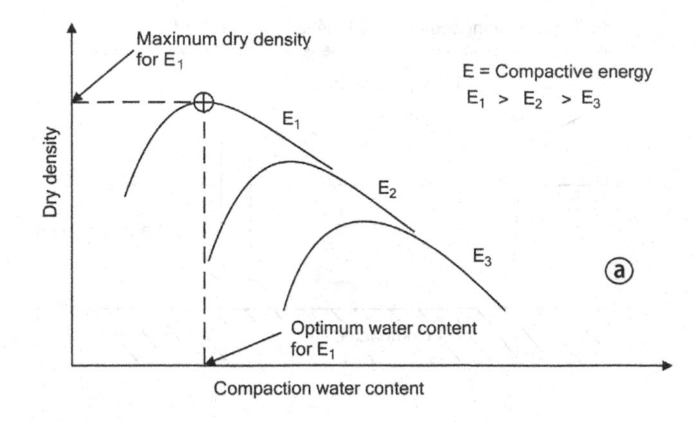

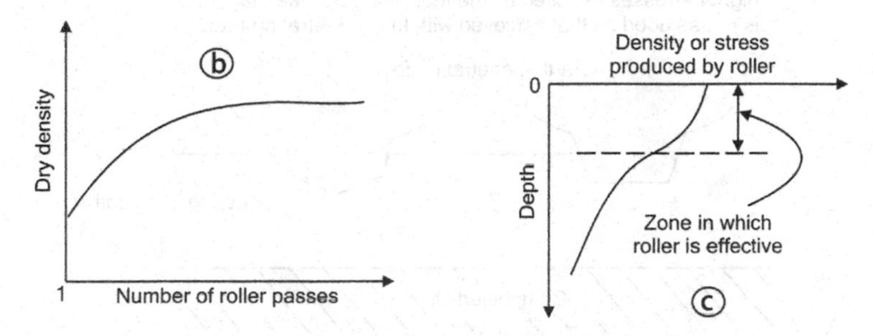

Figure 7.2 Relationship between: a: dry density, compaction water content and compactive energy, b: roller passes and dry density, c: depth and dry density.

The smooth surface promotes the close contact between soil and geomembrane that is essential for the functioning of a composite soil/geomembrane liner.

Laboratory compaction tests can be used to establish approximate values for the optimum water content and the maximum dry density for "light" or "heavy" compaction energies. Laboratory values will generally not agree with the fill characteristics for roller compaction as the method of applying the energy (usually by repeated blows of a drop hammer applied to thin layers of soil confined in a small rigid-walled mould) differs completely from that used in the field. Nevertheless, because design options usually have to be assessed by means of laboratory tests, it is also important to study laboratory compaction.

7.2 Uses of compaction in mine waste engineering

The main uses of compaction in mine waste engineering are as follows:

- The preliminary earthworks for tailings storages, i.e. toe walls and erosion catchment paddock walls for ring dyke storages and either the toe wall for a valley storage in which the retaining wall will be constructed of hydraulic fill tailings, or the complete retaining embankment if it is to be entirely constructed of earth.

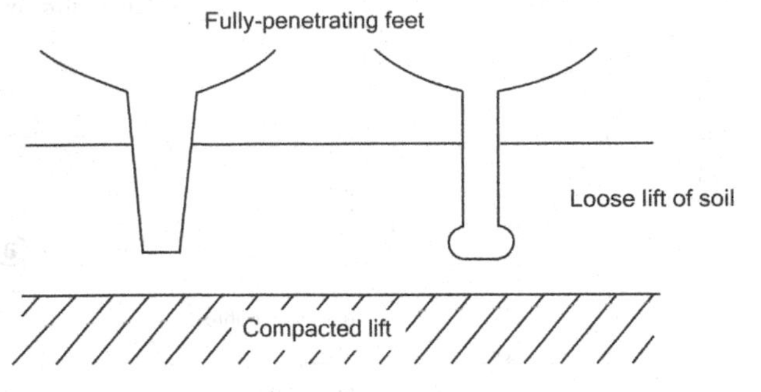

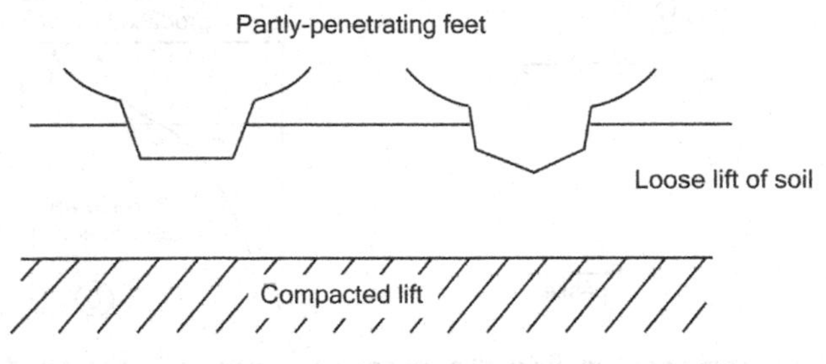

Figure 7.3 Two kinds of footed rollers on compaction equipment.

Return water reservoirs or dams are also usually constructed of compacted natural soil.

- Compaction is used to reduce permeability to air in the case of coal waste storages. Reducing permeability to air entry, and thus excluding oxygen, lessens the danger of spontaneous combustion and hence of the coal waste catching fire.
- Construction of impervious clay or composite clay/geomembrane liners. Special considerations apply to compaction of clay soils to construct relatively thin impervious liners that may cover extensive areas. These will be described in Sections 7.4 and 7.8.

7.3 The mechanisms of compaction

Essentially, compaction is a process whereby air is expelled from the pores of the soil, reducing the air-filled void volume and, in the process, forcing the solid particles closer together. It is the process of forcing more solid particles into a given volume that increases the dry density and hence increases the soil strength, reduces the compressibility and also reduces the permeability to air and water flow. However, to produce these effects, both shear and compressive stresses have to be applied to the soil, and that is what requires the expenditure of energy.

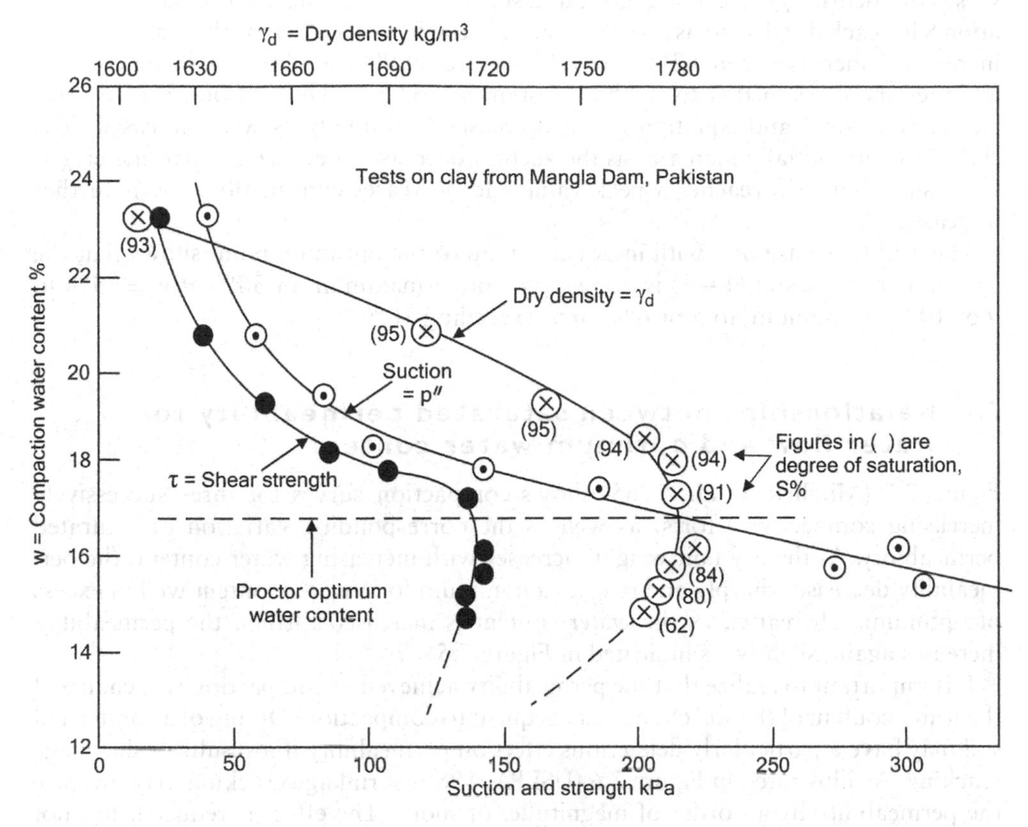

Figure 7.4 **Relationships between compaction water content w, dry density γ_d, suction p$$ and strength τ for compacted clay using standard Proctor compactive effort.**

At low water contents the resistance of the soil to compression and deformation is relatively high. The air-filled void spaces are interconnected and air can freely leave the soil. With a given expenditure of energy (or "compactive effort") only a relatively low compacted dry density can be achieved. As the water content is increased, the resistance of the soil to compaction decreases and higher dry densities result, but as the air-filled voids decrease, the resistance to the escape of air increases, until the air-filled voids become occluded, or sealed off from the atmosphere by the surrounding water-filled pore space. This point corresponds approximately with the maximum dry density and optimum water content for the particular compactive effort being used. From the optimum point onwards, as water is added to the soil, it occupies increasing space in the voids, while the air content remains almost constant. The result is that the dry density of the soil decreases progressively with increasing water content.

A set of relationships between water content, suction, strength and dry density is illustrated in Figure 7.4. It must be noted at the outset that the relationships apply after compaction. The relationships shown in Figure 7.4 apply to laboratory compaction, using standard Proctor compactive effort, of a clay derived from the weathering of shales at the site of the Mangla dam in Pakistan. They show compaction water content versus dry density γ_d, shear strength τ and suction $p^{||} = (u_a - u_w)$. The degree of saturation S for each dry density is given in brackets. It will be seen that as the water content increases, S increases from 62% at w = 15% to about 90% at optimum water content, and then increases further to reach a constant 94 to 95%. The suction $p^{||} = (u_a - u_w)$ (see section 4.4.3 and equation (4.9)a decreases continually as w is increased. The shear strength initially increases as the suction decreases, because the parameter χ is increasing, but then reaches a peak value and decreases continually as w is further increased.

The rapid increase of S with increasing w up to the optimum point shows that the air content of the soil (1 – S) is decreasing from a maximum of 38% at w = 15% to 9 or 10% at optimum, to 5 or 6% for w exceeding 18%.

7.4 Relationships between saturated permeability to water flow and optimum water content

Figure 7.5 (Mitchell, et al., 1965) shows compaction curves for three successively increasing compactive efforts, as well as the corresponding variation of saturated permeability. As the dry unit weight increases with increasing water content, the permeability decreases sharply and reaches a minimum for a water content well in excess of optimum. Thereafter, as the water content is increased further, the permeability increases again, slightly, as indicated in Figure 7.5.

It is important to realize that the permeability achieved at compaction will change if the water content of the soil changes subsequent to compaction. Drying of a compacted soil may have a particularly deleterious effect on permeability if it results in shrinkage cracking. As illustrated in Figure 7.6 (USEPA, 1989) shrinkage cracking may increase the permeability by an order of magnitude, or more. The effect is reduced, but not eliminated, if the cracks are forced to close by applied stress.

7.5 Laboratory compaction

Dry density versus compaction water content curves, such as those shown in Figures 7.2a, 7.4 and 7.5 are usually the result of compaction tests carried out in a laboratory. The equipment for laboratory compaction testing consists of a series of compaction moulds, (a typical one of which is illustrated in Figure 7.7) and a compaction hammer with a standard mass and dimensions that is designed to fall a standard distance in free fall. Thus the energy applied per hammer blow is the weight of the hammer multiplied by the height of fall. For the standard Proctor compaction hammer shown in Figure 7.7, the energy per blow is $25\,N \times 0.3\,m = 7.5\,Nm = 7.5\,J$, and the energy per m^3 of soil compacted in 3 layers with 27 blows per layer (i.e. standard Proctor compaction) is $7.5\,J \times 27$ blows $\times$ 3 layers $\times$ 1 000 = 607.5 kJ/m^3. Although the hand-held compaction hammer is a standard, most laboratories use mechanized hammers or compaction machines that distribute the selected number of hammer blows uniformly over the plan surface of the soil layer in the mould.

To establish a dry density versus water content curve, a number of specimens of loose soil are prepared at a series of increasing water contents. Each specimen is then compacted into a mould with the specified input energy, and weighed, after trimming

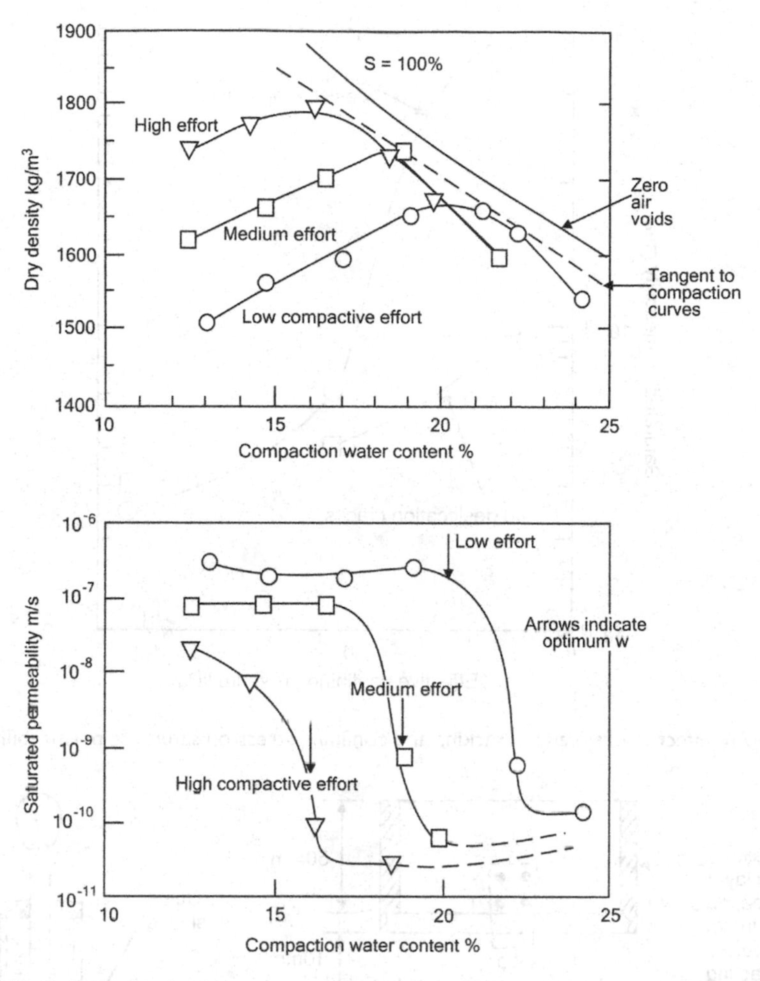

Figure 7.5 **Relationship between compaction water content, dry density and saturated permeability to water flow for low, medium and high compactive efforts.**

the soil surface flush with the top of the mould. The compacted soil is then removed from the mould and specimens of soil are taken to measure the water content.

7.6 Precautions to be taken with laboratory compaction

Bulk samples of soil taken for compaction testing are not usually sealed against loss of moisture, and often consist of an assemblage of clods, which, if allowed to dry, are extremely difficult to break down to achieve a homogeneous material for testing. Drying of a soil sample may have at least three consequences:

7.6.1 *Moisture mixed into the soil is not uniformly distributed:*

Most guides to compaction specify that specimens prepared for compaction should be stored in sealed containers overnight to allow the moisture content to equalize

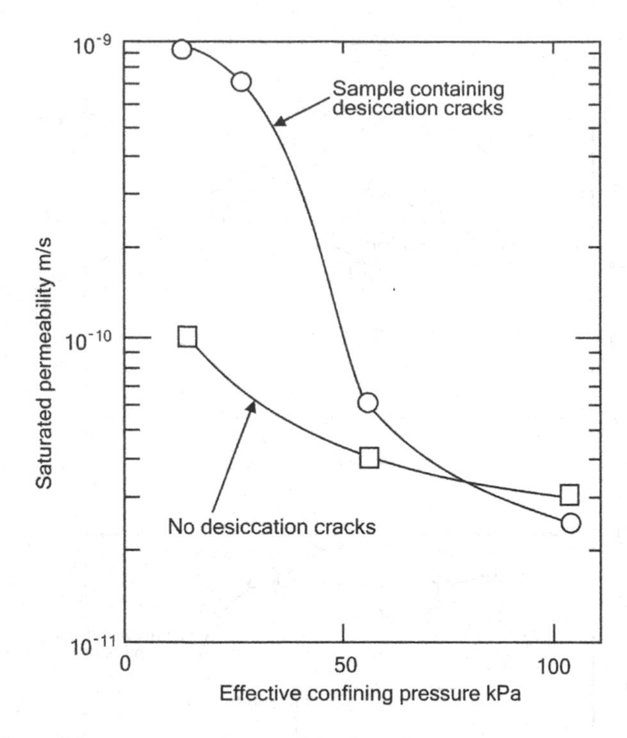

Figure 7.6 Effect of desiccation cracking and confining stress on saturated permeability.

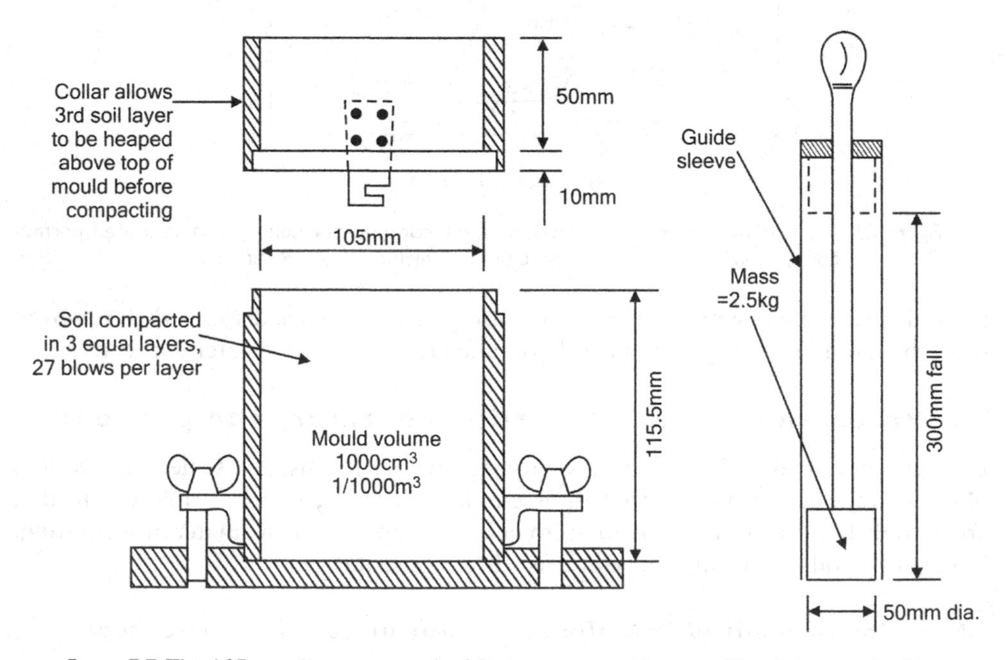

Figure 7.7 The 105 mm diameter standard Proctor compaction mould and the standard Proctor compaction hammer.

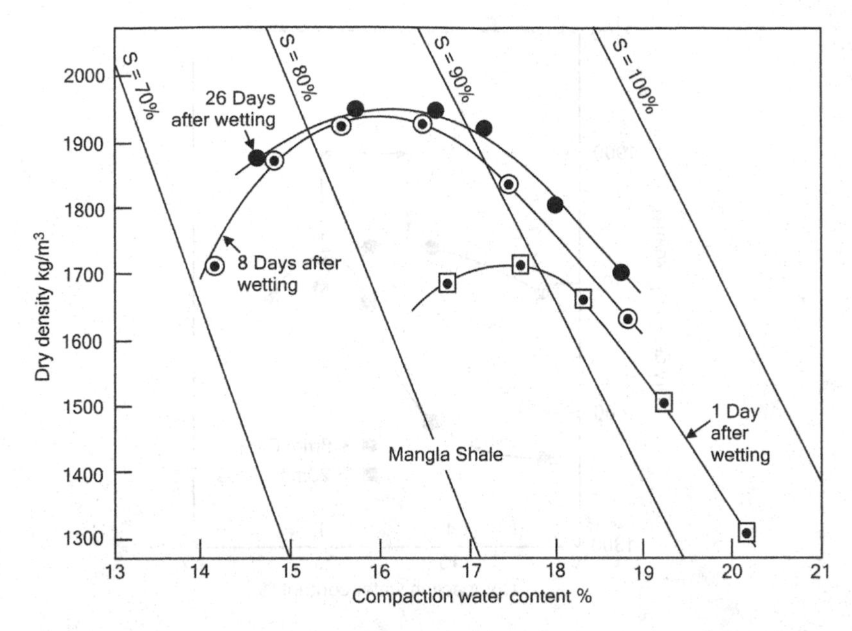

Figure 7.8 Time taken to achieve a stable, repeatable compaction curve after wetting an air-dry clayey soil.

throughout the specimen before compacting it. However, as shown by Figure 7.8, this moisture homogenization process takes much longer than a day, and at least a week should be allowed. The water not only has to be uniformly spread over all particles and soil aggregations, but also has to have time to diffuse into the soil aggregations.

7.6.2 *Soil aggregations or clods are not broken down*

Hard soil aggregations or clods affect dry density in the same way as stones do. The larger and more clods there are in the compaction specimen, the more the true compaction curve will be masked. Figure 7.9 (USEPA 1989) shows an example where larger (>20 mm) clods result in a falsely low density for compaction dry of optimum and require a larger optimum water content. A soil with smaller clods (<5 mm) shows a higher dry density over the whole water content range, and a 4% lower optimum water content. If the clods had been broken down completely, an even lower optimum water content would probably have resulted.

7.6.3 *Other treatments that will affect the laboratory compaction curve*

Figure 7.10 (Gidigasu and Dogbey, 1980) shows the compaction curve for a soil, taken from site at a water content of 8% and prepared for compaction by adding water,

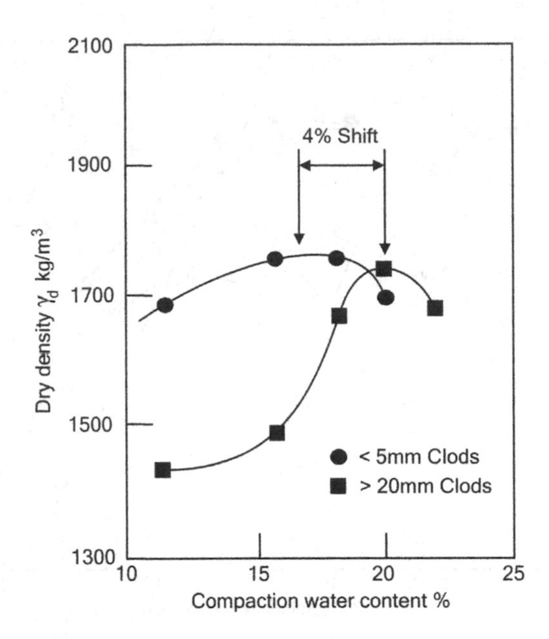

Figure 7.9 Effects of clod size on compaction curves.

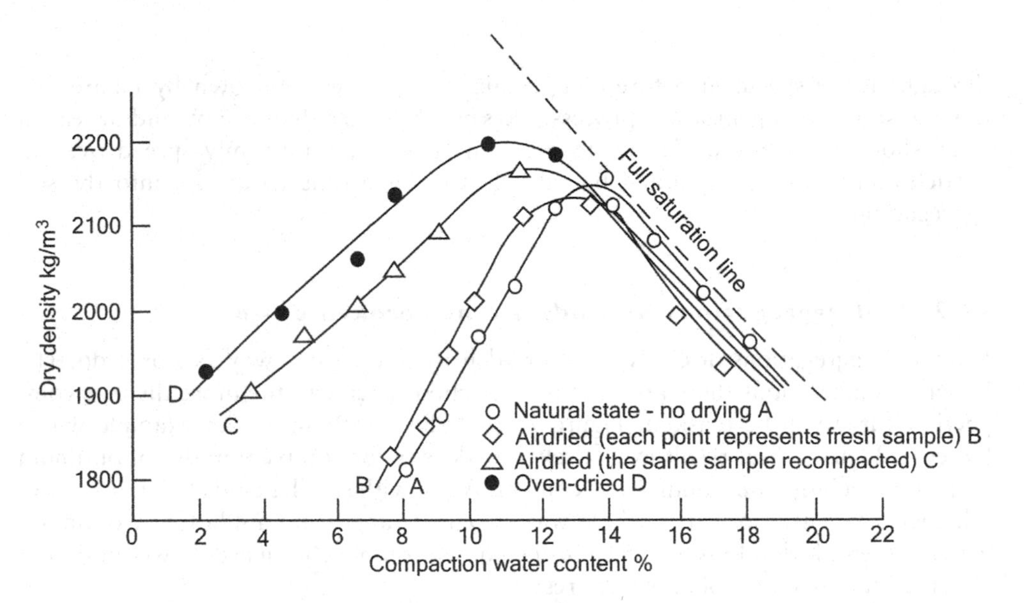

Figure 7.10 Effects of various treatments of a soil, before compaction, on the compaction curve.

allowing 7 days for dissemination of the added water throughout each sub-sample (the "natural state", A).

The same soil was air dried, rewetted and compacted using the same treatment as the "natural" sample, and the resultant compaction curve was very similar (B).

When, however, sub-samples that had received treatment (B) were re-compacted, a very different compaction curve resulted. As the soil contained some laterite, which pulverizes under compaction, it is likely that the change in compaction curve resulted from particle break-down caused by the first compaction. Finally after oven-drying and then rewetting and compacting, a fourth compaction curve resulted. This was probably because the drying temperature of 105°C had modified the clay minerals or metallic hydroxides contained by the soil.

It is therefore essential to avoid drying of a soil destined for compaction testing between taking the sample in the borrow pit and conditioning it in the laboratory for compaction testing. If the surface of the soil to be sampled is desiccated, the dried surface layer should be discarded and excluded from the sample. Samples should never be recompacted to produce a compaction curve. Use a fresh sample for each point on the curve.

7.7 Compaction in the field

Compaction is undertaken in the field using the best technology available, but by force of circumstances, not always the best possible technology. The compacted product requires compromises between energy and cost expended, and the value of the result obtained. Engineering design must recognize the reality of what can be achieved in the field, as compared with what can be achieved in the laboratory.

The in situ soil is usually variable in composition, particularly if it is a residual or alluvial soil, as the degree of weathering and yearly layers of deposition will usually be variable. Hence the selection of representative samples for testing can be a major problem. For the same reason good control of quality in compacted fills may be extremely difficult to achieve.

The compaction characteristics of soils may be very dependent on the method of applying the compactive energy. In particular, laboratory compaction curves may bear little resemblance to the compaction curve achievable in the field. This phenomenon is illustrated by Figure 7.11 which compares roller and laboratory compaction curves for a weathered granite pegmatite (Blight 1962). With soil A, it did not prove possible to achieve the required 100% of laboratory maximum dry density, until it was discovered that the optimum water content for laboratory compaction was 3% wet of that for roller compaction. At that time (1957), the link between permeability and compaction water content was not known. Nevertheless, an area of the fill compacted at roller optimum water content +2% was tested for permeability by excavating a shallow (150 mm deep) pond, filling it with water, covered with a film of engine oil to inhibit evaporation, and observing the rate of seepage. This showed that the permeability was acceptably low, and the compaction requirement was altered on site to "100% USBR laboratory maximum dry density at roller optimum water content +2%".

In the case of soil B, roller compaction was not able to achieve laboratory maximum dry density, although the roller and laboratory optimum water contents were almost the same. However, the roller maximum dry density was 98% of the laboratory maximum, and this was deemed sufficient to achieve the required strength. The in situ strength and permeability were checked and found to be satisfactory.

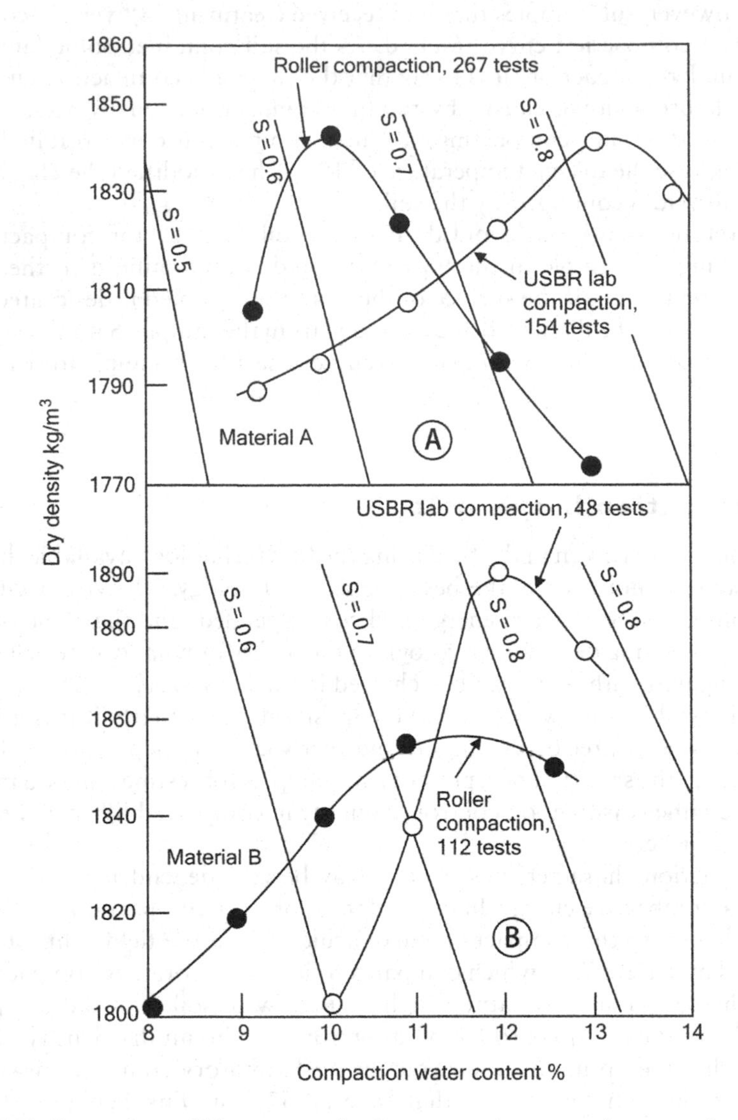

Figure 7.11 Comparison of laboratory and roller compaction curves for two soils.

7.8 Designing a compacted clay layer for permeability

The method for designing a clay layer so that it will have less than a specified maximum permeability is based on a series of tests like that illustrated by Figure 7.5. The procedure is as follows (USEPA, 1989):

Suppose that the specified maximum permeability is 10^{-9} m/s (30 mm/y). On a separate plot of dry density versus permeability, plot all the experimental points that correspond to a permeability of 10^{-9} m/s or less. An example of such a plot appears in Figure 7.12, and defines the zone of compaction water content and dry density within

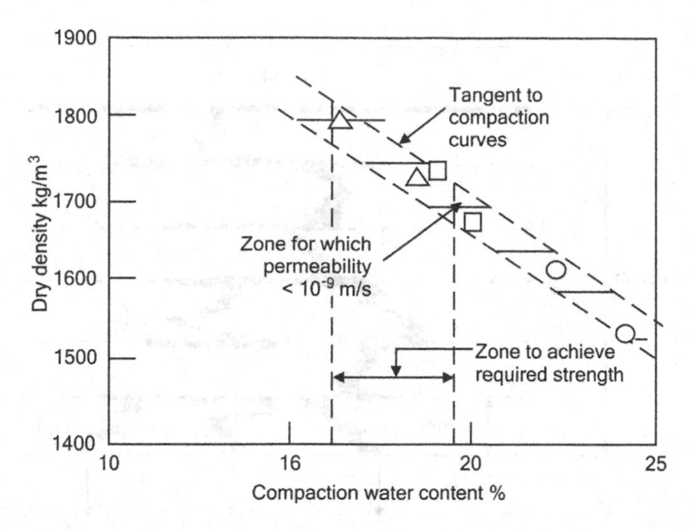

Figure 7.12 Zone for which permeability should be less than 10^{-9} m/s for compaction and permeability data shown in Figure 7.5.

which the permeability should meet the specified value. Strength requirements may also limit the compaction water content to a specific range, as indicated in Figure 7.12.

If it is not possible to reach the maximum permitted or designed level of permeability, it is usually possible to reduce the permeability of the natural soil by mixing in a proportion of a clay mineral such as bentonite. Bentonite is a highly expansive clay mineral, that occurs as two main types, sodium bentonite and calcium bentonite, where either sodium or calcium is the predominant exchangeable cation. Sodium bentonite is more expansive and less permeable than calcium bentonite. However, it should be noted that metallurgical extraction processes often produce and discharge a calcium-alkaline tailings, seepage from which will convert a sodium bentonite to a less expansive, more permeable calcium bentonite. If the use of bentonite is contemplated, therefore, the chemistry of the seepage water must be ascertained to ensure that the wrong type of bentonite is not used.

The bentonite is best mixed into the natural clay as a dry powder, using a mechanical mixer before placing, spreading and compacting the modified soil. If in situ mixing is to be used, a pulvi-mixer should be used. The quantity required is usually small (5 per cent of sodium bentonite added to a silty sand may reduce the permeability of the compacted soil by a factor of 10^{-4}.)

It is particularly important that any layer or sub-layer of a compacted soil mass, that is designed to be of low permeability, should be protected from desiccation by the sun after compaction. The usual method of protection is to spread the next layer of soil to be compacted over a compacted surface, as soon as compaction has been completed. When the soil will be exposed to the effects of desiccation over a long period, e.g. when used as a capping layer (see Figure 6.37), it must be provided with permanent protection against desiccation. For example, in Figure 6.37a the top filter layer acts as desiccation protection, and in Figure 6.37b the topsoil layer provides

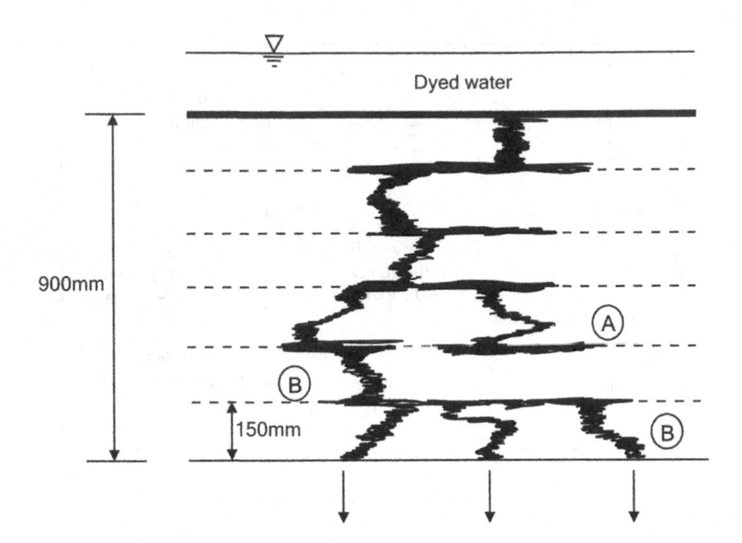

Figure 7.13 Liquid flow between compaction lift interfaces in a soil liner.

the protection. However, work recently reported by Benson, et al. (2007) indicate that even a thickness of 1 m of protection layers will not prevent desiccation of a compacted clay layer. (Also see Section 10.12.)

7.9 Seepage through field-compacted layers

The permeability of extensive field-compacted soil layers is very adversely affected by defects that inevitably occur. Examples of such defects are shrinkage cracks, poorly mixed or compacted or sandy zones in a layer and poor bond between layers. Figure 7.13, for example, shows the effect of shrinkage or tensile stress cracks, inter-linked by zones of poor inter-layer bond in a multi-layer compacted clay liner. The mechanism of leakage was observed by ponding dye on the surface of the liner and then excavating a hole to observe the seepage path taken by the dye (USEPA, 1989).

It would appear that a multi-layer compacted clay layer should be less pervious than a thin single layer because each layer would tend to interrupt and cut off flow through faults in contiguous layers. As shown by Figure 7.13, this does happen to a limited extent (A in Figure 7.13), but equally, may not happen (B in Figure 7.13).

Figure 7.14 shows measurements of the effect of liner thickness on overall permeability. Theoretically, because permeability is the ratio of flow velocity v to flow gradient i ($k = v/i$), and i would normally be unity in vertical seepage flow under gravity, permeability should be independent of liner thickness. In reality, because of defects such as those illustrated by Figure 7.13, permeability decreases as liner thickness increases. It will be noted from Figure 7.14 that what was judged to be "good and excellent construction" gave a slightly lesser decrease of k with increasing liner thickness, but the difference between "excellent" and "poor" construction was not very marked. Considering the lower bounding line, a 0.3 m thick liner achieved $k = 10^{-9}$ m/s, whereas a 1.5 m thick liner achieved $k = 4 \times 10^{-11}$, a 25-fold improvement for a 5-fold increase

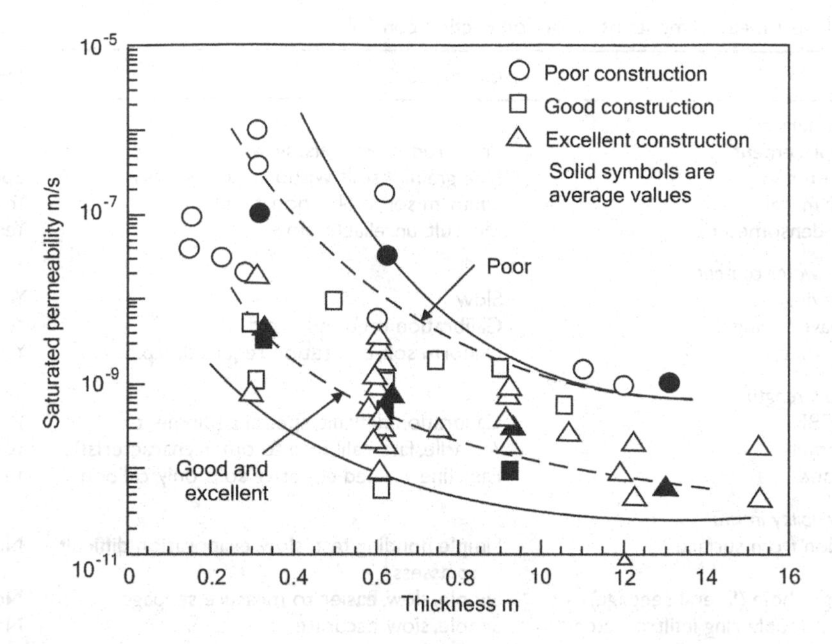

Figure 7.14 In situ measurements of permeability as a function of compacted layer thickness.

in thickness. Thus it has to be concluded that, given similar quality of construction, thicker clay liners considerably outperform thinner ones, but at the cost of providing the additional compacted clay. (USEPA, 1989).

7.10 Control of compaction in the field

In order to set appropriate control criteria, the function of the compacted fill must be understood, and a number of additional constraints must also be considered (Simmons and Blight, 1997).

- construction requirements related to weather and climate that will affect borrow pit and compaction water contents (limited time for completion, sunshine, rain, wind, low or high temperatures, etc.),
- moisture conditioning requirements for borrow pit material,
- availability and standards of resources for quality control,
- if available, performance data from field compaction trials either for a method specification or optimum compaction characteristics for equipment used,
- availability and standards of any reference laboratory testing on which specifications have been based.

Table 7.1 is a summary of tests that have been used for compaction control. These can be used as a guide to selection of appropriate testing. The frequency of testing (e.g. number of m^2 or m^3 per control test) should be related to a number of additional factors, and must be determined separately.

Table 7.1 Test measurements used for compaction control

Test	*Comments*	*Standard?*
1. In situ density		
Sand replacement	Preferred, most soils, slow	Yes
Core cutter	Fine grained soils without stones, slow	Special
Nuclear meter	Uniform soil, calibration, rapid	Yes
Balloon densometer	Difficult, unreliable, slow	Yes
2. In situ water content		
Oven drying	Slow	Yes
Microwave drying	Calibration required, rapid	Yes
Nuclear meter	Uniform soil, calibration required, rapid	Yes
3. In situ strength		
In situ CBR	Calibration difficult, special equipment	Yes
Penetrometer	Versatile, fast, calibrate to other characteristics	Yes
Shear vane	Fast, fine grained cohesive soils only, calibrate	Yes
4. Permeability in situ		
Infiltration from surface	Simple ponding test, slow, evaporation difficult to assess	No
Drill/auger hole (lateral seepage)	Simple, slow, easier to measure seepage	No
Covered double ring infiltrometer	Simple, slow, accurate	No
5. Laboratory (on undisturbed samples) Constant or falling head permeability	Constant head slow, falling head quicker. Representivity of field conditions doubtful	No
UU triaxial strength	Fast, select test conditions to suit purpose	Yes
CU triaxial strength	Slow, select test conditions to suit purpose	Yes

Table 7.2 Compaction control parameters

1	In situ dry density as compacted
2	In situ water content as compacted
3	In situ dry density within a range of water content as compacted
4	In situ strength as compacted
5	In situ permeability as compacted
6	Laboratory strength properties correlated to in situ measurements
7	Recipe specification

Table 7.2 lists five parameters that are commonly used for compaction control:

7.10.1 *In situ dry density as compacted*

In principle this is simple and direct. However, accurate field measurement of volume is time consuming and subject to procedural errors. Indirect measurements using nuclear moisture/density meters require careful calibration checks. Nuclear equipment may not be suitable for many remote sites, and the calibration can be affected by lightning occurring within 1 km from the site, as well as adjacent high voltage power lines.

The greatest disadvantage of density testing is that it offers only limited information on the fill properties that are really required for the function of the fill. The method of field compaction, and the particle sizes involved, may not correlate with the standard laboratory methods. At best, only inferences can be made about strength and permeability. Test results are compared to a laboratory value. Considerable material variability may occur in the field. Unless a corresponding laboratory reference test is performed for each field density measurement, there is a risk of field measurements being incorrectly interpreted.

In situ density has traditionally been the most popular method of measurement, due to its adoption and wide use over a long period of time.

7.10.2 *In situ water content as compacted*

Water content can be measured reasonably rapidly, and has the advantage that field variability can be assessed relatively easily by taking many measurements. Provided that the strength and permeability characteristics of the material are understood in relation to water content, water content can be used as an effective control parameter. However, the desired properties of strength and/or permeability are dependent on density as well as water content. The use of water content therefore should be supplemented by measurement of other properties. Alternatively, if following a field trial, the method of compaction may be specified so as better to achieve the requisite density. Water content could be used as the sole control parameter. However, this would apply only in unusual cases.

7.10.3 *In situ dry density within a range of water content as compacted*

In Section 7.8, a method was described for controlling the permeability of a compacted soil. This requires achieving a result within permissible ranges of both water content and permeability, as illustrated in Figure 7.12.

Figure 7.15 shows the statistics for the control of compaction water content and Proctor dry density that relate to a large soil compaction project. Each of the three stages of the histograms relates to approximately equal periods of time and, overall, the data show that control adequately met the set targets, and improved as the project progressed.

7.10.4 *In situ strength as compacted*

In principle, this is the most effective method of control, since strength is normally a direct indicator of performance. Strength should be measured with a rapid test which is not subject to significant interpretation problems. A variety of rapid strength measurements can be used, ranging from hand-held or hand-operated vanes or penetrometers to in situ CBR tests and larger heavier penetrometers. Difficulties may occur if large particles present in the soil compromise the performance and interpretation of the measurement technique. Technologically advanced compaction machinery or monitors, which use accelerometers to assess the response of a fill to vibration, must be calibrated to site conditions. Strength is also greatly dependent on water content and will change if water content changes subsequent to compaction. Thus water content must be used as a second control parameter.

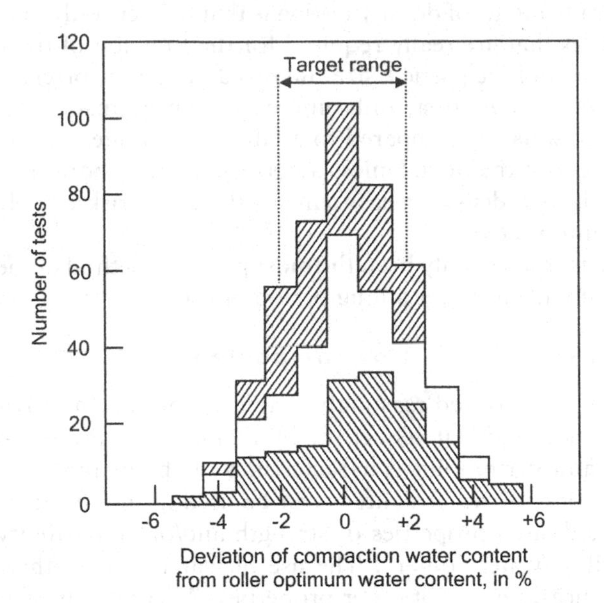

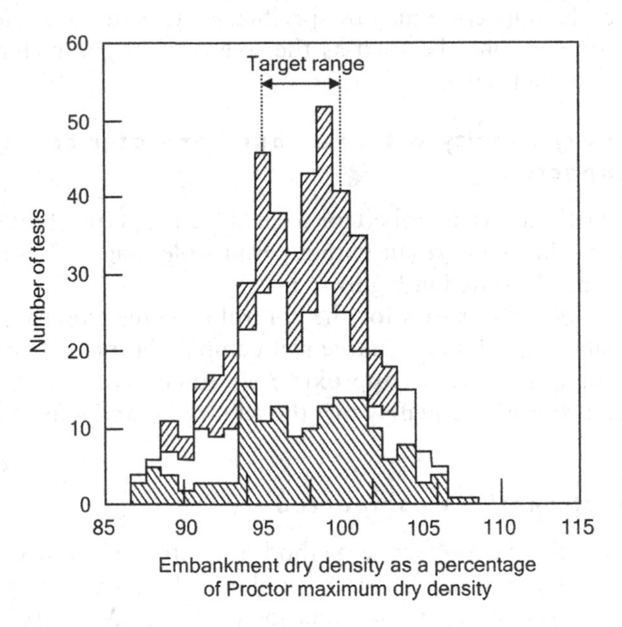

Figure 7.15 Progressive histograms for compaction water content and dry density, showing adequate and improving control as the project progressed.

7.10.5 *In situ permeability as compacted*

This is very effective where permeability is the most important characteristic for field performance. The greatest disadvantages are that field permeability testing is time consuming, very prone to errors, and it may not be possible to undertake a sufficient number of tests to give a statistically representative picture of field conditions.

Approximate field permeability tests, using simple infiltration methods, are easily performed provided that they can be done in areas which do not affect ongoing construction activities. In this case, permeability testing even if approximate, is a very useful guide.

Another disadvantage of permeability testing is that strength is usually also required. There are no effective correlations between strength and permeability, so that other forms of field tests are required anyway. Permeability cannot be used as a sole control measurement.

7.10.6 *Laboratory strength properties correlated to in situ measurements*

The advantages of laboratory testing relate to control and repeatability. Where this can be combined with an effective field measurement, very efficient compaction control can be achieved with a high degree of confidence.

The most widespread correlations are of laboratory strength related to field water content, for materials which have been adequately tested and whose variability is well understood. Thin-walled tube or core samples can be taken in the field, and tested at a site laboratory. The usual site laboratory strength test is the unconsolidated undrained (C_{uu} or 'Quick') triaxial test.

7.10.7 *Recipe specifications*

A recipe specification may call for compaction using a specified type and weight of roller, working within a specified range of water contents and using a specified minimum number of roller passes. In the case of footed rollers, passes may be substituted by a minimum coverage "C" where

$$C = A_f/A_d \times N \times 100\% \qquad (7.1)$$

A_f = area of foot, A_d = area of drum, and N = number of roller passes. A typical range of C would be 150–200%, the aim being to ensure that, other things being equal, the whole area of the layer being compacted is subjected to the higher "under-foot" pressure at least once.

(During the Vietnam war during the early 1960's, an article appeared in the American Society of Civil Engineers' magazine entitled "On the uselessness of elephants for compacting soils". The article told the story of a group of US army engineers who attempted to compact the surface of an area of soil by driving a herd of elephants across it. They discovered that not only is the foot contact pressure of an elephant very low, but also that all of the animals in the second and subsequent rows trod exactly in the footprints of their predecessors, so that coverage was minimised!).

7.11 Special considerations for work in climates with large rates of evaporation

In arid and semi-arid conditions it is usually planned to construct all, or the major part of the earth works during a single dry season. This expedient usually reduces the cost of, for example, diverting a river during construction and minimises delays due to wet weather, but limits the construction period to a maximum of about six months.

Placing and compaction operations often proceed on the basis of a 24-hour working day. Unless the working area is very large, up to three compacted layers, each 150 mm thick, may be deposited over the entire compaction area each working day. This precludes the usual method of compaction control by checking the embankment dry density and the compaction water content and rewatering and rerolling if either is not up to standard. Even if rapid methods of determining water content and density are used, the construction schedule will not usually allow time for recompaction of substandard layers.

Because of this, methods of control must be used which ensure that the soil is at the correct water content before compaction starts, and that the optimum number of roller passes is used to produce an adequate dry density.

Large water losses due to evaporation can take place during the dry season, and the in situ water content in the borrow pits decreases as the season progresses. As a result the water content of the soil usually has to be increased considerably before compaction can be undertaken.

The water content is sometimes increased by a sprinkler system in the borrow area. A few hours may elapse between spreading the soil in place and the start of compaction; in this time a considerable amount of water can be lost from the soil. The delay results from the practical necessity of keeping the slower moving compaction rollers and water tankers out of the way of the faster earth-moving equipment.

Typical water losses from two 200 mm thick layers of loose uncompacted soil on the surface of the embankment of a dam in Central Africa are shown in Figure 7.16. The shade and sun temperatures on the surface of the soil at the time are also shown. The water contents represent the average through the thickness of the layers. A total loss of water content of 6 percent took place in eight hours, 4 percent being lost within the first three hours. Because of these large potential evaporation losses, it is best, under such conditions, to add and mix in water to the soil on the embankment immediately before compaction.

Using the known capacity of the water tankers, a chart can be constructed relating the area of embankment to be treated and the required increase in water content to the number of full tanker loads to be discharged over a given area. An example of such a chart is shown in Figure 7.17.

Evaporation-time curves can be established for work on the day and night shifts. Twice in each shift, water content samples can be taken from the incoming soil and the amount of water to be added to bring it to the required compaction water content estimated, using the chart in conjunction with the appropriate evaporation-time curve.

7.12 Additional points for consideration

7.12.1 *Variability of borrow material*

Variability of borrow material may relate both to particle size distribution and to cohesive or frictional attributes. Compaction assessment should include monitoring of borrow materials to ensure that specification requirements can be met.

Field compaction trials are strongly recommended because they enable the work to be controlled as much as possible by factors that have been proven in the field. There is no problem with field trials being undertaken as part of the permanent works, but

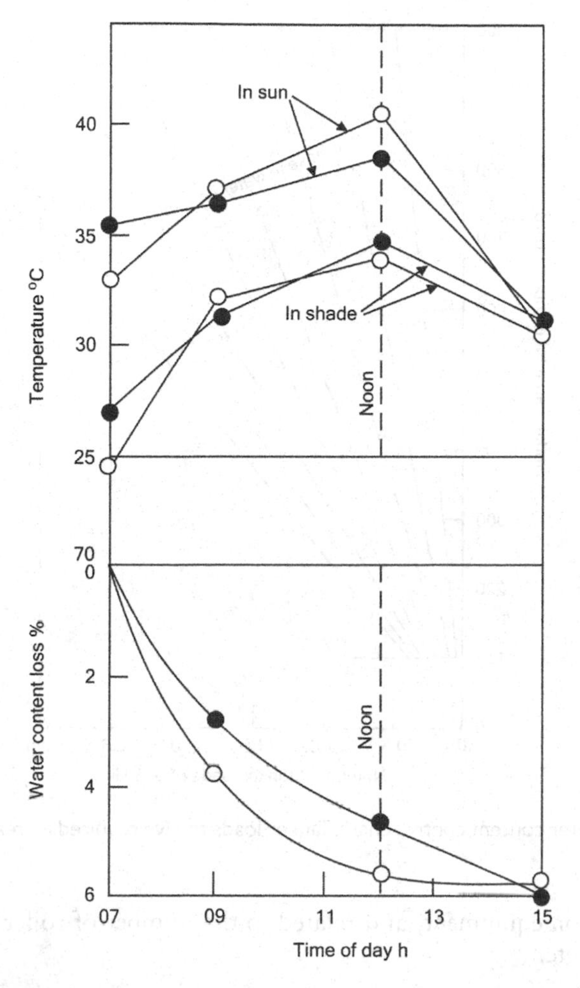

Figure 7.16 Water content losses from two 200 mm thick, loose, uncompacted soil layers in arid conditions. Losses are average through the entire layer thickness.

it is important to realize that trials can be slow and painstaking compared with full production. Both the supervisor and the constructor must recognize this and allow for it in their schedules and budgets.

The particular advantages of field trials are related to the practicalities of field conditions and to time constraints.

7.12.2 *Compactor performance*

Knowledge of compaction performance can be translated from laboratory test results, to results which directly reflect the performance of the compaction equipment. For example, field maximum dry density and optimum moisture content can be determined

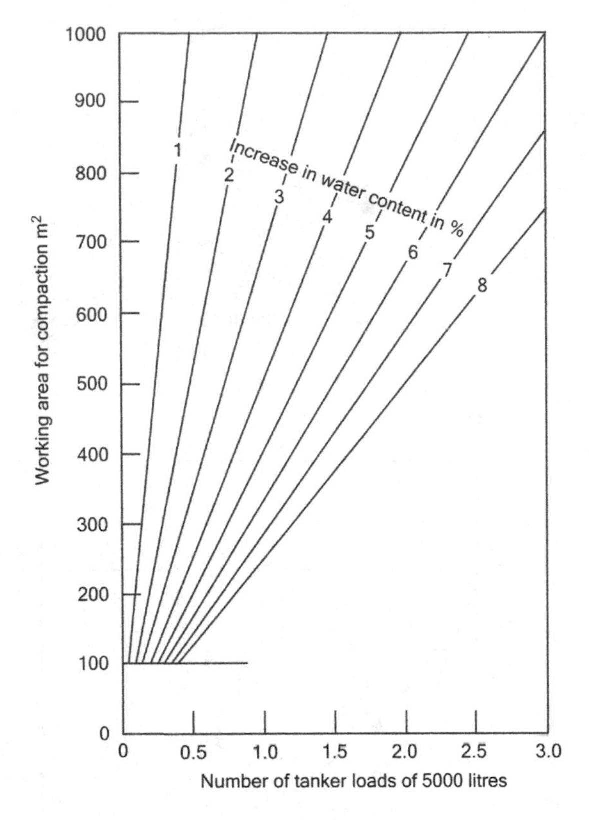

Figure 7.17 Water content control chart. Tanker loads to give required increase of water content.

for the compaction equipment, and related to the number of roller passes, thickness of placed layers, etc.

7.12.3 *Testing frequency*

The necessary testing frequency is very dependent on items 7.12.1 and 7.12.2 above. If materials are variable, or if compactor performance is poor or erratic, a higher testing frequency is required than if these items are satisfactory.

Selection of lot sizes (i.e. numbers of m^2 or m^3 per test) and testing frequency should be based on site conditions, and adjusted appropriately. It is generally accepted that sampling of a compacted layer for dry density and compaction water content should take place on a square grid pattern as this provides a method of choosing test locations that is independent of personal bias. It is, however, important to stagger the grid points from one layer to the next. The grid size or spacing is usually 10 to 15 m, corresponding to 100 to 44 tests per hectare, or 1 test per 15 to 34 m^3 of a 150 mm thick compacted layer.

Rapid field tests can be identified and selected so as to minimize the delays caused by waiting for laboratory test results.

References

Benson, C.H., Thorstad, P.A., Jo, H.-Y., & Rock, S.A.: Hydraulic Performance of Geosynthetic Clay Liners in a Landfill Final Cover. *J. Geotech. Geoenviron. Eng, ASCE 133*(7) (2007), pp. 814–827.

Blight, G.E.: Controlling Earth-Dam Compaction Under Arid Conditions. *ASCE Civil Eng.*, 1962, pp. 54–55.

Daniel, D.E.: Earthen Liners for Land Disposal Facilities. In: R.D. Woods (ed): *Geotechnical Practice for Waste Disposal '87*. Geotechnical Special Publication No. 13, New York, U.S.A.: American Society for Civil Engineers, 1987, pp. 21–39.

Gidigasu, M.D., & Dogbey, J.L.K.: The Importance of Strength Criterion in Selecting Some Residual Gravels for Pavement Construction. In: *Proc. 7th Regional Conf. Africa Soil Mech. Found. Eng.* Accra, Ghana. Vol. 1, 1980, pp. 300–317.

Mitchell, J.K., Hooper, D.R., & Campanella, R.G.: Permeability of Compacted Clay. *J. Soil Mech. Found. Div., ASCE. 91(SM4)* (1965), pp. 41–65.

Rodda, K.V., Perry, C.W., & Roberto Lara, E.: Coping with Dam Construction Problems in a Tropical Environment. *Engineering and Construction in Tropical and Residual Soils*. ASCE Geotech. Div. Spec. Conf., Honolulu, Hawaii, 1982, pp. 695–713.

Simmons, J.V., & Blight, G.E.: Compaction. *Mechanics of Residual soils*. Rotterdam, Netherlands: A.A. Balkema, 1997.

United States Environmental Protection Agency (USEPA): Requirements for Hazardous Waste Landfill Design, Construction and Closure. Seminar Publication, EPA 625 4-89/022, 1989.

Chapter 8

Methods for constructing impounding dykes for storing hydraulically transported tailings and other fine-grained wastes

8.1 Deposition methods and sequences

One of the most important, if not the most important feature of a storage for fine-grained wastes such as tailings, fly ash or waste gypsum is the impounding dyke that forms the outer wall of the storage. There is a variety of options for the method and materials to be used in constructing this bund, dam, dyke or retaining wall.

- Coarse wastes or waste rock which would normally be disposed of by dumping can be used to construct the impounding dyke and suitable drainage or sealing measures can be incorporated to ensure the dyke's stability and to retain any pollutants within the storage.

  Alternatively imported natural material from stockpiles or borrow pits, appropriately compacted, can be used to construct the dyke. This method is commonly used to form starter walls for the early stage of forming a storage, using fine waste.
- The storage can be constructed of the separated coarse fraction of the waste which is placed either mechanically or hydraulically.
- If the waste is fine-grained the storage can be constructed of run-of-mill waste by hydraulic-fill methods.

There are four basic methods (see Figure 8.1) of constructing an impoundment by hydraulic-fill means.

1. The downstream method in which the centre-line of the dyke moves progressively downstream. As Figure 8.1a shows, this method can provide a substantial embankment. The downstream method is now the preferred method of construction in many countries, and especially where seismic activity occurs, e.g. the United States of America, Canada and Chile.
2. The centre-line method in which the centre-line of the dyke remains in the same position throughout construction. As Figure 8.1b shows this method also provides a substantial outer dyke to the storage.

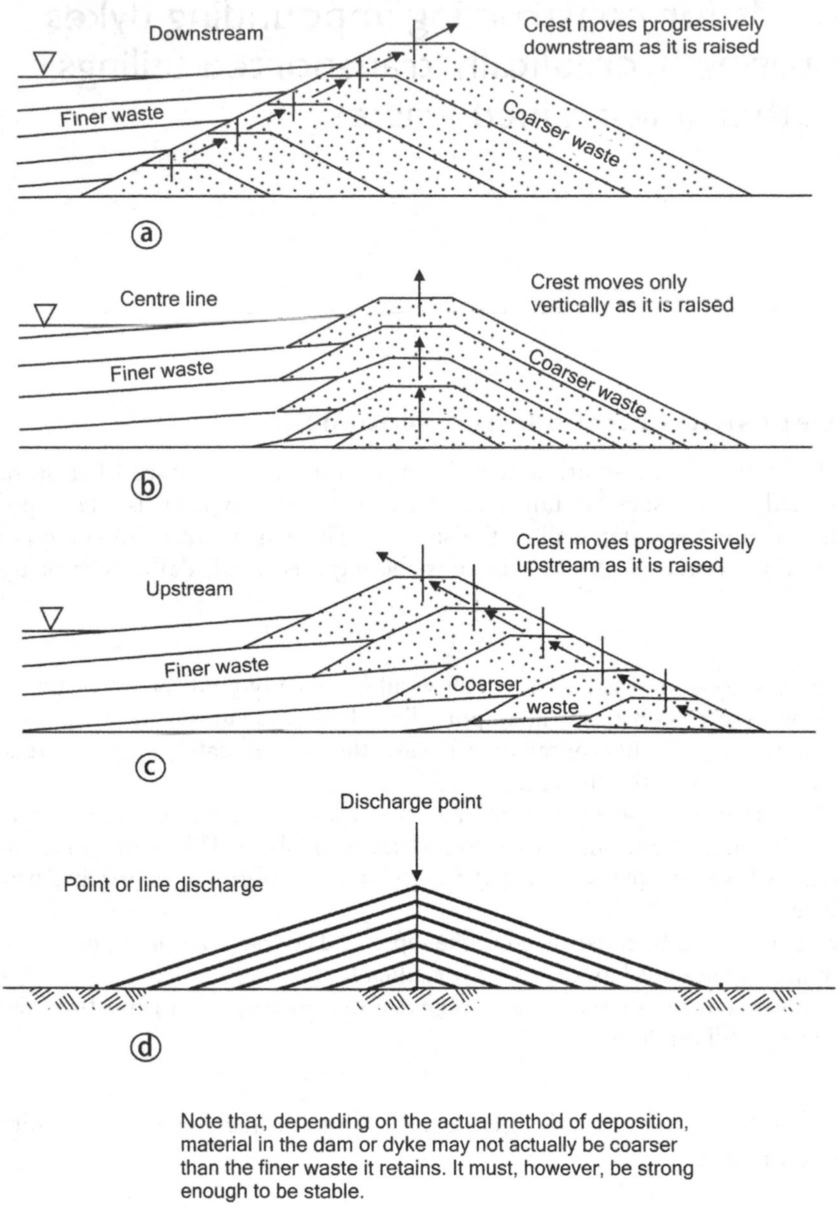
Downstream
Crest moves progressively downstream as it is raised
Finer waste
Coarser waste
a
Centre line
Crest moves only vertically as it is raised
Finer waste
Coarser waste
b
Upstream
Crest moves progressively upstream as it is raised
Finer waste
Coarser waste
c
Discharge point
Point or line discharge
d
Note that, depending on the actual method of deposition, material in the dam or dyke may not actually be coarser than the finer waste it retains. It must, however, be strong enough to be stable.

Figure 8.1 Various methods of constructing impounding dykes or dams: a: downstream construction. b: centre-line construction. c: upstream construction. d: point or line discharge construction.

3. The upstream method in which the centre-line of the dyke moves progressively upstream. As Figure 8.1c shows, this results in a less substantial outer shell for the storage. This particular method of construction is no longer used in many parts of the world, although it is still used in areas having an arid climate and no seismicity.

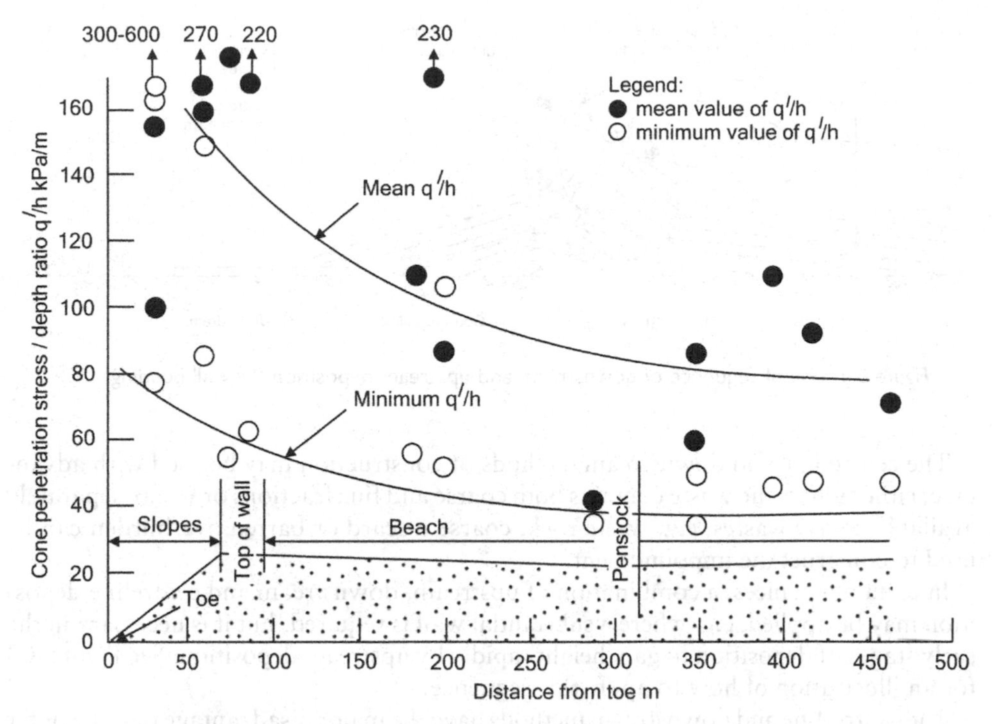

Figure 8.2 Variation of cone resistance with distance from toe for storage of gold tailings built by upstream method.

The reason is that if the climate is arid, with low rainfall and high evaporation, and if rates of rise are low (2 m per year or less), sun-drying of the fine waste blurs the distinction between the coarser outer retaining shell and the finer retained waste. This point is illustrated by Figure 8.2 which shows the results of a series of cone penetrometer tests made at increasing distances from the toe of a storage for gold tailings in South Africa built by the upstream procedure (at Merriespruit, see Section 11.3.2). As the diagram shows, there is no clearly defined outer retaining shell, but instead, a gradual reduction of the strength of the tailings occurs from the toe of the storage towards the decant point or penstock. Figures 8.1a to c are based on the concept that there is a complete separation between coarser and finer fractions of the total tailings material. In practice there is usually a relatively small difference between the particle size distribution of "coarser" and "finer" wastes.

4. As will be discussed in Section 8.4, if the tailings slurry is thickened to more than a certain relative density or solids concentration, gravitational particle size sorting becomes negligible and the slurry then behaves as a homogeneous viscous fluid. In the case of a thickened tailings, a method known as point or line discharge is usually used. The principle of this method is illustrated in Figure 8.1d which shows that a flat-sloping conical deposit is formed by discharging the thickened slurry from a progressively raised central point. For very large areas of deposition, multiple point discharges may be used to give a series of overlapping cones of deposition, or the tailings can be discharged along a line from, e.g., the top of a ramp.

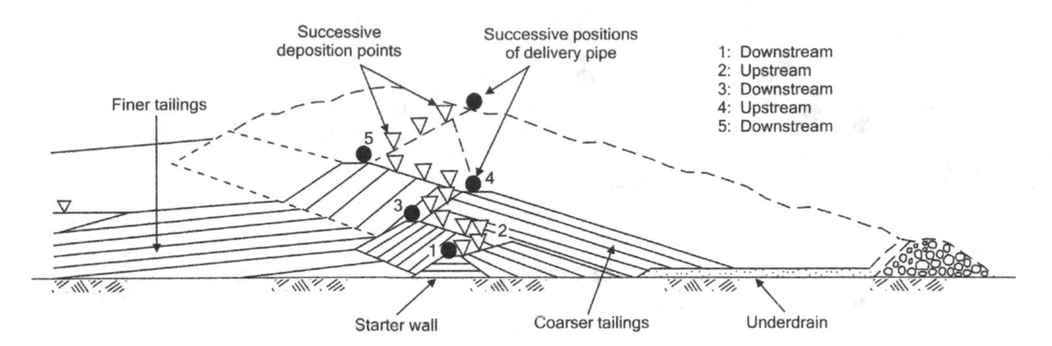

Figure 8.3 Use of sequence of downstream and upstream deposition for wall building.

The centre-line and downstream methods of construction may be used with advantage, either where the waste contains both coarse and fine fractions or where separately available coarse wastes, e.g. waste rock, coarse discard or barren overburden can be used to construct the impoundment.

In certain instances, a combination of upstream, downstream and centreline deposition may be applied, e.g. where a substantial wall is required, but it is necessary in the early stages of deposition to gain height rapidly by upstream deposition. See Figure 8.3 for an illustration of how to apply this sequence.

The centre-line and downstream methods have the major disadvantage that the outer toe of the dam moves out progressively as the dam is raised. This usually means that access roads and slurry delivery pipes have to be moved out progressively. Because the final outer slope of the storage can only be permanently protected against wind and water erosion once the storage is being decommissioned, temporary measures must continually be taken to mitigate water erosion and prevent the generation of blowing dust in dry windy weather. The upstream method of construction does not suffer from either of these two disadvantages. However, the same disadvantage applies to the point or line discharge method.

8.2 Beach formation in hydraulic deposition of fine-grained wastes

Tailings are transported to the discharge points on a tailings storage at a velocity sufficient to keep all particles in suspension. The power necessary to overcome the frictional resistance and maintain particles in suspension is applied to the system by the delivery pumps. On discharge, much of the power is dissipated and the available energy of the tailings stream reduces to a residual amount. The delivery line usually ends above the beach level and where the discharge stream impinges on the beach a "plunge pool" develops as illustrated in Figure 8.4. Within this plunge pool much of the energy of the discharge stream is dissipated. The slurry plunge pool and the geometry of the stream beyond the plunge pool down to the pond is regulated by the difference between the levels of the plunge pool and the tailings pond, the pond location and the rate at which tailings is deposited. As the level of tailings increases, the stream meanders down the beach and from time to time the position, width and

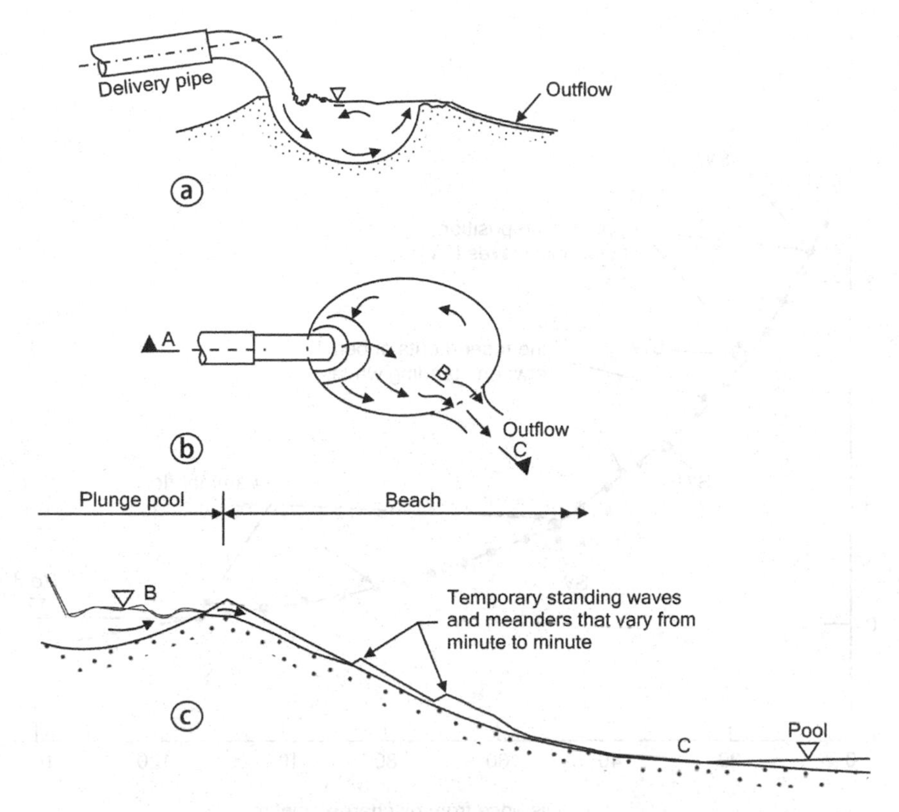

Figure 8.4 Schematic geometry of slurry plunge pool and tailings beach: a: section ABC through plunge pool formed at discharge point. b: plan of plunge pool. c: section B-C along outflow stream.

depth of the overflow from the plunge pool may change. Standing waves commonly develop in the tailings stream as it runs down the beach. (See Plates 8.1 and 8.2 for examples of meandering channels and standing waves during deposition on a tailings surface.)

At all times during the discharge period, deposition occurs in the plunge pool (which may also change size), at each point along the beach and in the pond (the pool of supernatant water that collects at the end of the beach). Hence the beach and the floor of the plunge pool aggrade. The loss of energy (kinetic-elevational) from one point on the beach to the next determines the mass of tailings deposited between those points and hence the rise in beach surface. Since large particles require more energy to be maintained in suspension than smaller particles, there is a greater probability that larger particles will be deposited close to the discharge point, while finer particles will transported further down the beach towards the pond.

Bentel (1981) observed that the standing waves form on the upper part of the beach where the velocity of flow is greatest and the flow is supercritical. As the energy of flow is dissipated by the standing waves, the velocity reduces and, as the pool is approached, becomes subcritical. Thereafter laminar sheet flow occurs. Bentel surveyed the profile

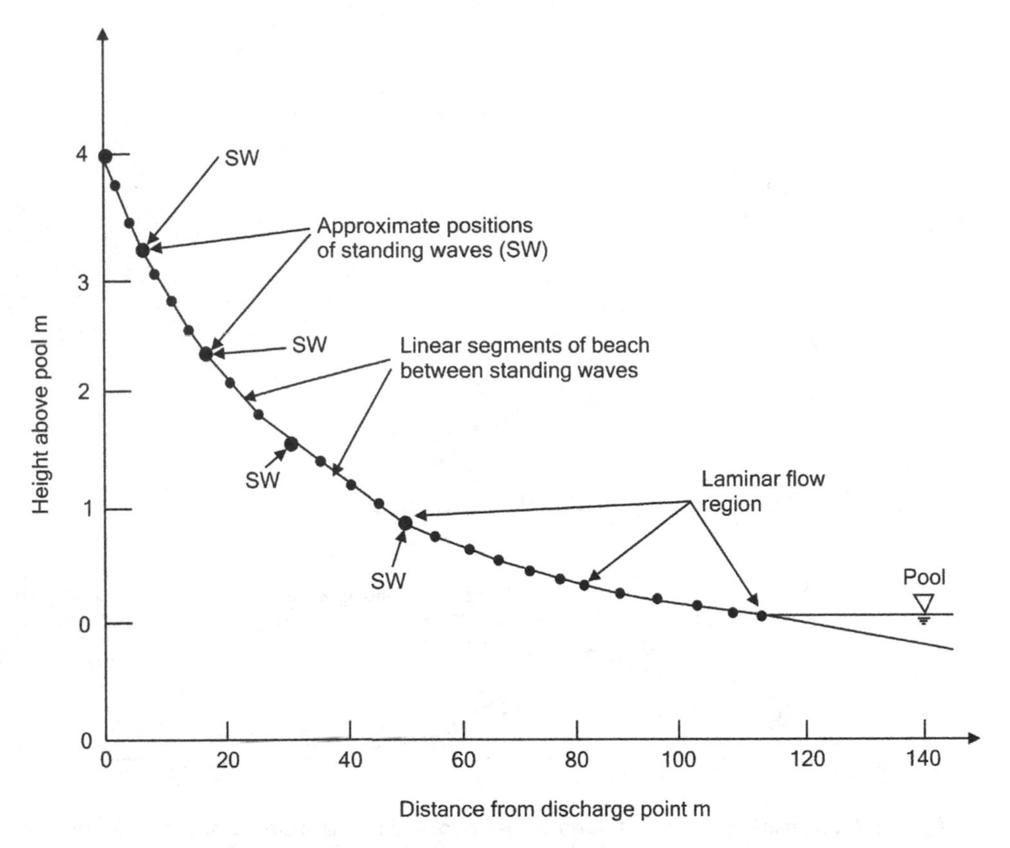

Figure 8.4d Measured profile showing linear sections of beach between standing waves and region of laminar sheet flow near pool.

of an unthickened platinum tailings beach (Figure 8.4d) and found that it consisted of a series of linear portions which flattened after the flow had passed each standing wave. Once the flow became subcritical, the profile became continuously curved. These features are shown in Figure 8.4d. Plate 8.1 shows meandering channels that occur on a typical hydraulic fill beach. It is tailings slurry spilling over the banks of these main flow channels that form the general beach surface. The positions of the channels change continually. Plate 8.2 shows a typical standing wave in a flow channel.

The mean surface profile of the beach is characteristic for a particular tailings and slurry water content. If the dimensions (length and elevation) of a tailings beach are non-dimensionalized, then all beaches of the same tailings, discharged at similar water contents, will adopt the same non-dimensional profile, regardless of the distance and difference in elevation between the point of deposition and the edge of the pool. This also applies to short "beaches" such as those formed by cyclone underflow deposits, as well as point or line discharge beaches. Beach shape is illustrated by Figure 8.5b

Plate 8.1 Meandering flow channels on a typical hydraulic fill beach.

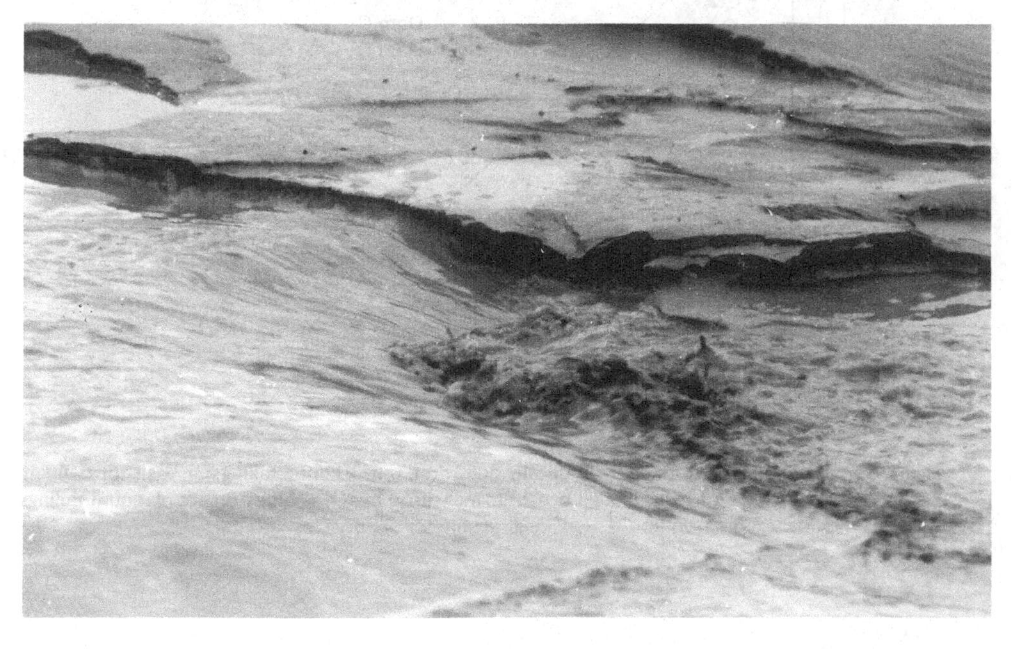

Plate 8.2 Typical standing wave in a flow channel.

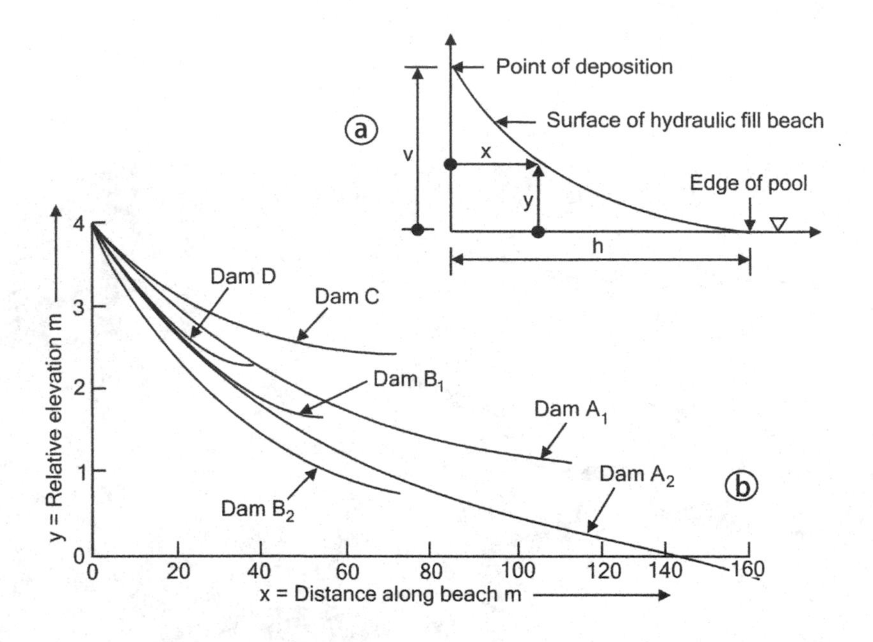

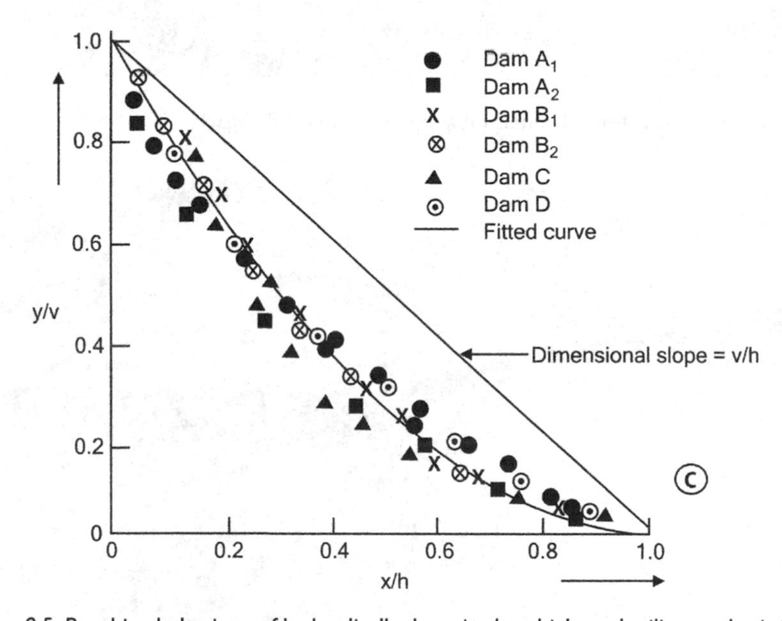

Figure 8.5 Beaching behaviour of hydraulically deposited unthickened tailings: a: basis for defining mean shape of hydraulic fill beach. b: measured beach profiles on six platinum tailings dams. c: common non-dimensional profile for six profiles in b.

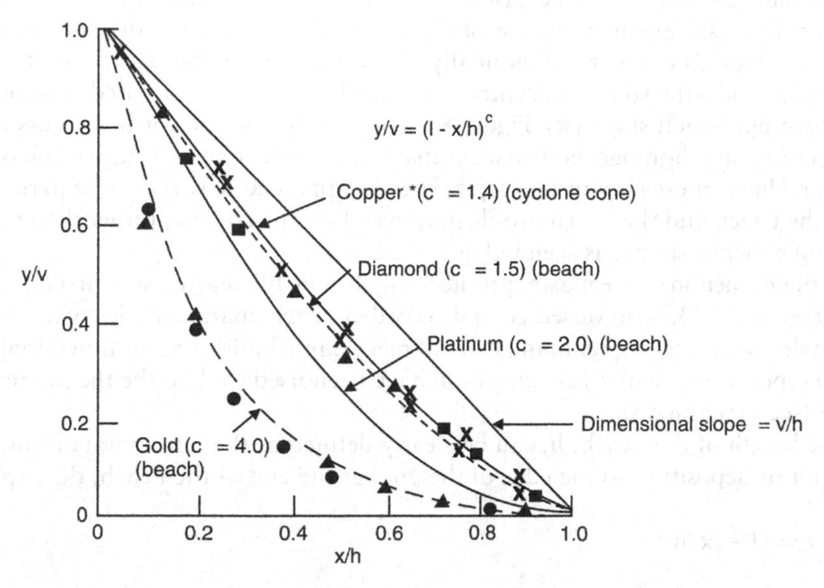

Figure 8.6 Dimensionless beach profiles for four different tailings materials (three beaches and one cyclone underflow cone).

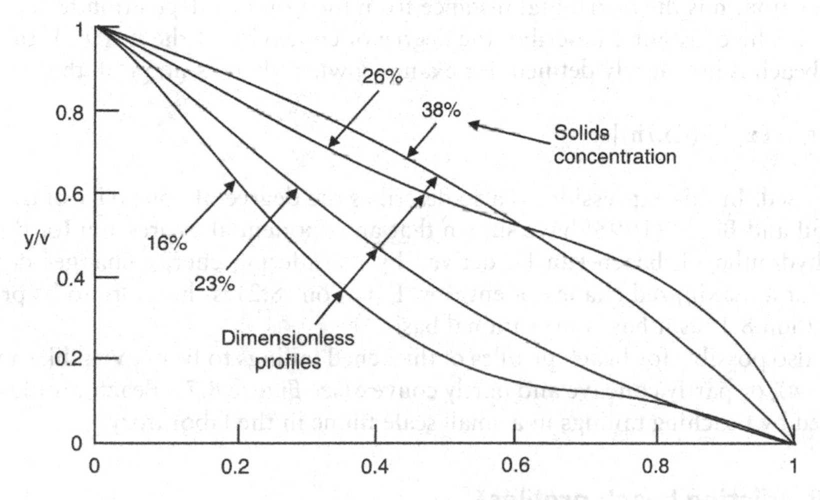

Figure 8.7 Change of shape of dimensionless profiles for model beaches of diamond tailings as solids concentration (mass of solids/total mass) increases.

which shows beach profiles observed on six different platinum tailings dams, and also the corresponding common non-dimensional "master" beach profile. Figure 8.6 shows dimensionless profiles for four different tailings, illustrating how the profile changes with tailings characteristics. The shape and overall gradient of the beach (v/h) depends on the solids concentration of the tailings, as shown by Figures 8.7 (shape) and 8.8 (gradient). Solids concentration = (mass of solids)/(total mass).

As the solids concentration increases, so does v/h up to a maximum. Thereafter, the gradient decreases again, but must obviously reach a minimum and start increasing again as the solids content is continually increased, and the tailings ceases to behave as a liquid. Once the solids concentration exceeds the value associated with the maximum average beach slope (see Figure 8.8), very little particle sorting occurs and the tailings flows as a homogeneous viscous material, rather than as a suspension of solids in water. Under these circumstances, there is no appreciable variation of particle sizes down the beach and the beach profile may even become convex, rather than the more common concave shape, as seen in Figure 8.7.

The phenomenon of the master profile for hydraulic fill beaches was first reported by Melent'ev, in 1973, who observed and recorded beach characteristics while working with hydraulic fill water retaining embankment dams built of natural materials.

Two types of mathematical expression have been used to describe the master beach profile (see Figure 8.5a):

If the length of the beach, h, can be clearly defined as the horizontal distance from the point of deposition to the edge of the pool at the end of the beach, the expression

$$y/v = (1 - x/h)^c \tag{8.1}$$

can be used. In this expression, y is the height of a point at a horizontal distance x from the point of deposition, and v is the height from the level of the pool to the point of deposition. h is the horizontal distance from the point of deposition to the edge of the pool. The constant c describes the degree of concavity of the beach. If the length of the beach is not clearly defined, for example, when there is no pool, the expression

$$y/v = \exp[-(Cx/h)] \tag{8.2}$$

can be used. In this expression. C also describes the degree of concavity of the beach. McPhail and Blight (1998) have shown that an exponential expression for the shape for a hydraulic fill beach can be derived by considering energy changes down the beach for a maximized change of entropy. Equation (8.2) is therefore to be preferred to equation 8.1, as it has some rational basis.

It is also possible for beach profiles of thickened tailings to be convex (like a viscous lava flow), or partly concave and partly convex (see Figure 8.7). Beach profiles can be modeled by beaching tailings in a small scale flume in the laboratory.

8.3 Predicting beach profiles

The concept of the master beach profile can be used to predict the profile of a full-size hydraulic-fill tailings beach on the basis of a small-scale laboratory model beach. However, the master profile cannot be used to predict the profile of a prototype beach unless the average or chord slope of the prototype (v/h in Figure 8.5a) can be determined, or predicted. Several methods have been developed to predict the slope of a prototype beach. As the slope must depend on the shear strength of the newly deposited tailings slurry, all of these methods depend on an analysis based on versions of the diagram shown in Figure 8.9a. This represents a two-dimensional element of a newly-deposited layer of tailings slurry that has been placed over a pre-existing tailings surface. The

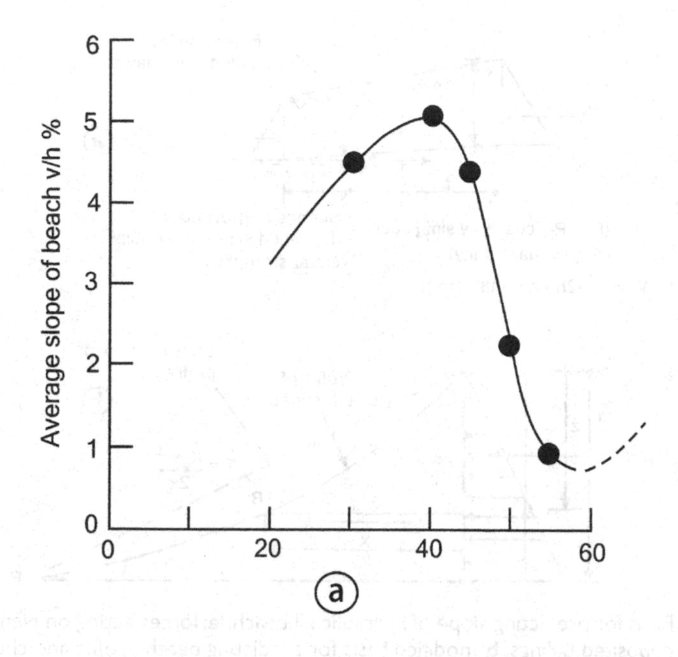

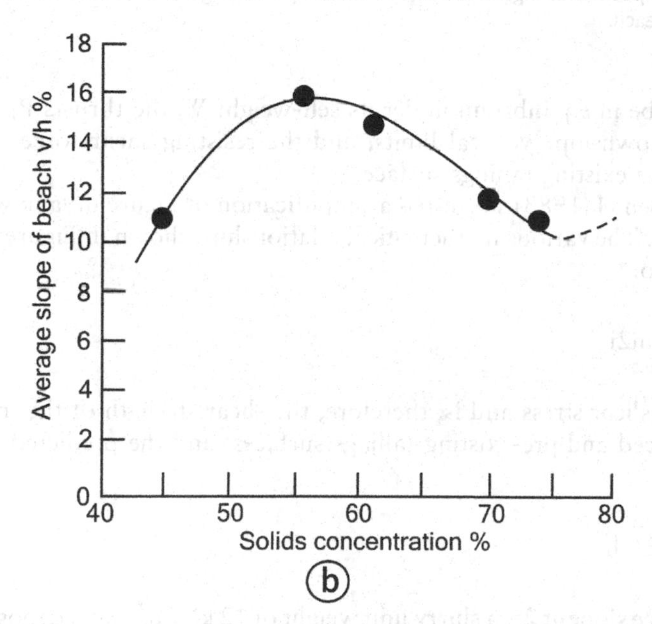

Figure 8.8 Variation of average or chord slope (v/h) of model hydraulic fill beaches with solids concentration for slurries of: a: power station fly ash. b: gold tailings.

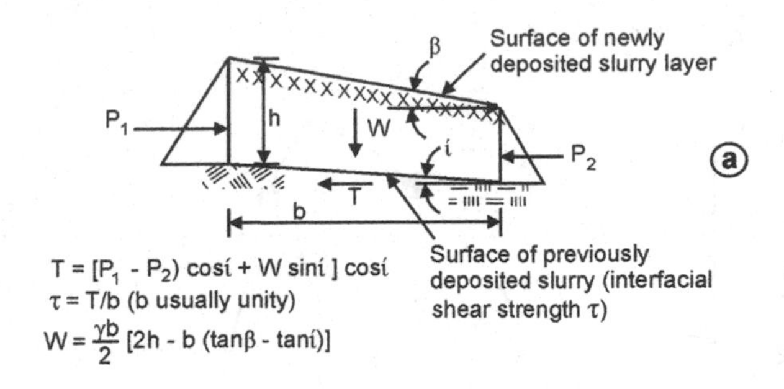

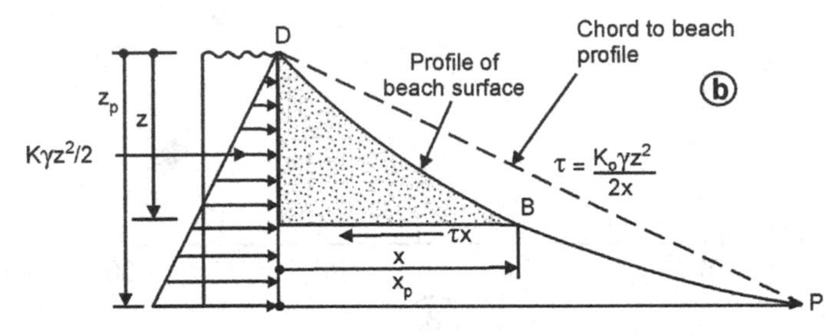

Figure 8.9 Basis for predicting slope of hydraulic fill beach: a: forces acting on element of recently deposited tailings. b: modified basis for predicting beach profile and chord slope (i_c) of beach.

element must be in equilibrium under its self-weight W, the thrusts P_1 and P_2 on its upslope and downslope vertical limits, and the resisting shear force T on its base, which is the pre-existing tailings surface.

Blight and Bentel (1983) suggested a simplification of Figure 8.9a in which $P_1 = P_2$, $\beta = i$ and $b = 1$. The various mathematical relationships shown in Figure 8.9a can then be simplified to:

$$\tau = ½.\gamma h \sin 2i \tag{8.3}$$

τ is a limiting shear stress and is, therefore, the shear strength of the interface between newly placed and pre-existing tailings surfaces, and the predicted slope angle is given by:

$$\sin 2i = 2\tau/\gamma h \tag{8.3a}$$

Taking a surface slope at 2°, a slurry unit weight of 12 kN/m^3 and a deposited thickness of 5 mm, from equation (8.3):

$$\tau = ½ \times 12 \times 10^3 \times 5 \times 10^{-3} \sin 4° = 2.1 \text{ Pa}$$

Thus the shear strengths involved are very small and fall in the range shown in Figure 4.48.

For beaches of unthickened tailings, as mentioned earlier, gravitational particle size separation occurs down the length of the beach, from the point of slurry discharge to the pool (see Figure 8.5a). The separation has the effect of causing the shear strength τ to vary down the length of the beach. This is one reason why beaches of unthickened tailings are usually concave, with steeper slopes adjacent to the point of discharge where the settled tailings is coarser and τ tends to be higher, and flatter slopes as the beach approaches the pool where the finer tailings has a lower shear strength. Thickened tailings, however, does not segregate in this way and therefore has a more uniform shear strength down the beach.

Figure 8.9b shows a basis for calculating the complete beach profile from a knowledge of the shear strength (τ) and the unit weight of the deposited tailings (γ). The method is essentially similar to that illustrated in Figure 8.9a, except that the shear stability is considered for progressively increasing lengths of beach (x) as indicated by Figure 8.9b. There is also no necessity to consider the depth or thickness of the deposited layers.

If a point on the beach surface (B), a distance x from the point of deposition (D), is just shear-stable along a horizontal line through B (as it must be for horizontal equilibrium), then the horizontal thrust acting over a depth $z(= v - y$ in Figure 8.5a) below point D must be balanced by the horizontal shear force over depth z, i.e.:

$$K\gamma z^2/2 = \tau x \tag{8.4}$$

K is the lateral pressure coefficient (σ_h/σ_v) for the newly deposited tailings.

If τ is constant, then point B on the beach profile will be located by the co-ordinates

$$\text{length } x, \text{with } z = (2\tau x/K\gamma)^{1/2} \tag{8.5}$$

and the complete beach profile can be calculated. If the values of z and x at the edge of the pool are z_p and x_p, then the chord slope, where DP is the chord, is given by

$$\tan i_c = z_p/x_p \tag{8.6}$$

Figure 8.10a shows (solid lines) a series of measured beach profiles for deposits of gold, platinum, uranium, diamond and zinc tailings. Of these profiles, the gold, platinum and uranium beaches were formed from unthickened tailings in which considerable particle size sorting took place on the beach. The deposits of diamond and zinc tailings were formed from thickened tailings in which little or no particle size sorting occurred.

The broken lines in Figure 8.10a (marked "fitted") represent equation 8.5 fitted to each measured beach profile at the point of deposition ($x = z = 0$) and the edge of the pool (x_p, z_p) for unthickened tailings beaches, or the lower extremity of the tailings slope (P in Figure 8.9b). The degree of fit is very satisfactory, particularly as the fitting process assumes a constant shear strength, whereas, as mentioned above, the shear strength for beaches of unthickened tailings reduces down the length of the beach as a result of particle size sorting. The shear strength of thickened tailings slurries are nearer to constant, but may increase slightly as the slurry runs down the beach and its

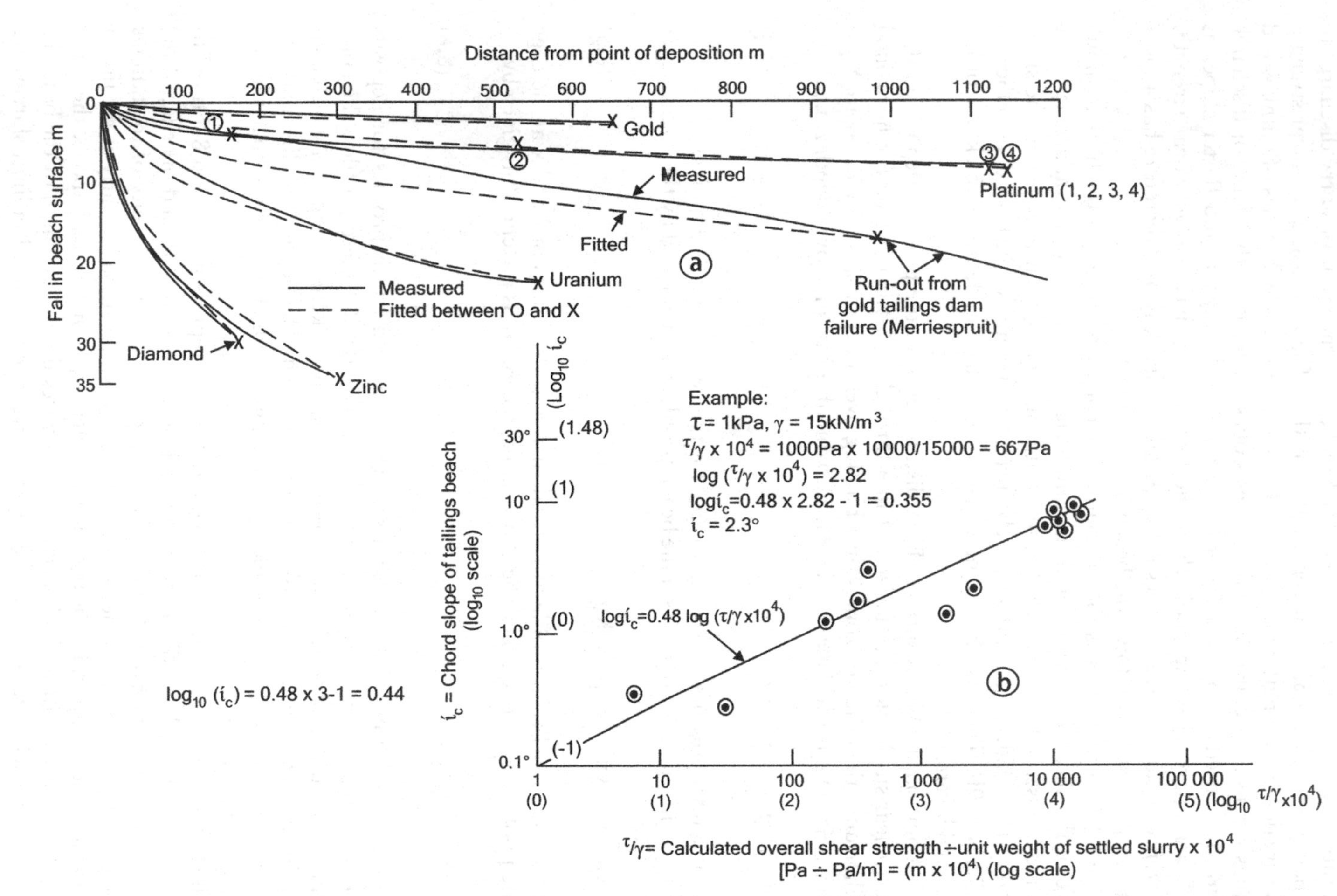

Figure 8.10 a: Measured beach profiles for various tailings (solid lines) fitted to equation (8.5). b: Relationship between chord slope i_c and strength to unit weight ratio, τ/γ.

water content is progressively reduced by evaporation and seepage into the existing beach surface.

Figure 8.10a also shows the measured profile of the tailings flood that escaped from the failure breach when the Merriespruit tailings dam failed in 1994 and the tailings ran through and devastated the village of Merriespruit (Fourie, Blight and Papageorgiou, 2001, Section 11.3.2). The measured profile was rather irregular, probably as a result of obstructions to the flow by houses, garden walls, etc., and the fit of equation (8.5) to the measured profile is not particularly good.

A consideration of equation (8.5) and Figure 8.10a will show that there is a relationship between the chord slope of a tailings beach and the ratio of shear strength τ to unit weight γ of the freshly deposited and settled slurry. Figure 8.10b has been constructed from chord slope angles (i_c), measured from Figure 8.10a, and the mean calculated shear strengths shown in Figure 8.11 for a unit weight of 16×10^3 Pa/m. The equation relating i_c (in degrees) to the ratio of τ (in Pa), to γ (in Pa/m, i.e. N/m^3) is

$$\log_{10} i_c = 0.48 \log_{10} (\tau/\gamma \times 10^4) - 1 \tag{8.7}$$

This can be used to calculate the chord slope as shown by the example in Figure 8.10b. Figure 8.11 presents profiles of shear strength down the various beaches, calculated for progressively increasing values of x and z for the measured beach profiles shown in Figure 8.10a. Calculated values for the thickened tailings beaches (diamond and zinc) were close to constant. For the beaches formed of unthickened tailings, however, the calculated shear strength increases with increasing x, in every case. (This is opposite to the expectation expressed earlier.)

Finally, it must be pointed out that there are many ambiguities in the type of calculations described above. Perhaps the most glaring is the assumption that the forces illustrated in Figure 8.9b, and hence the analysis resulting in equation (8.5), can be applied to cases (see Figure 8.10a) where the length of the tailings beach (x_p) can approach 1200 m and, simultaneously, z_p can be as small as 8 m. However, the results of this calculation appear to be sensible. A second ambiguity is illustrated by the contour plan for a ring dyke tailings storage with two pools that is shown in Figure 8.24 (Blight, 1994). The chord slope for beach CB is clearly flatter than that for beach AB, yet one would expect the shear strength of slurry newly deposited on both beaches to be very similar, as it comes from the same source. For beach CB, the chord slope $i_c = 0.26°$, and the average calculated shear strength is $\tau = 64$ Pa. For beach AB, $i_c = 0.48°$ and average $\tau = 81$ Pa. Concerning the master profiles for the two beaches, for CB, when x/h (Figure 1a) is 0.5, y/v is 0.46, and the corresponding value of y/h for beach AB is 0.5. Thus the two beaches have essentially the same master profile, but apparently different shear strengths. One must therefore view the concept of the master profile and calculations to predict beach profiles as only a simplified guide to the imperfectly understood, but much more complex processes that occur on a prototype tailings beach and control the beach profile.

8.4 Details of particle size sorting during hydraulic deposition

As already mentioned, in the process of beach formation, the coarser fraction of the tailings settles out closer to the point of deposition, while progressively finer material

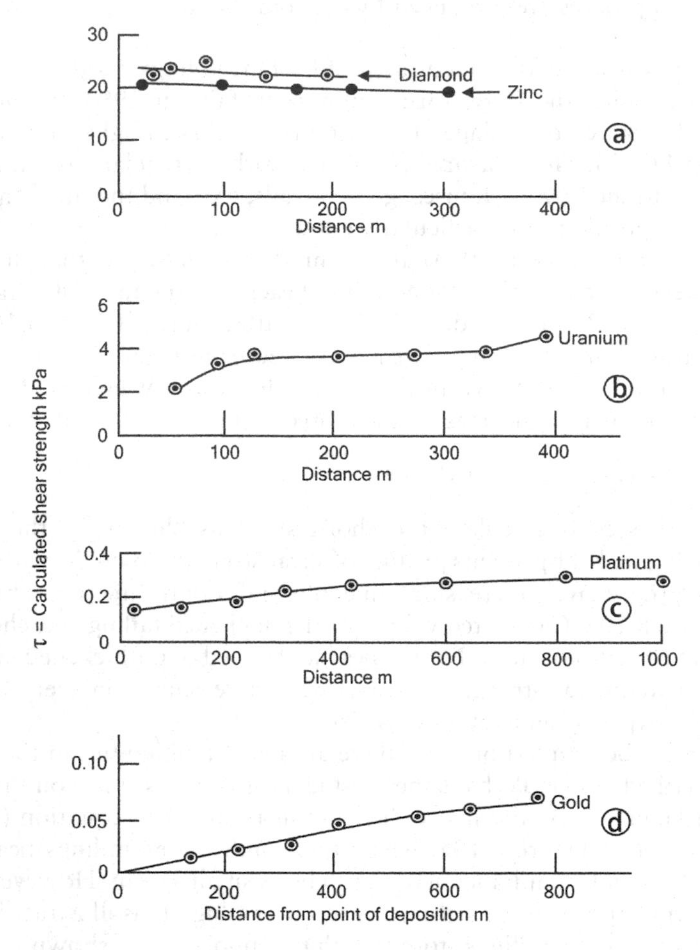

Figure 8.11 Profiles of shear strength τ versus distance x down the beaches of both unthickened and thickened tailings for profiles shown in Figure 8.10a.

settles out as the pool is approached. There is a discontinuity in the beach profile at the edge of the pool, with the slope of the underwater beach being much steeper than the beach deposited in air, because of the buoyancy provided by the water.

The first example of this type is given in Figure 8.12a which shows changes of particle size distribution on a laboratory model beach of diamond tailings having a solids concentration of 16%. Referring to Figure 8.7, the profile of the beach was concave, and the change in particle size distribution from $x/h = 0$ to $x/h = 0.9$ is pronounced. At a solids concentration of 38%, (Figure 8.12b) the beach profile was convex throughout and the variation of particle size distribution down the beach was negligible. Note that D_{50} in both diagrams a and b was the same. Figure 8.13 shows the variation of mean particle size D_{50} expressed as the ratio of D_{50} at any distance away from the point of deposition to D_{50} of the total tailings stream for a full size beach of diamond tailings. Note the general similarity of this distribution to the concave profiles in Figure 8.7.

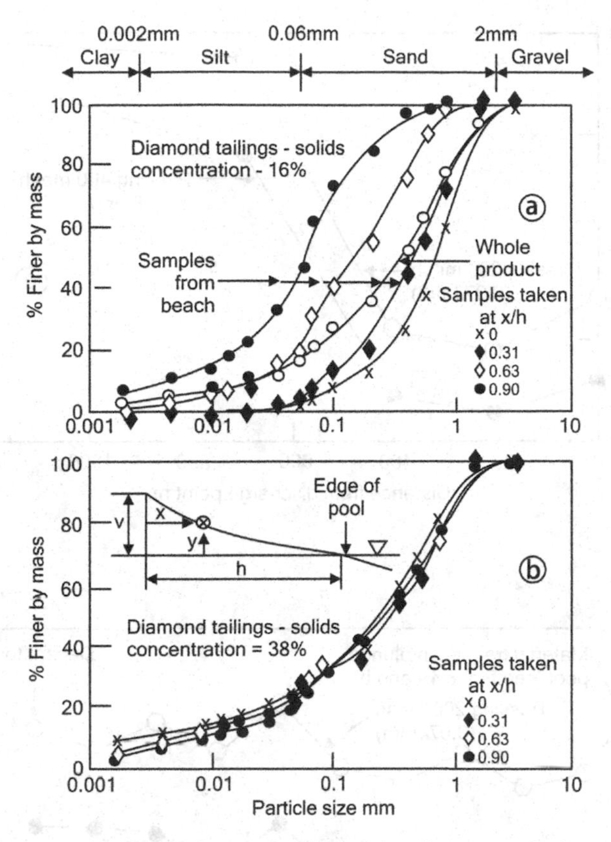

Figure 8.12 Comparison of particle size distributions: a: along an unthickened tailings beach, and b: a thickened tailings beach.

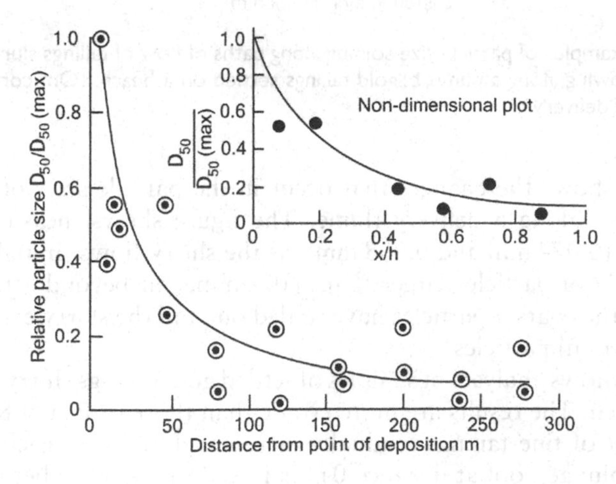

Figure 8.13 Particle size distribution measured on the beach of a diamond tailings storage.

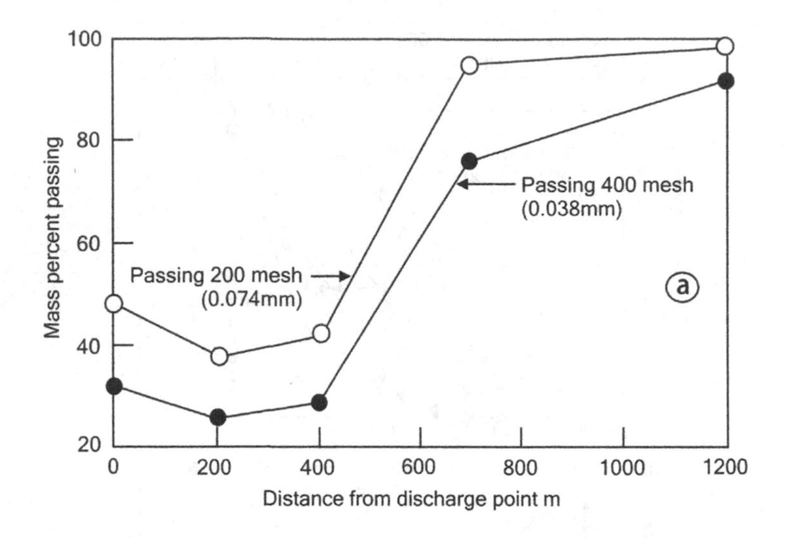

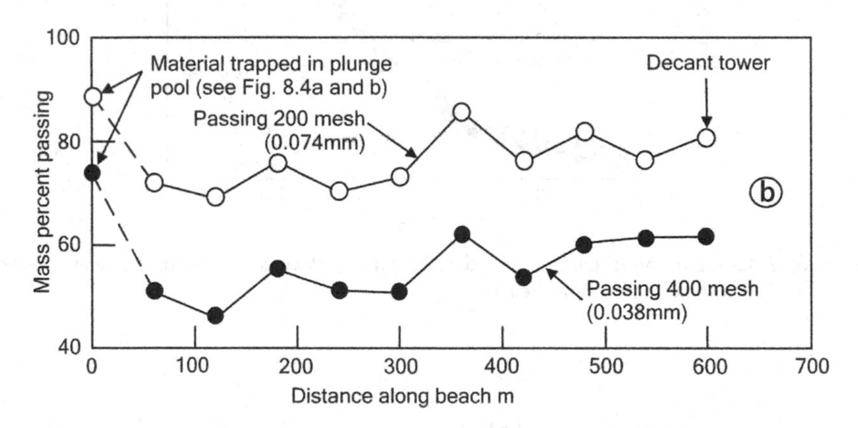

Figure 8.14 Examples of particle size sorting along paths of flow of tailings slurries: a: gold tailings flowing along a flume, b: gold tailings settled on a beach. (Om corresponds to point of delivery.)

Figure 8.14a shows the changes that occur in the particle sizes of a gold tailings slurry as it flows along a delivery flume. The figure shows measurements of two sizes of particle (0.074 mm and 0.038 mm), as the slurry flows. Initially, the slurry is composed mainly of particles larger than 0.074 mm, but beyond a flow distance of 400 m, most of the coarser particles have settled out and the slurry carries mainly fine (less than 0.038 mm) particles.

Figure 8.14b shows analyses of samples of settled gold tailings slurry taken down the surface of a beach. The results are more erratic than those of Figure 8.14a, but show how the content of fine tailings gradually increases down the beach. (The material trapped in the plunge pool at distance 0 m is finer (presumably) because the coarser particles had settled first and were now beneath the fines.

8.5 Effects of particle size sorting on permeability, water content and strength variation down a beach

8.5.1 *Permeability*

Particle size sorting has the considerable benefit that the coarser, more permeable material is deposited at the head of the beach, with the finer, weaker material in the interior of the impoundment, where it is contained by the coarser, stronger outer deposit. The combination of coarse material and a depressed phreatic surface means that the outer containing embankment is constructed of the strongest material in the whole deposit. The variation of permeability down the beach of a diamond tailings storage is shown in Figure 8.15a, while Figure 8.15b shows how the particle size sorting causes a depression of the phreatic surface. In arid and semi-arid climates the strength of the outer wall is augmented by capillary water stresses induced by evaporation from the surface of the wall and beach. In contrast, a beach of unsorted tailings will be uniform throughout its length and there will be no depression of the phreatic surface.

With tailings that contain appreciable proportions (i.e. more than 5 to 10%) of clay particles, for example kimberlite (diamond) tailings or tailings from a weathered, oxidized ore, particle size separation may have the unfortunate effect of separating out in the form of thin continuous layers of clay at the surface of each layer of deposition. In kimberlite tailings, the permeability of the clay may be as little as 10^{-5} (1/100 000) times that of the sandy material that forms most of the tailings. The result can be that k_v, the effective vertical permeability for the tailings body, is only 1/3000 of k_h, the effective horizontal permeability. A similar effect has been found with tailings from weathered, near surface platinum tailings. The result is, as illustrated in Figure 8.15c, that the phreatic surface is raised to a near horizontal position and exits the outer slope of the storage as a series of horizontal perched water tables, with an extremely dangerous effect for the stability and surface erosion of the slope. Possible remedies are to separate out the clay by cycloning, directing the clay-laden overflow into the basin of the storage, well away from the outer wall or to change the unthickened tailings system to a thickened tailings system in which the separation of the clay into layers will not occur.

Figure 8.15d shows the observed depression of the phreatic surface in a gold tailings storage that is constructed by upstream paddocking. The diagram is not as straightforward as it may seem. There is relatively little cross-wise particle size separation in the paddocks, but there is always a degree of sorting that results in the tailings being horizontally stratified in this area. In the beaching area, within the outer ring of paddocks, horizontal size separation would prevail with the deposited tailings becoming progressively finer as the pool area is approached. The two effects – vertical and horizontal particle size sorting – have opposite effects, lowering the phreatic surface in the beach area and raising it in the paddock area. However, in this case the lowering effect prevails.

8.5.2 *Water content profiles*

It would be expected from Figure 8.15a that a beach formed of unthickened tailings would, immediately after deposition and draining of excess water, have the lowest water content in the coarser material at the head of the beach and the highest in the finer material near the pool. On the other hand, a beach of unsegregated thickened tailings would be expected to have approximately the same water content throughout its length.

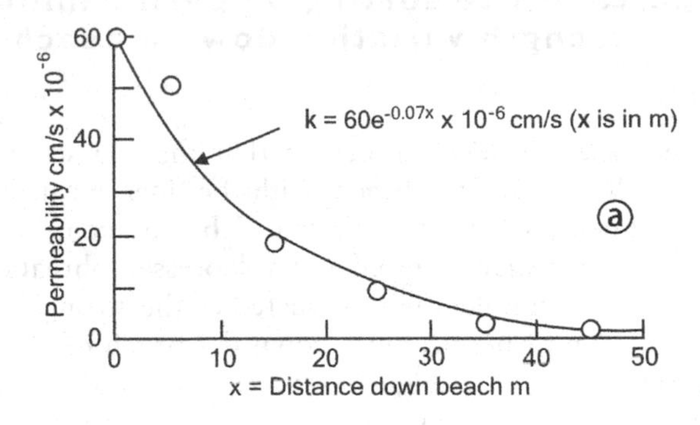

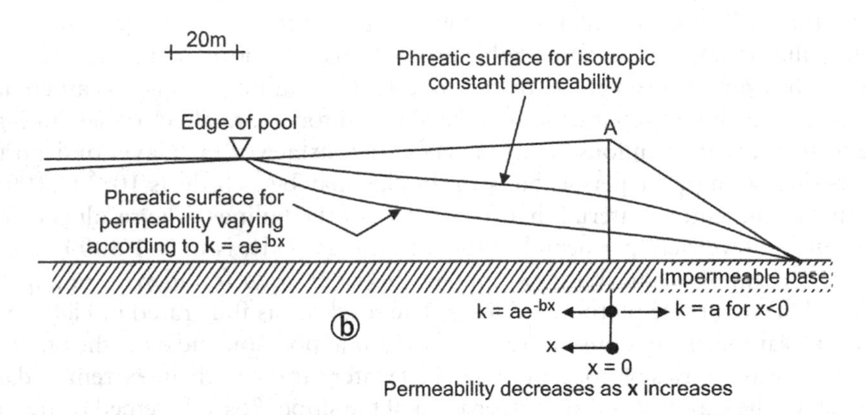

Figure 8.15 Effect of particle size sorting on permeability variation and consequent shape and position of phreatic surface: a: variation of permeability down a beach of diamond tailings, b: effect of varying permeability on elevation of phreatic surface.

Figure 8.16 compares water content profiles measured shortly after deposition on a full size beach of unthickened platinum tailings and a slope of thickened diamond tailings. The figure shows that the above expectation is fully realized, and that a beach of unthickened tailings increases in water content towards the pool, whereas thickened tailings has an almost constant water content.

8.5.3 *Shear strength profiles*

As is the case with water content profiles, one would expect the surface strength profile for an unthickened tailings beach shortly after deposition to have the highest strength at the point of deposition and that the strength would decline with distance down the beach. The surface strength of a thickened tailings slope, on the other hand, would be expected to be constant.

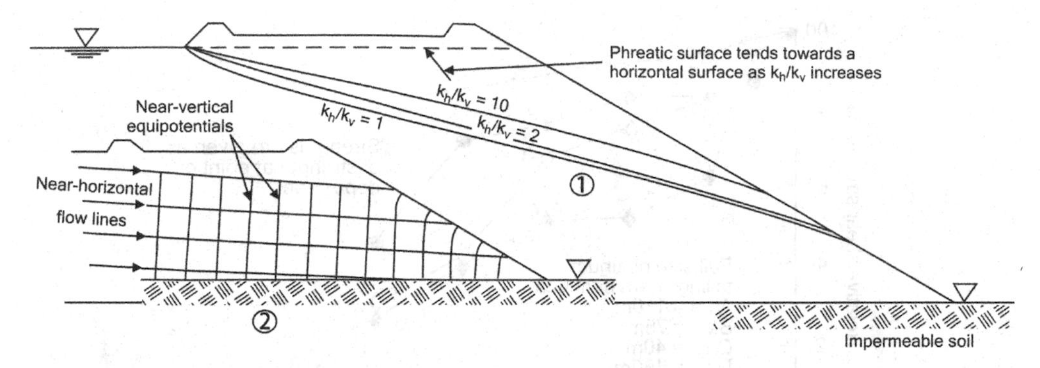

Figure 8.15c: Effect of vertical permeability k_v being small relative to horizontal permeability k_h: 1: Increasing k_h/k_v raises phreatic line. 2: Flow lines become flatter and equipotentials approach vertical. Pore pressure within storage increases with increasing k_h/k_v.

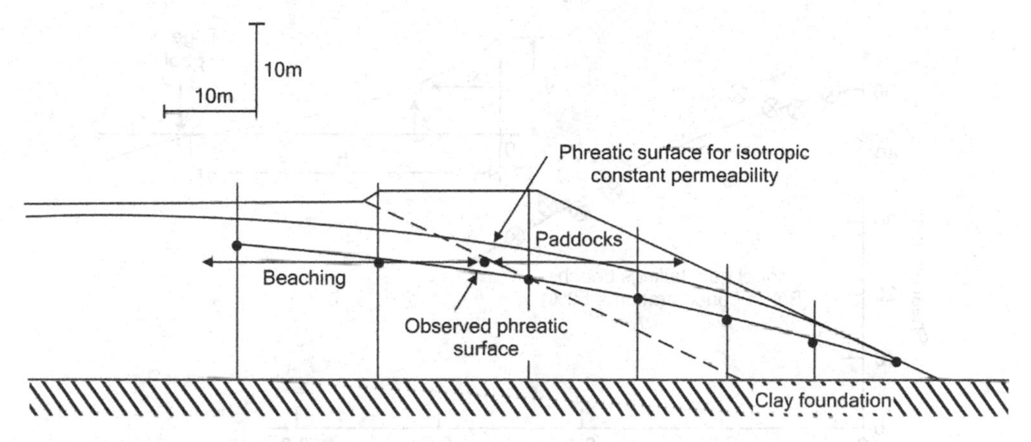

Figure 8.15d: Observed depression of phreatic surface as a result of particle size sorting on beaches of gold tailings storage.

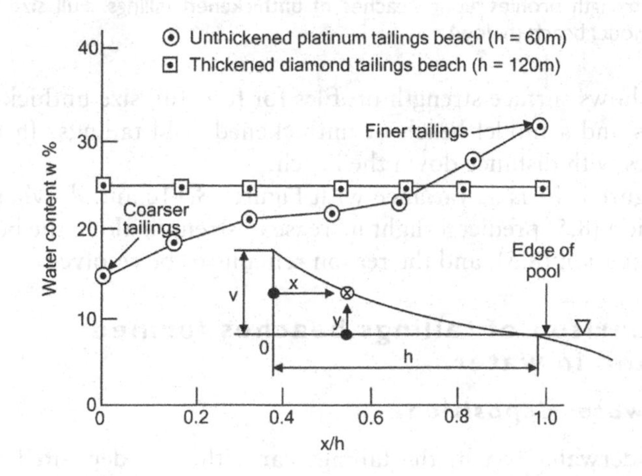

Figure 8.16 Water content profiles measured along unthickened and thickened tailings beaches.

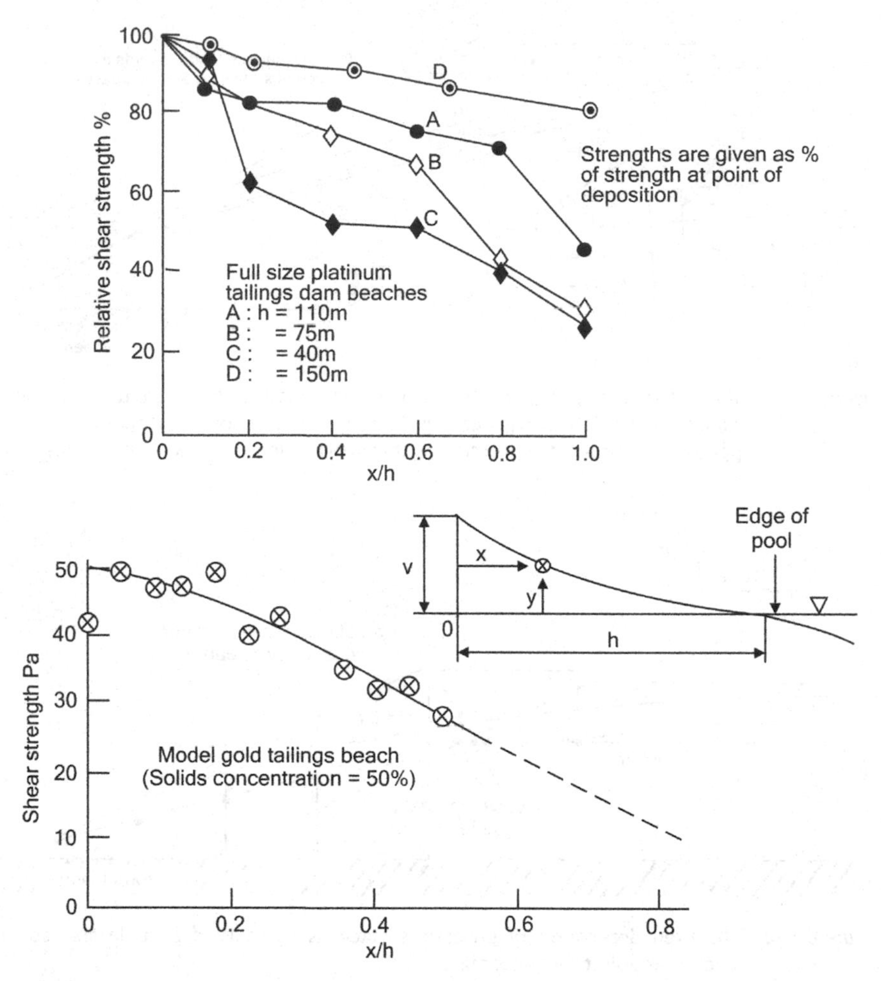

Figure 8.17 Strength profiles along beaches of unthickened tailings. Full size beaches (above), model beach (below).

Figure 8.17 shows surface strength profiles for four full size unthickened platinum tailings beaches and a model beach of unthickened gold tailings. In both cases the strength declines with distance down the beach.

However, Figure 8.17 is at variance with Figures 8.11c and d, where calculations based on equation (8.5) predict a slight increase in strength along the beach. The fault must lie with equation (8.5), and the reason remains to be resolved.

8.6 A comparison of tailings beaches formed in air and in water

8.6.1 *Underwater deposition*

To form an underwater beach, the tailings can either be deposited through water (i.e. allowed to fall freely through water from a deposition or discharge point at the

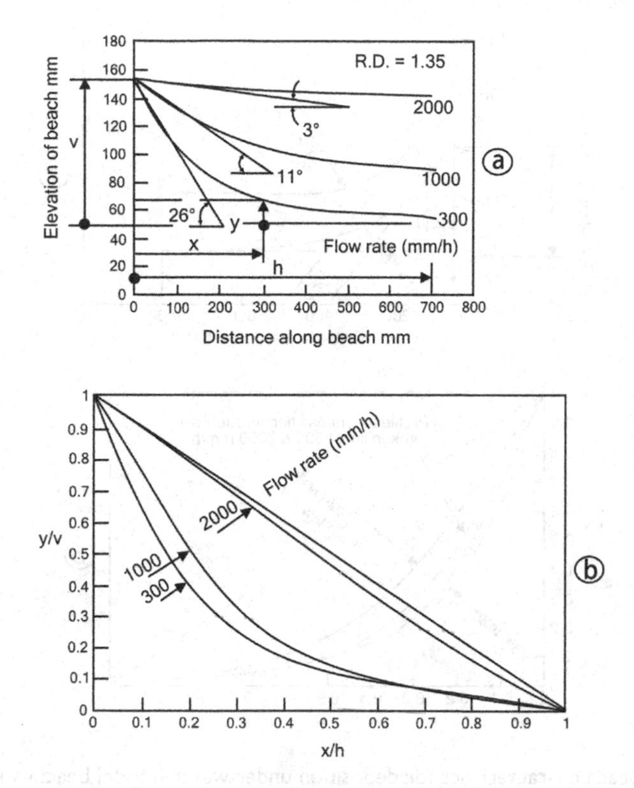

Figure 8.18 Beach characteristics for deposition through water. (Model beaches.)

water surface), or the deposition point can itself be underwater. The deposition through water (to be described in Section 8.9), took place from a spigot pipe supported at the surface of the water by means of a line of pontoons. The corresponding situation for underwater deposition would have had the spigots extended by means of flexible pipes that discharged the tailings onto the underwater tailings surface. This section will describe the results of a series of comparative model beaching tests on the deposition of a nickel tailings, both through and under water. These tailings had the particle size distribution of a sandy silt, containing 40% sand, 55% silt and 5% clay.

Figure 8.18 shows beach profiles formed by deposition through water at three flow rates. The flow rates are given in mm per hour of deposition as an average over the plan area of the laboratory flume in which the beach was formed. The dimensional beach profiles (Figure 8.18a) were found to flatten progressively as the rate of deposition was increased. At the slowest rate of deposition, 300 mm/h, the slope of the initial portion of the beach was 26° and this flattened to only 3° at the highest rate of deposition of 2000 mm/h. The non-dimensional beach profiles in Figure 8.18b show a progressive reduction in concavity as the flow rate was increased.

It appears from this rate effect that, at the slowest rate, the tailings were settling slowly enough for the material to be deposited in close to a fully consolidated state. As the flow rate increased, however, water was entrapped by the falling tailings which

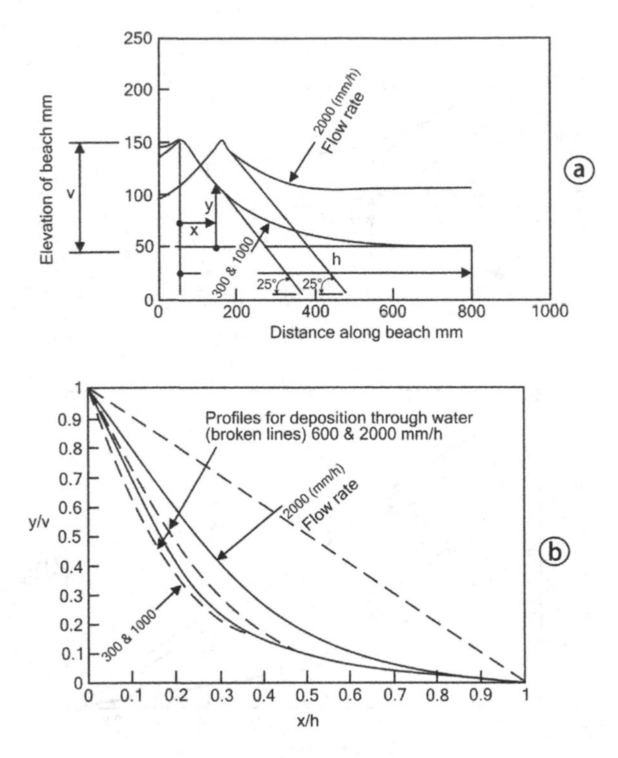

Figure 8.19 Beach characteristics for deposition under water (Model beaches.)

could not fully consolidate. Therefore the tailings had a progressively lower strength as they were deposited and consequently came to rest at progressively flatter slope angles.

Figure 8.19 shows the results of similar tests for deposition under water. When tailings are deposited in air, a crater-like plunge pool develops at the point of deposition. As described in Section 8.2, this pool causes a sharp reduction of the energy of the incoming tailings stream. The tailings beach then forms as the greatly de-energised tailings stream overflows the rim of the plunge pool and runs towards the pool. The plunge pools formed by underwater deposition are very similar in shape and dimensions to those formed in air. The dimensional beach profiles shown in Figure 8.19a all developed the same maximum angle (25°) at the head of the beach which was very similar to the maximum angle observed for deposition through water (26°). In this case, the beach profiles for rates of deposition of 600 and 2000 mm/h were almost identical. The dimensionless profiles shown in Figure 8.19b were also quite similar, and also very similar to the dimensionless profiles for deposition through water at rates of 300 and 1000 mm/h. It is obvious from Figure 8.19a that the size of the plunge pool depends on the rate of tailings flow and must be related to the reduction of energy caused in the tailings stream. Once the de-energized tailings stream overflows the rim of the plunge pool, it probably has a very similar energy per unit volume regardless of flow rate, and therefore forms

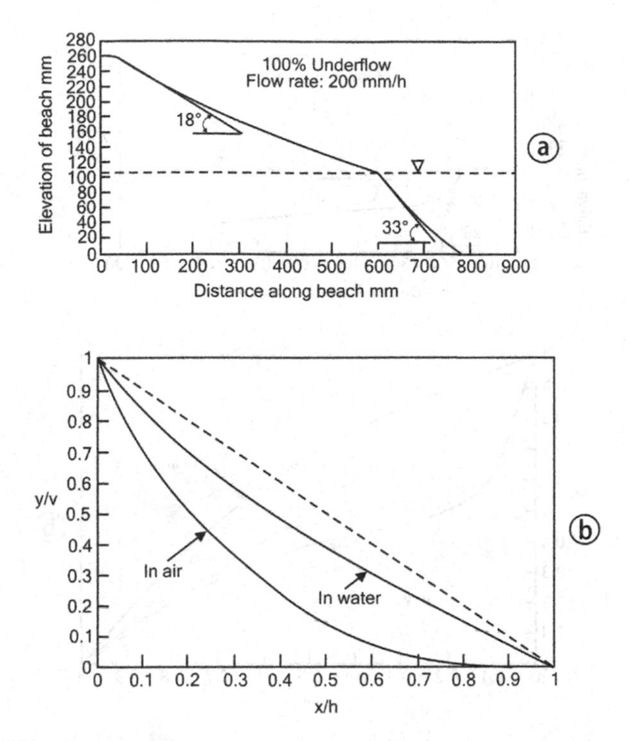

Figure 8.20 Beach characteristics for deposition in air, running into water – 100% underflow. (Model beach.)

beaches with very similar profiles. Because of the reduction of energy that occurs in the plunge pool, the progressive beach flattening with increased flow rate, that occurs with deposition through water, does not seem to occur with underwater deposition.

8.6.2 *Beaches in air that run into water*

It often occurs in hydraulic fill construction that material is deposited in air, forms a beach, and then runs into water. In normal tailings storage operation, the underwater portion of the beach is usually only a small part of its total length and (thus) unimportant. However in hydraulic fill land-building or land-reclaiming the underwater beach may be as important, or more important than the beach in air.

The experiments that will be described in this section were performed on a tailings material that had been split by cycloning into a sandy cyclone underflow and a silty overflow. (The principle of cyclone separation is illustrated in Figure 8.27). The particle size distributions of the split material consisted of:

- underflow – 95% sand, 5% silt,
- overflow – 60% sand, 37% silt, 3% clay

The overflow was thus not very much finer than the sand as it contained 60% of sand. It had only a small clay fraction of 3%. Despite its sandy

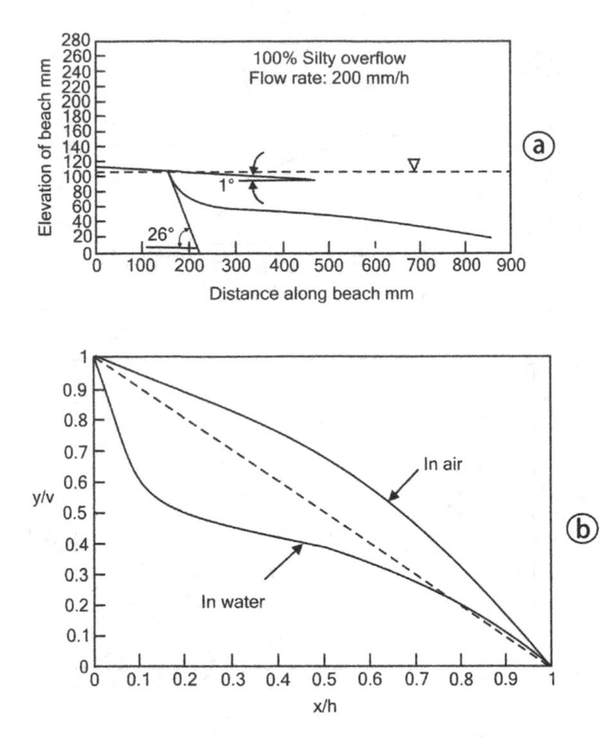

Figure 8.21 Beach characteristics for deposition in air, running into water, 100% overflow. (Model beach.)

nature, however, the slightly clayey overflow tailings behaved very differently to the sand from which it had been separated. Figure 8.20 shows the results of a model beaching test on the underflow sand at a relatively slow flow rate of 200 mm/h. This test was carried out with a slurry relative density of 1.5. Figure 8.20a shows that the beach had a sharp discontinuity where the tailings entered the water. The initial slope of the beach in air was 18°, while that of the underwater portion of the beach was 33° Figure 8.20b shows that the dimensionless underwater beach profile was much less concave than the profile of the beach in air.

Figure 8.21 shows a similar beaching test carried out with a slurry of the overflow also at a relative density of 1.5 and a deposition rate of 200 mm/h. In this case, Figure 8.21a shows that the beach in air had a very flat slope of only 1°. However, where the beach entered the water, the slope increased considerably to 26°. Figure 8.21b shows that the portion of the beach in air was convex, while that in water was initially concave and then became convex.

Figure 8.22 shows the shapes assumed by a beach of tailings consisting of a mixture of 80% of underflow and 20% of overflow, also deposited at a relative density of 1.5 and a deposition rate of 200mm/h. Figure 8.22a shows that the initial slope of the beach in air was intermediate (12°) between the corresponding slopes for underflow sand (18°) and overflow (1°). The initial slope of the underwater beach was 18°, less than that for either the underflow or the overflow. Figure 8.22b shows that the

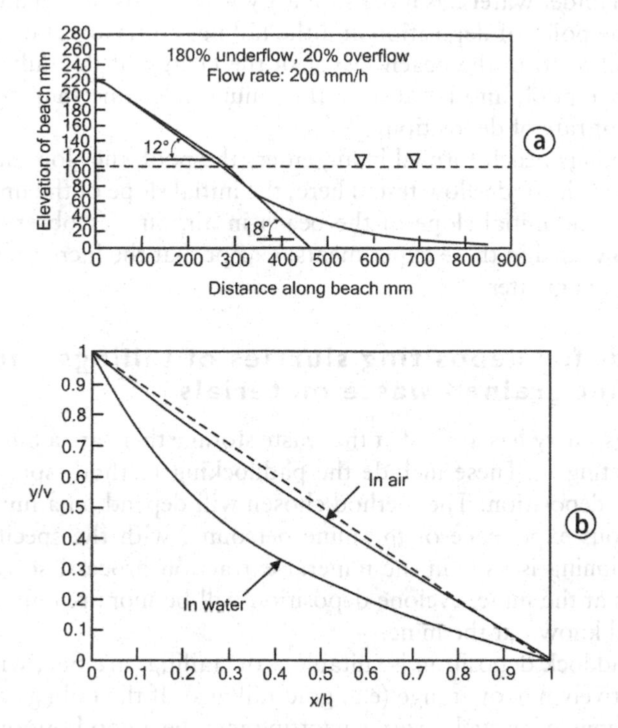

Figure 8.22 Beach characteristics for deposition in air, running into water, 80% underflow, 20% overflow. (Model beach.)

dimensionless profile for the beach in air was almost linear, while that for the beach in water was concave. Comparing Figures 8.22a and 8.22b shows that the shapes for the profiles in air and in water for the underflow/overflow mixture were the reverse of those for the underflow alone.

8.6.3 *Discussion*

1. The phenomena involved in hydraulic beach formation in air are not yet completely understood. Although a concave beach profile, for example, can be interpreted in terms of changes of energy and entropy (McPhail and Blight, 1998), this does not explain the formation of a convex or a concave/convex beach such as those illustrated in Figure 8.21b. It is obvious that deposition of tailings through or under water involves even more complex processes than the formation of beaches in air.
2. During the deposition of tailings through water, the initial slope of the beach depends very greatly on the rate of deposition. At slow rates the material settles on the beach in a consolidated state and the initial slope approximates an "angle of repose in water". As the rate of deposition increases, the tailings settle, still in slurry form, and, at very high rates, spread to form a near-horizontal beach surface.

3. Deposition under water has more similarity with deposition in air. A plunge pool forms at the point of deposition and the tailings slurry overflows the rim of the plunge pool to form the beach. Most of the energy of the tailings is dissipated in the plunge pool, and because of this, much the same shape of beach results regardless of rate of deposition.
4. When a tailings beach formed in air, enters the pool, the slope steepens sharply. In the case of the underflow tested here, the initial slope of the underwater beach was double the initial slope of the beach in air. Similar observations apply to the overflow used in these experiments, except that the increase in slope may be relatively even greater.

8.7 Methods for depositing slurries of tailings and other fine-grained waste materials

Once the tailings slurry has arrived at the waste storage there are a number of possible ways for depositing it. These include the paddocking method, spigotting or spray-bar and cyclone deposition. The method chosen will depend on a number of factors, including previous experience of the mine personnel with the specific tailings. For example, if cycloning is used in the mineral extraction process, so that expertise in cycloning exists at the mine, cyclone deposition will be more readily acceptable then if it is not a skill known at the mine.

In general, paddock deposition is suitable if the tailings are fine, with particle sizes falling in a relatively narrow range (e.g. gold tailings). If the tailings are less fine and cover a wider range of particle sizes, spigotting may be a good option (e.g. platinum tailings). Cycloning is usually applied to situations where spigotting would also be suitable. However, it has also been successfully applied to the deposition of gold tailings, where some particle size separation is produced together with a dewatering of the coarser material fraction.

The overall object of the deposition method chosen should be to use the mechanisms of beaching to best advantage for the particular tailings in hand. For example, if the tailings are a fine and/or virtually single sized product, little gravitational sorting will occur down a beach. On the other hand, if the material is deposited at intervals around the perimeter of a tailings storage and allowed to flow around the perimeter, a deposit will result that is segregated to a minimum degree. If the tailings have a wide range of particle sizes, beaching can be used to produce a storage with a coarse frictional outer shell with the finer material confined within the body of the deposit, and permeabilities that increase from inside of the storage to the outer wall.

8.7.1 *Paddock deposition*

In this system of deposition, tailings are deposited to form a "retaining" wall in a series of paddocks constructed by raising low bunds of previously deposited tailings. The paddock bunds are usually raised mechanically by means of a specially designed plough. 100 to 150 mm of tailings slurry are deposited in the paddock, and after settling, the supernatant water is drained off towards the pool. After a period of drying, the paddock bunds are raised again, and the cycle repeated. Plate 8.3 shows the outer paddock filled and with the bunds raised for the next deposition. Tailings

Plate 8.3 Surface of paddock wall on gold tailings storage. Paddock with sun-dried surface is to right of beach area.

deposited into the basin of the storage can be seen to the left of the paddock inner bund.

A chain of paddocks is constructed around the perimeter of the storage to form the impounding wall. Each paddock is used in turn for deposition with the balance of the material being deposited in the inner basin of the storage. The deposition cycle decides the rate of rise of the storage and is kept as long as possible to allow the previous layer of deposition to dry out by draining and evaporation before the next layer is deposited.

The outer paddock wall is usually built during daylight hours (the "day wall") and provides the freeboard to the storage. When the tailings delivery is not being used for wall-building, it is directed into the basin of the storage (the "night wall"). The tailings then beach from the delivery points towards the pool.

The success of the paddock method relies heavily on the densification of the day wall area and the adjacent beach areas as a result of desiccation by sun-drying and is best suited to an arid to semi-arid climate (see Figure 8.2).

The paddock system of construction can be used with either down-stream, upstream or centre-line construction methods as shown in Figure 8.23. It is particularly suitable for single particle-sized tailings. When used with graded particle-size tailings, vertical gravitational sorting of the particle sizes results in the formation of a series of fine horizontal impervious layers which have the effect of increasing

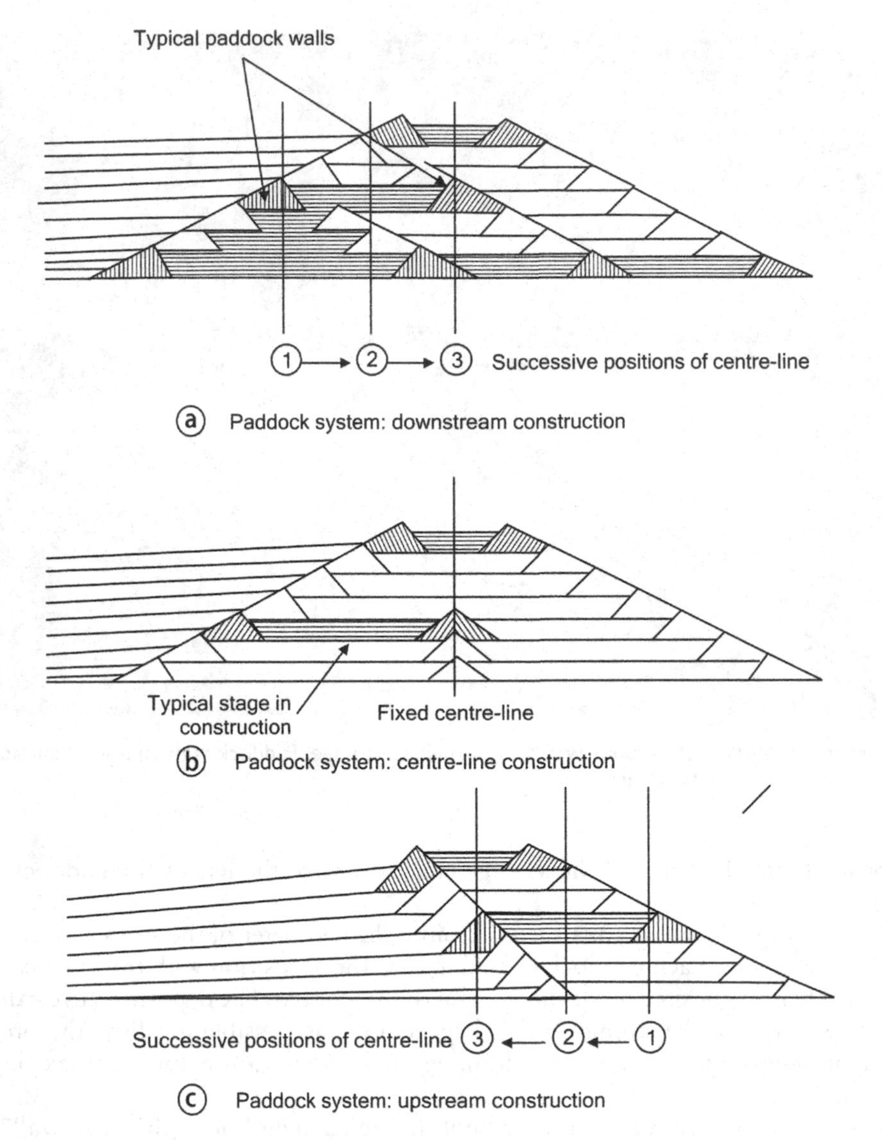

Figure 8.23 Application of paddock building to: a: downstream, b: centre-line, and c: upstream methods of construction.

the ratio of horizontal permeability to vertical permeability. The result may be a deposit with highly anisotropic properties, a high seepage surface and consequent problems with slope stability. A more serious result, however, is that the fine material may segregate at the low point of each paddock. As material is usually run in both directions from each delivery point, this results in a series of zones of both lower strength and lower freeboard than the zones adjacent to the delivery points.

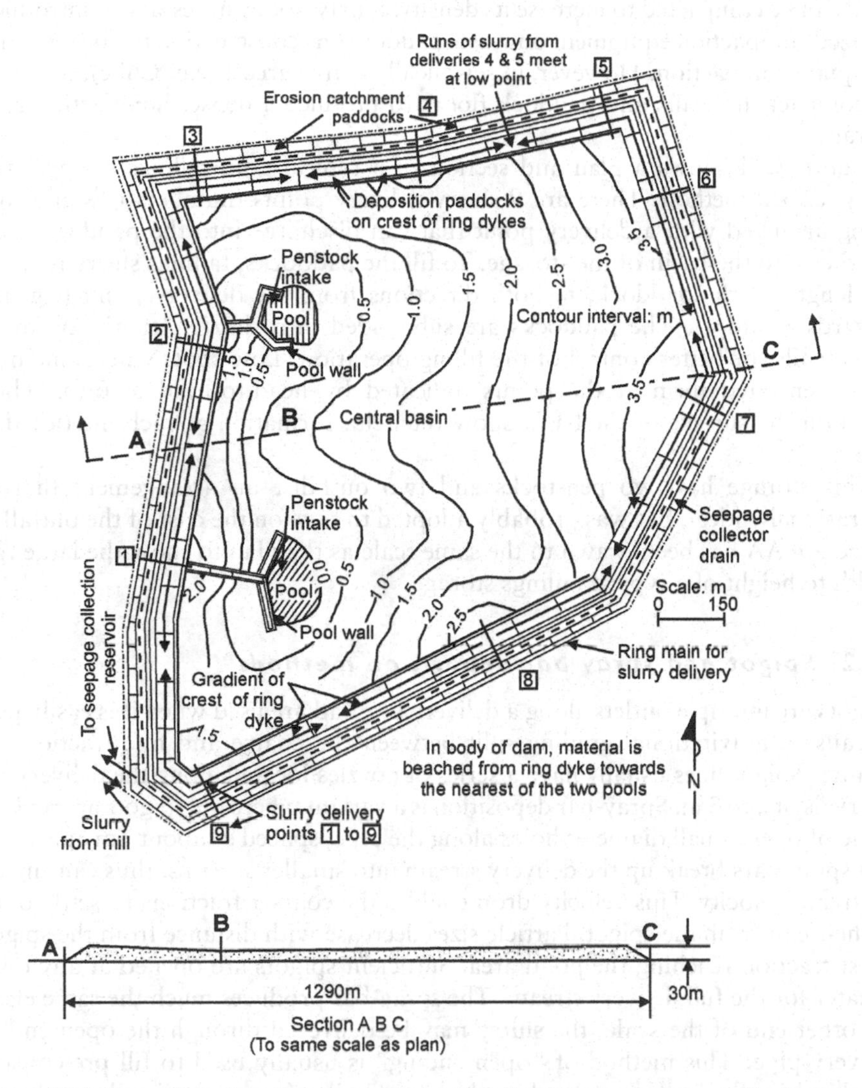

Figure 8.24 Plan and section of typical unthickened tailings storage built by paddock method.

Although the paddock system, as described, is very commonly used for gold, it has been adapted to other materials (particularly sands such as oil sands) on a much larger scale. Containment walls of up to 2 m depth are built round each paddock by mechanical means. The paddock is then filled by open-ending slurry into it. With sandy wastes, up to one metre of slurry may be deposited at a time. Wall building proceeds by scraping material up from the deposition area and placing it on the outer face of the storage by means of front-end loaders, bull-dozers or scrapers, to form the containment wall for future depositions. This operation is therefore carried out from inside the dam and can result in a loose uncompacted wall. Once the supernatant water has been drawn off, the sand can be compacted to increase its density. It may not be necessary to introduce specialized compaction equipment on waste paddocks as construction traffic may produce adequate compaction. However, in seismically active areas, (e.g. Chile), it is essential to compact the walls and paddock floors to produce a dense, liquefaction-resistant storage.

Figure 8.24 shows a plan and section of a typical gold tailings storage built by the paddock method. There are 9 slurry delivery points (numbers in squares), each being arranged with a delivery point that can discharge into the paddock area and another into the basin of the storage. To fill the paddocks, tailings slurry is run along the length of the paddock, in both directions from the delivery point (e.g. from 9 towards 8 and 1). The paddocks are subdivided into shorter lengths by means of cross-walls for better control of the filling operation. Low spots where fine material is concentrated form at the points indicated by head-to-head arrows. The contours on the surface of the basin show the delta formation at each interior delivery point.

This storage has two penstocks and two outfalls, an arrangement that is not operationally ideal, but was probably adopted to save on the cost of the outfall pipes.

Section AA has been drawn to the same scale as the plan to show the large ratio of width to height of a typical tailings storage.

8.7.2 *Spigot and spray bar deposition methods*

Spigots are multiple outlets along a delivery line and are used when it is easily possible to cause a gravitational grading split between the coarse and fine fractions of the tailings. Spigot lines usually have a series of nozzles located along the delivery pipe at intervals of 2 to 3 m. Spray-bar deposition is a variant where the spigots are replaced by a line of open, small diameter holes along the pipe, spaced at about 1m apart. Spigots and spray-bars break up the delivery stream into smaller streams, thus causing a drop in stream velocity. This velocity drop enables the coarser fractions to settle out close to their exit from the spigot. Particle sizes decrease with distance from the spigot, the finest fraction reaching the pond area. Sufficient spigots are opened at any one time to cater for the full delivery stream. The spray-bar produces much the same effect. At the other end of the scale, the slurry may be delivered through the open end of the delivery pipe. This method of "open ending" is usually used to fill pre-constructed ponds where the walls or dykes have been pre-built of compacted soil, overburden or tailings.

Wall building is a process requiring sections of the spigot lines to be raised in one section, while allowing uninterrupted deposition to continue on other sections. Bull-dozers, front-end loaders, backactors and scraper-loaders are used to build walls mechanically, while manual shovel packing may be used on small-scale operations.

In addition to the vertical freeboard provided by the wall construction, the spigotted beach develops basin or volume freeboard as the coarser fractions deposit at a steeper slope than the finer fractions. Figure 8.25 illustrates the spigot discharge deposition method. Plate 8.4 shows spigots in operation. The plunge pool formed by each spigot discharge should be noted. Plate 8.5 shows an unusually widely spaced spigot system in operation. Each spigot forms its own plunge pool and delta of deposition.

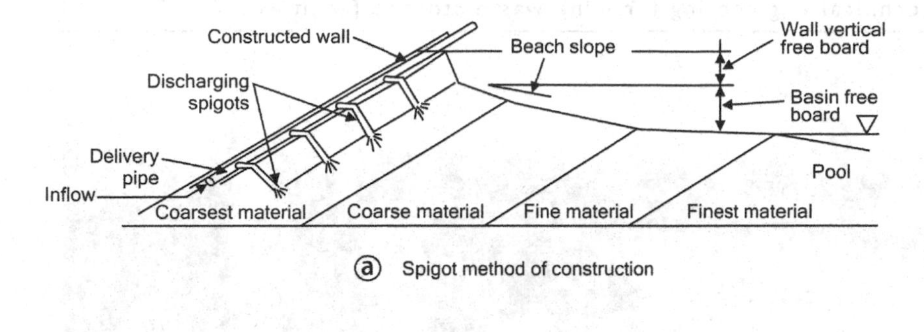

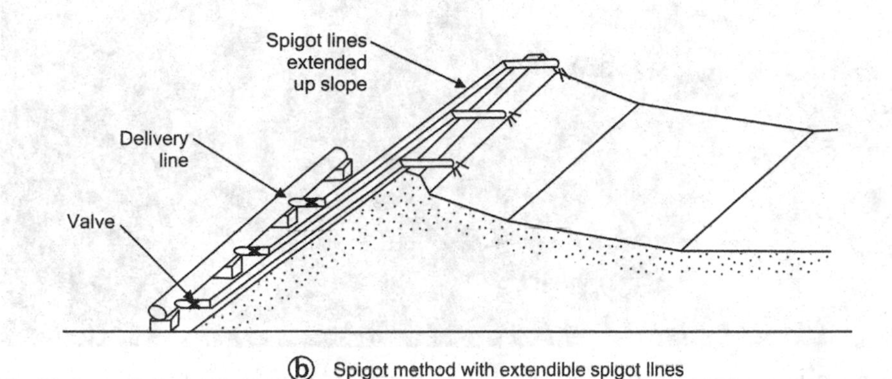

Figure 8.25 a: Principle of spigot deposition, b: typical method of extending spigot lines.

Plate 8.6 shows a spray-bar system being used to build the outer dyke of a down stream construction valley dam in Chile. Because of the seismic activity in the region, each lift of tailings is compacted before placing the next lift. The tracks of the compaction equipment are visible in the foreground. At the opposite extreme to the closely spaced openings in a spray-bar, Plate 8.7. shows red mud from a bauxite mining operation being open-ended into a tailings pond. Here, the delta of deposition extends more than half-way across the pre-constructed pond.

Spigotted dams often involve lower capital and operational costs than cycloned storages, but the particle size separation achieved is not nearly as controllable. A degree of horizontal layering is inevitable and this reduces vertical permeability and may result in near horizontal seepage in the tailings.

There are many methods of extending and raising the spigot lines as the height of the storage increases. Some of there are listed below:

- The main delivery line can be left in its initial position on the starter wall. Maintenance and inspections can then easily be carried out. This system, illustrated in Figure 8.25b, relies on extending the spigot outlet or branch pipes.

Plate 8.4 Spigot system of deposition in operation. (The operators are known, quaintly, as "mudguards", a term from the early 1900's.)

Plate 8.5 Unusually widely spaced spigots on a gold mine in Nevada.

Plate 8.6 A spray-bar system being used to construct a cross-valley dam in Chile. Each layer deposited is allowed to dry and is then mechanically compacted to reduce the void ratio.

- Branch lines can be laid from the main delivery line to the top of the storage. Here the branch line forms a secondary line inside and parallel to the crest of the storage, which has spigots in it at suitable intervals.
- When the spigot lines shown in Figure 8.25b become too long, the main delivery line can be lifted to a suitable new base elevation. To accomplish this lift, a suitable stepback is required in the slope profile of the storage.

Figure 8.26 shows a plan and section of a tailings storage that is being built by the spigotting method. Note the twin penstock towers and their separate outfall pipes. This ensures that should either of the penstocks or outfalls develop a fault that renders it inoperative, the storage will continue to operate safely without the pool encroaching on the freeboard. The large freeboard of 7 m above the penstock intake level should also be noted, as should the small alluvial fans opposite most of the delivery points. ("Small" means about 200 m radius.)

8.7.3 *Cyclone deposition methods*

The basis of the operation of a cyclone (or hydrocyclone) is illustrated in Figure 8.27. A cyclone is a very simple device, with no moving parts, that consists of a conical housing equipped with a feed pipe that enters the cone tangentially at its larger diameter closed end. A second pipe enters the cone on its axis and intrudes into the body of the cone. The slurry feed enters tangentially under pressure and is forced to swirl with a spiral motion towards the smaller open end. In the process, centrifugal force causes the larger

Plate 8.7 Open-ending bauxite red mud in Western Australia.

particles in the slurry to move down and away from the axis, towards the wall and the narrow exit of the cone, while the smaller particles concentrate on the axis of rotation and move, in the opposite direction, towards the axial pipe or "vortex finder". The net effect is that the finer particles and most of the water leave the cyclone through the vortex finder and form the "overflow", while the partially dewatered larger particles leave at the opposite end as the coarser "underflow".

Figure 8.27 shows examples of proportions for a typical "cyclone split" into finer overflow and coarser underflow. These proportions can be adjusted by changing the pressure of the slurry feed and the length and angle of the cone. The diagram also indicates how the coarser underflow can be used to build the outer wall of the storage while the finer, relatively less dense overflow is directed down the beach towards the pool.

Cyclone deposition can be applied to downstream, centerline and upstream building methods. In every case, the cyclone underflow is used to build the outer wall and gain freeboard rapidly. The overflow is deposited into the impoundment raised by means of overlapping underflow cones. Because the cyclone underflow is a relatively uniform and free-draining material, cyclone walls are generally stable even without the benefits of desiccation.

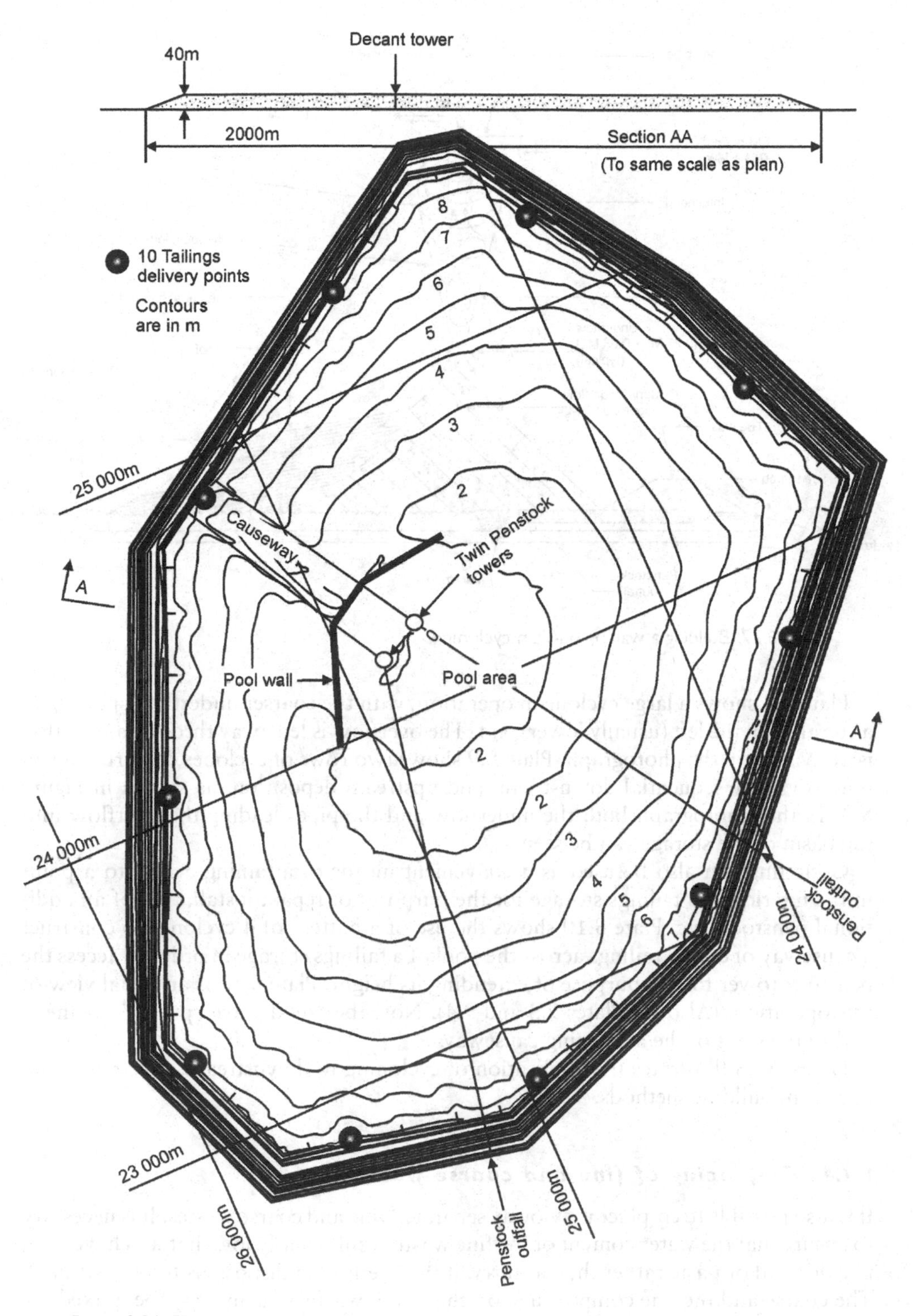

Figure 8.26 Plan and section of unthickened tailings storage built by spigotting method.

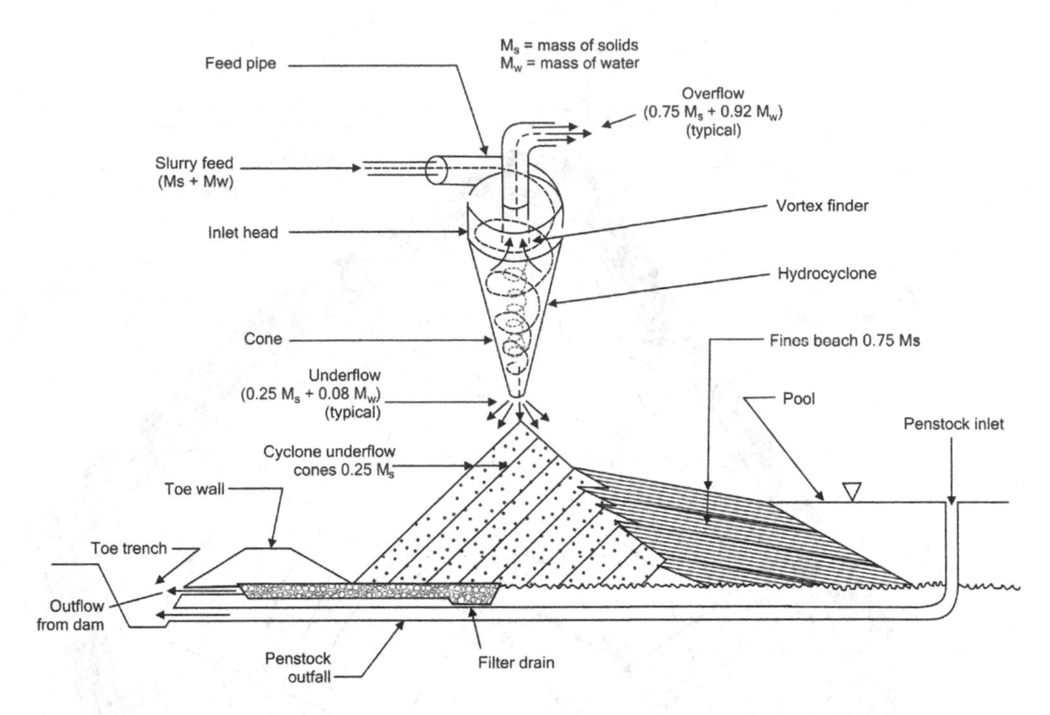

Figure 8.27 Building a wall by on-dam cycloning.

Plate 8.8 shows a large cyclone in operation, with the coarser underflow spewing in a spiral from the left (usually lower) end. The overflow is led away through a pipe that is not visible in the photograph. Plate 8.9 shows two rows of cyclones constructing an outer dyke by sequential downstream and upstream deposition, as shown in Figure 8.3. In this photograph both the underflow and the pipes leading the overflow into the basin of the storage can be seen.

Cycloning can also be used as a convenient means of attaining access to a point in the interior of a tailings storage for the purposes of repair, installation of an additional penstock, etc. Plate 8.10 shows the use of a battery of 4 cyclones to construct a causeway of coarse tailings across the pool of a tailings storage in order to access the penstock tower for the purpose of extending its height. Plate 8.11 is an aerial view of this operation. (Also see Plates 9.3 and 9.4). Note the "mud wave" pushed up ahead and to the sides of the advancing causeway.

Figure 8.28 illustrates the application of cycloning to downstream, centre-line and upstream building methods.

8.7.4 *Co-placing of fine and coarse wastes*

It is also possible to co-place previously separated fine and coarse wastes. It is necessary to ensure that the water content of the fine waste is sufficiently low that it behaves as a plastic solid or paste rather than a slurry, if this method of disposal is to be instituted. The coarse and the fine components of the total waste stream may be mixed by

Plate 8.8 A large cyclone in operation. Underflow is visible emerging as a spiral spray. Overflow is conducted away through a pipe.

Plate 8.9 Double row of cyclones building by sequential upstream and downstream deposition.

Plate 8.10 Battery of four cyclones building a causeway to gain access to a decant tower. Note mud wave ahead and to side of causeway.

discharging them onto a common conveyor belt that transports the waste to the dump. The mixing of the two components takes place by the vibration of the belt as it moves along. A co-placed deposit may be started from the top of a ramp constructed of compacted coarse waste. Once it falls from the end of the conveyor belt, the mixed coarse and fine waste flows out at a flattish angle to form a profile that has much in common with the beach profile of a hydraulically-deposited residue (see Figure 8.29). In Figure 8.29, the fine tailings was in the form of a non-segregating slurry (see Figure 8.13b). Plate 1.13 shows this method in operation. The ramp carrying the conveyor belt can be seen on the skyline to the left.

Alternatively, coarse and fine wastes may be co-placed by depositing a layer of coarse waste in an approximately horizontal or flatly sloping position, and then distributing the fines slurry over the surface of the coarse material to fill the void space in the coarse waste. In this method the fines slurry may need to be more fluid in order to flow into the interstices between the coarse particles.

This method of disposal has a number of advantages over the more usual separate disposal of coarse and fine wastes:

1. Less water is sent out to the storage as there is little or no free or decantable water in the co-placed waste.

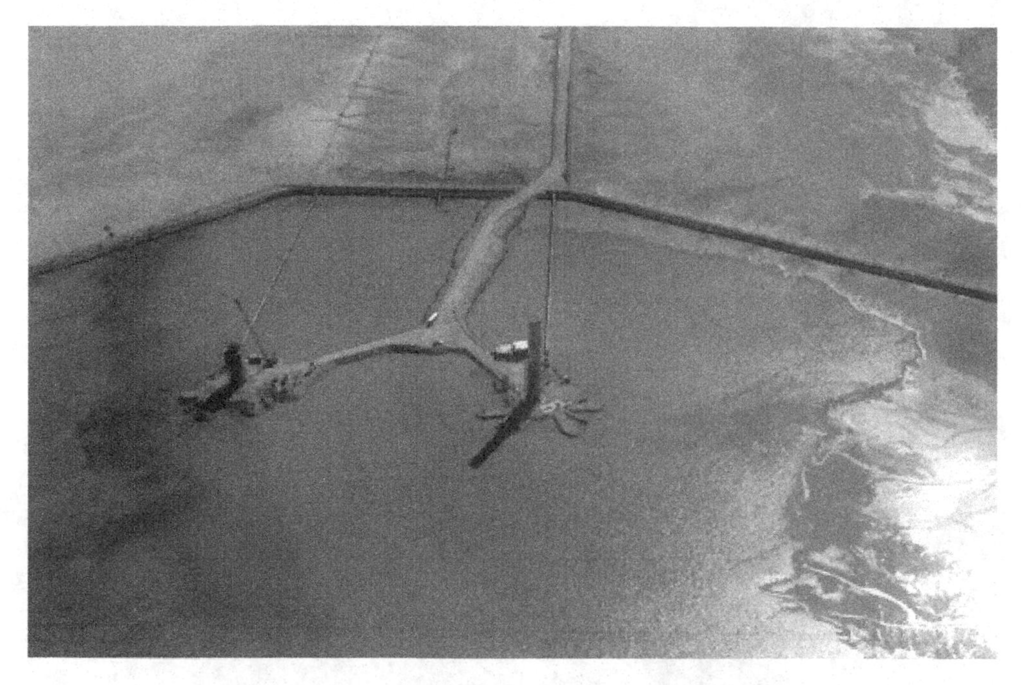

Plate 8.11 Aerial view of completed causeways. (Photo by Ljiljana Nedeljkovic,)

2. The lack of free water, combined, in arid and semi-arid climates with high evaporation rates, reduces the potential for ground-water pollution.
3. Because in the ramp method of placing, the mixed waste flows out under its own weight and adopts a slope that is stable for the wettest condition that is ever likely to occur in the material, there are few potential problems of slope stability. (This is, provided the strength of the foundation strata are adequate.
4. Because the fine waste fills the voids in the coarse, the fine waste is protected by the coarse particles from erosion by both wind and water. Hence overall erosion losses are reduced.
5. The air and water permeabilities of the co-placed waste are governed by the permeability of the fine waste. This reduces infiltration of rain water and also reduces the ingress of oxygen into the storage. In the case of combustible or pyritic wastes, therefore, the tendency to spontaneous combustion, as well as the rate of generation of acid from the oxidation of pyrite are inhibited.

However, there are also disadvantages:

6. Because of the flat beach angle assumed by the ramp co-placed waste, the length of run-out is great (10 to 15 times the height of the dump). Hence the method requires a large land area.
7. Although the surface roughness imposed by the coarse waste inhibits rain-water erosion, the longer slope length means that the slope collects more precipitation, which in turn tends to increase erosion. (Also see Section 8.8.4)

▼① ▼② ▼③ Successive positions of cyclones
Feed pipe gantry and line of cyclones
Overflow forms fines beach
Impoundment built of coarse underflow
④ ③ ② ①
Toe wall
Starter wall
ⓐ Centre-line cyclone wall building

① ② ③ ------▸ Successive slurry feed pipe positions
Wall crest
Overflow forms fines beach
④ ③ ② ①
Toe wall
Impoundment built of coarse underflow
Starter wall
ⓑ Downstream cyclone wall building

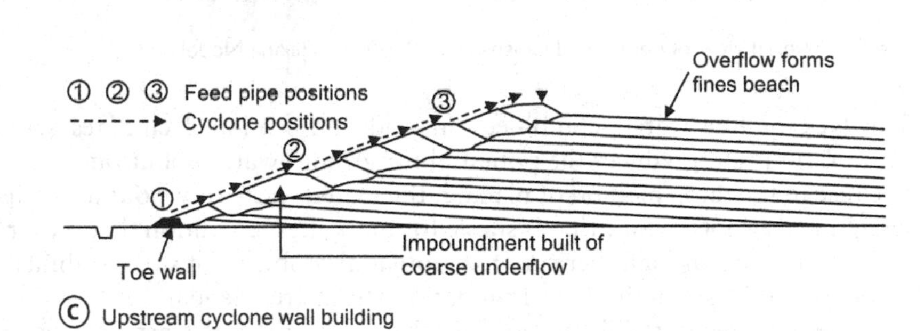

Figure 8.28 Various methods of wall building using cyclones.

8.8 Operational systems for tailings storages

The operational system for a tailings storage includes:

- The method of distribution and deposition of the tailings;
- The rate of deposition (which, taken in conjunction with the area of deposition, determines the rate of rise of the storage);
- The amount of water stored on the top surface of the storage and the length of time for which it is stored (which affects the position of the seepage surface within the outer walls of the storage, and the risk and consequences of failure. The amount of water stored also affects the position of the seepage surface within the storage

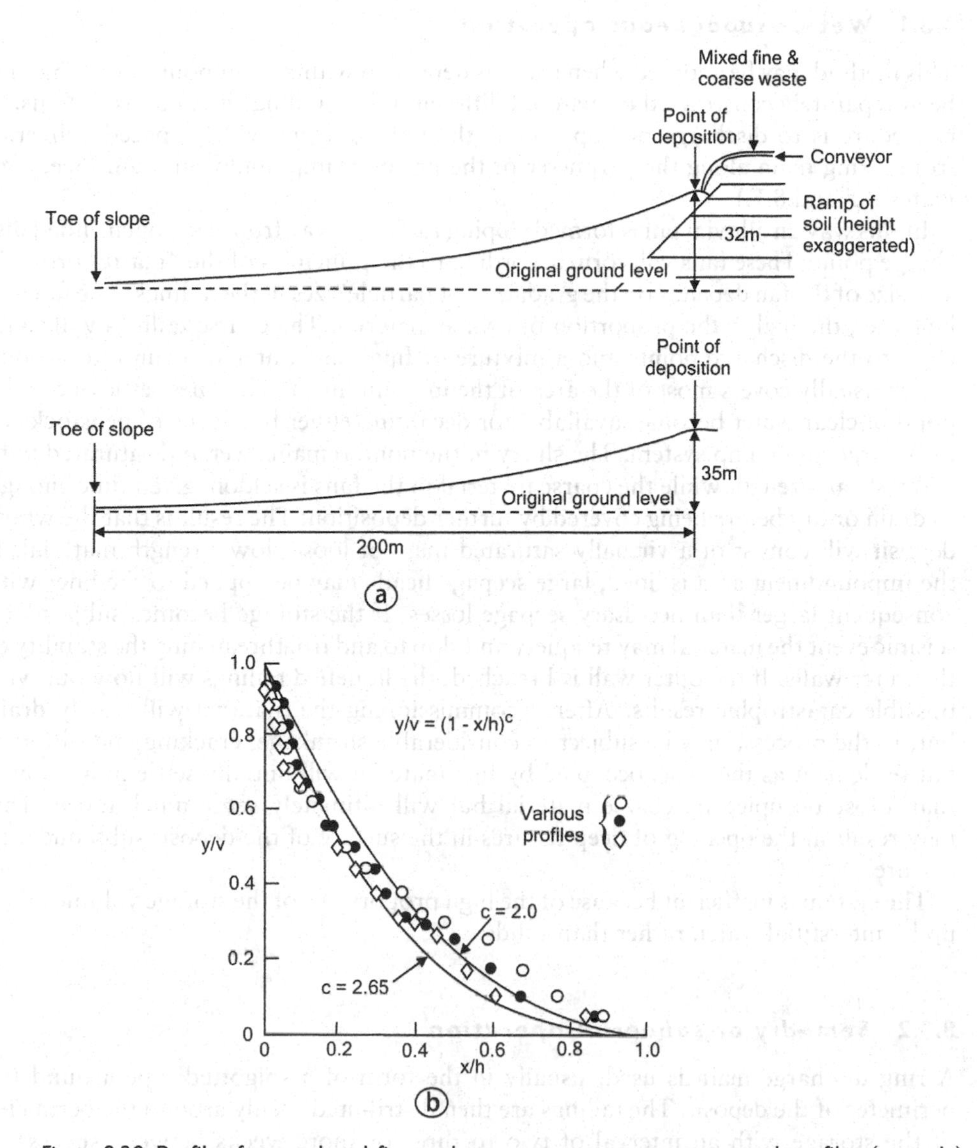

Figure 8.29 **Profiles for co-placed fine and coarse wastes: a: two measured profiles (to scale), b: corresponding non-dimensional profiles.**

at decommissioning, and therefore the rate at which the waste contained within the storage will drain after decommissioning).

With careful and rational operating procedures it is possible to control (within limits) the density, compressibility, permeability and shear strength of the tailings in various zones of the dam.

There are two recognized classes of operation for tailings dams: wet operation and semi-dry operation.

8.8.1 *Wet or subaqueous operation*

This method may be adopted when waste is deposited within an impoundment that has been separately constructed of material different to the tailings it contains. The usual procedure is to discharge or "open-end" the tailings from widely spaced deliveries from a ring main along the perimeter of the pre-built impoundment wall. (See, e.g., Plates 8.5 and 8.7.)

In this way an alluvial fan is formed sloping radially away from each open ended discharge point. These fans will form according to the principles of the "master profile". The size of the fan depends on the gradation of particle sizes in the tailings. The fan will be larger, the higher the proportion of coarse material. The coarse tailings will settle close to the discharge points and a mixture of fines and water will run into a pond which usually covers most of the area of the impoundment. The fines settle out and a pond of clear water becomes available for decanting either by means of penstocks or by a barge and pump system. The slurry in the pond remains wet and saturated with a low shear strength while the coarse material in the fans is seldom given time enough to drain or dry before being covered by further deposition. The result is that the whole deposit will consist of a virtually saturated mass of loose, low strength material. If the impoundment area is lined, large seepage heads may be applied to the liner with consequent larger than necessary seepage losses. If the storage becomes subject to a seismic event the material may reliquefy and slop to and fro, threatening the stability of the outer walls. If the outer wall is breached, the liquefied tailings will flow out with possible catastrophic results. After decommissioning the material will slowly drain but, in the process, may be subject to considerable shrinkage, cracking and differential settlement as the areas occupied by fine material will initially settle more slowly than those occupied by coarse material but will ultimately settle much more. This may result in the opening of deep fissures in the surface of the deposit subsequent to closure.

The system is inefficient because of the high proportions of the storage volume taken up by interstitial water, rather than solids.

8.8.2 *Semi-dry or subaerial operation*

A ring discharge main is used, usually in the form of a spigotted pipe around the perimeter of the deposit. The tailings are then distributed evenly around the perimeter of the storage with an interval of two to three or more weeks between successive depositions. The outer wall is raised by ploughing, or other mechanical means to build a dyke or paddocks around the wall area. Water released by settling out of the solids in the tailings is drawn off the storage as rapidly as possible and either recycled to the plant or held in the return water reservoirs. An absolute minimum of water is held on the top surface of the storage. As a result of this procedure most of the surface of a storage operated in this way is dry except when tailings are being deposited, or during rainy weather.

Resulting from the semi-dry method of operation, almost the entire surface of a tailings storage is maintained in a dry condition. This maintains the seepage surface at a low level which is beneficial from the point of view of the stability of the outer wall. It also means that the saturated pool area is always small and that the wall area

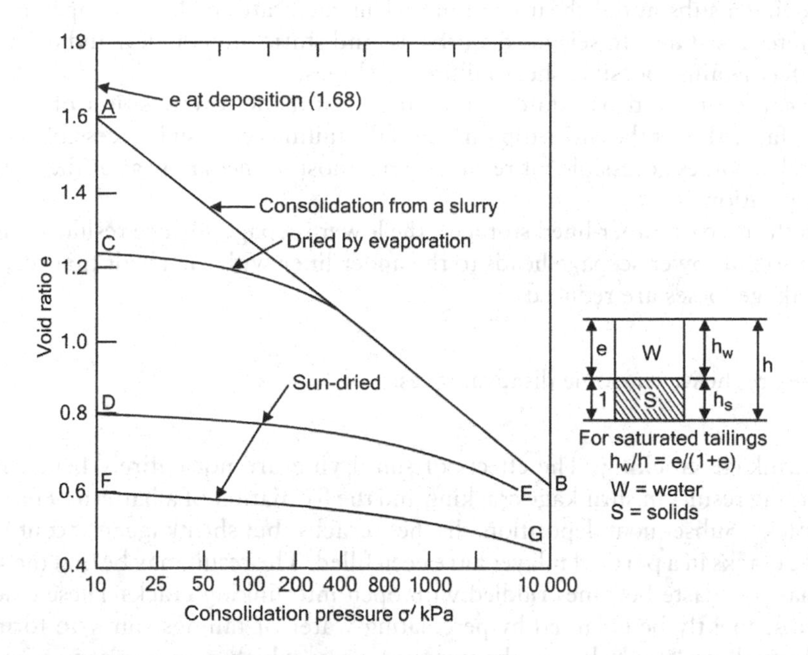

Figure 8.30 **Effect of sun-drying on void ratio and consolidation characteristics of gold tailings.**

of the storage is effectively far larger than the nominal upstream wall area, (see Figure 8.2). The semi-dry method of operation can also be applied when cycloning is used to separate the coarse fraction of the tailings from the fines, or when paddocking methods are used. In the latter case, water must be decanted from each paddock shortly after a deposition has occurred, and the deposited tailings must be allowed to dry before raising the walls of the paddock and depositing the next layer of tailings.

The important aspects of the method are:

1. A minimum of water is retained on the top surface of the storage.
2. The deposition of the tailings in thin layers at well-spaced time intervals allows as much time for drainage and sun-drying of successive depositions as possible.

The semi-dry method of operation avoids many of the disadvantages of the wet method of operation. Its principal advantages are:

3. Sun-drying causes a considerable shrinkage and compaction of the waste. As a result, a given mass of tailings occupies a smaller volume and the site is therefore used more efficiently. Figure 8.30 illustrates the considerable reduction in void ratio undergone by a gold tailings as a result of sun drying.
4. The compacting effect of the sun-drying considerably increases the shear strength of the material so that the deposit consists mainly of reasonably dense material

having a substantial shear resistance. This facilitates reclaiming operations, promotes resistance to seismic disturbance and static liquefaction and offers greater safety against possible shear failures of slopes.

5. Because of the dense state of the tailings and the low position of the seepage surface, almost the entire top surface of the tailings deposit is accessible at all times and becomes accessible for reclamation almost immediately after the cessation of deposition.
6. In the case of under-lined storages, the lower seepage surface results in the application of lower seepage heads to the under-liner with the result that seepage and leakage losses are reduced.

There are, however, some disadvantages:

7. Shrinkage cracking: The effects of sun-drying are not entirely beneficial. Sun-drying results in shrinkage cracking and the formation of a pattern of interlinking cracks. Subsequent deposition fills these cracks, but shrinkage may continue after the cracks in a particular layer have been filled. The result may be that the sundried mass of waste becomes riddled with open interlinking cracks These cracks may subsequently be enlarged by percolating water or tailings slurry to form piping channels or "rat holes" in the sundried waste which may result in serious gulley erosion and undermining of the outer slopes of the storage. Plate 8.12 shows an extreme example of piping occurring by flow through interlinked shrinkage cracks in a storage of a fine ash that was unusually shrinkable. In the foreground, a standpipe piezometer has been exposed in a piping channel. The man standing in the centre of the photograph is next to a large erosion gully that is in the process of extending towards the left. The paddock system of deposition is particularly vulnerable to the effects of shrinkage cracking at points where longitudinal runs of slurry end and overlap with runs coming from the opposite direction (see Figure 8.24). To prevent this, runs should be limited in length so that accumulations of fine tailings do not occur in specific areas of the wall of the storage.
8. If the tailings contain a wide range of particle sizes, the deposited waste may be stratified. This stratification can be undesirable, particularly if the ratio of the permeability parallel to layers to that across layers exceeds about 50. The large ratio between horizontal and vertical permeabilities may tend to concentrate flow in a horizontal direction through the coarser material with the effect that the seepage surface becomes raised or even perched. This may constitute a danger to the integrity of the dam walls as the upper part of the walls may rapidly wet up during periods when an unusually large quantity of water is held on the dam surface, e.g. during a period of prolonged heavy rainfall, resulting in a phreatic surface perched above the permanent phreatic surface.

7 and 8 above provide cogent reasons why water should never be held on the surface of a dam near to the outer wall.

It will be noted that in order to benefit fully from the effects of sun-drying, the rate of rise of a residue dam has to be limited to two to three metres per year.

Plate 8.12 Excessive shrinkage cracking of beach surface, resulting in piping erosion.

8.8.3 *In-pit disposal*

Using a pre-existing mining void to dispose of current waste production has the following advantages:

1. The major advantage of the in-pit disposal is that impoundment structures are minimal.
2. Because disposal occurs into an existing excavation, in-pit disposal produces no adverse visual impact.
3. It is suitable for both dumping and hydraulic fill methods of operation.
4. In-pit disposal may also be combined with dyke construction to enable the disposal of tailings to take place in a worked-out area of an operational pit.

However the method suffers from the following disadvantages:

5. Open-pit workings are usually deep enough to penetrate beyond the level of the local water table. It may therefore be necessary to store waste below the water table which may be undesirable if it contains any harmful pollutants and if the groundwater is subject to significant regional flow.

6. The sides of the pit may cut through faults or zones of preferential seepage which if they intersect the waste will cause the rapid introduction of any leachate or pollutants into the regional groundwater system.
7. Even if the waste is deposited above the present level of the water table, it is possible that this level may have been lowered by mining activities and will re-establish itself at a higher level once mining ceases. Also, it is possible that long term fluctuations in water level over periods of hundreds or thousands of years may ultimately mean the submergence of the waste below the water table.
8. Particular care must be taken not to store waste in the form of wet tailings above underground workings. Several disasters are on record of wet tailings breaking through into underground workings and causing the deaths of miners working there (See Table 11.1, entries (5), (18) and (19).)

8.8.4 *Thickened tailings and pastes*

Figures 8.7 and 8.12 show that once the water content of a tailings slurry has been reduced to below a given limit (and conversely, once the solids concentration has been increased to above a certain limit), gravitational sorting of particle sizes no longer occurs and the beach profile becomes planar to convex, instead of being concave. Tailings in this state are referred to as thickened tailings, and as their viscosity and shear strength increase further, as "paste tailings". As shown by Figure 8.8, even though the tailings are more viscous, the average beach slope (v/h) typically remains at less than 10% (in angular terms, less than about 5°).

The practice of thickening tailings before sending to storage is gaining popularity for a number of reasons, not all of which stand up to critical examination. Some of the claimed benefits of using paste and thickened tailings (P&TT) disposal practices are examined in what follows. (See Cincilla, et al., 1997, Williams and Seddon, 1999, Fourie, 2002 and Jewell, Fourie and Lord, 2002.)

All of these claims will not be examined, and indeed, the only way to validate most of them would be to carry out a full direct comparison of a number of methods of tailings disposal, as applied to a particular product produced at a particular site. However, three of the less scientifically based claims will briefly be examined.

CLAIM 1 Dry stacked tailings require a decreased footprint as the stacking height can be significantly increased.

The most popular configuration for a P&TT deposit is the single central point discharge in which the thickened tailings are discharged from a single elevated point and allowed to flow radially outwards to form a low conical pile or stack. A variant of this practice is the multiple point discharge system in which a series of overlapping cones are deposited from a pattern of discharge towers. Figure 8.31 compares the geometries of an unthickened tailings ring dyke deposit that occupies a circular area or footprint and a thickened tailings single point discharge cone that is confined to an identical footprint by a low circular bund. Figure 8.32 shows a plan and section of an actual single point discharge tailings storage (after Williams, 2002) while Figure 8.24 shows a ring dyke impoundment for unthickened tailings of a similar extent.

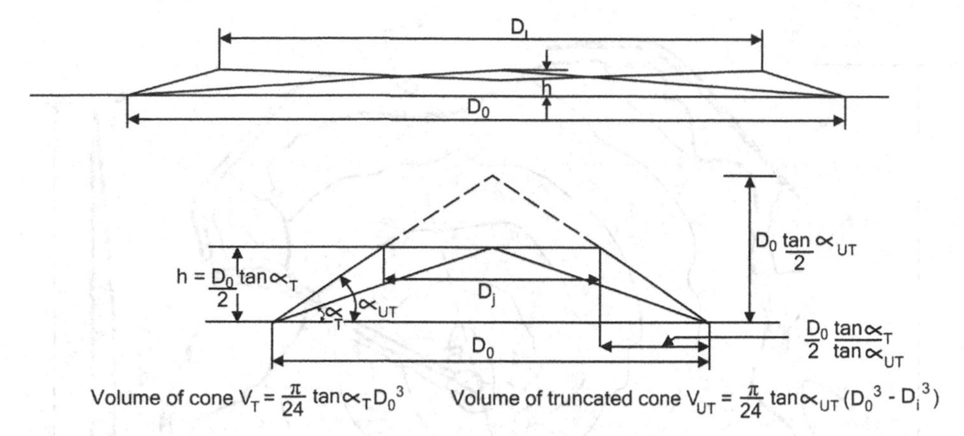

Figure 8.31 Comparison of geometries of central discharge thickened tailings deposit and unthickened tailings ring dyke impoundment.

The upper part of Figure 8.31 shows the geometries drawn to scale, the lower part exaggerates the height to make the details easier to show.

Expressions for the volumes of tailings deposited are as follows:

Volume of central discharge cone:

$$V_T = \pi/24.D_o^3.\tan\alpha_T \tag{8.1}$$

Height of cone:

$$h = D_o^2.\tan\alpha_T \tag{8.2}$$

Top diameter of truncated cone:

$$D_i = D_o(1 - \tan\alpha_T/\tan\alpha_{UT}) = \beta D_o \tag{8.3}$$

Volume of ring discharge truncated cone:

$$V_{UT} = \pi/24D_o^3[(1-\beta^3)\tan\alpha_{UT} - \beta^3\tan\alpha_B] \tag{8.4}$$

(where α_B is the slope of the unthickened tailings beach).

Ratio of volumes:

$$V_{UT}/V_T = [(1-\beta^3)\tan\alpha_{UT} - \beta^3\tan\alpha_B]\tan\alpha_T \tag{8.5}$$

As an example, the following numerical values have been used:

Slope of thickened tailings deposit:

$$\alpha_T = 4°(7\%)$$

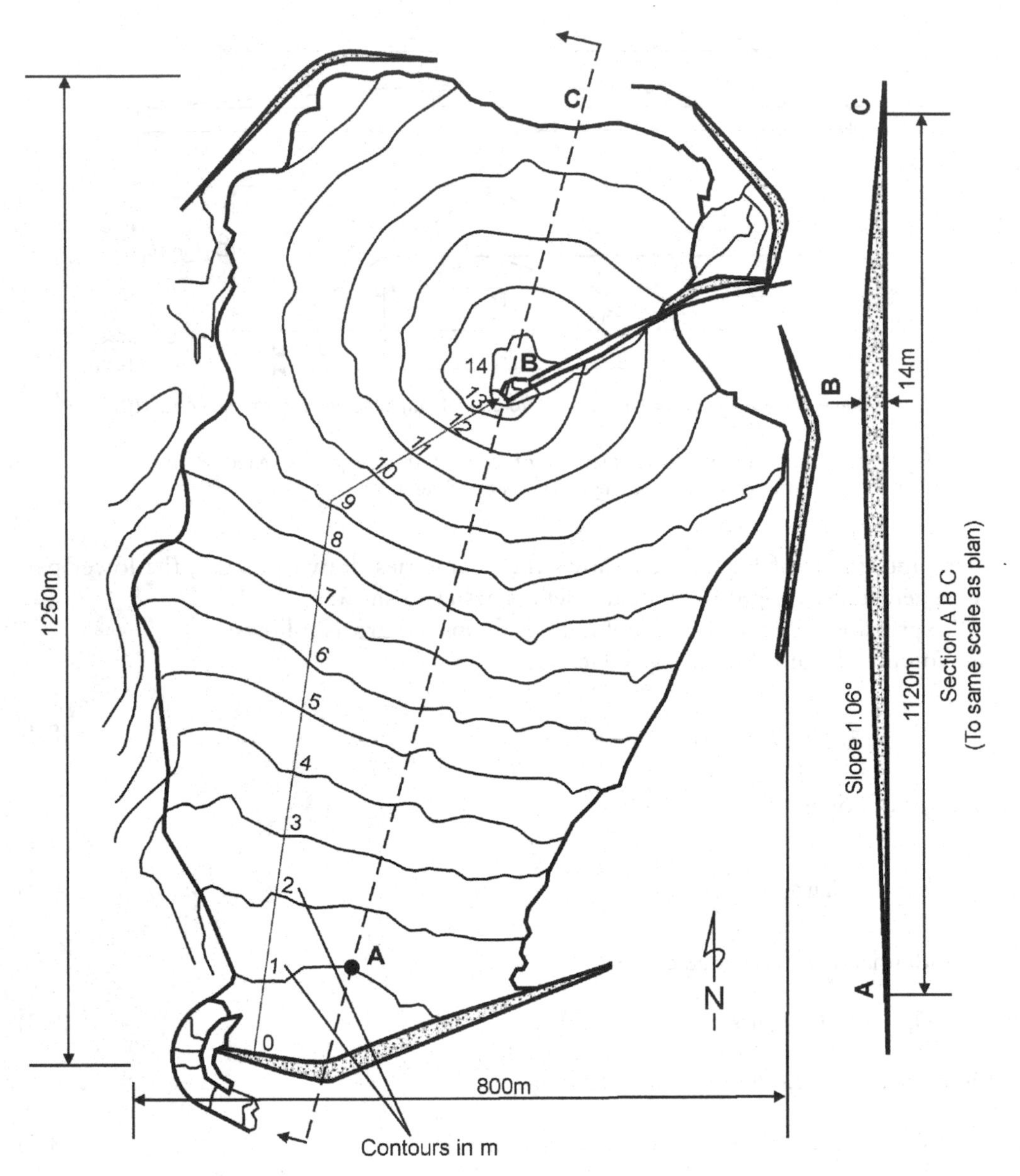

Figure 8.32 Plan and section of typical thickened tailings storage built by single point discharge method.

Outer slope of unthickened tailings deposit:

$$\alpha_{UT} = 15°(27\%)$$

Slope of beach of unthickened tailings:

$$\alpha_B = 2°(3.5\%)$$

The outer slope of 15° was chosen as it will allow the outer slopes of the unthickened tailings deposit to be built up of 18° segments with storm water control and access berms between them. Eighteen degrees is the maximum surface slope that can be worked by agricultural machinery (also see Section 12.4) and an overall slope of 15° (1 on 3.73) constructed by the upstream method at a low rate of rise will be completely stable against overall shear failure. For the above angles, $\beta = 0.74$, $\beta^3 = 0.40$ and $V_{UT}/V_T = 1.79$.

It appears from the data that have been given in Figures 8.16 and 8.17 that the average water content of a layer of unthickened tailings after deposition and drainage, but before drying, would be much the same as for a layer of the same tailings deposited as a thickened slurry. On this basis, for a given circular footprint on level ground, it is possible to store a mass of unthickened tailings that is 1.8 times that of the corresponding mass of thickened tailings, and still have very stable slopes for the retaining embankment.

It is also possible to combine the two methods by, for instance, depositing a thickened tailings cone on top of a previously constructed ring dyke. For the above example, the final total volume stored could be $(V_{UT} + V_T) = 1.56\ V_{UT}$, and $(V_{UT} + V_T)/V_{UT} = 2.8$.

CLAIM 2 Prompt creation of firm, well draining surface at completion.

This touches on an important difference between the upstream construction, unthickened tailings, beached tailings storage and both the single and multiple discharge thickened tailings storages. In the case of the former, it is possible to rehabilitate the outer (or downstream) slopes progressively as the storage grows in height, and hence progressively to protect them from both water and wind erosion. The interior basin of the storage is protected from wind erosion by the higher surrounding dyke or the valley dam and the sides of the valley, as the case may be. This cannot be done with either type of thickened tailings storage, or with centre-line or downstream construction with unthickened tailings. In all four cases, rehabilitation and hence protection against erosion and in particular, dust pollution as a result of wind erosion is impossible to effect until the storage is decommissioned.

CLAIM 3 Claim 3: Large potential reductions in water use.

This has been dealt with in detail in Section 10.5 and the reader is referred to this section for a comparison of the water-saving potentialities with unthickened and thickened tailings.

8.8.5 *The "best" tailings deposition system*

There is no best system for tailings deposition. There is only a best system for a particular tailings requiring storage at a site with a particular topography and given climatic conditions. It is the responsibility of the tailings storage engineer and is up to his ingenuity to find, compare, design and evolve the best system for each particular mine, and possibly for divisions of each mine.

8.9 An example of building an embankment by underwater deposition

Water decanted from diamond tailings, retained by a valley dam, was gradually being pushed up a side valley towards a small housing settlement, and it was decided to construct an embankment across the valley to provide a buffer zone between the tailings and the border of the settlement. The best way of doing this appeared to be to construct the embankment across the water-filled part of the valley by depositing tailings into the water.

8.9.1 *Laboratory feasibility investigation*

To investigate the possibility of successfully using this method, a laboratory experiment was carried out. A tank with transparent perspex sides, measuring 2 m long by 300 mm wide was filled with water, to a depth of 300 mm. Experiments were made with two methods of deposition. The first method was intended to simulate spray bar deposition from the surface of the water. The tailings stream, at a relative density of 1.3, was released at the water surface and the tailings were allowed to fall through water and settle on the bottom of the flume. The second method was intended to simulate tremieing the tailings through water. The tailings stream was released from an open-ended tube just above the bottom of the flume, and the pipe was gradually raised as the level of deposited solids rose.

Both methods worked successfully, and a steep-sided model embankment with slopes of up to 38° on either side was quickly deposited. During these experiments it was observed that much of the clay in the tailings dispersed into the water, so that the model embankment consisted primarily of sandy material.

One of the design concerns had been whether the downstream face of the embankment would be stable once the water on the downstream or land side had been drawn down, removing the buoyancy effect of the water and subjecting the downstream slope to seepage. This situation was simulated by siphoning off the water from one side of the model flume, leaving the model embankment to retain the remainder of the water. Only very minor slumping of the downstream slope occurred. This appeared to be the result of seepage erosion rather than shear instability.

Undisturbed samples were taken from the model embankments and tested for permeability. The spray bar and tremied model dykes were found to have much the same permeability of 1.5m/year, whereas the total tailings has a permeability of only 5×10^{-3} m/year.

8.9.2 *Full-scale implementation of underwater embankment*

Following on the promising laboratory tests, it was decided to proceed with the building of the full-scale embankment by depositing tailings by means of a spray bar from the surface of the water. A tailings delivery column was established from the plant to the site of the embankment, and a spray-bar column was extended across the water-filled valley on pontoons. Extensive use was made of helicopter lifting to overcome the difficulties posed by the rugged, rocky slopes of the valley. Plate 8.13 shows a pipe being helicoptered out to the spray-bar column visible as a line across the water in the middle distance.

Plate 8.13 Helicopter lifting pipes to form floating spray-bar column.

After only two weeks of deposition, the crest of the underwater embankment appeared at the water surface. Plate 8.14 shows the crest of the embankment shortly after it had emerged from the water. The spray-bar, now lying on the crest, is shown discharging upstream into the impoundment.

Once the crest of the embankment had emerged sufficiently to be easily accessible, a testing programme was carried out to establish the in-situ properties of the embankment and thus to determine whether it was safe to continue raising the embankment and dewatering the downstream area. The objectives of the testing were to establish the extent of consolidation, the in-situ strength and the coefficient of consolidation of the embankment. A series of piezocone probings was carried out from the top of the embankment, and the following properties were indicated.

1. The pore-water pressure versus depth relationships proved to be linear, with a unit water head versus depth gradient, showing that the pore pressure distribution in the embankment was hydrostatic, that is the embankment was fully consolidated.
2. Values of c_v measured within the embankment by means of the piezocones, were between 500 and 50 m^2/year (as compared with the 0.5 m^2/year of the total tailings). These values showed that the embankment consisted of silty or sandy

Plate 8.14 Embankment built by underwater deposition emerging at surface.

material which contained relatively little clay. Extensive seepage through the embankment could therefore be expected.

Values of c_v close to the base of the embankment were relatively low at 3-30 m^2/year. Hence the base was essentially impermeable. This reinforced the conclusion that the hydrostatic pore-water pressure profiles showed complete consolidation of the embankment.

3. The cone resistance values in the embankment showed that the angle of shearing resistance of the consolidated tailings varied from about 25–30°. As the interpretation assumed completely drained conditions in the tailings, and because the probing must have set up some excess pore-water pressures, actual angles of shearing resistance probably exceed these values. Although the angles of shearing resistance were not particularly high, they are typical of diamond fine tailings and were considered to be adequate for stability, especially as the embankment is not located in a seismically active area.

Profiles of the underwater embankment surveyed by sounding from a boat, indicated that its side slopes varied from 20–29°. Together with the observed state of consolidation and the measured angles of shearing resistance, this indicated that the underwater embankment would be stable under the prevailing conditions.

Although the piezocone results confirmed that the shear strength properties of diamond tailings deposited subaqueously were satisfactory, they also confirmed that the embankment was pervious (as predicted by the model tests). This caused concern that the phreatic surface in the embankment would be unacceptably high once the area downstream of the embankment had been dewatered. However, on the basis of the satisfactory stability considerations, the raising of the embankment in an upstream direction using conventional subaerial spray bar techniques was commenced. It was also considered that clay fines slowly settling out of the water on the upstream side of the embankment would provide some measure of seepage cut-off.

When it had been established that the embankment was fully consolidated and of adequate strength, the water on the downstream or landward side of the embankment was pumped back into the body of the impoundment.

However, it soon became evident that the high permeability and possible sub-horizontal layering of the embankment material was causing a phreatic surface that was unacceptably close to the downstream slope. The phreatic surface was found to be roughly parallel to the embankment slope and at a depth of about 0.5 m below the surface, with seepage exiting the surface approximately 20–30 m from the toe. It was therefore obvious that drainage of the downstream slope was essential for long-term slope stability and to minimize sloughing of material in the area of the toe of the slope. Because of the high phreatic surface it was also very obvious that any significant excavation into the downstream slope would be extremely difficult and hazardous.

After several alternatives with respect to drain installation had been considered, it was concluded that the safest and most cost-effective alternative that would satisfy the technical requirements of ensuring stability and minimizing sloughing, was immediately to install a series of shallow trench drains into the downstream slope parallel to the length of the embankment with provision for additional drains to be installed in the slope at a later stage, as conditions dictated. The drains were installed as deeply as possible in timbered trenches. Depths attained varied from 900 mm near the toe of the embankment to 1200 mm higher up the slope. The drains were successfully installed and flowed satisfactorily. Piezometers that had been installed in the embankment prior to the drain construction indicated a significant drop in phreatic level as a result.

On the basis of an assumed future phreatic surface, bearing in mind the drain depths and positions, calculations indicated that with a 1:3 (18.4°) side slope, the embankment can attain a final height of 20m with a factor of safety against slope failure of at least 1.3.

The slope of the downstream side of the embankment was successfully grassed, and the embankment is monitored by taking regular piezometer readings, observing the discharge from the drains, inspecting the toe for seepage and the top of the slope for signs of movement. Beacons have been installed on the embankment and these are regularly surveyed for line and level.

8.10 Pool control and decanting

8.10.1 *Pool control*

It is important to keep the pool well away from the walls of a storage to prevent a possible rise in elevation of the phreatic surface. Because the beach slopes are so flat (typically 0.5%), a small increase in elevation of a beach on one side of the pool will displace the pool horizontally in the opposite direction quite significantly. Thus, the pool position can be unstable particularly on an irregularly shaped dam which requires uneven deposition of the slurry to maintain a relatively even perimeter height. This can be a problem as the decant penstock or penstocks are fixed in position, and the only possible adjustment to centralize the pool is by way of varying the deposition around the perimeter of the storage.

To reduce this problem and stabilize the pool, a poolwall can be formed from one of the delivery positions on the perimeter towards the penstock – usually stopping some metres short of the penstock intake. Two wing walls, usually at about 90° to each other, are extended from the end of the pool wall on either side of the penstock. (See Plate 1.9). The pool wall and wing walls serve three purposes:

1. The pool wall provides access to the penstock.
2. The pool can be contained between the two wingwalls and its location is not as sensitive to variations in beach height.
3. The pool wall increases the flow path length from the deliveries closest to the penstock to the pond. Penstocks are often, although not ideally, placed eccentrically so as to be closer to the dam perimeter to save on the penstock outfall pipe length. Due to the shorter beach length, the deliveries in close proximity to the penstock cannot be utilized to the same extent as more distant ones. This area then becomes the low area on the outer wall. Increasing the beach length by deviating flow around the ends of the wingwalls allows greater utilization of those deliveries by increasing the flow path of the slurry and thus the general elevation of the area, as shown by Figure 8.33.

The pool, through judicious use of deliveries, should be kept within the included angle of the wingwalls. In particular, beaching must be used to prevent the pool from encroaching on the storage outer wall. Sharp angles in the pool and/or wingwalls should be avoided as these tend to become unfilled low spots that isolate pockets of water from the penstock.

8.10.2 *Decanting*

As the slurry runs down the beach, the solid particles settle out leaving relatively clear water and with careful deposition control, this water accumulates in the pool. The pool further helps settle the solids as the water is static, thereby reducing the velocity of inflowing beach streams which then drop their remaining solids. This supernatant water, predominantly derived from the deposition process but also from rainfall, must be decanted from the surface of the storage:

1. To conserve water (i.e. to recycle as much as possible).
2. To prevent accumulation of water, loss of freeboard and possible overtopping.
3. To reduce infiltration and potential phreatic surface rises as much as possible.
4. To reduce evaporation losses of the decantable water.

The rate of decanting water must be controlled. If it is allowed to occur too rapidly the velocity of flow of water through or across the pool area becomes too great and either does not allow the dropping of solid particles out of suspension or actually entrains solids into the flow. Wind-driven waves may also stir up previously settled solids. Water must therefore be decanted slowly and all penstock designs must allow

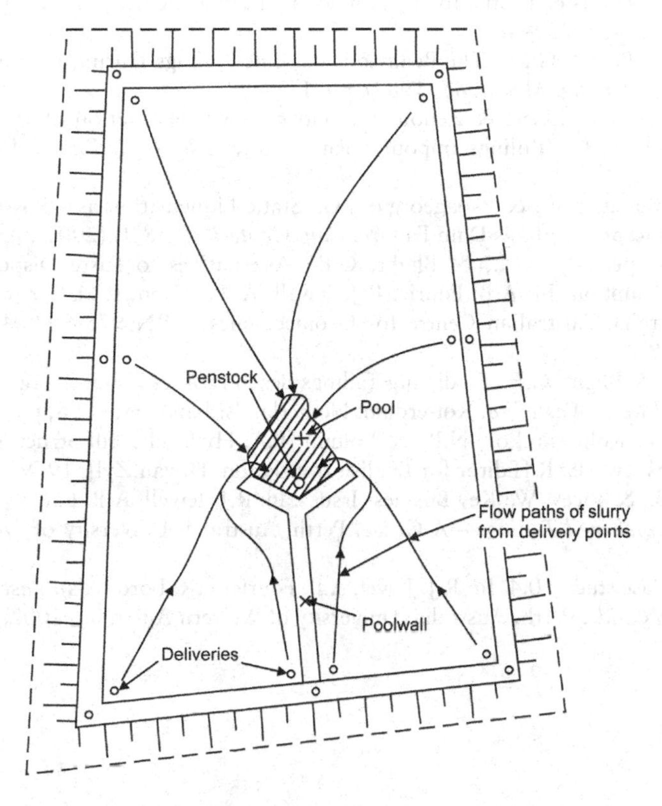

Figure 8.33 A pool wall and its effects on slurry flow path lengths.

for control of the crest level. Control implies observation, so decanting should only take place during the day. At night the penstock crest is usually raised to prevent decanting. The level of the inlet must, however, always be low enough to ensure adequate freeboard.

A pool of significant depth can develop overnight. The deeper pool allows the water to clear. During the next day, the pool can be decanted, removing a predetermined amount of the supernatant water. During windy conditions wave action may lift particles into suspension creating "dirty" water. (See Plate 9.11). If possible decanting should then be avoided – if there are no freeboard problems.

In the event of a large rainstorm the excess water should be decanted as quickly as possible as the occurrence of the next storm cannot be predicted. This storm decanting may require 24 hour per day operation. The size of the decant system must be adequate to decant the normal 24 hour slurry water within the hours of daylight and the design storm (usually a 1 in 100 year storm of 24 hour duration) within 3 to 4 days.

References

Bentel, G.M.: *Some Aspects of the Behaviour of Hydraulically Deposited Tailings*. MSc(Eng.) Thesis. Johannesburg, South Africa: University of the Witwatersrand, 1981.

Blight, G.E.: The Master Profile for Hydraulic Fill Tailings Beaches. *Proc., Inst. Civil Engrs, U.K., 107*, 1994, pp. 27–40.

Blight, G.E., & Bentel, G.M.: The Behaviour of Mine Tailings During Hydraulic Deposition. *J. S. Afr. Inst. Mining. Metall. 85* (1983), pp. 157–161.

Blight, G.E., Boswell, J.E.S., & Zenon, A.: Underwater Construction of an Embankment to Extend the Life of a Tailings Impoundment. *Proc., Inst. Civil Engrs, U.K., 113*, 1995, pp. 80–85.

Fourie, A.B., Blight, G.E., & Papageorgiou, G.: Static Liquefaction as a Possible Explanation for the Merriespruit tailings Dam Failure. *Can. Geotech. J. 38*(4) (2001), pp. 707–719.

Lyell, K.A., Copeland, A.M., & Blight, G.E.: Alternatives to Paste Disposal with Lower Water Consumption. In: A.B. Fourie, R.J. Jewell, A. Paterson, P. Slatter (eds): *Paste 2008*. Perth. Australia: Australian Centre for Geomechanics, ISBN 978-0-9804185-4-5., 2008, pp. 171–178.

McPhail, G.I., & Blight, G.E.: Predicting Tailings Beach Profiles Using Energy and Entropy. In: *Tailings and Mine Waste '98*. Rotterdam, Netherlands: Balkema, 1998, pp. 19–26.

Melent'ev, V.A., Kolpashnikov, N.P., & Volmin, B.A. Hydraulic Fill Structures (in Russian). *Energy*, Moscow, USSR [Editor for English translation: D. van Zyl], 1979.

Regensburg, B., & Tacey, W.: Key Business Issues. In: R.J. Jewell, A.B. Fourie, E.R. Lord (eds): *Paste and Thickened Tailings – A Guide*. Perth, Australia: University of Western Australia, 2000.

Williams, P.: Case study 10.4. In: R.J. Jewel, A.B. Fourie, E.R. Lord (eds): *Paste and Thickened Tailings – A Guide*. Perth, Australia: University of Western Australia, 2002, pp. 150–152.

Chapter 9

Water control and functional and safety monitoring for hydraulic fill tailings storages and dry dumps. Safety appraisal. Special considerations for carbonaceous and radioactive wastes

9.1 Basis of a water control system

A water control system for an hydraulic fill tailings storage is intended to fulfil the following functions:

Because of the differing elevations of a tailings storage and the terrain in which it is constructed, a number of different catchments will be created both on and surrounding the storage. Some of these catchments collect process water and precipitation that can be returned to the mineral extraction plant and re-used. Others collect water that has been in contact with the tailings, as well as solid material eroded from the slopes of the storage. Depending on the characteristics of the mineral being mined and of its ore, this "dirty" water may be contaminated with, as examples, acid drainage from the oxidation of sulphides, fluorine or dissolved uranium.

Water from "clean" catchments must be isolated from that collected in "dirty" catchments. Local statutory requirements vary. Most take a form similar to which that will now be set out, but the numerical requirements quoted below differ from jurisdiction to jurisdiction. The numbers given below should be regarded only as examples.

1 A system of storm water drains and diversion bunds must be designed and constructed to ensure that all water that falls outside the area of the waste storage is diverted clear of the storage.
2 All water that falls within the catchment area of the waste storage must be retained within that area. Provision must be made to store the maximum precipitation to be expected over a period of 24 hours with a probability of once in one hundred years. A freeboard of at least 1.0 metre must be provided throughout the system above the predicted maximum water level. This should apply to all waste deposits, including dry dumps and hydraulic fill storages.

Most waste storages may be subdivided into component catchments as illustrated in Figure 9.1 for the case of a ring dyke tailings storage.

Component catchments:

- The top area or basin of the storage together with any return water reservoirs which have been provided and which are connected to the basin of the storage by means of the outfall penstock. When considering cross-valley storages, the

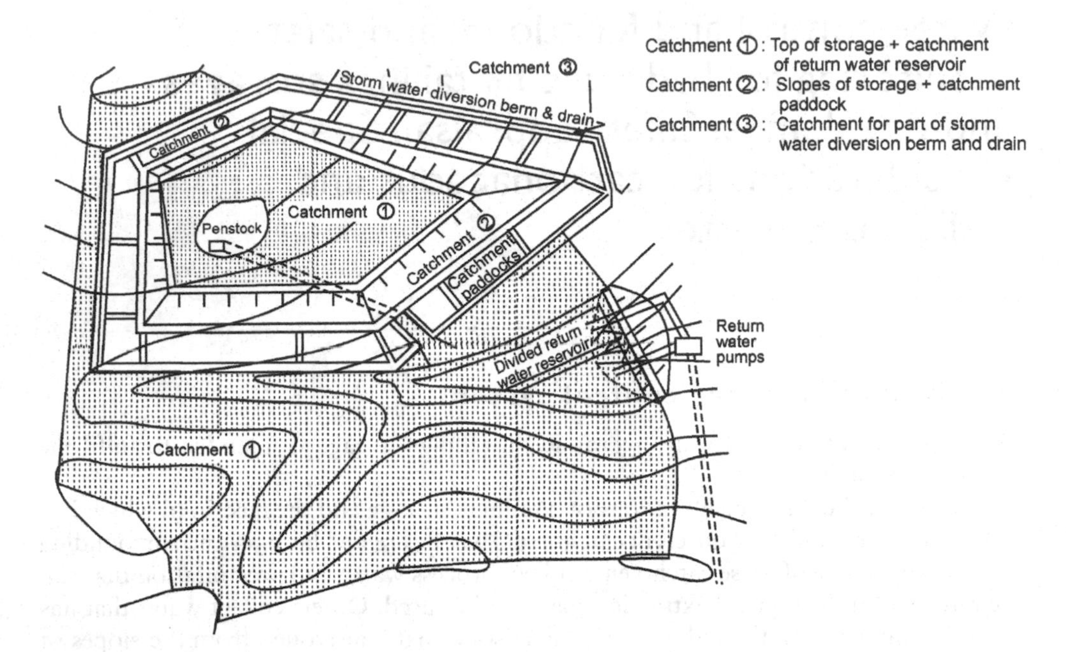

Figure 9.1 Diagrammatic representation of division of waste storage and surrounding area into separate catchments.

catchment of the valley upstream of the waste storage must be included with the top area of the tailings basin.

- The slopes of the waste storage together with the catchment paddocks provided to receive run-off and erosion from the slopes and any additional catchment dams associated with the slopes and catchment paddocks.

  The storage capacities for each component catchment must be sufficient to ensure a freeboard of at least 1.0 m above the maximum predicted water level, and should be based on the average monthly rainfall for the area concerned less the gross mean monthly evaporation in that area, plus the maximum precipitation to be expected over a period of 24 hours with a frequency of once in 100 years. Storage capacities for natural catchments should be calculated from accurately surveyed contours. The storage capacity of the basin of a tailings storage should be estimated by synthesizing the contours of the basin of the storage from relevant beaching profiles (see Section 8.3).

  For a waste dump or a thickened tailings storage, a storm water diversion drain will be required as well as a catchment reservoir for polluted runoff and seepage from the top surface and slopes of the dump or tailings mound.

3 For each of the hydrological sub-catchments a water balance should be set up, as described in Chapter 10.

Although a minimum freeboard of 1.0 m has been mentioned above, many hydraulic fill tailings storages are operated with minimum freeboards of much more than this, for example, up to 10 m on a large storage (500 ha and more in depositional area). Large freeboards are maintained particularly where available hydrological data is sparse or unreliable or if other special circumstances apply, e.g. if a breaching of the outer wall could result in severe consequences, including loss of life. Even if the design storm (e.g. the 24 hour storm of 1 in 100 years frequency) is specified by local regulations, it is incumbent on the designer to investigate and design for the most unfavourable likely storm. For example, a 1 in 20 year storm of half hour duration may be a critical event in a high rainfall year when available water storage may be close to full.

The runoff from the slopes of a waste storage is always polluted to some degree by soluble salts and fine solids eroded from the slopes. As indicated in Figure 9.1, this runoff should be captured in catchment paddocks that are appropriately sized to meet the freeboard requirements. Once in the catchment paddocks, the water can be allowed to evaporate, if the climate is suitable, or else it will have to be decanted by means of a penstock system. Because of the solids deposited in the paddocks, the freeboard is continually being reduced and the paddocks must be cleaned out at appropriate intervals, with the silt being returned to the inside basin of the storage. The erosion catchment paddocks also serve to trap spillage from burst or leaking delivery pipes, erosion caused by bursts, etc.

9.2 Penstocks or decant towers and spillways

In addition to specifying and designing to maintain a minimum freeboard on any waste storage, it is also important to reduce the level of the pool to its normal operating level as soon as possible.

Penstocks are usually designed to clear water from the design storm within a period of 48 to 72 hours. For maintenance of the stability of the storage and the limitation of seepage through its walls, the sooner excess water is cleared, the better. However, this general desideratum must be balanced against the cost of providing a penstock or penstocks that are larger than the minimum size required by regulation.

Penstocks can vary in form from very simple vertical towers to elaborate reinforced concrete structures with automated control of water outlet. In the case of valley storages, the penstock could consist of a culvert built of pre-cast sections up the side slope of the valley with discharge taking place through holes or slots in the culvert that are sealed or covered as the tailings level rises. With large storages, the penstock could consist of a tower with an outfall in the form of a tunnel. For ring dyke storages, a free-standing tower or towers connected to outfall pipe-lines are usual. Alternatively, a pump barge or a siphon intake moored in the pool and discharging the decant water through a floating pipe-line may be the most practical form of decanting arrangement.

9.2.1 *Penstock towers or shaft penstocks*

For smaller operations, penstocks are often constructed of nesting, interlocking reinforced concrete rings (see Figure 9.2). The crest height of this type of penstock is raised by adding rings as the level of the tailings and water rises, and subtracting rings to lower the crest level. The penstock functions as a sharp crested shaft or "morning

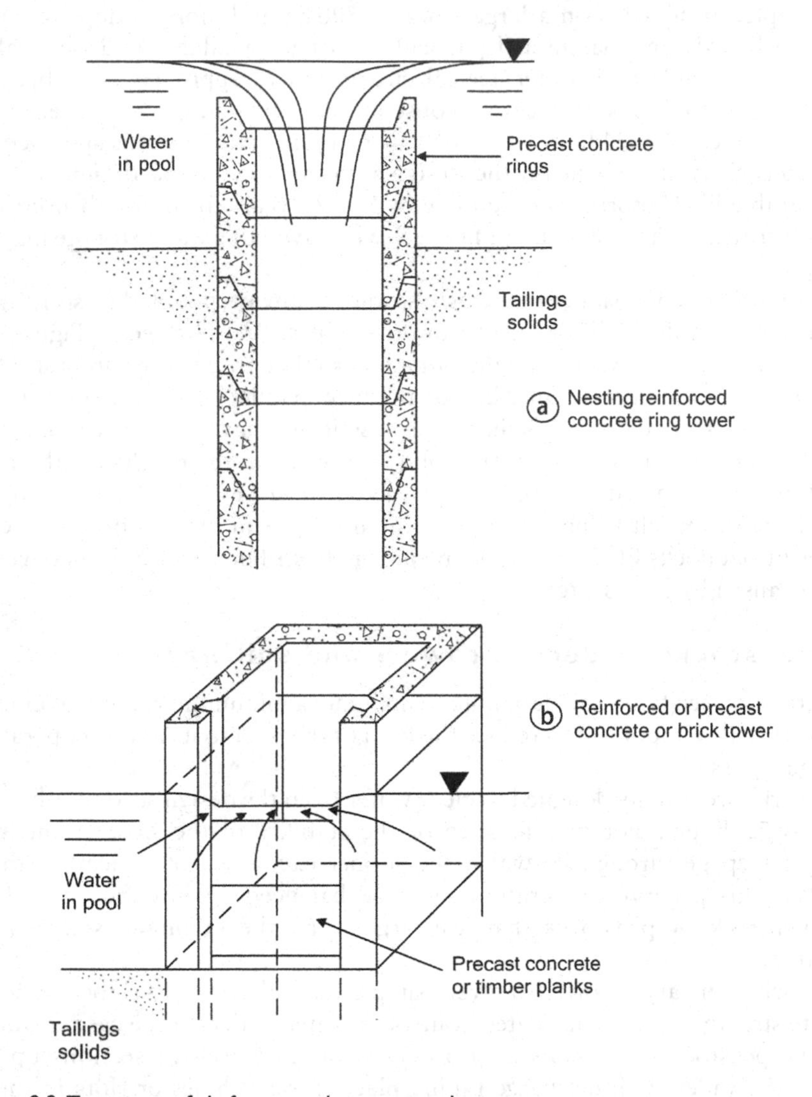

Figure 9.2 Two types of shaft penstock or penstock tower.

glory" spillway. Plate 9.1 is a view down a penstock shaft constructed of nesting concrete rings. Plate 9.2 shows a "lazy tong" device for safely placing or removing rings from the penstock shaft.

Less commonly, penstocks may be constructed as reinforced or precast concrete C-section (Figure 9.2b), timber, or brickwork towers. In this case the crest height is raised by adding or subtracting timber or precast concrete slats or planks to the slotted open side of the C-tower. This type of penstock usually functions as a sharp crested weir.

Concrete ring penstocks have the advantage of being cheap, with reasonably easily adjustable crest heights. They do, however, have some severe disadvantages. If the

Plate 9.1 Penstock shaft constructed of nesting concrete rings.

rings are not well made, with absolutely parallel sides, or if they are placed carelessly, the tower of stacked rings may go out of vertical, causing uneven stressing of the concrete. As a result, rings may crush or spall, resulting in piping and leakage of solids into the penstock shaft, or even complete collapse and blocking of the shaft. For this reason, penstock shafts for high storages are usually constructed monolithically of reinforced concrete and are designed for the frictional down-drag to which they are subjected by the consolidating tailings in contact with their outer perimeter.

The penstock shaft must be designed to resist both the lateral pressure of the tailings and the frictional down-drag caused by settlement of the tailings which surrounds the shaft (see Figure 9.3). It is recommended that the design lateral pressure be taken as 0.6 times the effective overburden pressure of the tailings plus the water pressure, while the down-drag be taken as the tangent of the angle of shearing resistance of the tailings times the effective lateral pressure.

If the depth of tailings is to exceed about 25 m, towers of nesting rings tend to become unstable and to go out of alignment. Also because of misalignment and down-drag, stresses in the lower rings may reach the crushing strength of the concrete, resulting in collapse of the penstock shaft. Hence for large depths of tailings a monolithic reinforced concrete or similar penstock shaft becomes necessary.

Plate 9.3 shows a circular reinforced concrete penstock shaft. The shaft has portholes in two vertical lines which act as overflow weirs. The overflow level can be adjusted by the sluice gates that hang on the sides of the tower. When the photograph was

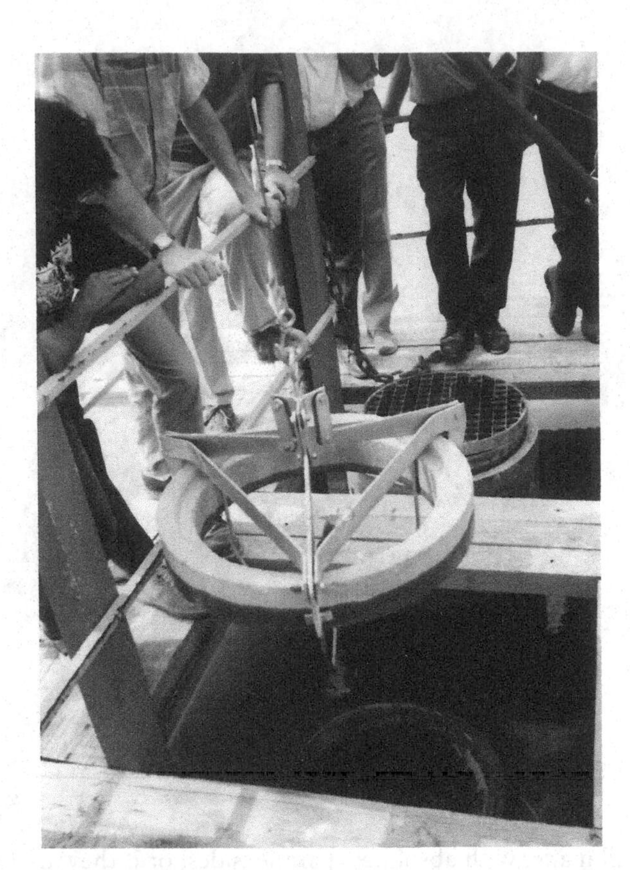

Plate 9.2 "Lazy tongs" device for placing or removing concrete penstock ring.

taken, the tower was buried almost to the top of its 40 m height and was about to be extended. Plate 9.4 shows the newly extended tower ready to continue in use. Plate 9.5 shows a similar system of penstock towers at a mine in the Andes of Chile. The tower on the left has almost reached the end of its operational life, and that to the right is ready to be taken into service. In this case, the outfall conduit is a tunnel through the rock, back to the valley in which the tailings storage has been built.

Plate 9.6 shows a pump barge in use on a large hydraulic fill tailings storage, while plate 9.7 shows a small pump barge serving the return water reservoir for a small tailings storage.

Referring to Figure 9.3, the increment of vertical frictional load on an increment of depth dh of penstock shaft per unit of perimeter will be:

$$dV = 0.6\gamma^{I} h \tan \varphi^{I}.dh$$

$$\text{i.e.} \quad V = 0.6\gamma^{I} \tan \varphi^{I} \int_{0}^{h} hdh$$

$$V = 0.3\gamma^{I} \tan \varphi^{I}.h^{2}$$

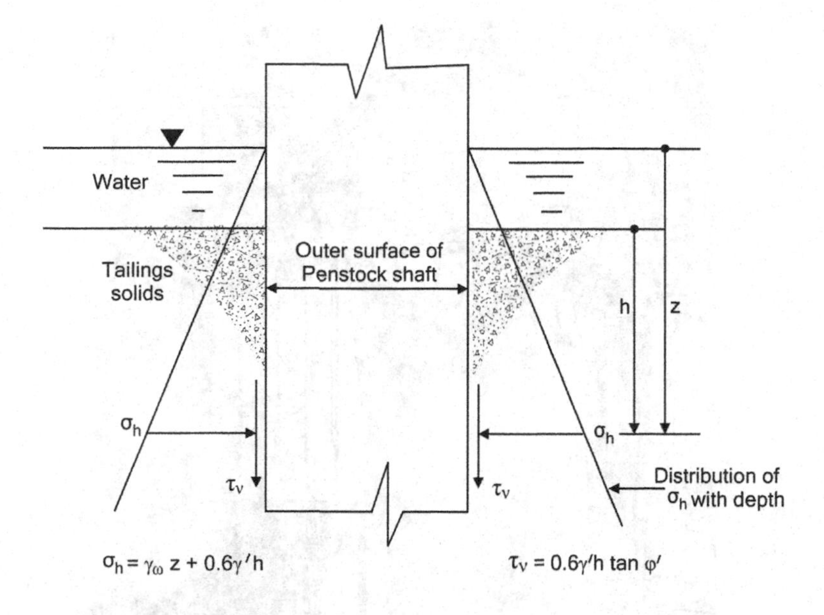

Figure 9.3 Stresses acting on a penstock shaft.

(This assumes that the pore pressure in the tailings is hydrostatic, i.e. the tailings underlying the pool is fully consolidated, which is the worst case for frictional downdrag.)

If the thickness of the wall is t, the compressive stress in the wall at depth h will be

$$\sigma_v = V/t = 0.3\gamma^I \tan\varphi^I h^2/t$$

As an example, suppose

$$\gamma^I = 8\,\text{kN/m}^3, \varphi^I = 30^O, h = 30\,\text{m} \quad \text{and} \quad t = 0.1\text{m}$$

$$\sigma_v = 12470\,\text{kN/m}^2 = 12470\,\text{kPa} = 12.5\,\text{MPa}$$

Thus if the tower is constructed perfectly vertically and the down-drag forces are perfectly symmetrical, σ_v would be an acceptable stress on a 30–40 MPa concrete tower.

Plate 9.3 Vertically adjustable stainless steel sluice gate hanging from free-standing reinforced concrete decant tower. Decanting takes place through pair of diametrically opposite 1m diameter holes in wall of circular tower.

9.2.2 *Flow capacity of shaft penstocks*

The quantity of discharge over a sharp crested weir such as that illustrated in Figure 9.2b is given by:

$$Q = CLH^{3/2} \quad \text{(US Bureau of Reclamation(1960 onwards))} \tag{9.1}$$

where

Q is the discharge in volume of water per unit time;
L is the length of the crest;
H is the head of water above the crest; and
C is the discharge coefficient, equal to 1.8 $m^{1/2}/s$

The same expression applies to a circular shaft spillway such as that illustrated in Figure 9.2a, provided the overflowing water can fall freely down the shaft without converging to form a single jet down the shaft axis. In this case L becomes the perimeter

Plate 9.4 Decant tower of Plate 9.3 after adding 40m extension. Original sluice gates were re-used.

of the shaft, $2\pi R$, where R is the inside radius. The above condition applies providing the ratio:

H/R does not exceed 0.45

If H/R exceeds 0.45 the same formula for Q can be used, but the value of C must be modified as given in Table 9.1.

The situation where H/R exceeds 0.45 should be avoided as it leads to pressure surges in the penstock shaft that may result in dislodgment of rings.

The units of the discharge coefficient can be established from the dimensions of equation (9.1):

$$[m^3/s\} = C[m.m^{3/2}] \therefore C = [m^3/s] \div [m^{5/2}] = [m^{1/2}/s](s = \text{second})$$

The approximate time required to empty a reservoir can be calculated as follows:

Suppose that the top dH of the depth of water over the penstock crest has an area A_1 over the whole of the reservoir. Then the time taken to discharge depth dH will be

$$A_1 dH/Q[m^2.m \div m^3/s = s].$$

Plate 9.5 Reinforced concrete decant towers at cross-valley tailings storage in Chilean Andes.

Table 9.1 Discharge coefficients C for sharp-crested circular weirs (H = head over crest, R = crest radius)

H/R	C $m^{1/2}/s$
0.45	1.8
0.70	1.6
1.0	1.1
1.3	0.9

dH would conveniently be the contour interval for the reservoir basin and A_1 the mean area between the contour for the water surface and the next contour down.

As an example, suppose that the penstock has a circular inlet of 1.2 m radius and that H/R will be limited to 0.45 (i.e. H is limited to $0.45 \times 1.2 = 0.54$ m.) Then

$$Q = 1.8 \times 2\pi \times 1.2 \times (0.54)^{3/2} = 5.38\,m^3/s$$

and when $H = 0.27$ m, $Q = 1.90\,m^3/s$

Plate 9.6 Large pump barge withdrawing water from pool of tailings storage.

If the mean of the initial water surface area and that when the pool has been drawn down by 0.27 m is 20 ha = 200 000 m^2

The time to half draw down the pool will be

$$t = 200\,000 \times 0.27/\frac{1}{2}(5.38 + 1.90) = 14835\,s = 4.12\,h(h = hour)$$

The time to draw the pool down to penstock crest level will be a further

$$200\,000 \times 0.27/0.47 = 31.9\,h$$

and the time for complete draw-down will be 36 h, provided there is no further inflow to the pool during this period. The calculation may be refined as much as is necessary by reducing dH, provided corresponding values of A can be estimated.

9.2.3 *Return water system*

The return water system should include silt traps, a return water reservoir, pumping system and return water pipeline. The objective of the return water system is to provide sufficient storage capacity to contain the design storm runoff from the entire catchment

Plate 9.7 Small pump barge serving return water reservoir in New Zealand.

(catchment 1 in Figure 9.1) and secondly to return all available water to the plant for re-use as required. The return water reservoir is usually situated at the lowest point of the site so that all water gravitates to it.

Silt traps

Water decanted from the pool of the storage will inevitably contain suspended solids. These solids will settle out in the return water reservoir under windless, calm conditions. Without a silt trap the silt will eventually accumulate to the extent that the reservoir no longer has sufficient capacity to fulfil its design function. Consequently, it is good practice to incorporate two silt traps in parallel upstream of the return water reservoir. The silt traps are small stilling basins where the solids can drop out before the return water enters the return water reservoir. The silt traps can be designed to be of manageable proportions and easy to clean on a periodic basis (typically every 2–3 years). Two silt traps are provided to allow one to be dried and cleaned whilst the other remains in operation. Splitter boxes upstream of the silt traps divert the inflow to one or the other. The easiest cleaning is usually by mechanical means. Removed silt should be returned the basin of the tailings storage.

The silt traps normally operate full of water as they must act as stilling basins. Consequently, concrete lined overflow spillways to the return water reservoirs are usually used. Depending upon the size of the traps they can be completely lined with concrete (to allow the entry of mechanical equipment) and/or have wave erosion protection along the water line.

Return water reservoir

The return water would spill from the silt traps into the return water reservoir. The return water dam should be designed as a conventional water retaining dam. If higher than 5 m and retaining more than 50 000 m^3, the dam would be classified as a large dam and subject to the national safety regulations applicable to large dams.

The return water reservoir must be able to retain the design storm volume, so under normal operating conditions, should be kept at a low level. For this reason the total return water impoundment area is often divided into compartments (Figure 9.1) where the smaller ones can be kept reasonably full to ensure a steady supply to the plant whilst the larger compartments are kept empty to accommodate storm precipitation. Frequently, return water reservoirs are compartmentalised to allow cleaning of one whilst the other is still in use and to provide greater flexibility to the operation of the system. Balancing spillways are provided between compartments. An emergency overflow spillway must also be provided to safely spill rainfall exceeding the design storm.

Return water pumping system

The return water pumping system is usually situated below the return water dam wall and can be fed by large diameter suction pipes passing through the dam wall into the reservoir basin. A protective grille or trash rack should be provided over the inlet to prevent the ingress of flotsam. The pumping system, including the return water pipeline, should be sized to deliver the maximum water demand of the plant. In areas of surplus water balance, additional pumping capacity must be provided to send the surplus water to a treatment plant prior to discharge.

It may be possible to link the return water pipe to the slurry delivery pipeline at strategic positions to facilitate flushing of the latter when necessary. The pumping facility may also be used to provide a high pressure system to flush tailings and sludge from the return water reservoir and deposit it back into the tailings storage. Pump stations are normally automated, and should be programmed to keep the return water reservoir level as low as possible.

9.2.4 Pressures on penstock outfall pipes or conduits

Penstock towers and outfall pipes are empty for most of their working life and the primary design loading which they must resist is the external pressure exerted by the tailings. As explained in Section 9.2.1, the lateral pressure on a penstock tower or outfall pipe at depth h below the surface of the waste and depth z below the seepage surface or water level is given by:

$$\sigma_h = 0.6(\gamma h - \gamma_w z) + \gamma_w z \qquad (9.2a)$$

in which γ is the saturated density of the tailings and γ_w is the density of water.

The corresponding vertical pressure on a penstock outfall pipe is

$$\sigma_v = \gamma h = (\gamma h - \gamma_w z) + \gamma_w z \tag{9.2b}$$

The reason for splitting the applied pressure into components is that the water pressure $\gamma_w z$ is equal in all directions and causes circumferential compression in the shaft or pipe wall, but no bending. The horizontal and vertical effective pressures are

$$\sigma_h^I = 0.6(\gamma h - \gamma_w z) \text{ and}$$
$$\sigma_v^I = \gamma h - \gamma_w z \tag{9.3}$$

This implies that $K_o = \sigma_h^I/\sigma_v^I = 0.6$, where K_o is the at rest pressure coefficient. Because they are unequal, σ_h^I and σ_v^I cause bending in the pipe wall. There are two potential failure modes of a cylinder or pipe subjected to outside compressive stress. The first is collapse by buckling instability of the wall, the second is failure by bending in the wall.

The uniform external pressure that will cause an empty pipe to collapse by buckling instability of the wall is given by

$$\sigma(\text{critical}) = 2.67E(t/D)^3 \tag{9.4}$$

where E is the elastic modulus of the wall material, t is the wall thickness and D the mean diameter of the pipe or shaft wall.

For example, for a concrete pipe with a mean diameter of 1m, a wall thickness of 100 mm and a value of E = 30 000 MPa, the critical external pressure will be

$$\sigma(\text{critical}) = 2.67 \times 30 \times 10^3 \times (0.1/1)^3 = 80\,\text{MPa}$$

and a safe value of external pressure would be 1/3 of this or 27×10^3 kPa. For a tailings having a saturated unit weight of 20 kN/m^3, the depth of the pipe to reach the limiting safe external pressure of 27 MPa would be 1350 m. Hence it is highly unlikely that buckling instability will ever become critical, although a combination of buckling and bending might result in collapse. It must also be remembered that straining and deflexion of the pipe wall will relieve the applied stresses. (Also see Section 6.14.3 and Figure 6.40.) Calculations according to equation (9.4) give an absolutely worst scenario that may not be realistic, except for thin-walled pipes. For example:

If the 1 m diameter pipe is of steel with E = 200 000 MPa and a 5 mm thick wall

$$\sigma(\text{critical}) = 2.67 \times 200 \times 10^6 \times (0.005)^3 = 67\,\text{kPa}$$

and the critical external collapse depth would be only 3.3 m. In this case collapse by inward buckling may become a reality and the pipe wall will need stiffening.

9.2.5 *Strength of penstock outfall*

The penstock outfall pipe would be designed to resist the full external water pressure and the horizontal and vertical components of the effective stress, σ_v^I and σ_h^I.

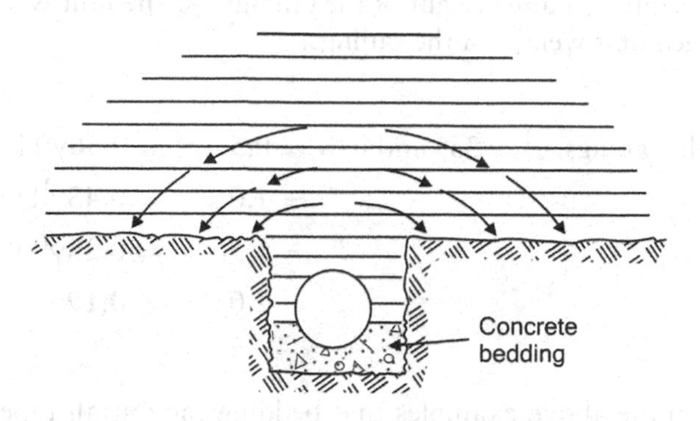

ⓐ Arching over trench reduces vertical stresses on pipe

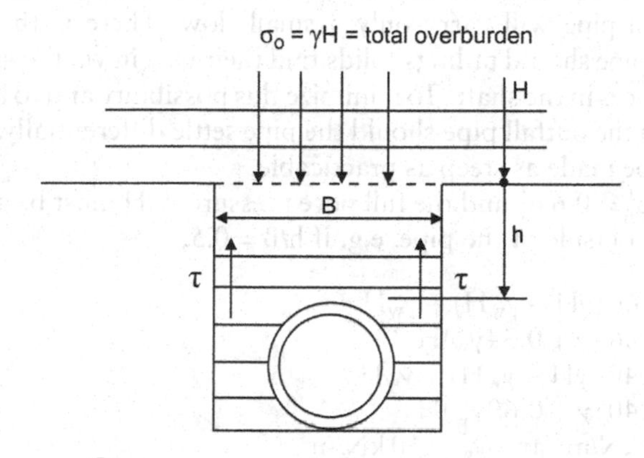

ⓑ Pressure relief by bedding pipe in a trench

Figure 9.4 Pressure reduction by laying pipes in trench.

As it is always likely that the outfall will be subjected to differential settlement, it should be provided with flexible couplings or joints to allow it to conform to the changing profile of its bed without cracking or rupturing. To reduce the effect of the overburden stress on the walls of the pipe, it is preferable to lay the outfall in a trench (see Figure 9.4) as the 'negative projection' achieved in this way reduces pressure on the pipe. This may be taken into account in the design. The pipe should never be supported on the ground surface or on pedestals clear of the ground as it will then project positively and also become subject to down-drag and (possible) bending between points of support. If the centre-line of the pipe is at depth h below the general ground surface which in turn is at depth H below the surface of the tailings (see Figure 9.4b), the pipe should be designed for an external effective vertical pressure of:

$$\sigma_v^I = (\gamma - \gamma_w)He^{-1.2\tan\varphi'.h/B} = \gamma^I He^{-1.2\tan\phi'.h/B} \tag{9.4}$$

in which γ is the saturated unit weight of the tailings, γ_w the unit weight of water and γ^I is the submerged unit weight of the tailings.

e.g. If, for the tailings, $\varphi^I = 35°$ and $h/B = 0.5 \quad \sigma_v^I = 0.66\gamma^I H$
$= 1.0 \quad = 0.43\gamma^I H$
$= 1.5 \quad = 0.28\gamma^I H$
$= 2.0 \quad = 0.19\gamma^I H$

It is obvious from the above examples that bedding the outfall pipe in a trench can have a considerable effect on reducing vertical effective stresses on the pipe. Because $\sigma_h^I = K_o\sigma_v^I$, horizontal effective stresses are also reduced. For most of the time the penstock outfall pipe will carry only a small flow. There is therefore a danger of siltation of the pipe should tailings solids find their way in via the crest of the penstock or through crevices in the shaft. To minimize this possibility and to avoid the formation of low points in the outfall pipe should the pipe settle differentially, the gradient of the outfall should be made as steep as practicable.

In each case $\sigma_h^I = 0.6\ \sigma_v^I$ and the full pore pressure $\gamma_w H$ must be added to get the full pressure on the outside of the pipe, e.g. if $h/B = 0.5$,

$\sigma_v = 0.66\,(\gamma H - \gamma_w H) + \gamma_w H$
i.e. $\sigma_v = 0.66\,(\gamma + 0.34\gamma_w)H$
$\sigma_h^I = 0.40\,(\gamma H - \gamma_w H) + \gamma_w H$
$\sigma_h = 0.40\,(\gamma + 0.60\gamma_w)H$
If $\gamma = 20\,kN/m^3$ and $\gamma_w = 10\,kN/m^3$
$\sigma_v = 15.44\,H\ kPa$, $\sigma_h = 10.40\,H\ kPa$ and $\sigma_h/\sigma_v = 0.67$.
i.e. σ_h/σ_v is approximately the same as σ_h^I/σ_v^I

The bending moments generated by the vertical and horizontal effective stresses σ_v^I and σ_h^I (equation 9.3) are summarised in Figure 9.5. As an example, suppose that a penstock outfall pipe has a diameter of 1.0 m and is buried under 40 m of tailings having a saturated unit weight of 20 kN/m^3. The phreatic surface is 10 m below the surface. At the level of the outfall:

$$\sigma_v^I = 20 \times 40 - 10 \times 30 = 500\,kPa$$
$$\sigma_h^I = 0.6 \times 500 = 300\,kPa$$

The bending moments at points A and B are as given in Figure 9.5:

$$M_A = 0.16 \times (0.5)^2 \times 500 = 20.0\,kNm/m$$
$$M_B = 0.025(0.5)^2 \times 500 = 3.1\,kNm/m$$

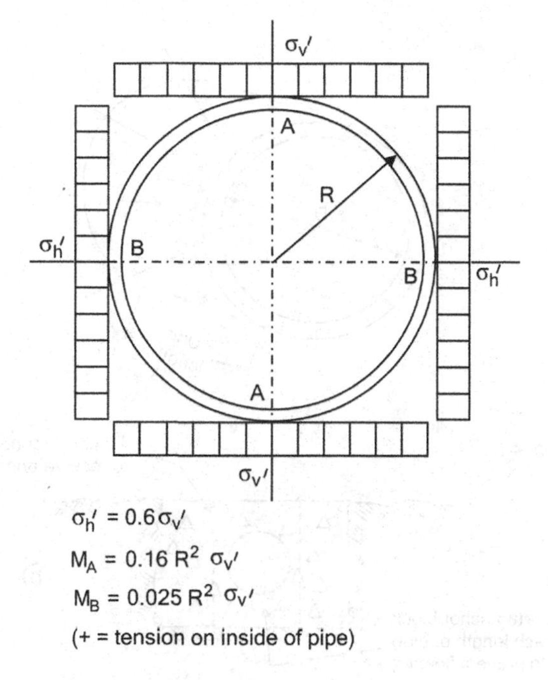

Figure 9.5 Bending moments in a pipe subjected to effective stresses σ_v^I and σ_h^I.

If the wall thickness of the pipe is 0.1 m, the second moment of area of the gross wall section will be:

$$I = 1/12 \times 1.0 \times (0.1)^3 = 8.3 \times 10^{-5} m^4/m$$

and the maximum bending stress will be (ignoring the effect of reinforcing):

$$\sigma_A = 0.05M_A/8.3 \times 10^{-5} = 12.0\,\text{MPa}$$
$$\sigma_B = 0.025/0.16 \times 12.0 = 1.9\,\text{MPa}$$

σ_A and σ_B will be tensile stresses. However, the effect of an all-round water pressure of 300 kPa will be to generate a compressive stress of:

$$\sigma = 300 \times 0.5/0.1 = 1500\,\text{kPa} = 1.5\,\text{MPa}$$

which will reduce σ_A to 10.5 MPa and σ_B to 0.4 MPa

However, as the short term tensile strength of the concrete is unlikely to be more than 5 MPa, (if the compressive strength is taken as 50 MPa,) the pipe wall will crack on the inside at positions A where the tension will have to be provided for by hoop reinforcing. The outside of the pipe will also be in tension, by 0.4 MPa, but with a concrete strength of 40–50 MPa, the pipe will be safe to use.

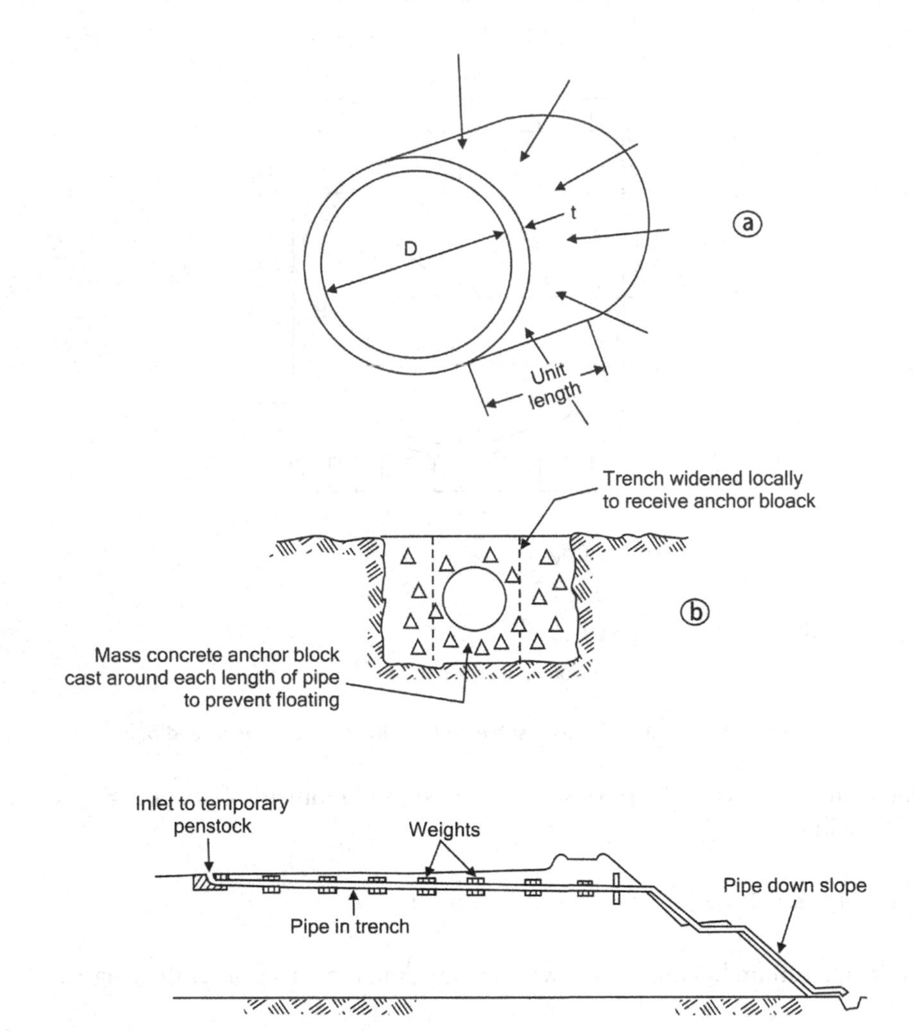

Figure 9.6 Floating of empty pipelines in trench.

9.2.6 *Possible floating of the outfall pipe*

Occasionally, there are reports or anecdotes of penstock outfall pipelines having floated out of the ground, displacing the tailings deposited above them. This section will examine the condition that results in "floating out" of buried pipelines and the way of preventing this.

Figure 9.6a shows unit length of an empty outfall pipe surrounded by tailings with a saturated unit weight of γ. The pipe is made of material with a unit weight of γ_p and has a bore diameter of D and a wall thickness of t.

The weight of the empty pipe per unit length is

$$\gamma_p \pi (D + t)t$$

and its overall volume per unit length is

$$\pi(D + 2t)^2/4$$

The weight of tailings displaced by the pipe per unit length is $\gamma\pi(D + 2t)^2/4$. Hence the buoyancy force is

$$\pi[\gamma(D + 2t)^2/4 - \gamma_p(D + t)t] \text{ per m length}$$

For the pipe to float, the buoyancy force must be positive, i.e.

$$\gamma(D + 2t)^2/4 \geq \gamma_p(D + t)t, \text{ or}$$

$$\gamma/\gamma_p(D + 2t)^2/(D + t)t \geq 4 \qquad (9.5)$$

For a reinforced concrete pipe of 1.0 m bore and wall thickness 0.15 m with $\gamma_p = 24\,kN/m^3$ in a tailings having $\gamma = 18\,kN/m^3$, the pipe will float, because:

$18/24\,(1.3)^2/(1.15) \times 0.15 = 7.35$, which is obviously >4. Because the wall thickness is unlikely to exceed 0.15m, all practically dimensioned concrete pipes are likely to float in tailings.

Alternatively, Weight of pipe $= 24\pi(1.15) \times 0.15 = 13\,kN/m$

Weight of tailings displaced $= 18\pi(1.3)^2/4 = 24\,kN/m$

$24/13 = 1.84$ and $7.35/4 = 1.84$which checks equation (9.5).

To prevent floating, the pipe must be weighted by the buoyancy force, increased by a factor of safety F. For a length of pipe L, the net buoyancy force will be

$$\pi L[\gamma(D + 2t)^2/4 - \gamma_p(D + t)t] \qquad (9.6)$$

As an example, if $D = 1.0\,m$, $t = 0.15\,m$, $\gamma = 18\,kN/m^3$ and $\gamma_p = 24\,kN/m^3$, each 3 m length of pipe must be weighted by

$$\pi 3[18(1.3)^2/4 - 24(1.15)(0.15)] = 32.6\,kN$$

If $F = 1.2$, the weight per pipe must be 39 kN, or say, a 1.6 m^3 concrete saddle, such as that indicated in Figure 9.6b.

A good example of an emergency situation where it is essential to remember the floatation propensities of an empty pipeline is that which arises when a penstock tower or outfall fails or becomes blocked and has to be replaced, in a hurry, by a temporary penstock and outfall, as illustrated in Figure 9.6c. Because of the urgency, it is quite likely that steel or even plastic pipes will be used, and it is absolutely essential to weight these correctly. As temporary installations all too often tend to become permanent, all the other usual design criteria should be applied, especially allowance for differential settlement between pipes and between the new emergency penstock and the outfall pipeline.

If the difference in surface elevation between one side of a site and the other is more than a few metres, it may be necessary, until the entire area is covered with tailings, to operate two or more penstocks in two or more sections of what will eventually become a single tailings storage. In a case like this, either two or more permanent penstock systems could be established, or one permanent penstock and outfall together with as many temporary penstocks as required to operate the separate sections. The temporary penstock systems could be blocked and abandoned when no longer required, or temporary towers could be used with access to a single permanent outfall pipeline serving the permanent penstock.

9.3 Monitoring systems for waste storages

It is essential to implement an ongoing system of instrumentation and monitoring that is designed to guard against the most likely causes of failure that are listed in Figure 2.2. The results of the monitoring measurements must be reported at regular intervals (e.g. quarterly) to a safety committee of knowledgeable people appointed to receive and study the reports and who accept the responsibility for safety and good operation of the storage.

Even with this committee system in place and running smoothly, complacency can erode vigilance. Care must be taken that making and recording the measurements and producing the quarterly safety reports does not become an end in itself, instead of a means of guarding against safety-reducing trends or incidents. The best way of doing this appears to be to periodically rotate the membership of the safety committee, regularly inviting and introducing new-comers who are encouraged to question current practices and study the quarterly reports with unjaded eyes and uncomplacent minds. If possible, the newcomers should not have previous acquaintance with the operating system and the particular waste storage concerned, but should come in as experienced but unfamiliar outsiders.

9.3.1 *Drive or walk-over visual inspections*

Regular three-monthly inspections of the condition of the slopes and perimeter installations of waste storages form an essential part of a monitoring programme. During these close visual inspections by the safety committee particular attention should be paid to the following:

- the presence of cracks parallel to or transverse to the crest of the slopes or the presence of similarly oriented cracks on the slopes themselves;
- the presence of cracks in the soil at the toe of the slope and any visible displacement or distortion (either horizontally or vertically) of drainage trenches, channels or pipes at the toes of the slopes;
- any visible sagging of the crest of the slope or bulging at the toe of the slope;
- the visible emergence of seepage at the toe of the slope as indicated by visible wetness (usually a darker colour) of the surface, local concentrations of vegetation growth or local excessive erosion of the slope surface. Depending on the quality of the seepage, a local die-back of the vegetation may occur, rather than a concentration of plant growth. A dark brown (tea coloured) seepage is indicative of the emergence of acid drainage. The appearance of any of these warning signs is

Plate 9.8 Cracks behind crest of outer dyke of paddock-built tailings storage.

a strong indication that the slope may be unstable and that the advice of a professional geotechnical engineer is required. (Here, it should be noted that the point of apparent emergence of the seepage surface from a slope usually defines the top of the capillary fringe above the seepage surface, and not the seepage surface itself. However, it is conservative to assume that this point represents the emergence of the seepage surface.) The close inspections can, with advantage, be supplemented by periodic (6 monthly) helicopter inspections to get an overall view of the storage.

- In addition to the geotechnical inspection, the safety committee should also note and report any potentially dangerous working practices that may have developed.

Plate 9.8 shows a series of ominous cracks behind the crest of the outer dyke of a tailings storage. The cracks warned of an imminent failure of the slope. The cause of the problem was that the slope was being prepared for vegetating by leaching the acid and soluble salts close to the tailings surface down into the deeper, alkaline tailings by the use of the water sprays that can be seen on the slope (supported by wooden stakes). This had caused an excessive rise in the phreatic surface that had not been monitored. In this state, even if the irrigation was stopped, a rainfall that could fill the cracks with water would be sufficient to cause failure. The slope failed the next night. The inspection came too late to save it.

Plate 9.9 shows clear evidence of the emergence of the phreatic surface high up the slope at approximately the level where the vegetation stops and the severe erosion

Plate 9.9 Raised phreatic surface in dyke of abandoned gold tailings storage resulting in severe erosion of toe and seepage of acid drainage (pool in foreground).

begins. The hole in the foreground of the picture is filled with dark brown acid leachate that has emerged from below the slope. The slope needs urgent attention in the form of a drained buttress and a cut-off drain to remove the acid seepage.

9.3.2 *Drainage and seepage*

The functioning of the under-drainage of a dam wall can be monitored by measuring the rate of discharge of individual drains. The total recorded drainage should agree reasonably well with the calculated quantity of discharge. (Also see Chapter 10). Monitor wells around the perimeter of a waste storage (See Figure 9.7) penetrating the water table are used to monitor possible pollution of the groundwater. Both the effluent from under-drains and samples taken from monitor wells should be analysed to ensure that the quality of the ground water is not being affected by the presence of the waste storage. If for some reason the monitoring procedure shows that unacceptable pollution is occurring, it may be necessary to isolate the storage by constructing an impervious cut-off wall or else a cut-off drain around the waste storage. The installation of such measures will require the complete reappraisal of the stability of the storage, as, particularly in the case of a cut-off wall, the water levels in the foundation strata may be significantly affected.

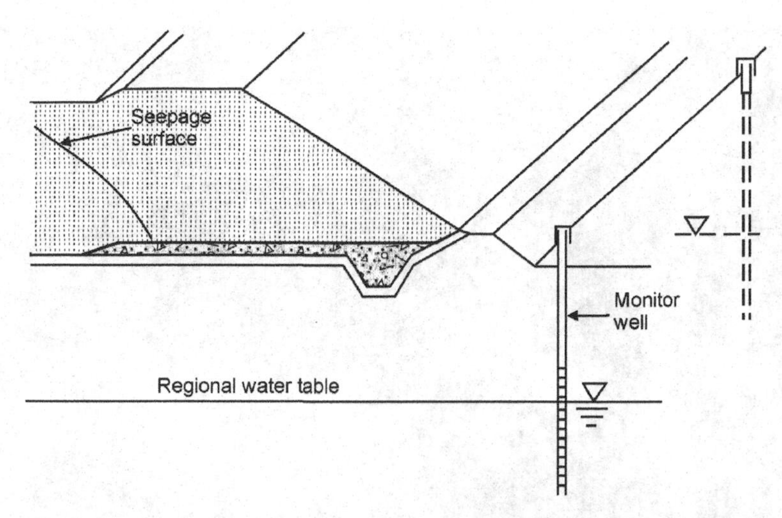

Figure 9.7 Monitor wells installed around perimeter of tailings storage to monitor for pollution of regional ground water.

Plate 9.10 shows the outlets of two adjacent under-drains. One is flowing freely, but the other appears to have partially blocked and needs to be cleared by jetting with water.

Plate 9.11 shows a clear stream of seepage water from the slope under-drains passing over a measuring weir and joining the penstock outflow, which is slightly dis-coloured by fine particles, probably stirred up from the bottom of the pool by wind-driven waves.

9.3.3 *Monitoring by survey*

Survey methods may also be used to monitor waste storages especially if visual observations indicate possible slope instability. The simplest form of survey system consists of a line of beacons along the toe of the slope. These beacons are aligned or co-ordinated on installation and also levelled. Thereafter, periodic observations for line and level are made. The upward or outward, or a combination of upward and outward movement of any of the line of toe beacons may indicate the development of shear instability in the slope. A similar line of beacons could be installed on berms part way up the slopes of a waste storage. Surveying of beacons can be undertaken by GPS (Geographic Positioning System.)

Slope indicators can be installed in slopes to show if deep-seated movement is taking place. (A slope indicator consists of a lined borehole in which the lining is equipped with two sets of longitudinal guiding tracks at right-angles. The slope indicator is lowered down the borehole and continuously records the verticality of the lining in two directions at right angles. These directions are usually parallel to and at right-angles to the length of the slope. By integrating the recorded slopes of the borehole lining, the movement at any depth relative to the borehole collar can be obtained. The

Plate 9.10 Outlets from toe drain of tailings storage. Outlet on left is blocked, that on right is flowing freely.

position of the collar is surveyed in separately.) Plate 9.12 shows a slope indicator guide tube being installed near the crest of the outer dyke of a tailings storage.

Slope cross-sections, crest levels and available freeboard for a tailings dam should also be checked by survey on a regular basis. (Also see Section 10.4.)

9.3.4 *Phreatic surfaces and penstock outfalls*

The height and position of phreatic surfaces may be observed by means of stand-pipe piezometers, such as those shown in Figure 9.8. It is usually convenient to install stand-pipe piezometers once the storage has reached a height of about 3m above base level. The stand-pipes can then be extended as necessary by adding sections of tubing at the surface to increase their height. Stand-pipes are usually installed along a section normal to the slope of the dam at critical points for stability. A sufficient number of stand-pipes must be installed to enable the phreatic surface to be accurately determined (see, e.g. Figure 6.19) and to pick up certain critical points such as the pore pressure adjacent to under-drains (in order to check if the drains are working effectively), etc.

When using a seepage surface to analyse a slope for shear stability, it is important to note that the pore water pressure at any point is not given by the depth of the point below the phreatic surface. (See Figure 6.19.) In order to estimate pore pressures, it

Plate 9.11 Clear drainage from toe drains joins murky outflow from penstock. Note measuring weir on toe drain channel. Similar weir on penstock channel is upstream (to right) of channel junction.

is necessary to construct a complete flow net for the slope, taking any anisotropy of waste permeability into account and estimating pore pressures from the resulting equi-potentials.

Figures 6.23 and 8.15c illustrate the effect of anisotropic permeability on the position of the phreatic surface. It will be noted that the larger the ratio of the permeabilities k_h/k_v, the higher does the phreatic surface rise, other things being equal.

Only if a piezometer is isolated from points above and below the porous tip will it record the pore pressure at that point. The usual unsealed standpipe records only the position of the seepage surface, and pore pressures estimated from the elevation of the seepage surface will over estimate actual pore pressures. (See Figures 6.19 and 6.22.) As an alternative a piezocone probe may be used to directly measure a series of static pore pressure profiles in a slope. From these profiles, contours of pore pressure can be constructed.

To report the results of routine monitoring, the measured piezometric levels should be plotted to scale on a section of the slope. If they have changed appreciably since the previous set of measurements, or if they are unrealistic, this will immediately become apparent.

Plate 9.12 Inclinometer casing installed in wall of cross-valley dam.

The penstock tower and the outfall pipe or conduit should also be inspected at regular intervals. Inspection of small diameter penstocks which are not very deep can be done visually from the top, using a flood-light, or even a mirror to direct sunlight down the shaft. It is possible for a man to descend a shaft on a bosun's chair if the diameter is 600mm or more, but obviously only once the crest has been raised to exclude the water, and all the necessary safety precautions are in place.

If the outfall pipe is of large enough bore diameter (larger than 1m diameter) it can be inspected by a man, but for smaller diameters, a television camera must be used. Plate 9.13 shows an inspection of a 1.25 m diameter outfall pipe by two men on a pedal-driven trolley (one of whom was the author). The large "growth" to the right of the photograph consists of calcium carbonate leached from the concrete by water infiltrating a joint between two pipes. The dark colour is iron oxide staining from corrosion of the steel reinforcing in the pipe wall. The pipe had been in place for 27 years.

Plate 9.14 is from a video taken inside a 450 mm diameter reinforced concrete outfall pipe that had been in place for 32 years and apart from a few steps in the invert, one of which is shown in Plate 9.14, was in perfect condition.

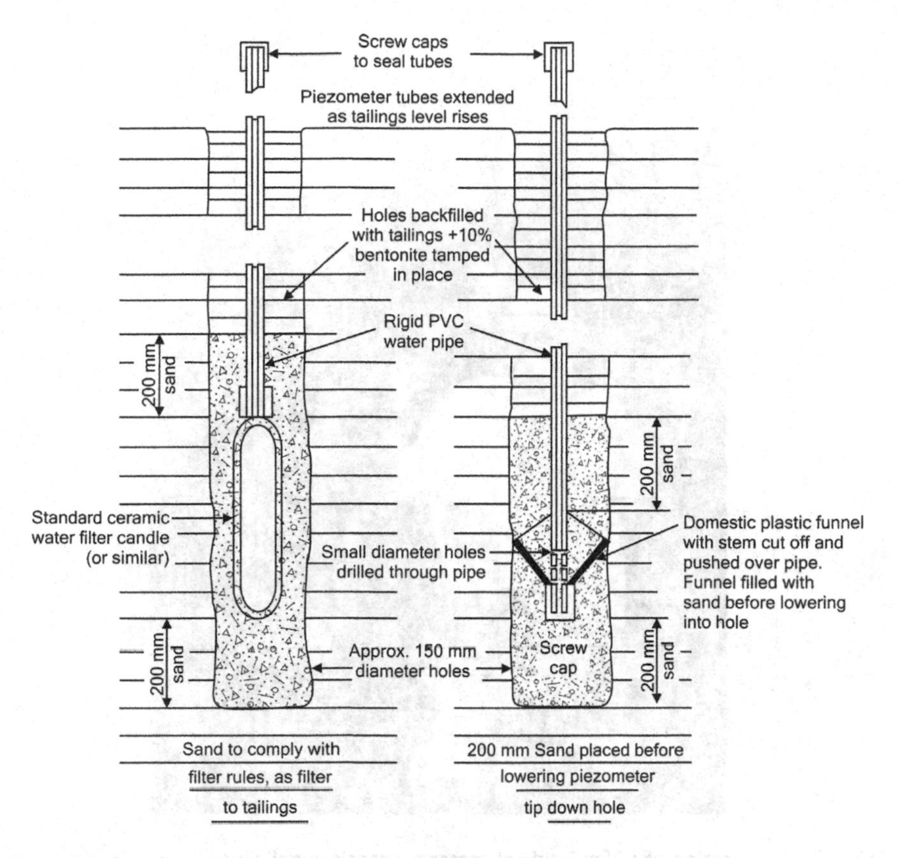

Figure 9.8 Standpipe piezometers suitable for installing in tailings storages.

9.3.5 *Strength profile*

The strength profile of the material constituting a storage outer wall is best monitored by performing in situ vane shear or piezocone tests after three to five metres of material have accumulated in the deposit. The results of these in situ tests can then be compared with conditions assumed for the design and the slope of the tailings storage can, if necessary, be amended accordingly.

9.4 Appraisal of safety for waste storages

It may prove necessary to carry out an appraisal of the safety of an existing tailings storage. This may arise either because the storage was not originally adequately designed or because a change in the usage of the storage (e.g. an increase in the rate of rise) or the characteristics of the tailings has occurred. Alternatively, a change in the deposition or the management system, size or geometry of the storage may be contemplated.

9.4.1 *Inspection*

In the case of such a re-appraisal, the first step should be to carry out a careful inspection of the storage in accordance with the guidelines set out in Section 9.3.

Plate 9.13 Inspection of 1.25m inside diameter penstock outfall pipe.

If appropriate, reference to published slope stability charts may be made to get a preliminary estimate of the stability of the slopes from the knowledge of their height, angle and rate of rise. If this preliminary stability survey does not prove reassuring further measures to investigate the stability must be taken.

9.4.2 *Measurements*

The first essentials are to locate the phreatic surface in the storage (if any) and to assess the shear strength of the waste. Measurements of in situ shear strength profiles for the waste may be combined with the operation of installing piezometers.

Strength profiles are first measured at strategic points using the vane shear apparatus, the pressuremeter, or the piezocone, after which stand-pipe piezometers may be installed in the shear strength holes. The strength profiles should extend through the full height of the waste storage and as far as possible into the foundation stratum. Because of the slope of the sides, a vertical hole will penetrate material that has been deposited further and further from the slope as the depth increases. This should be born in mind when analysing the results of a vertical profile of strength measurements. Strength profiles should also be measured and piezometers installed in the foundation

Plate 9.14 Video inspection of 460mm diameter penstock outfall pipe. Note step in pipe invert.

stratum at the toe of the slope. For dry dumps of coarse waste, it is usually only possible to measure the shear strength profile of the foundation soil at the toe of the dump, and the shear strength of the coarse waste must be estimated from the angle of repose of the slopes, if end-tipped, or by cutting a slope in the waste to allow the waste to assume its angle of repose.

9.4.3 *Analysis and remedial measures*

Once the results of these measurements have become available a full stability analysis of the storage in its present condition can be performed and the effect of changes in future use of the storage, extensions of its height, an increase in the rate of rise, etc. investigated and designed for. In order to effect such extensions and changes with safety, it may be necessary to install additional drainage measures on the existing top surface of the storage. This can be done by conventional means, bearing in mind that the existing surface will settle when additional material is deposited over it. Alternatively, it may prove necessary to install additional drainage at the base of the slopes. This may be done in a variety of ways:

- by pushing or jetting perforated pipe drains into the existing outer slope (see Figure 9.9a);

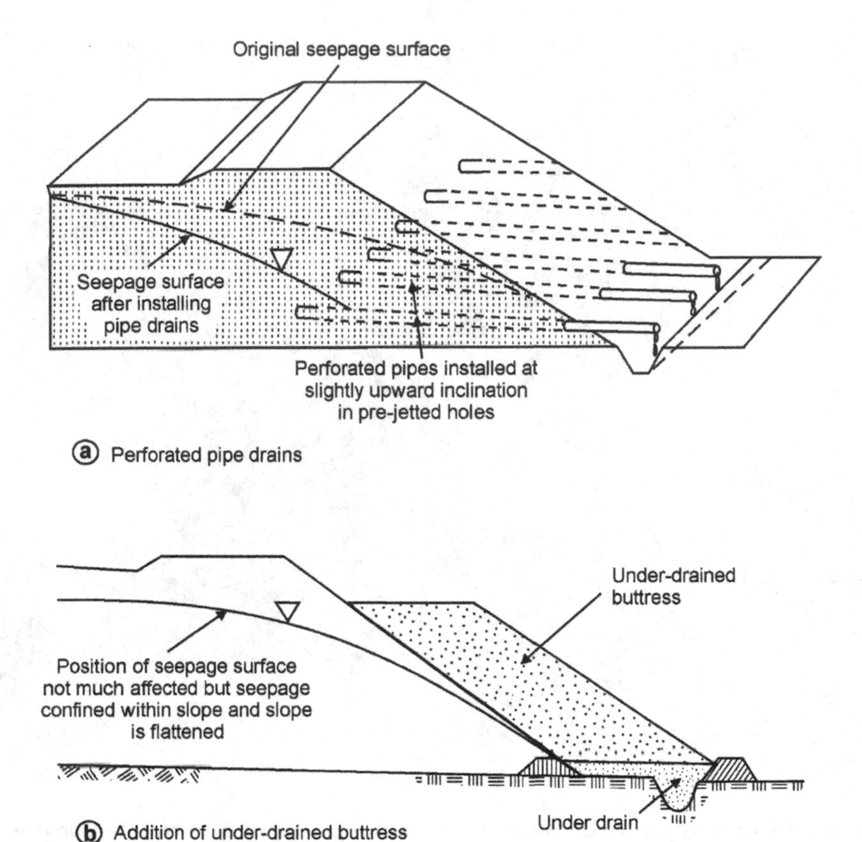

Figure 9.9 Methods of augmenting or improving the drainage of slopes.

- by constructing additional drains in front of the existing toe and then reconstructing the slope of the waste storage on top of the new drainage measures (see Figure 9.9b.)

  A variety of other options is available.

It may also be necessary to flatten the slopes of the storage by stepping back in the form of a berm and then continuing at either the original or at a reduced slope. Another possible remedial measure is the construction either of an under-drained buttress of tailings, or a buttress of rock or other permeable waste (e.g. cyclone underflow tailings) at the toe of the existing storage.

9.5 Special considerations for carbonaceous wastes

Colliery wastes are carbonaceous and often contain pyrite. Both carbon and pyrite are subject to oxidation on exposure to air and the oxidation process is accompanied by the evolution of heat. The process of oxidation is also vastly accelerated by the presence of sulphur oxidising bacteria (e.g. Thiobacillus thiooxidans). Oxidation proceeds slowly at ambient temperatures, but accelerates if the temperature increases as heat is evolved. Oxidation and the consequent build-up of heat can only occur if the waste has access to oxygen. The ultimate result of the oxidation process may be ignition and burning of the waste. (See Plate 4.1). Adverse environmental effects include the emission of smoke and noxious gases and the generation of acid and sulphate-bearing runoff and seepage. (Also see more detailed note in Section 9.5.5 as well as Section 12.7.)

9.5.1 *Dumps of coal waste*

It has been noted above that heating and spontaneous combustion of coal wastes pose major problems and that the only way to diminish or prevent the problem is to exclude air from the pores of the waste.

The most effective means of excluding air from dumps of coal waste is by means of compaction and all the principles of compaction set out in Chapter 7 apply.

- The waste should generally be spread and compacted on the dump in layers not exceeding 200 mm in loose thickness. These will compact to a thickness of about 150 mm. Depending on the available compaction equipment, it has been found that adequate compaction of layers up to 400 mm thick can be obtained by using heavy rollers (particularly vibrating and impact rollers, see Figure 7.1) and giving attention to the compaction water content.
- When a load of waste is end-tipped, segregation of the particle sizes occurs with the larger particles tending to roll down the tip face and form a coarse layer at the bottom of the pile. In end-tipped dumps this segregation can be a major cause of combustion as the large particles represent a pervious layer under the dump that admits the oxygen necessary for oxidation and heating. A similar situation will exist if a dump is built up by end-tipping in layers of excessive thickness, e.g. 2 or 3 metres. Even if the surface of these thick layers is heavily compacted, the effects of the compaction will not extend through the thickness of the layer (see Figure 7.2c). In addition, the compacted surface of each thick layer will be immediately overlain by a continuous coarse segregated layer that will admit oxygen relatively freely, thus negating the effect of the compaction.
- The best methods of deposition are either to dump truck-loads at a regular spacing across the previously compacted surface of the dump and then to spread by dozer or grader prior to compacting; or preferably to spread directly in a 200 mm thick layer from a moving bottom dump truck.
- If the dump is formed with side slopes of less than 18° to the horizontal (1 vertical on 3 horizontal) the side slopes can be compacted to seal the surface and be further sealed against ingress of air by placing a layer of soil, which can be compacted, over the slopes of the dump surface.

9.5.2 *Stability of coal waste dumps*

The shear stability of coal waste dumps can be calculated in exactly the same way as for dumps of other material. However, as Figure 4.29b shows, coal waste may differ from other wastes in that the angle of shearing resistance φ^I may be less than the angle of repose β. Not enough is known of the shear strength properties of coal wastes to be able to state whether or not this is always the case. It appears that wastes having less than a certain minimum carbon content may have φ^I at least equal to β whereas the phenomenon of φ^I less than β may apply to varying degrees for more carbonaceous material.

9.5.3 *Monitoring of coal waste dumps*

Section 9.3 describes monitoring procedures designed to detect signs of instability in waste storages. With coal waste storages there is an additional requirement, that of detecting actual or incipient burning.

Inspections

Signs of actual or incipient burning may be detected by sight, smell or temperature.
Sight: Burning is often indicated by:

- thin wisps of smoke issuing from surface cracks;
- plumes of steam or water vapour issuing during or after rain;
- heat haze or shimmer above hot areas;
- deposits of yellow flowers of sulphur on the sides of surface cracks;
- glowing patches visible at night when burning is actually occurring. These are usually not visible in day light.

Smell: The distinctive acrid smells of sulphur dioxide, and hydrogen sulphide indicate that combustion is occurring.
Temperature: Metal rods (e.g. discarded drill steels) can be driven into the waste to a depth of 1½ to 2 m below the surface. If the protruding end of the rod feels hot, this may indicate a dangerous increase of temperature below the surface.
Remote sensing by means of thermal imagery from the air is a useful way of monitoring the overall temperature of a coal waste storage.

Temperature measurements

Temperatures may be monitored by driving small diameter steel pipes or tubes, closed at the remote end, into the waste, and then measuring temperatures inside the tube. Temperatures may be measured by lowering a maximum thermometer down the pipe and leaving it for some time to equilibrate before withdrawing and reading it. A better method is to equip the pipe with thermocouples which remain permanently in place and are read by a portable thermocouple bridge. These could also be read by data logger and recorded remotely.

Any tendency for the temperature to increase above 45°C and particularly for the rate of temperature increase to accelerate, indicates the development of burning in the vicinity of the temperature probe.

It is important not to use open ended pipes for the probes as they may admit oxygen to the waste at depth and thus promote a rise in temperature. It is also not advisable to install temperature probe pipes by excavating holes and backfilling by placing waste around them, unless special precautions are taken to hand-compact the waste tightly around the pipe. If this is not done, the zone of loose backfilled waste may act as a chimney and admit oxygen to the waste at depth.

Gas analysis

The gas in the pores of a waste dump can be sampled by driving a small diameter closed ended pipe into the waste. If a few small-diameter perforations have been provided adjacent to the closed tip of the pipe, the gas at that level can be withdrawn, sampled and analysed. Any presence of the products of combustion, i.e. carbon monoxide, sulphur dioxide or hydrogen sulphide, is a warning sign.

9.5.4 *Extinguishing a burning coal waste dump*

This is a very difficult and highly specialized subject which depends very much on local conditions and experience. The basic principle is to smother the fire by excluding oxygen from it and then permanently to exclude oxygen from the coal waste by constructing an impervious capping layer over the storage. It is not intended to deal with the subject in any detail in this section. The material that has been included is intended to inform the reader of the complexities and not to act as an operational guideline.

No attempt should ever be made to douse a burning dump by applying water. The application of water may prove to be dangerous, as the addition of water to glowing waste will generate water gas and steam which may explode if mixed with air and ignited by coming into contact with other burning waste.

The capping layer is usually fairly complex, consisting possibly (from bottom to surface) of a compacted impervious insulation and air exclusion layer, a drainage and capillary break layer, a neutralization layer and a growing layer.

The air exclusion and insulation layer is intended to exclude air from the hot and possibly burning waste beneath it, and also to insulate the overlying capping layers from the hot material below. The drainage and capillary break layer consists of a layer of gravel to intercept water infiltrating from above, and to prevent this infiltration from seeping into the air exclusion layer and thence into the waste. It also traps potential acid-bearing water from below and prevents it from rising by capillary action into the overlying layers.

The neutralization layer is a lime-rich layer intended to neutralize any acid rising by capillarity that eludes and bypasses the capillary break layer.

The procedure of capping is not easy to carry out, as there are many hazards attendant to working on a burning waste dump.

In the case of operating mines with sufficient life ahead of them, the best way of blanketing a burning dump appears to be to surround it with a properly constructed compacted waste dump. The compacted waste can be placed and compacted right against the burning waste to be smothered. More usually, an insulating layer, commonly of inert soil is placed and compacted against the surface of the burning or burned out waste, and the compacted coarse waste is placed outside this soil blanket. Plate 9.15 shows this process, with the burning dump in the background and the compacted waste encasement in the foreground.

Plate 9.15 **Burning coal waste dump in process of being blanketed with outer layer of well-compacted coal waste.**

Noxious gases

These will probably be present in the work area. A programme of regular sampling and analysis must be set up to ensure that the gases are not present in dangerous concentrations.

- The work must be supervised and the supervisors equipped with breathing apparatus and trained in its use and in rescue procedures.
- Machine operators must be acquainted with the dangers and symptoms of gas poisoning, particularly those due to carbon monoxide. They must be instructed to leave the work area and move upwind immediately they notice any ill effects.
- Machines breaking down in the work area must be abandoned by their operators and towed out of the work area before attempting to repair them.
- A cell phone or radio link should be maintained between the work area and the mine first aid post.
- Machinery should be removed from the work area overnight.

Explosions

The danger of explosions caused by water gas has already been mentioned. There is also a danger of coal-laden dust, stirred up by the work, exploding if accidentally ignited. Dust can be diminished by watering, but care must be taken to apply only enough water to lay the dust and not so much that it percolates into the waste, where it might form water gas.

Cavities and unsafe ground

- Combustion of the waste may produce cavities or soft areas into which men or machines may fall. Wherever possible, working should be done remotely.
- If quantities of smouldering waste have to be moved, they should be pushed downhill by dozer, using gravity to further move and spread the waste or (preferably) be pulled downhill by dragline or cable operated scraper.
- Men and machines should move only by routes probed and known to be safe. When a new route has to be established, it should be proved by probing with a long, small diameter steel rod. For heavy machinery the route should be tested by walking a crane over it, dropping a heavy weight ahead to test the ground over the width of the proposed route.

9.5.5 *More detailed consideration of burning of coal waste*

Causes of fires

- Coal waste storages may be accidentally ignited by grass fires, by the kindling of small fires for heating purposes or by tipping hot boiler ashes on the dump. The most common cause of burning, however, is spontaneous combustion resulting from the natural oxidation by atmospheric oxygen of carbon and pyrite contained by the waste.
- The oxidation process is both exothermic and temperature dependent i.e. it is accompanied by the evolution of heat and proceeds faster at higher temperatures.
- If the heat is not dissipated as rapidly as it evolves, and if the oxygen is not limiting, the temperature rises, the rate of oxidation doubling for each 10°C rise in temperature.
- Coal ignites at a temperature of 350°C, but if other combustible material is present in the waste, e.g. timber, cotton waste, paper, cardboard or domestic refuse, this will ignite at lower temperatures (e.g. domestic refuse ignites at 260°C).

Factors affecting spontaneous combustion and burvning

Coal rank Wastes from lower rank coals are more reactive and hence more susceptible to burning than wastes from higher rank coals. Hence bituminous coal wastes (low rank) require more care in disposal then wastes from anthracite (high rank) if combustion is to be prevented.

Pyrite content Pyrite, an iron sulphide, combines with oxygen in the presence of water to produce iron sulphate and sulphuric acid with the evolution of heat. The presence of pyrite therefore exacerbates the tendency of coal waste to ignite. Oxidation of pyrite is accelerated by the presence of sulphur oxidizing bacteria, but it is not known how important this effect is in causing ignition.

Once ignition has occurred, pyrite burns to sulphur dioxide, but if insufficient oxygen is present, hydrogen sulphide will be given off. The characteristic pungent smells of sulphur dioxide and hydrogen sulphide may provide a warning that combustion is occurring within a waste storage.

Moisture

Free moisture is essential to the oxidation of pyrite while the condensation of water on the surface of carbonaceous material, giving up its heat of vaporisation, can cause an appreciable rise of temperature. The presence of moisture in a coal waste therefore makes a positive contribution to heating. In a dump of coal waste, therefore, heating effects are exacerbated during warm weather after soaking rains have fallen.

Once combustion is occurring, the addition of water results in the formation of steam, and water gas (a highly explosive mixture of carbon monoxide and hydrogen) may also be produced.

If combustion is localized, the steam will tend to migrate into adjacent areas of cooler waste, giving up its latent heat of vaporisation as it condenses, and raising the temperature of the cooler material. The increase in temperature will create more favourable conditions for the spread of combustion.

There is also a danger that a mixture of water gas and air will explode, possibly opening up the burning residue to air and causing the fire to spread.

Particle size and particle size distribution

The smaller the particles of waste, the larger will be their specific surface area. If oxygen is available, a mass of smaller particles will heat faster than the same mass of large particles.

However, heating is absolutely dependent on the presence of and continued access to oxygen. The permeability of a granular material is approximately proportional to the square of the particle size. Hence permeability to air decreases rapidly with decreasing particle size. Since oxygen is usually only available at the free surface of the mass of waste, and will only flow in if there is a pressure gradient into the mass, the net effect of decreasing the particle size of a coal waste is to inhibit heating.

Material most susceptible to heating is large single sized waste through which air can move relatively easily under low pressure gradients caused by wind and variations in barometric pressure, i.e. effects are worse on the windward side of a dump. If, however, the material is well graded so that the voids between the large particles are filled with finer material, the permeability to air will essentially be that of the fine material and the tendency towards heating will be less.

Noxious gases produced by combustion

Mention has already been made of the generation of *sulphur dioxide* and *hydrogen sulphide* as well as, in the presence of water, *water gas* and *carbon monoxide*. In addition to these four gases, *carbon dioxide* may also be produced by coal waste fires.

The sulphur gases are readily detectable by smell, taste and the irritation they cause to eyes and throat, long before they reach dangerous or lethal proportions. They are therefore primarily a nuisance and an indication that burning is present, rather than a danger.

Carbon dioxide is not poisonous, but being an inert gas and heavier than air, it is possible to drown in a hole or hollow filled with the gas. (This may be the cause of death of night-shift workers sleeping on duty in sheltered corners.)

Table 9.2 Characteristics of gases released by burning coal waste

Gas	Danger Threshold Concentration by Volume of Air	Physiological Effect
Sulphur dioxide	200 ppm	Eye, nose and throat irritation
Hydrogen sulphide	200 ppm	Eye, nose and throat irritation, headache
Carbon dioxide	50 000 ppm	Breathing laboured
Carbon monoxide	400 ppm	Headache and discomfort. Possible collapse after exposure of 2 hours at rest or 45 minutes exertion

Carbon monoxide is the most dangerous gas as it cannot be detected by smell, taste or irritation even when present in lethal concentrations. It is also the most common gas found in the vicinity of burning coal deposits.

Water gas is a mixture of carbon monoxide and hydrogen which is highly explosive when mixed with air.

"Danger threshold" values for concentrations of the various gases and the corresponding physiological effects at these concentrations are summarized in Table 9.2.

Water pollutants from coal mining wastes

Water pollutants arise directly from the oxidation and combustion of pyrite in the residue. These include:

Sulphuric acid
Iron sulphate and
Ammonium sulphate

Water pollution must be controlled by the water control measures described in Section 9.1, and if necessary, by providing drained impermeable linings beneath new installations.

9.5.5 *Coal washery slurry ponds*

Coal washery slurry consists of finely divided carbonaceous material that may be of saleable quality. The slurry is usually deposited hydraulically into a system of ponds where the solids settle out and the water is recovered for re-circulation to the washing plant. At smaller collieries where the rate of production of slurry is not very large, it is feasible to excavate the slurry from a pond, once it has been dewatered, and to transfer it to a waste dump, if it is unsaleable. For this reason, slurry ponds are often designed for reasonably rapid dewatering of their contents and for access by earth-moving machinery to empty the ponds. A sufficient number of separate ponds must be provided to enable a sequence of filling, draining and emptying to be established.

As rates of production of washed coal have increased, there has been a tendency to move away from the traditional slurry pond to a co-disposal system in which an outer dyke is built of compacted coarse waste to contain a pond of hydraulic-fill washery sludge. This method thus combines the compacted coarse discard dump with the traditional slurry pond. Obviously, the volume of coarse material deposited in the outer dyke can be designed so that the slurry production can be accommodated while retaining a satisfactory freeboard.

Design of coal slurry ponds

The design of coal slurry ponds has many features in common with the design of return water reservoirs. There are, however, important differences:

- A return water reservoir is designed to contain and store water whereas a slurry pond is designed to store solids and to separate the solids from the water.
- The desilting of a return water reservoir constitutes necessary maintenance. The emptying of slurry pond may be part of the functional operation of the pond.

Coal slurry is not generally considered a suitable material for constructing the outer walls of ponds as it erodes very badly. Hence the walls of slurry ponds are usually constructed of earth fill. In certain cases, however, it may be possible to construct the slurry pond walls of compacted coarse coal waste, blanketed on the outside with a layer of compacted soil, to prevent spontaneous combustion.

Coal slurry has also been disposed of by building a conventional hydraulic fill tailings storage of the slurry. One problem with this method is that the coal slurry is very susceptible to erosion. It also has a low density and is therefore more susceptible to the effects of a high phreatic surface than is compacted soil.

DESIGN FOR CAPACITY

If it is not proposed to reclaim the slurry either for sale or for depositing on a waste dump, a pond can be designed for a specified life at a given rate of production of slurry. This applies particularly to the composite coarse discard-slurry retaining structures mentioned above. Otherwise, the pond can be designed for a specified time for the cycle of filling, dewatering and drying and excavating.

In any case, tests must be performed on the slurry to establish:

- the density to which it settles and the relationship between density and effective overburden stress and;
- the rate of drainage or consolidation of the material.

If it is proposed to excavate the slurry periodically, the maximum depth of settled material should not usually exceed 3 m. This recommendation is important both from the aspect of draining the slurry prior to excavation and considering the safety of men engaged in excavating the drained slurry. Greater depths than 3 m can be used if investigation proves that the proposed depth is not excessive for both drainage and safety.

HYDROLOGICAL DESIGN

Slurry ponds should be hydrologically isolated from their surroundings by means of a stormwater diversion drain designed in accordance with Section 9.1. Ponds would normally be provided with a penstock leading to a return water or polluted water reservoir. The catchment of the pond within the storm water diversion drain as well as the return water dam should also be designed to Section 9.1.

PENSTOCK DESIGN

If it is not planned to empty the pond of its contents periodically, no special measures apply to the construction of the penstock and a conventional penstock tower e.g. of nesting rings, as described in Figure 9.2a would be appropriate. If, however, it is intended to excavate the pond contents, the penstock should be specially designed and protected against accidental damage by earth-moving machinery. As the penstock will now be a permanent re-useable structure, the type illustrated in Figure 9.2b and constructed in brickwork or reinforced concrete would be preferable.

WALL OR EMBANKMENT DESIGN

Because the object of a slurry pond is to store coal solids and not water, the walls of a slurry pond do not need to retain water. At the same time, water passing through pond walls, being polluted, needs to be controlled. For this reason, the "filter-dams" shown in Figure 9.10 are useful, as they retain solids while allowing the passage of water in a controlled way. Polluted seepage or drainage water can then be intercepted and stored, for the purpose of treatment, in a seepage-proof reservoir immediately downstream of the slurry pond.

As in the case of return water reservoirs, careful attention must be given to the foundations. While it is not always necessary to remove permeable layers, weak layers must be removed and it may be necessary to remove loose permeable layers and replace them with a series of compacted layers.

9.6 A brief note on characteristics of radioactive wastes

Uranium tailings and other radioactive wastes usually contain traces of long-lived radio-nucleides, in particular:

Thorium 230 (80 000 year half life)

Radium 226 (1 620 year half life)

and their radioactive decay products. Of these

Radon 222 (3.8 day half life)

is particularly difficult to deal with because of its gaseous nature.

The two main routes whereby radionucleides are released into the environment are via surface and ground water pollution and air pollution.

Surface and ground water pollution may be controlled by the hydrological and hydraulic measures described in Section 9. There is for example a possibility that

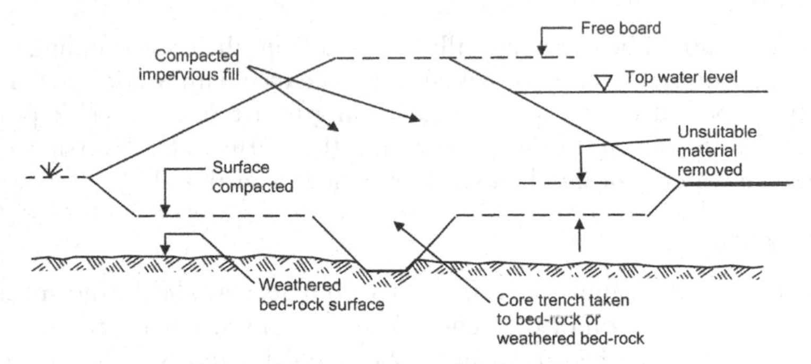

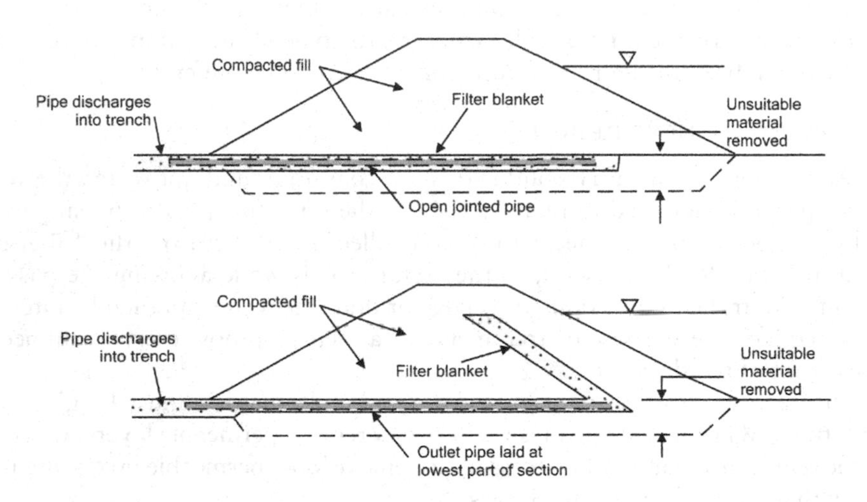

Figure 9.10 Typical sections for water control dams.

radium 226 may be leached from a uranium tailings storage, but the impact of such leaching is likely to be significant only on a localized scale (e.g. individuals or animals drinking the contaminated water). This is an additional reason why waste storages should be securely fenced.

Air pollution may arise by direct emission of radon into the atmosphere or by the release of radioactive particulate material (i.e. dust). Measurements made in the USA have indicated that radon concentrations above the natural level may be detectable for 1 to 2 km downwind of a uranium tailings storage. In addition, decomposition products of radon, particularly Lead 210 and Polonium 210 may be deposited on crops and subsequently be ingested by man or animals. (It is correct that both atomic weights are 210.)

Wind-generated dust may result in contamination by

Direct inhalation
Contamination of vegetation and soil surfaces
Subsequent contamination of rainwater runoff

The radionucleides of particular concern are:

Thorium 230
Radium 226
Polonium 210

However, because dust is wind supported, downwind contamination drops off more rapidly with distance than in the case of radon. Severe contamination via wind-borne dust may extend to some hundreds of metres from a waste storage, while significant contamination has been recorded up to 1 km away. Particular care should therefore be taken to provide any storage of radioactive waste with a buffer zone having a width of at least 1 km, and also to pay particular attention to dust control, especially for the sake of persons working on or close to the deposit for protracted periods of time.

Because of Radon's short half life, one method of control is to cover the surface of a Radon-emitting storage with a 1 to 2 m thickness of non-radioactive waste rock. This provides a layer of stagnant pore air space above the emitting surface, from which the Radon can only escape by diffusion. As diffusion is a relatively slow process, the Radon should be well along its decay curve before it diffuses into the air above the waste rock cover, where it can be swept away by wind.

Chapter 12 advises on the minimization of erosion by both water and wind and therefore the minimization of the removal of radioactive material from a tailings storage.

As an example of potential emissions from a uranium tailings storage, a tailings storage at a uranium mine in Canada was abandoned in 1959 and rehabilitated in 1996 (Lee, et al., 2000). Rehabilitation consisted of removal of radioactive material from the waste rock dump and spreading of this rock over the tailings storages, then "thick-capping" the tailings storages with silty clay from a site nearby.

A subsequent audit of health risks from the rehabilitated storage, completed in 1999 used estimated exposure-route-specific incremental doses to conclude that the ranking of potential daily doses by various pathways was as follows:

external exposure to gamma radiation (highest);
ingestion of surface water from nearby water bodies;
inhalation of radon;
inhalation of dust. (lowest)

The numbers of consecutive days of exposure to reach a 1 mSv annual incremental dose limit for the various exposure pathways were estimated as:

External exposure to gamma radiation 26 days
Ingrestion of surface water 29 days
Inhalation of radon 83 days
Inhalation of dust 11400 days

It was estimated that simultaneous exposure to all pathways would require 17 days to reach the 1 mSv/year limiting dose. (The calculations were statistical and the figures are for a 95th percentile exposure to a particular route.)

References

Ashby, R.J., & Chilver, A.H.: Problems in Engineering Structures. London, U.K.: Arnold, 1965.

Chamber of Mines of South Africa: Guidelines for Environmental Protection: The Engineering Design, Operation and Closure of Metalliferous, Diamond and Coal Residue Deposits. Volume 1/1979 (Revised 1983 and 1995), (1996).

Lee, R.C., Robinson, R., Kennedy, M., & Swanson, S.: Identification of Long-term Environmental Monitoring Needs at a Former Uranium Mine. In: Tailings and Mine Waste '00. Rotterdam, Netherlands: Balkema, 2000, pp.361–368.

Pippard, J.J.S., & Baker, J.: The Analysis of Engineering Structures. London, U.K.: Arnold, 1976.

U.S. Department of the Interior, Bureau of Reclamation (1960 onwards). Design of Small Dams. Washington, U.S.A.: US Government Printing Office, 1960.

Chapter 10

Water balances for tailings storage facilities and dry waste dumps

10.1 Water balances in general

This chapter deals with the basic principles of assessing water balances in the design and operational phases as well as the closure and aftercare phases for hydraulic fill tailings storages. The water balances for the closure and aftercare phases for dry dumps are very similar to those corresponding to tailings storages. The examples are drawn from research on hydraulic fill ash dams, dry ash dumps and municipal solid waste landfills as well as metalliferous tailings storages. This demonstrates the generality of the principles – they can be applied to any mine waste deposit of any type.

It will become abundantly clear from the practical examples of water balances for both operational hydraulic fill tailings storages and for a closed tailings storage or a dry dump that water balances cannot be calculated with reliable accuracy if the necessary waste properties are not known. A "generic" knowledge of these properties is insufficient, the properties must be measured for the specific waste and the specific method of operation of the storage. It must also be borne in mind that the properties of the waste being deposited will be variable from hour to hour, day to day, year to year and decade to decade. It may be, and often is necessary to use generic properties for the waste at the design stage. However, once the storage has been commissioned, the actual parameters for the waste must be determined. Measurements must not only be made once when the waste storage is commissioned, but monitoring of properties and adjustment of the storage operation must be ongoing throughout its operating life and when it is finally being rehabilitated.

Water balances for tailings storage facilities are important at four stages and scales:

(1) Design and operational, when the overall water balance for the storage is considered, mainly as a water management tool. Main inputs are tailings slurry water and rainfall. Outputs are decant and drainage water (representing return water) and evaporation from the pool and beaches. Inputs and outputs for the side slopes are usually ignored. Seepage through the base or foot print of the dam is usually not accurately determinable, but can be estimated from a knowledge of the overall permeability of the foundation strata, as well as from the closure water surplus for the water balance.

(2) Rehabilitation and aftercare. For this stage the inputs are rainfall infiltration and (possibly) irrigation water used to establish surface vegetation. The outputs are surface run-off, surface evapotranspiration and possible seepage through the base. Depending on climate, if infiltration exceeds evapotranspiration, and the

water balance for the storage is positive, leachate may form and exit the base of the rehabilitated storage.

(3) For most storages on which operations have ceased, there will be a stage intermediate between a) and b) when the phreatic surface is subsiding, but seepage still exits the footprint. Depending on the height of the storage and the permeabilities of the tailings and foundation strata, this phase may last several years, and possibly as long as two or more decades.

(4) It is also important to realize that "dry" dumps of mining waste (e.g. waste rock, soil and rock overburden, colliery discards) may also emit leachate from their base, if their water balance is positive.
If the waste material contains pyrite, as many waste rocks as well as colliery discards do, the dump may emit leachate that is either unacceptably acid or saline. Predicting the occurrence of such emissions therefore also involves a careful consideration of the water balance for the dump, which will usually fit into category (2) above.

Whether the water balance is positive or negative is mainly a function of the balance between infiltration (rainfall plus irrigation minus run-off) and evaporation (or evapotranspiration). To a major extent, this is controlled by the local climate, modified by factors such as the efficiency of decanting and storm-water drainage, the presence of an underliner or an impervious base to the impoundment, etc.

During operation, inputs are usually minimized and return water maximized to save water. After closure, and during aftercare, the slopes and surface should (if possible in the prevailing climate) be kept in a negative water balance state (compatible with requirements of vegetative growth) to minimize seepage and maximize slope stability. It is essential to realize that both of the above phases, or stages of life are important: The operational phase of a waste storage may extend over 30 to 50 years, i.e. 1 to 1½ human generations, but the aftercare period will extend beyond the foreseeable future. Already, periods of aftercare of centuries and possibly millennia are being spoken of. This is not impossible to consider as realistic. In China's amazingly long continuous civilization, ancient tombs and tumuli have been maintained without interruption for several millennia.

Section 10.2 of the chapter lists, describes and considers the data required, and 10.3 describes operational hydraulic fill tailings storage and presents the corresponding water balance equation. Section 10.4 illustrates (by means of an example) the difficulties, both of obtaining good information and of the innate variability of available data. Because the slurry density represents the quantities of both solids and water being deposited, it is truly the driving variable of the water balance. A satisfactory representation of the water balance is not possible without accurate data on slurry density.

Recession of the phreatic surface after ceasing deposition (Section 10.5) takes place mainly by seepage into the foundation strata. The rate of recession can be estimated by applying D'arcy's law to the system of storage plus foundation. Interstitial water will also reduce as the recession occurs, but fine grained tailings will remain saturated until the suction in the tailings exceeds the air entry value.

Section 10.7 points out how, once operations have ceased, the important components of the water balance change, and the climatic variables of rain infiltration, evaporation or evapotranspiration from the surface and recharge, i.e. seepage into the foundation, become important. The chapter describes methods for measuring infiltration and for estimating and also measuring evaporation. A further example of transferability of principles from one type of waste deposit to another is the recommended use of the method of estimating evaporation from A-pan measurements, which is now in use for municipal solid waste landfills.

10.2 Required data

Data required for the operational stage relate mainly to:

- the rate of deposition of tailings expressed in terms of tons of dry solids;
- slurry density or relative density on arrival at the storage facility;
- intended rate of decant from the pool;
- storage of water within the pores or interstices of the tailings (not usually well established, but can be judged from data on profiles of void ratio for the storage and the position of the phreatic surface);
- evaporation from the surface of the storage, including the pool and wet and dry beaches;
- rainfall precipitation on the surface of the facility; and
- seepage from the foot-print (difficult to estimate from first principles because of uncertain knowledge of permeability parameters, but can also be estimated from the water balance itself).

Data required for the recession of the phreatic surface, rehabilitation and aftercare stages are:

- net recharge or percolation of rainfall (infiltration) as well as seepage through the footprint to enter the ground water system (long-term deep seepage);
- rainfall infiltration on slopes and top surfaces; and
- evaporation or evapotranspiration from slopes and top surfaces.

Both slopes and top surfaces may be vegetated, or armoured to reduce surface erosion. Both treatments will affect infiltration and evaporation (or evapotranspiration).

It is not known if the period for subsidence of the phreatic surface has been investigated, or is known for any tailings storages, but the rate of recession can be calculated as shown later.

10.3 Components of the water balance for an operational tailings storage

The water balance for the top area of an operational hydraulic fill ring-dyke tailings storage impoundment and its associated return water storage and evaporation reservoir (if used) would have the following components (see Figure 10.1a and b):

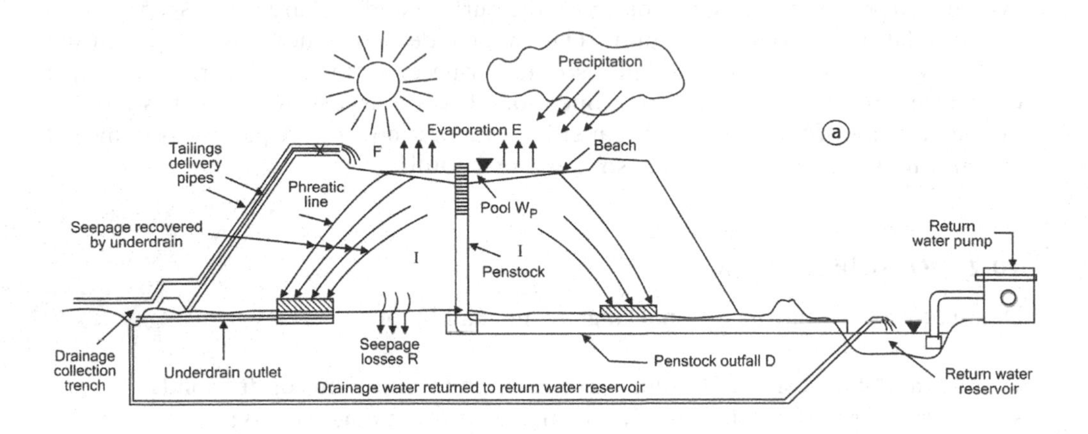

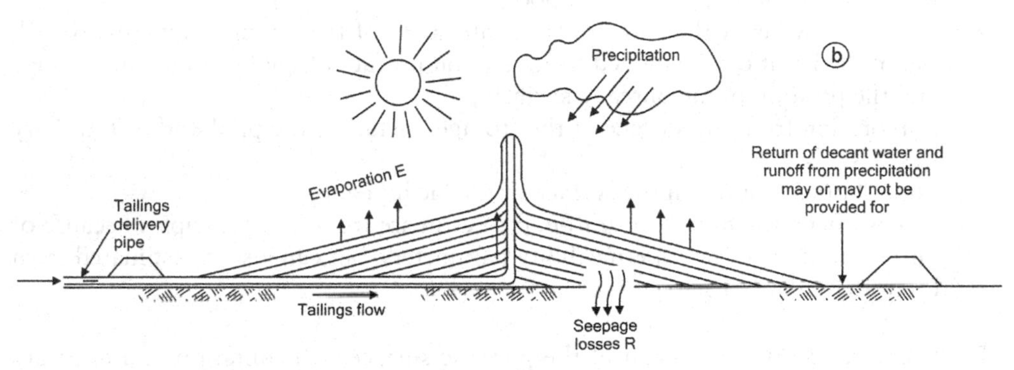

Figure 10.1 Diagrammatic representation of components of water balance for: a: beached unthickened tailings storage, b: point or line discharge thickened or paste tailings storage.

10.3.1 *The water input is based on the mass or tonnage of tailings deposited, expressed in terms of dry solids (T) and would consist of:*

- Water contained in the feed from the tailings slurry delivery line (F). This comprises interstitial water (I) that remains in the pores of the settled tailings, and supernatant water (S) that bleeds out of the surface of the settling tailings slurry and collects in the pool.
- Rainfall precipitating on the surface of the storage and in the catchment of the return water reservoir (P).

10.3.2 *The water output would consist of:*

- The discharge or decant water (D) from the penstock to the return water reservoir.

- The seepage into the underdrainage system (if any) which surrounds the tailings storage and which is fed back to the return water reservoir (d). Unless separated from the penstock outflow, this forms part of D, the decant water.
- Seepage losses into the foundation of the storage and return water reservoir that are irrecoverable and constitute a recharge (R) of the ground water system.
- Evaporation (E) from the surface area of the tailings storage and the return water reservoir.

(Guidance on the accurate measurement of flows D and d is given in Section 10.21.)

10.3.3 *The water storage consists of*

- Water retained in the voids or interstices of the settled tailings by capillary action (I). The proportion of water retained in this way will depend on the void ratio, particle size analysis of the tailings and the way in which the deposition is arranged, i.e. whether or not there is a coarse-from-fine separation on the tailings storage either gravitationally or by cycloning.
- The water (W_p) stored in the pool.

Over any interval of time during the life of the tailings storage, the change in water storage will equal the inflow minus the outflow while the integral of the changes in water storage may be used to predict the water level at any time, both on the top of the tailings storage and in the return water reservoir. At no time may the calculated water level encroach on the minimum required freeboard height.

The equation for the water balance can therefore be written as follows:

$$\Sigma(F + P) - \Sigma(D + d + E + I + W_p) = \Sigma R + \text{losses} \tag{10.1}$$

The losses arise mainly from a lack of accuracy in estimating the remaining terms in the equation, which, in turn, usually arises from inaccurate measurement, unwarranted assumptions about certain quantities and poor record keeping. Equation (10.1) shows all the components of the water balance, but does not show their relative magnitudes and therefore importance.

10.4 Examples of water balances for operating hydraulic fill tailings storage impoundments

The first example of a water balance for an operating tailings impoundment is a synthetic example that uses the deposition quantities for one operational tailings storage along with variations in ρ_s(the mass density of the slurry stream) measured on another (different) storage. It is designed to show the difficulties that can arise in calculating a water balance if excellent data are not available. The example illustrates the relative magnitudes of the water balance components and the type of information required, and also illustrates the often severe difficulty of making a satisfactory water balance calculation because of lack of information, or variability of tailings properties and tailings slurry density, or all three.

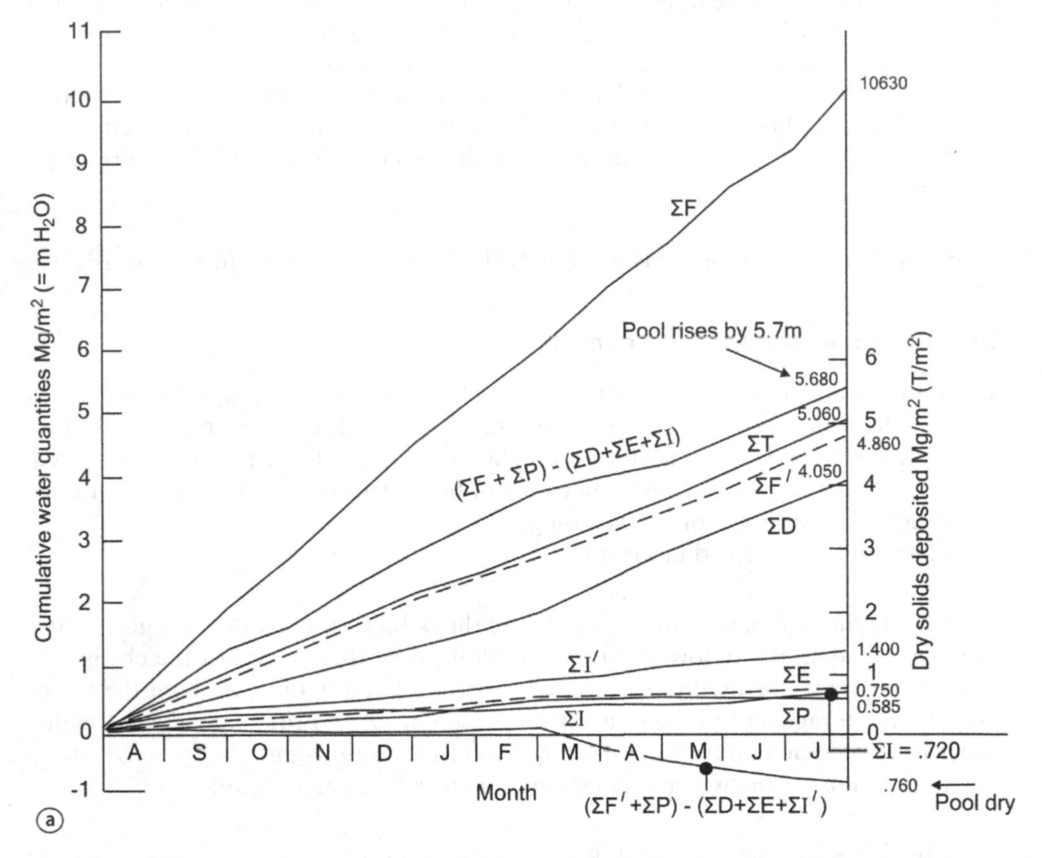

Figure 10.2a: Examples of water balance calculations: a: designed to show effects of variations in void ratio, slurry density and particle relative density on water balance for a hydraulic fill tailings storage.

In the example, the data available are the monthly values of

- T the mass rate of deposition of dry tailings solids in kg per m^2 of impoundment surface area per month (kg/m^2/month);
- ρ_s the saturated mass density of the slurry stream in kg/m^3,
- G the solids relative density (dimensionless);
- P the monthly precipitation in kg/m^2/month (=mm/month, with 1 Mg/m^2/month = 1 m/month);
- D the monthly rate of water decanting together with drain flow in kg/m^2/month = mm/month over the surface area of the impoundment).

Note that all quantities of solids and water have been expressed in the same units (kg/m^2/month or Mg/m^2/month) so that comparison between their various magnitudes is facilitated.

Figure 10.2a shows graphs of cumulative (Σ) T, D, P and E. The interstitial water, I can be calculated from a knowledge of G and the void ratio e_o of newly deposited and settled tailings. This is the first of the potential problems with data, because, although

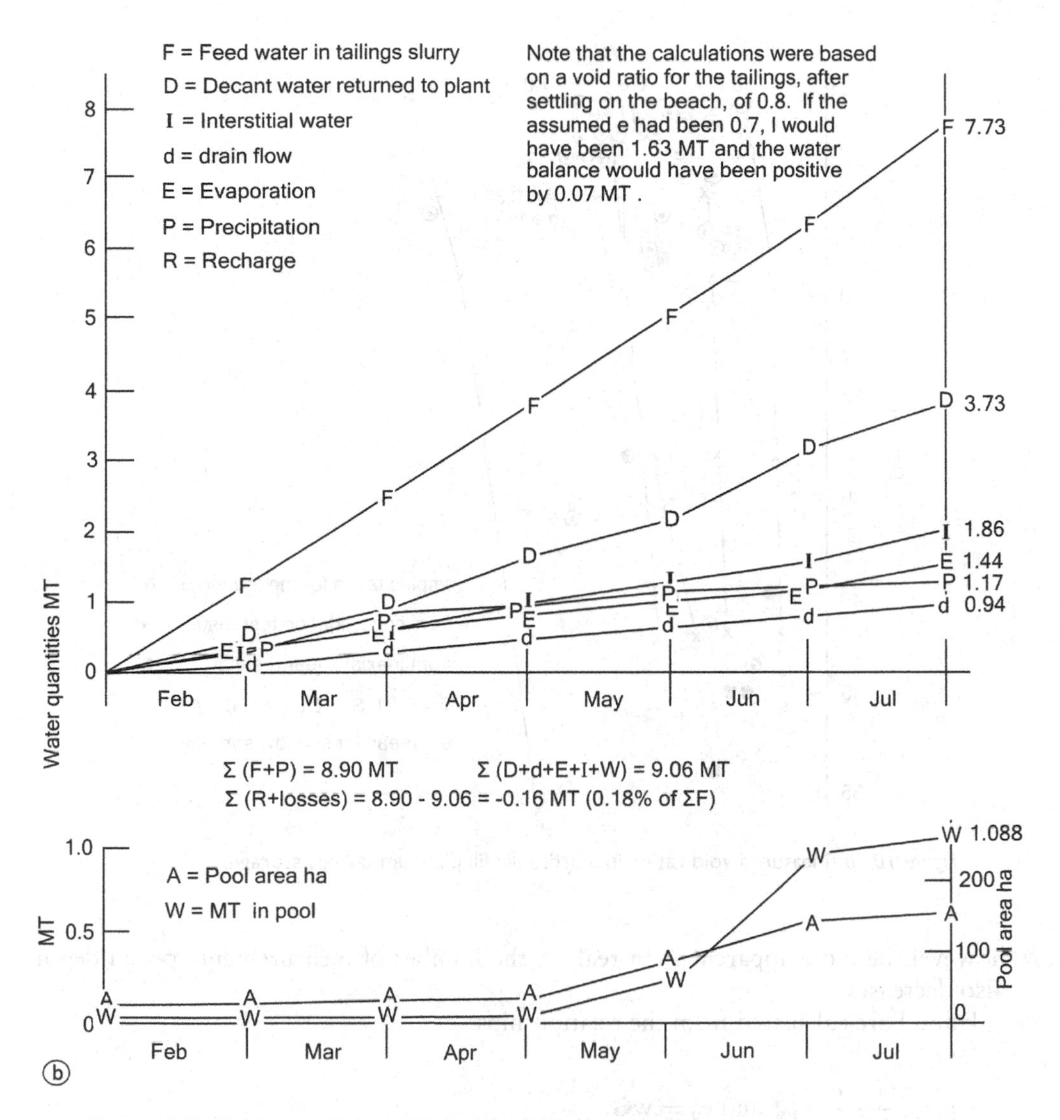

Figure 10.2b: Real water balance for hydraulic fill tailings storage.

measurements of void ratio e on the actual storage may exist, as shown in Figure 10.3a, the void ratio can vary widely from 0.58 to 1.25 (in Figure 10.3a). The data in Figure 10.3a were obtained from samples taken specifically to measure e_o in freshly deposited tailings, from samples taken near-surface or at depth and subjected to compression in oedometer tests, and from samples consolidated and sheared undrained in triaxial compression tests. The figure displays 70 measurements of e.

Figure 10.3b shows a second void ratio profile, in this case measured on undisturbed samples. In this case, the void ratio varies from less than 0.5 to 2.2 in the first 15 m, but the variation appears to decrease with depth. The decrease in variability may,

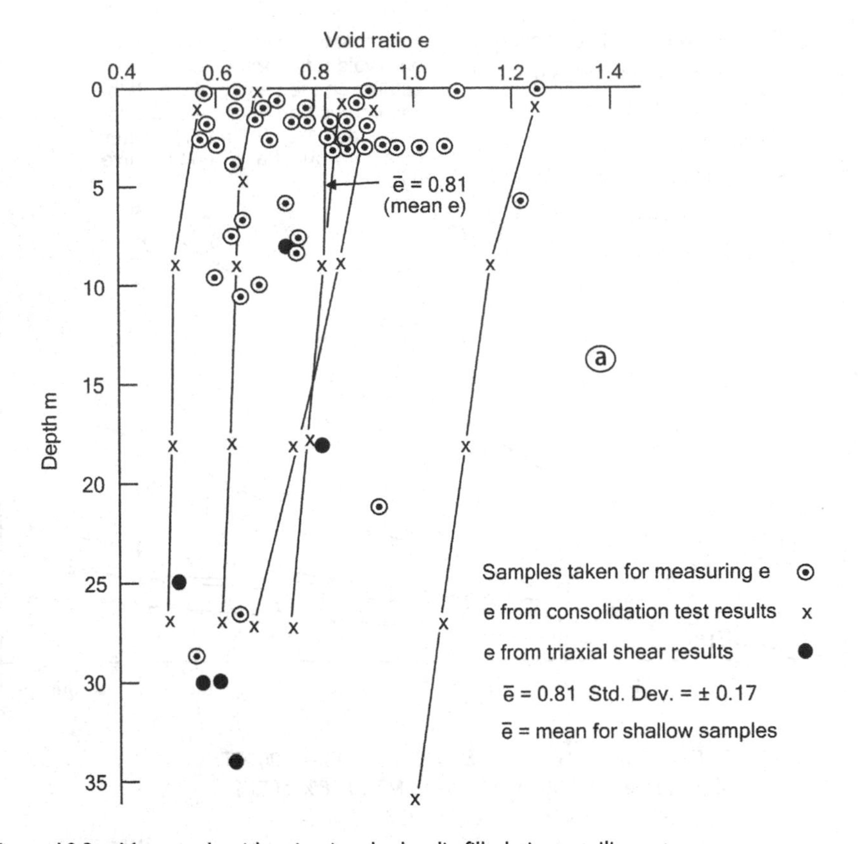

Figure 10.3a: Measured void ratios in a hydraulic fill platinum tailings storage.

however, be more apparent than real, as the number of measurements per m depth also decreases.

F and I are calculated from the relationships

$$\rho_s = \frac{G + e_o}{1 + e_o} \cdot \rho_w \text{ and } e_o = wG.$$

$$\text{Hence } w = \frac{G\rho_w - \rho_s}{G(\rho_s - \rho_w)}.$$

and F = wT

Also, I = eT/G

In the above, ρ_s is the slurry density, ρ_w is the density of water, G is the particle (or solids) relative density, and w is the water content. ρ_s can be measured very simply and quickly by means of a mud balance, a spring balance graduated in units of slurry relative density (ρ_s/ρ_w or RD). A calibrated bucket or measuring cylinder is filled with one litre of slurry and weighed on the mud balance. The calibrations of the balance give

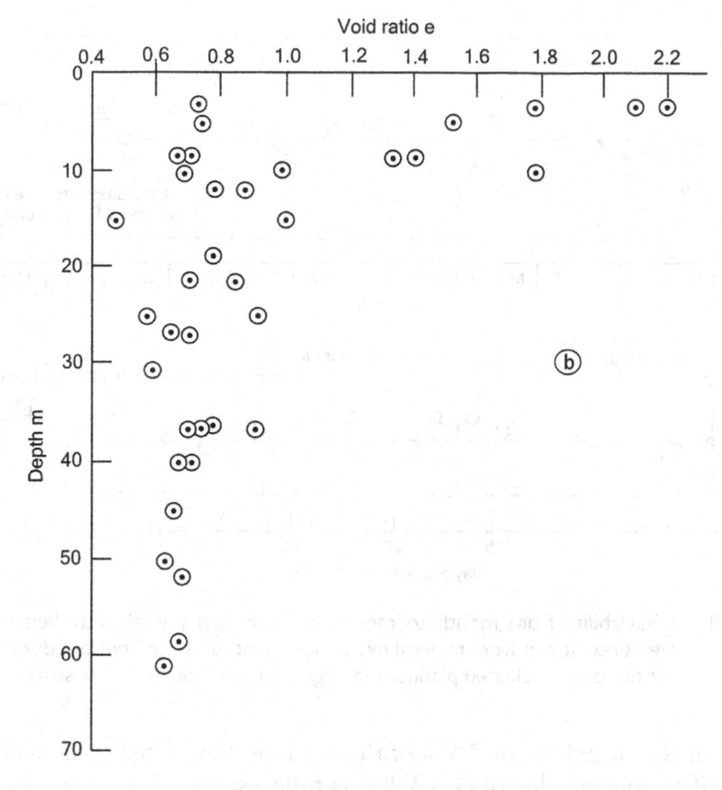

Figure 10.3b: Measured void ratios in a hydraulic fill copper tailings storage.

RD (numerically equal to ρ_s in kg/litre if $\rho_w = 1$ kg/litre) as well as the corresponding w for various values of G. G has to be measured separately in a soil laboratory, by a standard method.

The problem here, is that ρ_s for the slurry stream is often a "reputed" value, not a measured value, and is also variable. ρ_s for the slurry can vary quite widely, as the mineral extraction plant cannot always control the slurry density to keep it constant. Ideally, ρ_s for the slurry should be measured frequently and recorded as it arrives on the tailings deposit itself. It is also the variability of the slurry stream that is partly responsible for the variability of e shown in Figure 10.3. The other cause of variability is that e depends on the particle size distribution of the tailings, which varies as particle size sorting takes place down the length of the tailings beach. The measurements in Figure 10.3 do not record the plan position on the beach from which the sample was taken, but this would not help unless the effect of distance down the beach on particle size distribution is known for the particular tailings storage.

Figure 10.4a shows variations in RD measured in a mineral extraction plant by means of a single measurement made near the beginning of each month. If it is assumed that the variation of RD from hour to hour, from day to day and from week to week is similar to that from month to month, shown in Figure 10.4a, it will be seen that

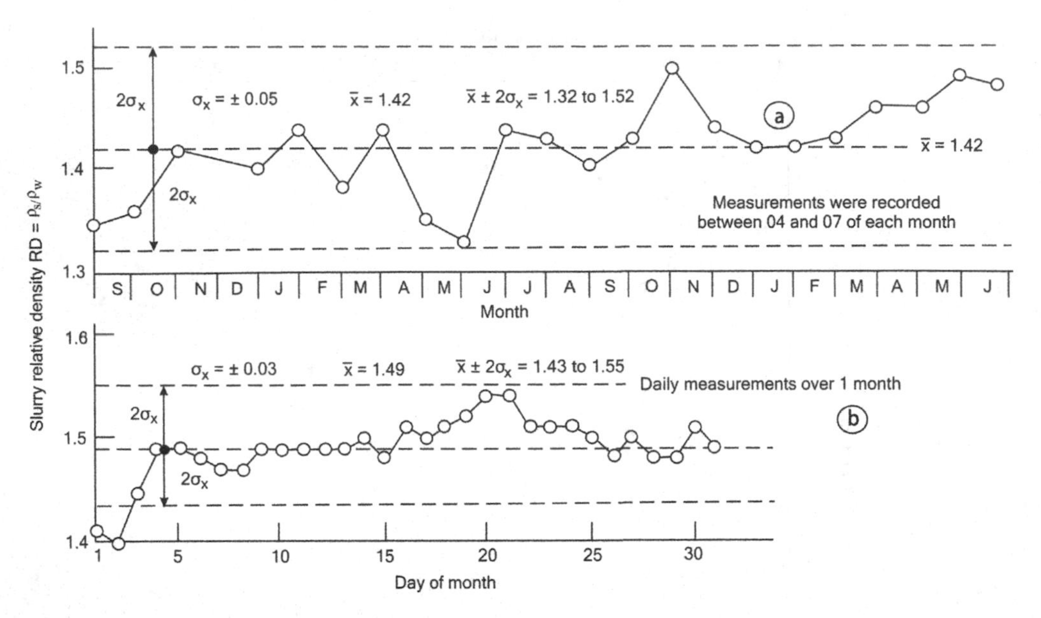

Figure 10.4 **a: Variability from month to month in slurry density of unthickened gold tailings measured at exit from mineral extraction plant. b: Variability from day to day in slurry density of unthickened platinum tailings measured on arrival at storage.**

a variation in RD of ±0.05, or 7.5% of the minimum expected value will occur. The effect on w if the mean value of G is 3.2 is as follows:

If RD = 1.32, w = 183%
If RD = 1.52, w = 101%

The variation is thus 81%

The third source of variation is G which in this case is reputedly 3.2 but almost certainly varies from this value. Calculations of the effect of variations in G on the water content w show that allowing G to vary by ±10% (i.e. from 2.88 to 3.52) has the following effect:

If RD = 1.32, w varies from 169 to 195%, i.e. a 26% variation.
If RD = 1.52, w varies from 91% to 109%, an 18% variation.

Therefore, if the slurry density (or relative density RD) is high, variations in G have a lesser effect than if the slurry density is low.

Hence the effect of variations in slurry density ρ_s far outweigh that of variations in G.

Figure 10.4b shows a similar set of data to Figure 10.4a consisting of daily measurements of ρ_s as the slurry arrives at the tailings storage. Probably as expected, the daily variation over a month is slightly less than the monthly variation over a year, but still very significant, as the results of calculations based on the measurements in Figures 10.3 and 10.4a and recorded in Figure 10.2 will show.

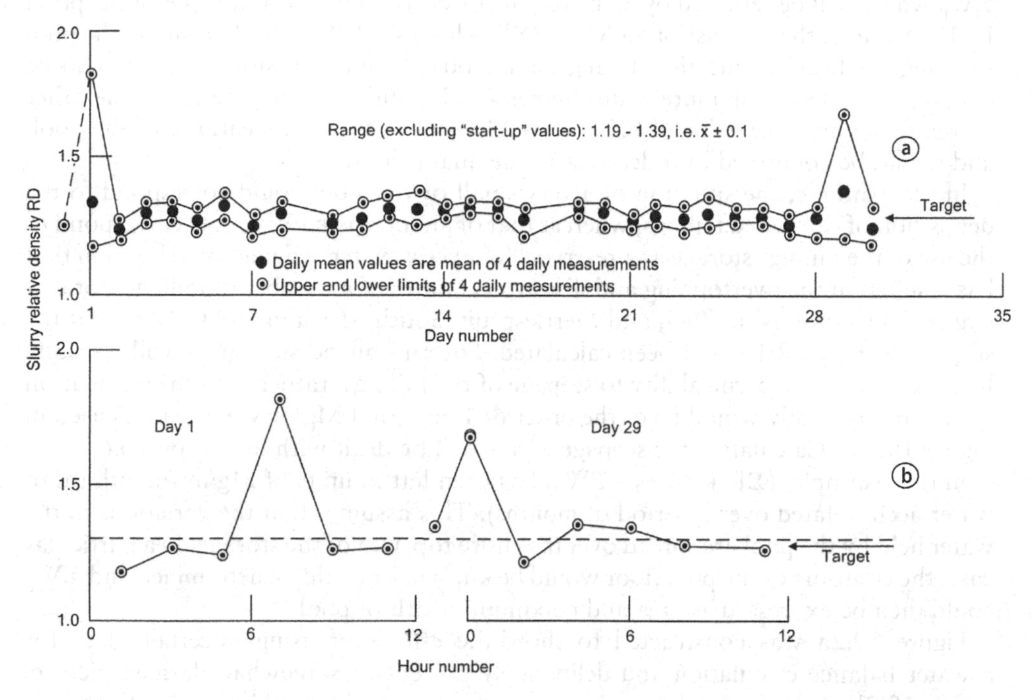

Figure 10.5 a: Variability from day to day in slurry density of unthickened diamond tailings measured on arrival at storage. b: High values resulting from settling of solids in pipes during interruption of flow.

Figure 10.5a shows a second set of on-storage measurements of relative density made over a period of a month. These measurements were specially made at the author's request. Four measurements were made each day with 3 intervals of 4 hours between measurements. Figure 10.5a shows the greatest, least and mean values for each day. Although at first sight the data look better than that in Figures 10.4a and b, they are actually worse, with a range of (mean $\pm$ 0.1), as compared with (mean $\pm$ 0.05) for 10.4a, and (mean $\pm$ 0.03) for 10.4b. Twice during the month, stoppages resulted in a plug of high density tailings coming through, but control was re-established very soon afterwards, as shown by Figure 10.5b. It is obviously very difficult to control ρ_s more consistently than to $\pm$ 0.1.

In Figure 10.2a, the driving quantity is T, the dry solids disposed, while the largest quantity of water has to be F, given by F = wT, with w, in turn, dependent on ρ_s and G. Figure 10.2a shows two versions for F, namely, F, calculated from one extreme set of values for ρ_s and G, and F^l, calculated for the other extreme set of values. The same applies to the quantities of interstitial water, I and I^l. Values of D shown in Figure 10.2a are measured quantities of combined decant and drainage water. It will be noted that ΣF is 2.2 times larger than ΣF^l, showing that apparently quite small variations in ρ_s and G can result in large differences in water quantities.

Also, the difference between $\{(\Sigma F + \Sigma P) - (\Sigma D + \Sigma E + \Sigma I)\}$ and the corresponding values calculated for F^l and I^l are very large, and in the latter case become negative. With any calculated value of $(\Sigma R + \text{losses} - \Sigma W_p)$, the major quantum will consist of

ΣW_p which will be reflected by an increasing or decreasing size and depth of the pool. In the example, the increase of 5.68 m in (ΣR + losses − ΣW_p) would result in the pool flooding the beaches and threatening the freeboard. In a real situation, it would be countered by decanting more water (increasing D) and/or increasing ρ_s. At the other extreme, the negative value of 0.76 m would result in the disappearance of the pool, and would be countered by a decrease in the quantities decanted.

In the example, the situation of a very small or dry pool could correspond to the deposition of thickened tailings, whereas that of an increasing W_p could correspond to the use of the tailings storage as a reservoir for excess water, a dangerous situation that has resulted in the overtopping and disastrous failure of a number of tailings storages (e.g. at Stava in Italy in 1985, and Merriespruit, South Africa in 1994). Note that the seepage recharge R has not been calculated. For an unlined storage, R will typically be governed by the permeability to seepage of the tailings, rather than the foundation strata, and typically would be of the order of 1 m/y (or 1 Mg/m^2y in the units used in Figure 10.2a). Calculating the seepage loss R will be dealt with in Section 10.6.

In this example, (ΣR + losses− ΣW_p) has been left in units of Mg/m^2/month (m of water accumulated over a period of months). This assumes that the variations in the water held by the pool are spread over the entire top area of the storage. In a particular case, the contours of the pool floor would be known, or could be determined and ΣW_p could then be expressed as area and maximum depth of pool.

Figure 10.2a was constructed to show the effects of using uncertain data for a water balance calculation and deliberately presents a somewhat alarmist picture. Figure 10.2b represents real measurements for an operating unthickened tailings storage. The example was chosen because during the six months that it covers, extensions were being made to the decant towers and this resulted in difficulties in regularly decanting water. As a result, the pool considerably increased in area. The water stored in the pool and the pool area are shown in the lower diagram in Figure 2b. Because this example is worked on the basis of total quantities and not quantities per m^2, the water quantities have been given in MT (mega tons). It is interesting that the balance closed to within 0.18% of ΣF. This must be at least partly fortuitous, as the individual quantities are not known to such a fine accuracy.

10.5 The possibilities for saving water

The possibility for effecting savings of water consumption is an important consideration at every mine that is situated in an arid or semi-arid climate, or does not have a plentiful and cheap source of water. Considering Figure 10.1a, showing a beached unthickened tailings storage, of the total water input, including precipitation, only the decant water that collects in the pool and the relatively small amount of water collected by the underdrainage system can be returned to the plant. All evaporation is lost, whether from the beaches, the pool or the return water reservoirs, and so is all seepage into the foundation strata. There are only a limited number of options for saving water in the case of the unthickened tailings storage. These are:

1. Improve the housekeeping, which includes:
    - keeping the pool to a minimum size within the constraints of allowing enough retention time for solids to settle out, and the decant water to clarify; and

- eliminating, as far as possible, leakage or seepage and evaporation from return water reservoirs or tanks. This requires deep, minimum surface area return water reservoirs, lined with impervious geomembranes and also the covering of all return water storage tanks to minimize evaporation.

2. Increasing the rate of rise of the storage. In theory, this will increase the ratio of decant water to evaporation, but also brings with it the danger of raising the phreatic surface and decreasing the shear stability of the outer wall of the storage. Interstitial water per ton of deposition will be largely unchanged and drain flows should increase because of the elevated phreatic surface.
3. Reducing the tonnage of tailings sent to storage. This could be achieved by more selective mining, reducing ore dilution by barren waste rock and also, by this means, reducing the cost of hauling barren rock to the surface, milling it and sending it through the mineral extraction process, prior to deposition in the tailings storage.

In the case of thickened tailings or paste storages (see Figure 10.1b), the intention is to save water by thickening and thus eliminating decant water. Losses of interstitial water cannot be eliminated and evaporation from the surface will continue to occur, as will seepage from the base of the storage, if unlined. The assumption is usually made that by eliminating decant water, the overall use of water must be less than in the case of a beached, unthickened tailings storage. (See Section 8.8.4.). This is only so, however if the void ratio or water content of the thickened or paste tailings, as deposited, is less than that which occurs on a comparable beach of unthickened tailings after all decant and bleed water has run off. Referring to Figure 10.3 for example, the mean void ratio of the platinum tailings after beach deposition is 0.81. If G is taken as 3.2, the corresponding in situ relative density RD is

$$RD = (G + e)/(1 + e) = 2.22$$

It is only if the thickened or paste tailings can be thickened and placed at that density or higher that, given good housekeeping, water will actually be saved over and above that expended in forming a conventional unthickened tailings beach. Similarly, for the gold tailings described in Figure 8.30, if the void ratio as deposited on the beach is 1.6, and $G = 2.67$, the corresponding slurry relative density would be 1.64, and the tailings would need to be thickened to more than a relative density of 1.64 before real savings of water could be achieved.

Figure 8.16 shows a constant water content of 25% down a beach of thickened diamond (kimberlite) tailings. G for kimberlite is 2.8 and because, in a saturated tailings, $e = wG$, the void ratio of the deposited tailings is $e = 0.25 \times 2.8 = 0.7$ and $RD = (2.8 + 0.7)/(1 + 0.7) = 2.06$.

It would appear from this comparison that the average relative density on the beach of an unthickened slurry storage would have to exceed 2.06 in order to be more sparing in water expenditure than the thickened tailings process. However, the comparison is deceiving and invalid for the following reason: Recently published data on the thickening of kimberlite tailings from Botswana (Ntshabele, et al., 2008) shows that great difficulty is experienced in achieving slurry relative densities of 1.8 and that most tailings are very difficult to thicken to above 1.5. The discrepancy between the relative densities reported above and the 2.06 measured on the beach arose from evaporation and absorption of water from the thickened tailings by the sun-dried surface of the previous deposition, as the most recent layer was placed. In other words, the relative density of the thickened slurry as it reaches the point of deposition was unlikely to have been more than 1.5 and the water that makes the difference between 1.5 and 2.06 was lost by seepage and evaporation. It was not saved.

This point is supported by the data shown in Figure 10.6 which shows that the relative density for a thickened kimberlite tailings from a different mine to that mentioned above, averaged no more than 1.5 as it left the thickener to be transported to the thickened tailings deposit.

A question that remains to be addressed is: by what degree does the relative density of an unthickened tailings slurry change as a result of particle size sorting and drainage after it has been deposited on a beach? This is almost impossible to answer, as the whole

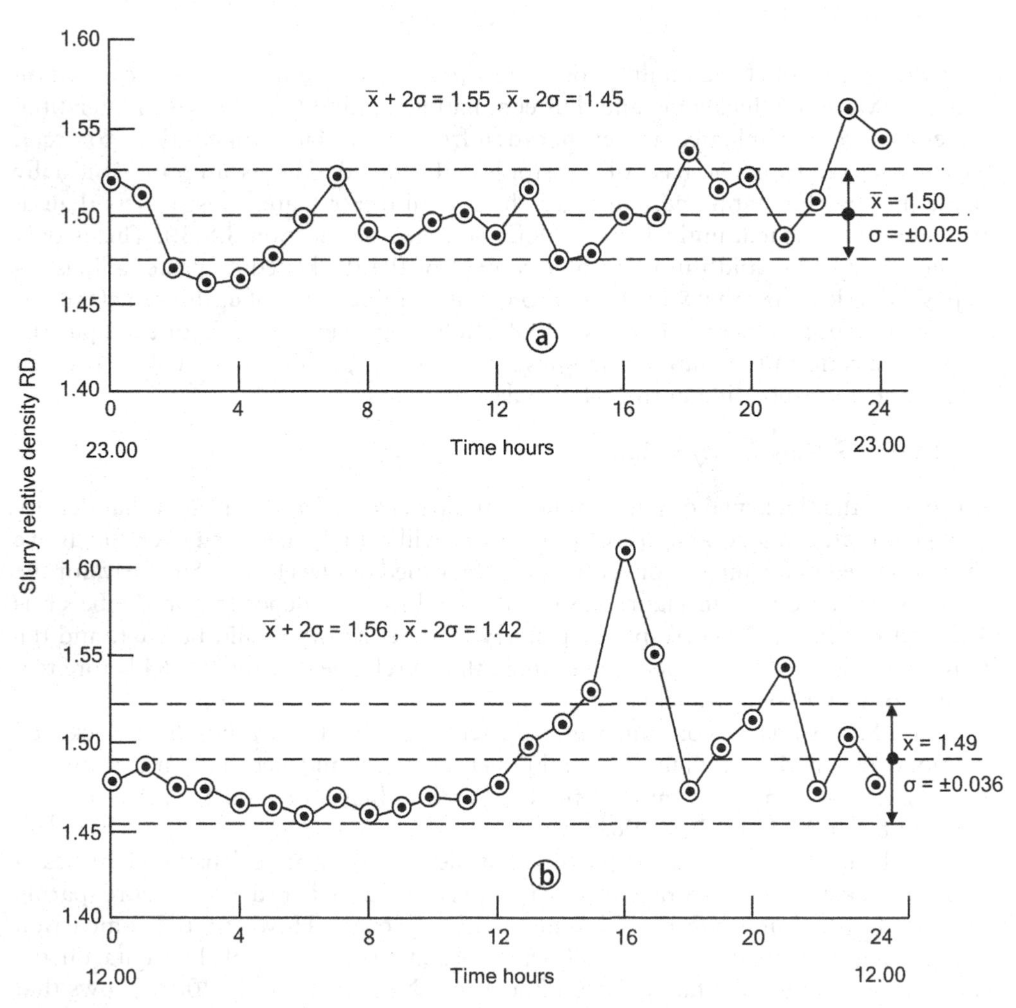

Figure 10.6 Hourly measurements of relative density of thickener underflow for diamond tailings at tailings plant taken over two randomly selected 24 hour periods.

length of the beach, including the underwater slope into the pool would have to be sampled, evaporation accounted for, and the resulting weighted mean RD compared with the RD of the slurry at discharge.

To obtain a preliminary answer to the question, the following experiment was performed on slurries of a fine, almost single sized sand with G = 2.65. The single-sized sand was chosen to eliminate, as far as possible, the effect of particle size sorting. The slurries, prepared at a series of RDs were poured into an open-topped cylinder measuring 100 mm diameter by 100 mm deep to simulate the placing of a 100 mm thick layer of slurry on a beach. The base of the cylinder was closed by a plate perforated with closely spaced 1 mm diameter holes and lined with a disc of geofabric to allow water to drain out of the slurry, while retaining the solids. The cylinder, standing in an empty basin, was weighed immediately after filling and the basin and cylinder were then enclosed in a large plastic bag to prevent evaporation losses. After a drainage time of 8 hours, the basin and cylinder were re-weighed as a check against losses and the cylinder and (basin plus drainage water) were weighed separately to establish the drained mass of the slurry. The drained sand was then sampled for water content, two samples being taken, one from the upper and one from the lower half of the cylinder. There was a small difference between upper and lower water contents, typical top and bottom values being 19.1 and 21.2% (i.e. 2.1% or 10% of 21.2%).

The results of the experiment are summarized in Figure 10.7 which shows average relative densities and void ratios, measured after draining, plotted against the corresponding initial values. All specimens deposited at RDs of less than 2.08 (void

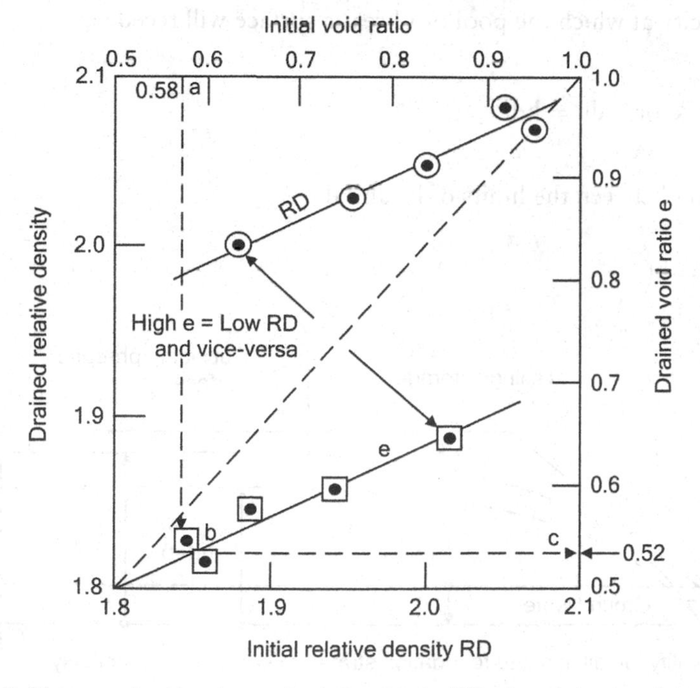

Figure 10.7 Changes of void ratio e and relative density RD of sand slurries after deposition and draining.

ratios more than 1.85) lost water by drainage and therefore increased in RD. However, the final, drained RDs did not reach the "non-draining RD" of 2.08. The preliminary conclusion is that an unthickened tailings slurry deposited on a beach will probably not thicken by drainage alone to reach a "non draining RD". But similarly, a tailings will need to be thickened to a "non draining RD" or beyond if it is not to lose water by draining after deposition.

Following path abc in Figure 10.7, a void ratio e of 0.58 corresponds to a water content (w) of $w = e/G_s = 21.9\%$ while $e = 0.52$ corresponds to $w = 19.6\%$. For these tests, the difference in water content (2.3%) represents the maximum possible saving of water by a thickened (non-draining) deposition system over an unthickened beaching system in this experiment. But a large part of the difference can be saved by returning the decant and drainage water (say 67% of 2.3%) leaving a net possible saving of perhaps 0.8%.

10.6 Seepage from the tailings storage into the foundation strata and the recession of the phreatic surface following cessation of operations

Figure 10.8 shows a section through a hydraulic fill tailings impoundment on which deposition has recently ceased. It will initially be assumed that the coefficient of permeability (k) of the tailings equals that of the foundation stratum (both for vertical flow). The height of the pool in the operational storage, or the phreatic surface in the tailings at time $t = 0$ is h_o and at $t = t$, the height is h. The head driving vertical water flow is h_o over a distance h_o, i.e. the flow gradient is $i = 1$.

The velocity at which the pool or phreatic surface will recede is

$$\frac{-dh}{dt} = \text{k, or} - \text{dh} = \text{kdt}$$

Integrating between the limits of h_o and d

$$h_o - d = kt$$

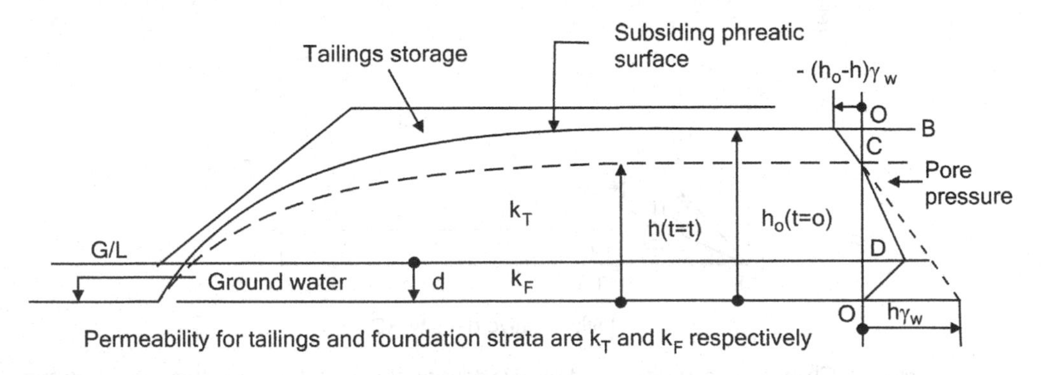

Figure 10.8 Recession of water table in tailings storage after cessation of operations.

Thus the recession time is

$$t = \frac{h_{0-d}}{k} \tag{10.2}$$

As an example, if $h_o = 35$ m, $d = 5$ m and $k = 1$m/y, $t = 30$ years.

k also represents the flow rate of water from the tailings into the foundation stratum, because the flow gradient will be approximately 1.0. The units of flow rate will be $m^3/m^2/y$ and kt will be the flow into the foundation in m^3/m^2 over a time period of t.

If the coefficient of permeability of the foundation stratum (k_F) is greater than that of the tailings (k_T), the recession time will be unaffected as it is governed by k_T. If k_F is less than k_T, the outflow is governed by k_F, and the recession time will be

$$t = \frac{h_0 - d}{k_F}$$

Supposing that $k_F = 0.1$ m/y, $t = \frac{30}{0.1} = 300$ years.

Alternatively, and probably more realistically, the recession of the phreatic surface can be analysed as a falling head process. Re-arranging equation (6.2)b and setting $a = A$, and $L = \frac{1}{2}(h_o + d)$, the recession at time t after cessation of deposition can be calculated from

$$t = 2.3\frac{L}{k}\log_{10}(h_0/d) \tag{6.2c}$$

Taking $h_o = 35$ m, $d = 5$ m and $k = 1$ m/y, as before

$$t = 2.3(35 + 5)\frac{1}{2}\log_{10}\left(\frac{35}{5}\right) = 68 \text{ years}$$

Whereas, if $k = 0.1$ m/y, $t = 680$ years.

The two approaches give different time predictions, but of the two, the falling head approach is the more realistic.

This simple analysis considers only one-dimensional vertical flow. This is justifiable as a first approximation, as the width of a ring dyke tailings storage would usually be 20 or more times the depth of tailings, (see, e.g., Figures 8.24 and 8.26) and away from the edges, lateral flow towards the outer edge of the storage would be small in comparison with vertical flow.

The analysis also considers a very simple section for the tailings storage. An example of a more complex section in which the tailings storage is underlined with a semi-impermeable liner, is shown in Figure 6.29.

10.7 Drainage of interstitial water as the phreatic surface recedes

In the case considered above, the pore pressure at the height of the phreatic surface was zero, that at the initial height h_o was $-(h_o - h)\gamma_w$ and that at the ground water

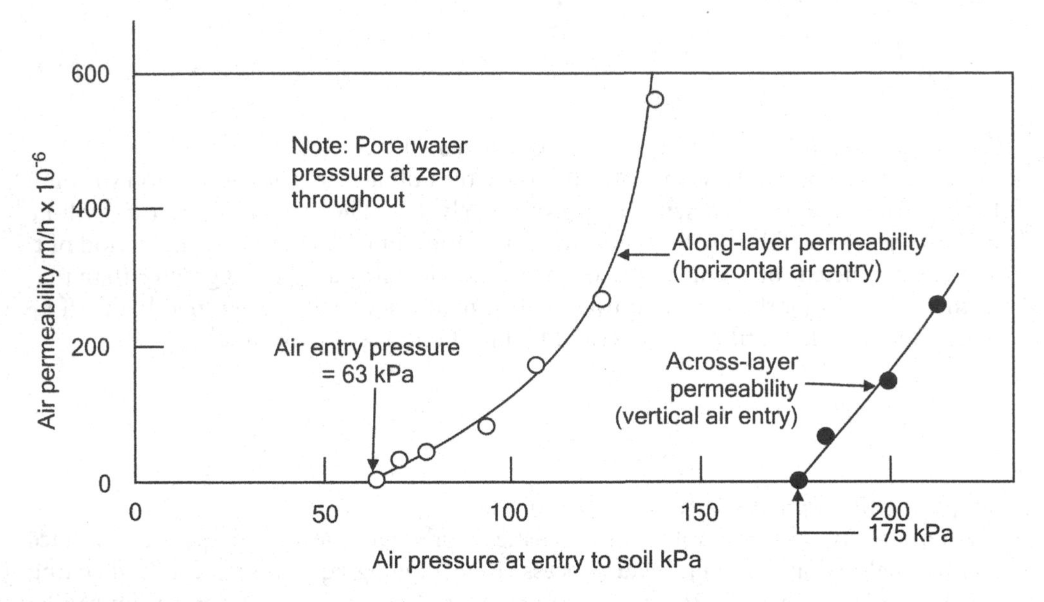

Figure 10.9 The process of air entry into an undisturbed sample of saturated gold tailings.

level was $h\gamma_w$. As the phreatic surface receded and $(h_o - h)$ increased, the pore pressure would have become more and more negative. Expressed differently, in the terms set out in equation (4.9), the suction $(h_o - h)\gamma_w$ increased, and therefore the effective stress increased and the tailings would have compressed. This compression would have released additional interstitial water. As the suction in a saturated particulate material increases, a suction is reached at which air starts to enter the larger pore spaces, displacing water. This is known as the air entry pressure or the air entry suction. Figure 10.9 shows air permeabilities measured on an undisturbed sample of saturated gold tailings, both parallel to the layering (i.e. horizontal, in situ) and across the layering (vertical, in situ). The air pressure at which the air permeability becomes measureably greater than zero represents the air entry pressure. In this case, the air entry pressure was 175 kPa for vertical flow in the field. This represents a suction head of $175\,kN/m^2 \div 10\,kN/m^3 = 17.5\,m$ of water. In other words, referring to Figure 10.8, until $(h_o - h)$ exceeds 17.5 m for this gold tailings, the tailings below the original level of the phreatic surface will remain saturated.

Figure 10.10a shows the effect of air entry on the water content of a gold tailings (a coarser material than that represented in Figure 10.9). The water content was initially 32%. As the applied pore air pressure u_a was increased, keeping the pore water pressure u_w at zero (gauge) and the confining stress σ at 400 kPa, the increased suction expelled 6% of the interstitial water. Once air had entered, an additional 14% of water was expelled. After the air pressure exceeded 250 kPa, no more water was expelled.

Figure 10.10b shows the effect of drainage by air entry on the shear strength of the gold tailings. For these tests, $(\sigma - u_a)$ was kept constant at 100 kPa, the pore water pressure u_w was constant at zero and hence the suction $(u_a - u_w)$ numerically equaled the pore air pressure. The shear strength increased until $(u_a - u_w) = 250\,kPa$.

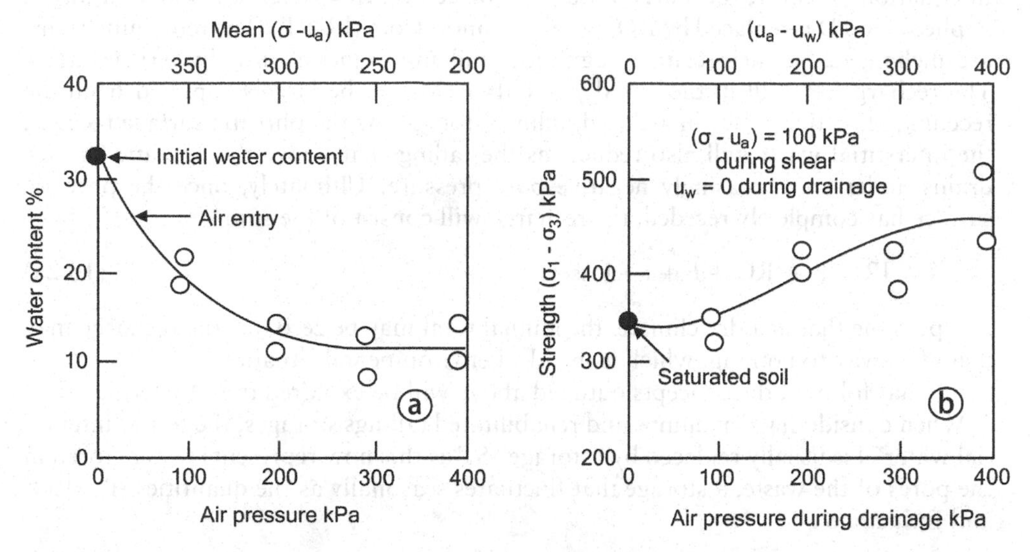

Figure 10.10 a: Drainage of water from initially saturated gold tailings. b: Effect of increasing suction $(u_a - u_w)$ keeping $(\sigma - u_a)$ constant.

Thereafter, because the water content was longer changing, even though the suction was increased, no change in shear strength occurred. In other words, referring to the effective stress equation

$$\sigma^{l} = (\sigma - u_a) + \chi(u_a - u_w)$$

the increase in $(u_a - u_w)$ was compensated by a decrease in the effectiveness parameter χ.

To investigate the likely annual quantity of interstitial water draining as a result of recession of the phreatic surface, suppose that the phreatic surface in a saturated tailings is receding at 1 m/year, that the average void ratio e is 0.8 and that the void ratio changes by $\Delta e/(1+e) = 0.01/0.8 = 0.0125$. In a column of saturated tailings 30 m high, this would amount to a compression of 0.0125 x 30 m, or 375 mm/year. Thus the quantity of interstitial water released into the ground water would be 375 mm/year, and therefore relatively small. If the tailings is sufficiently coarse to desaturate, the quantity of interstitial water would be slightly larger.

10.8 The water balance for a "dry" dump or a closed and rehabilitated tailings storage

If the dump has been deposited mechanically i.e. "dry" or if a closed and rehabilitated hydraulic fill tailings storage is under consideration, certain terms in equation (10.1) either fall away, or must be modified. The modified equation is:

$$IR + P = RO + E + I + (R + \text{losses}) \qquad (10.2)$$

In equation (10.2), F, the slurry water, is replaced by IR = irrigation water (if any is applied), and D is replaced by RO = surface runoff from (IR + P). The remaining terms are unchanged, but their relative magnitudes and importance are usually very different. The recharge, R, will in the initial years after closure, be largely supplied from the receding phreatic surface in a closed tailings storage. As the phreatic surface recedes, the interstitial water will also reduce, as the tailings left above the phreatic surface drains under an increasingly negative pore pressure. Ultimately, once the phreatic surface has completely receded, the recharge will consist of the sum

$$R = IR + P - (RO + E + I + \text{losses}) \tag{10.2a}$$

It is possible that in a dry climate, the annual total may be zero, i.e. the recharge may dwindle away to nothing, which is the ideal environmental situation.

In what follows, the concepts outlined above will be explored term by term.

When considering dry dumps and rehabilitated tailings storages, the term "interstitial water" I is usually replaced by "storage" ST as this now represents water stored in the pores of the waste, a storage that fluctuates seasonally as the quantities (P – RO) and E fluctuate.

10.9 Measuring potential infiltration and runoff

The mechanics of infiltration and runoff were investigated comprehensively 30–40 years ago in the field of soil physics. The results of this research are well summarized by Hillel (1980). The work that will now be described was undertaken to confirm the transferability of the concepts to the specific circumstances being considered here, namely the surfaces of waste storages, which differ somewhat from those for the earlier applications, in agriculture.

Precipitation can be partitioned into infiltration, that portion of precipitation that enters through the surface of the waste, and runoff, the remainder of the precipitation. In more detail:

$$\text{infiltration} = \text{precipitation} - (\text{interception} + \text{surface evaporation} + \text{runoff})$$

$$IN = P - (INT + E + RO) \tag{10.3}$$

For a vegetated surface, interception is that part of the rainfall that is intercepted by the plant cover or anti-erosion armouring and evaporates without reaching the surface. If a surface has a higher temperature than the rain, some of the precipitation will evaporate either as it reaches the surface, or just above it. Depending on the amount of precipitation, the nature of the surface and the type of vegetative cover, interception and surface evaporation can vary from a small to a major proportion of precipitation. For example, if a small amount of rain falls on a hot paved surface, or a densely forested area, the entire precipitation can be either evaporated or intercepted, with no infiltration or runoff. On the other hand, a similar amount of rain falling on a ridge-ploughed surface or a paddocked surface may all infiltrate.

It could be argued that to find a limiting infiltration rate it is only necessary to carry out a double-ring infiltrometer test or some other saturated in situ permeability test on the soil surface and measure the limiting rate of infiltration. Any rain falling

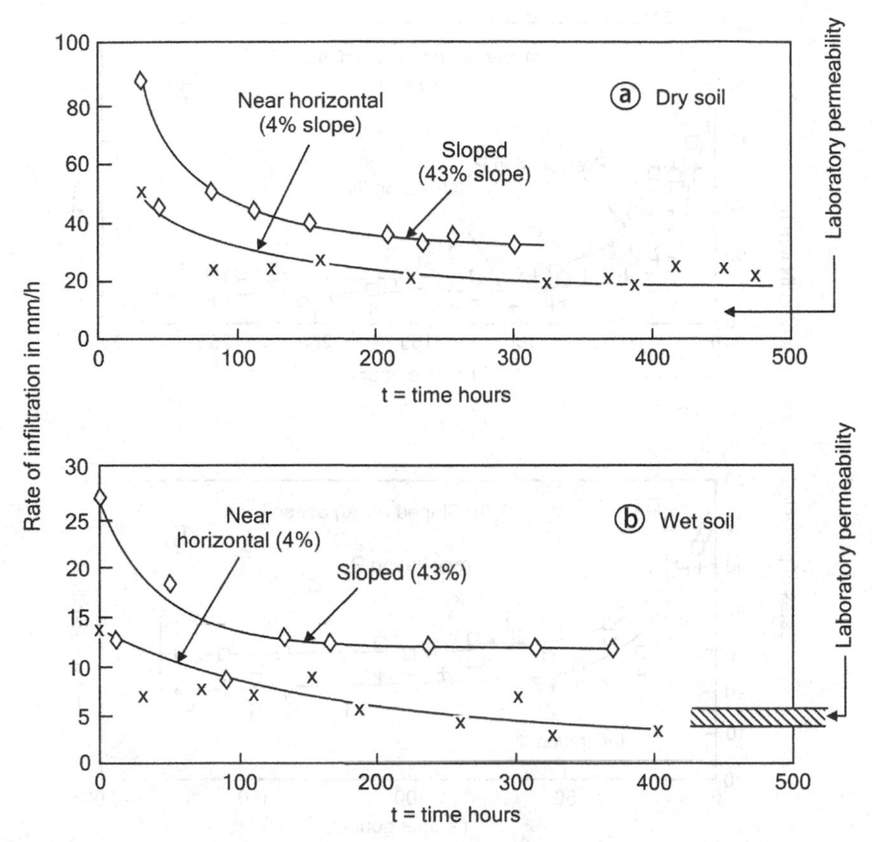

Figure 10.11 Double-ring infiltrometer tests before and after sprinkler irrigation of a soil surface: a: dry soil, b: wet soil.

at an intensity less than this rate will infiltrate, and if the rainfall intensity exceeds the limiting infiltration rate, the excess over this rate will run off. Unfortunately, it is not as simple as this, even though the principle is correct. Not only does the limiting infiltration rate depend on the slope angle and state of compaction of the surface, but also on the moisture content of the material underlying the surface, i.e. on the degree to which suction in the underlying unsaturated material draws water in. This in turn depends on rainfall or irrigation that took place in the recent past (antecedent precipitation). This is illustrated in Figure 10.11, which compares the results of double-ring infiltrometer tests on the surface of a power station ash dump (a) starting with a nominally dry surface, and (b) starting after the surface had been thoroughly wetted by sprinkler irrigation. It can be seen that the limiting infiltration rate into a near-horizontal surface (4% or 2.3° slope) was in this case less than that into a sloped surface (43% or 23° slope). The reason for this is that the surface of the ash on the slope was less compact and therefore accepted water more readily than that on the near-horizontal surface. Also, the limiting infiltration rate into the initially dry surface averaged about 30 mm/h for the two sites after 300 h, whereas that into the pre-wetted

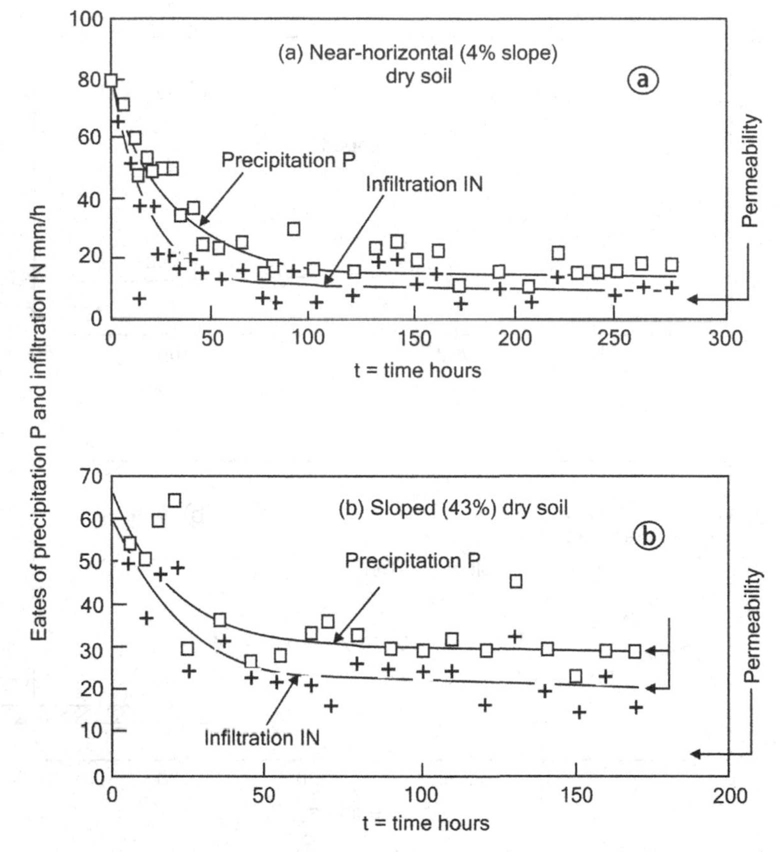

Figure 10.12 Sprinkler infiltration tests to determine limiting infiltration rates for soil surfaces. Precipitation and infiltration versus time: a: near-horizontal (4% slope), dry soil, b: sloped (43%), dry soil.

surface at exactly the same spots averaged about 10 mm/h. Figure 10.11 also shows the saturated permeability for the ash as measured in the laboratory. This corresponds to the limiting infiltration rate for the wet, near-horizontal site.

The reason for the discrepancy between the initially dry and wet sites is that, because the double-ring infiltration test affects a relatively small volume of ash, even after 300 h (nearly 2 weeks) of infiltration into the dry ash, there were still significant suction gradients augmenting the infiltration rate at the boundary of the zone affected by the infiltration. A steady state had not yet been reached, and nor, whatever the reason, had a truly limiting infiltration rate.

Figure 10.12 shows the results of two sets of sprinkler irrigation tests on the same site. A test plot measuring 9 m × 9 m was surrounded by bunds and gutters to prevent run-on from adjacent areas and to collect runoff from the test plot (see Plate 10.1). Rain was then simulated using a sprinkler system having a series of sprinklers with an overlapping deposition pattern. The precipitation on, and runoff from the test plot were measured for various rates and sequences of precipitation.

Plate 10.1 Collecting gutters and measuring container for sprinkler infiltration test on surface of waste storage. Note rain gauge in left background.

Plate 10.1 shows the arrangement of gutters and measuring bucket at the low point of the test plot. One of the rain gauges can be seen in the background.

Figure 10.12 shows the results of two tests starting with nominally dry soil surfaces. The test was designed to apply precipitation at a rate just large enough to produce runoff, and began with high rates of precipitation (80 mm/h for the 4% slope and 55 mm/h for the 43% slope). The corresponding rates of runoff were measured and the rates of precipitation were adjusted progressively so that, while runoff still occurred, its rate was minimized and thus approximated a limiting infiltration rate. As Figure 10.12 shows, a final limiting rate of infiltration of about 15 mm/h was measured for the 4% slope and that for the 43% slope was about 20 mm/h. These rates were slightly less than the limiting double-ring infiltration rates for the dry sites shown in Figure 10.11, but the limiting rate was reached in much less time (2.5 h for the sprinkler tests, as compared with 250 h for the double-ring infiltrometer). The smaller time to equilibrium probably arises because, in contrast with the double-ring infiltrometer, the sprinkler infiltration tests affected a relatively large volume of soil and therefore reduced the suction in the soil more rapidly.

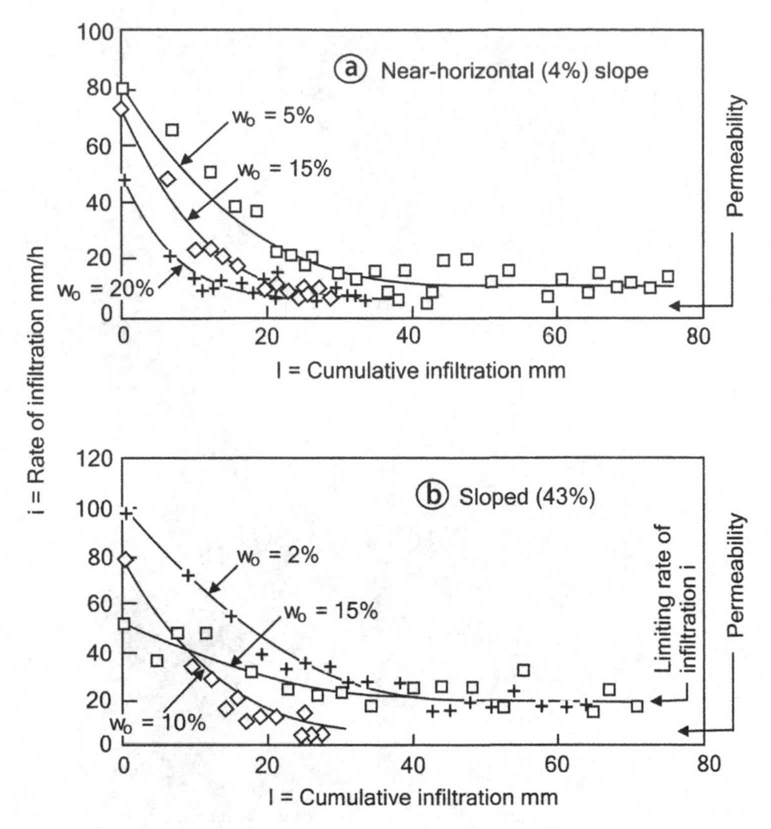

Figure 10.13 Sprinkler infiltration tests to determine limiting infiltration rates for a soil surface. Infiltration rate versus cumulative infiltration: a: near-horizontal (4%) slope, b: sloped (43%).

Figure 10.13 shows the limiting rates of infiltration from Figure 10.12 ($w = 2\%$ and 5%) together with the results of two sets of similar tests, plotted against cumulative infiltration into the surface. Each of the sets of three measurements corresponds to a different initial water content w_o, and the results illustrate the effect of higher soil suctions at the lower water contents on increasing the initial infiltration rate. The results for $w_o = 10\%$ at the sloped site are anomalous, but no explanation has been found. From Figure 10.13 it appears that for the surface considered, a fairly steady rate of infiltration was reached after a cumulative infiltration into a "dry" surface of 40–60 mm had occurred. Thus the rate of infiltration is affected by the permeability of the soil, its surface gradient and its water content or suction.

If i is the rate of infiltration (mm/h) and t hours of precipitation occurs at rate r (mm/h), i_o is the infiltration rate at $t = 0$ and i_L is the limiting rate of infiltration after a long time, then the infiltration curves in Figure 10.12 can be described by an expression due to Horton (1940):

$$i = i_L + (i_o - i_L)e^{-kt} \tag{10.4}$$

where k is a characterizing constant. Thus, at $t=0$, $i=i_o$, and at $t=\infty$, $i=i_L$, and the value of k decides on how rapidly i approaches i_L. For example, for the lower curve in Figure 10.12a, $i_o = 80$ mm/h, $i_L = 10$ mm/h and k – 1.76/h.

Equation (10.4) is an empirical expression that fits the shape of the infiltration curve. Alternative expressions have been proposed by Philip (1957) and Holtan (1961). Similarly, the infiltration rate i can be related to the cumulative infiltration I (see Figure 10.13) by the expression

$$i = i_L + (i_o - i_L)e^{-KI} \tag{10.5}$$

For the $w_o = 5\%$ curve in Figure 10.13(a), K = 0.078/mm.

The difficulty with applying equations (10.4) and (10.5) is that the four characteristic parameters i_o, i_L, k and K have to be determined experimentally. Philip's expression,

$$i = i_L + \tfrac{1}{2}St^{-\frac{1}{2}} \tag{10.6}$$

has the advantage over Horton's expression that it requires the evaluation of only two characteristic parameters, i_L and S.

The data shown in Figures 10.11 to 10.13 illustrate the potential for infiltration. However, actual infiltration will depend on the rainfall intensity of a storm as well as the suction and permeability characteristics of the surface. The intensity of rainfall in a natural storm is very variable. Figure 10.14a shows an intensity-time relationship for a natural storm that was recorded at the site for the sprinkler tests. In this storm of 26.5 mm total rainfall, the highest intensity reached nearly 100 mm/h for 5 min (8 mm of precipitation). A curve of potential infiltration rate versus time has been superimposed on Figure 10.14a. Any (rainfall intensity × time) exceeding this curve would have resulted in runoff, and any intensity less than the curve would have infiltrated. It is obvious that, in this case, only the excess peak intensity (60 mm/h for 5 min, or 5 mm) resulted in runoff. The measured runoff for the whole storm was 5.3 mm, or 20% of the total precipitation.

Figure 10.14b shows a second intensity-time relationship for natural rainfall. In this case the rainfall intensity was lower, but the duration greater. The entire storm fell below the potential infiltration curve, with the exception of the two short-duration peaks. Hence this event resulted in zero runoff. For this surface it appears that runoff from rainfall events of up to 30 mm must be very low. Appreciable runoff would only occur from rainfall events exceeding 30 mm in 24 h.

Hence for assessing infiltration into the surface of a "dry" dump or closed tailings storage, infiltration measurements and a knowledge of rainfall patterns at the site are imperative.

10.10 Estimating evaporation or evapotranspiration (E)

The determination of evapotranspiration has been of concern to agriculturalists and hydrologists for over a century (e.g. Penman, 1963). Many methods for assessing potential evapotranspiration (the evapotranspiration when the availability of water is not a limiting factor) have been developed. Among these are methods by Blaney & Criddle, (1950), Thornthwaite (1954), and Penman (1963).

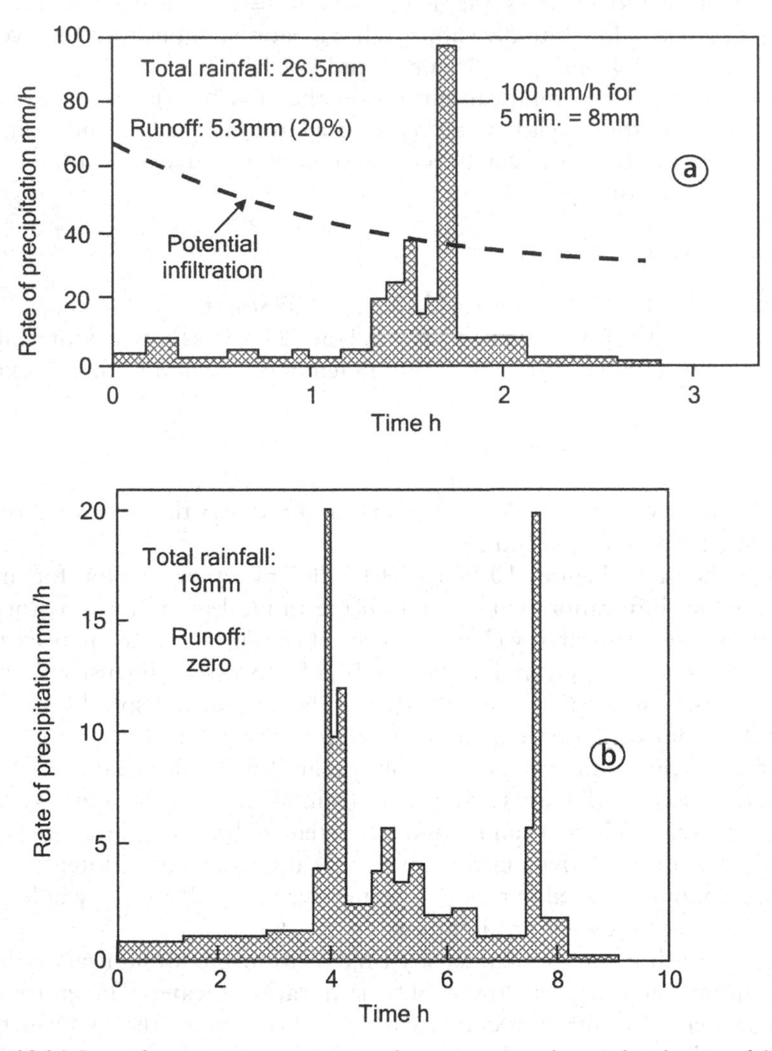

Figure 10.14 Rate of precipitation versus time for two storms observed at the site of the sprinkler infiltrometer tests: a: total rainfall 26.5 mm, run-off 5.3 mm (20%), b: total rainfall 19 mm, run-off 0 mm.

All these methods were evolved for irrigated fields, and for climates with a positive annual water balance, where the actual evapotranspiration will usually be close to the potential evapotranspiration. However, for arid, and semi-arid, non-irrigated areas, which are often those of interest to the geotechnical engineer concerned with mine waste storage, the availability of water may limit evapotranspiration, and actual evapotranspiration may drop well below the potential figure. As soil suction increases, increasing quantities of energy are required to draw soil water into the atmosphere. Soil suction is expressed in units of pressure (Pa or N/m^2). These are equivalent to units of energy per unit volume of soil (Nm/m^3 or J/m^3). Thus soil suction represents the energy necessary to extract a unit volume of water from the soil. The energy demand is, however, not as important as is often supposed. A simple calculation will show that, even at the wilting point of vegetation, where the suction is of the order of 1500 kPa, the energy needed to extract unit mass of water from soil is about 1700 times smaller than that required to evaporate the same mass of water (1.5 kJ/kg as compared with 2450 kJ/kg). Hence, if there is enough energy to evaporate water from the soil surface, there will be enough energy to draw it up through the soil. However, as suction increases, soil permeability to water flow declines and this may have a major effect on inhibiting water loss from the soil.

Many attempts have been made to predict actual evaporation from climatological data. Most researchers in this area have derived empirical or semi-empirical equations. For example, Thornthwaite (1954) produced an empirical equation and tables relating actual evapotranspiration to potential evapotranspiration and soil moisture deficit. Much of Thornthwaite's work was carried out in the humid north-east of North America and thus may not be suitable for extrapolation to more arid conditions. Thornthwaite's empirical equation is:

$$E_m = 16.2\{(10t_m)/\Sigma i\}^a \qquad (10.7)$$

in which E_m is the monthly evaporation (mm), t_m is the mean monthly temperature (°C), i is $(T/5)^{1.5}$, where T is the mean daily temperature, and a has a limiting value of 0.5 for all practical temperatures.

Turc (1955) produced one of the earliest attempts to take a limiting water supply into account. He did this by introducing a term for precipitation into his empirical equation for evapotranspiration. The equation was, however, developed for wet climate conditions, and has the limitation that evaporation is expressed as a fraction of precipitation which, in periods of drought, is not correct. Turc's empirical equation is as follows:

$$E_m = \frac{P_m}{0.9 + (\frac{P_m}{L})^2} \qquad (10.8)$$

in which E_m is the monthly evaporation (mm), P_m is the monthly precipitation (mm), and

$$L = 300 + 25T_m + 0.05T_m^3$$

where T_m is the mean monthly temperature (°C).

The most rational approach to calculating potential evaporation is due to Penman (1963). He used many of the concepts that follow, and produced a semi-rational equation, based on the energy balance at the soil surface:

$$E_p = \frac{\frac{\Delta Rn}{\lambda} + PE_a}{\Delta + P} \qquad (10.9)$$

in which E_p is the potential evaporation, Δ is the slope of the temperature versus saturated water vapour pressure curve at the prevailing air temperature, R_n is the net incoming solar energy, λ is the latent heat for vaporization of water, P is the psychrometric constant (66 Pa/°C), $E_a = 0.165(e_{sat} - e_a)(0.8 + u_2/100)$ mm/day, e_{sat} is

Plate 10.2 Evaporation pan (A-pan) and rain gauge set up near the decant of a tailings storage.

the saturated water vapour pressure in air (mbar), e_a is the actual water vapour pressure in air (mbar) and u_2 is the wind speed at a height of 2 m (km/day).

As Δ and P both have units of Pa/°C, if R_n has units of J/m^2 per day and λ has units of J/kg, the units of E_a are kg/m^2 or, taking the density of water as 1000 kg/m^3, the equivalent of mm/day of water.

The potential evaporation rate from a soil or waste surface is often assumed to be equal to the evaporation rate of water from an American standard "A-pan" (Class A evaporation pan). Many attempts have been made to find modification factors that will convert A-pan evaporation to actual evaporation from cropped fields. For instance, Penman (1956) found ratios of E_p/E_A varying from 0.8 for summer to 0.6 for winter to 0.7 for equinoctial months for Western Europe. Others (quoted by Fenn et al. (1975)) found values for E_p/E_A varying in the range 0.5–0.9. Evaporation from open fresh water is usually assumed to equal 0.8 E_A. (Plate 10.2 shows a standard A-pan set up with a rain gauge over the pool of a tailings storage.)

Work in South Africa (Blight, 2006) has shown that the A-pan can greatly exaggerate evaporation rates from water, because the water not only absorbs solar energy through its surface, but also through the sides and bottom of the A-pan. Figure 10.15 shows a typical result comparing A-pan evaporation with evaporation measured from the surface of a landfill in South Africa by means of the solar energy balance method (to follow in Section 10.11). The figure shows that the ratio of evaporation by energy balance (E_B) to mean A-pan evaporation ($\bar{E}_A$) can vary seasonally from 0.2 to 0.7,

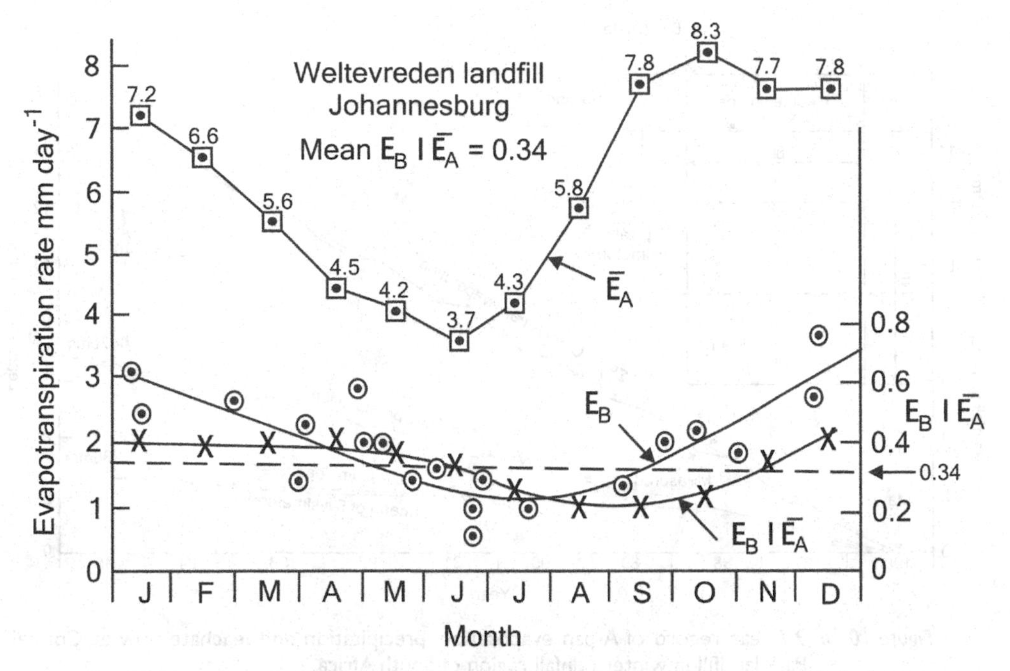

Figure 10.15 Ratio of evaporation from a landfill surface E_B (solar energy balance) to mean American A-pan evaporation $\bar{E}_A$ for a site in South Africa.

with an average value of 0.34. Table 10.1 shows the results of sets of measurements made at six separate sites in South Africa, including areas having winter or summer rainfall. The mean value of $E_B/\bar{E}_A$ is 0.39 and the standard deviation ± 0.05.

Thus, taking a conservative approach, the value of $E_B/\bar{E}_A$ could be taken as $(0.39 - 2 \times 0.05) = 0.38$ with a probability of less than 3% that the value would be less than 0.38.

In support of the six one year sets of observations, Figure 10.16 shows a 9.7 year record of measurements of precipitation, P and leachate flow L measured by lysimeters at the Coastal Park Landfill (see Table 10.1). The cover material on this landfill is a clean beach sand and runoff of precipitation is negligible. The net infiltration (P – L) is therefore a close approximation to the actual evaporation E from the landfill surface. As recorded on the figure, the ratio of $\Sigma(P - L)/\Sigma E_A$ was 0.40 over the 9.7 year period, as compared with the value of $E_B/\Sigma\bar{E}_A$ of 0.46 for the 1 year series of solar energy balance observations (Table 10.1, Coastal Park). ($P =$ precipitation, $L =$ leachate flow, $E_A =$ A pan evaporation.)

10.11 Measuring evaporation by solar energy balance

10.11.1 *Solar radiation*

The evaporative process requires the supply of energy, which in the case of the surfaces of soils or waste deposits is supplied by the sun. If the amount of energy consumed

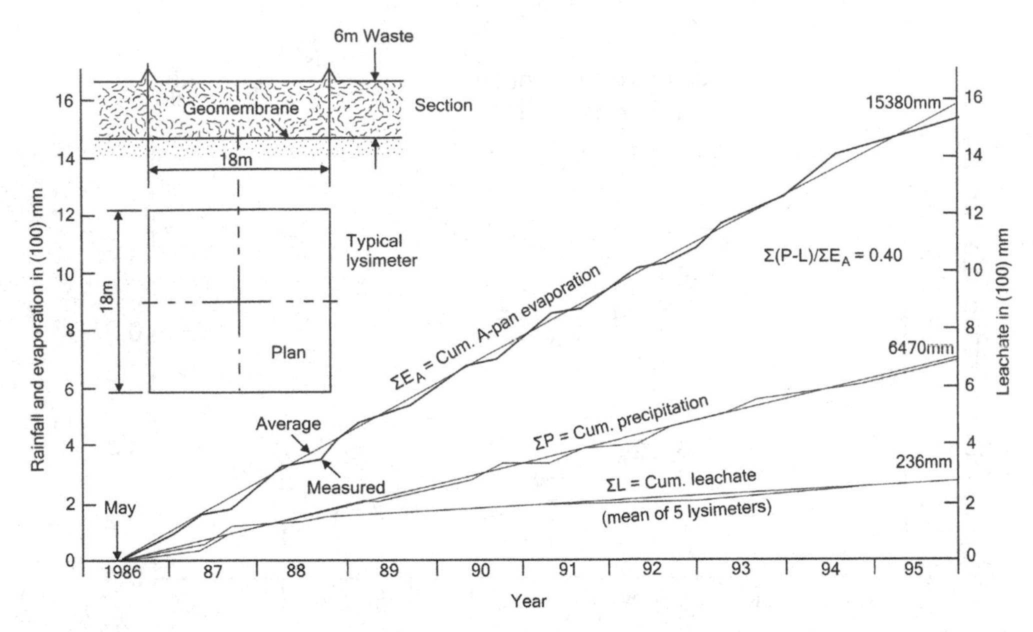

Figure 10.16 9.7 year record of A-pan evaporation, precipitation and leachate flow at Coastal Park landfill in winter rainfall region of South Africa.

Table 10.1 Evaporation and evapotranspiration measured at six sites in South Africa using the solar energy balance technique

Area (province)	E_B *(mm/y)*	*P (mm/y)*	$\bar{E}_A$ *(mm/y)*	$E_B/\bar{E}_A$
Clarens (Free State)	720	862	1930	0.37
Coastal Park Landfill (Western Cape)	823	690	1790	0.46
Holfontein Landfill (Mpumalanga)	762	732	2228	0.34
Johannesburg (grassed site) (Gauteng)	1000	940	2228	0.45
Nylsvley (Limpopo)	835	555	2154	0.39
Weltevreden Landfill (Gauteng)	768	732	2228	0.34
Mean				0.39
Standard deviation of sample				±0.05

by evaporation can be measured, the corresponding mass of water evaporated can be deduced.

The quantum of solar energy arriving at the earth's surface depends on the

- solar constant
- latitude of the location being considered and the time of year
- influence of the atmosphere (dust and clouds)
- albedo of the earth's surface
- elevation of the location

The solar constant is the quantity of solar energy at normal incident angles to the atmosphere at the mean sun-earth distance. Some of this incident radiation is reflected

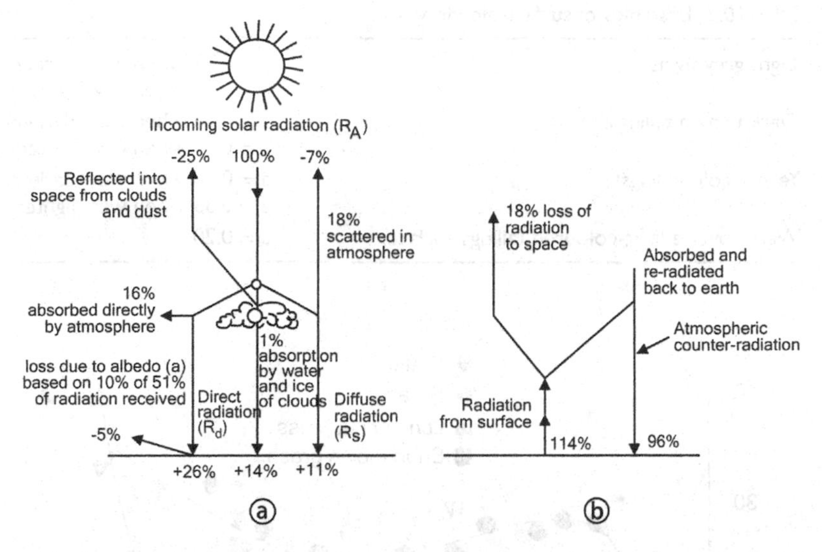

Figure 10.17 Annual radiation balance for Earth: a: short wave length solar radiation, b: long wave length terrestrial re-radiation.

or absorbed by the atmosphere and does not reach the earth's surface. Most of the radiation reaching the earth's surface is confined to short wavelengths, in the band $0.3 - 3\,\mu$m. On reaching the ground some of the short-wave radiation is also reflected back by the surface.

Figure 10.17 (after Flohn 1969) shows diagrammatically how incoming solar radiation is split, with some being reflected or absorbed by the atmosphere, some being reflected by the ground surface, and some being absorbed at ground level. The diagram represents average annual conditions for the whole earth.

The overall radiation balance between space and the earth's surface may be described by the equations:

$$(R_d + R_s) = R_A(1 - \propto) \tag{10.10a}$$

$$R_n = (R_d + R_s)(1 - a) \tag{10.10b}$$

Here, R_A is the direct solar radiation received at the outer limit of the atmosphere and $\propto$ is the planetary albedo or reflectivity, the proportion of solar radiation reflected by the atmosphere. The planetary albedo $\propto$ varies with cloud cover and latitude, but in the earth's inhabited latitudes of 60°N to 45°S, lies in the range from 0.3 to 0.5 with 0.4 being a reasonable annual average value (Robinson 1966).

R_n is the net radiation, R_d is the amount of incident direct solar radiation, R_s is the amount of incident scattered sky radiation and a is the surface albedo or reflectivity of the ground surface.

The surface albedo a varies throughout the day and depends on the colour and texture of the earth's surface Table 10.2, for example, gives average daily albedo

Table 10.2 Examples of surface albedo values

Light grey fly ash:	a = 0.22 when wet (darker) a = 0.33 when dry (lighter)
Dark brown tailings:	a = 0.06 when wet (darker) a = 0.14 when dry (lighter)
Yellow gold tailings:	a = 0.16 when wet (darker) a = 0.33 when wet (lighter)
Water over a light-coloured tailings surface:	a = 0.20

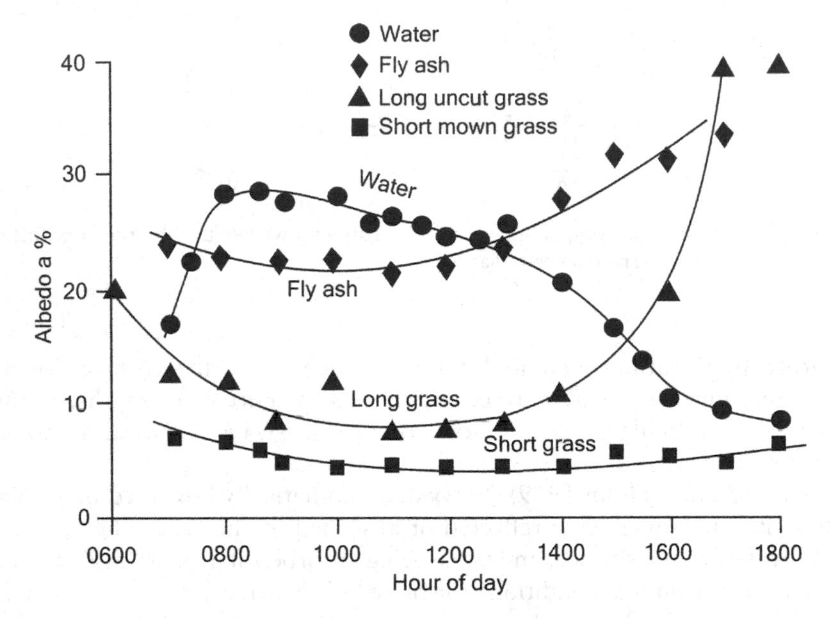

Figure 10.18 Variation of surface albedo during the day for a water surface and three land surfaces at 26°S latitude.

values for various mineral tailings surfaces as well as for shallow water over a tailings surface.

Figure 10.18 shows how the albedo varies throughout a typical day for four surfaces: Shallow water over a light-coloured tailings surface, soil surfaces covered by long (0.5 m) grass and short (mown) grass and light grey fly ash. For solid surfaces, the albedo is greatest in the early morning and evening, when the sun's rays strike the surface at a low angle, and least at midday. For water the reverse applies, presumably because of the influence of ripples on the surface.

Calculated direct solar radiations received at the outer limit of the atmosphere are given in Table 10.3 (taking the planetary albedo into account), as values of $(1 - \alpha)R_A$.

Note that the annual value of R_A at the latitude of Stockholm (60°N) is 57% of that at the equator, while at the latitude of New York and Madrid (40°N) the annual value of R_A is 79% of that at the equator. Hence useful quanta of solar energy are available even at relatively high latitudes.

Table 10.3 Values of $(1-\alpha)R_A$, solar radiation at outer limit of earth's atmosphere

Latitude	*Jan*	*Feb*	*Mar*	*April*	*May*	*June*	*July*	*Aug*	*Sept*	*Oct*	*Nov*	*Dec*	*Annual*
60°N	112	274	550	863	1124	1235	1158	927	620	335	142	71	7411
40°N	464	630	860	1063	1207	1257	1221	1094	903	685	499	412	10294
20°N	819	932	1065	1148	1183	1189	1183	1151	1075	960	836	777	12317
0	1095	1129	1140	1100	1042	1008	1028	1064	1119	1124	1096	1076	13021
20°S	1259	1196	1080	925	789	728	763	882	1030	1158	1238	1269	12317
40°S	1295	1129	890	646	464	387	432	588	814	1060	1248	1340	10294

Units are MJ/m^2/month (final column MJ/m^2/year)
After Angot, quoted by Wilson 1970

Apart from the planetary albedo α, the values of $(R_d + R_s)$ actually received at the earth's surface depend on cloud cover, elevation or altitude and the number of hours of sunshine. Examples of empirical relationships between $(1 - \alpha)R_A$ and $(R_d + R_s)$ are as follows (Penman 1963):

For southern England (latitude 52° N):

$$R_d + R_s = (1 - \alpha)R_A(0.18 + 0.55\,n/D) \qquad (10.11a)$$

For Canberra, Australia (latitude 36° S)

$$R_d + R_s = (1 - \alpha)R_A(0.25 + 0.54\,n/D)$$

(n = number of hours of sunshine, D = number of hours of daylight.)

Hence even on completely overclouded days (n/D = 0) about 20 to 25% of solar radiation reaches the earth's surface while on cloudless days 75 to 80% arrives (see also Figure 10.17). On an average day in Stockholm, the value of $(1 - \alpha)R_A$ (from Table 3) will be 20 MJ/m^2 and if the day is cloudless, with an albedo of 0.3, the energy reaching the surface will (by equation (10.11(a) be about 10 MJ/m^2. In Madrid, the corresponding energy would be about 14 MJ/m^2.

10.11.2 *Surface energy balance*

At the surface of a waste storage, the net incident radiation R_n is converted into L_e, the latent heat of evaporation, in accordance with the surface energy balance.

$$R_n = G + H + L_e \qquad (10.12)$$

$$\text{or } L_e = R_n - (G + H)$$

This is the situation during the day. At night, when R_n is zero, the heat energy stored in the soil, G is converted into L_e.

The value of the evaporation from the soil that corresponds to a given cumulative value of L_e is obtained by dividing L_e by the heat of vaporization of water λ which is 2.45 MJ/kg, to give evaporation in kg of water per m^2, equivalent to mm, of water depth.

R_n can be measured directly and G can be estimated from changes in the temperature and temperature gradient below the surface together with the specific heat capacity for the waste. H, the sensible heat may be calculated from temperature and relative humidity gradients above the surface. In principle, the calculation of the subdivision of R_n into G, H and L_e is simple, but in practice it is difficult because the equations tend to be ill-conditioned, especially when temperature and humidity gradients are small.

Data for energy balance calculations can be collected automatically by monitoring and recording the outputs of a radiometer, heat flux plates or thermocouples buried in the soil or waste and psychrometers, used to measure air temperature and relative humidity at two heights above the surface (500 mm and 1500 mm would be typical heights). Readings taken at two minute intervals, and averaged over longer periods (15–20 minutes) are recommended for best results. Hand-held portable instruments can also be used to measure the energy balance at significantly lower instrument cost and with much greater flexibility (but obviously with greater labour costs). The manual method is generally to be preferred if it is wanted to monitor several sites with less detail.

An A-pan should be monitored at every measurement site, for comparison to the quantity of evapotranspiration calculated from the energy balance. Wind speed is also measured, as advection of heat and water vapour due to gusts and eddies of wind may give rise to anomalies in the measured energy balance. (If the test site is located in the centre of a large uniform area, this effect is minimized). Precipitation should also be measured at each site, preferably both daily quantity, as well as time and intensity, but quantity is a minimum requirement.

Figure 10.19 shows an automatic weather station that can continuously record the measurements necessary to evaluate equation (10.12) and Plate 10.3 shows such a weather station set up in the field, in this case, on a landfill with an experimental geomembrane cover. Figure 10.20 shows typical measurements made by an automatic weather station and Figure 10.22 shows a typical set of daytime measurements made with hand-held instruments.

The data for calculating the sensible heat H is particularly difficult to measure and the method of calculation usually used, the Bowen ratio method (Bowen, 1926) is inaccurate because it depends on small differences between large measured quantities. Its results are not always credible, as the following will show.

In Figure 10.20, values of H, calculated by the Bowen method have peak values that far exceed corresponding values of G, the soil heat. Both H and G can alternatively be calculated by the same completely basic method, i.e.

$$\text{Heat energy absorbed} = \text{volume of substance heated} \times \text{density} \\ \times \text{average increase in temperature} \\ \times \text{specific heat of substance.}$$

Considering a 1 m^2 area of either air or soil,

$$\text{Heat energy absorbed} = z\rho\Delta\overline{T}C_h$$

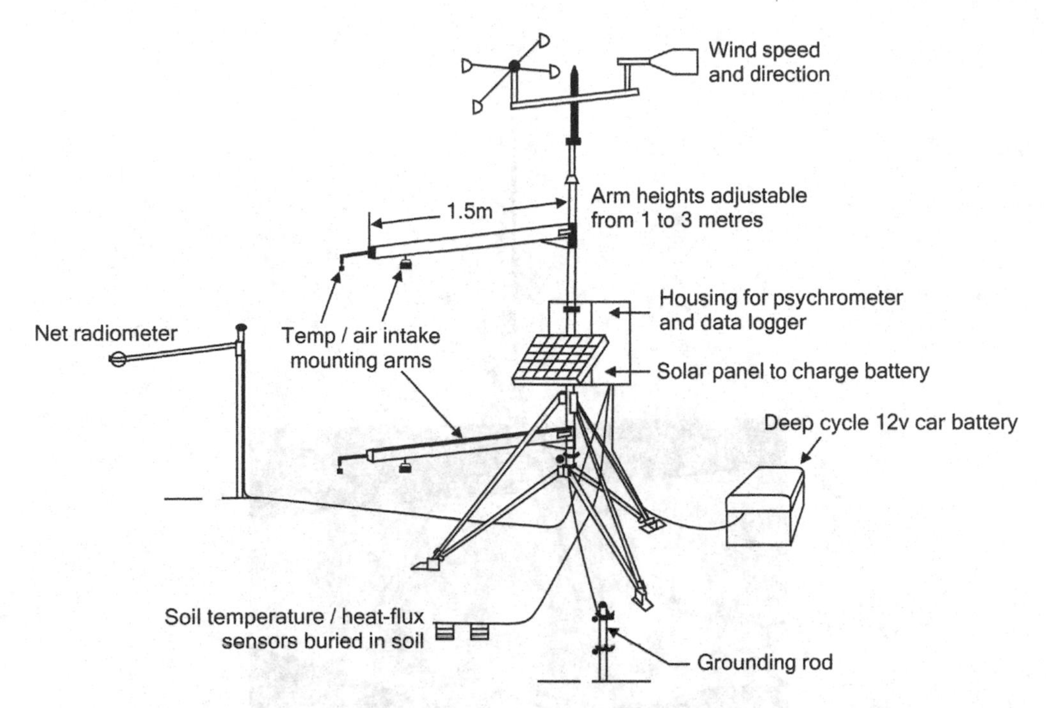

Figure 10.19 Automatic weather station for recording data used to evaluate surface (or solar) energy balance, by Bowen ratio method. (Campbell Scientific Instruments.)

where z is the depth of material [m], ρ is the density [kg/m³], $\Delta\overline{T}$ is the average increase in temperature [°C] and C_h is the specific heat [kJ/kg°C]. The units of heat absorbed are thus kJ/m².

Densities and specific heats for air at various temperatures are given in Table 10.4.

The specific heat of a soil or mine waste can be calculated by the rule of mixtures as:

$$C_G = (C_{Gd} + wC_w)/(1 + w) \tag{10.13}$$

in which
C_{Gd} = specific heat of the dry soil particles;
C_w = specific heat of water; and
w = gravimetric water content of the soil (mass of water/mass of solids).
C_{Gd} has a value of about 0.85 kJ/kg°C
C_w has a value of 4.19 kJ/kg°C and is identical with Joule's mechanical equivalent of heat.

Because C_w is so much larger than C_{Gd}, the water content of the soil is very important and should be measured. It may also be necessary to allow for variations in water content during the course of a set of measurements, especially if the measurements extend over a period of several days or weeks.

Plate 10.3 Automatic weather station mounted on surface of waste storage.

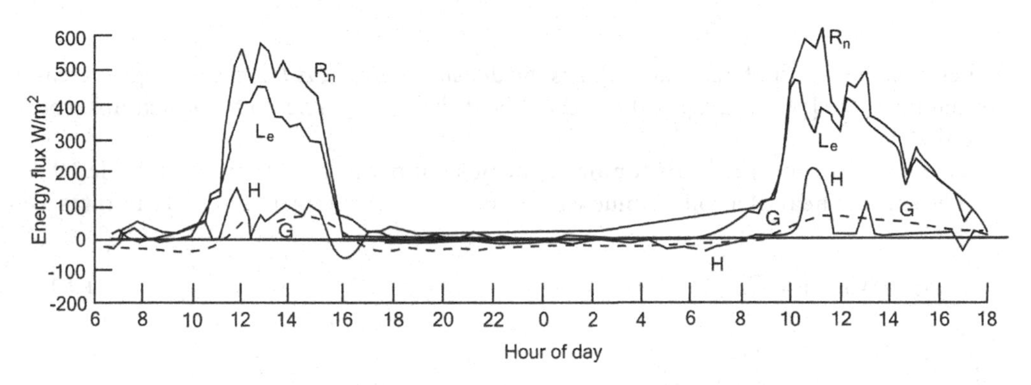

Figure 10.20 Measurements of components of R_n, G and H of the surface energy balance made by an automatic weather station.

As an example for H, Take $z = 2$ m, $\rho = 1.28$ kg/m^3 at $T = 0°C$, $\Delta = 10°C$ and $C_h = 1.004$ kJ/kg°C. Then $H = 25.7$ kJ/m^2. For G, take $z = 0.2$ m, $\rho = 1500$ kg/m^3, $\Delta T = 10°C$ and $C_h = 2$ kJ/kg°C. Then $G = 6000$ kJ/m^2. Thus the ratio $G/H = 6000/25.7 = 233$. Even if the depth of air that affects heating of the surface is taken as 4 m, $G/H = 116$. In other words H is negligible in comparison with G and could be neglected with little effect on L_e. The only situation in which it should be considered is in temperate maritime climates where diurnal temperature variations are small and therefore G

Table 10.4 Density and specific heat of air

Temperature °*C*	*Density* ρ_a *kg/m²*	*Specific heat* C_a *kJ/kg°C*
0	1.28	1.004
25	1.18	1.005
50	1.09	1.006

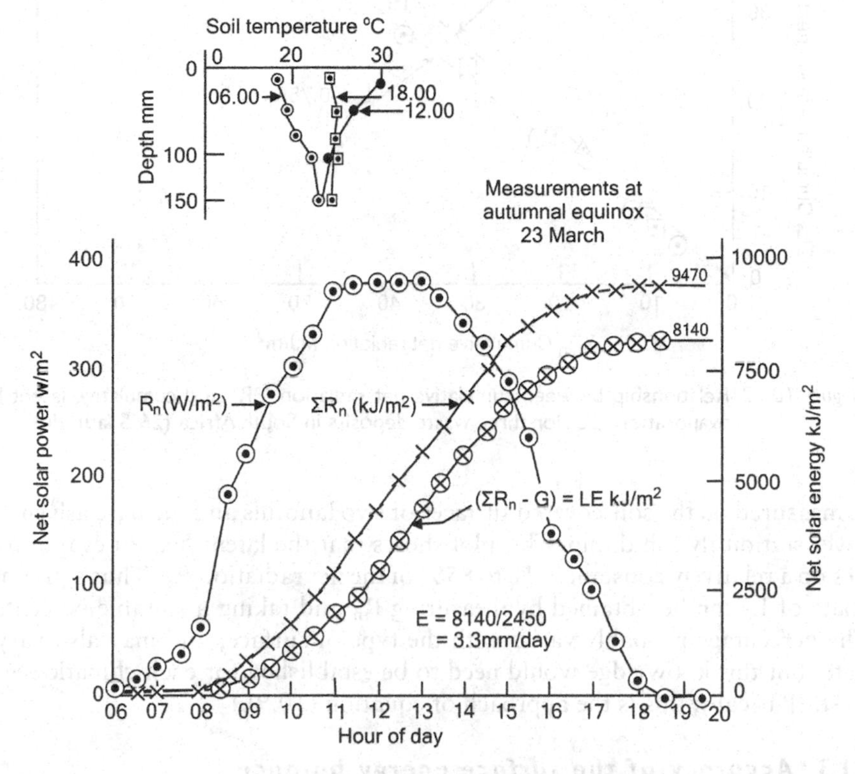

Figure 10.21 Measurements of net solar power R_n (W/m²) and cumulative net solar energy ΣR_n (kJ/m²) measured by hand-held instrument. Inset shows near-surface soil temperature profiles at sunrise (06.00), noon and sunset (18.00). Note: Depth to which soil temperature changes significantly is 150–200 mm.

is small. At the same time winds can move large amounts of warm or cold air onshore or offshore, thus significantly changing and possibly enlarging H.

Ordinates in Figures 10.20 and 10.21 are energy fluxes, i.e. rates or solar power in W/m² (1 Watt = 1 Joule/second). Areas under the curves represent energies in J/m²/day. There is relatively little interchange of energy during the night, hence it is sufficient to take hand measurements only during daylight hours. The figures show that G makes up a relatively minor part of the net radiation R_n. Figure 10.22 shows the cumulative latent heat of evaporation (ΣL_e) plotted against the cumulative net radiation (ΣR_n) for the surfaces of three waste storages in South Africa over a period of a year. These

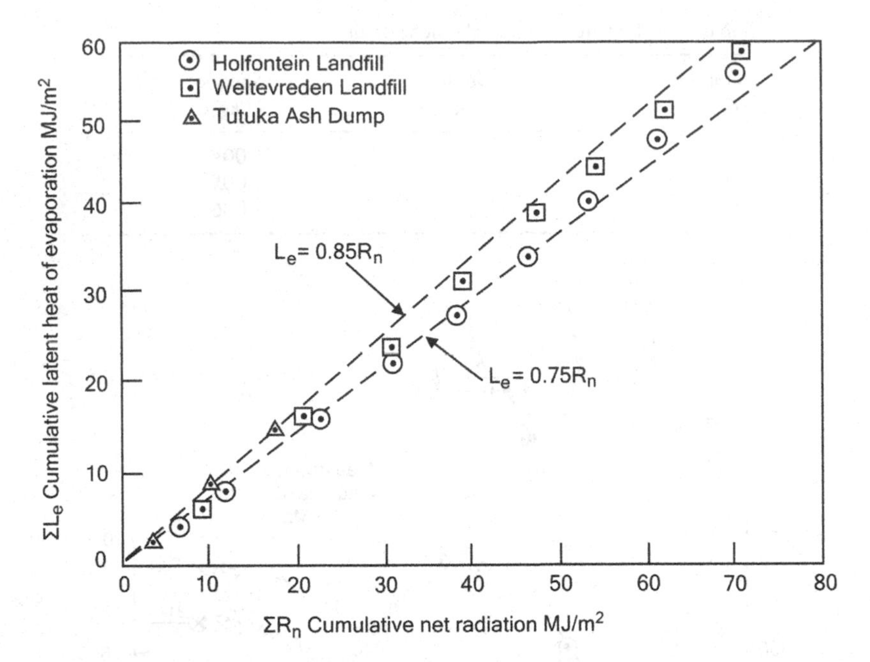

Figure 10.22 Relationship between cumulative net radiation ΣR_n and cumulative latent heat of evaporation ΣL_e, for three waste deposits in South Africa (26°S latitude).

were measured on the soil-covered surfaces of two landfills and the bare ash surface of a power station fly ash dump. This plot shows that the latent heat of evaporation L_e makes up a relatively constant 75% to 85% of the net radiation R_n. Thus a preliminary estimate of L_e can be obtained by measuring R_n and taking a suitable percentage of it. The percentage probably varies with the type of surface, and may also vary with climate, but this knowledge would need to be established for each climatic region of interest. (Basically, this is the approach of equation (10.9).)

10.11.3 *Accuracy of the surface energy balance*

Checking the accuracy of evaporation measured by means of the surface energy balance is difficult because of the difficulty of establishing an absolute basis for comparison. Two examples of accuracy checks will be given here, one for laboratory measurements and one for field measurements.

Blight and Lufu (2000) describe a series of tests in which evaporation from trays of wet mine tailings 200 mm deep kept in a greenhouse (to eliminate the effects of wind and rain) was estimated by the surface energy balance and checked by direct weighing. The results of these measurements, for three different types of tailings, are shown in Figure 10.23. Evaporation found by energy balance (ΣE_{EB}) has been plotted against evaporation from mass loss (ΣE_M) of the trays. The comparison was excellent for the heavy minerals tailings, but less so for the gold tailings (maximum error an underestimate of 2 mm in 23 or 9%) and the fly ash (maximum error an overestimate of 2 mm in 13 or 15%). Errors were found to increase as the tailings became drier,

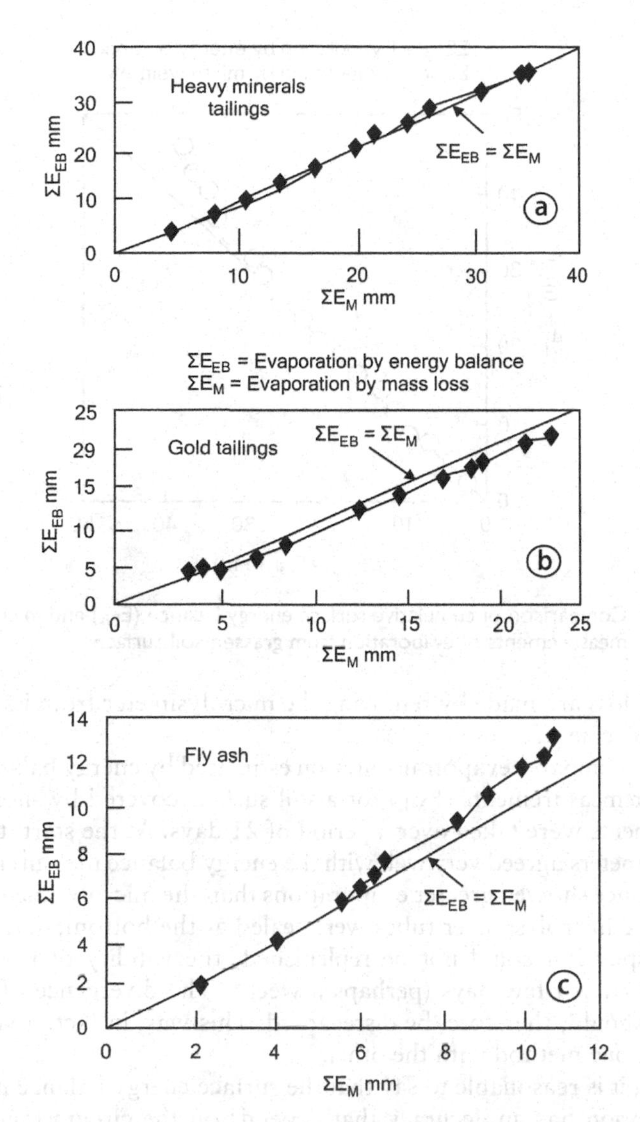

Figure 10.23 Assessment of accuracy of surface energy balance measurements applied to measuring evaporation from tailings surfaces in the laboratory

probably because of inaccuracies in estimating the specific heat of the tailings as this is heavily influenced by the water content.

The second example relates to Figure 10.24 where an independent check on evaporation was made using the microlysimeter technique (Boast and Robertson, 1982). A microlysimeter consists of an open-ended coring tube (in this case 150 mm in diameter by 150 mm long) that is used to remove an undisturbed core of soil. The core is left in the tube, and its base is sealed using wax or a plastic sheet. The core plus coring tube is then replaced in the hole from which it was taken. Daily measurements

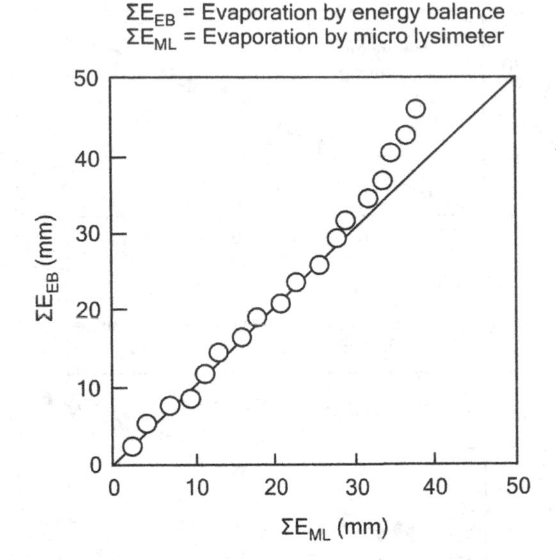

Figure 10.24 Comparison of cumulative surface energy balance (E_{EB}) and micro lysimeter (E_{ML}) measurements of evaporation from grassed soil surface.

of evaporative loss are made by removing the microlysimeter from its hole, weighing it and then replacing it.

Figure 10.24 compares evapotranspiration estimated by energy balance (ΣE_{EB}) with microlysimeter measurements (E_{ML}) for a soil surface covered by short mown grass. The measurements were taken over a period of 21 days. At the start, the mean of the four microlysimeters agreed very well with the energy balance measurements, but later the energy balance showed greater evaporations than the microlysimeters. Because the soil cores in the microlysimeter tubes were sealed at the bottom, and water depleted by evapotranspiration could not be replenished, the validity of the measurements only extended over a few days (perhaps a week). The divergence of measurements at later times should, therefore, be disregarded. This was, in fact, a very satisfactory comparison of one method with the other.

In summary, it is reasonable to say that the surface energy balance method of measuring evaporation has an accuracy that depends on the circumstances and way in which it is used. Accuracies in the range of ±10% appear to be attainable, but the accuracy may be much worse, especially when only a few sets of measurements are available and when dealing with dry or progressively drying soils for which water contents are not accurately known.

10.11.4 *Calculating E from L_e*

ΣLE in kJ/m^2 can be used to calculate the quantity of daily evapotranspiration by dividing the daily total ΣLE by the latent heat of vaporization λ in kJ/kg, i.e. Evapotranspiration $= \Sigma LE/\lambda$ in kg/m^2/day, the equivalent of mm of water/day. For water, λ varies almost linearly with temperature (e.g. Calder, 1990) from 2477 kJ/kg at 10°C to 2417 kJ/kg at 35°C, i.e. by 2.4 kJ/kg per °C. In an unsaturated soil, evaporation could occur throughout the depth of heating of the soil (see Figure 10.21). Hence it is probably fair to use a value for λ that corresponds to the mean temperature in the depth of heated soil. In fact, over the whole range of temperatures given above (10–35°C) the value of λ only changes by 2.5% of its value at 35°C and it is sufficiently accurate to use a mean value of 2450 kJ/kg, as shown in Figure 10.21.

Although an estimation of the daily quantum of evapotranspiration can be made, the way in which the water moves up to the surface and escapes as vapour is obscure. It is known that evapotranspiration can draw water to the surface from depths of more than 10 m (see later), thus the reservoir of water available for transpiration may be large. It will be seen from Figures 10.20 and 10.21 that the solar power input is negligible between sunset and sunrise, so that the major impetus for upward water flow to the soil surface prevails during the daylight hours, and is replaced at night by the conversion of G into L_e.

10.11.5 *Evaporation from wet and dry hydraulic fill beaches*

Ring dyke hydraulic fill tailings storages are usually operated on a cycle of deposition that, depending on the length of perimeter and rate deposition, may take up to 3 or 4 months to complete. The current area of deposition is visibly wet, (see Plates 1.9 to 1.13), while most of the remaining area usually appears dry although it remains moist to a varying degree just below the surface. When calculating the water balance for the impoundment, the question often arises as to whether the evaporation rate for a wet beach surface differs from that for an apparently dry surface.

Possible reasons for differing evaporation rates are:

- differing albedos or energy reflectances for wet and dry surfaces:
- differing permeabilities of the immediate surface layers to the passage of water; and
- differing soil suctions of the surface layers.

Figure 10.25 compares two sets of simultaneous day time surface energy measurements, one made for a visibly wet surface, the other for an apparently dry surface. The measurements have been shown as cumulative energy in MJ/m^2 (i.e. the time-integrated form of Figures 10.20 or 10.21). The material was a power station fly ash, the wet surface had been deposited on the previous day and the dry surface a week previously. The two sites were about 200 m apart.

The lower diagram shows that the albedo ($a = R_o/R_i$ where R_i is the incoming radiation and R_o the outgoing reflected energy) for the dry area was slightly higher than that for the wet area throughout the day, because of its lighter colour. This resulted in ΣR_n for the wet surface being slightly larger than for the dry surface. The wet near-surface ash also heated up more and its ΣG was larger. This resulted in the day time ΣL_e for the wet surface being less than for the dry surface. The dry surface evaporated 1.7 mm (by radiation absorption) as compared with 1.5 mm for the wet surface.

Hence the comparison shows that evaporation from wet and dry beaches will be practically the same, provided the relative humidity above the dry surface is close to 100%, and the beach has not dried out to the extent that the near-surface permeability has been significantly reduced, thus limiting evaporative outflow.

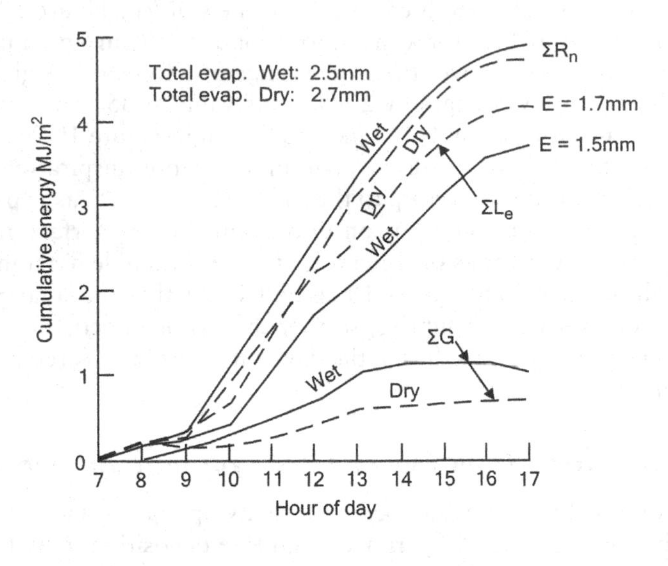

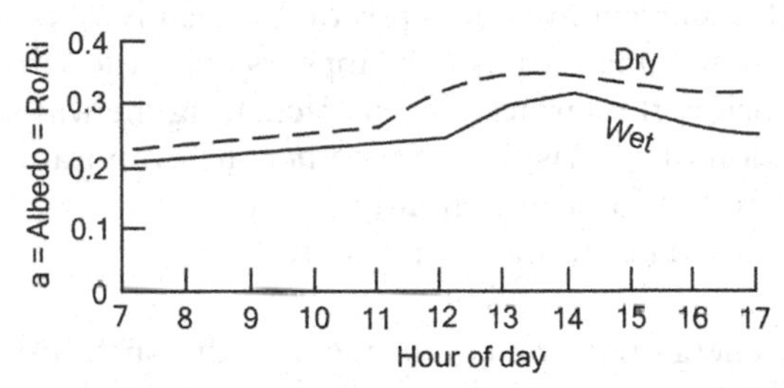

Figure 10.25 Set of comparative daytime measurements for assessing evaporation from visibly wet and apparently dry hydraulic fill tailings beaches.

This set of measurements was backed up by sampling a wet beach to measure its water content profile immediately after deposition on the beach had been completed. The sampling was repeated at regular intervals up to 16 days after deposition had ceased.

The resultant water content profiles are shown in Figure 10.26a. The profiles were close to a series of vertical lines, with the water content at the end of the first day being about 41%. The ash appeared to dry out and drain fairly uniformly over the top 300 mm, although there is a tendency for the measurements from 250 to 300 mm to be slightly wetter. This seems to show that the depth to which short-term drying or drainage extended was about 300 mm, probably with some downward percolation also occurring below this depth. The field moisture capacity of the ash was also measured at about 40% which agrees with the data of Figure 10.26a. However, if deeper drainage into the ash occurred, a suction would be set up that would cause drainage to occur to below the field capacity.

Figure 10.26b shows the water content measurements for all depths plotted against the time in days since deposition. As this figure shows, the average gravimetric water content of the ash decreased from about 41% to about 24% during the

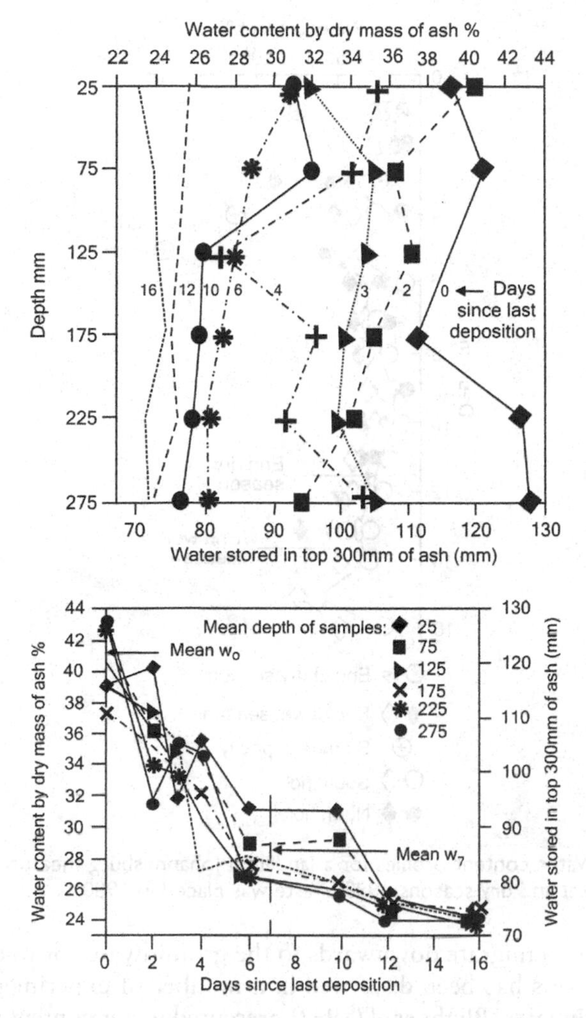

Figure 10.26 Results of water content sampling of fly ash beach for 16 days: a: water content profiles, b: decline of water content with time.

period of 16 days. Over the first seven days, the water content reduced almost linearly with time, at about 1.8% per day, which was very close to the daily evaporation measured by energy balance, of 1.7%. Thus this series of measurements showed by two methods that the rate of evaporation from wet and dry beaches, certainly for the first week of drying, are near-identical. Figure 10.26b also shows that, at least down to a depth of 300 mm, the loss of water content was reasonably constant with depth.

10.12 Depth to which evaporation extends

There is a common misconception that if water penetrates a soil or waste surface to deeper than 1 m or so, the water will move beyond the influence of evaporative forces,

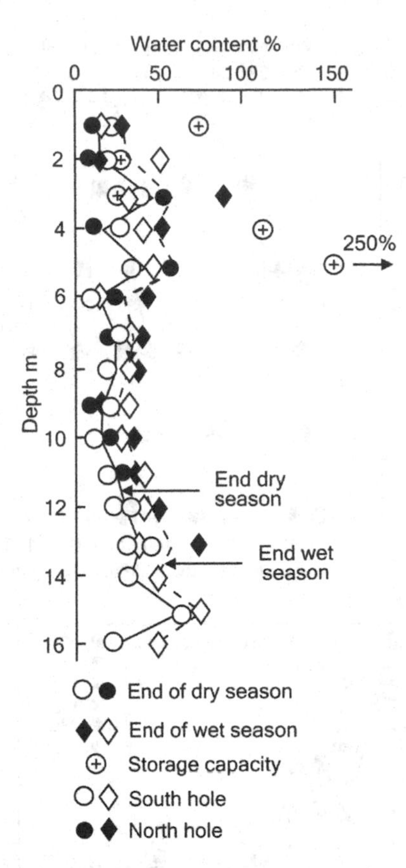

Figure 10.27 Water content profiles for a landfill in Johannesburg measured at the ends of the wet and dry seasons, 1988. Waste was placed in 1980.

and will continue to migrate downwards to the groundwater or to exit at the base of the waste body. This has been disproved in a number of experiments on landfills in water-deficient climates. Blight et al (1992) presented water content profiles measured at the ends of the wet and dry seasons in landfills situated in both Cape Town and Johannesburg. The Cape Town landfill was temporarily capped with 300 mm of clean beach sand, while the Johannesburg landfill was temporarily capped with 300 mm of a semi-pervious silty sand residual from the decomposition of granite. The water content measurements showed that in Cape Town, waste seasonally dried by evaporation to a depth of 7.5 m (the full depth of the waste). In Johannesburg similar profiles showed seasonal drying to a depth of 16 m. One of the pairs of profiles from Johannesburg is shown in Figure 10.27. In a separate experiment, (originally set up to test the results of the water content profile observations), Roussev (1995, quoted by Blight, 1997) constructed two pairs of identical lysimeters in a landfill in Johannesburg. Each lysimeter measured 4.5 m square in plan, one pair of lysimeters was 3 m deep, the other 5.5 m deep. The dimensions are illustrated in Figure 10.28. The 4 sides and the base of each lysimeter were sealed by means of sheets of geomembrane welded to form an impervious box. Each lysimeter was equipped with a drainage layer

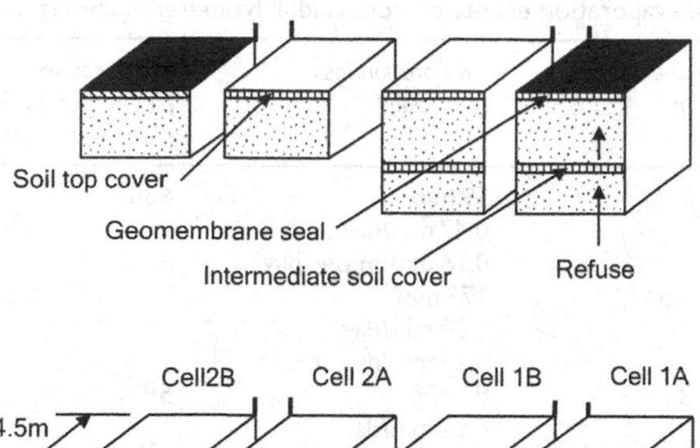

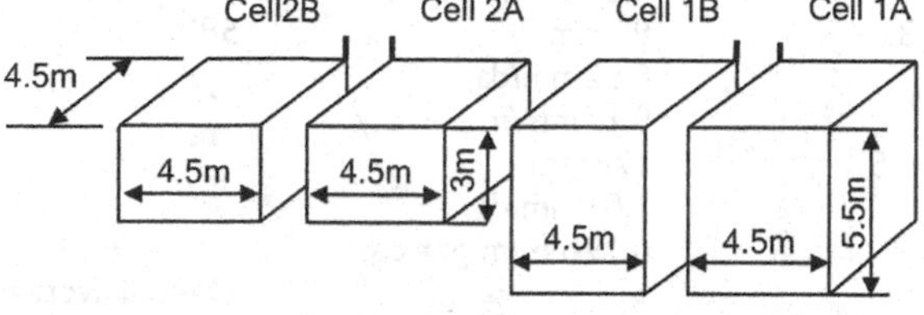

Figure 10.28 Lysimeter experiment to test depth to which influence of evaporation at surface penetrates into a landfill.

at the base and a 100 mm diameter observation well and was filled with compacted waste. The top surfaces of two of the lysimeters (3 m and 5.5 m deep) were then sealed with a geomembrane, while the surfaces of the other two lysimeters were left open to the atmosphere. All four lysimeters were brought to their water storage capacity by irrigating their surfaces until leachate appeared in the observation wells. By lowering a submersible pump down each observation well, the leachate was then pumped out until no more collected. At this stage the waste in the lysimeters was at its water storage capacity. The lysimeters were then left for a period of 4 months and measured quantities of water were then slowly added at the surface until leachate again appeared in each observation well. The difference between the water added to the open-topped and closed-topped lysimeters was then taken to be the evaporative water loss from the open-topped lysimeters. The experiment was then repeated over a period of 6 months. It was found that the 3 m deep lysimeter lost an average of 0.17 mm per m depth of waste per day while the 5.5 m deep one lost 0.22 mm/m/d. This was taken as a demonstration that evaporative losses can occur from waste at least to depths of 5.5 m and supported the evidence of the earlier water content profiles. The detailed measurements are shown in Table 10.5.

The water lost seasonally from the profile shown in Figure 10.27 can be calculated to be equivalent to 720 mm, as compared with the rainfall for that year of 730 mm. Hence if the entire annual rainfall infiltrates into the landfill it can be completely lost again by evaporation. The average annual A-pan evaporation for the area is 1860 mm. Hence for this landfill the equivalent to the ratio $E_B/\bar{E}_A$ from Figure 10.15 is 730/1860 = 0.39. This is larger than but very similar to the ratio of $E_B/\bar{E}_A$ shown in Table 10.1 for Weltevreden landfill in the same region (0.34), but larger than the average of E/E_A for the lysimeter measurements (Table 10.5). This shows that E/E_A is not an intrinsic constant, but depends on how much water is available to be evaporated.

Table 10.5 Results of evaporation estimates from landfill lysimeters (see Figure 10.28)

Drying period: days	*Cell depth: m*	*Evaporation loss from cell, E*	*Evaporation from free water surface (A-pan), E_A: mm*	E/E_A
124	3	59 mm 0.47 mm/day 0.16 mm/m per day	630	0.09 (0.03/m)
	5.5	173 mm 1.39 mm/day 0.25 mm/m per day		0.27 (0.05/m)
171	3	89 mm 0.52 mm/day 0.17 mm/m per day	580	0.15 (0.05/m)
	5.5	178 mm 1.05 mm/day 0.19 mm/m per day		0.31 (0.06/m)
			Overall average E/E_A	0.21

The upper limit would be the ratio P/E_A where P and E_A are annual quantities. The seasonally variable availability of water for evaporation is also the main reason for the seasonal variation of $E_B/\bar{E}_A$ shown in Figure 10.15.

10.13 The effects of slope angle and orientation on solar radiation received by slopes of waste storages

In the northern hemisphere a north facing slope receives less solar radiation than does a south facing one. In the southern hemisphere south facing slopes receive less radiation than north facing ones. In both hemispheres, east and west facing slopes receive less radiation than slopes that face towards the sun. A steep east-facing slope loses the sunshine earlier in the day than a flat slope. The object of this section is to show the effects of slope angle and orientation on received radiation and hence on potential evapotranspiration from the surface.

At noon at a particular spot on the Earth's surface, the solar power reaching the outer limit of the Earth's atmosphere is given by:

$$R_A = S_o(1 - \alpha)(\sin\varphi \sin\delta + \cos\varphi \cos\delta) \tag{10.14}$$

Where R_A = net incoming radiation above the limits of the atmosphere (incoming minus reflected); S_o = solar constant (=1380 W/m^2); α = planetary albedo or the reflectance of the Earth. α varies from 0.3 to 0.5 with 0.4 being a reasonable average; φ = latitude of the place under consideration; δ = declination of the sun which varies from 0° at the equinoxes to 23½° at the solstices. Assuming $\alpha = 0.4$, $S_o(1-\alpha) = 828$ W/m^2 which is the net incoming solar power at noon, on the Equator, at the equinoxes, or at the appropriate tropic (either Cancer or Capricorn) at the solstices.

The solar power can be converted into daily solar energy in MJ/m^2 by integrating the power over the length of the day from dawn to dusk. This can be done approximately by assuming that the power at a particular point varies through the day from sunrise to sunset in a parabolic fashion. For example the daily net solar energy at the outer limit of the atmosphere at the Equator, at the equinox would be

$$2/3(828\,\text{W/m}^2 \times 12\,\text{h} \times 3600\,\text{s/h}) = 23.85\,\text{MJ/m}^2$$

(Also see Section 10.11 and Table 10.3. This value is less than given by Table 10.3.)

However, what is required for practical use is not the power or energy at the outer limit of the atmosphere, but the quantity available at the surface of the Earth. The intensity of solar radiation will be reduced considerably as a result of absorption by the atmosphere, reflection from high clouds and dust particles and reflection from the surface of the Earth. These losses amount to about 50% for nominally cloudless days, leaving the balance available for heating the near surface air and soil and for evaporating water from the soil.

Figure 10.29 shows the effect of surface orientation on solar radiation normal to five surfaces throughout the day at the southern hemisphere spring equinox, as measured at latitude 26°S, altitude 1700 m. The surfaces were oriented as follows: horizontal (O), and facing north (N), east (E), south (S) and west (W). Apart from the horizontal surfaces, the other four surfaces were all inclined at 30° to the horizontal. The curves in the upper diagram in Figure 10.29 show hourly values of solar power (W/m^2) received normal to the five surfaces. Although measured on a cloudless day, it is apparent that more power was received before noon than after noon. (Compare the 11h00 peak for the east facing slope with the 14h00 peak for the west facing slope. Measurements of power for the two orientations should have peaked at the same number of hours before or after noon and have been the same.) The discrepancy is probably a result of varying concentrations of water vapour and air-borne dust in the atmosphere. The lower diagram in Figure 10.29 gives the integrals of the power curves above (in MJ/m^2). It is clear that for this latitude and season of the year, the north-facing slope received the most energy. The numbers in parentheses to the right of the energy curves are the relative values of energy received over the day. The south-facing slope received less than 60% of the energy received by the north-facing slope.

The figures to the immediate right of the cumulative curves in Figure 10.29 are evaporation values in mm/d assuming the albedo $a = 0.1$, and dividing L_e by 2.45 MJ/kg (Section 10.11). The rates of evaporation may not sound very impressive, but a north-facing 30° slope 30 m high (slope length $2 \times 30 = 60$ m) and 1 km long could be evaporating

$$3.26\ell/\text{m}^2\text{d} \times 2 \times 30 \times 1000 = 0.2\text{M}\ell/\text{d} \text{ (approximately)},$$

while a 1 km × 1 km top surface could, on the same day, evaporate

$$3.01\ell/\text{m}^2\text{d} \times 1000 \times 1000 = 3\text{M/d} \text{ (approximately)}.$$

Figure 10.30 shows (above) the solar power curves for a 30° slope on a day at the summer solstice at Johannesburg and (below) the solar power curves for a day at the winter solstice. The power curves for the autumn equinox should be very similar to those for the spring equinox (Figure 10.29) and were not measured. Interestingly, the solar

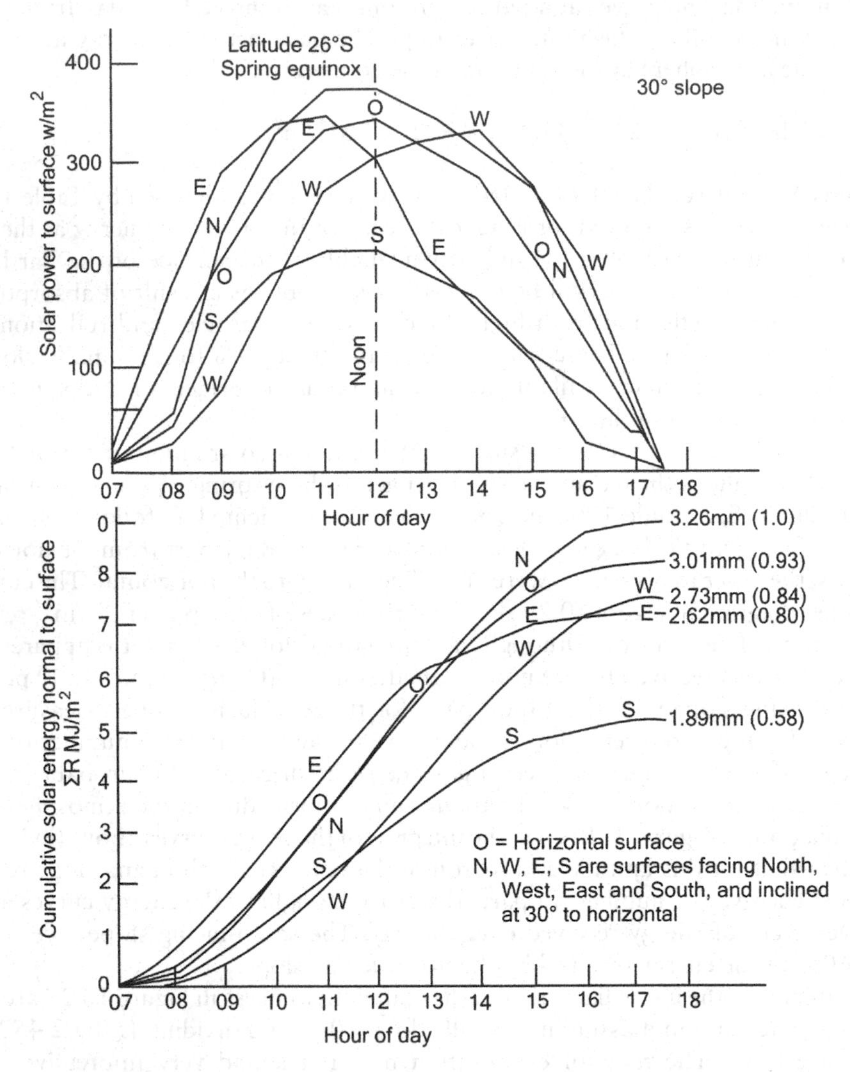

Figure 10.29 Solar power and daily energy received at Earth's surface at 26°S at spring equinox on 30°slope.

power received at noon at the summer solstice by a horizontal surface (~500 W/m^2) is about 100% larger than the peak solar power at the winter solstice (~250 W/m^2), and the same relationship applies to the solar energy, even though the length of day varies.

Figure 10.31 shows the principle for calculating the solar power incident to a slope at noon:
z, the zenith angle at noon is given by

$$\cos z = \sin\varphi \sin\delta + \cos\varphi \cos\delta \qquad (10.15)$$

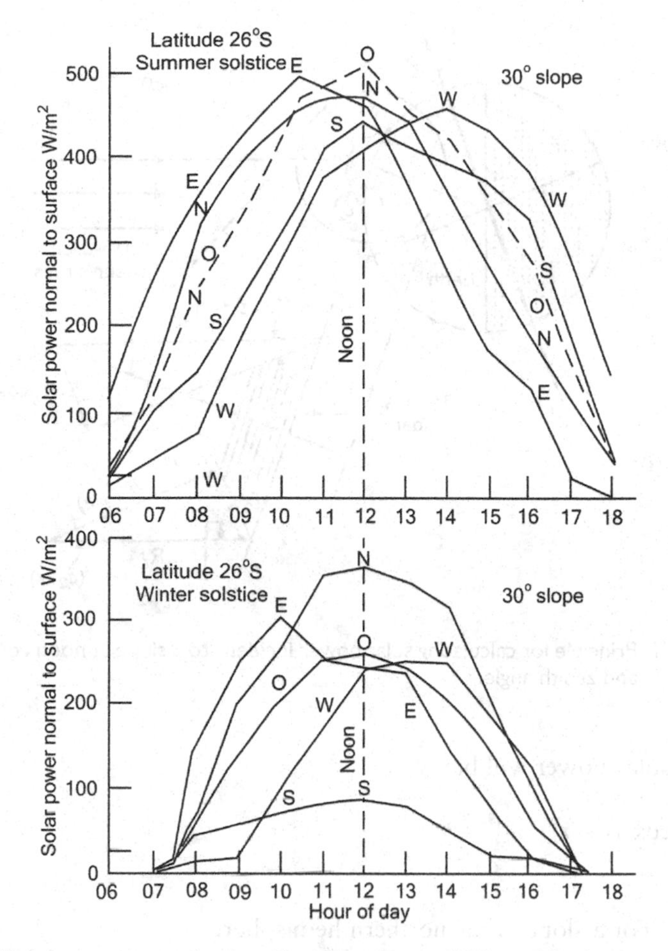

Figure 10.30 Solar power received at Earth's surface at 26°S on 30° slope at summer solstice (above) and winter solstice (below).

For cloudless weather R_i is assessed as 50% of R_A, and for cloudy weather a suitable reduction can be made. Referring to Figure 10.31 for a slope where the sun's rays at noon strike the surface at an angle of

$$90^\circ - (i_1 + z)$$

the incident solar power will be

$$R_s = R_i \cos(i_1 + z) \tag{10.16}$$

And for a slope where the noonday sun's rays strike the surface at

$$90^\circ - (i_2 - z)$$

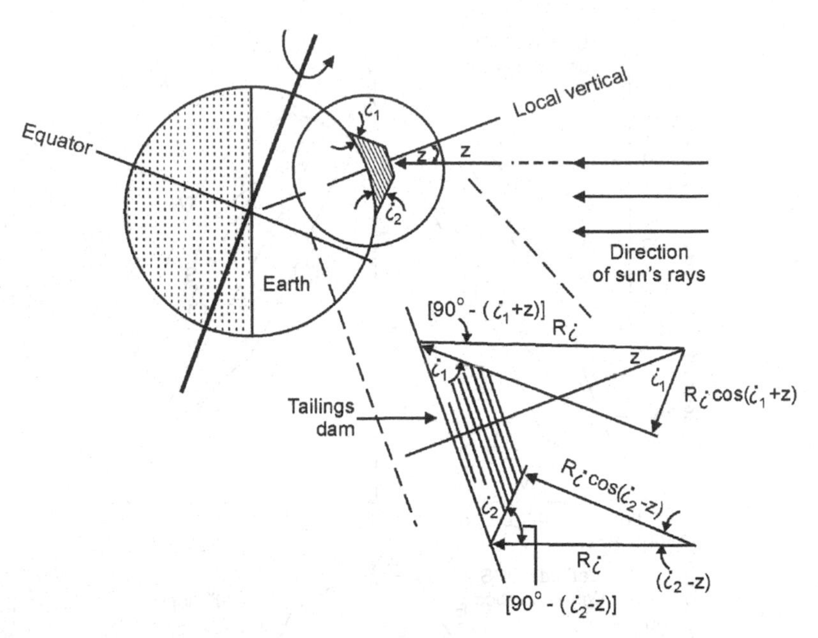

Figure 10.31 Principle for calculating solar power incident to a slope at noon considering latitude and zenith angle.

the incident solar power will be

$$R_s = R_i \cos(i_2 - z) \tag{10.16a}$$

EXAMPLE For a slope in the northern hemisphere:

$\varphi = 35°N, \delta = 0, i_1 = 30°$(slope to N), $i_2 = 20°$ (slope to S).
Then $\cos z = \cos\varphi$ and $z = \varphi$
$90° - (z + i_1) = 25°$ and $R_s = R_i\cos(65°) = 0.42R_i$;
$90° - (i_2 - z) = 105°$ and $R_s = R_i\cos(-15°) = 0.97R_i$

Thus the south-facing slope receives more than twice the power received by the northern slope. The power can then be integrated over the length of the day to give the daily incident energy.

10.14 Water balances for "Infiltrate, Store, Evapotranspire"(ISE) covers and for impervious cover layers on mine waste storages

Impervious cover layers are increasingly being used to isolate the contents of mine waste storages from the surrounding environment. The reasons are to exclude oxygen from pyritic wastes or wastes that might be subject to spontaneous combustion, or

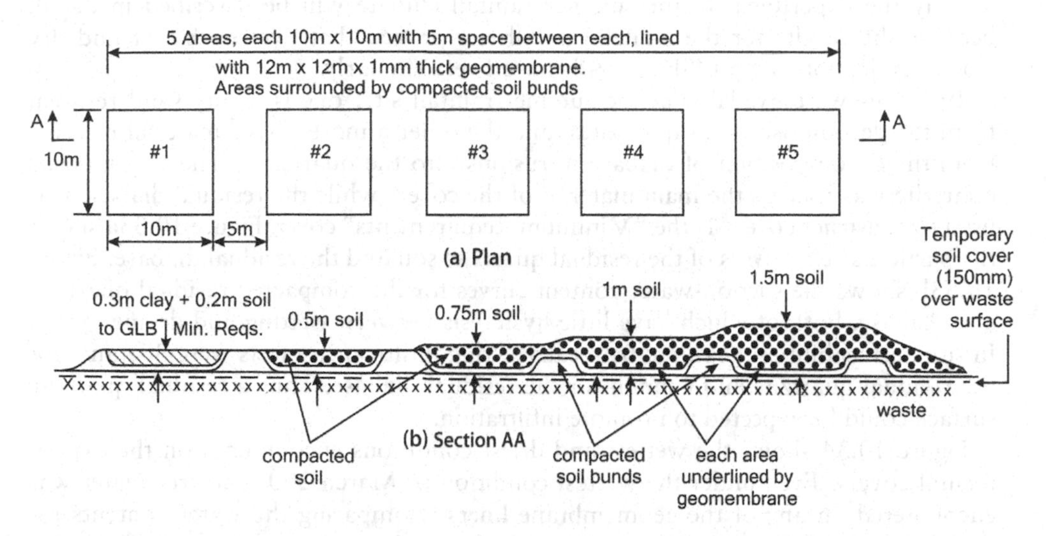

Figure 10.32 Layout of experimental ISE covers.

to prevent the ingress of water that might percolate through the waste and enter and pollute the underlying ground water. The latter may and does occur in areas that have a water surplus, but in water deficit areas, covers can be used that allow controlled infiltration during the wet season, but lose the stored infiltration by evapotranspiration during the ensuing dry season, so that no leachate ever accumulates in the waste, or exits the base of the storage.

To illustrate the concept of the "infiltrate, store, evapotranspire" or ISE cover, two large-scale experiments were set up, one in a summer rainfall, dry winter climate, the other in a winter rainfall, dry summer climate. The layout of both experiments was similar, and is illustrated in Figure 10.32.

The geomembrane lining mentioned in Figure 10.32 (plan) defined the base of each trial cover. Any infiltration in excess of the water storage capacity of the soil would be intercepted and trapped by the geomembrane basin where it could be detected and measured, but would still be available to evaporate out through the soil cover at a later time. This gave an indication of the quantity of water available to infiltrate the underlying waste. The first cover (#1 on the left of Figure 10.32) consisted of a 300 mm clay layer covered by 200 mm of soil. This satisfied the then current South African statutory landfill Minimum Requirements (1998) specification for a GLB^- landfill (i.e. a Large landfill receiving General waste and located in a B^-, i.e. water-deficient climate) and allowed the behaviour of a cover incorporating a clay layer to be studied. Thicknesses of compacted soil cover (silty sand) increasing from 0.5 to 0.75, to 1.0 to 1.5 m were provided in covers 2 to 5. Each cover area was separated from its neighbours by a 0.5 m high compacted soil bund, square in plan, within which the geomembrane formed an impervious basin. The possible accumulation of excess free water within each basin could be detected and measured by means of four perforated vertical standpipes built into the corners of each basin. Surface bunds, 0.3 m high, were constructed to prevent run-on of water from the 1.5 m-thick cover to the 1 m cover and from the 1 m to the 0.75 m cover.

Only the experiment in the summer rainfall climate will be described in detail, because the results for the winter rainfall test proved that over each wet and dry season cycle, both sets of ISE covers behaved very similarly.

Two soils were available at the summer rainfall site, one is a silty sand residual from the decomposition of quartzites, and the other a more clayey material residual from the decomposition of diabase intrusions into the quartzites. The decomposed quartzite was used as the main material of the covers while the residual diabase was used to construct cover #1, the "Minimum Requirements" cover. Figure 10.33a shows the particle size analyses of the residual quartzite soil and the residual diabase. Figure 10.33b shows the suction-water content curves for the compacted residual quartzite and diabase, both of which have little hysteresis between wetting and drying cycles. In situ permeabilities measured by means of ring infiltrometer tests were 35 mm/h for the residual quartzite and 6 mm/h for the residual diabase. Thus a residual quartzite surface could be expected to promote infiltration.

Figure 10.34 shows the wettest and driest conditions encountered on the experimental covers. Even under the wettest condition of March 2004, no free water was encountered on any of the geomembrane liners. Comparing the water contents for the wettest condition (Figure 10.34a) with the suction-water content curves shown in Figure 10.33b, there appeared to be a small wave of moisture moving down through cap #5, but the cap as a whole had not yet reached its field moisture storage capacity.

Figure 10.35 shows the water balance for the experimental covers from October 2002 to March 2004. The water balance can be expressed as

$$P - RO = \Delta I + E \tag{10.3a}$$

Where P = precipitation, RO = runoff, ΔI = change in interstitial water, i.e. in water storage, and E = evaporation from surface.

Figure 10.35 shows the cumulative precipitation, ΣP, evaporation rates found by solar energy balance measurements on the landfill surface, $\dot{E}_B$, and the storage in the five covers, I, expressed in mm of water. (The energy balance method used to measure $\dot{E}_B$ has been fully described in Sections 10.11.1 to 10.11.4.) The figure shows the results of every set of measurements made in the first 18 months of the experiment. The numbers on the curves of storage I, cumulative rain, ΣP, and cumulative evaporation, ΣE_B, indicate when each measurement was made.

The upper diagram in Figure 10.35 (which is on the same time scale as the rest of the figure) shows rates of evaporation measured in situ on the landfill between December 2002 and October 2003. For periods in which there is either very little rain, or for which runoff can be regarded as negligible, the water balance equation can be rearranged as

$$E = P - \Delta I \tag{10.3b}$$

and actual evaporation rates in mm/d can be estimated ($\dot{E}$ in Figure 10.35).

Because of the relatively permeable nature of the residual quartzite, it had been assumed, in the early stages of the experiment, that no run-off was occurring and that all rainfall infiltrated. When after the five month wet spell from mid-October

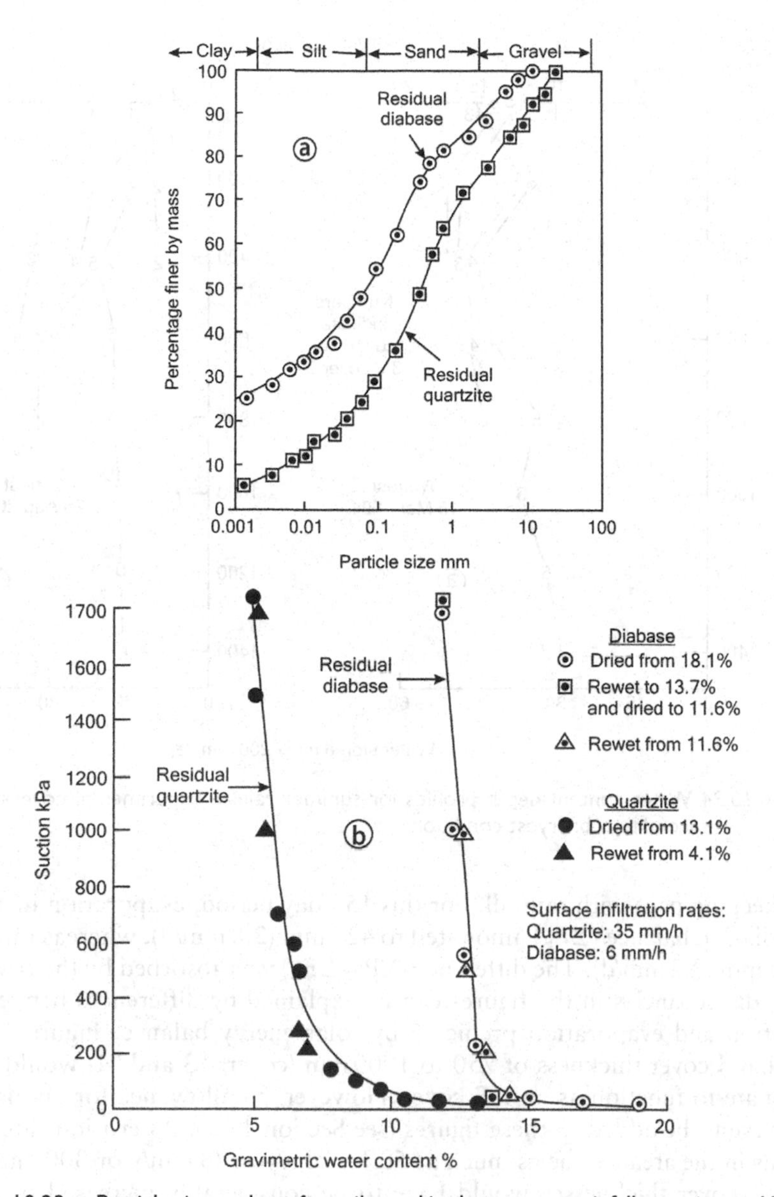

Figure 10.33 **a: Particle size analyses for soils used in the summer rainfall experimental ISE covers. b: Suction: water content relationships for soils used in summer rainfall experimental ISE covers.**

2000 to mid-March 2004, it was found that no free water had accumulated on any of the geomembranes underlying the caps, it was at first thought that run-off was the reason for this. However, as the calculations on Figure 10.35 show (assuming that evaporation was occurring in accordance with measured $\dot{E}_B$) run-off either did not occur (negative figures for RO) or was very small (19 mm for cover #1 and 2 mm for cover #4). It then became apparent that evaporation from the surface must have been

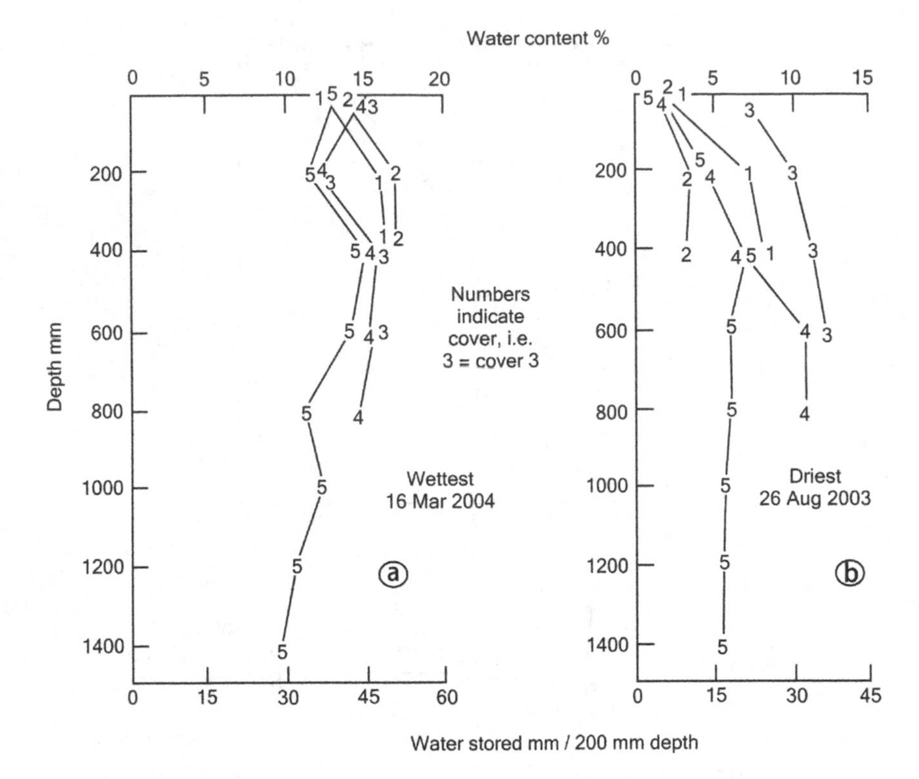

Figure 10.34 Water content-depth profiles for summer rainfall experimental covers: a: wettest condition, b: dryest condition.

almost keeping pace with rainfall. For this 150 day period, evaporation measured by solar radiation balance (ΣE_B) amounted to 420 mm (2.8 mm/d), whereas rainfall (ΣP) was 485 mm (3.2 mm/d). The difference ($\Sigma P - \Sigma E_B$) was absorbed by the cover layers, and any discrepancies in the figures can be explained by differences between actual evaporation and evaporation predicted by solar energy balance. Figure 10.35 also shows that a cover thickness of 750 to 1000 mm (covers #3 and #4) would probably be adequate to function as an ISE cover. However, an allowance for annual erosion would have to be added to these figures (see Section 12.2). As erosion rates for silty sand soils in the area can be as much as 50 Tons/ha/year (3 mm/y or 300 mm/century) the above cover thicknesses would have to be considerably increased to allow for long-term erosion.

The second example given in this section is that of a cover designed to be impervious to infiltration by rain. The measurements were taken from Heerten and Reuter (2006) and represent one of the few published examples of water balance observations for impervious covers that were made over a period of several years. The cover consisted of 500 mm of soil over a single geosynthetic clay liner (GCL) (see Section 6.13.1). Heerten and Reuter do not give details of how the measurements were made, but presumably, separate measurements were made of rainfall, run-off and drainage off the top surface of the GCL. It is also possible that seepage through the GCL was measured directly.

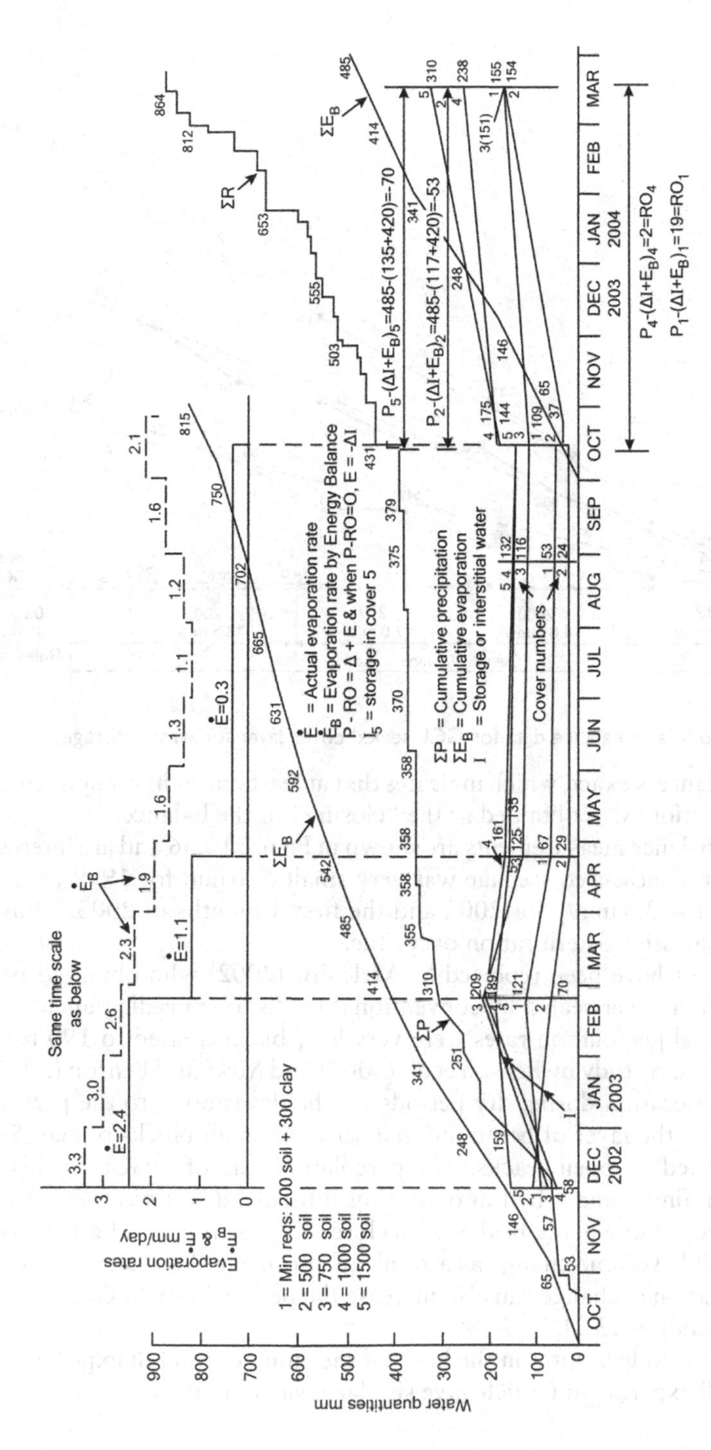

Figure 10.35 Water balance data for experimental ISE covers in summer rainfall climate, October 2002 to March 2004.

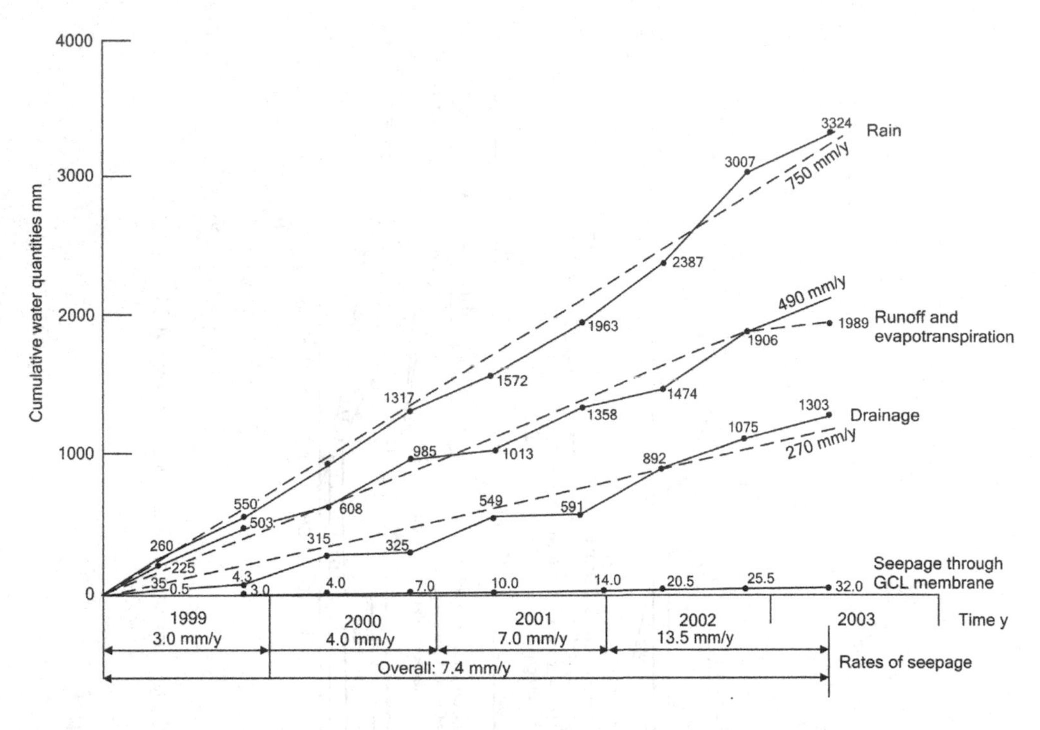

Figure 10.36 Water balance data for GCL-sealed cover layer for waste storage.

The water balance is exact, which indicates that at least one of its components (possibly evapotranspiration) was obtained as the "closure" of the balance.

The water balance measurements are shown in Figure 10.36 and are interesting from two aspects: the measured seepage was very small (3.0 mm for 1999), but increased progressively to 13.5 mm/y for 2002 and the first 4 months of 2003. Thus the GCL showed a progressive deterioration over time.

Similar results have been reported by Melchior (2002) who observed two 100 m^2 test sections of a cover sealed by conventional GCLs, one needle punched, the other stitched. Annual percolation rates were very low, but increased to 190 to 220 mm/y after a few years. A study by Benson et al. (2007) and Meer and Benson (2007) showed that severe desiccation during dry periods can be devastating to the performance of GCLs, reducing the layer of bentonite to a series of small blocks of clay 5 to 10 mm in size, separated by open cracks. The percolation rate of a GCL in this condition is effectively infinite, and rehydration during subsequent wet weather may cause the cracks to narrow, but will not heal or even close them. Bentonite, like all soils is subject to non-reversible volume change as a result of shrinkage. In addition to the effect of desiccation, cation exchange can also increase the permeability of GCLs to a dramatic extent (see Section 6.13.2).

It may be concluded, both on the basis of the summer rainfall experiment, and the winter rainfall experiment (which gave similar results), that:

- Covers for waste storages are usually designed to exclude infiltration into the waste by maximizing run-off and storing any infiltration within the soil layers of the cover pending its re-evaporation or transpiration. Any infiltration into the waste body is usually regarded as a leakage, and therefore a short-coming of the cover.

- The ISE cover is designed to allow limited infiltration into the waste during the wet season and then facilitate any excess infiltration to be re-evaporated during the ensuing dry season, without allowing any leachate to exit the base of the waste storage.
- The experimental covers have demonstrated that the ISE concept will work in semi-arid climates with summer rainfall. They have also demonstrated that in drier years, an ISE cover will store all infiltration within the cover and attenuate infiltration before it reaches the waste. In wetter years, or during exceptionally wet periods, the ISE cover will function as designed.
- The observations of Melchior (2002), Heerten and Reuter (2006) and Meer and Benson (2007) have shown that covers, intended to be impervious, should not incorporate GCLs unless there is certainty that neither severe desiccation nor exchange of the sodium cations in the bentonite for calcium or magnesium can occur.

10.15 The Water Balance for a Dry Ash Dump

The concepts that have been explored above will now be further illustrated by an investigation of the water balance for a 30 m high dry power station ash dump. The dump was the site of the infiltration and runoff tests illustrated by Figures 10.11, 10.12, 10.13 and 10.14.

The dump has been placed on the surface of a backfilled open-cast coal mine. The water table is at a depth of about 10 m below the rather irregular surface of the mine backfill. The section of the ash dump where the measurements were made was placed in 1987, and the tests to be described below were performed in 1993 and 1994. The ash had been placed at a nominal water content of 15% (for dust control), but the actual water content at placing was not well controlled. For current ash placement it can fall well below 15% in dry windy weather and be well over 15% in wet weather.

To predict the interstitial water or water storage curves for the ash dump, two vertical columns consisting of 200 mm bore rigid plastic piping were set up. One column was 6 m high and the second was 12 m high. The columns were filled with a paste of the ash at a water content of 50%. A free water surface was maintained at the base of each column, in a plastic tray of water fed by an upturned plastic bucket filled with water that maintained the water level in the tray at a constant level, on the principle of a pets' drinking fountain, and evaporation was allowed to occur from the open upper end of the column. After allowing 3 months for the columns to reach moisture equilibrium, and hence their equilibrium water storage capacities, they were dismantled and sampled for moisture content. While the columns were equilibrating, the suction-water content relationship for the ash was determined in the laboratory by subjecting a sample of ash to an increasing suction in a triaxial cell. The water pressure at the base of the ash specimen was maintained at atmospheric pressure while an air pressure was applied at the top of the triaxial specimen. The suction applied was thus $(u_a - u_w)$. At any height h above the water level in the ash column, $u_w = -\gamma_w h$ and hence the suction $(u_a - u_w) = u_a + \gamma_w h$, or $+\gamma_w h$ if u_a is taken as a zero datum.

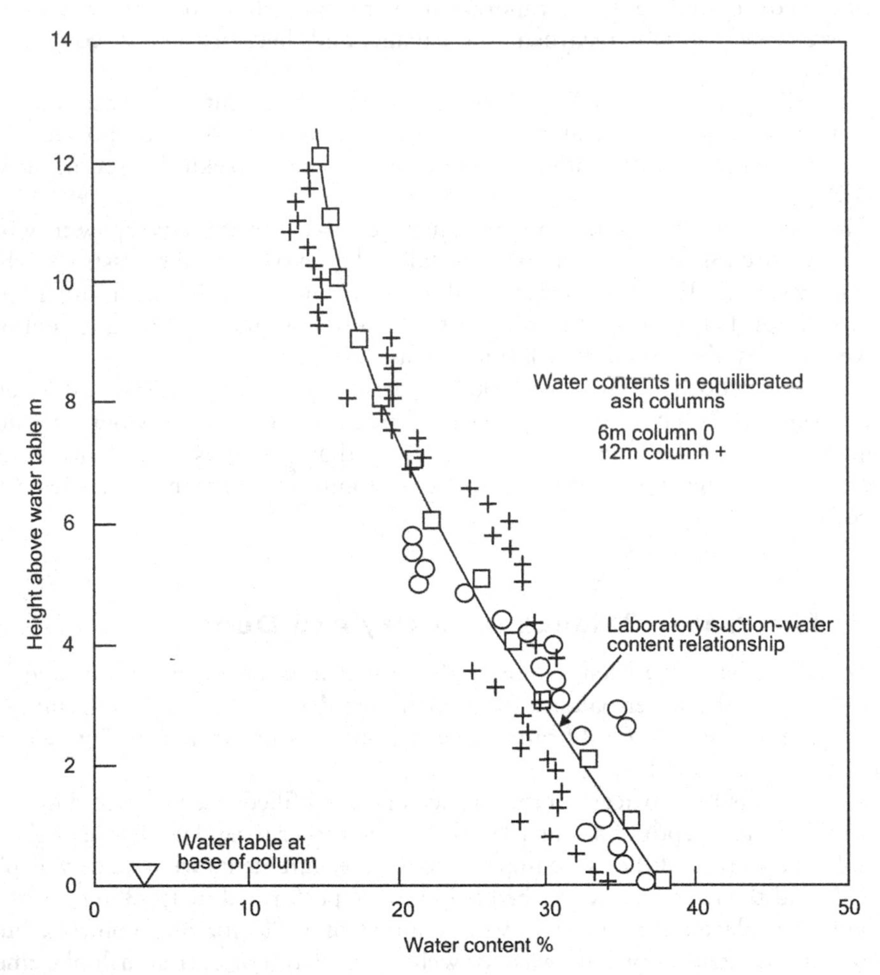

Figure 10.37 Results of experiments to determine storage capacity-depth relationship for power station ash.

The equilibrium water content at each value of $\gamma_w h$ was tracked, in the laboratory specimen, by measuring the outflow of water through the base of the specimen.

The moisture content profiles for the ash-filled columns are shown in Figure 10.37. The laboratory suction-water content curve for the ash, with the suction expressed as metres of water head is also shown in the figure. The suction-water content relationship was determined for a constant value of $(\sigma_3 - u_a) = 50\,kPa$ and would be expected to give a slightly higher water content than the column tests, where the height of the ash column would have approximated to equivalent overburden stresses. The two types of curve should, however, coincide at suctions corresponding to the tops of the ash columns.

As Figure 10.37 shows, the laboratory suction-water content relationship almost coincides with the equilibrium water contents in the columns, and can thus be used to determine the water storage capacity-depth relationship for the ash storage.

The principle of this observation could be extended to any waste storage. However, as discussed above, to obtain a water storage capacity versus depth curve, the confining stress should also be varied. The ash column tests used in this experiment were also not ideal because of the reduction of overburden stress caused by friction between the ash and the walls of the column. Ideally, triaxial tests, in which the suction, overburden and lateral stresses can accurately be controlled, should be used to define the water storage capacity versus depth relationship.

To investigate the actual water content profile in the ash dump and to compare it with the water storage capacity as established from the column and laboratory tests, holes were drilled by dry augering through the full height of the dump and into the soil below. The holes were drilled in October 1993, at the end of the dry season and again in April 1994, at the end of the wet season.

The water content profiles in Figure 10.38 show a very variable water content, with a slight trend to increase with depth. If it is assumed that the water content was essentially constant with depth, the mean water contents, plus or minus one standard deviation (SD), are as follows:

Date sampled	Mean + 1 SD	Mean − 1 SD
Oct. 1993	15.3%	11.7%
Apr.1994	16.4%	11.4%

On this basis there was little change in the water content from the end of the dry season to the end of the wet season.

Because of the mining activities adjacent to the ash dump, the water table has been drawn down to 40 m below the top of the ash dump. Hence, it may be assumed that a depth of 40 m below the ash dump surface marks a level of zero suction. The laboratory suction-water content curve, on the assumption of zero suction at a depth of 40 m, has been superimposed on Figure 10.38. It lies on the dry side of the water content profiles and, therefore, if it correctly represents the water storage capacity of the ash, there is a water surplus which is slowly draining downwards. The amount of surplus water stored in the ash amounts to about 1450 mm and, based on permeability measurements for the ash (Fourie et al, 1995), it can be calculated that this surplus moisture is recharging the water table at the very slow rate of about 20 mm/year.

10.16 Disposal of industrial waste liquids by evaporation and capillary storage in waste

In semi-arid and arid regions the atmosphere has a considerable capacity to evaporate liquid industrial wastes such as brines. The evaporation capacity may be as high as 2000 mm/year, which means that 20 megalitres of liquid can potentially be evaporated per year per hectare. In the past, lined evaporation ponds were used to evaporate waste liquids. Recently, the tendency has been to spray-irrigate waste liquids over the top surfaces of storages of solid waste, e.g. landfills, ash dumps and mine waste dumps. To be environmentally acceptable, the rate of irrigation must be related to the overall water balance of the waste body, so that all waste liquid that infiltrates the waste is either re-evaporated or permanently held within the pores of the waste by capillarity. No liquid can be allowed to escape from the often unlined base of the waste body to become a potential or actual source of groundwater pollution.

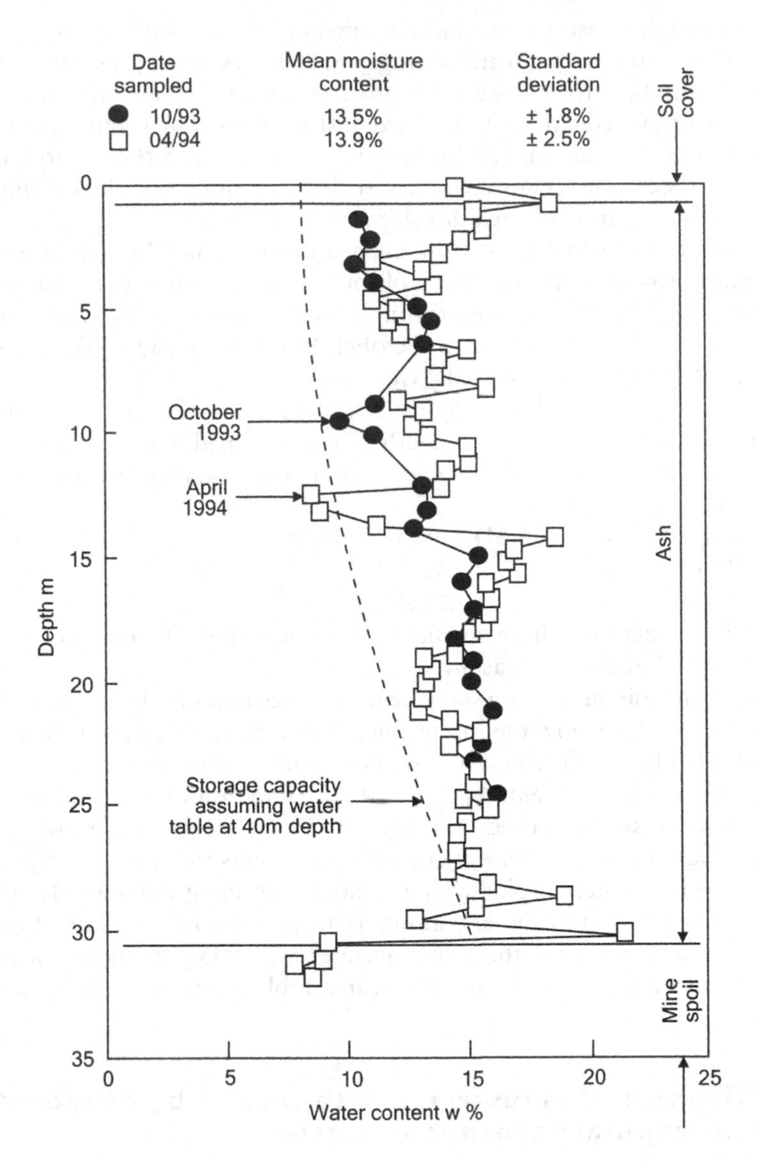

Figure 10.38 Water content profiles measured in ash dump in 1993 and 1994.

This section will describe how to establish acceptable rates of irrigation as well as irrigation cycles. Also, how to establish the quantity of water it is possible safely to absorb in the waste body so that nothing will leach into the ground water even under the most adverse weather conditions.

10.16.1 *Operating a waste storage to dispose of surplus (waste) liquids*

To operate an evaporative waste water disposal system optimally, the application rates for the water should be matched to the evaporation rates. If too much water is applied, the limiting moisture storage capacity of the waste in the body of the storage will be exceeded and waste water will start leaching out at the base of the storage.

Over any 24 hour period, part of the storage surface will usually be sprayed and the remainder will not. For the part that is sprayed, suppose that the spraying takes place for h hours in 24. If the total surface area available for irrigation is A and the area that is sprayed on a typical day is A_s, the water balance for the area that is being sprayed will be:

$$S.h/24A_s = (E_a.h/24 + E_{ss} + I + P_o)A_s - RA_s, \text{ hence}$$
$$(I + P_o) = (S - E_a)h/24 - E_{ss} + R \quad (10.17)$$

In equation (10.17) all parameters are measured in rates of mm per 24 hour day, i.e. mm/d

S = application of waste water by spraying;
E_a = evaporation of water into the air above the waste surface over the area that is being sprayed;
E_{ss} = evaporation rate from the saturated surface that is being sprayed (E_{ss} is the average rate for the day on which A_s is sprayed;
I = infiltration of waste water into the waste;
P_o = ponding depth of waste water on the waste surface;
E_s = evaporation of water from the waste surface that has not been sprayed on a particular day;
R = rainfall for the day.
$(I + P_o)$ forms the input moisture for the areas $(A - A_s)$ of waste that is not being sprayed during the $(A/A_s - 1)$ days left in the spraying cycle. For this area the water balance is

$$(I + P_o) = (E_s - \check{R})(A/A_s - 1) \quad (10.18)$$

where $(A/A_s - 1)$ is the number of days in each rotation when water is not being sprayed over an area, and $\check{R}$ is the average daily rainfall for days on which no spraying took place. (Note that R is the rainfall for the day on which spraying took place).

Thus for an exact water balance for the entire surface being spray-irrigated, the input on spraying days must equal the output on non-spraying days.

$$(S - E_a)h/24 = E_{ss} - R + (E_s - \check{R})(A/A_s - 1) \quad (10.19)$$

For example, suppose $A/A_s = 5$, $E_a = 3$ mm/d, $E_{ss} = 4$ mm/d, $E_s = 2$ mm/d, $R = 0$, $\check{R} = 0$ and $h = 10$ hours

Then $S = 31.8$ mm/d

In other words, if 20% of the waste surface area is sprayed every day on a 5 day rotation, with no rain, $31.8 \times 10/24 = 13.25$ mm of water could be applied each day without causing an overall increase of the water content of the waste body.

Clearly, if it rains by more than a value given by

$$R + \check{R}(A/A_s - 1) = E_{ss} + E_s(A/A_s - 1) = 12\,\text{mm/d} = R + 4\check{R} \qquad (10.20)$$

spraying will not be possible without upsetting the water balance, and if $\check{R}$ has been 3 mm/d or more, spraying is also not permissible on day 5 of a 5 day cycle.

10.16.2 *Measuring the water balance parameters*

The water balance parameters that need to be measured are E_a, E_{ss} and E_s. These parameters all vary seasonally and with the weather and considerable effort is required to assess them and their variation and limits. S, the rate of application of waste water by sprinkler, also needs to be established in terms of the performance characteristics of the sprinklers and pumping system being used.

The overall limiting moisture storage capacity also needs to be established by performing a suction-water content test on the waste, as described in Section 10.15.

Aerial evaporation, E_a

This can be assessed by comparing the rate of application measured by means of a flowmeter in the pipe feeding a sprinkler, with measurements made by a number of rain gauges placed just above the surface, within the area covered by the sprinkler. The difference between water supplied and water actually reaching the waste will be a measure of E_a, while the water reaching the waste will be S.

Surface evaporation, E_s and saturated surface evaporation, E_{ss}

These can be estimated by means of radiation balance measurements or by means of microlysimeters, as previously set out in Section 10.11.

10.16.3 *Spray irrigation tests to estimate E_a*

Figure 10.39 shows the results of a spray irrigation test to measure E_a, the aerial evaporation. The figure shows measured contours of irrigation water reaching the surface of the ash, expressed in mm (ℓ/m^2) of water over the 6.5 h duration of the test. The total volume of irrigation was found by calculating the volume under the contoured surface ($m^2 \times \ell/m^2 = \ell$) and then subtracting this from the total volume of water pumped through the sprayer in the test period. The difference represented E_a, the aerial evaporation. The average volume of E_a was found by dividing the total volume of evaporation by the area enclosed by the zero irrigation contour in Figure 10.39. This gave an average value for E_a of 3.5 mm for the 10 h of daylight during which sprinkling continued. Because of heat stored in the surface layers of the waste, about a quarter of the total evaporation ($E_a + E_{ss}$) occurs during the hours of darkness, giving a total value for E_a over 24 h of $3.5 \times 4/3 = 44.7$ mm. This is equivalent to 47 kℓ per ha of waste surface.

10.16.4 *Summary and conclusions*

Evaporation by controlled spray-irrigation is a promising way of disposing of waste water without causing pollution. There are, however, a number of aspects that need to

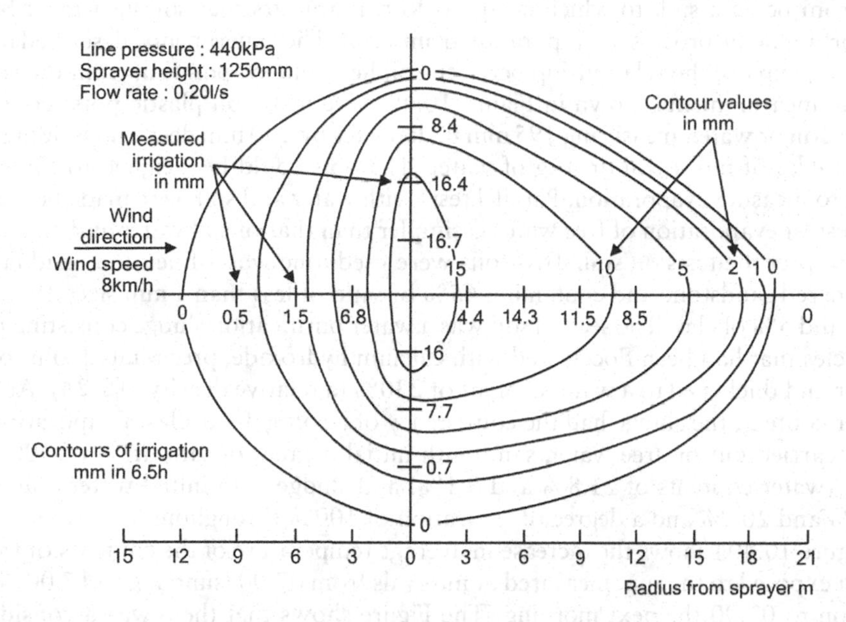

Figure 10.39 Contours of spray irrigation water reaching waste surface in test to measure aerial evaporation E_a.

be investigated before such a system can be designed with confidence. These include the behaviour of the wetting front produced by a period of a spray irrigation and the drying front that will develop during the ensuring period of drying. Provision must also be made to store water during periods of rain when spray irrigation has to be stopped. The possible problem of pollution of surrounding land by saline spray blown off the waste deposit by high wind also needs investigation.

10.17 The role of soil heat G in evaporation of water from a soil

In his classical book on evaporation from soil, Penman (1963) does not include soil heat as a term in his much-used equation (10.9) for the prediction of evaporation. Calder (1990) in his book "Evaporation in the uplands" states that the photo-synthetic energy consumption is negligible, and so is the soil heat G. In warm dry climates, however, measured values of G are usually almost equal to the vertical incoming radiation ΣR_n until noon, and then decline to less than ΣR_n as the day passes. The maximum value of G may be as much as 50 to 70% of ΣR_n for the daylight hours. It is usually assumed (see Figure 10.20) that as the incoming solar radiation is negligible during the hours of darkness, evaporation also does not occur at night. It is taken that the soil surface is the "surface of reaction" at which net solar energy is converted into L_e, the latent heat of evaporation (equation10.12). There is no consideration that water may be evaporating within the pores of an unsaturated soil.

In order to evaporate from a partly saturated soil, soil water must be drawn up towards the soil surface, and heat energy must be supplied to cause vaporisation. The most likely source of heat energy for evaporation of soil water is G and therefore, far from being a sink to which energy is lost, it appears that soil heat must be very important in the process of evaporation from a soil. The experiments, described in what follows, explore the relationship between soil heat and evaporation from the soil.

The measurements shown in Figure 10.40 were made on plastic pots, containing either soil or water, measuring 195 mm in diameter by 200 mm deep and holding either about 4 kg of moist soil or 3 kg of water. The pots could be weighed to the nearest 0.5 g to measure evaporation. Parallel tests with water and soil were made because the process of evaporation of free water is simpler than that of soil water and can be used to interpret the tests on soil. Two soils were used, one was a fine sand residual from weathered sandstone and containing 65% of sand of less than 1 mm size, 30% of silt sizes and 5% of clay. The second soil was a water purification sludge consisting of clay particles that had been flocculated with calcium hydroxide, precipitated out of raw water and thickened to a water content of 230% (a relative density of 1.24). At 230% water content, the sludge had the consistency of a soft gel-like clay. Evaporation tests were carried out on free water, sand with initial degrees of saturation of 100% and 24% (water contents of 21.8% and 9.1%) and sludge with initial water contents of 230% and 209% and a degree of saturation of 100% throughout both tests.

Figure 10.40a shows the increase in average temperature of the contents of the pots when exposed to the sun, measured at intervals from 07.00 (sunrise), to 17.00 (sunset) and on to 07.00 the next morning. The Figure shows that there was a considerable amount of heat stored in the water, sand and sludge at sunset (17.00) (86% of the maximum at 15.00 for water, 64% for the saturated sand, 77% for the moist sand and 90% for the sludge) and that this heat was gradually lost during the hours of darkness, between 17h00 and 07.00.

Figure 10.40b shows the rates of evaporation, with the two initially saturated soil specimens evaporating at maximum rates of just over 0.8 mm/h, water at just under 0.7 mm/h and the unsaturated sand, which reached the highest temperature, at the lowest maximum evaporation rate of 0.3 mm/h.

Figure 10.40c records the cumulative evaporation over the 24 h day and shows that evaporation continues through the night and that night-time evaporation can amount to 20% to 30% of day-time evaporation. As the only source of energy for the night-time evaporation can be the heat stored in the air, the water or the soil, and as the heat stored in the air is small compared with that stored in the water or soil, this shows that stored water heat or soil heat is the main source of energy for evaporation.

Figure 10.41a shows the results of the five tests plotted in equivalent energy terms. The horizontal axis represents the measured cumulative evaporation expressed as g of water and the vertical axis the corresponding water heat (W) or soil heat (G) expressed as potential evaporation in g of water, i.e. as W/λ or G/λ (λ being the latent heat for vaporisation of water).

Figures 10.40 and 10.41a are labeled “Thinly insulated”. The “thin” insulation around the sides and base of the pots consisted of 4 layers of plastic “bubble wrap”. Figure 10.41b shows the effect of using additional insulation consisting of 50 mm of dry wood shavings. With the thick insulation, heat losses from the sides and bases of the pots were reduced, and the maxima for W and G increased. However, actual cumulative evaporation was almost unchanged, which probably indicates that the water surface or a saturated soil surface is indeed the surface at which evaporation occurs. It has to be the reaction surface for water and saturated soil, but it remains

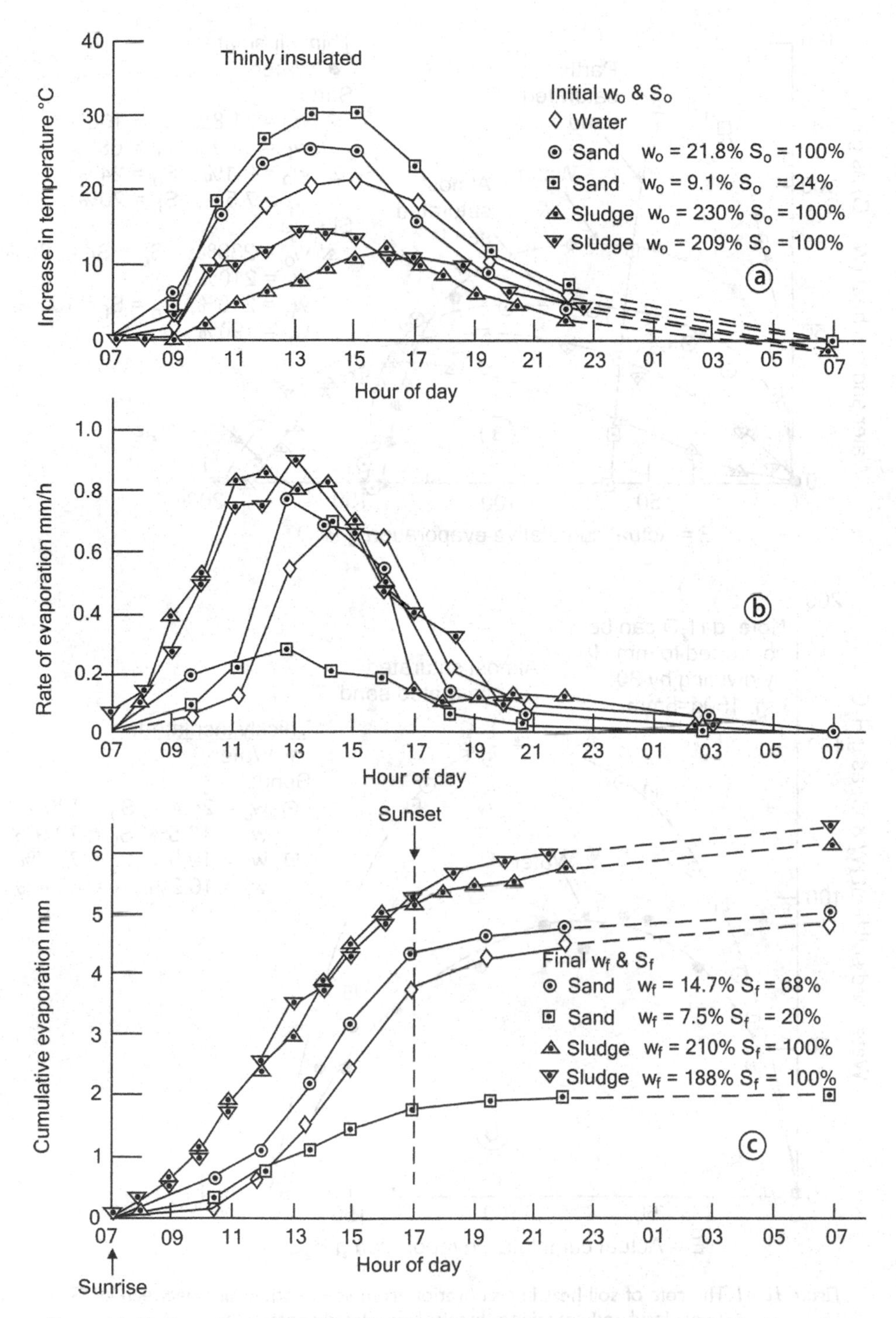

Figure 10.40 Experiments on the role of soil heat in evaporation from soil with time: a: increase in mean temperature, b: rate of evaporation, c: cumulative evaporation.

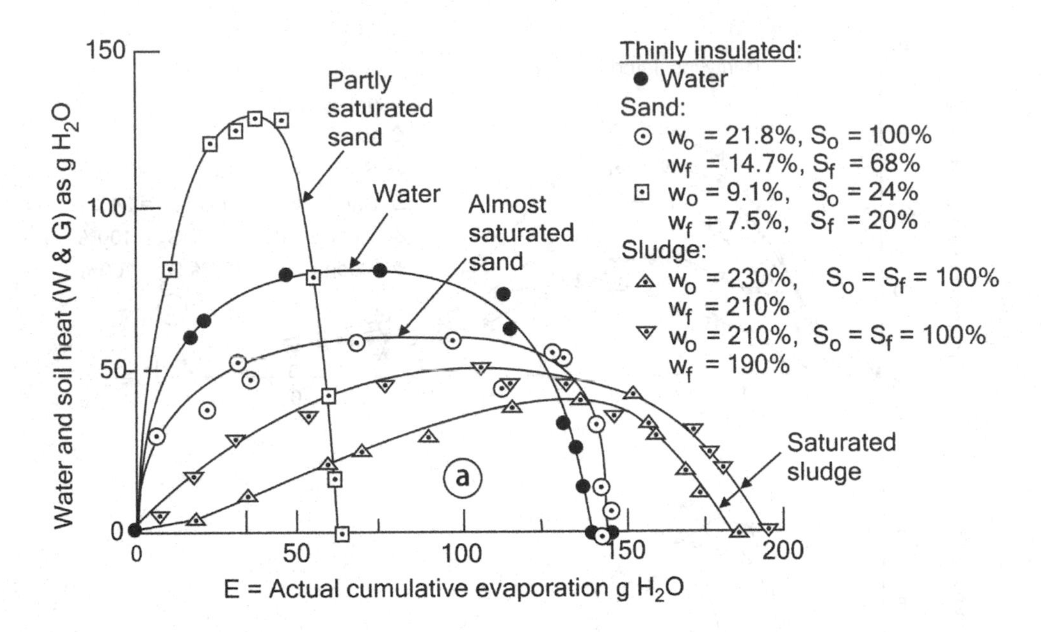

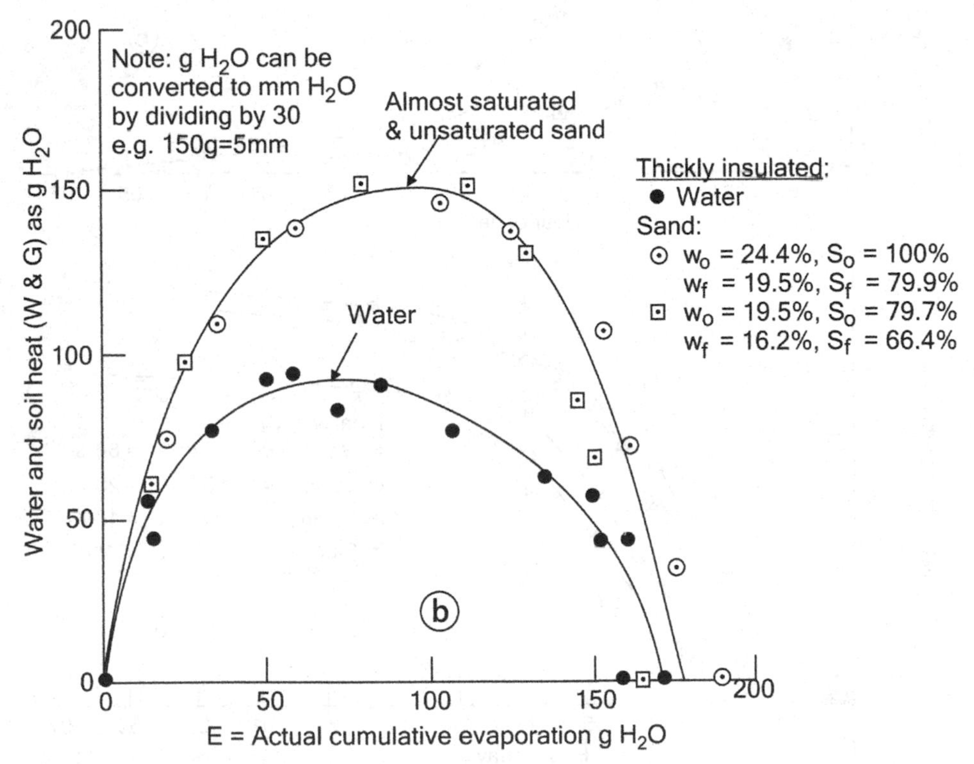

Figure 10.41 The role of soil heat in evaporation from soil-relationships between W, G and E: a: thinly insulated containers, b: thickly insulated containers.

likely that in an unsaturated soil, evaporation could occur within unsaturated pores, the water escaping as vapour from the soil surface. When the degree of saturation is fairly low (as with the partly saturated soil of Figure 10.41a), water probably moves to the surface as vapour rather than the liquid phase and this is why relatively little evaporation occurs for quite large values of G.

The positions and shape of the loops in Figure 10.41a and b give an indication of the efficiency of conversion of incoming solar energy into latent heat of evaporation. The horizontal E-axis represents 100% efficiency of conversion of W or G into E with no storage of heat, while the W and G axis represents zero conversion of incoming and stored energy into E, i.e., all storage and no conversion. Thus in Figure 10.41a, the most efficient conversion of G into E occurred with the saturated sludge and the least efficient conversion occurred in the partly saturated sand. The almost saturated sand was more efficient as an energy converter than the water, because it reflected less of the incoming solar energy and water was present at the surface to evaporate.

Increasing the effectiveness of the insulation (Figure 10.41b) appears to decrease the efficiency of conversion into evaporation, but this is more apparent than real, because the total evaporation increased slightly, and the increased storage of W and G was actually caused by a reduction in heat losses through the sides and bases of the containers. It is also interesting to see that improving the insulation appears to reduce the efficiency of energy conversion in the sand, as compared with the water.

10.18 Further points to consider

10.18.1 *General*

This section has dealt with the basic principles of assessing water balances in the design and operational phases as well as the closure and aftercare phases for hydraulic fill tailings storages. The water balances for the closure and aftercare phases for dry dumps are very similar to those corresponding to tailings storages. The examples are drawn from research on hydraulic fill ash dams, dry ash dumps and municipal solid waste landfills as well as metalliferous tailings storages. This demonstrates the generality of the principles – they can be applied to any mine or industrial waste deposit of any type.

It is abundantly clear from the practical examples of water balances for both operational hydraulic fill tailings storages and for a closed tailings storage or a dry dump that water balances cannot be calculated with reliable accuracy if the necessary waste properties are not known. A "generic" knowledge of these properties is insufficient, the properties must be measured for the specific waste and the specific method of operation of the storage. It must also be borne in mind that the properties of the waste being deposited will be variable from hour to hour, day to day, year to year and decade to decade. Measurements must not only be made once, when the waste storage is commissioned, but monitoring of properties and adjustment of the storage operation must be ongoing throughout its operating life.

The same basic water balance equation applies to all phases, but the relative magnitudes and importance of the various inputs and outputs, as well as the objective of establishing the water balance will change as the waste storage progresses through its life cycle.

Water balance is a generic concept that is equally applicable to wastes of any mineralogy or origin that have been produced by any process. For example the water balance for a gold tailings dam is assessed in exactly the same way as that for a power station ash dam or a municipal solid waste landfill.

10.18.2 *Rate of deposition of tailings (T):*

In the design phase, a rate of production is targeted that will result in a financially viable venture. Almost inevitably, if the first few years of operation are successful, the rate of production will be increased, and T will also increase. This will have three important consequences:

- more water (F) will resort to the tailings storage;
- the rate of rise of the tailings storage will increase; (see Fig.12.8 in Chapter 12);
- the deposition area will initially increase at a greater rate, and after the entire footprint has been covered by tailings, the area will decrease at a faster rate.

These consequences will, in turn, affect the overall stability of the slopes, the quantities of decant water D, the pool area and depth (Wp), evaporation losses E, and seepage losses into the foundation (R). All of these will need to be re-checked.

10.18.3 *Water contained in the tailings feed (F)*

In the early stages of commissioning and operation of a new plant, the slurry density ρ_s tends to be poorly controlled and erratic. Control of the deposition process at the tailings storage may also be poor and erratic. For example, if deposition is by means of cycloning, the designed particle size split between underflow and overflow may not be achieved, resulting in potentially unstable outer slopes and difficulty in gaining freeboard. It is therefore critical to measure incoming slurry density regularly at the storage, as well as the slurry density and particle size analysis of the cyclone underflow and overflow. It is only in this way that abnormal conditions can be detected and recorded and corrective measures taken.

10.18.4 *Interstitial water in tailings on the beach (I)*

Void ratios (e) on the beach tend to be erratic on the best managed tailings storages and must therefore be measured regularly to build up a good statistical knowledge for the storage. Ten or more undisturbed samples should be taken from the dry beaches (which are dry enough to be accessible on foot) at least six-monthly. This is a simple operation using a thin-walled sampling tube pushed in by hand. The sampling tubes should then be sealed against moisture loss (and therefore shrinkage) and taken to a soil mechanics laboratory where the void ratios and particle relative densities G of the samples are immediately measured and recorded.

10.18.5 *Precipitation on basin of storage (P)*

Rainfall is notoriously variable over a large area such as a tailings storage, and rain gauges are cheap to install and simple to read. Tailings storages with footprint areas of up to 100 ha should be equipped with 5 rain gauges, one at the penstock, where an

"A" evaporation pan (or A-pan) should also be mounted, plus 4 gauges spread around the perimeter of the deposition area. For storages larger than 100 ha, an additional rain gauge should be installed for each additional 100 ha, giving, e.g. 8 gauges for a 400 ha footprint.

10.18.6 *Decant and drainage water (D)*

Discharge points from underdrains must be inspected regularly to ensure that they are discharging water, and that the water is clear. Any cloudiness in the water may indicate that a piping erosion problem is developing. Any drain that is not discharging must be investigated to find and clear the cause of the blockage. Drain flows should be measured separately from decant flows, and both types of flow should be measured by means of a carefully calibrated measuring weir. (See Section 10.19.)

10.18.7 *Evaporation from the surface area of the storage (E)*

Ideally, evaporation from the surface of the storage should be measured by the solar energy balance method for each storage, but as this has to be done over a period of at least a year and requires specialized equipment, it is suggested that the equation $E = 0.4E_A$ (Blight, 2006) be used in water deficient climates, where E is the surface evaporation and E_A is the corresponding standard American A-pan evaporation. In water surplus climates, take $E = 0.7E_A$ (Penman, 1963).

Each tailings storage should be equipped with at least 2 standard A-pans, one at the penstock over the pool, and one on the perimeter of the storage. Both must be adjacent to a rain gauge so that water level changes in the evaporation pan caused by rain can be accounted for. An obvious point that is sometimes overlooked, is that each A-pan must be topped up regularly once a week to replenish the water lost by evaporation. Also, if the pan overflows as a result of rain, the water level must be lowered to the "full" level of the A-pan. In each case the date and time at which the water level was adjusted must be recorded, and the new water level must be measured.

10.18.8 *Water stored in pool (Wp)*

The water level in the pool should be observed and recorded daily, and especially after rain. If the volume of water in the pool is required, it can be estimated from the current contours of the basin profile. At the design stage, the basin profile can be obtained by predicting the beach profile. Once the tailings storage is operational, the contours can be measured periodically by a combination of aerial survey, soundings taken from the access catwalk to the penstock and/or from a boat, and distances measured by, e.g. a range-finder.

10.18.9 *Seepage losses into foundation (R)*

During the design and operational phases these can be estimated by the methods described in Section 6.10 and confirmed by means of the water balance calculation. To do this successfully for an unlined footprint, detailed measurements of the permeability of the footprint surface are required that are best obtained by means of double ring infiltrometer tests and should have been obtained during the site investigation and/or design phases. The original geotechnical site investigation will assist in pinpointing the areas of the footprint likely to have differing permeabilities and several sets of double ring infiltrometer tests should be made in each geotechnically distinct area.

10.18.10 *Irrigation and precipitation on storage surface (IR+P) during closure, rehabilitation and aftercare*

Precipitation (P) can be measured on a similar basis to that described above. The quantity of water delivered by an irrigation system for a particular pressure requires a different technique. A test plot should be marked out on the irrigated area. Convenient dimensions are 10 m × 10 m (100 m^2). Rain gauges should be mounted 1 m above the surface on short stakes and should be arranged in a regular pattern with a minimum of 5 gauges. (e.g. one gauge at the quarter-length point of each diagonal (4 gauges) plus one at the intersection of the diagonals). Rain gauges are used rather than measuring the flow through the feeder pipe or pipes because evaporation of water in the air can be significant, and so can drift of the spray caused by the wind. Once any vegetation on the test plot is wet, interception is negligible and can be ignored.

10.18.11 *Runoff from surface (RO)*

This can be measured in the way described in Section 10.9, by enclosing a test plot with collector trenches and bunds to exclude run-on.

10.18.12 *Storage of water in pores of waste (ST)*

As indicated in Section 10.3 and Figure 10.3, storage of water in the pores of the waste and any cover layers (interstitial water) is best measured by sampling. In fine grained uniform material such as tailings, a 50 mm diameter hand auger can be used to sample down to a depth of 2 m which is as deep as necessary. If the material is coarse grained (i.e. contains gravel and boulder sizes), test holes have to be dug, either by hand or using a small back-hoe.

10.19 Principles of the measuring weir

The accuracy of an operational water balance depends to a great degree on the accuracy with which the decant flow D and the drain flow d can be measured. In their journey from the storage to the return water reservoir, both flows usually pass along open channels for some distance. The ideal way of measuring the quantity of flow is to equip these channels with measuring weirs. The principle of a measuring weir is illustrated by Figure 10.42a, a section along the centre-line of a flow channel that is intercepted by a sharp-crested measuring weir. The discharge over the weir can be determined by integrating the velocity of flow over the depth of flow above the crest. The assumptions made are as follows:

- there is no vertical contraction over the crest (i.e. depth h remote from the weir is the depth of flow over the crest);
- The pressure is atmospheric right across the nappe (i.e. the flow does not cling to the downstream surface of the weir);
- there is no energy loss in passing over the weir (i.e. $V_0^2/2g + h = H =$ constant).

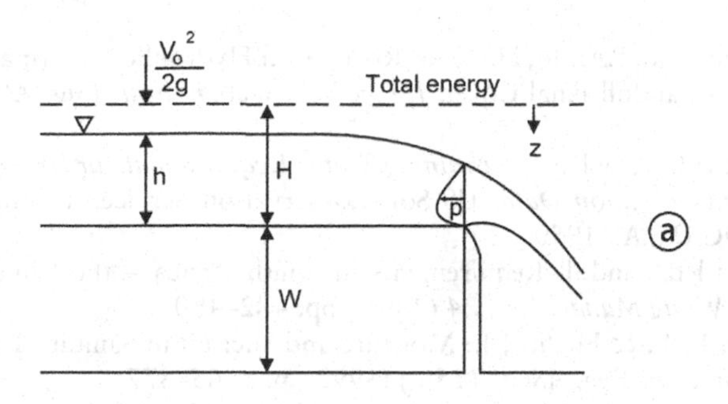

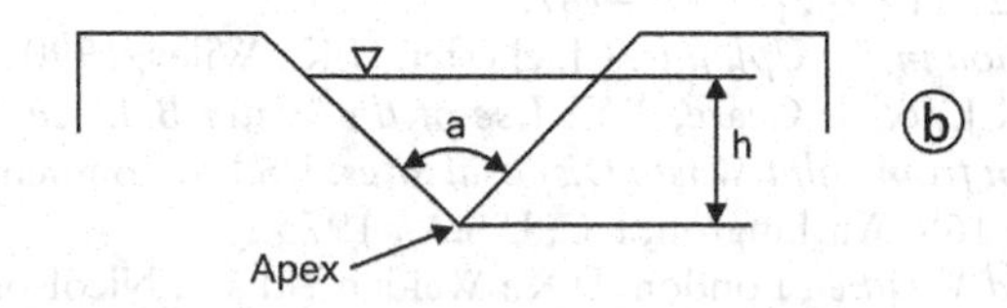

Figure 10.42 Measuring weirs: a: two-dimensional flow over sharp-crested weir, b: triangular or V-notch weir.

The triangular, or V-notch weir illustrated in Figure 10.42b is the most useful for measuring flow in an open channel, because it can measure accurately over a large range of values of h. For the V-notch weir, the discharge Q is given by

$$Q = 8/15 \cdot C_d \cdot \tan(\alpha/2) \cdot \sqrt{}(2g) \cdot h^{5/2} \qquad (10.21)$$

Usually, $\alpha = 90°$ and therefore $\tan\alpha/2 = 1$

C_d is a dimensionless empirical discharge factor, $C_d = 0.585$ and hence

$$Q = 1.382h^{5/2} \qquad (10.21a)$$

If h is in m and g in m/s², Q has units of $\sqrt{}[ms^{-2}]m^{5/2}$, i.e. Q is in m³/s.

The V-notch weir has been in use for centuries, and it is interesting to note that in Rankine's "Manual of Civil Engineering", published in 1862, the value of C_d is given as 0.595, less than 2% different from the currently accepted value.

The beauty of the method is that a single measurement of h, made remote from the weir, gives the discharge quantity. In practice, if the width of the approach channel is b, h should be measured at a distance of 2b upstream of the weir. It is important to ensure that the datum for measuring h is exactly equal to the level of the apex of the notch.

References

Benson, C.H., Thorstad, P.A., Jo, H.-Y., & Rock, S.A.: Hydraulic Performance of Geosynthetic Clay Liners in a Landfill Final Cover. *J. Geotech. Geoenviron. Eng. ASCE, 133*(7) (2007), pp. 814–827.

Blaney, H.F., & Criddle, W.D.: *Determining Water Requirements in Irrigated Areas from Climatological and Irrigation Data.* US Soil Conservation Service. Technical Publication 96. Washington, DC U.S.A., 1950.

Blight, G.E.: Graded Landfill Requirements in South Africa – the Climatic Water Balance Classification. *Waste Manag. Res. 24* (2006), pp. 482–490.

Blight, G.E., Ball, J.M., & Blight, J.J.: Moisture and Suction in Sanitary Landfills in Semi-arid Areas. *J. Geoenviron. Eng. ASCE 118*(6) (1992), pp. 865–877.

Blight, G.E., & Kreuiter, A.: Disposal of Industrial Waste Liquids by Evaporation and Capillary Storage in Waste Dumps. In: *Tailings and Mine Waste '00.* A.A. Balkema: Rotterdam, Netherlands, 2000, pp. 141–148.

Bowen, I.S.: The Ratio of Heat Losses by Conduction and by Evaporation from Any Water Surface. *Phys. Rev. 27* (1926), pp. 779–787.

Calder, I.R.: *Evaporation in the Uplands.* Chichester, U.K.: Wiley, 1990.

Fenn, D.G., Hanley, K.J., & de Geare, T.Y.: *Use of the Water Balance Method for Predicting Leachate Generation from Solid Waste Disposal Sites.* US Environmental Protection Agency. Report EPA/530/SW168. Washington, DC, U.S.A., 1975.

Flohn, H.: *Climate and Weather.* London, U.K.: Weidenfeld and Nicolson, 1969.

Heerten, G., & Reuter, E.: Is the European Regulatory Capping Design of Landfills Equivalent to Geosynthetic Equivalents? In:13th *Danube Conf. Geotech. Eng.* Ljubljana, Rotterdam, Netherlands: A.A. Balkema, 2006, p.8.

Hillel, D.: *Applications of Soil Physics.* New York, U.S.A.: Academic Press., 1980.

Meer, S.R., & Benson, C.H.: Hydraulic Conductivity of Geosynthetic Clay Liners Exhumed from Landfill Final Covers. *J. Geotech. Geoenviron. Eng., ASCE 133*(5) (2007), pp. 550–563.

Melchior, S.: Field Studies and Excavations of Geosynthetic Clay Barriers in Landfill Covers. In: H. Zantziger, R. Koerner, E. Gartung (eds): *Clay Geosynthetic Barriers.* Lisse, Netherlands: Swets and Zeitlinger, 2002, pp. 321–330.

Ntshabele, K., Cooks, M., Busani, B., & Dode, J.: The Effects of Water Quality on Dewatering Properties of Debswana kimberlitic Ores. In: A.B. Fourie, R.J. Jewell, A. Paterson, P. Slatter, (eds): *Paste 2008.* Perth, Australia: Australian Centre for Geomechanics, 2008, pp. 113–123.

Penman, H.L.: *Vegetation and Hydrology.* Commonwealth Agricultural Bureaux Technical Communication No. 53. Harpenden, U.K., 1963.

South African Department of Water Affairs and Forestry. *Minimum Requirements for Waste Disposal by Landfill.* The Department, Pretoria, South Africa, 1998.

Thornthwaite, C.W.: A Re-Examination of the Concept and Measurement of Potential Transpiration. In: J.R. Mather (ed.): *The Measurement of Potential Evapo-Transpiration.* Publications in Climatology. New Jersey, U.S.A.: Seabrook, 1954, pp. 200–209.

Turc, L.: Le bilan d'eau des sols. Relation entre les prècipitations, l'èvaporation et l'ècoulement. *Ann. Agron. 6* (1955), pp. 5–131.

Chapter 11

Failures of Mine Waste Storages

11.1 Failures: causes, consequences, characteristics

Mine waste storages are very large structures, easily visible from space, that have very long operating lives and have often not been properly planned or carefully operated. They are under construction for the whole of their operating lives and are operated by a succession of people, not all of whom are dedicated to carrying out their assigned tasks to the best of their abilities, not all of whom are properly trained and not all of whom understand why they have to undertake certain tasks and what the consequences of negligence may be. Not all of the workers can recognize that a dangerous situation may be developing, and not all of them know the correct course of action to be taken in an emergency. The foregoing statement may not be flattering, but it is a reality.

Most important of all, the operation of a mine's waste storage is a net cost to the mine, a deduction from the bottom line and a deduction that may make the difference between a profitable and a loss-making operation. A mine is operated for the benefit of its share-holders and there are few mine managers who, to safeguard their livelihood, will not view the waste storage operation with an ungenerous eye and attitude.

At the same time, a failure of the waste storage operation may negatively affect, severely impede or cause a complete temporary cessation of mining operations and consequently cause a reduction, or completely wipe out the all-important profitable bottom line.

Then there are natural hazards, severe rain storms, earthquakes, undetected adverse geological or ground water conditions, human errors such as well-intentioned, but faulty design, theft of vital components, warning systems that fail at the crucial time, and finally, "Acts of God".

It is no wonder that the list of mine waste failures is a long and well-populated one, and that mine waste storages are regarded as dangerous, and they can be dangerous, if they are not respected and treated as such. Above all, it should be remembered that:

"Todays act of God is yesterday's criminal act of negligence"

(United Nations Department of Humanitarian Affairs, Geneva, 1996).

Failures of tailings and coarse waste storages can take many forms, the most dangerous and destructive of which are those in which the waste loses strength, becomes mobile and flows as a viscous fluid in which the supporting fluid can either be air or more commonly, water. Table 11.1 lists 19 major flow failures that have occurred

Table 11.1 19 Flow failures of mine waste dumps, tailings storages and industrial waste dumps that have resulted in deaths, major environmental damage, or major damage to structures and infrastructure.

Year and number	*Location*	*Waste stored (height where known)*	*Cause of failure*	*Volume of flow*	*Consequences*
1928 (1)	Barahona, Chile	copper tailings	8.2 Richter* earthquake	3×10^6 m^3 fine tailings	environmental devastation
1961 (2)	Jupille, Belgium	fly ash	removal of toe support of dump	100–150 $\times 10^3$ m^3 fly ash	11 deaths, houses destroyed
1965 (3)	El Cobre (2 impoundments)	copper tailings	7.5 Richter earthquake	(1) 1.9×10^6 m^3, (2) 0.5×10^6 m^3 fine tailings	300 deaths, village buried in tailings
1966 (4)	Aberfan, UK	coal waste	dumping of waste over spring	108×10^3 m^3 waste	144 deaths, 116 children, extensive damage to property
1970 (5)	Mufulira, Zambia	copper tailings	collapse of tailings dam into workings		89 miners killed underground
1972 (6)	Buffalo Creek, USA	coal waste (15 m)	overtopping of waste impoundment	500×10^3 m^3 water + waste	118 deaths, 4000 homeless, US$50 million damage
1974 (7)	Bafokeng, South Africa	platinum tailings (20 m)	overtopping of tailings dam	3×10^6 m^3 fine tailings	13 deaths, extensive damage to mine installation and environment
1978 (8)	Mochikoshi, Japan	gold tailings	7.0 Mercalli** earthquake	80×10^3 m^3 fine tailings	environmental devastation
1985 (9)	Stava, Italy	fluorite tailings (25 and 30 m)	shear failure of retaining dyke	190×10^3 m^3 fine tailings	268 deaths, extensive damage to property and environment
1985 (10)	Quintette Marmot, BC, Canada	coal waste	pore pressure resulting from collapse settlement	2.5×10^6 m^3	environmental damage – river valley filled with waste for 2.5 km

1993 (11)	Saaiplaas, South Africa (3 failures in 3 days)	gold tailings (28 m)	high phreatic surface in ring dyke	140×10^3 m^3 (slides 1 and 2) 140×10^3 m^3 (slide 3)	minimal environmental damage. Not reported by news media
1994 (12)	Merriespruit, South Africa	gold tailings (31 m)	overtopping of tailings dam	600×10^3 m^3 fine tailings	17 deaths, extensive damage to housing and environment
1995 (13)	Omai, Guyana	gold tailings	piping erosion of retaining dyke	4.2×10^6 m^3 slurry	80 km of river devastated
1995 (14)	Surigao del Norte, Philippines	gold	dyke failure	50×10^3 m^3	12 deaths, environmental devastation
1996 (15)	Sgurigrad, Bulgaria	lead, zinc, copper	overtopping of retaining dyke	220×10^3 m^3	107 deaths, environmental devastation
1998 (16)	Los Frailes, Spain	lead, zinc, copper (25 m)	foundation failure of tailings dam	4×10^6 m^3 slurry	environmental devastation
1999 (17)	Surigao del Norte, Philippines	gold	tailings slurry escaping from burst pipe	700×10^3 m^3	17 houses destroyed, agricultural land devastated
2000 (18)	Inez, Kentucky, USA	coal wastes	tailings dam failure from collapse of underground workings	950×10^3 m^3	120 km of rivers devastated by slurry
2000 (19)	Star Diamonds, South Africa	diamond	collapse of under-ground workings resulting in flow of tailings into workings	100×10^3 m^3	4 deaths of miners underground

Note: Entries have been selected, list is not comprehensive. Table was drawn from a number of sources, most of which appear in the reference list.

* Richter magnitude $M = \log_{10} A$, A = Maximum amplitude (microns) recorded by standard seismometer at 100 km distant from epicentre.

** Mercalli magnitude 7.0: Damage negligible in well designed and constructed buildings, moderate in well-built structures, considerable in poorly built structures.

Table 11.2 Damage to Japanese tailings storages more than 15 m in height

Number of the damaged dams.									
Location of dam	*Sapporo*	*Sendai*	*Tokyo*	*Nagoya*	*Osaka*	*Hiroshima*	*Shikoku*	*Fukuoka*	*Total*
Up to 1950	5	24	3	3	5	8	0	19	67
1951–1955	3	9	3	1	5	5	1	17	44
1956–1960	4	25	8	8	9	0	3	2	59
1961–1965	13	5	6	0	6	6	4	2	42
1966–1972	4	7	5	1	6	5	1	3	32
Total	29	70	25	13	31	24	9	43	244

Causes of damage:
Violent rainfall 59% Earthquake 3%
Prolonged rainfall 3% Snow or ice 2%
Flood 2% Landslide 1%
Erosion 1% Unknown 29% 100%

between 1928 and 2000 and together have caused at least 1080 deaths of which 1065 occurred between 1965 and 1996, an average of over 34 deaths per year. Deaths from failures of other forms of waste storage do not lag behind. Deaths caused by only three failures of municipal solid waste dumps between 1993 and 2005 alone totaled 464, an average of more than 38 deaths per year (Blight, 2008). However, to keep these numbers in perspective it must be remembered that they are miniscule in comparison with the yearly death toll on the roads of the world (see Section 2.7) and with other preventable causes of death, such as AIDS.

It is also interesting to note that many of the failures listed in Table 11.1 occurred in small waste storages at operations that were being run on a "shoe-string", or at mines threatened by closure. The failures at Mochikoshi, Stava, Saaiplaas and Star Diamonds were of this ilk. Also, by no means all mine waste "incidents or accidents" are reported in the widely read or viewed news media nor do all lead to disasters. For example, Table 11.2 shows incidences of damaged tailings storages greater than 15m high reported in Japan up to 1950 and from 1950 to 1972 (Muramoto, et al., 1986). None of these seem to have been major failures, but the number involved, in a country that is not usually regarded as a major mining centre, is startling.

A more optimistic, probably more realistic way of viewing the data in Table 11.2 is that tailings dams, are subject to many mishaps that cause damage. However, these seldom cause major damage or set off chains of events that lead to major failures. In many cases, emergency repairs may be carried out that save the storage from complete failure. In other cases, it is possible to cordon off the failure, temporarily cease deposition in the affected area, and operate on the remainder of the deposition area while the failure is repaired. (See Section 11.9). In the case of small waste storages, this may not be possible as the failure may have displaced almost all of the waste (e.g. at Stava where 65 per cent of the deposited tailings were displaced). In cases where the failure has resulted in multiple deaths, public opinion may prevent the reinstatement and re-use of the facility. After the flow failure at Merriespruit in 1994 that killed 17 residents in the nearby Merriespruit village, extensive discussions and consultations took place between the mine owners, the local population and government agencies.

The residents agreed to having the breach in the storage repaired and the storage recommissioned, on condition that a monitoring committee with representatives of the community, government agencies, the mine manager and the consulting engineers was set up and carried out an inspection every three months (Brink, 1998). In the case of the Aberfan disaster that killed 144, the waste dump was removed from the hillside above the Aberfan village, and a memorial garden was created in the area where the flow side had demolished the village primary school, killing 116 children.

The object of this chapter is to make the reader familiar with the characteristics of failures in both hydraulic fill storages of tailings and in dumps of waste material. This will be done by describing a number of case histories of failures, including what is known of the chain of events that lead up to the failure, both in terms of natural events, human failings, and geotechnical processes. This will include an examination of the characteristics of the seldom described post-failure profiles and the bearing that these characteristics have on improving design and preventing future failures. Finally, the analysis of potential failures by slope stability theory will be considered.

11.2 Failures of hydraulic fill tailings storages caused by seismic events

Common features of failures in tailings dams that have failed by seismic action are (Blight et al., 2000):

- the presence of a large pond in the impoundment that may have encroached on the outer impoundment dyke;
- an outer dyke formed of loose, poorly compacted or uncompacted tailings sand that is contractive and potentially liquefiable when subjected to shear stress (See Sections 4.5.8 to 4.5.10 and 5.4);
- poor separation of the sand used to build the retaining dyke from the silts stored within the impoundment, possibly with weak lenses of silt included in the dyke; and
- dykes usually built (at least partially) by upstream deposition over the impounded silts.

When an earthquake of sufficient magnitude occurs, a failure typically develops as follows:

1. the shear strains and the corresponding shear stresses imposed by the earthquake cause the weaker, fine, possibly partly consolidated tailings in the basin of the impoundment to strain-soften (e.g., Figure 4.40). If the shear strength falls to a low enough value,
2. liquefied tailings and ponded water will be agitated by the earthquake accelerations into forming waves, alternately drawing down and overtopping the upstream slope and crest of the confining dyke;
3. the upstream slope of the dyke may slide into the impoundment when drawn down and the dyke may crack;

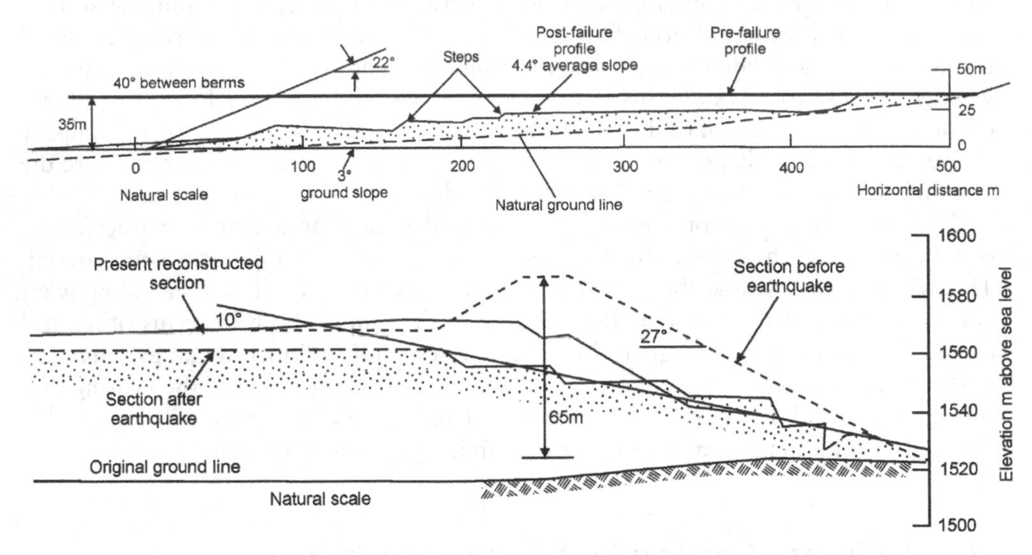

Figure 11.1 a: Pre-and post-failure longitudinal profiles of El Cobre old dam. b: Pre and post-failure sections of Barahona dam.

4. when the wave of water and liquid tailings returns, it may overtop the failed section of the dyke, eroding it and forming a beach, while water and liquid tailings may flow into and through cracks in the dyke, eroding and enlarging them.
5. the downstream slope of the dyke may fail in shear, as a result of strain-softening accompanied by erosion;
6. as the breach in the dyke rapidly enlarges, the contents of the impoundment flow out of the breach starting the tailings flood, which is sustained by retrogressive liquefaction of the tailings within the impoundment;
7. the failure process and flow of tailings cease once the shear strains imposed by the earthquake diminish and a stable surface profile is developed by the breached dyke and the tailings flood that has escaped from the impoundment.

11.2.1 *The El Cobre flow failure*

The El Cobre (Antiguo) failure (Table 11.1(3)) is a good example of a flow failure caused by an earthquake (Dobry and Alvares, 1967). Figure 11.1 shows cross-sections through the side-hill storage before and after failure. The storage was commissioned in 1930, but after the Nuevo (new) dam (Table 11.1(3) was constructed in 1963, the Antiguo (old) dam was used only periodically as a standby. The dyke had been built by upstream hydraulic filling, and the downstream slope of the dyke was 35 m high at the time of the failure. The epicenter of the 7.5 Richter magnitude La Ligua earthquake that resulted in the failure was 70 km from the dam with a focal point at a depth of 61 km.

During the quake a huge cloud of dust arose from the dried surface of the, only periodically used, impoundment. The flow failure continued for 20 minutes after the quake had ended, as 1.9×10^6 m^3 of a total storage of 4.25×10^6 m^3 of tailings flowed down a dry valley for a distance of 12 km. A small town in the path of the flow was annihilated with 300 deaths occurring.

As shown by Figure 11.1, the storage was constructed on ground with a slope angle of 3° and the average slope of the post-failure profile through the breach was only 4.4°. The flow was reported to have covered its 12 km course in a few minutes. This is too imprecise to allow the speed of the flow to be estimated, but it must have been about 20 km/h (see Section 11.8).

It should be noted that after a failure, the flow of liquefied tailings from the impoundment will continue until a surface profile compatible with the reduced strength of the tailings has developed. Once this stable surface has formed, loss of tailings from the impoundment will cease. In the case of El Cobre (Antiguo) the average stable slope was 4.4° under static conditions because the quaking had stopped. It is possible that aftershocks could have resulted in further flattening of the profile, and further loss of tailings.

The presence of occasional steps in the post failure profile (Figure 11.2) should be noted. They mark points at which the post-failure surface cut through layers of denser tailings that had probably been densified by desiccation during periodic dry spells or intermissions in the tailings deposition.

11.2.2 *The Barahona failure*

The Barahona dyke (Table 11.1 (1)) is an example of the failure of an upstream dam that failed and distorted severely, but did not flow. (Dobry and Alvares, 1967). Figure 11.1b shows sections of the dyke before and after the 8.2 Richter magnitude earthquake that caused it to fail. Originally constructed with a slope of 27°, the distorted post-failure slope flattened after the 1928 earthquake to an average slope of 10°. The profile was reconstructed in 1929 as shown in Figure 11.1b.

In 1991 (63 years later) Troncoso et al (1993) carried out cone penetration tests in the tailings below the level of the disturbance caused by the failure. After all this time, the minimum strength from the cone resistance (q_c), using $N_c = 20$ (see Section 5.3) was only about 50 kPa while the minimum sleeve resistance was 20kPa. (The corresponding maxima were 300kPa and 100kPa respectively.) This illustrates the extreme slowness of the consolidation of large depths of fine tailings.

11.3 Flow failures caused by overtopping

It is relatively seldom that the management decisions and mistakes that lead to a failure come publicly known. In the cases of two flow failures of tailings storages in South Africa, caused by overtopping, a lot of detail of the decisions and actions that eventually resulted in the failures are on record because they became known when mine officials and contractors who operated the storages were cross-examined at the judicial inquests that followed the failures. The two failures took place at Bafokeng in 1974 and Merriespruit in 1994 and both will be described here. However, it must be emphasised that in both cases overtopping represented the trigger action that set the failure in motion. In both cases the state of stability of the affected storages was marginal, and the root cause of both failures was poor management, resulting from ignorance of the principles of soil mechanics, poor training of staff, negligence, and in the second case, a measure of dishonesty.

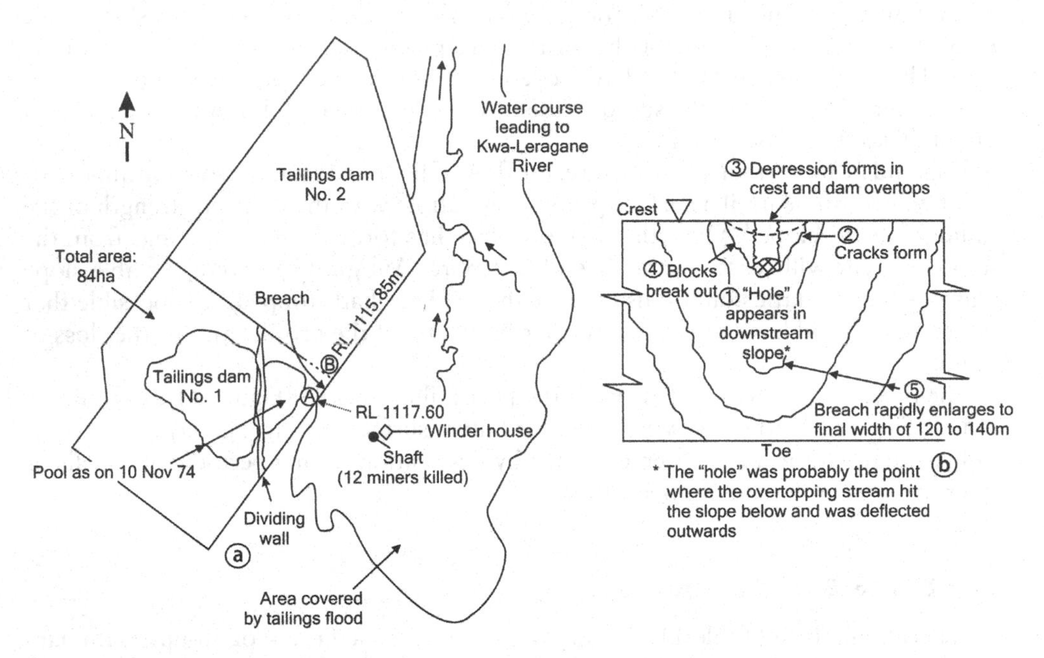

Figure 11.2 a: Plan of Bafokeng tailings dams showing position of breach, course of flow failure and extent of pools prior to failure. b: Stages of failure of Bafokeng dam wall.

Other note-worthy failures that were caused by overtopping include those at Buffalo Creek, USA (6 in Table 11.1) and Sgurigrad, Bulgaria (15). The worst failure of all, in terms of death-toll, that at Stava, Italy, is listed as a shear failure, but the author believes that all the available information points to the triggering action being an overtopping during the spring thaw.

11.3.1 *The tailings storage flow failure at Bafokeng, 1974 (Table 11.1 (7))*

On the morning of November 11, 1974, the southwestern wall of the No. 1 tailings dam at Bafokeng Platinum Mine, near the town of Rustenburg, South Africa, failed with disastrous results. Before the failure, the storage (which was 85 ha in area and 20 m high) contained about 17×10^6 m^3 of tailings. Approximately 3×10^6 m^3 of liquefied tailings slurry flowed through the breach in the wall, engulfed a vertical shaft of the mine, and flowed into and on down the valley of the Kwa-Leragane River, a seasonal stream which was flowing at the time.

The slurry demolished or damaged many surface structures at the shaft, and the flood also carried away with it vehicles and items of equipment waiting to be taken underground. A large quantity of slurry flowed down the shaft, trapping some workers underground and tearing loose certain shaft equipment. Twelve died by drowning underground in the disaster. Miraculously, although several people were caught in the surface flood of tailings, none was drowned and no one was killed on surface.

Events leading to failure at Bafokeng

The events leading to the failure of the No. 1 tailings dam at Bafokeng have been well reported by Midgley (1979) as follows:

During the early morning of Monday, November 11, 1974, heavy rain fell in the Rustenburg area. Rain gauges in the vicinity differed widely, but it was estimated that at the Bafokeng Mine 77 mm of rain fell between 02h00 and 04h00 on that day. At the time of the accident, as may be seen from Figure 11.2a, the impoundment of Tailings Dam No. 1 was divided into two portions by a diagonal wall extending from the southeastern to the northeastern side. The area of the storage behind the dividing wall was about 73 ha and that of the triangular downstream portion about 11 ha. The peripheries of the ponds on top of the storage, as on the day before the failure, are indicated in Figure 11.2a. The approximate water surface areas in the upper and lower compartments were 20 ha and 4.5 ha, respectively.

Adjacent to dam No.1 was dam No.2, its crest at a substantially lower elevation. Workers had cut a track (at B on Figure 11.2) through the dividing wall between dams 1 and 2, near the eastern corner, to bring a front-end loader onto dam No. 1, reportedly for the purpose of pushing up barriers of dry tailings to drive the water back from the crest of the dam near A.

At about 08h45, workers noticed water overtopping the diagonal dividing wall, which soon breached, releasing water from the upper pool to the smaller lower one in which the water was judged to have been standing about 2m lower. There was a sudden rise of level in the lower pool and water started to flow through the cutting B from dam No. 1 into dam No. 2. Flow was not so strong, however, as to cause substantial erosional down-cutting. The elevation of the crest of the overflow (B) section was subsequently measured as RL 1115.85 m compared with RL 1117.60 m at point A (the lowest part of the outer crest of what remained of tailings dam No. 1).

From these elevations, Midgley concluded that the failure could not have been initiated by overtopping.

The process of failure appears to have been as follows:

- At about 10h15, water appeared on the surface of the southeastern dyke of the storage, appearing to onlookers to be a jet of water emerging about two thirds of the way up the slope and falling on the slope below.
- The hole or breach from which water was emerging quickly grew and the jet became a large stream pouring down the slope.
- Above the stream of water two or more cracks developed in the slope, extending upward at fairly steep angles toward the crest, to form a wedge with its apex down.
- Blocks of material collapsed out of this wedge, which fell as solid masses into the now rapidly increasing flood of water, tailings slurry and blocks of tailings.
- The resulting gulley widened rapidly to about 130 m as three million cubic metres of liquefied tailings poured out. After spreading over the fairly level ground surface adjacent to the breach and destroying the shaft and its winder house, the escaping slurry found its way into a seasonally dry watercourse (which was flowing as a result of the early morning rain-storm) and thence into the Kwa-Leragane River.

These stages of the failure are illustrated in Figure 11.2b, an elevation of the wall of the dam at the breach.

Plate 11.1 Aerial view of flow failure at Bafokeng showing breach and failure scar.

At a distance of 4 km from the breach in the dam, the flood of slurry had spread to a width of 0.8 km and was 10 m deep. The flood continued down the Kwa-Leragane River into the Elands River, and an estimated 2×10^6 m^3 of tailings eventually flowed into the reservoir of the Vaalkop (water storage) dam 45 km downstream of Bafokeng.

Plate 11.1 is an aerial view of the Bafokeng failure, with the breach in the right foreground and the tailings flow going towards the lower right corner of the plate. Plate 11.2 shows the breach at centre-left, with the head-gear of the shaft, down which 13 miners were killed, in the centre of the picture. The stream of escaped tailings can be seen winding its way into the distance. The mine hostel and hospital in the foreground were, very fortunately, up-slope from the breach and escaped damage. Plate 11.3 shows the damage caused to the housing of the shaft winding engine by the tailings flood.

The platinum tailings at Bafokeng are coarse (see Figure 1.3) and have a silt and clay tail. The dam was built by subaerial spigotting from the perimeter toward the center of the storage. Hence, the dyke would have been loose, though consolidated, but the impounded fines probably were not completely consolidated. It also appears that the mobility of the tailings that left the breach was very dependent on the condition of the ground surface over which the material was moving. If the ground surface was already covered by a sheet of flowing storm-water, the viscosity of the tailings at the interface between the tailings and the ground surface would have been much reduced, and this would have facilitated the flow of the tailings. This is probably why the tailings

Plate 11.2 Aerial view of Bafokeng failure showing course taken by fugitive tailings. Hospital and workers' hostel in foreground that escaped the tailings flow. Shaft headgear in centre of view.

Plate 11.3 Damage to shaft winder house caused by tailings flow.

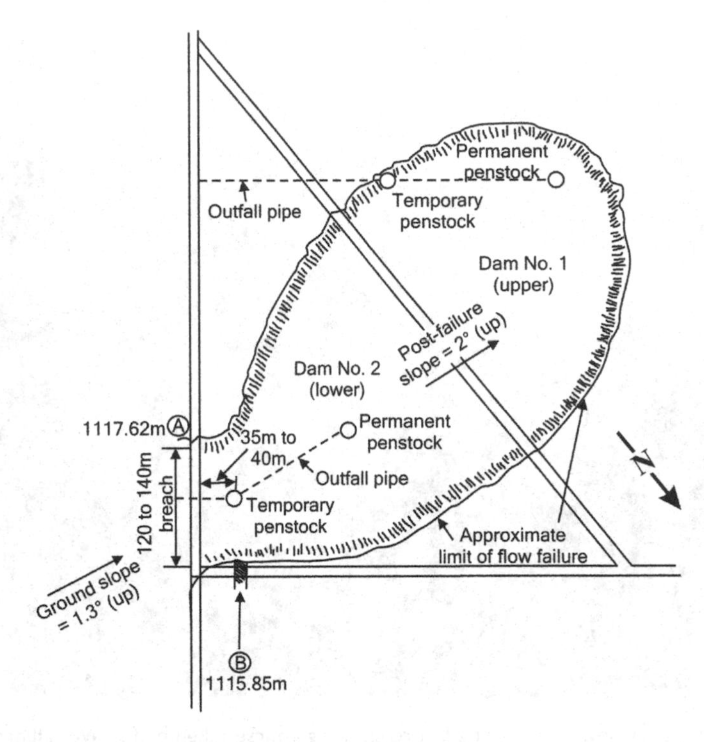

Figure 11.3 Plan showing positions of penstocks and limits of flow failure in Bafokeng dam No.1.

from Bafokeng diluted by water, both on the ground surface and in the river channel, flowed so far down the Kwa-Leragane-Elands river valleys.

Possible causes of failure at Bafokeng

It appears that at first sight that the dyke did not fail by conventional overtopping. Eyewitness accounts all point to a failure by piping erosion. However, a satisfactory explanation of how the initial hole formed in the wall was never reached. The following quotes from Jennings (1979) point to what are believed to be serious possibilities:

- "The slimes dam foreman reported that he was on the dam when the dividing wall between the upper and lower ponds failed but that nevertheless he continued for some time thereafter to work with his front-end loader in the area of the crest of the dam where the breach subsequently occurred."
- "....The foreman on the dam was not in a position to see water jetting from the wall. To him the failure would have appeared to be due to a deepening of a gulley which appeared in a place where he had been trying to push the water back with his front-end loader."

If the foreman was "trying to push the water back," it would appear that the crest of the dam must have been in imminent danger of overtopping, even though the quoted crest levels indicate a freeboard of 1.75 m or so. A vibrating machine, even a light, wheeled front-end loader, working on a loose saturated sand could possibly cause the sand to liquefy or settle at least locally. The gulley reported by the foreman and later the hole that eye-witnesses reported in the wall could have been one and the same and have resulted from the liquefaction.

It is very likely that a hole did not form, but that what the eye-witnesses actually saw was the nappe of water shooting out of the erosion gulley and falling onto the slope below. This was reported in a later failure at Merriespruit (Wagener, et al., 1977). Settlement and surface erosion would have resulted in a deepening depression forming in the crest, followed by large-scale overtopping and breaching of the wall. Hence the foremen's desperate attempt to save the dyke may have caused its destruction.

Only one of the papers that describes the failure even mentions the outlet penstocks. Midgley (1979) referred to attempts to lower the water level on the dam by lowering the crests of the decant penstocks, but gave no further details. Aerial photographs of the failure scar show no signs of any penstock shafts or towers. However, Figure 11.3 (taken from an unpublished report on the failure) shows that the storage was equipped with four penstock towers and two outfall pipes, all of which disappeared during the course of the failure. The temporary penstock in the triangular portion of the storage was less than 40 m from the crest of the dyke at the breach. If this penstock was operational and if, in their hurry to lower the water level, too high a water head was allowed over the penstock crest, the penstock could have started to flow full. The shaft of the tower was built of stacked, precast concrete rings. It has long been known (e.g., US Bureau of Reclamation 1960) that if the head of water over the crest of a shaft spillway exceeds 0.45 times the shaft radius, the shaft will tend to flow full and the flow may surge. The resulting pressure fluctuations have been known to dislodge rings and destroy similar penstock shafts. The vibrations from the surging could also have been the primary cause of liquefaction of the saturated sand, which resulted in the breach. Alternatively, if the penstock rings were dislodged, a cavity would have been eroded around the penstock shaft, and that cavity could have rapidly enlarged until it emerged through the downstream slope, causing the breach.

In summary, the dyke must have failed by overtopping that was triggered by a form of erosion failure caused by either or both

- localized liquefaction caused by vibrations set up by the front-end loader, or
- dislodgement of rings from the temporary penstock shaft caused by surging flow and resulting in the formation of a cavity.

Both of these possible causes could have resulted from over-zealous and uninformed action by the contractor's staff in their attempts to save the dyke.

Operational and management failings at Bafokeng

The inquest into the deaths of the 13 miners found that the failure was an Act of God and therefore that no one was to blame for the disaster. In almost every case of failure of an engineering structure, some blame attaches to the designers, constructors, supervisors, or operators of the structure. In the case of Bafokeng, at a distance in time of more than 35 years, the failings that resulted in the disaster are not very clear:

- One of the basic rules of tailings storage operation is that the pool should be kept as small as possible and be located as far as possible from the outer walls. Figure 11.2 shows that the pool was dangerously large on the afternoon prior to the failure, and the lower pool was only about 30m from the wall that breached. There was evidence that the mine had a surplus of water at the time and that the tailings storage was deliberately being used to store the surplus.
- The storage was not equipped with piezometers, so the position of the phreatic surface was not known.

For whatever reason, some person or persons were responsible for the decisions to hold an excessive amount of water in the storage and not to install piezometers.

- It also appears that the operation of the storage was not subjected to regular examinations and supervision by a knowledgeable geotechnical engineer.

There was thus a lack of concern for and caution about the stability of the storage. Some blame for this should have been attached to the management of the mine.

To summarize the management failures,

- There was complacency about the state of safety of the storage.
- There was a lack of knowledge and understanding of how tailings storages function and what affects their safety. The decision to store large volumes of water on top of the storage probably arose from this ignorance.
- There were no regular inspections or assessments of the storage's state of safety. In this regard, the lack of piezometers was a major shortcoming.
- There was obviously no emergency action plan that the operators of the storage were aware of. They were poorly trained and nonplussed when the emergency confronted them.

The aftermath of the Bafokeng failure

Engineering has a history of advancing via a series of disasters, and tailings dam engineering is no exception. For example, mine tailings disposal was largely unregulated in Britain until the Aberfan disaster occurred (Anonymous, 1967). Prior to the failure of the Bafokeng storage, there was some rudimentary guidance available on tailings storage construction in South Africa. This was based on research carried out under the direction of J.E. Jennings and issued as an internal document by the Chamber of Mines in 1959. After the Bafokeng disaster, the Chamber of Mines commissioned the preparation by the present author of a comprehensive technical guideline to tailings storage design and operation (1979). This was voluntarily adopted by the whole gold mining industry. It was revised in 1983 and was again in the process of revision in 1993 and 1994 when the failures at Saaiplaas and Merriespruit occurred. During the inquest on the Merriespruit failure, it emerged that the responsible staff on the mine were quite unaware of the existence of this publication, as were the site staff of the contractor who was operating the tailings storage. This made it quite clear that the mining industry in South Africa was not able to regulate itself and, in the interest of safety, regulation and coercion by the State were necessary.

Thus it appears that the lessons that should have been learned from the Bafokeng failure had largely been forgotten 20 years later when two further incidents, one resulting in many fatalities, occurred.

11.3.2 *The tailings dam failure at Merriespruit, 1994 (Table 11.1(12))*

On the night of February 22, 1994, a 31 m high tailings dyke upslope of the village of Merriespruit, South Africa, failed with disastrous consequences. The dyke breached a few hours after 30 to 55 mm of rain fell in approximately 30 minutes during a late afternoon thunderstorm. The failure resulted in some 600 000 m^3 of liquefied tailings flowing through the town, causing the death of 17 people, widespread devastation of property, and environmental damage. The tailings flowed for a distance of about 4 km, with some material reaching a small tributary of the (usually dry) Sand River, which runs on the northern boundary of Merriespruit.

The wave of tailings was about 2.5 m high when it reached the first row of brick-built single-storey houses, about 300 m downslope of the dam. Some of these houses were swept off their foundation slabs, while walls were ripped off and roofs collapsed on others. The mudflow continued down the main street narrowly missing a home for the aged and lapping at the wall of a mine hostel housing several hundred mine workers. It was miraculous that only 17 people died.

From eyewitness reports, a strong stream of water had entered the top end of the village at dusk (about 17h00). The stream flowed down streets and through gardens. One person reported seeing water cascading over the top of the tailings storage dyke. The mining company and contractor were informed, but when their representatives arrived on site it was already dark. An employee of the contractor rushed to the stacked ring penstock decant tower inlets and found water lapping the top ring but not overflowing. He removed several rings from the two penstock inlets. A second employee was below the dam near the overtopped area and, in the half-light, saw blocks of tailings toppling from a recently constructed tailings buttress against the north wall of the storage. There was an attempt to raise an alarm, but before the inhabitants could be warned a loud bang was heard, followed by the wave of tailings that engulfed the village.

Condition of the Merriespruit tailings storage prior to failure

The northern wall of the Merriespruit tailings storage, only 300 m upslope of the village, had been showing distress for a number of years in the form of seepage and minor sloughing near the toe. Two years before the failure, a drained tailings buttress was constructed along a 90 m length to stabilize the wall. Continued sloughing resulted in a decision by the mine and the contractor to discontinue tailings deposition in the storage about 15 months prior to the disaster. According to the contractor, the freeboard on the storage was at least an acceptable 1.0 m at the time when the decision was made to retire the storage.

Sloughing at the toe continued after deposition had supposedly stopped. shortly before the disaster, small slips occurred on the lower slope of the dyke immediately above the buttress. The slips were repaired by dozing in dry tailings from the surface of the wall adjacent to the slips.

After the instruction to stop operating the storage, raising of the outer wall ceased. However, excess water was still illicitly delivered to and stored in the impoundment by the mine, and this water must often have contained tailings. The stored water was decanted through the penstock when required by the plant, but the tailings slowly accumulated and filled the available freeboard. Shortly before the failure, the freeboard was a mere 0.3 m. What was even more serious was the fact that the pool became displaced from the penstock and pool training wall by the uncontrolled tailings deposition and moved towards the crest of the outer dyke. To a large extent, this situation arose because, for reasons of cost saving, there was no return-water pond to which excess water could be decanted and stored.

Merriespruit was covered by the orbits of a Landsat satellite that transited the area every 16 days. The nearest date for which an image was available prior to the failure was February 1, 1994, because clouds masked the site on the next pass. Several points emerged from a study of a series of the satellite photographs.

- The unauthorized deposition of water and tailings had not only reduced the freeboard to a dangerous and illegal level but had also pushed the pool away from the penstock causing it to be located next to the northern wall just prior to the failure, as shown in Figure 11.4.
- Wet conditions could be seen below some parts of the north wall that eventually breached.
- Free water lay impounded near the north wall for a considerable period of time. Even in October 1993, there was a large pool of water very close to the north wall and well away from the penstock inlet.
- By superimposing the outlines of the storage and the water surface on the contour plan, investigators were able to estimate the water level on February 1, 1994, to have been about 0.45 m below the crest of the dyke.

Plate 11.4 is a view of the Merriespruit failure, with the devastated village in the foreground and the tailings storage in the background. Plate 11.5 is a close-up aerial view of the failure scar. Plate 11.6 is an oblique view of the Merriespruit failure that gives an idea of the proportion of the top area of the storage occupied by the failure. Plate 11.7 is a close-up of the breach in the Merriespruit dyke showing a tailings delivery pipe bridging the chasm. Plate 11.8 is a satellite photo showing the pool (black area) lapping against the crest of the dyke three weeks before the failure.

Most likely mode of failure

The Merriespruit dam was equipped with piezometers at several sections, including a line of piezometers close to the breach. The contractor had carried out a stability analysis of the dyke a few months previously, using an angle of shearing resistance φ^I of 35° and pore pressures based on the piezometer measurements. The conclusion was that the minimum factor of safety F against general shear failure was 1.34. After the failure, the profile of the dyke on either side of the breach was surveyed, and the shear stability re-analysed on the basis of laboratory tests on undisturbed specimens taken on either side of the breach and the latest available piezometer measurements. The values of the shear strength parameters were $c^I = 0$, $\varphi^I = 33°$ and the calculated factor of safety

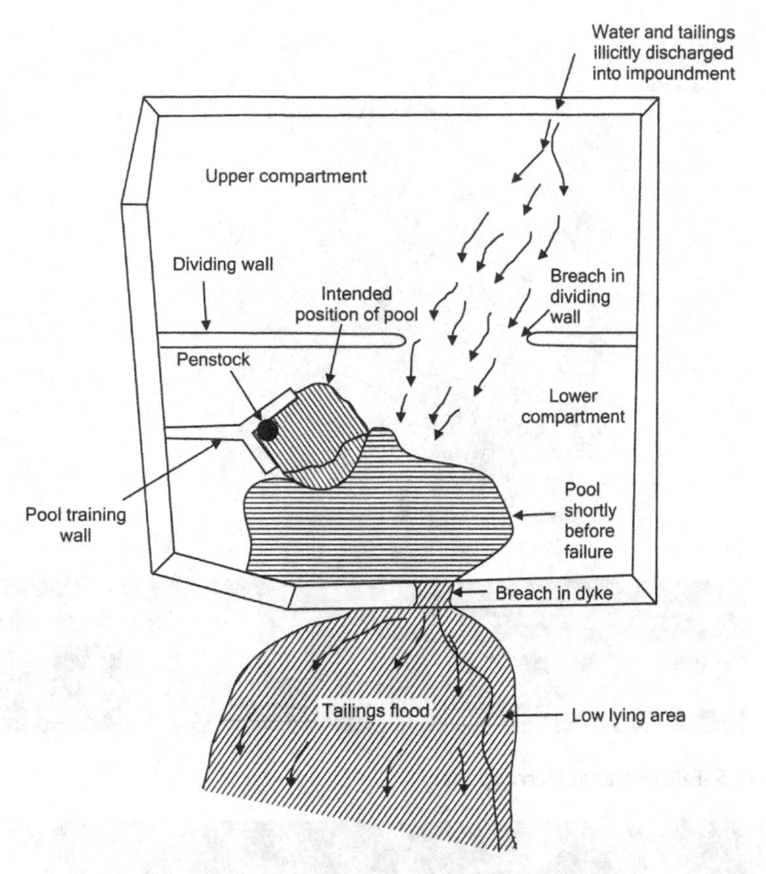

Figure 11.4 Plan of Merriespruit tailings storage showing position of pool at time of failure, intended position of pool, breach in dyke and path of tailings flood.

Plate 11.4 Flow failure at Merriespruit showing course of flow.

Plate 11.5 Failure scar at Merriespruit.

Plate 11.6 View of failure showing size of scar relative to area of storage.

Plate 11.7 View of breach in outer dyke at Merriespruit.

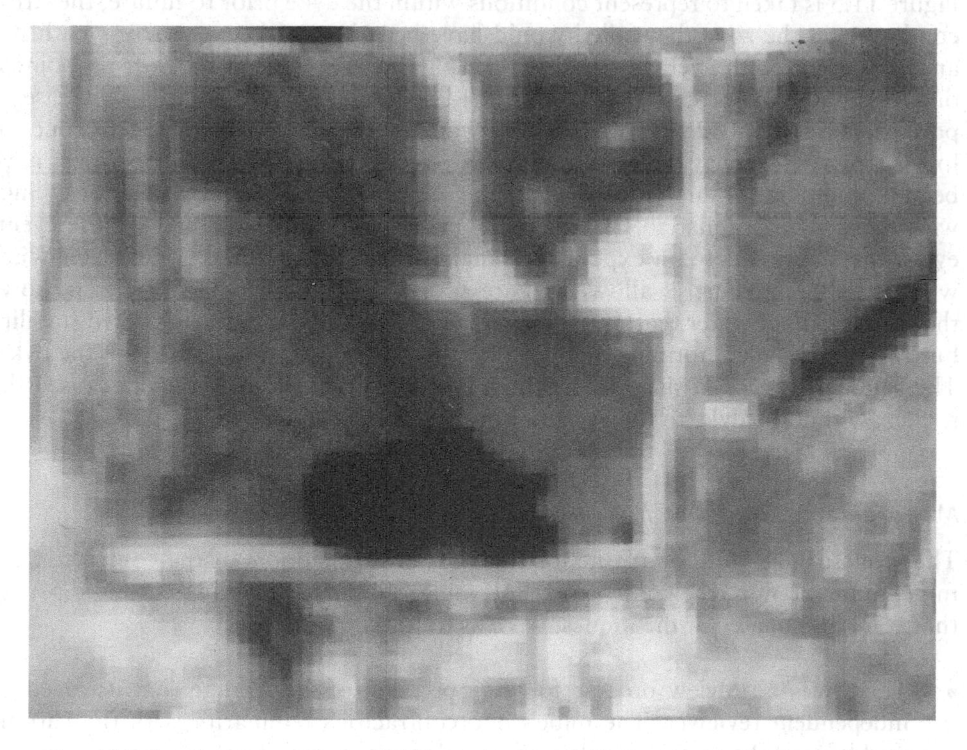

Plate 11.8 Satellite photo showing position of pool at Merriespruit two weeks before failure.

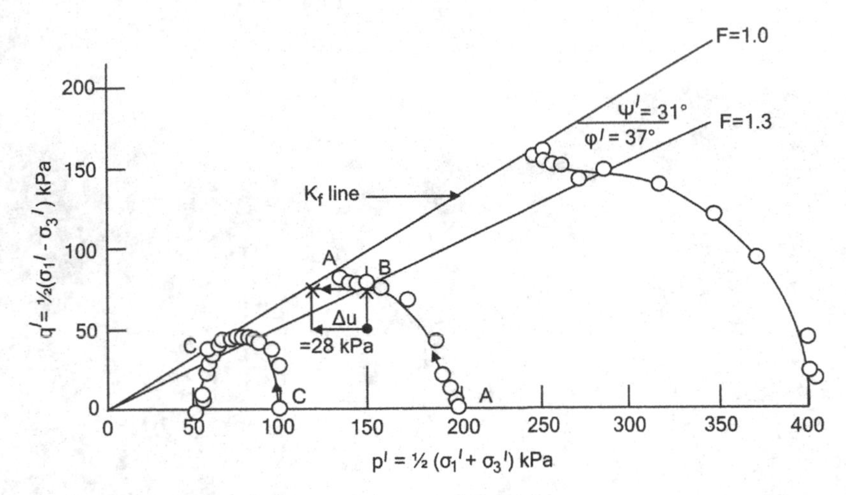

Figure 11.5 Deduced effect of movement of pool at Merriespruit towards crest of outer retaining dyke.

was $F = 1.31$. However, the analysis stopped short of taking account of the increase of pore pressure that must have occurred because of the slow progressive movement of the pool towards the crest of the dyke. Figure 11.5 is a repeat of Figure 4.34a, a set of post-failure triaxial shear test results for Merriespruit for which $\varphi^{I} = 37°$. If Figure 11.5 is taken to represent conditions within the dyke prior to failure, the stress conditions in the zone that failed would have been located between stress path AA and the origin, with average conditions on the line $F = 1.3$. To move from point B on stress path AA to the $F = 1$ (i.e. the K_f) line would have required an increase of pore pressure of only 28 kPa, or 2.8 m of water, and proportionately less in zones of lower stress. Hence, at the instant before overtopping started, the slope must have been on the point of failing in shear along a circular arc. Also, any layers of tailings with stress paths similar to CC in Figure 11.5 would have liquefied. In the event, eye-witness accounts to an experienced engineering geologist (the late Dr Ben Wiid) who was a talented artist, allowed Figure 11.6 to be constructed. The sketch shows that the stream of water that overtopped the crest eroded the recently-placed stability buttress which failed, forming a gulley that rapidly eroded its way through the dyke. The sudden increase in shear stress caused by local removal of the dyke caused the tailings within the dyke to liquefy and flow out of the breach.

Management failures at Merriespruit

The inquest judge laid the blame for the disaster at the doors of the contractor, the mine, and certain of the contractor's and mine's employees. Failings of these parties that were illuminated at the inquest were as follows:

- There was no review process for the operation of the storage that involved an independent reviewer. The mine's and contractor's familiarity with the chronic problems of the storage resulted in complacency about their seriousness.

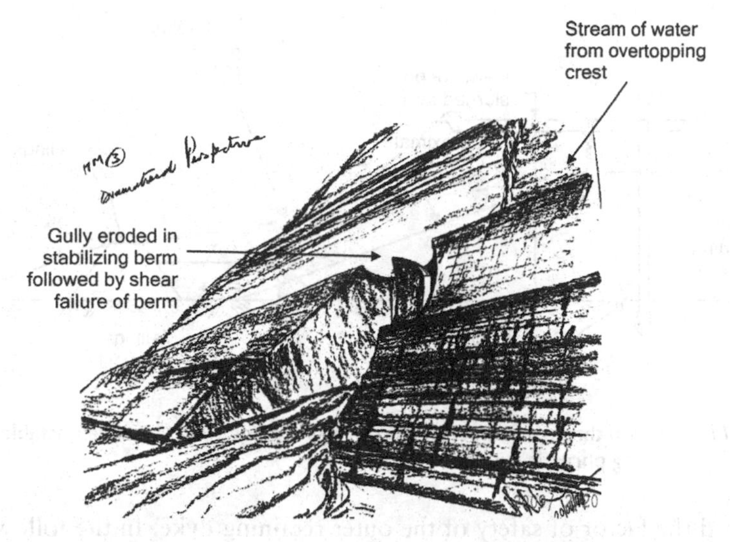

Figure 11.6 Sketch showing development of failure as seen by eye-witnesses. (Sketch by Dr. B.L. Wiid.)

- The only involvement of a trained geotechnical engineer in the problems of the storage was that of an employee of the contractor, who became involved occasionally, only by request, and whose roles and responsibilities were ill defined.
- There were regular meetings between the mine and the contractor. However, decisions were poorly recorded, which led to confusion about responsibilities and agreed actions.
- The contractor's office at the mine did not keep the head office adequately informed of happenings at the storage. The head office was ignorant of problems and potential problems at the site and could thus not take corrective action.
- The contractor's local office was aware that water was being stored in the storage by the mine, but it took no action and did not inform either head office or seek the advice of its geotechnical engineer.
- Although the contractor had operated the storage since its inception, he had never been requested to upgrade the facilities of the storage and so bring it in line with acceptable practice, as spelled out in the industry guideline. (Chamber of Mines of South Africa 1979, 1983). Thus, the storage continued to be operated without a return-water pond. This necessitated storing water in the storage.
- Remedial measures taken to restore the stability of the northern wall were ad hoc and not the result of an adequate geotechnical investigation and design.

11.4 Failure caused by increasing pore pressure

To a certain extent, both of the overtopping failures described in Section 11.3 were caused by increasing pore pressure. At Merriespruit, by the pool moving across the basin of the storage towards the crest of the outer dyke, raising the phreatic surface as it did so, and at Bafokeng by the large pools of water stored on top of the two sections of Dam No.1. In both cases the rising phreatic surface caused the shear strength to fall

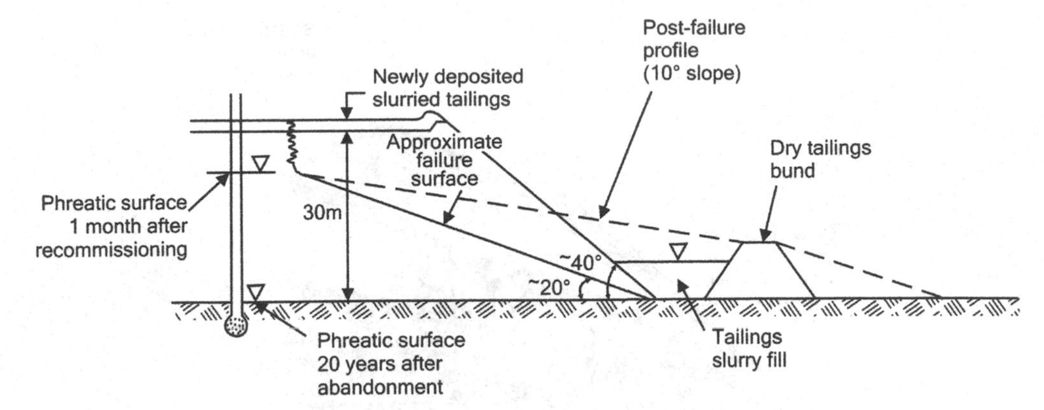

Figure 11.7 Section through Simmergo gold tailings storage which became unstable as a result of a rising phreatic surface.

and reduced the factor of safety of the outer retaining dyke. In the following case, a perfectly stable storage was caused to fail, almost entirely because of a rising phreatic surface.

11.4.1 *The Simmergo dam failure*

Figure 11.7 shows a section through a gold tailings storage, 30 m high, with a 40° outer slope. The tailings storage had been abandoned 20 years previously and was completely stable. As part of a tailings re-mining project, it was decided to re-commission the tailings storage to store re-mined tailings slurry. A piezometer installed in the tailings showed that the water table had subsided to the level of the soil surface underlying the tailings. The storage was taken back into use, but a month afterwards, it was found that the phreatic surface had risen by 20 m. As a safety measure, it was decided to buttress the toe of the ring dyke. This was done by constructing a bund of compacted dry tailings at the toe and filling the space between the bund and the pre-existing dyke with re-mined tailings slurry. The result was an almost immediate failure of the dyke which is illustrated in Plate 11.9.

Although the phreatic surface in the storage had subsided since abandonment, the tailings remained saturated, though subjected to a hydrostatic suction, as demonstrated by Figure 5.2. Very little addition of water was necessary to cause the loss of the suction and the re-establishment of a high phreatic surface. (See Figure 4.59a, for example, where a change of water content of 2% was sufficient to reduce the suction from 1000 to 100 kPa.)

Figure 5.2 shows measurements of vane shear strength made in a deposit of gold tailings in which the water table was at 17 m below surface. The figure shows that the in situ strength can reasonably be represented by the equation:

$$\tau = \{\gamma h + \gamma_w(h_w - h)] \tan \varphi^I \qquad (11.1)$$

If $h_w = 0$, this becomes

$$\tau = h(\gamma - \gamma_w) \tan \varphi^I \qquad (11.2)$$

Plate 11.9 Failure at Simmergo, showing liquid tailings splashed over compacted tailings bund.

As shown by Figure 5.2 this could represent a very considerable loss of shear strength.

The Simmergo failure (Figure 11.7) occurred at 10 a.m. on a Saturday when deposition was not in progress. A maintenance workman was standing on the section of the wall that failed. According to him, there was a tremor and a loud rumbling. He clung to the tailings delivery pipe that is visible on either side of the failure (Plate 11.9) and was let down gently as this subsided into the gap. Fortunately, no one was injured.

As the failure took place into the still-liquid tailings slurry retained by the dry tailings bund, quantities of liquid slurry as well as fragments of dried tailings from the original slope splashed over the bund and can be seen in the foreground of Plate 11.9.

11.5 Failures caused by excessive rate of rise

The subject of rate of rise and its effect on shear strength has been discussed in Section 6.11.2. The case that follows illustrates how an excessive rate of rise, or an increase in the rate of rise can cause destabilization of the outer dyke of an hydraulic fill tailings storage.

11.5.1 *The Saaiplaas tailings dam failure*

The Saaiplaas No. 5A Dam is a ring-dyke gold tailings storage located near the town of Virginia, South Africa. It was built by upstream paddocking with subaerial deposition

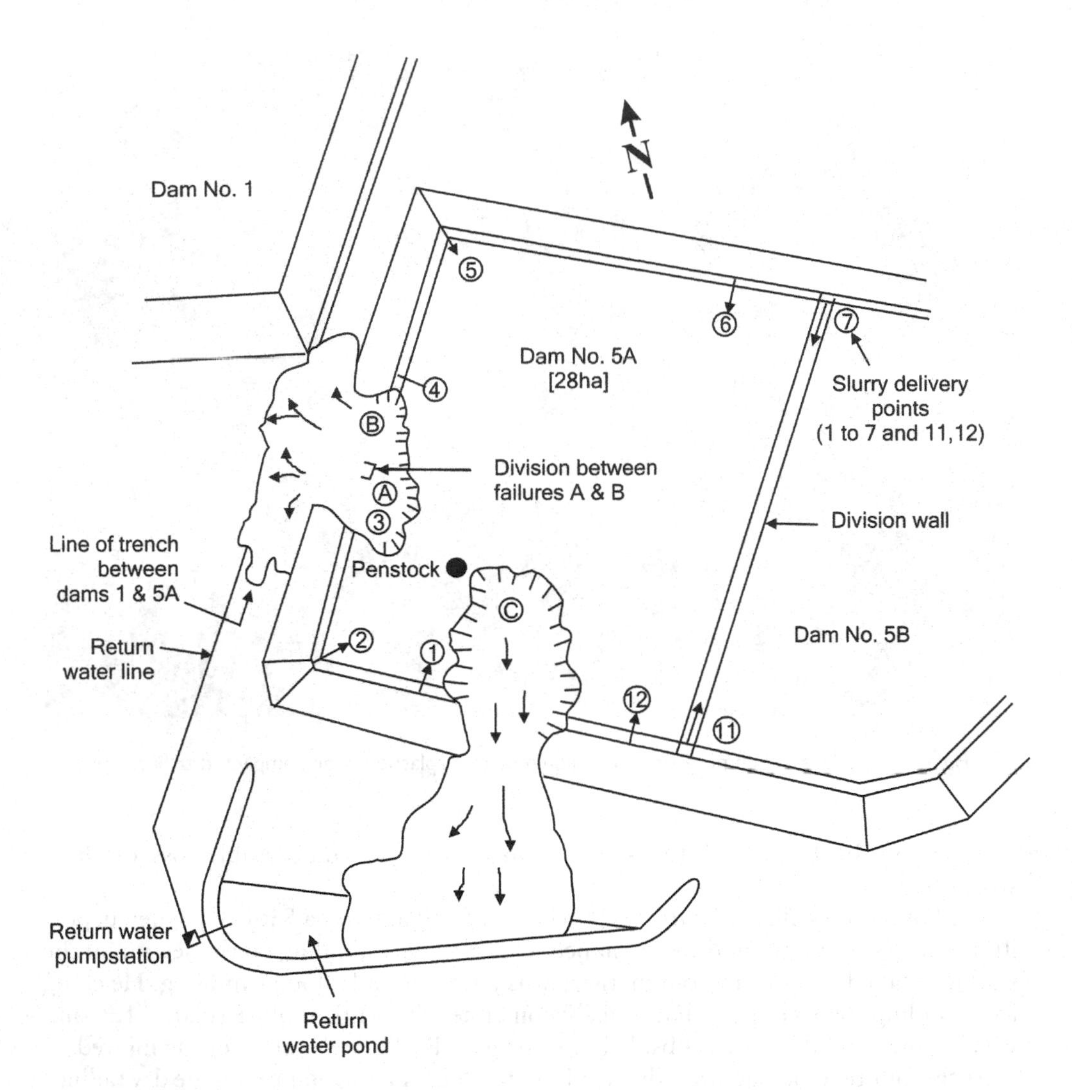

Figure 11.8 Plan of Saaiplaas Dam 5A showing locations of failures.

from several points off a ring-feed main into the impoundment. In March 1993, the storage was 28 m high with an average slope angle of 21°. At 12 h15 on March 18, 1993, a rotational slide occurred in the western side of storage dyke. This was followed on the night of March 19 by a second similar failure, separate but immediately adjacent to the first. At 07 h30 on March 22, these two failures were followed by a third failure on the southern side of the ring dyke. Figure 11.8 shows the layout of No.5 Dam and the location of the three failures. Plate 11.10 shows all three failures that occurred at Saaiplaas, A and B in the foreground and C at centre-left. Plate 11.11 is a closer view of failures A and B at Saaiplaas showing the narrow section of the outer dyke that separated the two failures.

Plate 11.10 View of three failures at Saaiplaas.

Plate 11.11 Close-up view of two separate failures A and B.

Fortunately, no one was injured, possibly because very little water was ponded around the penstock and the tailings flow was contained by adjacent structures and did not travel very far. Failure C was the most mobile of the three failures, but fortunately it took place into the return water pond, which dammed the tailings flow.

Background to the failures

No 5A Dam had been an old tailings deposit that was re-mined for reprocessing in the late 1970s. In 1981, the 28ha storage was re-commissioned. At that time, Saaiplaas Gold Mine was considered to be on the brink of closure and the re-commissioning was intended only to accommodate tailings for the few remaining months of the mine's operation. To save costs, deposition recommenced with no underdrains under the containing dyke, and no piezometers were installed to monitor the phreatic surface. To further save costs, the decant penstock was placed close to one corner of the rectangular impoundment instead of near the geometric center where it would have kept the pool further from the wall, thus depressing the phreatic surface.

Two of the required operating conditions set for the re-commissioned dam were that the rate of rise be limited to 1.5 m/y and that the height be limited to 20 m. A 1.5 m/y rate of rise should have resulted in a well-consolidated tailings deposit with adequate slope safety at the design overall slope angle of 26°.

The mine, however, did not close. Dam 5A continued to be operated, and by the end of 1992, was 28 m high and had risen at an average rate of 2.33 m/y. Over the previous year, however, the rate of rise had averaged 2.57 m/y, and in February 1993 it was 2.83 m/y. For a number of years prior to the failure, the contractors who operated Dam 5A on behalf of the mine had pleaded for the installation of piezometers and had also insisted on the need to move the penstock away from the retaining dyke. The plea for piezometers fell on deaf ears, but the request to relocate the penstock was eventually acceded to, but only because the contractor had repeatedly drawn attention to the visibly wet condition of the southern slope of the dam. A further request to step back the wall, forming a berm and thus flattening the overall slope of the dam, was refused.

If the coefficient of consolidation of the Saaiplaas tailings were to be taken at $200\,m^2/y$ (see Figure 4.42b) and the rate of rise averaged 2.33m/y over 12y, then according to Gibson's theory (Figure 6.34) for an impervious foundation, the time factor T would have been $T = (2.33)^2 \times 12/200 = 0.32$ and the degree of consolidation would have been about 0.8 (80%). However, the tailings had been re-mined and re-milled and were therefore finer than average and perhaps a $c_v = 50\,m^2/y$ would be more appropriate. This would have changed T to 1.3 and reduced the degree of consolidation to about 0.65 (65%). Gibson's (one-dimensional) theory predicts the time taken for the pore pressure in the accumulating mass of tailings to reduce from its initial value to the hydrostatic value (see Figure 6.35). The initial value of the pore pressure if the tailings had been deposited instantaneously would have been $u_o = RD \cdot \gamma_w h$ where RD is the relative density of the tailings after settling on the beach. This is likely to have been about 1.5. Thus the excess pore pressure u(max) at the base of the mass of tailings would have been (1–0.65) (1.5–1) $\gamma_w h$ where $\gamma_w = 10\,kN/m^3$ and $h = 28$ m, i.e. u(max) = 91 kPa to which the hydrostatic pore pressure of $\gamma_w h = 280$ kPa has to be added, giving 371 kPa. The shear strength would have been $\tau(max) = (15 \times 28 - 371) \tan \varphi^I = 49 \tan \varphi^I$. Taking $\varphi^I = 33°$, $\tau(max) = 32$ kPa.

If, however, the tailings had been deposited on a pervious base, the pore pressure at the base of the outer dyke would have been zero, and the shear strength $\tau(\max) = 1.5\gamma_w h\tan\varphi^I = 273\,\text{kPa}$. Hence if the outer dyke had been properly underdrained, even the excessive rate of rise of 2.83 m/y (a degree of consolidation of about 80% at mid-height of the dam and 100% at the base) would probably not have been too rapid for safe operation. The fault therefore lay with the combination of a high rate of rise combined with the absence of underdrainage.

If the originally specified maximum rate of rise of 1.5 m/y had been adhered to, taking $c_v = 200\,\text{m}^2/\text{y}$, $T = 0.135$, and the degree of consolidation would have been about 95% for an impervious base. The corresponding $\tau(\max)$ would have been $28[1.5-(0.05 \times 0.5 + 1)] \times 10\tan 33° = 86\,\text{kPa}$. This is a considerable improvement on 32 kPa, but probably still not sufficient for stability, even though the outer dyke would also have consolidated by flow to its free outer slope surface.

There was no overtopping involved in these failures. The high rate of rise and lack of under-drainage had resulted in an insufficiently consolidated wall, the shear strength of the tailings was deficient, and the wall failed in shear. Nevertheless the three failures must also be classified as flow failures. However, because failures A and B took place into a narrow passage between Dams No.1 and 5A (see Figure 11.8) and were further constrained by a deep trench between the two dykes, they did not travel far. Failure C ran into the return water pond and was contained by the pond wall. Nevertheless, the failures show that flow failures do not necessarily have to be triggered by overtopping.

Aftermath of the Saaiplaas failures

Clearly, the profit motive was dominant in the minds of mine management who, in order to save costs, refused to entertain any warnings that the dam was unsafe.

Because the failures did not result in injury or death, were confined to the boundaries of mine property, and damaged only mine property, they were successfully hushed up and not reported in local newspapers, let alone radio or television. Hence, these major failures went virtually unnoticed, and no lessons were learned from the incident by the gold mining industry. Nor were any alarms raised. Less than a year later, the failure at Saaiplaas was followed by the Merriespruit tragedy, only 5km away. If the Saaiplaas failure had been publicized by a public enquiry, it is possible that the Merriespruit failure could have been avoided.

11.6 Failure caused by poor control of slurry relative density

Sections 10.3 and 10.4 showed how critical it is for the tailings slurry to be sent to the storage with a well-controlled relative density. A well-controlled RD is particularly important when a new plant and tailings storage are being commissioned together. Figure 12.8 shows that for a typical storage, deposition initially takes place on a small area, at the lowest point, against the toe wall and hence the initial rate of rise will be relatively high. Unless the toe of the storage is well drained, a pocket of tailings may form that will take many years to consolidate and gain in strength. This weak zone will always detract from the overall shear stability of the tailings storage at the section that will probably have the greatest height. The example that follows will illustrate what might happen in a case like this:

A relatively small alluvial mining activity operated three separate mineral extraction plants, each having its own tailings storage. In the interests of efficiency and overall cost saving, it was decided to construct a new central plant with a new tailings storage to replace the scattered small plants. The new storage had a starter wall of compacted alluvial overburden which was 8m high at its highest point. It was planned to deposit the tailings upstream from the starter wall by means of a spraybar system designed

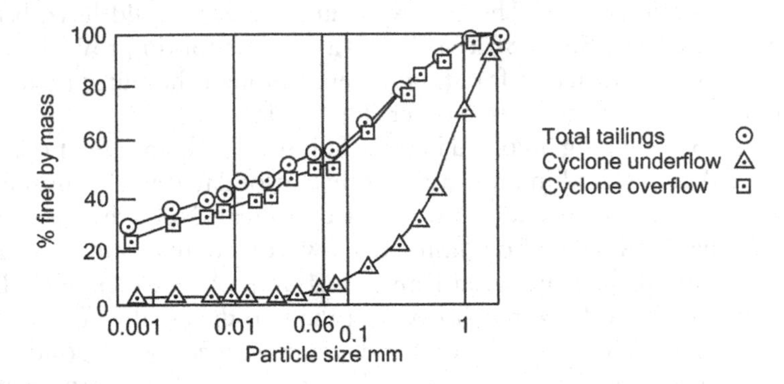

Figure 11.9 Particle size distributions for failure described in Section 11.6.

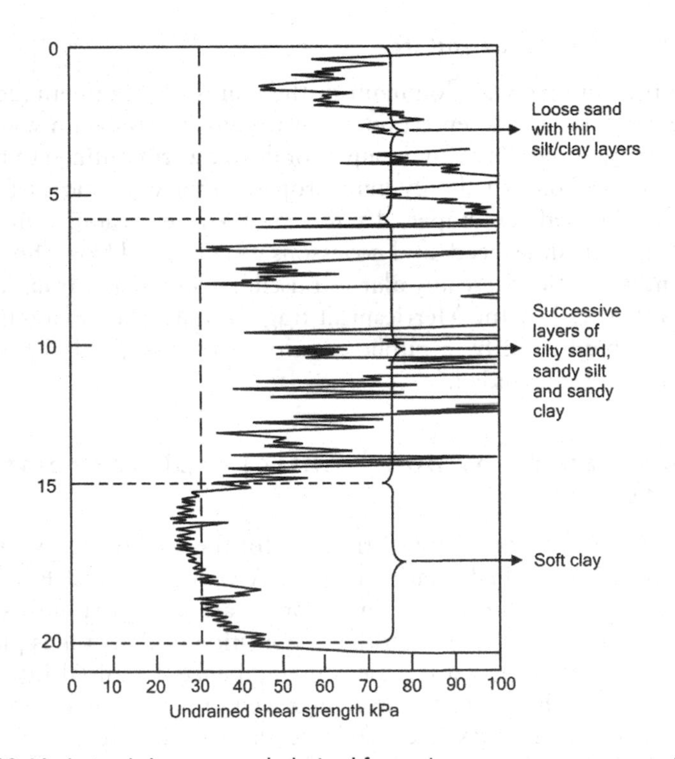

Figure 11.10 Undrained shear strength derived from piezocone measurements ($N_c = 20$).

to produce a substantial outer wall of the sand coarse fraction of the tailings. This coarse zone was intended to act as a pervious conduit to the toe drain on the inside of the starter wall. However, as a result of commissioning problems at the new plant, the first tailings to be delivered had too high a relative density to allow of particle segregation as the slurry was beached away from the starter wall. The net result was that the first 5m thickness of unsegregated tailings deposited against the starter wall had a clay fraction of over 30% and a coefficient of consolidation c_v of only $5\,m^2/y$, instead of (possibly) $500\,m^2/y$.

When this was discovered, the spraybar system was abandoned and cyclones were introduced to produce the required outer shell of coarse pervious material. Figure 11.9 shows the particle size distributions of the almost unsegregated total tailings deposited by spraybar, the cyclone underflow and the cyclone overflow. It is obvious that the cyclones were able to perform superbly giving a sand fraction (<2 mm>0.06 mm) with a minimum tail of finer material. The overflow was deposited with the remainder of the total tailings to give a material very similar to the tailings previously deposited by spraybar.

When the height of the dyke had reached 15m above the top of the starter wall, two separate rotational failures took place, 4 days apart. All deposition had been stopped when the first failure occurred, and the second failure took place 300 m from the first. This indicated that the shear stability of the dyke was critical over much of its length. The sections of the dyke at the two failures were superficially very similar, and their composition was explored by means of piezocone probes from the outside of the dyke. Figure 11.10 shows one of the strength profiles interpreted from the probing, based on a value of $N_c = 20$. Equilibrium pore pressures (u_e) from the piezocone showed that pore pressures were approximately hydrostatic between depths of 7 and 14m below the level of the third bench (see Figure 11.11), but below this, dropped to close to zero at the level of the base of the starter wall. Hence the toe drain was effective in draining the initial layer of unsegregated tailings, but the rate of drainage was impeded by the low coefficient of consolidation of the tailings. A slip circle analysis based on the observed movements showed that an average shear strength of 30 kPa was required for a factor of safety of 1.0 on the circular slip shown in Figure 11.11. The required shear strength matched the strength of the soft clay almost perfectly and constituted good evidence that the assumed mechanism of failure was correct.

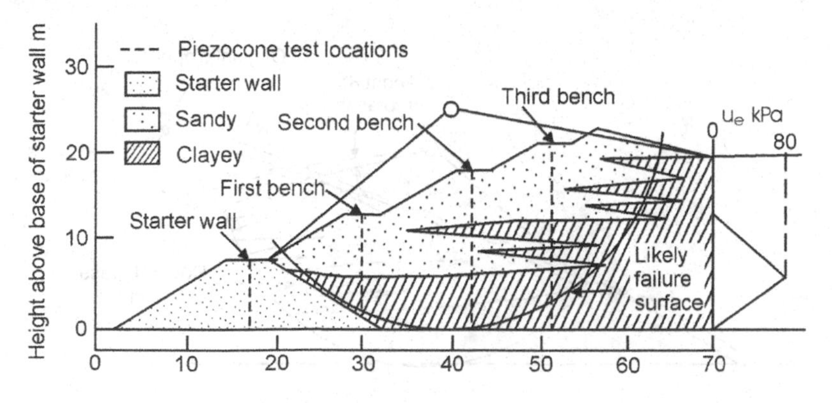

Figure 11.11 Interpreted cross-sectional profile and locations of piezocone tests.

11.7 Post-failure profiles of hydraulic fill tailings storages

11.7.1 *Failures that did not flow extensively*

Essentially, a slope fails because it had been built too steeply to be stable above a certain height, or has become too steep and high to be stable, as a result of conditions that were not foreseen, such as a rise in the phreatic surface. The profile assumed by the failed slope will be close to the equilibrium profile for the conditions that resulted in the failure. As a first example, Figure 11.7 shows the post-failure profile observed for the Simmergo failure. The run-out of the tailings was interrupted by the presence of the dry tailings bund and much of the wedge of tailings that came to rest beyond the bund, consisted of slurry that was splashed over from behind the bund. Whereas the original outer slope of the storage had been 40°, the post-failure slope was 10°. The vertical step at the crest of the slope was stable because, being above the elevated phreatic surface, it remained subject to negative (capillary) pore pressure. The height of the slope was also effectively reduced by the presence of the dry tailings bund.

In the failure illustrated by Figure 11.11, relatively little slope flattening occurred. The post failure slope profile, shown in Figure 11.12, was only 3° flatter than the pre-failure slope, and the height of the slope was reduced by 2 m, the vertical distance between A and A^{I}, the angular rotation amounted only to about 8°.

Figure 11.13 shows pre- and post-failure sections of a rotational failure that took place in the north east ring dyke of the Bafokeng No. 2 tailings dam (see Figure 11.2). Plate 11.12 shows the failure seen from above. In this case the average post-failure slope was slightly steeper than the pre-failure slope, but the height of the slope had been reduced by about 5m from the original 15m and the slope was buttressed by the run-out of tailings debris. In each of the above cases, the profile and height of the slope adjusted itself to restore stability.

11.7.1.1 *Failures that led to extensive flow*

Figure 11.1 shows the post failure profile of the El Cobre old dam which ended up with an average slope of 4.4°. The actual slope profile was not uniform, but consisted of a series of near-horizontal surfaces joined by near-vertical steps of up to 5 m in height.

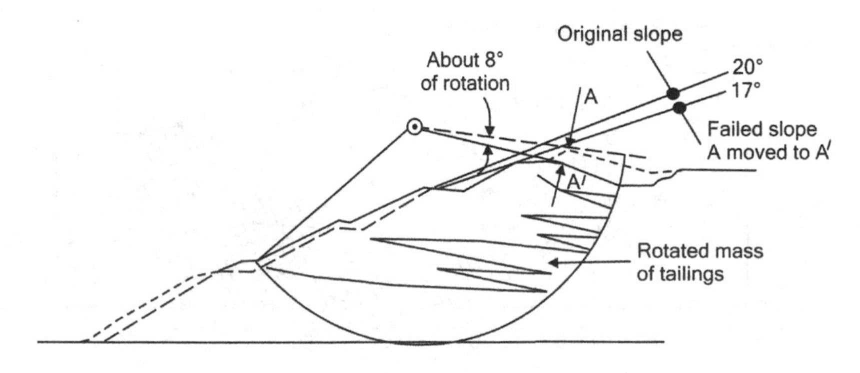

Figure 11.12 Post-failure profile of failure illustrated in Figure 11.11.

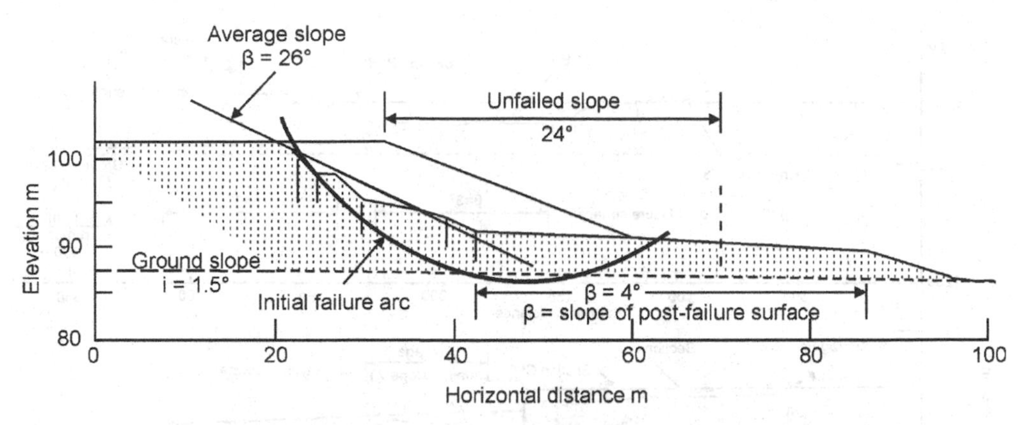

Figure 11.13 Pre-and post failure profiles of a rotational failure that took place in north-east side of No.2 tailings dam at Bafokeng, South Africa. (See Plate 11.12.)

Plate 11.12 Rotational slide in Bafokeng No.2 dam wall.

The steps appear to have formed as a result of the presence of denser layers, possibly where the tailings had been heavily overconsolidated by sun-drying while the storage was not in use.

In Figure 11.14, section BB is a section through failure C at Saaiplaas (see Figure 11.8), and section B′B′ is a section of the unfailed dyke. The tongue of the slide

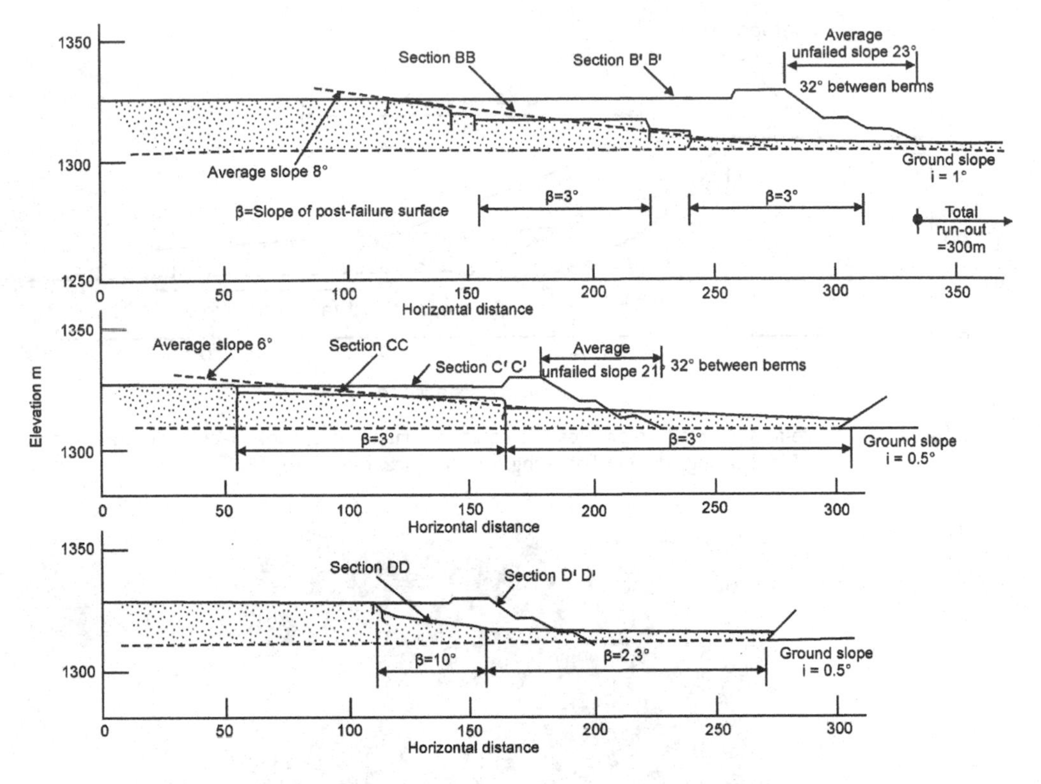

Figure 11.14 Sections through Saaiplaas failures.

and portions of the failed surface had a slope of 3°, while the average slope of the upper part of the failed surface was only 8°, compared with the average slope of the unfailed dyke of 23° (with sections between berms at 32°). Sections CC and C′C′, DD and D′D′ show similar information for the twin failures A and B at Saaiplaas. The slopes of the slide tongues were very similar (2.3° and 3°), as were the average slopes of the upper parts of the failed surfaces. For all three sections, it will be noted that a number of small near-vertical steps occurred in the post-failure surface, similar to those at El Cobre.

Figure 11.15, shows sections of the post-failure surface for the Merriespruit tailings dyke. Section E′E′ is the pre-failure section normal to the wall and EE is a section through the overtopping breach. Section FF runs at right angles to EE, and GG runs at 45° to EE. The intersections of FF and GG with EE are marked in Figure 11.15. The slope of the tongue of escaped tailings was 2°, which is very similar to the slopes of large portions of FF and GG. In other words, the post-failure surface had flattened to a slope of 2–3°, with some portions around the perimeter of the failure scar being as steep as 10–20°. Because of the disturbance caused by the failure, it is very difficult to know from what depth in the impoundment the material that composes the post-failure surface originated. The surface was too soft to be accessible after the failure until a drying crust had formed. Hence it was not possible to sample the post-failure surface for water content straight after the failure to help identify its depth of origin.

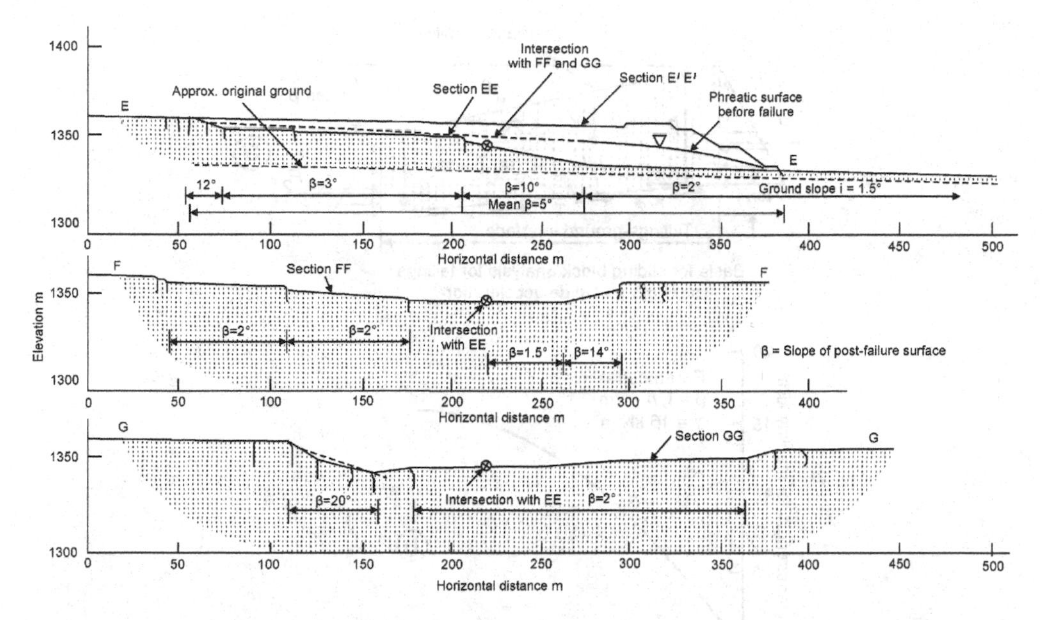

Figure 11.15 Sections through Merriespruit failure.

It seems likely, however, that the tailings that moved out of the breach consisted of the upper more recently deposited layers, and that the post-failure surface consisted of relatively undisturbed deeper layers exposed as the slope was flattened by the outward flow of the tailings.

Figure 11.16a shows the basis for a simple sliding block analysis to calculate the relationship between the post-failure slope, β, of a tongue of escaped tailings, the slope, i, of the ground surface and the interfacial shear strength, τ, between the ground surface and the tailings. The analysis assumes that the pore-water pressure within the tailings is hydrostatic throughout the sliding block, once it comes to rest. The shear strength τ is given by

$$\tau = [(P_1 - P_2)\cos i + W \sin i] \cos i / L \tag{11.1}$$

where

$$P_1 - P_2 = (K\gamma^{i} + \gamma_w)[h^2 - (H^I)^2]/2 \tag{11.2}$$

in which K will probably have a value equal to K_A, the active pressure coefficient, and

$$H^I = h - L(\tan\beta - \tan i) \tag{11.3}$$

and

$$W = \gamma L[2h - L(\tan\beta - \tan i)]/2 \tag{11.4}$$

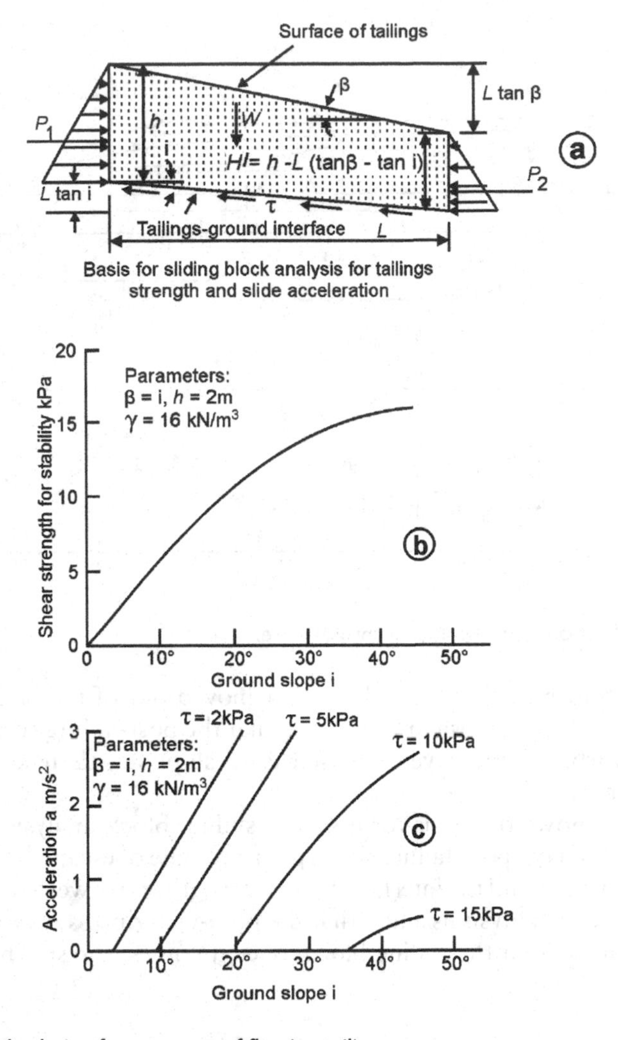

Figure 11.16 Analysis of movement of flowing tailings.

In equations (11.2) and (11.4), γ is the bulk unit weight of the tailings, γ^I is the effective unit weight, and γ_w is the unit weight of water. The analysis has been applied to the post-failure surface slopes recorded for the Bafokeng (Figures 11.3 and 11.13), Saaiplaas (Figure 11.14) and Merriespruit (Figure 11.15) failures. The results of these analyses are summarized in Table 11.3. The table shows that the interfacial shear strength must have been extremely low for the observed post-failure surface angles to have developed. Figure 11.16b shows a relationship between the shear strength τ for stability, and ground slope i for a tailings flow that is 2 m deep.

Reference to Figure 5.8a shows that for gold tailings, the remoulded φ^I may be as low as 5°. The interfacial angle of shearing resistance between the gold tailings (Saaiplaas and Merriespruit) and the underlying ground was obviously about half of this and was probably affected by ground surface being wet, although the Saaiplaas data do not

Table 11.3 Summary of observed post-failure surface slopes and corresponding ground/tailings interfacial shear strengths.

Tailings dam	*Post-failure surface slope β*	*Ground slope i*	*Interfacial shear strength τ(kPa)*
Bafokeng (Figure 11.3)	4°	1.5°	5.2
Bafokeng (Figure 11.3)	2°	1.3°	1.6
Saaiplaas			
(No rain)	2.3°	−0.5°	3.4
(No rain)	3°	−0.5°	3.6
(After rain)*	3°	1°	2.3
Merriespruit			
(After rain)			
(Flow slide)	2°	1.5°	1.0
(Failure basin)	2°	0	1.8

* Failures A and B at Saaiplaas took place 2 days after 4 mm of rain had fallen.
Failure C took place after 19mm of rain had fallen.

show this. For the failures at Bafokeng, the calculated interfacial shear strengths are not consistent with the 1–4 Pa range of measurements shown for platinum tailings in Figure 4.48.

It could be argued that the extremely flat angles of 2°–4° taken up by the post-failure surfaces (e.g. Figures 11.14 and 11.15) were at least partly as a result of erosive and momentum forces that acted as the liquefied tailings flowed out of the breach. This view is probably at least partly valid. If one were to use the post-failure profiles within the failure bowls as a guide to the post-liquefaction (or residual, or steady-state) angle of shearing resistance of gold tailings, a range of 8°–10° would probably be a fairer estimate. This range is in agreement with the result of a study to determine the residual strength of a natural sand (Byrne, et al., 1994) which was that φ^I (residual) = 11° using triaxial compression tests, whereas using triaxial extension tests, residual shear strengths could be as low as φ^I (residual) = 6°. Both of these values are compatible with the 8°-10° recorded above. Work done by Troncoso (see Blight, et al., 2000) measuring undrained vane shear strengths concluded that peak total strength values of φ as high as 16° could correspond to residual (or remoulded) strengths as low as 2°.

11.8 Analysis of the motion of flow failures

A simple extension of the block analysis illustrated in Figure 11.16a allows the motion of a flow slide to be analysed (Blight, et al., 1981). Instead of calculating the shear strength to maintain static equilibrium, as in equation (11.1) the full equation of motion is considered, as follows:

For motion of the sliding block illustrated in Figure 11.16a, downstream forces-upstream forces = mass of block × acceleration, i.e.

$$(P_1 - P_2)\cos i + W\sin i - \tau L/\cos i = W \cdot a/g \qquad (11.5)$$

The symbols are defined in Figure 11.16 and a = acceleration of the block, g = gravimetric acceleration. From Equation (11.5)

$$a = [(P_1 - P_2)\cos i + W\sin i - \tau L/\cos i]g/W \quad (11.5a)$$

If the block of material comes to rest, a = 0 and

$$\tau = [(P_1 - P_2)\cos i + W\sin i]\cos i/L \quad (11.1)$$

If the block is accelerating, its increase in velocity after time Δt will be

$$\Delta v = a\Delta t \quad (11.6)$$

In Equation (11.5)

$$W = [2h - L(\tan\beta - \tan i]\gamma L/2 \quad (11.4)$$

and

$$h = H^I + L(\tan\beta - \tan i) \quad (11.3)$$

where γ is the bulk unit weight of the material in the block.

If the surface of the flow (i.e., of the block) is parallel to the ground surface, $\beta = i$ and $P_1 = P_2$, $h = H^I$.

If the pore water pressure in the block is taken as hydrostatic with free water at the surface of the slide,

$$(P_1 - P_2) = (K\gamma^I + \gamma_w)[h^2 - (H^I)^2]/2 \quad (11.2)$$

where γ^I is the effective unit weight, γ_w is the unit weight of water and K is the active lateral pressure coefficient, K_A.

It was noted from Table 11.3 that if a liquefied waste flow debouches onto wet ground, e.g., when failure follows heavy or prolonged rain, the interfacial shear strength will be reduced by the water already at the wet waste-to-ground interface, and the flow will be more mobile than if the ground surface had been dry.

The data in Figure 11.16c show relationships between the acceleration a and ground slope i for a depth of flow of 2 m and various interfacial shear strengths τ. The graphs correspond to a simple case in which the surface of the flowing waste is parallel to the ground surface, but via Equation (11.6), give some idea of the speed with which a flow slide can move. For example, if the acceleration from rest is only 0.1 m/s^2 and this is maintained for 1 minute, the flow will accelerate to 6 m/s or 20 km/h in this period. This is faster than many humans can run. The consequences of higher rates of acceleration are frightening. In the flow failure at Bafokeng (Jennings, 1979), the flow velocity a short distance after leaving the breach in the impoundment was estimated from stagnation flow heights on damaged buildings (by equating the potential energy of the stagnation height against the building to the kinetic energy of adjacent unimpeded flow. (Blight et al., 1981) to have been 10 m/s or 36 km/h, even though the ground surface was almost level. A flow moving at this speed would be impossible to escape from by running ahead of it.

A similar approach to estimating flow velocity can be applied in cases where a downhill flow crosses a valley and stagnates at a given elevation on the opposite slope, as in a flow failure of a municipal solid waste dump in Istanbul (Blight, 2008). Here, the flow reached stagnation at an elevation of 15m above the bottom of the valley. Assuming the bulk unit weight γ of the liquefied waste to have been 10 kN/m^3, an approximate energy balance per m^3 of waste would be:

$$1/2\gamma v^2 = \gamma g\Delta h \text{ or } v = (2g\Delta h)^{1/2} \quad (11.7)$$

where v is the velocity of flow at the bottom of the valley and Δh is the stagnation height above the bottom of the valley. For the Istanbul case, $\Delta h = 15$ m and the (minimum) $v = 17$ m/s, or 60 km/h. This ignores energy consumed in overcoming shear at the interface of the hillside and the flowing waste. Applying similar reasoning to the flow at Aberfan, (see Section 11.10.2), the speed of the flow was estimated to have been between 10 and 20 miles per hour (4.5 to 9 m/s) by Hutchinson (1986). This gives a stagnation height of between 1 and 4 m. Where the debris flow crossed a disused railway embankment before entering Aberfan village and devastating it, the embankment was only 1.3 m above the natural ground surface on the uphill side. If the embankment had been 3 m higher, it would have stopped the flow and saved the village.

The basis of the sliding block analysis can therefore be used to design protection measures such as deflection dykes and safety platforms to protect installations from the effects of waste flows (e.g., Blight et al., 1981; Miao et al., 2001). This has been put into effect at a number of mines in South Africa. The design procedure is to apply the block sliding procedure step-wise along the steepest line from an assumed position for a breach towards the installation (e.g. a shaft head) that is to be protected (see Blight, et al., 1981). The object is to predict the flow velocity and depth of flow when a hypothetical waste flow reaches the installation and hence to calculate the stagnation height and build the protection works to above that level. (In the 2005 New Year tsunamis in South East Asia, it was reported that in one case the "tidal wave" reached a depth of 20 m when it reached a village several kilometres inland. What was actually observed was more likely to have been the marks showing the stagnation height of 20 m for a 3 m deep wave travelling at about 18 m/s (65 km/h).)

11.9 The effects of failure geometry on insurance rates

Damage to outside installations and compensation for loss of life and injury aside, the insurance premium for a tailings storage is very dependent on the expected loss of revenue if the mine has to close for a period of time because of a failure in the tailings storage. This is certainly something that needs consideration during the design of any new tailings storage. If the mine has more than one tailings storage, e.g. at Bafokeng (Figure 11.2) or Saaiplaas (Figure 11.8), any disused or "spare" storage or storages should be capable of taking the deposition that was going to the failed storage, and should be maintained so that if any of the complex of operational storages should be put out of action for any reason, the tailings flow can be diverted to another without any effect on the mine's production.

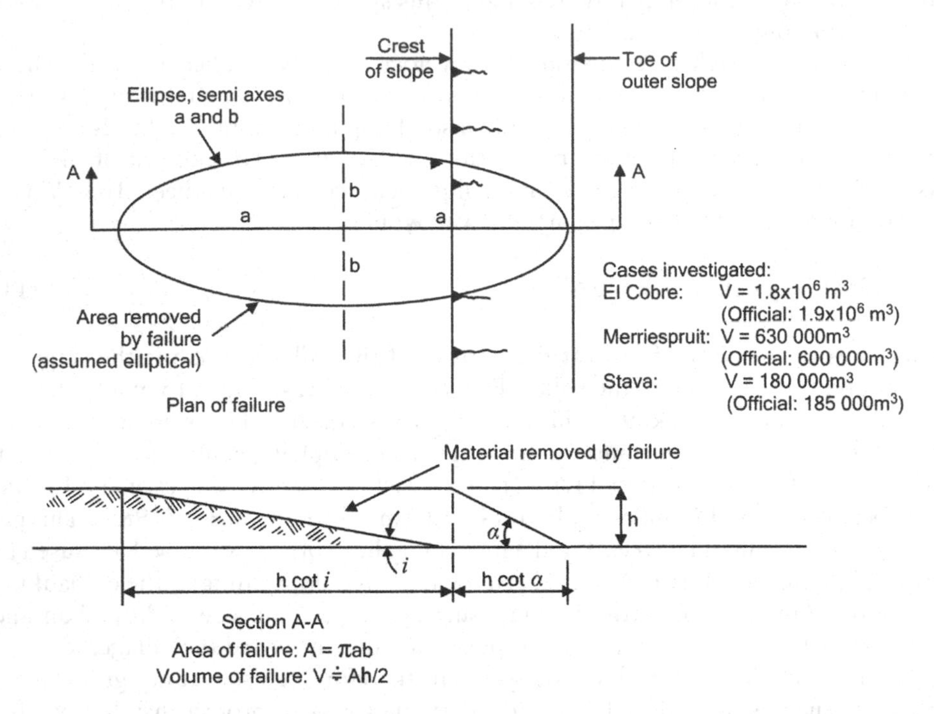

Figure 11.17 Simple estimation of area and volume involved in a failure.

Alternatively, if the storage is large enough to leave, say 60–90% of the deposition area still available for deposition, and if the rate of rise is moderate, and the stability of the rest of the storage dyke is not in question, the storage could continue in use. This is what happened after the Simmergo failure. A low bund was constructed around the failure, and deposition continued with very little interruption of activity. Even at Merriespruit, viewing Plate 11.6, it would have been technically possible to isolate the failure breach and continue depositing on the remainder of the area, if stability analyses had shown it to be safe to do so.

The likely dimensions of a failure scar can be estimated from the height of the storage and the likely slope angle of the floor of the failure basin, as indicated in Figure 11.17. Using the assumption of an elliptically shaped failure scar enables the effect of the ratio of length (a) to breadth (b) on areas and volumes of failure to be investigated easily. As shown in Figure 11.17, the calculation can give a very realistic result.

11.10 Failures of dumps of coarse waste

This section will describe the mode of failure of dry dumps of coarse waste such as waste rock from the development of underground mines, discards from coal mines, ash from power stations or coal gasification plants, metallurgical slags, waste gypsum, etc. These behave in a very different way to the hydraulic fill storages of fine waste described up to now. It should be noted that certain fine wastes, e.g. power station fly ash and waste gypsum can be disposed of either by hydraulic filling or dry dumping,

Plate 11.13 Failure in waste rock dump showing wall of earth pushed up in failure.

with dry dumping tending to replace hydraulic filling in areas where there is a shortage of water. The failure behaviour of the two types of deposit may, however, be completely different. Flow failures may and have also occurred in dumps (see failures (2), (4) and (10) in Table 11.1) and a flow failure in a dump will also be described.

11.10.1 *Failures of mine development waste rock*

The rock dumps that will be considered are similar to those illustrated in Figure 1.8 and Plate 1.4. In all cases, the waste rock was end-tipped from the top of a ramp or from the edge of a dump built up from a lower level. Specifically, referring to Figure 1.8 they are heaped, or side-hill dumps. Plates 11.13, 11.15 and 11.16 all illustrate dumps of this type that have failed in shear. Plate 11.13 is a frontal view of a waste quartzite dump, which is also illustrated in Figures 11.18 and 11.19. The wall of earth in the left foreground was pushed forwards and upwards by failure of the waste rock behind it. The darker-toned slope to the right reached the same height, but did not fail. Plate 11.14 is a view of the top of the same dump showing the 6m high displacement caused by sliding of the wedge of waste rock that caused displacement of the wall of soil shown in Plate 11.13.

Plate 11.15 shows a dump of coal gasifier ash that has failed in much the same way, pushing up a wedge of ash and a low wall of earth in front of it. The two swiveling

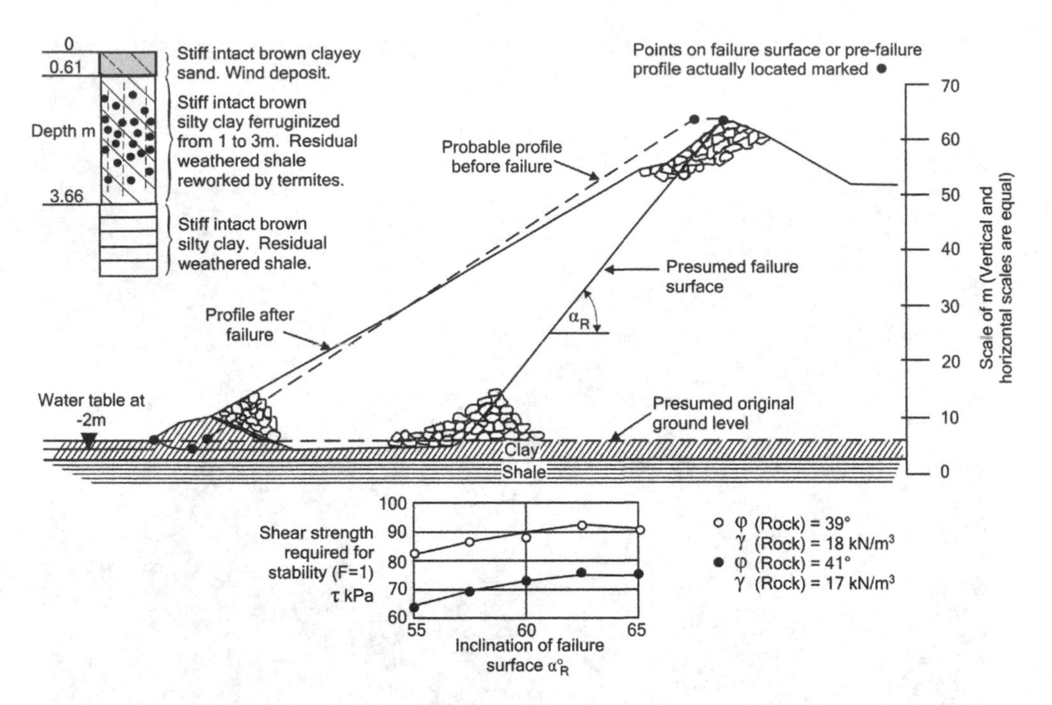

Figure 11.18 Detail of a first-time failure in a waste rock dump on a thin clay foundation stratum.

conveyors used to end-dump the ash on the sides and leading edge of the dump are visible on the skyline.

Finally, Plate 4.1 shows a dump of colliery discards that has not only caught fire by spontaneous combustion, but has failed, pushing up a wall of soil ahead of it (on the right). The failures shown in Plates 11.13 and 11.14 and Figures 11.18 and 11.19 were described and analysed in detail by Blight (1969, 1981).

Both of the dumps shown in Figures 11.18 and 11.19 consisted of run-of mine waste rock varying in size from 300 mm down to dust. The grading curve on the conventional percentage finer versus log particle size basis was close to a straight line with the ratio of D_{90} to D_{10} being about 100. The density of the dumped rock averaged between 17 to 18 kN/m^3 and the angle of repose of the rock varied from 37° to 40°. The angle of shearing resistance of the rock confined within the body of the dumps was slightly larger and appeared to be about 41°.

All of the dumps in the area of those studied were formed on approximately level ground by end-tipping from a ramp, the height of the dumps being continually increased. It was only if foundation conditions were poor that failures occurred at dump heights of less than 75 m. If a dump failed in shear, it was then extended horizontally at the highest elevation it was possible to maintain.

In each case the foundation of the dumps consisted of a stratum of stiff fissured clay which was thin in relation to the height of the dump (typically about 10% of the maximum dump height). The clay stratum overlay rock, and when failures occurred the failure surfaces through the clay foundation were constrained to follow a horizontal plane.

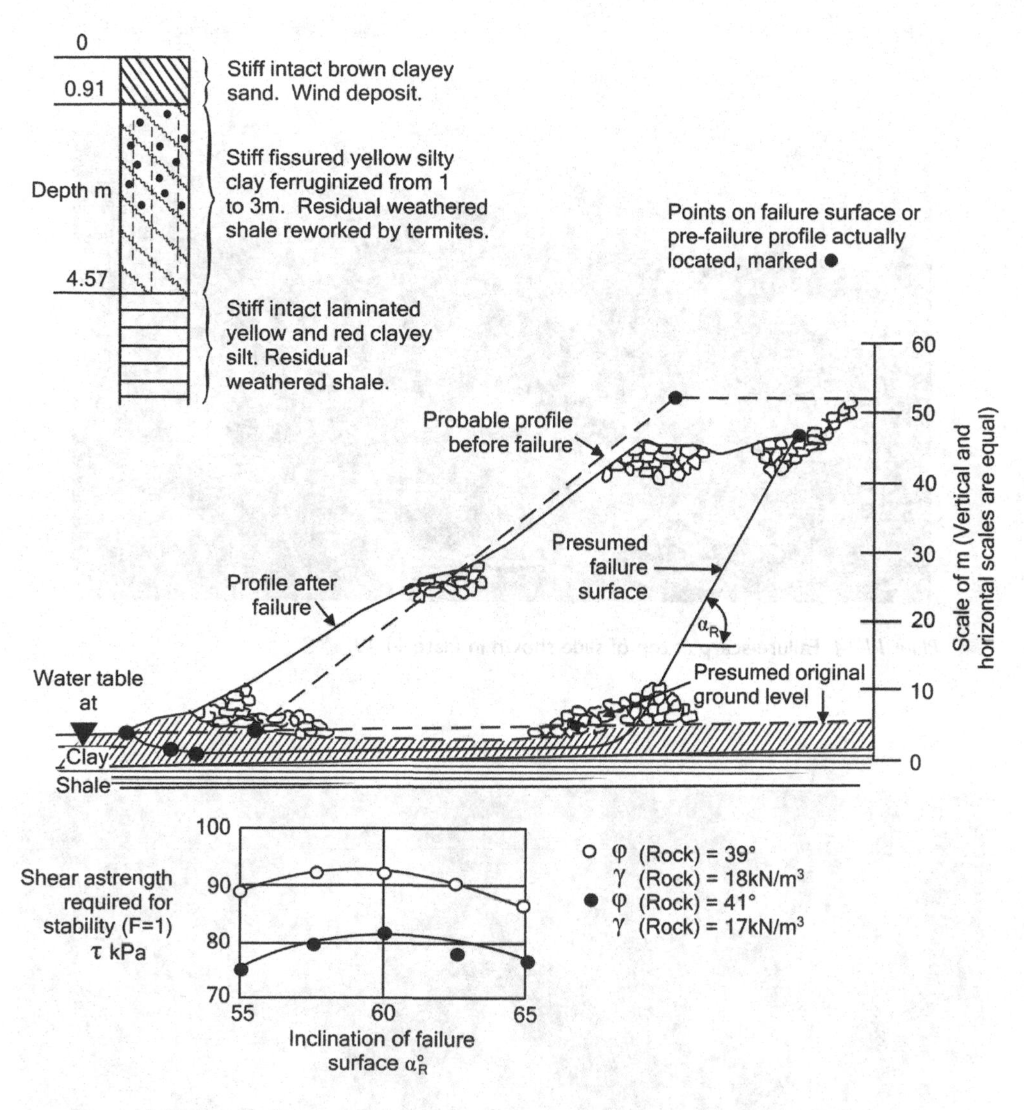

Figure 11.19 Detail of a second failure in a waste rock dump on a thin clay foundation stratum.

Figures 11.18 and 11.19 show surveyed sections through two failed rock dumps. Figure 11.18 shows the failure of the side of a ramp and illustrates the form that these failures take when they occur for the first time. To construct the figure, points on the pre-failure profile were located by lining up the failed section with adjacent unfailed sections. Points on the failure surface were either visible outcrops, or were located by hand-augering into the soil scarp at the toe of the slide and carefully examining the soil profile. The failure surface could readily be located by noting changes in consistency and colour of the profile. The profile above the failure surface consisted of loose, broken, disoriented blocks of clay, whereas the soil underlying the failure surface was intact, horizontally stratified and considerably stiffer.

Plate 11.14 Failure scarp at top of slide shown in Plate 11.13.

Plate 11.15 Failure in dry ash dump.

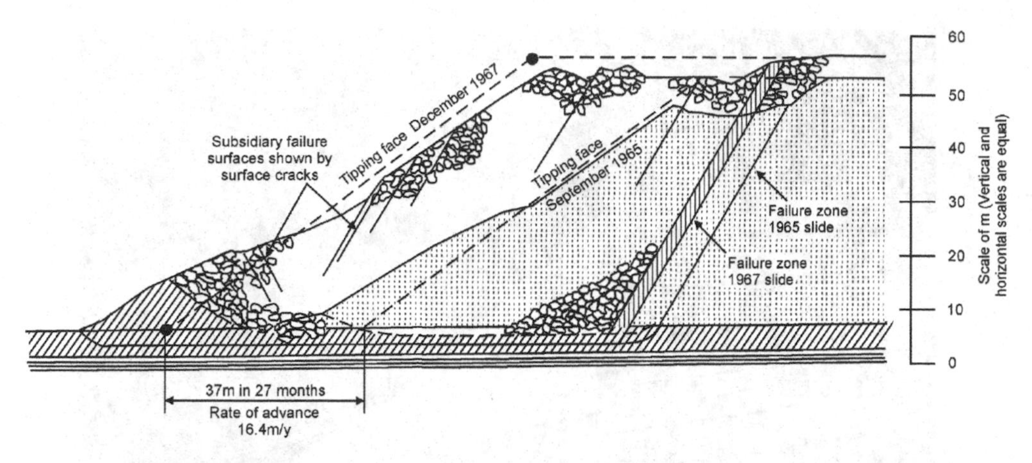

Figure 11.20 Comparison of successive failures in a waste rock dump. Note that failure zone of September 1965 slide almost coincided with that for December 1967 slide.

When a dump is advanced over a failed section, it generally fails again. Observations of the surveyed section of a dump that was failing repeatedly (Figure 11.20) indicated that failure through the rockfill tends to recur in the same position as the tip face advances. Hence the profile of a failed dump changes from the type illustrated by Figure 11.18 to that illustrated by Figure 11.19. Figures 11.19 and 11.20 are both sections of the dump illustrated in Plates 11.13 and 11.14, observed approximately 2 years apart.

The form taken by the failure surface (actually a complex of shear surfaces) within the rockfill was explored by means of laboratory models, photographs of two of which are shown in Plates 11.16 and 11.17. The models were built of a crusher sand produced from the waste quartzite of the full-scale dumps and having a similarly shaped grading curve and a similar angle of repose. In order to build and then induce failure of the model slopes, they had to be built on a foundation stratum that would model the characteristics of the foundation strata on the full-scale dumps, and at the same time permit model completion before it failed.

Studies of the rate of horizontal advance of the prototype dumps (16 to 61 m/y) in relation to the consolidation characteristics of the foundation clay showed that failure through the clay was occurring in an essentially unconsolidated-undrained state. Hence the foundation stratum should be modeled by a viscous $\tau = c$, $\varphi = 0$ material that would progressively lose strength with time under load. It was found that a 80–100 penetration bitumen produced the desired effect at the ambient laboratory temperatures of 25°C.

Plate 11.16 shows a model failure that simulates a "first-time" failure of a prototype rock dump. As indicated diagrammatically in Figure 11.21a the failure occurred by the formation of an active (or apex-down) wedge bounded by the slope face, and two shear surfaces inclined at steep angles to the horizontal. The active wedge displaced a passive (or apex-up) wedge ahead of it. The horizontal displacement of the passive wedge took place by shearing through the foundation in the direction of the arrow. The values of the angles indicated on Figure 11.21 are typical measured values.

Plate 11.16 First-time type failure in model rock dump.

Plate 11.17 Subsequent-type failure in model rock dump.

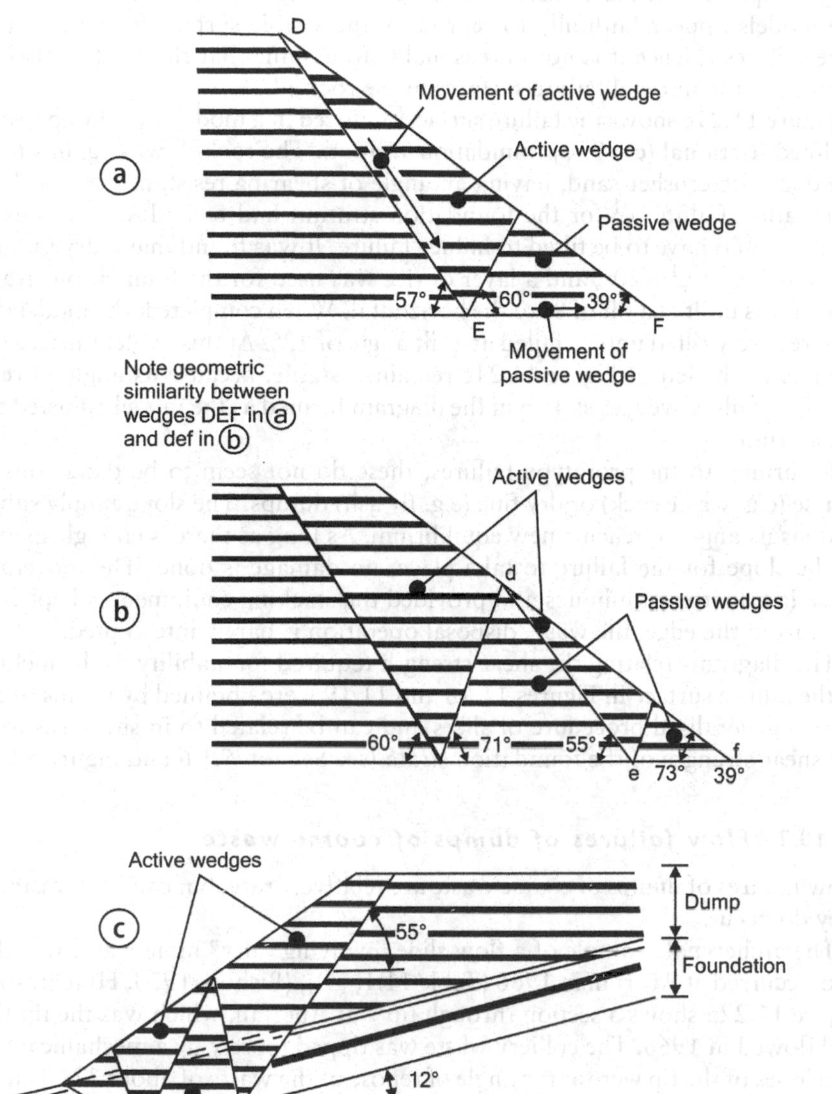

Figure 11.21 Active and passive wedges formed in failure of model rock dumps on a: and b: horizontal foundation strata and c: inclined foundation stratum. a: was a first-time failure, b: was a subsequent-type failure. a: and b: were on c, $\varphi = 0°$ foundation strata, c: was on a φ, c = 0 foundation stratum.

By adjusting the thickness of the model foundation stratum, it proved possible to cause failures simulating the condition of repeated failure of rock dumps. One such model failure is shown in Plate 11.17 and, in diagrammatic form, in Figure 11.21b. With this type of failure, at least two active wedges form which horizontally displace the corresponding pair of passive wedges.

A comparison of Plates 11.16 and 11.17 with Figures 11.18 and 11.19 shows that the models appear faithfully to reproduce the visible surface features of the prototype failures. Hence it is not unreasonable to assume that the models also faithfully reproduce the internal failure surfaces in the rockfill.

Figure 11.21c shows the failure surfaces induced in a model rock dump resting on an inclined frictional ($c = 0$, φ) foundation stratum. The rockfill was again simulated by the quartzite crusher sand, having an angle of shearing resistance $\varphi^I = 35°$. To cause foundation failure, φ^I for the foundation stratum had to be less than this, and the model would have to be tilted to induce failure. It was found that a dry unhusked rice had a value of $\varphi^I = 20°$, and a layer of rice was used for the foundation stratum. The model was built at a tilt of 8° to thc horizontal. When completed, the model dump was progressively tilted until it failed at a tilt angle of 12°. At this angle, the free surface of the rice to the left of Figure 11.21c remained stable, because its angle of repose was 20°. The failure wedges shown in the diagram formed as the model adjusted to its new equilibrium.

Returning to the prototype failures, these do not seem to be dangerous either in coarse (e.g. waste rock) or dry fine (e.g. fly ash) dumps. The slope simply subsides and flattens its angle to reach a new equilibrium. As long as there is enough space in front of the slope for the failure to take place, no damage is done. The movement takes place in a matter of minutes and provided the stacking equipment is kept far enough back from the edge, the waste disposal operation is barely interrupted.

The diagrams relating the shear strength required for stability to the inclination α_R of the failure surface in Figures 11.18 and 11.19 were obtained by means of the Janbu (1973) generalised procedure of slices and can be related to in situ measurements of the shear strength of the foundation strata (see Section 5.1.6 and Figure 5.9).

11.10.2 *Flow failures of dumps of coarse waste*

Flow failures of dumps of coarse waste are relatively rare, but can be devastating when they do occur.

The archetypal example of a flow slide involving "dry" mine waste was the failure that occurred at Aberfan in 1966 (Table 11.1, (4)), (Bishop, 1973, Hutchinson, 1986). Figure 11.22a shows a section through tip 7 at Aberfan, which was the tip that failed and flowed in 1966. The colliery waste was tipped loosely by a mechanical tipper and the slopes of the tip were at the angle of repose of the waste of about 37°. Under the toe of tip 7 was a spring, fed by water in the underlying sandstone under artesian pressure between the uppermost coal seam and the surface layer of alluvial boulder clay, which acted as lower and upper aquicludes. The height of the tip when the failure occurred was about 67m from toe to crest.

The first slide in tip 7 had occurred in 1963, but had been dumped over. The dump continued to advance down-hill and grow ever higher. Between 1963 and 1966 a number of slides had occurred. The spring had eroded fines from the waste which covered the slopes below the dump as a sheet of slurry. The failure was probably initiated by this series of slides which were triggered by the artesian pressure of the spring and exacerbated by contraction of the loose, bulked waste (see section 4.10) as it became saturated by upward seepage from the spring. Figure 11.22b shows a plan of the slide that records changes of surface level that occurred as a result of the failure.

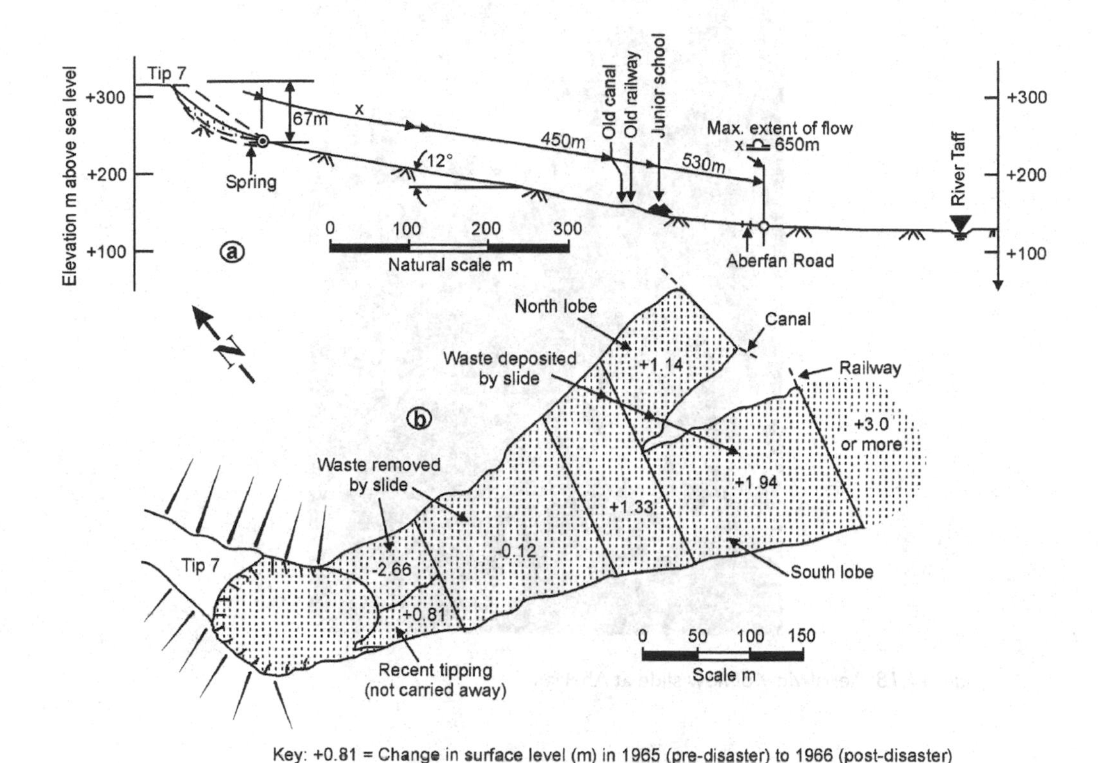

Figure 11.22 a: Section to natural scale through course of south lobe of waste flow at Aberfan. b: Plan of waste flow showing bifurcation of flow.

At 07.30 on the morning of the failure, the tipping gang found that the crest of tip 7 had moved downwards by 3m over a distance of 10–12 m from the edge. By 08.30 this displacement had increased to 6m. At 09.10 the toe of the tip started moving down the 12° hillside at walking speed, pushing the sheet of dried, crumbling slurry ahead of it. Within a few minutes the flow down the hillside had accelerated, splitting into two lobes, as shown in the plan view. The north lobe was halted by a low embankment alongside an old canal, but the south lobe crossed the canal as well as an abandoned railway embankment behind it and entered the village. The flow travelled 530 m before reaching the houses and school which it destroyed, and came to a halt 120 m further downhill. (See Section 11.8 for an analysis of the motion of the flow.)

This was the first disaster of such magnitude that had occurred in Britain and caused a serious reconsideration of current practice in the disposal of mine and quarry wastes. This resulted in the adoption of a new set of regulations for the control of the disposal of mine waste in Britain (Bishop, 1973).

Plate 11.18 is an aerial view of the flow, with the north lobe towards the right, and the more mobile south lobe towards the left. Plate 11.19 is a view 30 years later. The light-coloured square building in the centre of the view is the re-built junior school. Much of the coal waste has been removed from the hillside and what remains has been rehabilitated. The site of the demolished school is now a garden of remembrance.

Plate 11.18 Aerial view of flow slide at Aberfan.

Plate 11.19 Aberfan village 40 years after failure.

The failure at Aberfan is not the only flow failure on record for dumped dry waste. Table 11.1 records two more, number (2) at Jupille in Belgium in 1961 and number (10) at Quintette Marmot, Canada. The failure in Canada was ascribed to pore pressure (possibly including pore air pressure) caused by collapse settlement of the loosely dumped waste. The slide in Belgium involved dry-dumped fly ash which was apparently mobilised by pore air pressure and flowed downhill as a dry air-borne fluid, demolishing or filling houses with fluid ash and killing 11 people (Bishop, 1973).

11.11 Failures caused by collapse of tailings storages into subterranean caverns or underground workings

In the 1960's, a group of new gold mines was being developed in South Africa to the west of the original Witwatersrand gold mining area. Great difficulties were being experienced with in-rushes of water from the karstic near-surface dolomite rock strata that overly the gold-bearing Witwatersrand series reefs. Because the dolomites were divided into water-tight compartments by a series of near-vertical dolerite dykes, the mining companies decided jointly to dewater the compartments within which most of the mines were located. The water table was lowered by as much as 500 m in places. This alleviated the underground water problem but because the bouyancy of the water had been removed, surface subsidences, and particularly sink-holes began to appear. Plate 11.20 shows the result of a sinkhole forming beneath a tailings storage

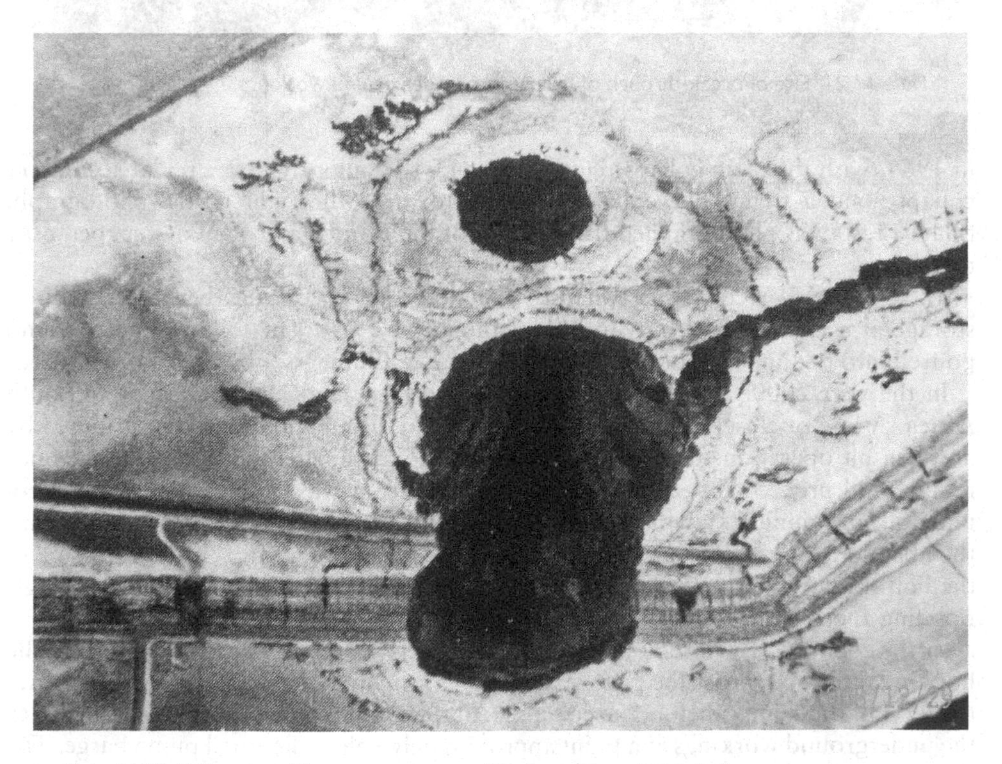

Plate 11.20 Failure in tailings storage caused by formation of sink hole.

Plate 11.21 Site of break-through of tailings into underground workings.

in 1964, with the main collapse cutting through the outer dyke, a circular secondary collapse and a trench-like collapse running towards the right of the photograph. (Plate 11.20 was photographed from a copy of the Rand Daily Mail newspaper, now defunct.)

Table 11.1 (5) records the collapse of underground workings beneath a tailings storage at a copper mine in Mufulira, Zambia, in 1970. The tailings liquefied and poured into the underground workings, drowning 89 miners.

In the year 2000, a similar collapse occurred at a small diamond mine in South Africa. A line of closely spaced diamond-bearing volcanic pipes was being mined by an open pit operation. Because of difficulty in controlling seepage into the pit once mining had proceeded deeper than the water table, it was decided to change to an underground mining system. This was done, and the tailings were open-ended into the abandoned pit. Very unfortunately, the mine workings approached too closely to the bottom of the pit and the tailings broke through into the underground workings, flooding them with liquid mud and drowning 4 miners. Plate 11.21 shows the pit after the collapse, which took place beneath the pool of water in the photograph. The horizontal terrace across the picture to the right, towards the back of the view shows the level of the tailings before the collapse. All the tailings below this level rushed into the underground workings at a point approximately below the small pump barge. The shaft tower can be seen on the horizon.

11.12 Failures of impervious linings installed on steep slopes

It happens, not infrequently, that an impervious lining incorporating a geomembrane and a drainage layer has to be constructed on a slope. This is particularly the case with in-pit storages and can also occur when an impervious lining is required on the upstream face of a cross-valley storage.

11.12.1 *The design concept and construction method*

Three methods of design for impervious linings on slopes are in common use. These are the so-called "limit equilibrium" method described by Giroud & Beech (1989) and Koerner & Hwu (1991), the closely related "limit" method proposed by Richardson & Koerner (1987) and Koerner (1990) and to a lesser extent, the "equal strain or composite column" method proposed by Long et al. (1994). The critical period in the life of the lining is during and immediately after construction, when the weight of the geomembrane plus its protective layer and the drainage layer (less friction at the lining-to-foundation interface) have to be hung from an anchorage at the top of the slope.

Equilibrium parallel to the slope requires the disturbing force, arising from the down-slope components of the self-weight of the lining and its protection and drainage layers to be balanced by a system of resisting forces. The latter consist of interlayer shear forces resulting in tensile forces transmitted to a tension anchorage at the top of the slope.

The operation of such a multi-force system is demonstrated by the results of a series of large-scale, inclined ramp tests (1.9 m long × 0.47 m wide) by Palmeira et al. (2002 and 2003), which are illustrated by Figure 11.23. The layers of the liner system consisted of a soil layer overlying a composite geotextile and geogrid layer that in turn rested on a geomembrane in contact with the surface of the inclinable ramp. In Figure 11.23, the driving shear stress (D) represents the self-weight component of the soil layer parallel to the ramp surface. As the ramp inclination was increased from 0 to 10°, D was almost entirely balanced by the shear stress on the ramp surface (B). As the inclination was increased from 10 to 15°, B passed its maximum value and then slowly declined as the soil layer tended to slide down the ramp, but was restrained by increasing tensions transferred by interlayer friction to the geomembrane (GM) and the geotextile and geogrid (GT/GG). At all inclinations, the sum of the stresses (B + GM + GT/GG) must, for equilibrium, add up to stress D, and, as shown by the measurements for an inclination of 24°, they did balance very accurately.

A failure occurred in a prototype lining for which the components are illustrated in Figure 11.24. The failure is shown in Plate 11.23. In the prototype lining, the counterpart of shear stress B in Figure 11.23 would be the shear stress between the under-side of the geomembrane and the geofabric-covered sand-cement geomembrane base (see Figure 11.24). The counterpart of shear stress GM would be the shear stress on the upper side of the geomembrane, which would be partly balanced by B and partly transmitted by tension in the geomembrane to the anchorage at the top of the slope. The counterpart of GT/GG would be shear transmitted from the sand-cement protection layer, partly balanced by shear on the upper side of the geomembrane and partly directed as a tension in the geotextile/geogrid to the top anchorage.

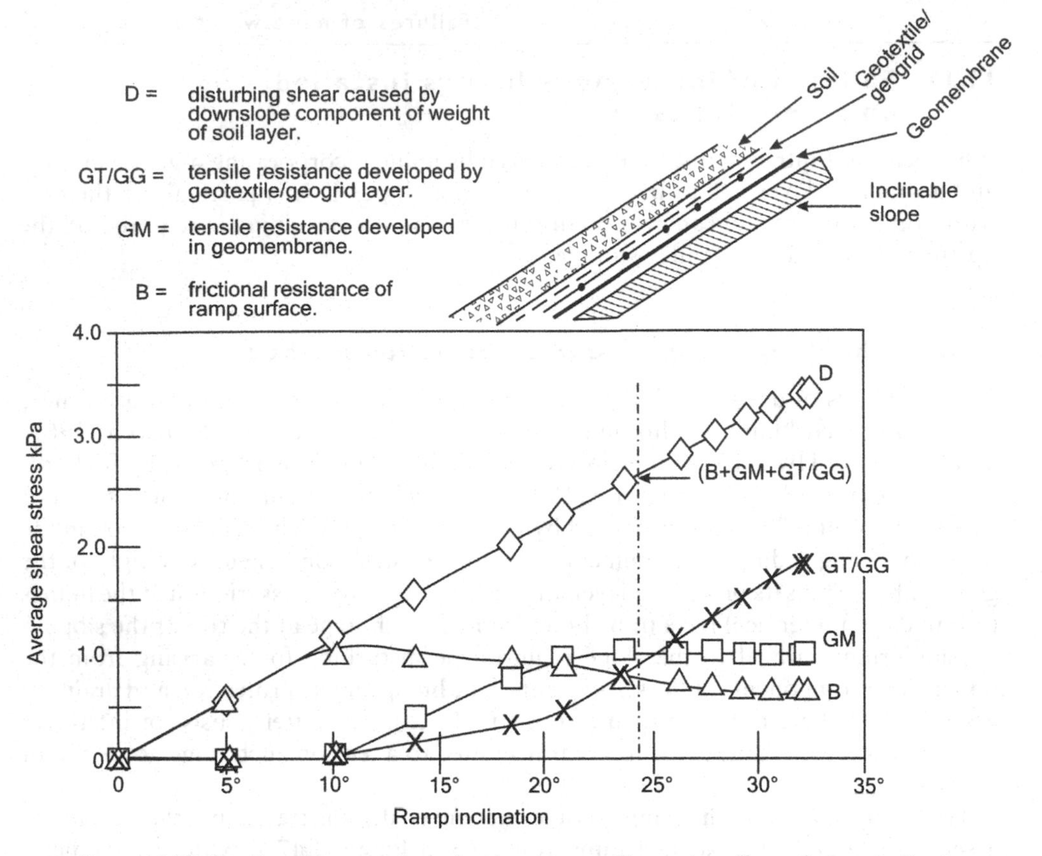

Figure 11.23 Results of large-scale inclined ramp test illustrating effect of ramp inclination on driving force D and resisting forces B, GM and GT/GG.

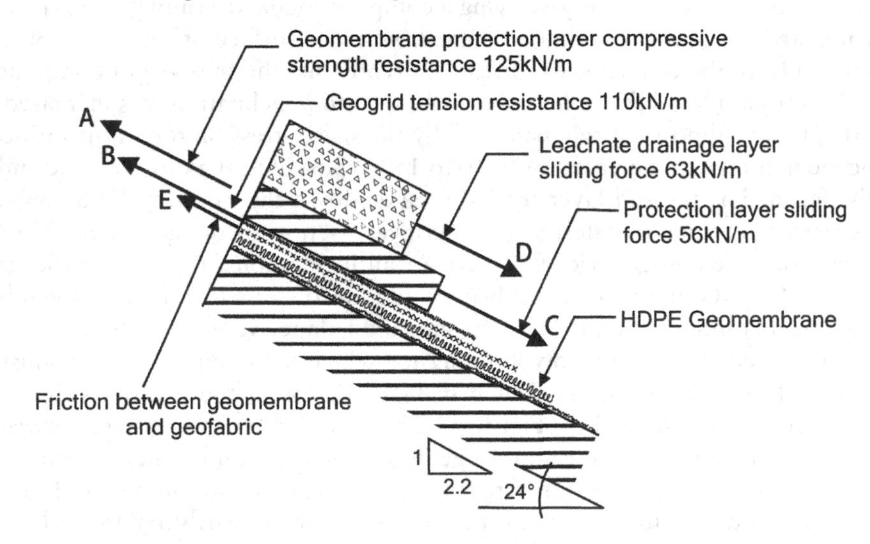

Figure 11.24 Components of prototype lining and forces assumed in design.

Plate 11.22 Failure of impervious lining during construction on a slope.

As the designers intended that the 125mm thick sand-cement geomembrane protection layer would be constructed upwards from the bottom of the slope, they assumed that the tension in the geotextile/geogrid would be small, and that resistance to the downslope forces, arising mainly from the self-weight of the sand-cement and drainage layers, would be resisted by compression in the sand-cement, with the maximum compression developing at the toe of the slope.

The geomembrane-to-geofabric contact was intended to provide a low friction interface that, in the event of sliding of the geogrid and the layers (and possibly waste) above it, would minimize force GM and protect the geomembrane from disruption and damage. The design assumed that negligible friction would develop on this interface. As the coefficient of friction between the smooth HDPE and the geofabric was measured by the author at 0.14, this was a conservative assumption for estimating GT/GG, but not for estimating GM. (However, as can be seen in Plate 11.22, the geomembrane appears undamaged by the subsequent failure, and close examination after the failure, confirmed that the geomembrane was indeed undamaged. Note the wrinkles in the geomembrane caused by expansion in the sun's heat.)

Figure 11.24 shows the force system, in the plane of the slope, that was assumed by the designers, A being a maximum at the toe of the slope, and B and E being maxima at the top anchorage. The sand-cement protection layer was intended to be constructed in sub-horizontal panels, working from the bottom of the slope upwards, so that successive panels would be supported by compressive resistance provided by the panels below them.

The contractor, however, decided that it was too difficult to build the protection layer upwards from the bottom of the slope and requested to build it downwards from the top. The designers acquiesced, although they must have realized that force A in Figure 11.24 would now disappear and have to be replaced by force B, the tensile resistance of the geogrid. Almost the entire downslope disturbing force would now hang from the top anchorage. During construction the disturbing force would also be increased by downslope stresses induced by the small crawler dozer used to spread and compact the sand-cement and drainage layers.

The designers also did not realize that the nominal strength of the geogrid actually represented the maximum strength in a short term test under ideal conditions. The geogrid was also likely to suffer some damage during installation and would be subjected to long term sustained loading (see Section 6.14.3). As a result, its actual strength was unlikely to exceed half of its nominal strength.

11.12.2 *The failures and reconstruction*

The first failure of a section of lining took place one night when 30 mm of rain fell in a little more than an hour. The protection layer was complete, but placing of the leachate drainage layer had not yet started, when a strip 14 m wide and extending over the whole height of the slope slid to the bottom. The bottom panel of the soil-cement protection layer for this strip had been placed earlier that day and consisted of soft, wet uncured material. Areas of the protection layer on either side of the slide were slightly damaged and cracked, but remained in place. The area of this slide can be seen under repair to the left of Plate 11.22, which is a view of the two failures.

Eighteen days later, without any additional rainfall having occurred, the second slide took place. A strip of completed protection layer, covered to the full 200 mm depth with its stone drainage layer, about 24 m wide and extending over the whole height of the slide collapsed, as shown (right of centre) in Plate 11.22. This slide was separated from the first by a 15 m wide strip of protection layer, partly covered with the drainage layer. After removing and replacing the geosynthetic materials damaged by the two slides, the protection layer was reconstructed from the bottom up, as originally required. The stone drainage layer was placed from the top down, as before. The repair was completely successful and the lining remained stable.

11.12.3 *Summary of causes of failure*

- Changing the sequence of construction from bottom-up to top-down completely changed the structural functioning of the elements in the geomembrane protection system. The sand-cement layer (unsupported from below) could not provide any compressive resistance and the entire downslope component of the weight of sand-cement and drainage layer now depended upon the geogrid (apart from the unreliable friction between geofabric and geomembrane). If the designers had insisted on the original construction sequence being adhered to, the failure would probably not have happened, as demonstrated by the successful reconstruction.

- The designers did not realize that the nominal strength of the geogrid was a figure of merit and not a true strength, in much the same way as the specified minimum 125 kN/m strength based on laboratory cube strengths would not have been a true measure of the in situ compressive strength of the soil-cement layer. They appeared unaware that the strength of every material used in construction depends on the time under load, on the extent to which the material is damaged during installation and also on the in situ variability of dimensions and material quality.
- If a design depends on a particular construction sequence, that sequence should not be varied without a very careful consideration of the possible consequences. Designers should also clearly understand the properties, behaviour and limitations of the materials they intend using.

11.13 Methods for analysis of the stability of slopes

The objective of this chapter has been to describe failures, the events and errors that led to their occurrence, the effects of the failures on geotechnical practice for mine waste storage, and the stable (or meta stable) slope surfaces that are left after failures have occurred. It has also been intended to show that not all failures are dangerous or disastrous. In fact, the disastrous failures are those that occur in disastrously located positions, e.g. at Stava, where the tailings storage was perched on high, steep ground above a well-populated valley, at Merriespruit, where the tailings storage was located on the outskirts and uphill of a village and at Aberfan where the coal discard dump that failed loomed on the skyline above the village. Other failures, where the site could contain the run-out material, such as those at Saaiplaas, were never reported to the public at all and therefore the geotechnical profession could have missed the benefit of the experience. This has also applied in the case of failures of municipal solid waste landfills (e.g. Blight, 2008). Hence the siting of a waste deposit of any kind is all-important, not only for public safety, but also to limit environmental damage. (An additional irony of the Merriespruit disaster is that Merriespruit village was originally set out on the site of the tailings storage which was to be sited downhill of the village. To save costs of road construction, the two sites were interchanged without considering the consequences of a failure of the tailings storage.)

The geotechnical literature has always concentrated on the mechanics of a failure, and hardly at all on the human causes. The human errors that actually caused all of the failures listed in Table 11.1 are almost routinely ignored, covered up or, what amounts to the same thing, attributed to "acts of God". (The author was once castigated by his peers for mentioning that the engineer who, for the past two years, had been responsible for a landfill that failed with a terrible death toll, had never visited the landfill and did not know where it was until it failed.) Recording and analysis of conditions after the failure to assess what would have been a safe geometry has also been done very rarely. Environmental damage is also seldom mentioned. Its remediation and the extent to which there has been irreversible environmental damage is usually ignored. (We call ourselves environmental engineers but actually concern ourselves very little with the environment, its protection or its remediation after damage.)

However, in order to design a slope in the first place, it is essential to analyse its shear stability. The methods for analysis are widely available in the geotechnical literature, and computer packages for slope stability analysis are easily available and used

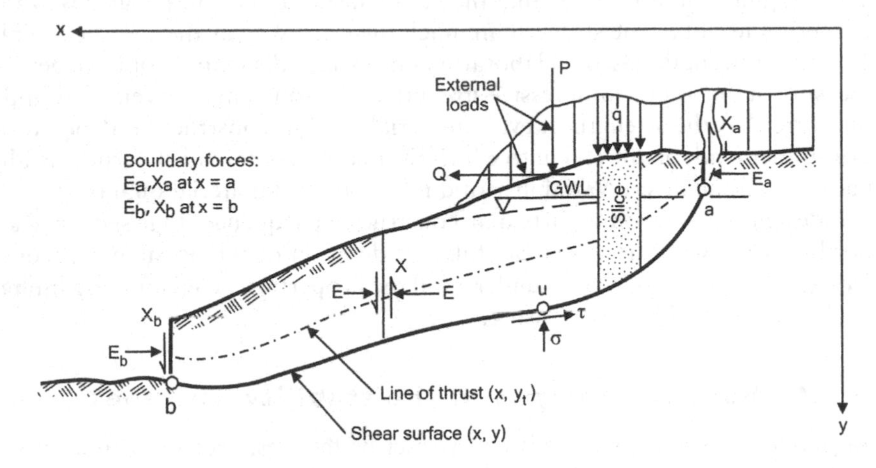

Figure 11.25 Definitions and symbols used for Janbu's generalized method of slices.

routinely (e.g. PCSTABL, SLOPW and PLAXIS). It is essential to understand the basis for the various methods of analysis and carefully to assess the design soil and waste parameters to be used, even though the detailed calculations can be left to the computer. This chapter will therefore examine the basis for three methods of analysis, all variants of the same basic approach, but will not dwell on the details of their application.

11.13.1 *Janbu's generalised method of slices*

When analysing the stability of a slope, the surface geometry of the slope is known, or is one of a series of trial geometries from which the most suitable must be selected. The strata composing the slope will have been decided by the function of the slope, if the slope is being designed, or otherwise must be established by soil exploration techniques. Hence, by testing, the unit weights and shear strengths of the soils composing the slope will be known. Similarly, the pore pressure distribution in the slope can be found either by prediction or by measurement in situ.

In Figure 11.25 the shear surface ab is one of many possible surfaces, and the problems to be solved are which surface will give a minimum factor of safety against sliding and what is the minimum factor of safety. If there is a surface for which the factor of safety is unity, it will be the probable surface of failure.

The mass of soil above a shear surface to be investigated is approximated by dividing it into a series of vertically sided slices of which the slice in Figure 11.26 is typical. The resultants of the horizontal and vertical interslice forces are E and X; ΔS and ΔN are the resultants of the shear stress τ and normal stress σ acting on the base of the slice, and $\Delta N = \sigma \Delta l$ and $\Delta S = \tau \Delta l$. In general, E, X, ΔN and ΔS are unknown. The basic assumptions are:

- The problem can be regarded as two-dimensional.
- The equilibrium shear stress along the shear surface is $\tau = \tau_f/F$ where τ_f is the shear strength and F is the factor of safety.

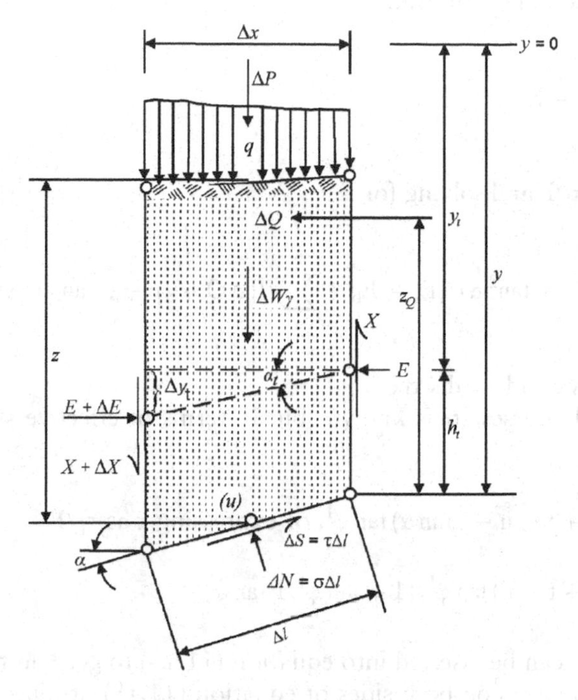

Figure 11.26 Forces acting on the boundaries of a single slice.

- The resultant ΔN acts where $\Delta W_\gamma + q\Delta x + \Delta P$ intersects the base of the slice.
- The line of thrust for the side forces E is known.

The average total vertical stress on the base of the slice is

$$p = \Delta W/\Delta x = \gamma z + q + \Delta P/\Delta x [\text{in kN/m}^2] \tag{11.12}$$

Horizontal forces applied at ground level or within the mass (e.g. earthquake forces) are represented by ΔQ acting at z_Q above the base of the slice. h_t is the height of the line of thrust and α_t its inclination. For cohesionless soils, h_t would be close to z/3 and for cohesive soils h_t would be more than z/3 towards the toe of the slope and less than z/3 near the crest.

The equations that must be solved simultaneously are:

$$\tau = c^I + (\sigma - u)\tan\varphi^I \tag{11.13}$$

$$\sigma = \Delta W/\Delta x + dX/dx - \tau\tan\alpha \tag{11.14}$$

$$\Delta E = \Delta Q + (\Delta W/\Delta x + dX/dx)\Delta x\tan\alpha - \tau\Delta x(1 + \tan^2\alpha) \tag{11.15}$$

$$X = -E\tan\alpha_t + h_t \cdot dE/dx - zQ \cdot dQ/dx \tag{11.16}$$

Equation (11.13) defines shear equilibrium,
equation (11.14) defines vertical equilibrium for each slice,
equation (11.15) combines horizontal and vertical equilibrium for each slice, and
equation (11.16) defines moment equilibrium.

For overall horizontal equilibrium

$$\sum_{a}^{b} \Delta E = E_b - E_a \tag{11.17}$$

Introducing $\tau = \tau_f/F$ and solving for F

$$F = \sum_{a}^{b} \tau_f \Delta \times (1 + \tan^2 \alpha)/\{E_a - E_b + \sum_{a}^{b} [([\Delta Q + (p + t) \tan \propto \Delta x]\}] \tag{11.18}$$

where $p = \Delta W/\Delta x$ and $t = dX/dx$

In terms of total stresses, τ_f is known, but in terms of effective stresses τ_f must be written as

$$\tau_f = c^I + (p + t - u - \tau \tan \alpha) \tan \varphi^I, \text{ or expressing } \tau \text{ as } \tau_f/F$$

$$\tau_f = c^I + (p + t - u) \tan \varphi^I/[1 + \tan \varphi^I/F \tan \alpha] \tag{11.19}$$

Equation (11.19) can be inserted into equation (11.18) to get F in terms of c^I and φ^I. Because F then appears on both sides of equation (11.18), it must be solved by trial and error.

Janbu's theory is very useful because it can be used for either circular or non-circular shear surfaces. The shear surfaces in Figures 11.18 and 11.19 were analysed using the Janbu method.

11.13.2 *Bishop's method of slices for circular shear surfaces*

Figure 11.27 shows a typical slice of the soil above a circular shear surface, together with the vector diagram for the forces acting through the midpoint of the base of the slice. For moment equilibrium about the centre of the circle:

$$R\Sigma W \sin \alpha = R\Sigma T, \text{or}$$

$$\Sigma W \sin \alpha = \Sigma T = \Sigma(c^I l/F + P^I \tan \phi^I), \text{ with } P^I = P - ul$$

If the lateral forces do not differ by much on each slice, their net moment about the centre of rotation can be ignored and

$$F = \Sigma(c^I l + P^I \tan \varphi^I)/\Sigma W \sin \alpha \tag{11.20}$$

Also, resolving forces on the slice vertically:

$$W = P \cos \alpha + T \sin \alpha \tag{11.21}$$

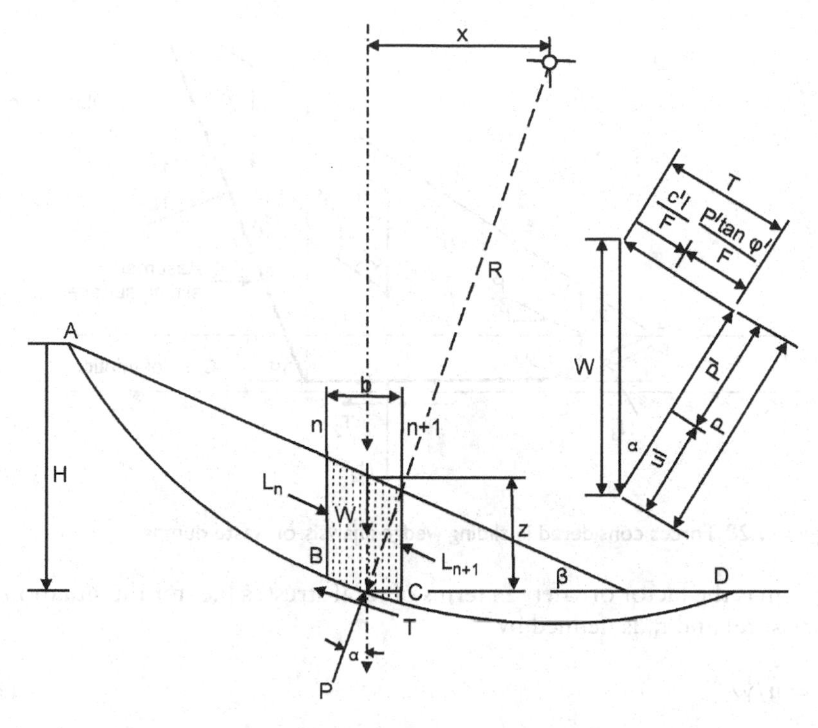

Figure 11.27 Basis of Bishop's method of slices for circular shear surfaces.

Hence

$$(P^I + ul) = W \sec\alpha - [c^I l/F + P^I \tan\varphi^I/F] \tan\alpha$$

and

$$P^I = [W - l(u\cos\alpha + c^I/F \sin\alpha)]/[\cos\alpha + \tan\phi^I \sin\alpha/F] \qquad (11.22)$$

Substitute in equation (11.20) to get:

$$F = 1/\Sigma W \sin\alpha \Sigma \cdot \{[c^I l + (W - ul\cos\alpha - c^I l \sin\alpha/F)\tan\varphi^I]/[\cos\alpha + \tan\varphi^I \sin\alpha/F]\} \qquad (11.23)$$

Because F appears on both sides of the equation, the solution requires a trial and error approach.

A series of stability charts has been prepared, based on equation (11.23) (Bishop and Morgenstern, 1960). These charts (which are reproduced in most text books on soil mechanics) use the expression for F:

$$F = m - nr_u \qquad (11.23a)$$

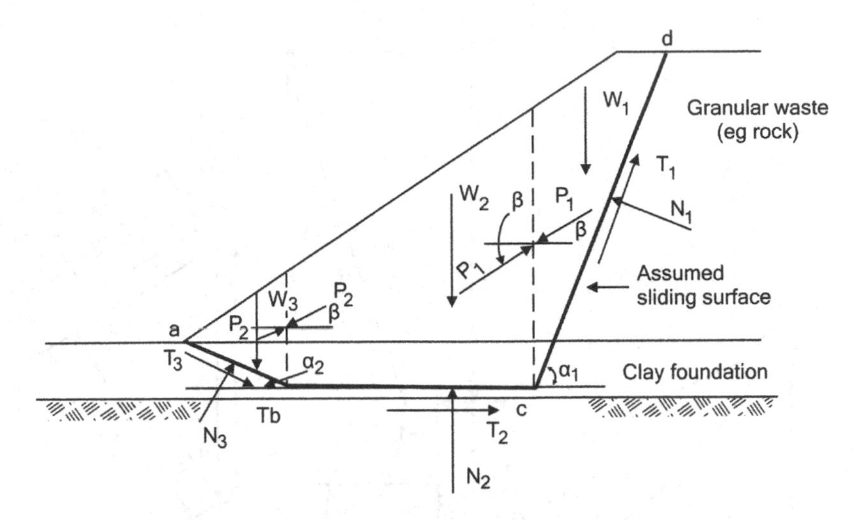

Figure 11.28 Forces considered in sliding wedge analysis of waste dumps.

In which m is the factor of safety in terms of total stresses (i.e. for the situation of zero pore pressure) and r_u is defined by

$$r_u = u/\gamma z \tag{11.23b}$$

r_u may range in value from 0 to γ_w/γ. For example, if $\gamma = 15\,\text{kN/m}^3$, the maximum value of $r_u = 10/15 = 0.67$, if $\gamma = 18\,\text{kN/m}^3$ the maximum value is $10/18 = 0.56$. The stability charts give values of m and n for a range of slope angles β and angles of shearing resistance in effective stress terms, φ^I. The charts vary with the ratio $c^I/\gamma H$ where H is the overall height of the slope.

11.13.3 *The sliding wedge analysis*

The sliding wedge analysis which is often applied to rock dumps, is illustrated in Figure 11.28 and is essentially a Janbu analysis with only three slices.

The equations for vertical equilibrium are:

$$W_1 = P_1 \sin\beta + N_1 \cos\alpha_1 + N_1 \sin\alpha_1 \tan\varphi_R^I/F \tag{11.24a}$$

$$W_2 = N_2 + (P_2 - P_1)\sin\beta \tag{11.24b}$$

$$W_3 = -P_2 \sin\beta + N_3 \cos\alpha_2 - [(N_3 - u \cdot ab)\tan\varphi_f^I/F + c^I \cdot ab/F]\sin\alpha_2 \tag{11.24c}$$

(The subscripts R and f refer to "rock" and "foundation", respectively.)

For horizontal equilibrium:

$$P_1 \cos\beta + N_1 \tan\varphi_R^I/F \cdot \cos\alpha_1 = N_1 \sin\alpha_1 \tag{11.25a}$$

$$(P_1 - P_2)\cos\beta = (N_2 - u \cdot bc)\tan\varphi_f^I/F + c^I \cdot bc \tag{11.25b}$$

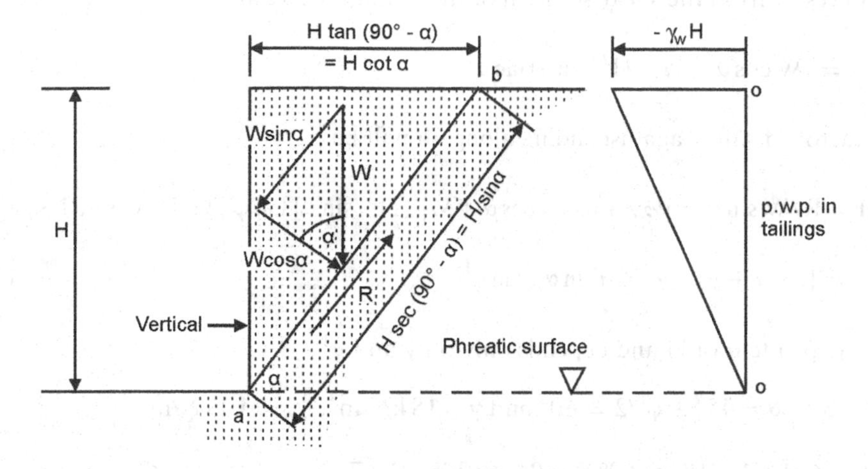

Figure 11.29 Basis for analysis of stability of vertical face of tailings.

$$P_2 \cos\beta = N_3 \sin\alpha_2 + [(N_3 - u \cdot ab)\tan\varphi^I/F + c^I \cdot ab]\cos\alpha_2 \qquad (11.25c)$$

The unknowns in the six equations are:

$P_1, P_2, N_1, N_2, N_3, \beta$ and F (7 unknowns). The solution is usually found by assuming values for β and then finding the minimum value for F. (This requires the assumptions that F for the granular waste is the same as that for the foundation material and that the length of surface cd cutting through the foundation is small in comparison with the length cutting through the granular or rock waste. The angles α_1 and α_2 must also be assumed.)

Referring to Figure 11.21, the angle α for F = 1 will be given approximately by

$$\alpha_1 = 45° + \varphi^I_R/2, \text{ e.g. if } \varphi^I_R = 30°,\ \alpha_1 = 60°, \text{ if } \varphi^I_R = 40°,\ \alpha_1 = 65°.$$

If the foundation stratum is thin in comparison with the height of the dump, the slice at the toe of the dump can be ignored, thus removing unknowns P_2 and N_3, and reducing the problem to one of 5 unknowns and 4 equations.

11.13.4 *Analysis of vertical slopes*

Vertical slopes can occur in tailings storages above the phreatic surface. For example, the post-failure profile of the Simmergo slope (Figure 11.7) had a vertical face above the phreatic surface. Vertical faces are also frequently formed when a tailings storage is remined for re-processing. The following analysis examines the stability of such a face.

Figure 11.29 shows the forces that will act on a potential plane failure surface inclined at α to the horizontal. The phreatic surface is at the toe of the face and the tailings are assumed to remain saturated.

$$W = \tfrac{1}{2}H^2 \tan(90° - \alpha) \cdot \gamma = \tfrac{1}{2}\gamma H^2 \cot\alpha$$

R will result from the shear strength of the tailings along ab.

$$R = (W\cos\alpha + \tfrac{1}{2}\gamma_w H^2/\sin\alpha)\tan\varphi^I$$

The factor of safety against sliding along ab will be

$$F = R/W\sin\alpha = [\tfrac{1}{2}\gamma H^2 \cot\alpha\cos\alpha + \tfrac{1}{2}\gamma_w H^2/\sin\alpha]\tan\varphi^I/\tfrac{1}{2}\gamma H^2\cos\alpha, \text{ i.e.,}$$

$$F = [\cot\alpha + \gamma_w/(\gamma\cos\alpha\sin\alpha)]\tan\varphi^I \tag{11.25}$$

F is independent of H and depends only on γ and φ^I

If $\varphi^I = 30°$, $\alpha = 45° + \varphi^I/2 = 60°$ and $\gamma = 18\,\text{kN/m}^3$, $\gamma_w = 10\,\text{kN/m}^3$

$F = (\cot 60° + 10/18\cos 60°\sin 60°)\tan 30° = 1.07$

If $\varphi^I = 28°$, and γ remains unchanged

$F = (\cot 59° + 10/18\cos 59°\sin 59°)\tan 28° = 0.99$

It is obvious that if $\gamma = 18\,\text{kN/m}^3\varphi^I$, cannot be less than 30°, or the face will fail, but this applies regardless of the height of the slope, H.

11.14 Further points regarding the failure of slopes

- It has long been recognized in seismically active areas of the world that liquefaction caused by cyclic loading is one of the most important factors leading to the failure of tailings dams or retaining dykes. It is only relatively recently that it has been realized that static liquefaction can also occur and has played an important role in failures in seismically stable areas. Static liquefaction is now known to be closely related and to be caused by similar factors to liquefaction under cyclic loading (see Figure 4.40).
- One of the most important factors in preventing failure is therefore to deposit the tailings in such a way that they will not liquefy. This means that they should be deposited at a slow enough rate to allow of full consolidation and at a relative density such that when fully consolidated the void ratio is below the critical value (see Figures 4.33 and 4.34).
- It has also long been realized that in water surplus areas of the world (where annual rainfall exceeds annual evapotranspiration) upstream building is not a safe method of construction, because it results in fine tailings with a generally low coefficient of consolidation and high void ratio, being deposited close to the outer slope of the dyke. Upstream building has been widely and successfully used in non-seismically active water deficit areas (where annual potential evapotranspiration exceeds annual rainfall) together with slow rates of rise. This is because a slow rate of rise allows full consolidation to take place, with the added benefit of desiccation of the beaches, during the relatively long period between depositions, to reduce the void ratio.

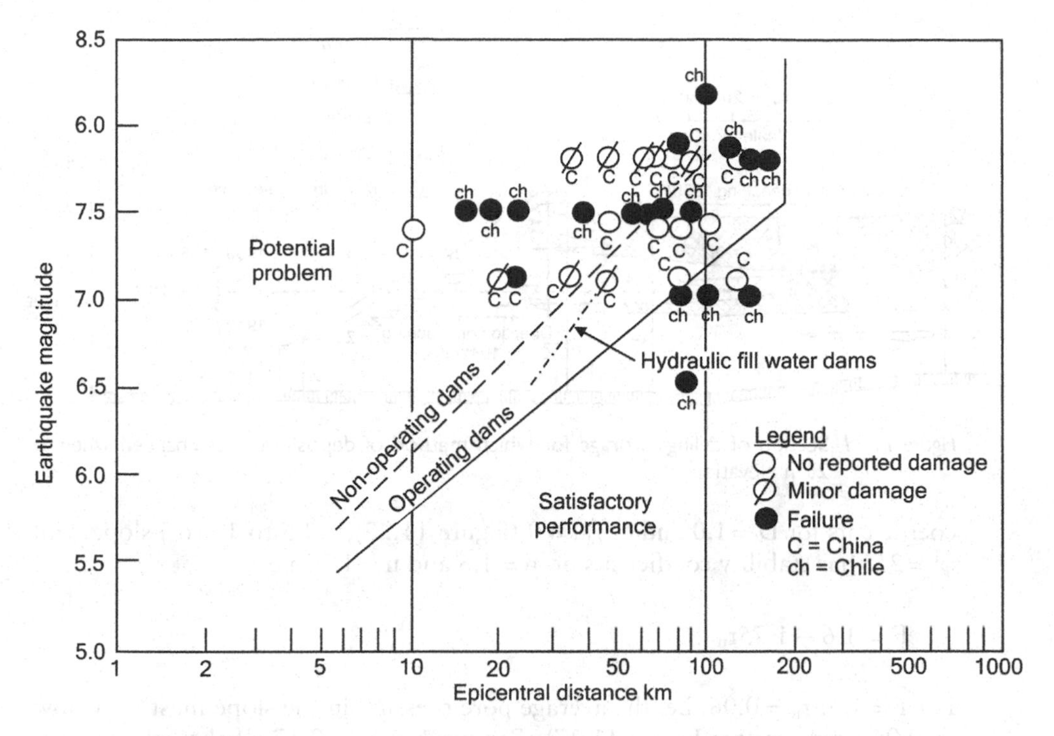

Figure 11.30 Susceptibility of upstream tailings dams to earthquake-induced damage.

- The susceptibility of upstream tailings dams to earthquake-induced damage is well illustrated by Figure 11.30. This shows a susceptibility chart by Conlin (1987) that relates the susceptibility of tailings slopes built by upstreaming to earthquake magnitude and the distance of the epicentre from the slope concerned. Events falling below the "operating tailings impoundment" line should not cause failure or damage, whereas those falling above can be expected to cause failure or damage.

  The points plotted on the diagram and identified by ch (Chile) were collected by Troncoso (Blight et al., 2000), whereas those identified by C (China) were collected by Xin, et al. (1992). It will be seen that 16 Chilean dams failed (actually 17, as in one case, 2 dams failed at the same mine). Of the 17, 4 plot within the "satisfactory performance" area. Of the 19 Chinese dams, only 2 failed, 8 reported damage and the remaining 9 dams were undamaged, even though 8 of these plot within the "potential problem" area. The difference appears to be that Chinese dams are constructed with relatively flat slopes of 11°–14°, whereas the Chilean dams were much steeper with the 2 steepest slopes being at 40° (1.2 to 1), 6 slopes were at 33.5° (1.5 to 1), 5 were at 30° (1.7 to 1), 1 at 26.5° (2 to 1) and the flattest slope was at 15° (3.7 to 1). (The last did not fail.)
- Considerations related to erosion of slopes and their rehabilitation (see Section 12.4.6) converge on a range of slope angles of 12° to 18° as being suitable for long term surface stability. Applying the Bishop and Morgenstern (1960) stability

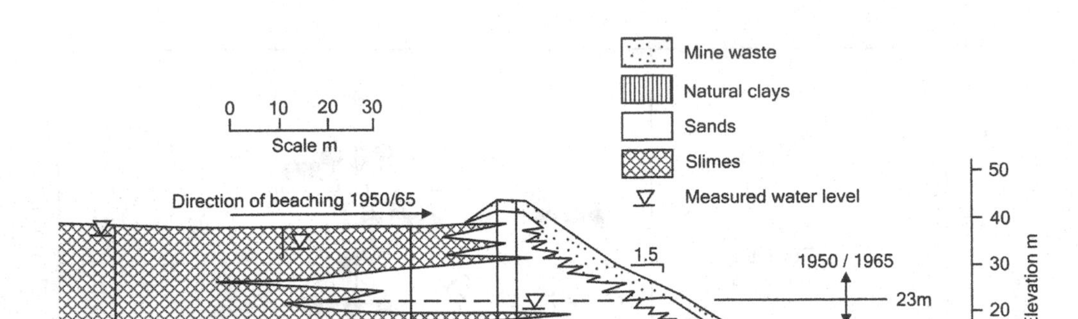

Figure 11.31 Section of tailings storage for which method of deposition was changed when at 23 m elevation.

coefficients for $D = 1.0$ and $c^I/\gamma H = 0$ (Figure 11.27), a 3.5 to 1 (16°) slope with $\varphi^I = 25°$ has stability coefficients of $m = 1.6$ and $n = 1.75$ i.e.

$$F = 1.6 - 1.75r_u$$

For $F = 1.5$, $r_u = 0.06$. i.e. the average pore pressure in the slope must be as low as 0.06 times γz (see Figure 11.27). For $F = 1.3$, $r_u = 0.17$. If the value of γ is 16 kN/m^3, the value of r_u for a phreatic surface at half of the depth z of tailings above a potential failure surface would be $u = \frac{1}{2} \times 10/16 = 0.31$. This emphasises the importance not only of a flat slope angle, but also of maintaining a well-depressed phreatic surface by the use of drains under the slope to augment stability.

- The stability and safety of any structure can be severely affected by constructing it in a manner different to that for which it was designed. Section 11.12 gives a good example of this, and the dictum also applies to dams or dykes built of tailings. One example is the description in Section 11.6, where failure occurred because particle size separation to give a coarse zone over the toe drain was not achieved. A second example relates to a tailings storage in Missouri, U.S.A. (Vick, et al., 1993). The structure, no longer operational, was checked for seismic stability. An idealized section is shown in Figure 11.31. From 1942 to 1950 the upstream construction proceeded by spigotting to form a substantial partial ring dyke of sand with the pool distant from the dyke. In 1950 the deposition method was changed. The perimeter dyke was raised upstream using coarse mine waste instead of spigotted tailings and the tailings were discharged from a point over 3 km upstream, towards the dyke. The result was that the fine tailings and the pool ended up against an insubstantial dyke of mine waste. When the stability of the structure against seismic forces was checked, it was found that the stability was controlled by the mass of fine saturated tailings against the upper part of the dyke, and in particular, by the finger of fine material that reaches towards the dyke at about mid-height.

- The final point relates to stability analysis of slopes. It is common when analysing the stability of a slope by one of the methods discussed in Section 11.13 to use an effective stress analysis with pore pressures derived from piezometer measurements and peak strength parameters in terms of effective stresses. Referring to Figure 4.35, it will be seen that if this "drained" approach is used, it will not give a realistic result for a contractant tailings, because the predicted strengths will be far too high. It is also extremely difficult, if not impossible to predict failure pore pressures. It is therefore suggested that a total stress approach be adopted, using total stress strength parameters identified as shown in Figure 4.35.

References

Anonymous: *Report of the Tribunal Appointed to Inquire into the Disaster at Aberfan*. London, U.K.: H.M.S.O., 1967.

Bishop, A.W.: The Stability of Tips and Spoil Heaps. *Q. J. Eng. Geol. 6(3 and 4)* (1973), pp.335–376.

Bishop, A.W., & Morgenstern, N.R.: Stability Coefficients for Earth Slopes. *Geotechnique 10* (1960).

Blight, G.E.: Foundation Failures of Four Rockfill Slopes. *J. Soil Mech. Found. Div., A.S.C.E., 95(SM 3)* (1969), pp.743–767.

Blight, G.E.: On the Failure of Waste Rock Dumps. In: *Symposium on Non-Impounding Mine Waste Dumps*. Denver, Colorado, U.S.A.: American Institute of Mining Engineers (1981), 7pp. [Papers numbered individually].

Blight, G.E.: Destructive Mudflows as a Consequence of Tailings Dyke Failures. *Proc. Inst. Civil Engrs, Geotech. Eng. 125(Jan.)* (1997), pp.9–18.

Blight, G.E.: Slope Failures in Municipal Solid Waste Dumps and Landfills: A Review. *Waste Manage. Res. 26* (2008), pp.448–463.

Blight, G.E., Robinson, M.J., & Diering, J.A.C.: The Flow of Slurry from a Breached Tailings Dam. *J. S. Afr. Inst. Mining Metall. Jan.* (1981), pp.1–8.

Blight, G.E., Troncoso, J.H., Fourie, A.B., & Wolski, W.: Issues in the Geotechnics of Mining Wastes and Tailings. *GeoEng. 2000*, Invited Papers, Lancaster, P.A., U.S.A.: Technomic Publishing, 2000, pp. 1253–1285.

Brink, D.: The Long-term Repair of the Merriespruit Tailings Dam. *Tailings and Mine Waste '98*. Rotterdam, Netherlands: Balkema, 1998, pp. 953–957.

Byrne, P.M., Imrie, A.S., & Morgenstern, N.R.: Results and Implications of Seismic Performance Studies, Duncan dam. *Can. Geotech. J.* (1994), pp.979–988.

Chamber of Mines of South Africa. The Design, Operation and Closure of Metalliferous and Coal Residue Deposits. *Handbook of Guidelines for Environmental Protection, Vol. 1*. Johannesburg, South Africa: The Chamber (1979, revised 1983 and 1986).

Conlin, B.H.: A Review of the Performance of Mine Tailings Impoundments Under Earthquake Loading Conditions. In: *Vancouver Geotech. Soc. Symp.*, Vancouver, Canada, 1987.

Dobry, R., & Alvares, L.: Seismic Failures in Chilean Tailings Dams. *J. Soil Mech. Found. Eng. Div., A.S.C.E. 93(SM 6)*, 1967, pp.237–260.

Giroud, J.P., & Beech, J.F.: Stability of Soil Layers on Geosynthetic Lining Systems. In: *Geosynthetics '89*. Industrial Fabrics Associates International, 1989, pp.35–46.

Hutchinson, J.N.: A Sliding-Consolidation Model for Flow Slides. *Can. Geotech. J. 23*(2) (1986), pp.115–126.

Janbu, N.: Slope Stability Computations. In: R.C. Hirschfeld, S.J. Poulos (eds): *Embankment Dam Engineering*. New York, U.S.A.: Wiley (1973), pp.47–88.

Jennings, J.E.: The Failure of a Slimes Dam at Bafokeng. Mechanisms of Failure and Associated Design Considerations. *Civil Engnr S. Afr. 21* (6) (1979), pp.135–150.

Koerner, R.M.: *Designing with Geosynthetics*. Englewood Cliffs, NJ., U.S.A.: Prentice-Hall, 1990.

Koerner, R.M., & Hwu, B.L.: Stability and Tension Considerations Regarding Cover Soils on Geomembrane Lined Slopes. *Geotext. Geomembr. 10* (1991), pp.335–355.

Lo, R.C., & Klohn, E.J.: Design Against tailings Dam failure. In: *Int. Symp. Seismic Environ. Aspects Dams Des.*, Santiago, Chile, Vol. 1, (1996), pp.35–50.

Long, J.H., Gilbert, R.B., & Daly, J.J.: Geosynthetic Loads in Landfill Slopes: Displacement Compatibility. *J. Geotech. Eng.-ASCE. 120* (1994), pp. 2009–2025.

Miao, T., Liu, Z., & Niu, Y.: A Sliding Block Model for Run-Out Prediction of High-Speed Landslides. *Can. Geotech. J. 38*(2) (2001), pp.217–226.

Midgely, D.C.: Hydrological Aspects and a Barrier to Further Escape of Slimes. *Civil Eng. S. Afr. 21*(6) (1979), pp.151–159.

Muramoto, Y., Uno, T., & Takahashi, T.: Investigation on the Collapse of the Tailings Dam at Stava in Northern Italy. In: G. Tosatti (ed) (2003), *A Review of Scientific Contributions on the Stava Valley Disaster (Eastern Italian Alps), 19 July 1985*. Bologna, Italy: Consiglio Nazionale delle Ricerche, Pitagora Editrice. ISBN 88-371-01405-2, 1986, pp.13–46.

Palmeira, E.M., Lima, N.R., & Mello, L.G.R.: Interaction between Soils and Geosynthetic Layers in Large-Scale Ramp Tests. *Geosynth. Int. 9* (2002), pp.149–187.

Palmeira, E.M., & Viana, H.N.L.: Effectiveness of Geogrids as Inclusions in Cover Soils of Slopes of Waste Disposal Areas. *Geotext. Geomembr. 21* (2003), pp.317–337.

Qian, X., Koerner, R.M., & Gray, D.H.: Translational Failure Analysis of Landfills. *J. Geotech. Geoenviron. Eng., ASCE 129* (2003), pp.506–519.

Richardson, G.N. & Koerner, R.M.: *Geosynthetic Design Guidance for Hazardous Waste Landfill Cells and Surface Impoundments*, EPA-600/2-87-097. US EPA, Cincinnati, OH., U.S.A.: Hazardous Waste Engineering Research Laboratory, 1987.

Soroush, A., Morgenstern, N.R., Robertson, P.K., & Chan, D.: Post-Earthquake Deformation Analysis of the Upper San Fernando Dam. *Int. Symp. Seismic Environ. Aspects Dams Des.*, Santiago, Chile, Vol. 1, 1996, pp. 405–418.

Tosatti, G. (ed): *A Review of Scientific Contributions on the Stava Valley Disaster (Eastern Italian Alps), 19 July 1985*. Consiglio Nazionale delle Ricerche, Pitagora Editrice, Bologna, Italy. ISBN 88-371-01405-2, 2003, p. 243.

United Nations Department of Humanitarian Affairs: *Mudflows – Experience and Lessons Learned from the Management of Disasters*, Geneva, Switzerland: United Nations, 1996.

U.S. Bureau of Reclamation. *Design of Small Dams*. Washington D.C., U.S.A.: US Government Printing Office, 1960.

Wagener, F., Strydom, K., Craig, H., & Blight G.E.: The Tailings Dam Flow Failure at Merriespruit, South Africa – Causes and Consequences. In: *Tailings and Mine Waste, 1997*. Rotterdam, Netherlands: Balkema, 1997, pp. 657–666.

Xin, H.B., Finn, W.D.L., & Wang, Y.Q.: Lessons from Seismic Performances of Chinese Tailings Dams. Internal Report quoted by W.D.L. Finn (1996). Seismic Design and Evaluation of Tailings Dams. *Int. Symp. Seismic Environ. Aspects Dams Des.* Santiago, Chile, 1992, pp. 7–34.

Chapter 12

Surface stability of tailings storages slopes – erosion rates, slope geometry and engineered erosion protection

12.1 Past practice for slope angles of tailings storages

Conventional practice in many parts of the world has been to construct tailings storages at slopes that would be considered extremely steep in comparison with slopes either cut or constructed of natural soils. This practice has also been followed principally in mining areas having semi-arid to arid climates, e.g. Western Australia, South Africa and Chile.

The use of very steep slope angles for the outer slopes of tailings storages in water-deficient climates arose because, when tailings storages were first introduced in the early 1900's, the outer slopes were built by hand and it was convenient to adopt angles of 1 on 1.5 (34°), 1 on 1.25 (39°), etc. These steep angles became accepted practice which continued until the early 1990's when (in South Africa) regulations were introduced requiring permanent rehabilitation of tailings impoundments before closure certificates were issued, that relieved the mine owners of responsibility for the waste deposit. It has slowly become accepted by mine owners that erosion from steep slopes by both water and wind can be very large and that permanently rehabilitating excessively steep slopes for a maintenance-free closure becomes almost impossible.

Plate 12.1 shows an example of the extremely steep angles that have been built to in the past. The lower slopes were built at an angle of 40°, while the upper slopes are at 35°.

It is becoming common in some countries (see, e.g. Smith et al., 1997) to require that the rehabilitation of a tailings dam or other waste deposit be designed to be maintenance free for periods as long as 500 to 1 000 years. This seems quite unrealistic when one thinks of the changes only 100 years can cause to a landscape (e.g. the areas now occupied by the cities, freeways and waste deposits of many cities in the "new world" were untouched countryside just over a century ago). However, very ancient man-made mounds exist in many parts of the world with slopes that have resisted 1 000 or more years of erosion and still exist in good condition. Figure 12.1 shows the profiles of several ancient man-made funerary mounds in China. All of these are accurately dated and most are well over 1000 years old. The mounds around Xi'an are all constructed of fine sandy silts from river alluvium and loess, which are intrinsically fairly erodible, and the climate is not dissimilar to that of the mining districts in Australia, Chile and South Africa. There are distinct wet and dry seasons, rainfall is between 500 and 1 000 mm/y, and there is an annual water deficit. The mound near

Plate 12.1 Excessively steep slopes to which outer dykes of tailings storages were built during 20th century.

Yinchuan is also of loess but the area has a desert climate. It would have been expected that over a period of 1 000 years, the slopes would all have eroded to similar and very flat slope angles. But this has not occurred. As shown in Figure 12.1, the slopes vary from 16° to 28°. Hence one cannot conclude that if a slope has less than a certain limiting angle, it will not erode, or vice versa. The contrast between the mounds near Xi'an and those near Yinchuan shows, however, that climate plays a role in determining the maximum angle of an erosion-free slope. It is also quite possible that these slopes have been periodically maintained and are thus not completely unaffected by erosion.

12.2 Acceptable erosion rates for slopes

When in doubt about the best way to achieve a geotechnical objective, it is often useful to study analogous situations or processes in nature (Smith et al, 1997). What are the characteristics of naturally erosion-resistant slopes? Figure 12.2(a) shows the profile of a natural slope of transported soil in Johannesburg that has been derived from the weathering of greenstone (an ancient metamorphic schist). The slope surface is covered by vegetation of various types (mainly grass and bushes) and is scattered with gravel and boulders. It appears to be very stable, with little evidence that erosion is occurring. Measurements of erosion on this slope (made by observing, over a period of 3 years, changes of the ground surface level against the tops of two parallel lines of steel pegs

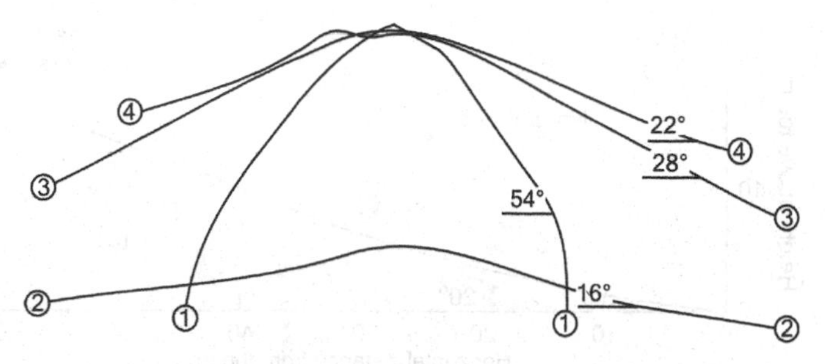

Figure 12.1 Profiles of ancient earth mounds in China: 1. Xia Kingdom, 900-700y BP near Yinchuan (desert climate). 2. Qin Shi Huan, 2000y BP, said originally to have been 116 m high, now 52 m high. 3. Liu Che, 2100y BP, now 46 m high. 4. Gaozong, 1320y BP. (2, 3 and 4 semi-arid climate, near Xi'an.) (BP = Before Present.)

driven into exposed soil patches on the slope) are shown in Figure 12.2(b). The slope is eroding near its head and accreting near its toe. The erosion and accretion shown in Figure 12.2(b) are probably exaggerated and more than the average for the slope because they were measured on patches of bare soil on the slope. Erosion is much less under the clumps of vegetation and is zero under the pebbles and boulders, unless one of them is undermined by erosion of the soil, and rolls downhill. Erosion by loss of soil from small soil pockets between boulders that locally concentrate the flow of runoff down the slope will also exaggerate erosion. The slope surface is thus probably as stable as is possible under natural conditions. Other similar natural slopes in other climatic regions of South Africa have shown similar erosion or accretion rates and distributions.

The upper part of the slope is eroding at rates approaching 50T/ha/y at the top, reducing to zero at about half the slope length. The lower part is accreting at much the same maximum rate. The area under the erosion rate curve (in Tm/ha/y) is slightly less than that under the accretion rate curve, but erosion and accretion are close to being in balance. The measured difference of 150Tm/ha/y in 1300Tm/ha/y (11.5%) probably arose because erosion occurring above the boulders was not recorded, but contributed to the accretion. Figure 12.3 shows the profile and rates of erosion or accretion of a second natural soil slope of transported soil derived from weathering of quartzites. This slope is slightly steeper than the slope described by Figure 12.2, but has a similar surface cover of vegetation and scattered gravel and small boulders. The slope also ends in large boulders at its upper end. The average erosion rate (observed over 2.5 years) averages only 5T/ha/y with occasional local values of 20T/ha/y. Erosion takes place over most of the slope, with accretion occurring near to the toe. Similar rates of erosion and accretion have also been observed on a third slope in the Witwatersrand area. Similar patterns of erosion/accretion occur on the slopes of tailings storages, if they are not protected against erosion, although rates of erosion/accretion of tailings slopes may be as high as 500T/ha/y (see Figure 12.4 and Table 12 to follow) rather than the 5-50T/ha/y for natural slopes.

Figure 12.4 shows the slope and erosion rate profiles of an unprotected gold tailings storage at Doornkop, near Johannesburg. In Figure 12.4, the 94/96 measurements

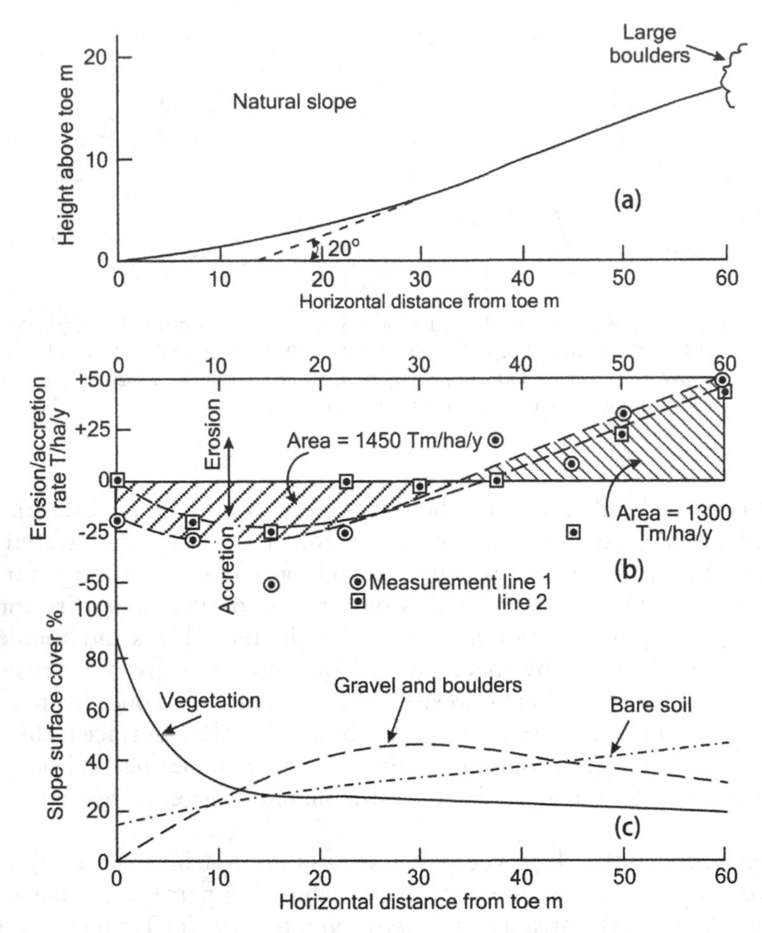

Figure 12.2 Erosion of a natural soil slope in residual greenstones: a: profile of slope, b: distribution of erosion/accretion rates, c: surface cover of slope.

extended from March 1994 to March 1996 and the 96/98 measurements ended in July 1998. The slope has a relatively flat angle (for gold tailings storages) of 16°, with a steeper 28° slope above it. Water and eroded solids from the slopes above were cut off by means of a sheet steel barrier set into the berm above the slope. Plate 12.4 (see page 537) shows the eroded surface of the unprotected surface referred to in Figure 12.4 (central section). The sections of slope an either side have been protected from erosion. The measurements for 96/98 could not be made on the upper slope because the measurement pegs in use at that stage had eroded out and been lost in the 4.3 years since installation. Also, the pattern of accretion at the toe observed in Figures 12.2 and 12.3 did not occur here because the eroded material was regularly removed from the toe of the slope during 94/96 and weighed to determine the erosion loss.

The most obvious difference between Figures 12.2 and 12.3 and Figure 12.4 lies in the erosion rates: up to 530T/ha/y for the upper slope, falling to less than 50 close

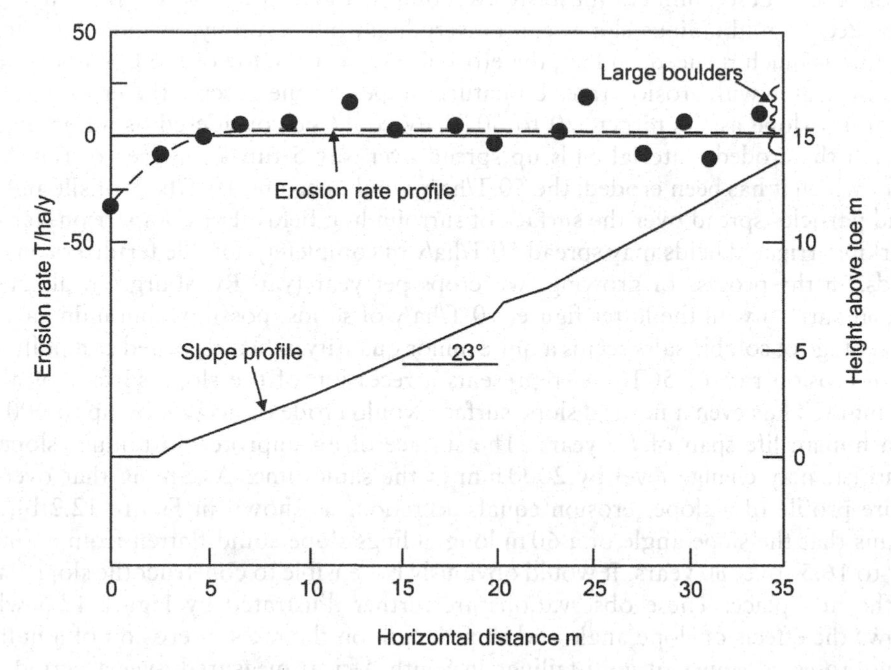

Figure 12.3 Erosion of a natural soil slope in residual quartzite, slope profile and distribution of erosion/accretion rates.

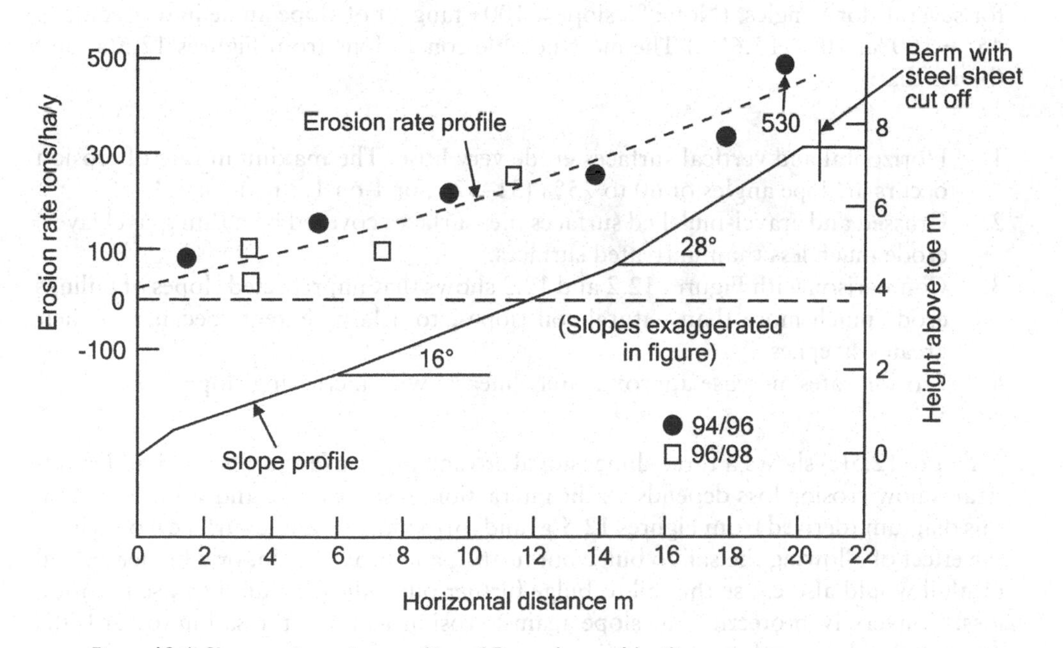

Figure 12.4 Slope and erosion profiles of Doornkop gold tailings storage.

to the toe. These numbers illustrate two things: firstly, the erosion rates for steep unprotected gold tailings slopes can be very high, but secondly, at moderate slopes, the rate is much reduced. In fact, the erosion rate near the toe of the Doornkop slope is comparable with erosion rates on natural slopes. If one accepts the erosion rate of a natural slope as the target, 20 to 50T/ha/y could be considered as an acceptable rate. If the eroded material ends up spread over, say 5 times the area of the slopes from which it has been eroded, the 50 T/ha/y would become 10 T/ha/y of silt and fine sand particles spread over the surface of surrounding fields. For comparison, farmers working irrigated fields may spread 50 T/ha/y of completely soluble fertilizers on their fields, in the process of growing two crops per year (van Rensburg, et. al. 2008). In comparison with the latter figure, 10 T/ha/y of solids, possibly containing a small percentage of soluble salts seems a quite minor quantity, when regarded as a pollutant.

An erosion rate of 50T/ha/y represents a recession of the slope surface of about 2.5 mm/y. Thus even a natural slope surface would erode or accrete by up to 200 mm in a human life span of 80 years. The surface of an unprotected tailings slope, in contrast, may change level by 2000 mm in the same time. Assuming that over the entire profile of a slope, erosion equals accretion, as shown in Figure 12.2(b), this means that the slope angle of a 60 m long tailings slope could flatten from an initial 20° to 16.5° over 80 years. It would obviously be sensible to construct the slope flatter in the first place. These observations are further illustrated by Figure 12.5 which shows the effects of slope angle and slope length on the rates of erosion of a number of unprotected slopes of gold tailings in South Africa, measured over a period of 4 years (Blight, 1989). Much of the scatter of observations arises because slope erosion rates are affected by both angle and length. In Figure 12.5(a), for any one slope angle (e.g. the most popular, 33°) data are shown for slopes of various lengths. Similarly, for the most popular slope length between berms (25 m in Figure 12.5(b)), data are shown for several slope angles. (Note: % slope = 100x tangent of slope angle in degrees, e.g. 45° = 100%, 10° = 17.6%.) The most notable conclusions from Figures 12.5(a) and (b) are that:

1. Horizontal and vertical surfaces erode very little. The maximum rate of erosion occurs at slope angles of 60 to 75% (31–37°), or 1 on 1.7 to 1 on 1.3).
2. Grassed and gravel-mulched surfaces (i.e. surfaces covered by a thin gravel layer) erode much less than untreated surfaces.
3. Comparison with Figures 12.2 and 12.3 shows that unprotected slopes of tailings erode much more than natural soil slopes, to a large extent, because of their greater steepness.
4. Erosion rates increase approximately linearly with increasing slope length.

Figure 12.5(c) shows a three-dimensional erosion rate surface or "sail" that demonstrates how erosion loss depends on the interaction of slope angle and slope length. In this diagram,(derived from Figures 12.5(a) and (b)) a more erodible surface would have the effect of allowing the sail to bulge out further, and heavier rain, or a higher annual rainfall would also cause the sail to bulge further out indicating an increased erosion loss. Conversely, protecting the slope against erosion will pull the sail in towards the "mast", the slope angle axis.

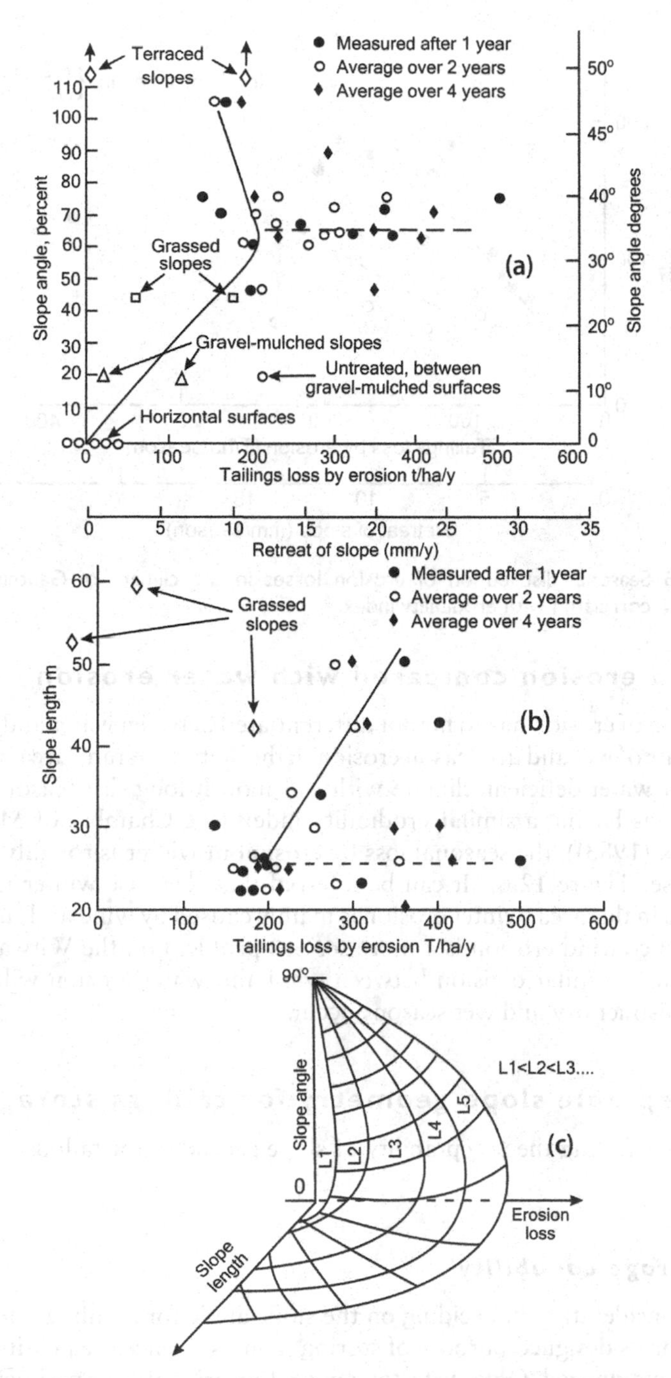

Figure 12.5 Rates of erosion of surfaces of gold tailings storages in South Africa: Effect on tailing loss of: a: slope angle, b: slope length. c: Three dimensional erosion rate surface or "sail".

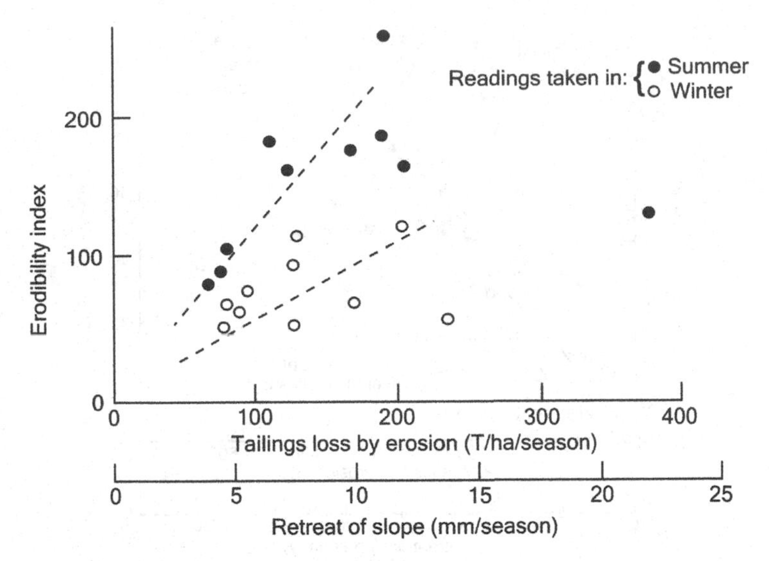

Figure 12.6 Seasonal distribution of erosion losses in the climate of Gauteng, South Africa, correlated with erodibility index.

12.3 Wind erosion compared with water erosion

The discussion of erosion has so far not differentiated between water and wind erosion. Measurements of wet and dry season erosion in the Witwatersrand area of South Africa (which has a water-deficient climate with a 5 month long dry season) have shown that for tailings having a similar erodibility index (the Chamber of Mines COMET erosion index (1983)), the seasonal loss by erosion in winter is roughly twice the loss in summer (see Figure 12.6). It can be inferred that, because winter is a dry, dusty, windy period in this area, winter erosion is mainly caused by wind and summer erosion by water. Hence wind erosion is a more serious problem on the Witwatersrand than water erosion. A similar division between wind and water erosion will occur in any area where distinct dry and wet seasons occur.

12.4 Acceptable slope geometry for tailings storages

Some factors affecting the acceptability of slope geometry for tailings storages are as follows:

12.4.1 *Storage capability*

A primary consideration in deciding on the slope angle for a tailings storage is that it must be fit for its designed purpose of storing tailings. Hence the possible storage volume must be maximized. Obviously, for a given footprint area or perimeter, the steeper the slope angle, the more tailings can be stored. The effect of slope angle β on deposition area A is illustrated in Figure 12.7 where A has been plotted against β for storages that are level and square in plan (L × L) and of either 100 ha or 400 ha in extent.

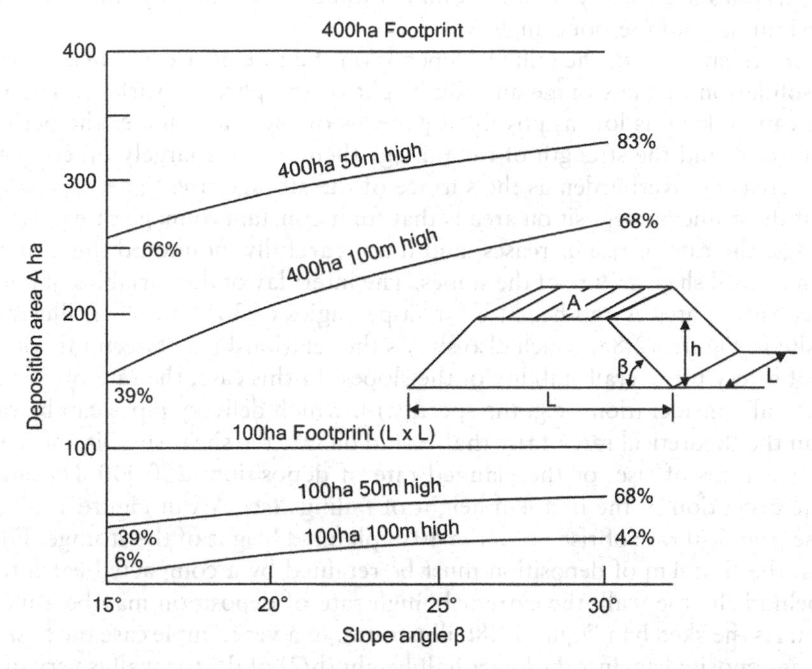

Figure 12.7 Effect of slope angle β on deposition area A for tailings storages with square and level footprints L × L and areas of 100 and 400 ha.

Two heights have been considered for each size of storage: h of 50 m or 100 m. The relationship between A, h, L and β is:

$$A = (L - 2h\cot\beta)^2 \tag{12.1}$$

For a relatively small footprint area of 100 ha and 50 m height, A is 68% of footprint area for $\beta = 30°$, but only 39% for $\beta = 15°$. For the corresponding 400ha footprint, A is 83% of footprint area for $\beta = 30°$ and 66% for $\beta = 15°$. A small relatively high storage (footprint = 100ha and h = 100 m) becomes ridiculously inefficient at full height with A being only 42% of footprint area at $\beta = 30°$ and 6% at $\beta = 15°$. In fact, with $\beta = 15°$ the storage effectively becomes a pyramid. It is clear that storage capability is less affected by slope area as the footprint area increases.

12.4.2 *Overall stability*

The overall shear stability of an hydraulic fill tailings slope of height h (more details are given in Chapter 11) depends on the following:

- the weight per unit area of the slope, γh, where γ is the weight of tailings per m^3,
- the slope angle β; and
- the in situ shear strength of the tailings, τ.

These variables are usually used in combination as the "stability number" $N = \tau/\gamma h$ (dimensionless) and the slope angle β.

The in situ strength of the tailings depends on the rate of rise and hence the degree of consolidation of the storage and the height of the phreatic surface. The phreatic surface can be kept as low as possible by means of toe drains under the perimeter of the footprint, and the strength of the tailings then depends largely on consolidation under increasing overburden as the surface of the storage rises. The most important effect of the reducing deposition area is that for a constant tonnage rate of deposition of tailings, the rate of rise increases, and if not carefully monitored this can possibly result in overall shear failure of the slopes. The interplay of the variables of deposition rate, deposition area A and height h for slope angles of 30° and 20° is illustrated for a real site in Figure 12.8a, which also shows the relationship between rate of rise and factor of safety for overall stability of the slopes. In this case, the rate of rise dictated by practical considerations, e.g. the speed with which delivery pipes can be raised, is less than the theoretical rate of rise that would dictate the shear stability of the slopes. The actual rates of rise for the planned rate of deposition (250 000 T/month) will, with the exception of the first 4 m height of tailings (see AA in Figure 12.8), be less than the practical rate of rise up to the 40 m planned height of the storage. For safety reasons, the first 4 m of deposition must be retained by a compacted earth toe wall. Even behind the toe wall, the extremely high rate of deposition may be a matter for concern. As the sketch in Figure 12.8b illustrates, in a very simple case the first volume V_1 that is deposited against the lower half height (h/2) of the toe wall is very much less than the volume required to raise the tailings level a further h/2 to the top of the toe wall. In the simple illustrative example, the rate of rise for the lower h/2 is 6 times that for the upper h/2, for a constant rate of delivery. Referring to Figure 6.34, dealing with Gibson's theory for consolidation, this means that the time factor T for consolidation of this first deposition V_1 would be $(6)^2 = 36$ times that for V_2. Hence, if the toe is not effectively drained, a pocket of partly consolidated tailings could exist at the low point of the storage for years to come, which could compromise the slope stability of the storage (see, e.g. Section 11.6).

Basically, the flatter the angle of the outer slope, the faster the deposition area reduces and the greater becomes the overall rate of rise. Hence the less stable the slope becomes. Obviously the primary demands for efficient use of the footprint area and slope stability during the operational phase must be reconciled with the long term (possibly 1000 year) demands of surface stability and environmental acceptability.

12.4.3 *Berms, ramps and step-ins*

Berms and ramps are necessary to provide access to the slopes and top surface of a tailings dam during its operating life and also for purposes of maintenance once it has been decommissioned and rehabilitated. Step-ins are sometimes used either as a pre-planned method of reducing the overall slope angle to increase the factor of safety against overall shear failure, or as an unplanned emergency measure to achieve the same aim. However, it must be remembered that each berm, step-in or ramp will receive the runoff from the slope above, as well as the solids eroded from the slope. Berms, step-ins and ramps must therefore be designed to safely both intercept and drain runoff and to intercept and retain eroded solids. This, in turn means that berms,

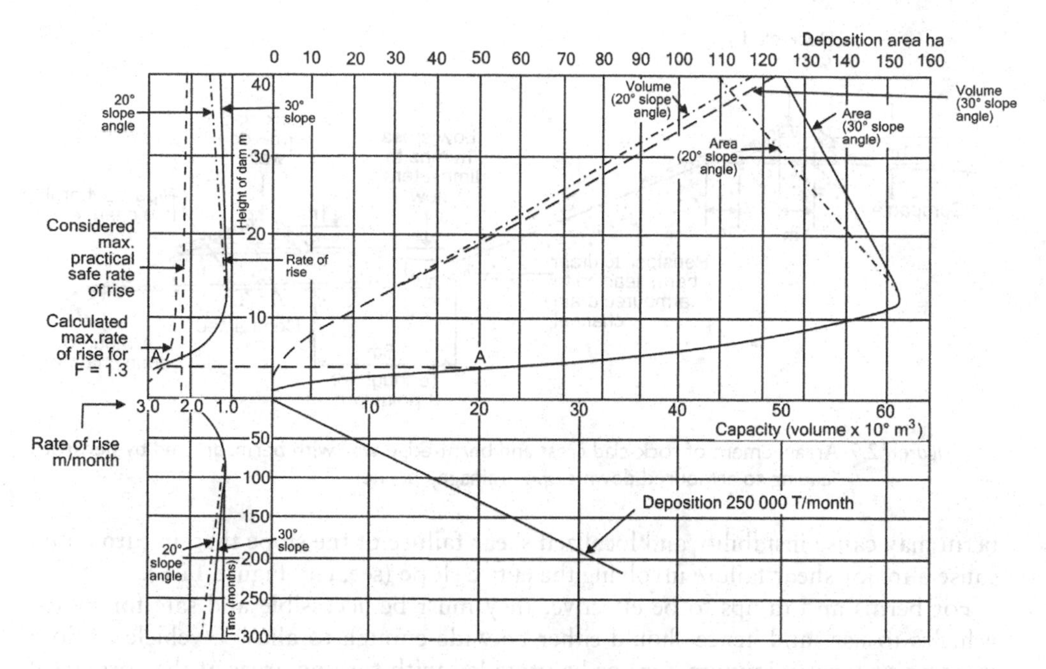

Figure 12.8a: **Interaction between deposition area and rate of rise for constant tonnage rate of deposition.**

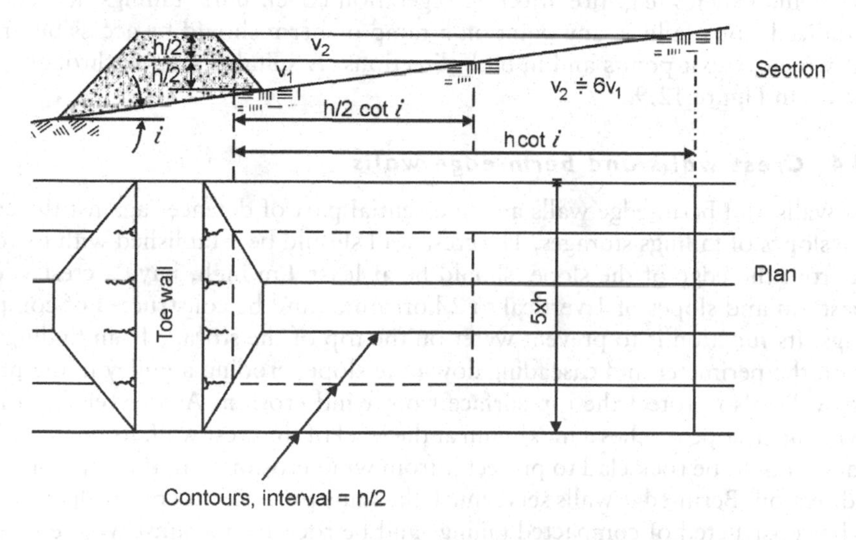

Figure 12.8b: **Sketch showing extremely high rates of rise possible at start of new tailings storage.**

step-ins and ramps must be maintained during the operational life of the storage and beyond rehabilitation and decommissioning. An alternative strategy is to design berms to retain eroded solids and runoff and evaporate the runoff on the berm. This is not a safe procedure, as the berm may overtop at a local low spot, causing an erosion gulley to form. In addition, holding a pool of runoff water at the top of the slope below the

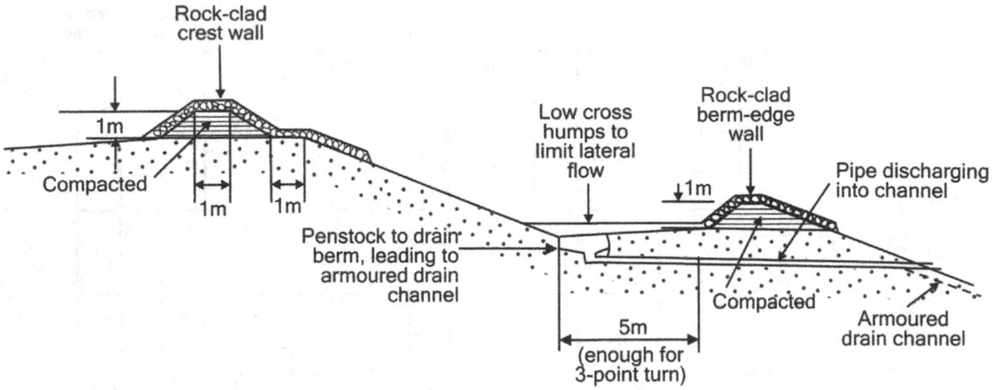

Figure 12.9 Arrangement of rock-clad crest and berm-edge wall with berm, drained by penstock leading to armoured down-slope drainage channel.

berm may cause instability and localized shear failure of the slope that, in turn could cause a major shear failure involving the entire slope (see, e.g. Figure 11.6).

For berms and ramps to be effective, they must be accessible and safe for motor vehicles to use, and hence should either be wide enough to allow a vehicle to do a three point turn (minimum 5 m) or be provided with turning areas at the corners of the dam and the tops of access ramps. Because ramps and berms may have to be used in emergencies (e.g. fire affecting vegetation cover, burst tailings delivery pipes or localized slope failure, any point on a ramp or berm should be accessible from at least two entry/exit points and in both directions. A suitable section through a berm is shown in Figure 12.9.

12.4.4 *Crest walls and berm-edge walls*

Crest walls and berm-edge walls are an essential part of defences against the erosion of the slopes of tailings storages. The crest wall should be established with its toe 1 m back from the edge of the slope, should be at least 1 m high, have a crest width of at least 1m and slopes of 1 vertical to 2 horizontal and be constructed of compacted tailings. Its function is to prevent water on the top of the storage from finding a low spot on the perimeter and cascading down the slope, eroding a gulley in the process. Crest walls also protect the top surface from wind erosion. As the velocity of wind blowing up a slope reaches a maximum at the level of the crest wall, its outer and inner surfaces should be rock clad to protect it from wind erosion with the wind blowing in any direction. Berm-edge walls serve much the same purposes as crest walls and should also be constructed of compacted tailings and be rock clad against wind erosion. To reduce overall berm width, the outer slope can form a continuation of the slope below and the inner slope can be the same as the outer slope.

12.4.5 *Height between berms*

Opinions differ on the maximum height between berms. As indicated in Figure 12.5(b), a popular slope length between berms is 25 to 30 m, which at the most popular angle of 33°, gives an average vertical height between berms of 15 m. However, some authorities believe that the vertical spacing of berms should be as little as 7 m. If the line defining the relationship between slope length and erosion rate in Figure 12.5b is extrapolated to an erosion rate of 100T/ha/y the corresponding slope length is 9 m. In Figure 12.5a, a slope angle of 18° also corresponds to a rate of erosion of 100T/ha/y (both for unprotected slopes of gold tailings). The height of an 18° slope, 9 m long is only 2.8 m. However 7 m between berms at 18° gives a slope length of 22.7 m, corresponding to an erosion loss of 175T/ha/y for an unprotected slope. Hence 7 m seems to be a good compromise height between berms, but the slopes will need erosion protection to reduce rates of erosion to acceptable limits.

12.4.6 *A suitable slope angle*

In the light of the above discussion as well as the information contained in Chapter 11, what would be a suitable compromise slope angle? The answer appears to lie between 12° and 18°. A slope of 1 vertical to 3 horizontal, or 18° is often advocated, as this is accepted as the steepest slope angle on which agricultural machinery can work safely and effectively. As will be seen from Figure 12.5(a), unprotected slopes in gold tailings can be expected to erode at a rate of 100 + T/ha/year, but it will be shown later that with protection against erosion, the rate can be reduced to values similar to those of natural slopes. Even in seismically active slopes, the information given in Figure 11.30 indicates that a slope of 12° to 18°, combined with good under-slope drainage, would be safe against most earthquakes.

12.5 Protection of slopes against erosion by geotechnical means

A number of methods of protecting tailings slopes against erosion by geotechnical means have been evolved and experimented with. The following briefly describes these:

12.5.1 *Gravel mulching*

It was mentioned above that moderate, i.e. 20% (10°) or less tailings slopes can be protected from erosion by covering them with a thin gravel layer or "gravel mulch". One experiment that used a single particle thickness layer of 19 mm concrete stone on a 12° slope (see Figure 12.5a) and was laid in 1980, initially eroded at an average of less than 100T/ha/y and looks much the same at present, 28 years later. Plate 12.2 shows a gravel-mulched crest wall that was established in 1979. The photograph was taken in 1982. Plate 12.3 shows the same scene from the opposite direction in 2007. The gravel mulched surfaces have had no maintenance in the intervening 25 years. This experiment led to the concept of "rock cladding" (the entire surface covered by a layer of waste rock) while the measurements on natural slopes (where the surface is covered by a discontinuous layer of rock and gravel) led to the concept of "rock armouring".

12.5.2 *Rock cladding*

Two major experiments have been undertaken to explore the concept of rock cladding. One of these was started in March, 1994 at Doornkop gold tailings storage, near Johannesburg, the other at FS S6 gold tailings storage in the Free State province.

Plate 12.2 Gravel mulch on gentle slope 3 years after placing gravel.

The Doornkop experiment was set out on the slope shown in Figure 12.4, which was chosen because, initially, rock cladding was thought likely to be suitable only for moderately steep slopes, and this slope had two sections, one of 16° and the other of 28° slope angle. The experiment at FS S6 was started in October 1998. The slope angle there is 30°.

At Doornkop the slope was divided into 12 panels, each measuring 20 m long (upslope) by 10 m wide. Each panel is isolated from its neighbors by means of 0.5 m-high metal sheets partly dug into the tailings surface to form a low vertical dividing wall. Each panel was originally equipped with a sprinkler irrigation system to simulate rain, 3 rain gauges, and a set of 10 surface level pegs embedded 0.5 m into the prepared tailings surface. The toe of each panel terminates in a catchment paddock to capture and hold solids removed from the slope by water erosion. Simulated rain was used to obtain initial results (Phase 1) but for the next 6 years the slopes were exposed to natural weather. Figure 12.10 shows a typical plan of part of the experiment. The various surface treatments have been summarized in Table 12.1a, which also summarizes the experimental results from March 1994 to March 2000 (Phases 2 to 5).

Plate 12.4 shows the unprotected control section in the Doornkop experiment (Panel 11 in Table 12.1a) with the simulated natural (or analogue) slope (Panel 0) to the right and Panel 10 to the left. Plate 12.5 shows two rock clad sections (Panel 3 to the left and Panel 4 to the right). Both of these photographs were taken in April 2008, 14 years after the experiment was started. No maintenance of any sort had been carried out on the test slopes since the experiment was constructed.

Plate 12.3 Similar slopes on same storage 28 years after placing gravel.

Tables 12.1a and 12.1b record the relative cost, of each treatment, relative erosion rates and a cost-effectiveness number represented by the product of relative cost (C) and relative erosion (E). It should be noted that the erosion rates were measured differently for phases 1 to 2 and 3 to 5 of the experiment. For phases 1 and 2 the material eroded off each panel was caught in the catchment paddocks at the toe of the slope, then collected and weighed. During phase 3, an unusually heavy rainstorm caused tailings to be washed onto the test slopes from above, thus rendering the origin of the mass of caught material questionable. For this reason, erosion for phases 3 to 5 was assessed by measuring the retreat of the slope surface against the surface level pegs that had been installed at the start of the experiment and, up to the time of the storm damage, had been used to check on the weighed erosion losses. With hindsight, this is a more reliable form of measurement and also shows the distribution of the erosion loss from the slope.

On the basis of cost-effectiveness for phase 2, conventional grassing ranked just above no treatment at all, and for both phases 2 and 3, a soil layer covered with grass sods (Panel 10) rated top. However, the grass sods quickly deteriorated, most of the grass died, and the remaining soil has eroded away. Only the presence of the grass roots maintained the effectiveness of the treatment for a year or two. On Panel 9, which had a thicker soil layer and was expected to outperform Panel 10, the grass died and the panel's rating dropped from 2 in Phase 2 to 8 in Phases 3 to 5.

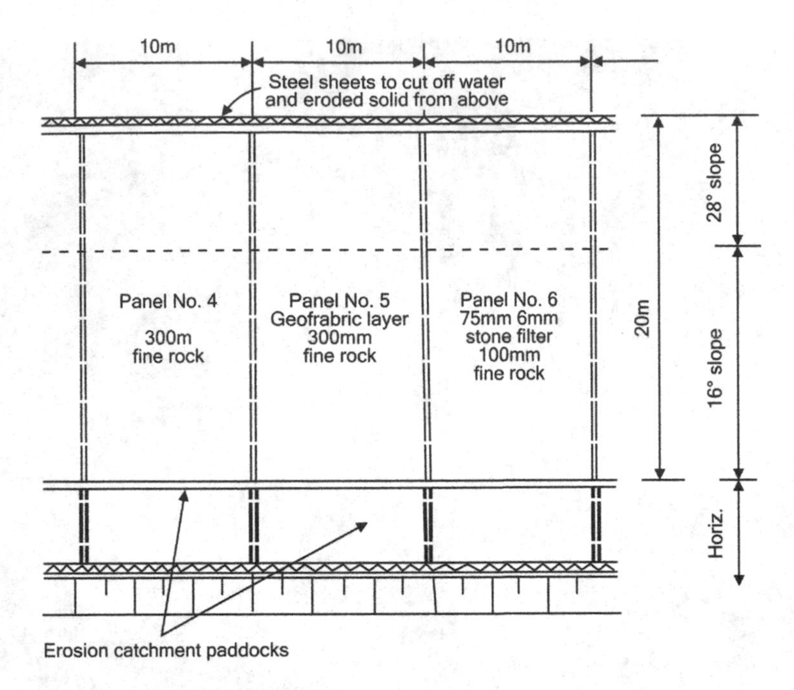

Figure 12.10 Part layout of large-scale erosion protection experiment at Doornkop.

Tables 12.1a and 12.1b show that geotechnical treatments occupy 8 of the first 9 places in the ranking and should therefore be seriously considered for use. However, the study of natural slopes (see above) has shown that the slope surface probably does not have to be completely covered with rock or gravel, which would reduce the cost of the treatment with little or no loss in effectiveness. An additional simulated "natural analogue slope" or "rock armoured" panel was prepared in 1996 (Panel 0 in Table 12.1a) to test this hypothesis. The simulated natural slope surface was vegetated to the extent of about 33% cover with tufts of grass and another 33% was covered with random sized rock and gravel particles. The remaining exposed area of tailings was left unprotected. Quantitative measurements are not available for this panel as the steel cut-off sheet at the top of the slope was vandalized and measurements showed the whole panel to be accreting with tailings washed down from the slope above.

However, 12 years later, in 2008 the panel was completely covered with luxuriant grass and shows no visible signs of erosion damage. (See right side of Plate 12.4.)

It is important to note from Table 12.1b that a protective treatment for a slope surface takes a number of years to show its true value. The efficiency of a protective treatment cannot be assessed on the basis of its performance in a single wet season, and certainly not on the basis of applying artificial rain by spray-irrigation for a limited period. The point is further illustrated by Figure 12.11 which shows how the performance of five of the panels changed with time over the first 6 years, with the performance of some remaining unchanged, some improving and others deteriorating with time.

Table 12.1a Doornkop experiment – Description of panels, relative costs and erosion rates.

Panel number	Treatment	Level & compact	Level only	Relative cost/ha C%	Erosion rate (tons/ha/year)			
					Phase 2	Phase 3	Phase 4	Phase 5
0	Simulated natural slope	√	–	100	–	–	–	*ACC
1	Conventional grassing	√	–	100	164	164	164	ACC
2	100 mm ballast(50 mm size)	√	–	67	105	32	35	23
3	300 mm coarse rock	√	–	62	170	12	23	23
4	300 mm fine rock	√	–	62	38	96	70	63
5	Geofabric + 300 mm fine rock	√	–	120	22	15	70	65
6	75 mm 6 mm stone + 100 mm fine rock	√	–	66	42	82	61	69
7	300 mm fine rock	–	√	54	118	75	67	53
8	250 mm open pit overburden	–	√	64	203	161	175	175
9	250 mm soil + Ag Lime + grass sods	–	√	120	19	72	78	74
10	100 mm soil + Ag Lime + grass sods	–	√	96	21	15	36	10
11	Control (no treatment)	–	–	0	276	257	316	261

*ACC = Accretion, no net erosion

Table 12.1b Doornkop experiment – Relative erosion rates and cost-effectiveness evaluation of slope protection methods.

Panel no.	*Relative erosion rate E(%)*				*Cost-effectiveness C X E (%)*				*Cost-effectiveness ranking*			
	Phase 2	Phase 3	Phase 4	Phase 5	Phase 2	Phase 3	Phase 4	Phase 5	Phase 2	Phase 3	Phase 4	Phase 5
0	–	–	–	–	–	–	–	–	–	–	–	–
1	59	64	52	–	59	64	52	–	10	10	10	–
2	38	12	11	9	25	8	7	6	7	4	2	2
3	62	5	7	9	38	3	5	6	8	1	1	2
4	14	37	22	24	9	23	14	15	3	7	6	5
5	8	6	22	25	10	7	27	30	4	3	7	7
6	15	32	19	26	10	21	13	17	4	6	5	6
7	43	29	21	20	23	16	11	11	6	5	3	4
8	74	63	55	67	47	40	35	43	9	9	9	9
9	7	28	25	28	8	34	30	34	2	8	8	8
10	8	6	11	4	7	6	11	4	1	2	3	1
11	100	100	100	100	0	0	0	0	11	11	11	10

Phase 2: cumulative for 2 wet seasons (94/95 to 95/96)
Phase 3: cumulative for 4 wet seasons (94/95 to 97/98)
Phase 4: cumulative for 5 wet seasons (94/95 to 98/99)
Phase 5: cumulative for 6 wet seasons (94/95 to 99/2000)

Plate 12.4 Erosion of unprotected tailings slope surface compared with protected surfaces on either side.

At the time of writing, 2009, the experimental panels at Doornkop had been in place for 15 years. Panels 0, 1 and 2 to 7 were in excellent condition, except for small erosion rills on panel 1 and some slippage of the rock cover on the underlying geofabric on panel 5. Panels 8 to 10 are now badly eroded and the tailings surface is exposed on the upper 28° section of the slope. Erosion has transformed the untreated control panel 11 into a 10 m wide trench about 1m deep.

Measurements were stopped after Phase 5 (6 years after construction) because tailings eroded from the slopes on the opposite (south west) side of the deposit (more than 1km away) by the dry-season winds were being deposited on the surface of the experimental area and the slopes above it. This is the reason for the "accretion" entries in Table 12.1b for panels 0 and 1 for Phase 5. Measurements of surface movement against the measuring pegs became meaningless after year 6. When the experiment was laid out, the tops of the panels were at the top of the tailings dam, the highest part of the wall, and ground was still in the process of being covered to the south-west. It was only in year 5 that the deposit reached a height of 5 or 6m on the opposite side.

The wind rose for the area is shown in Figure 12.12. The winds from the east-north-east blow onto the experimental section and do not affect it, but winds from the south-west that mostly blow during the end of the winter dry season (August to September) cause the deposition of air-borne solids on the experimental slope.

Plate 12.5 Examples of rock-clad slopes 14 years after placing.

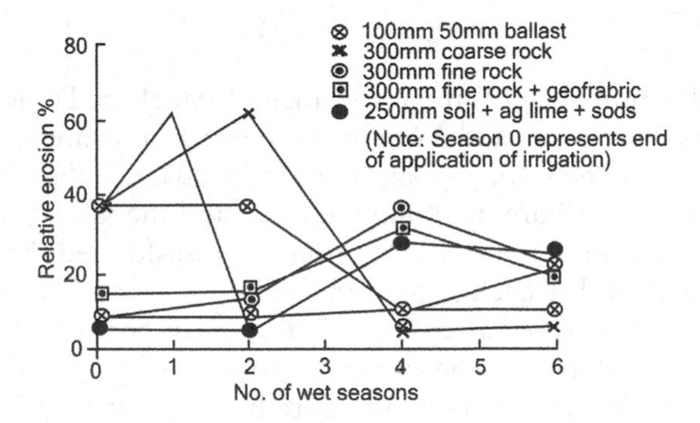

Figure 12.11 Erosion protection experiment at Doornkop tailings storage: Relative erosion versus number of wet seasons. (100% = erosion of unprotected control panel.)

Measurements taken in the FS S6 experiment have been tabulated in Table 12.2. The measurements of soil loss were taken over only one month and have arbitrarily been multiplied by 4 to obtain an equivalent annual wet season loss. (This multiplier could equally well have been 3.) Soil losses were very low, except for the dry land (i.e. un-irrigated) vegetation plot and plot 6 where topsoil was placed over the waste rock and eroded badly (as could be expected).Unfortunately, no soil loss measurement was made on the untreated control area.

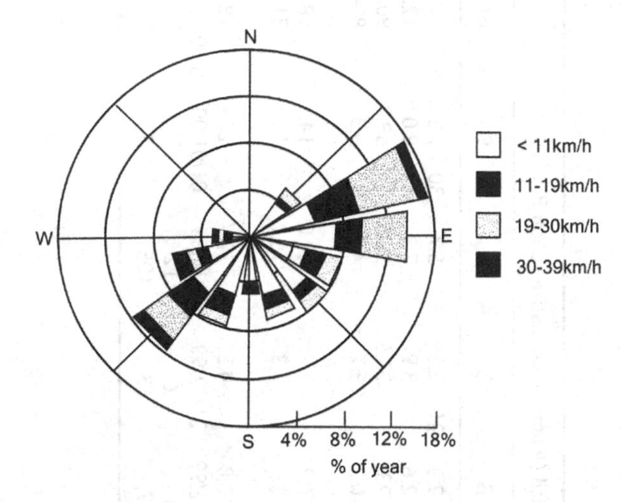

Figure 12.12 Wind-rose for area in which Doornkop is situated.

The emphasis of the experiment was mostly on runoff from the underlying tailings surface. Measurements were made on 5 different occasions, in December 1999 and January 2000, spaced 5 to 6 days apart. It is not clear if the quoted rainfall fell on several occasions during each period, but this seems likely. It would be unusual to have such relatively heavy falls at such regular intervals.

Figure 12.13 shows the relationships between cumulative rainfall and cumulative runoff for the various panels in the experiment. The three rock clad panels (1, 2 and 4) had very low and similar runoffs of 5% of rainfall, whereas the runoff for the control area (7) was 25%. The highest runoff of 47% came from the dry land vegetation area (vegetation planted directly into tailings) (3), with the irrigated vegetation (5) giving a runoff of 9%. The topsoil-covered rock cladding (6) gave the third highest runoff of 15%. The low 5% rate of runoff of the various rock clad panels (1, 2 and 4) in comparison with the 25% runoff from the control panel is the probable reason for the effectiveness of rock cladding in reducing rates of erosion.

12.5.3 *Full-scale field trials of rock cladding and rock armouring*

A full-scale field trial of rock cladding was carried out at the Cooke L1 tailings storage over a period of two years. The complete outer slopes of the 100 ha tailings storage were rock clad with a 300 mm layer of run-of-dump rock on an unprepared surface. The results of one of the 6 different profiles of erosion measurements are shown in Figure 12.14. There were localized high erosion rates (e.g. over 400T/ha/y near the top of the slope in Figure 12.14), but the average was 80T/ha/y. It is thought that some of the measured retreat of the surface was caused by compression of the 300 mm rock layer as the particles settled into place and became embedded into the tailings after the initial measurements. Comparison of the measured erosion rate with that tabled for the Doornkop panels 3 and 4 shows that from Phase 3 onwards, the cladding at

Table 12.2 Results of experiment at FS 6[a]

Panel no.	Treatment	Soil loss T/ha/y[b]	Rainfall (A) mm Cumulative runoff B									
			A	B	A	B	A	B	A	B	A	B
1	Compacted surface (pH 4.4) + 300 mm rock	75	47	2.3	21	2.3	37	5.7	30	7.5	60	11.5
2	Compacted surface (pH 5.8) + geotextile + 300 mm rock	0		2.6		3.9		7.3	?	8.0		12.1
3	Dry land vegetation (pH 4.1)	580		23		28.5		56.5		59.7		90.7
4	Uncompacted surface + lime + fertilizer + compost + 200 mm rock (pH 6.9)	0		0.2		3.2		4.8		5.3		8.4
5	Vegetation with irrigation (pH 7.5)	0		6.3		7.7		12.0		14.3		18.7
6	Uncompacted surface + lime + 200 mm rock + topsoil + lime + fertilizer + seeding over rock (pH 7.6)	630		0		1.3		16.2		16.7		31.1
7	Control area-no treatment			2.8		4.2		22.7		27.1		48.1
	Date of measurement	Mar2001	15/12/99		21/12/99		27/12/99		01/01/00		06/01/00	
	Days since 15/12/99		1		6		12		17		22	
	Cumulative rain since 15/12/99		47		68		105		135		195	

a Experiment started October 1998

b Loss for March multiplied by 4 to get tabulated equivalent T/ha/y

? Rainfall value doubtful

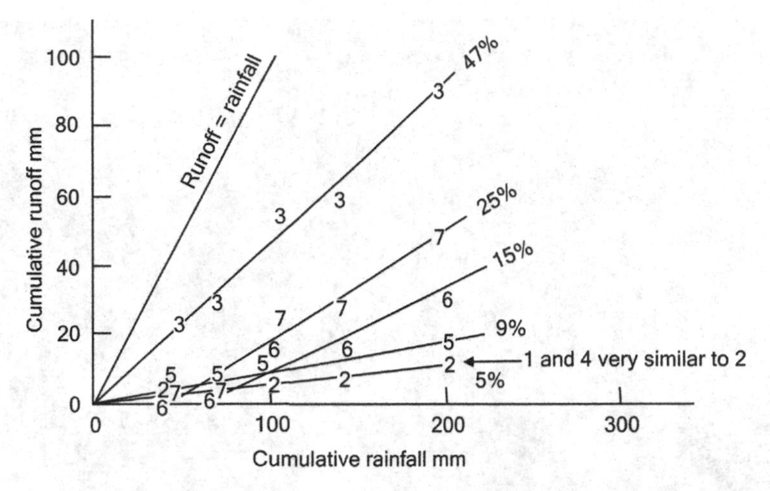

Figure 12.13 Cumulative runoff versus cumulative rainfall for FS6. experiment (numbers refer to panels described in Table 12.2.)

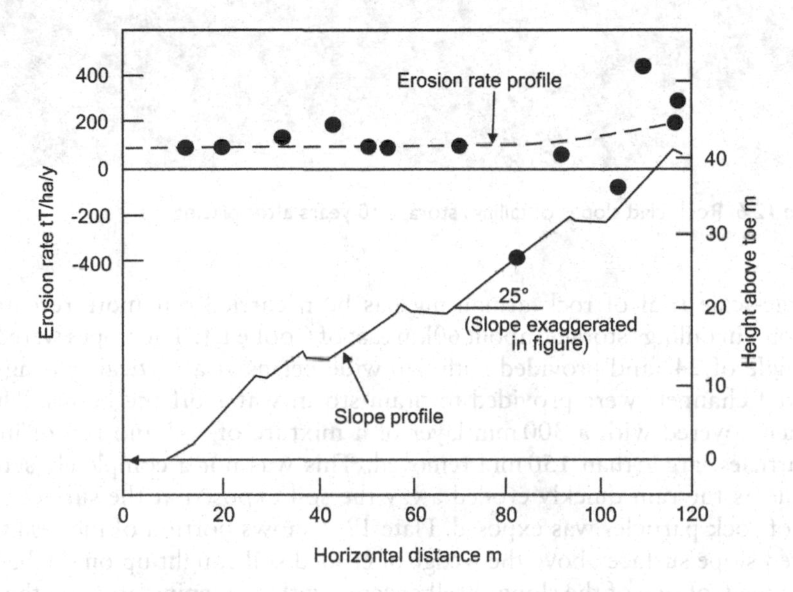

Figure 12.14 Profile and erosion rate measurements for rock-clad Cooke LT gold tailings storage.

Cooke behaved very similarly to that at Doornkop, where much of the test area was 9° flatter (16° as compared with 25°). Hence the full-scale field trial thus far, had behaved very satisfactorily.

Plate 12.6 shows a section of the slopes of Cooke L1 photographed 10 years after the cladding had been placed. No maintenance had been carried out on the cladding since it was placed, despite several rainstorms that exceeded 100 mm in an hour having been recorded on the slopes. Similar rainstorms also fell at Doornkop which is less than 5 km away from Cooke L1.

Plate 12.6 Rock-clad slopes of tailings storage 10 years after placing.

A large-scale trial of rock armouring has been carried out more recently at the Daggafontein tailings storage, about 60km east of Cooke L1. The slopes were flattened to an angle of 24° and provided with 4 m wide berms at a vertical spacing of 15m. Armoured channels were provided to drain storm water off the berms. The slopes were then covered with a 300 mm layer of a mixture of soil and run of mine rock with particles larger than 150 mm removed. This was not a completely satisfactory technique as the rain quickly eroded away the soil exposed at the surface, until the matrix of rock particles was exposed. Plate 12.7 shows portion of the resultant rock armoured slope surface above the wedge of eroded soil caught up on the berm. Plate 12.8 is a view of one of the slopes, well covered with a creeping grass on the left, and with the rock layer still partly exposed on the right. These photographs were taken four years after the slopes were rehabilitated and with no intervening maintenance. The treatment appears very successful, but unfortunately, no measurements of erosion have been made and there is no way of telling, quantitatively, to what extent erosion is continuing.

12.5.4 *Stabilization of slope surfaces using cement or lime*

Work by Blight and Caldwell (1984) showed that stabilizing a 75mm thick layer of tailings on the surface of a tailings dam using either cement or lime was a promising

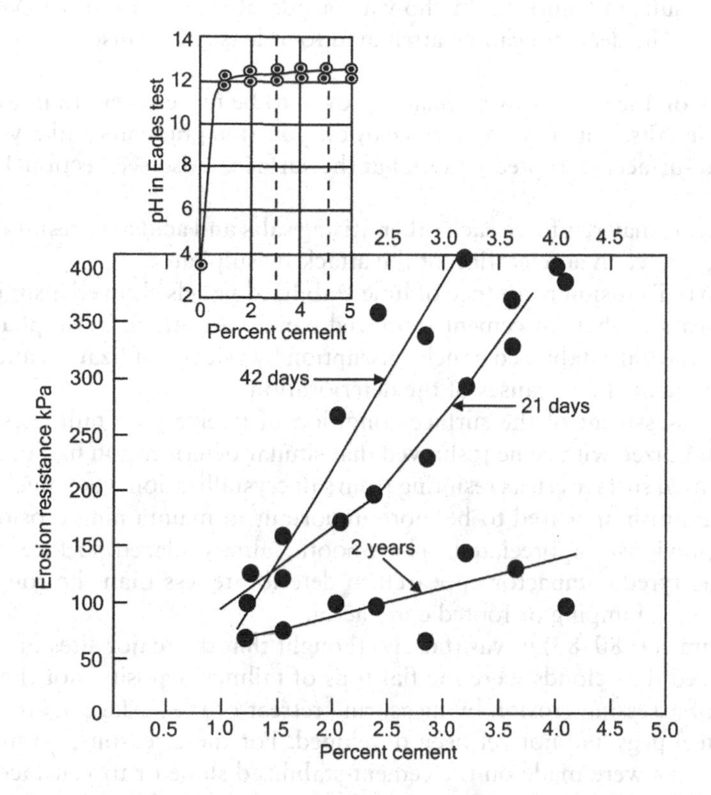

Figure 12.15 **Results of tests on stabilization of gold tailings with Portland cement.**

way of hardening a slope surface to resist erosion. Sufficient cement or lime has to be added to raise the pH of the tailings to 12.5 or above before the cementitious reaction occurs, but for oxidized gold tailings, this requires only about 1% of cement or lime by mass (Figure 12.15).

Figure 12.15 shows the results of a series of field measurements to relate the cement content of layers of oxidized cement-stabilized and compacted gold tailings to the erosion resistance measured by means of the COMET erosion tester. This directs a jet of water 0.8 mm in diameter at the surface of the slope from a distance of 25 mm. The pressure behind the jet is increased at a steady rate until the surface breaks up. The pressure at which the disruption occurs is recorded as a measure of the erosion resistance.

Figure 12.15 shows the measured erosion resistance plotted against percentage cement at times of 21 days, 42 days, and 2 years after stabilizing. It became apparent from these results that (at least in the early ages) cement was more effective than lime, and it was decided that the top surfaces of a number of tailings dams should be stabilized with cement (3 per cent) to a depth of 75mm, and that the performance of these surfaces should be observed over a number of years. Twelve tailings deposits were stabilized in this way.

The test results in Figure 12.15 show a considerable decrease in COMET readings after 2 years. The decrease can be attributed to at least two causes:

(i) attack on the cement by sulphates proved to be more severe than expected; and
(ii) soluble salts, drawn to the surface by evaporation gradients and crystallizing out at the surface, disrupted material at the surface. (Also see Section 13.8.2).

These effects may reinforce each other. Rising salts and acid may result in a lowering of pH and, hence, an acceleration of the attack by sulphates.

The COMET erosion resistance of lime-stabilized panels showed a similar decrease after two years to that for cement-stabilized panels. As attack by sulphates could be ruled out in the lime-stabilized panels, disruption by salt crystallization and acid attack appeared to be the main causes of the deterioration.

A visual assessment of the surface condition of twelve gold tailings storages that had been stabilized with cement showed that similar deterioration had occurred there. "Fluffy", raised surface crusts resulting from salt crystallization were much in evidence. The surface finish appeared to be more important in maintaining erosion resistance than was previously appreciated. The smooth, almost-glazed surface produced by a pneumatic tyred compactor appeared to deteriorate less than the rough, indented surface left by a tamping or footed compactor.

At that time (1980–83) it was (falsely) thought that the major sites of wind erosion that produced dust clouds were the flat tops of tailings deposits, not the slopes. The technique of assessing erosion by measuring retreat of the surface against the tops of a grid of steel pegs had not yet been developed. For these reasons, no measurements of erosion rates were made on the cement-stabilized slope or top surfaces. However, hydraulic fill fly ash deposits develop a cemented outer crust by carbonation of the lime hydroxide in the ash (Fourie, Blight and Barnard, 1999). Compressive strengths of as much as 800 kPa have been measured in laboratory tests. Erosion measurements were made on several ash dams in the late 1990s (Amponsah-DaCosta, 2000) and these give the best available indication of the erosion resistance of slopes with hardened surfaces. Another possibility, where waste rock is not available, shown feasible by recent research, is to produce blocks or bricks of cement-stabilized tailings to use in rock-cladding or rock armouring.

Figure 12.16 shows a profile of the slope and erosion rates for the Duvha #2 hydraulic fill ash storage. It will be noted that the lower slopes accreted by about 50T/ha/y, whereas the upper slopes which were in the process of cementing by carbonation, but were not yet fully hardened, eroded by more than 500T/ha/y in the '97/'98 year and by 200T/ha/y in the 98/99 year. The accretion seemingly originated from material eroded from the upper slopes which washed down and was deposited on the lower slopes. The fully hardened lower slopes, however, did not suffer erosion.

The flyash slopes at Duvha were experimentally protected against erosion by covering them with agricultural shade cloth. This was claimed to be very effective against both wind and water-erosion, but unfortunately comparative measurements of erosion rates of covered and uncovered slopes were not made and hence the claim has not been substantiated. Plate 12.9 shows the experimental shade cloth-clad slope.

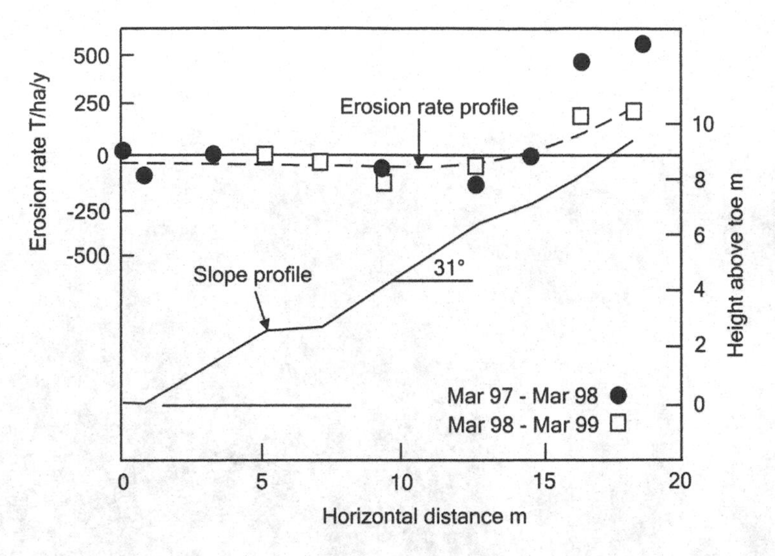

Figure 12.16 Profile and erosion rate measurements for Duvha #2 hydraulic fill ash storage.

Plate 12.7 Rock armoured surface: erosion of soil from between rock particles (below) to leave rock matrix at surface (above).

Plate 12.8 View of rock armoured slope 4 years after rehabilitation.

12.6 Special considerations applying to badly eroded abandoned or neglected tailings storages

12.6.1 *Erosion gulleys*

Erosion gulleys in the outer slopes of tailings storages are usually caused by water collecting on the top surface and cascading down the slope at a low point, or by enlargement of a hole formed by piping erosion through a shrinkage crack in the tailings surface, or through an abandoned tailings delivery pipe that has rusted away.

The formation of gulleys by cascading water can be prevented by the following:

- The top surface of the abandoned deposit should be divided into a series of paddocks by means of dividing walls built of compacted tailings (Figure 12.17). The object is to prevent water from accumulating to form large pools that could overtop the edge of the deposit. A substantial crest wall is an important and integral part of such a system. The walls must be designed to hold the precipitation of the 1 in 100 year storm of 24 hour duration without being overtopped, and should be laid out in such a way as to avoid holding a pool of water near to the crest wall to prevent piping failures from occurring through shrinkage cracks in the tailings.

Plate 12.9 Experimental use of shade cloth to protect fly ash slope against wind erosion.

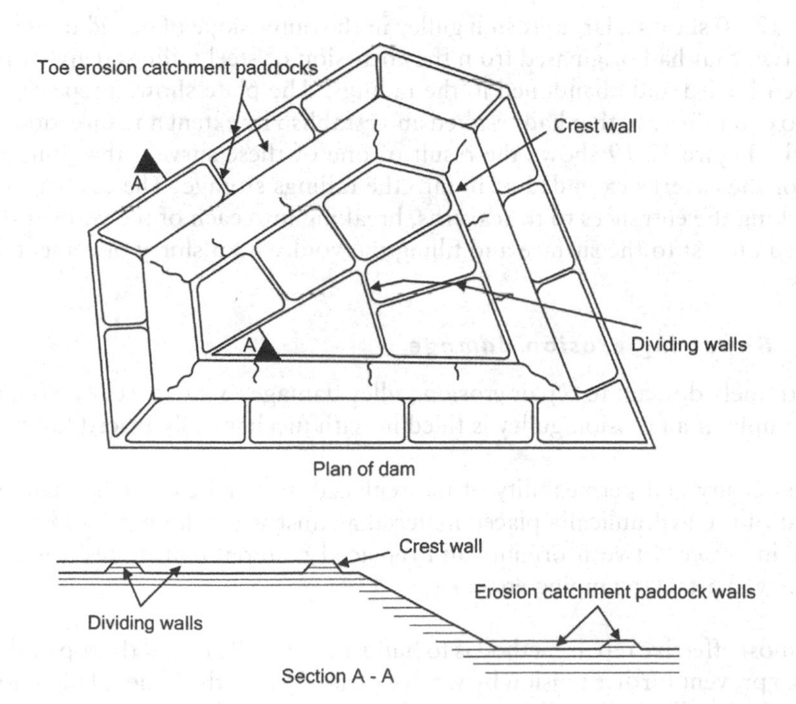

Figure 12.17 Principle of dividing surface of abandoned or decommissioned tailings storage into paddocks (with substantial crest wall) to prevent large accumulations of water.

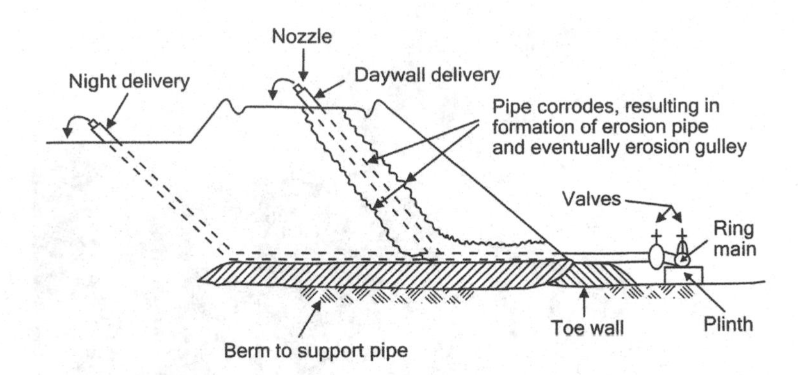

Figure 12.18 "In wall" tailings delivery pipes that corrode away after being abandoned.

- It is common to embed steel tailings-delivery pipes in the outer walls of a storage, as illustrated by the much used detail shown in Figure 12.18. The pipes are eventually abandoned, and if not grouted up with alkaline cement grout, corrode away resulting in the formation of an erosion pipe that is enlarged by water entering and eroding the channel away until it collapses to form an erosion gulley that can be the full height of the deposit. Either this detail should not be used, or the pipes should be grouted up when abandoned.

Plate 12.10 shows a large erosion gulley in the outer slope of a gold tailings storage, one of two that had originated from the corrosion of steel tailings delivery pipes that had been buried and abandoned in the tailings. The plate shows preparations being made to enter the cave that had resulted and establish its extent and direction by survey methods. Figure 12.19 shows the result of one of these surveys that found that the larger of the caverns extended 46 m into the tailings storage. The caverns were filled by blocking the entrances to the caverns, breaking into each of the vertical shafts that extended almost to the surface and filling the void with a slurry of cement-stabilised tailings.

12.6.2 *Repairing erosion damage*

It is extremely difficult to repair erosion gulley damage to the slopes of a tailings dam. For example, if an erosion gulley is filled in with mechanically placed tailings, then:

- the density and permeability of the replaced material cannot be made to match that of the hydraulically placed material against which it abuts, and
- the interface between original and replaced material constitutes a weakness for renewed attack by piping erosion.

The most effective repair method is to build a crest wall around the top of the erosion gulley to prevent further erosion by water cascading into the gulley. This stabilizes the near-vertical walls of the gulley and halts further erosion damage. A wall built across the exit of the gulley will entrap any material eroded from the walls of the gulley. Once

Plate 12.10 Large gulley eroded in slope of abandoned gold tailings storage.

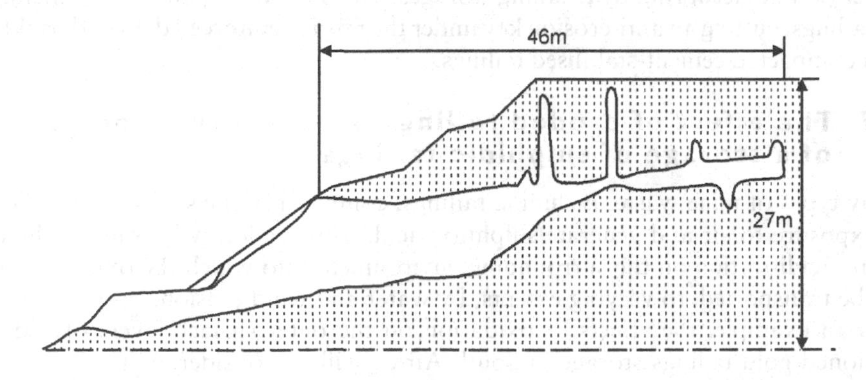

Figure 12.19 Section through large cavern eroded in slope of abandoned tailings storage.

Plate 12.11 "Pipe" eroded under silt retention dam.

these defences have been provided, the gulley can be filled in, compacting the fill in layers from the bottom of the gulley towards the top. (Filling the gulley may only be necessary for cosmetic reasons).

Plate 12.11 shows a piping failure through the tailings foundation stratum of a fabric-reinforced dam wall that was built to contain material eroded from the slopes of two closely adjacent ring-dyke tailing storages. The pipe was repaired by tunneling out the tailings, cutting an anti-erosion key under the fabric-reinforced dam and backfilling with compacted cement-stabilised tailings.

12.7 The effect of eroded tailings on the surroundings of a storage of sulphidic tailings

Many types of mine waste, including tailings, contain metallic sulphides that oxidize on exposure to air and generate sulphuric acid. This section will examine the effect of this acidification on the surrounding environment into which the oxidized tailings will be transported, to varying extents, by water and wind erosion.

As an example, the effects of oxidation on the outer surface layers of a decommissioned gold tailings storage in South Africa will be considered. The storage had been decommissioned about 10 years previously. Figure 12.20 shows profiles of total

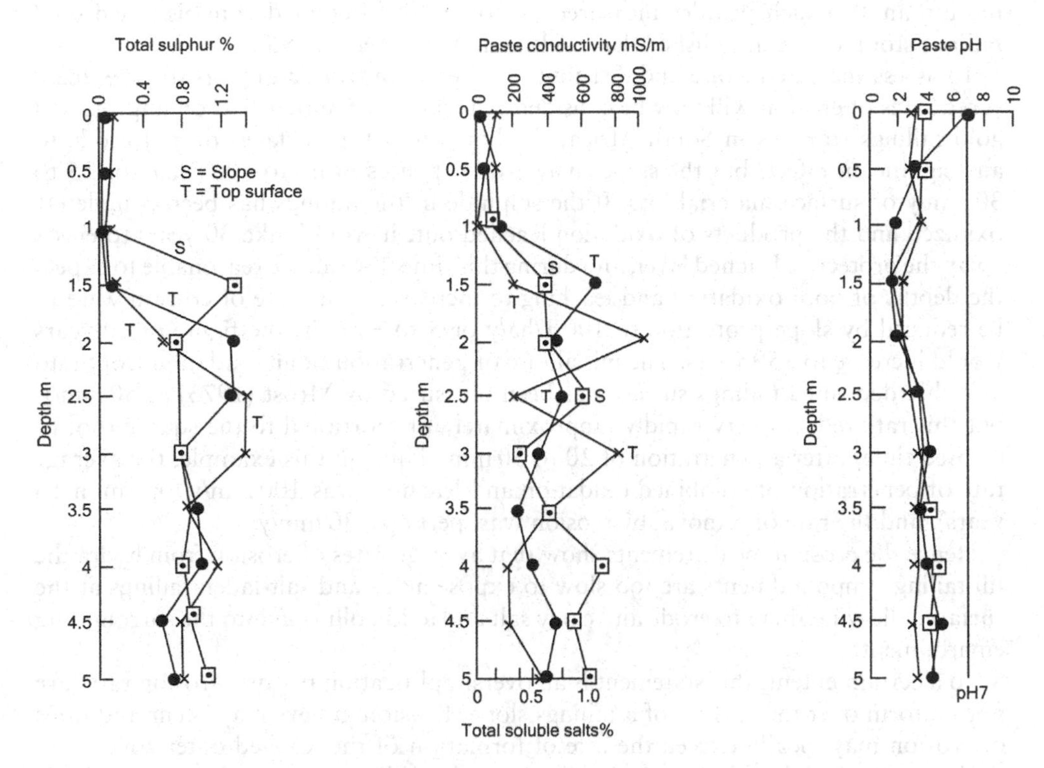

Figure 12.20 Profiles of sulfur, conductivity and pH beneath surfaces of gold tailings storage.

sulphur, paste conductivity and paste pH beneath the surfaces of a slope (S) and the top (T) of the tailings storage. The profiles were measured by augering holes into the surface of the tailings and sampling every 0.5m. The pH and conductivity were measured immediately in the field and the sulphur content was measured in the laboratory, subsequent to sampling. The tailings had been deposited in an alkaline condition at a pH of 9 to 10.

Under the slope, the total sulphur content indicates that the tailings have almost fully oxidized to a depth of about 1m whereas the depth of oxidation under the top surface appears to be 1.5 m. The conductivity profiles indicate that the upper 1m of tailings has been leached of soluble salts. (The soluble salt content is directly proportional to the conductivity, and 100 mS/m corresponds to a soluble salt content of 0.1% by dry mass of tailings, Blight, 1976). The pH profiles show either downward leaching of acid into the tailings, or continuing partial oxidation of the sulphide content, thus reducing the pH. Both mechanisms probably occur. Thus the acid- and salt-laden tailings in the deposit are separated from the atmosphere by a 1 to 1.5 m thick outer layer of oxidized, leached, relatively innocuous material. Unless this layer is eroded away, acid and salt pollution by the dispersion of acid- and salt-laden particles into the surrounding environment cannot occur. Even if the leached layer erodes more slowly than the downward progression of leaching, significant pollution cannot escape via salt- and acid-laden erosion. (It must be emphasized that Figure 12.20 is typical of more than 100 such profiles measured on some 20 different decommissioned gold tailings storages). (Unpublished data collected by du Preez, 1995).

To assess the extent, rate and distribution of erosion from a tailings storage, measured rates of erosion will now be considered. Figure 12.5 shows that on unprotected gold tailings storages in South Africa, the horizontal top surfaces do not erode to any significant effect, but the slopes may erode at rates of up to 500T/ha/y or 25 to 30 mm/y of surface material loss. If the sulphide in the tailings has been completely oxidized and the products of oxidation leached out, it would take 30 years to erode away the protective leached layer, and during this time it would be reasonable to expect the depths of both oxidation and leaching to increase. If the rate of erosion were to be reduced by slope protection to 100 T/ha/y or 5 to 6 mm/y, the figure of 30 years would increase to 150 years. The initial rate of penetration of an oxidation front into a freshly deposited tailings surface has been measured by Mrost (1973) as 50 mm/y, but this rate reduces very rapidly (approximately proportional to the square root of elapsed time) after a penetration of 20 to 30 mm. Thus, in this example, the average rate of penetration of combined oxidation and leaching was 100 mm/y (or 1m in 10 years), and the rate of removal by erosion was, perhaps, 30 mm/y.

Hence the erosion measurements show that average rates of erosion from hydraulic fill tailings impoundments are too slow to expose acid- and salt-laden tailings at the surface, allowing them to erode and carry salt and acid pollution into the surrounding environment.

To a certain extent, this statement is an oversimplification because erosion rates are not uniform over the surface of a tailings slope. Erosion gulleys may occur and rates of erosion may locally exceed the rate of formation of the leached outer zone. Also during prolonged dry spells, evaporation from the tailings surface may cause soluble salts to rise into the leached zone. Subsequent rainfall runoff could then erode this salt-enriched layer.

However, it appears unlikely that erosion from the surfaces of hydraulic fill tailings impoundments is a major source of acid or salt pollution.

12.8 Wind speed profiles, amplification factors and wind erosion

12.8.1 *Variation of wind speed with height above ground level – the wind speed/height profile*

Wind arises from the movements of air masses across the surface of the Earth, which are powered by the solar energy received by the atmosphere. As an air mass moves across the Earth's surface, its motion is retarded by friction between the moving air and the Earth's surface, taken to be stationary relative to the wind. As a result the wind speed increases with height above the ground surface, according to the equation (e.g. Oke, 1978)

$$u = u_* \log_e z/z_o \tag{12.2}$$

where u is the wind speed at height z above the surface, u_* is the wind speed when $\log_e z/z_o = 1$(i.e. $z/z_o = 2.72$) and z_o is a reference height.

Figure 12.21 shows two examples of wind speed profiles calculated from equation (12.2) for $u_* = 5$ km/h and $z_o = 0.2$ m and for $u_* = 5$ km/h and $z_o = 0.5$ m. Wind speeds are conventionally measured at a height of 2 m above the surface. Considering the profile for $z_o = 0.2$ m, the figure shows that if the wind speed at $z = 2$ m is 11.5 km/h, it will be 27.5 km/h at a height of 50 m. The ratio of these two speeds $(27.5/11.5 = 2.4)$ is called the wind speed amplification factor or simply the wind amplification factor (with the amplification factor at $z = 2$ m being 1.0). Figure 12.21 also shows the wind amplification factor curve for $z_o = 0.2$ m.

12.8.2 *Erosion and transportation by wind.*

The ability of the wind to erode and transport particles increases with wind speed. According to the classical work of Bagnold (1941) the rate of movement of material increases with the cube of the wind speed at the surface. His equation is:

$$R = B(u - u_t)^3 \tag{12.3}$$

In equation (12.3)

R is the rate of transport of material along the surface by the wind, in kg.h/km^3;
B is a constant of proportionality with units of kg.h^4/km^6;
u is the wind speed in km/h;
u_t is a threshold wind speed at which the surface material loses stability and starts to be transported.

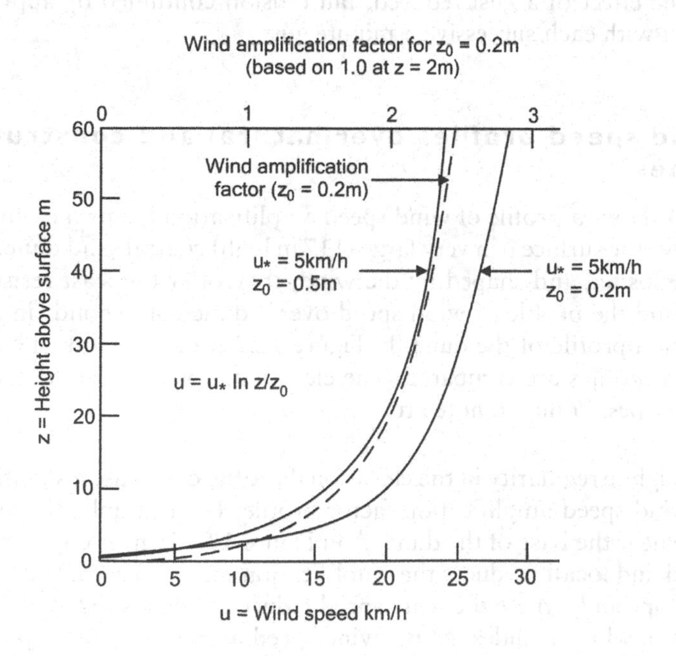

Figure 12.21 Wind speed profile over flat terrain.

B and u_t vary with the particle size distribution of the material being eroded and transported.

Figure 12.22 shows examples (to the same scale) of wind transport curves (a) for gold tailings and (b) for uranium tailings (coarser overall than the gold tailings). Figure 12.22c shows how the threshold speed u_t tends to increase as the material becomes coarser. In this figure, D_{60} is the particle size such that 60% by mass of particles are finer than D_{60}.

The data shown in Figures 12.22a, b and c were for dry material, and it is usually dry material that is involved in serious wind erosion. Figures 12.22d and e show some slightly different experimental data on the wind erosion characteristics of a phosphogypsum waste The gypsum consists of aggregations of needle-shaped crystals with $D_{60} = 0.015$ mm. It is therefore finer than the gold tailings referred to in Figures 12.22a and c. Figure 12.22d shows that the threshold velocity u_t for the gypsum is 20 km/h, and that it behaves in a similar way to a typical dune sand. (The sand referred to was sampled from the large dune shown in Figure 12.23. The gypsum is hygroscopic, and when dried in air at a relative humidity of 35%, it retains a water content of 16%. As shown in Figure 12.22d, the cumulative material loss at a wind velocity of 54 km/h (15 m/s) was 3.73 kg/m^2, but the loss diminished rapidly as the water content was increased, being only 0.35 kg/m^2 at a water content of 31%. It is common experience that the capillary water forces at the surface of a damp granular material hold the particles together and either prevent wind-erosion from occurring, or else reduce it considerably.

Figure 12.22e shows the effect of exposing a gypsum surface to repeated 5 minute gusts of wind at a velocity of 15m/s. The largest loss occurred during the first gust. Thereafter the effect of a gust reduced, but erosion continued by approximately the same amount with each successive 5 minute gust.

12.9 Wind speed profiles over natural and constructed slopes

Figure 12.23 shows a profile of wind speed amplification factor measured at a height of 0.1 m above the surface of a very large (117 m high) coastal sand dune. Because sand dunes are deposited and shaped by the wind, they offer the least resistance to wind movement, and the profile of wind speed over a dune corresponds in a general way to the elevation profile of the dune. In Figure 12.23, the elevational and wind speed amplification profiles are compared. The elevational profile has the same horizontal and vertical scales. Points to note are:

- Even a slight irregularity in the elevational profile can cause a significant response in the wind speed amplification factor profile. For example, the lee of the small slip face near the base of the dune (A in Figure 12.23) gives complete shelter from the wind and locally reduces the amplification factor to zero. Sand is transported up the slope and, where the wind speed reduces, is deposited at the top of the slip face, down which it slides. As the wind speed increases further up the slope, sand is once again picked up and transported towards the crest of the dune, ultimately

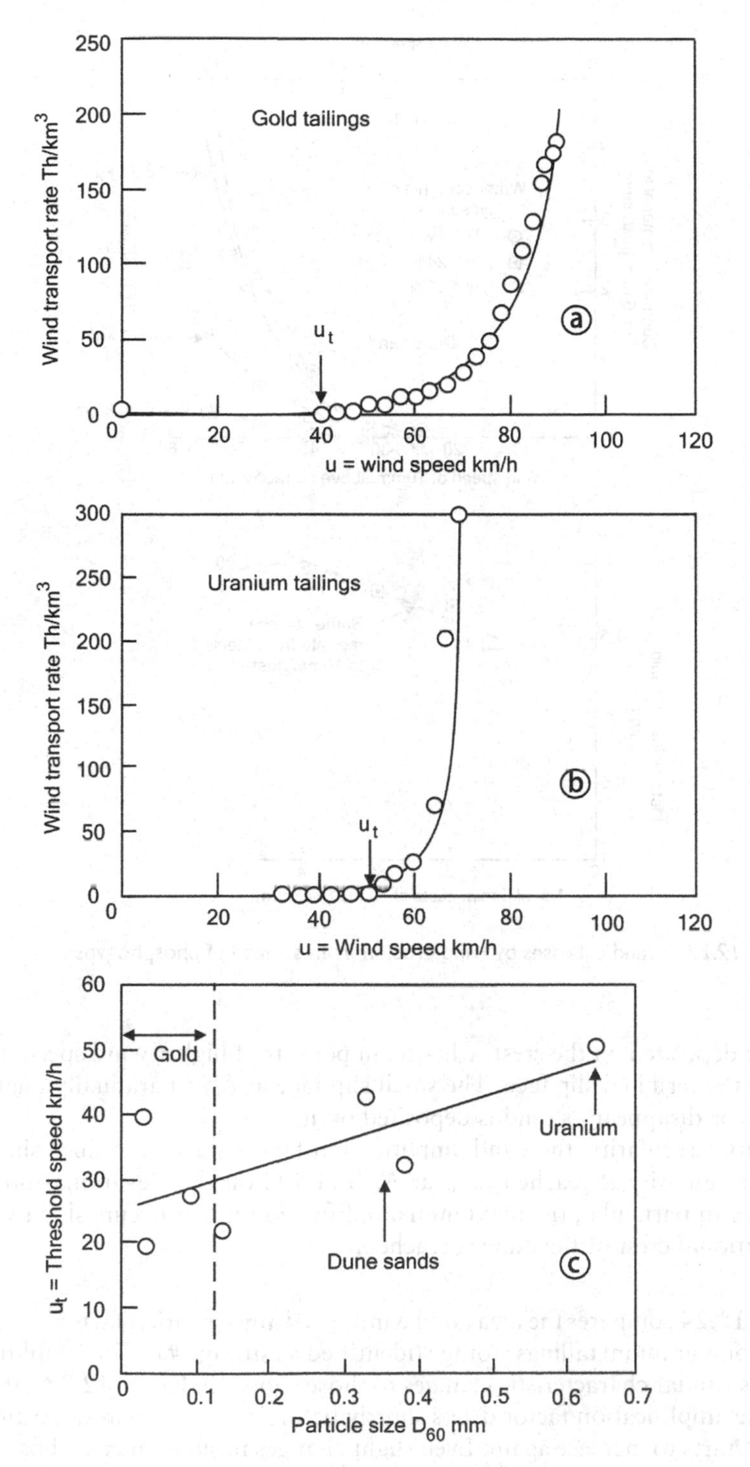

Figure 12.22 Material transport by wind: a: and b: wind transport curves for gold and uranium tailings, respectively, c: effect of D_{60} particle size on threshold wind speed u_t.

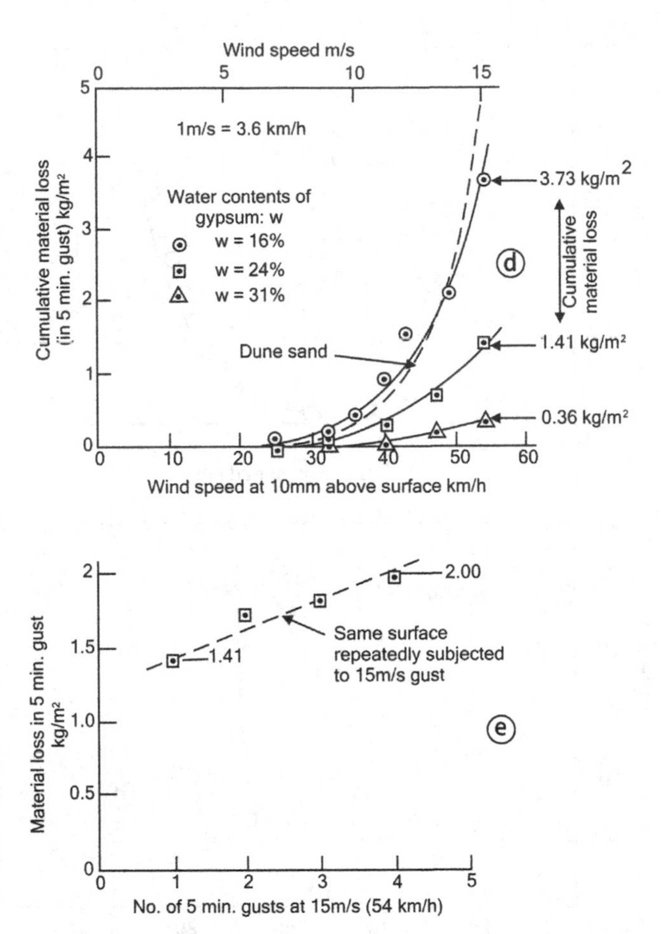

Figure 12.22 d: and e: Losses by wind erosion from surface of phosphogypsum.

to be deposited on the crest, whence, in periods of higher wind speed, it is carried onto the terminal slip face. The small slip face at A will gradually migrate up the slope or disappear as sand is deposited on it.

- At any irregularity, the wind amplification factor starts to reduce shortly before the irregularity is reached (e.g. at A, B and C on the elevational profile of the dune). In particular, the maximum amplification factor occurs slightly before the elevational crest of the dune is reached.

Figure 12.24 compares the measured wind speed amplification factor and elevational profiles for a uranium tailings storage (identified as storage #1). The amplification factor shows similar characteristic changes to those shown in Figure 12.23. At A, B, C, D and E, the amplification factor drops shortly before a reduction in slope angle occurs, and then starts to increase again. Even slight changes in slope such as those at C and F, reduce the amplification factor. A line of scrapped earth-moving tyres experimentally

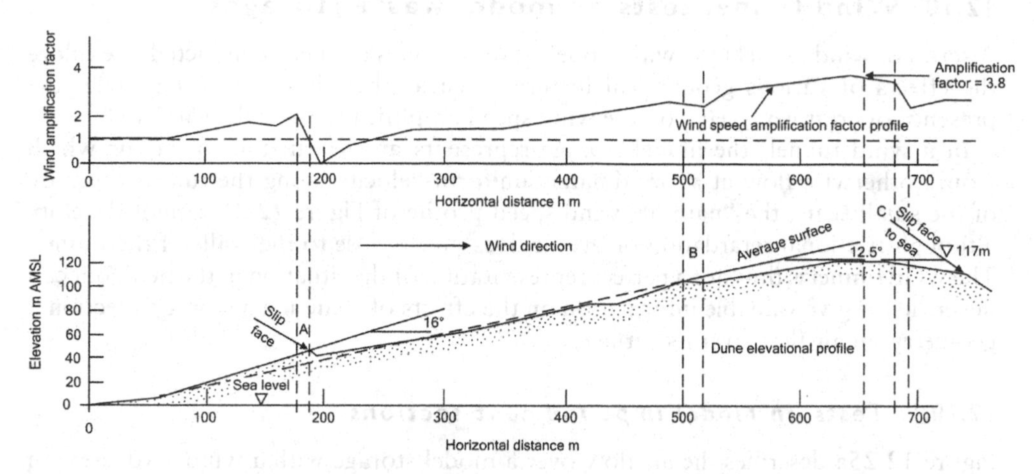

Figure 12.23 Profiles of elevation and wind speed amplification factor (measured 0.1 m above dune surface) on large natural sand dune.

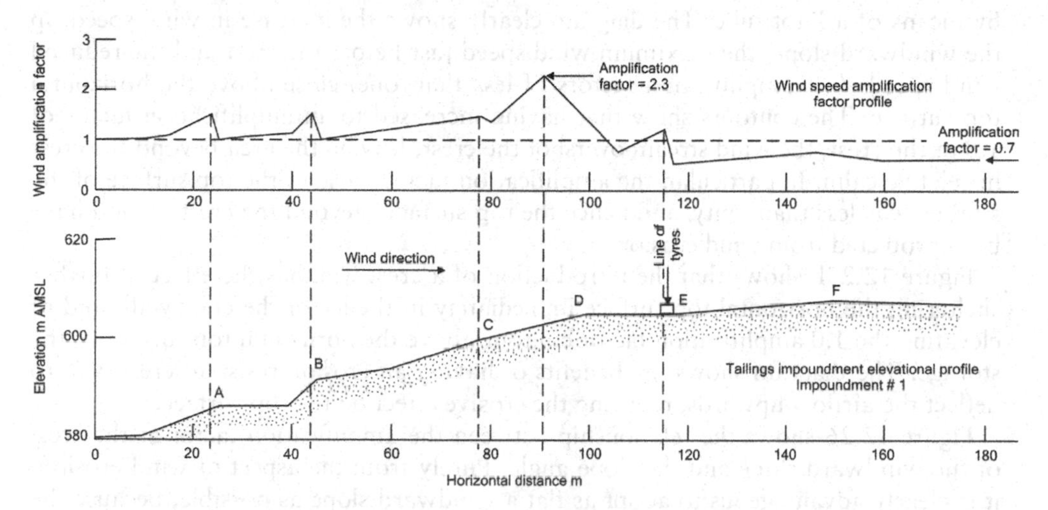

Figure 12.24 Profiles of elevation and wind speed amplification factor (measured 0.1 m above surface) for tailings storage slope. (Storage #1).

placed horizontally near the crest of the slope at E reduces the amplification factor significantly. (However, the tyres would have been more effective if they had been placed at, or even upwind of the crest at D.) Once the crest at F has been passed, the amplification factor becomes almost constant at 0.7 (compared with the maximum of 2.3). As with the natural dune, one would expect erosion of the surface to occur on the slopes before points A and B and, particularly, close upwind of D. Depending on the actual wind speed in relation to the threshold speed u_t (Figure 12.21), deposition (or a lesser rate of erosion) would be expected shortly beyond A, B, D and possibly E. Because of the low amplification factor over the top surface of the storage, wind erosion of the top surface would be much less than of the windward slope.

12.10 Wind tunnel tests on model waste storages

A series of wind tunnel tests with model waste storages has been conducted to explore the effects of various geometrical features such as the windward slope angle, the presence of crest walls, etc., on the wind speed amplification profile generated.

In a wind tunnel, the model storage represents an obstruction to the air which would otherwise flow at approximately uniform velocity along the tunnel. Because of the small scale, the "natural" wind speed profile of Figure 12.21 cannot develop, although frictional retardation of the air does occur close to the walls of the tunnel. The wind tunnel is thus not a perfect representation of the situation in the field, but can nevertheless give valuable information on the effects of changing a storage's sectional geometry on air flow over its surface.

12.10.1 *Tests on model impoundment sections*

Figure 12.25a describes the air flow over a model storage with a windward slope of 45° and no crest wall. The model was truncated by a vertical leeward slope. The figure shows contours of amplification factor measured in the air above the model profile by means of a Pitot tube. The diagram clearly shows the increase in wind speed up the windward slope, the maximum wind speed just before the crest and the reduced wind speeds (i.e. amplification factors of less than one) close above the horizontal top surface. The contours show that having increased to an amplification factor of 1.8 at the crest, the wind stream overshot the crest, leaving the area beyond the crest in relative calm. In particular, the amplification factor close to the top surface of the storage was less than unity, and hence the top surface, beyond the crest, would have been protected from wind erosion.

Figure 12.25b shows that the introduction of a crest wall has the effect of further sheltering the horizontal top surface immediately in the lee of the crest wall, and of elevating the 1.0 amplification factor contour above the horizontal top surface of the storage. The diagram shows the benefits of having an erosion-resistant crest wall to deflect the airflow upwards, reducing the erosive effect on the top surface.

Figure 12.26 shows the relationship between the amplification factor at the crest of the windward slope and the slope angle. Purely from the aspect of wind erosion, it is clearly advantageous to adopt as flat a windward slope as possible, because the maximum amplification factor for the slope increases linearly with slope angle. For example, in Figure 12.26 increasing the slope angle from 20° to 35° increased the amplification factor at the crest from 1.25 to 1.6, an increase of 28%. Even more importantly, for an approach wind of 30 km/h, the speed of the wind striking the crest of the slope would increase from 38 to 48 km/h.

12.10.2 *Wind flow over the top surface of a storage*

Figure 12.25 demonstrates that the upward deflection of the wind as it reaches the crest of a slope provides shelter for the horizontal or near horizontal surface of an impoundment. However, as most impoundments are several hundreds or even thousands of metres across, it can be argued that this sheltering effect may be relatively localized. Alternatively, even if the approach wind speed is reduced by a factor of 0.8 or 0.7, winds may still be sufficiently high over the top surface of an impoundment to

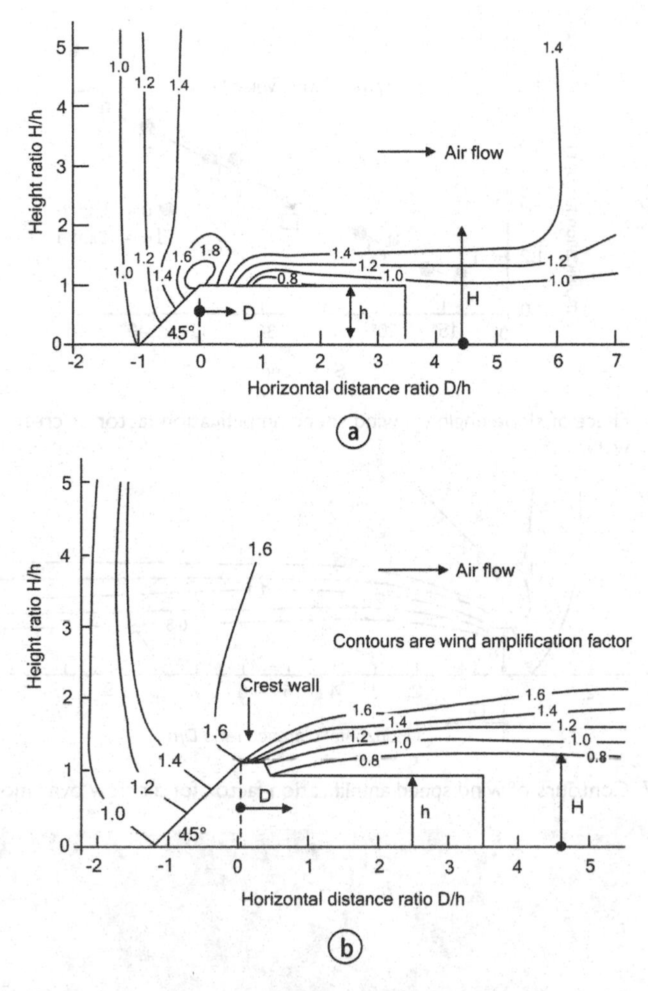

Figure 12.25 **Wind speed amplification factors observed in wind tunnel tests on model storage sections: a: 45° slope without crest wall, b: 45° slope with crest wall.**

exceed the threshold speed and cause wind erosion from the top. Low (1m) windrows of waste rock across the direction of the dry season prevailing wind or even ridges formed by ploughing are often used as a measure intended to limit erosion of top surfaces.

Figure 12.27 shows contours of wind speed amplification factor for air flow over two adjacent model windrows. The leading windrow could represent the crest wall of an impoundment or the first of a series of parallel windrows. The spacing in Figure 12.27 of 8.5 times the windrow height represents the maximum that was possible in the wind tunnel used for the model tests.

The contours show very similar characteristics to those illustrated by Figure 12.25. The contours were close to parallel to the surface and the wind speed near to the horizontal surface was small. Figure 12.27 also shows that windrows do not have to

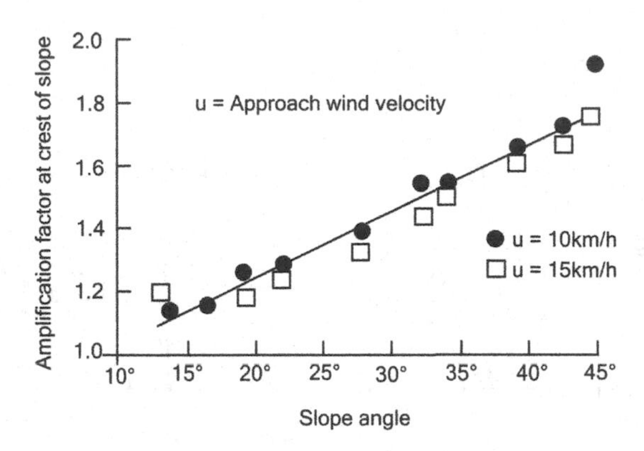

Figure 12.26 Effect of slope angle on wind speed amplification factor at crest of slope (no crest wall).

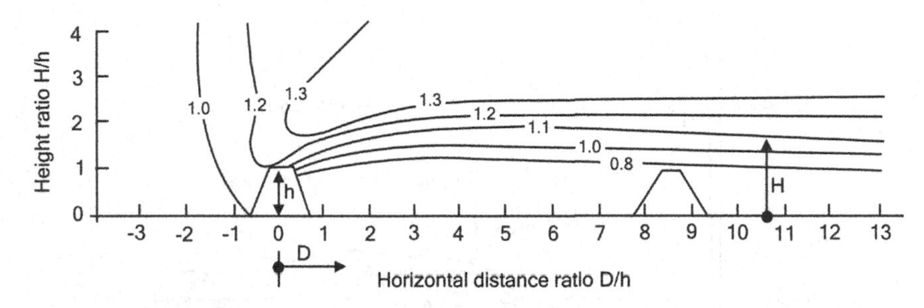

Figure 12.27 Contours of wind speed amplification factor for air flow over model windrows.

Plate 12.12 Dust clouds blowing over leeward crest of tailings storage.

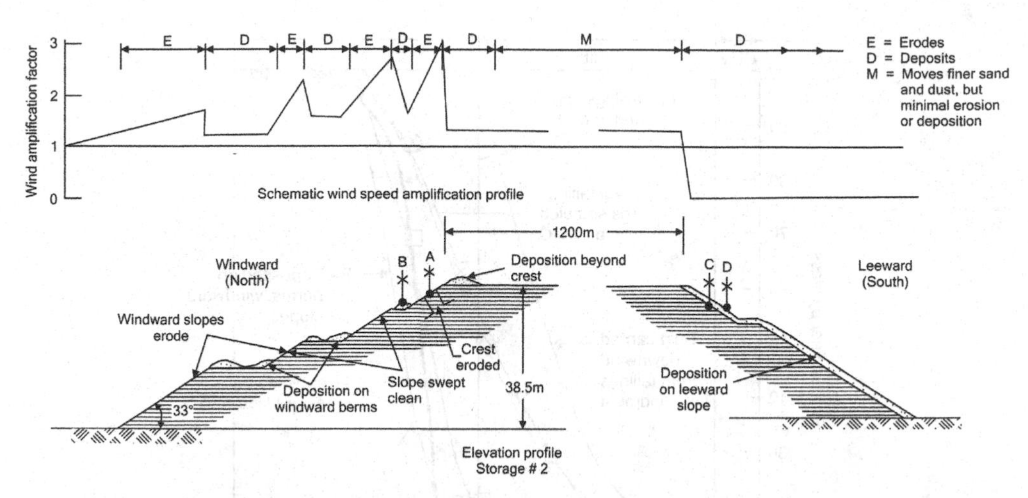

Figure 12.28 Observed patterns of erosion and deposition on gold tailings storage (#2) after period of dry windy weather. (Wind amplification factor schematic, not measured.)

be closely spaced to protect a horizontal surface from wind erosion. The maximum spacing has not been established, but spacings of up to 10 windrow heights will clearly give good protection. Because major wind directions vary from season to season, in practice, windrows should be zig-zagged to provide protection for a range of wind directions. The angles for the zig-zagging are best established from a wind rose for local wind directions. (see, e.g., Figure 12.12). This information is available from meteorological offices.

Hence, within their limitations, the wind tunnel tests, as well as the data given in Figures 12.28 to 12.31 support the contention that wind erosion of the top surfaces of waste storages will be small, except (possibly) in conditions of unusually high wind. Erosion will be greatest from the windward slopes.

12.11 Erosion and deposition by wind on full size waste storages

Because clouds of dust are often observed billowing off the top surfaces of tailings storages, as seen in Plate 12.12, it is commonly believed that the dust has been eroded from the top surface by the wind. Plate 12.12 shows dust blowing off the top of a large platinum tailings storage. The horizon to the right of the picture is the crest of the storage, which extends horizontally across to the left of the centre of view. However, as will now be argued, the dust is actually eroded from the windward slopes and does not originate from the storage basin formed by the beaches of the storage.

Figure 11.28 shows the patterns of erosion and deposition observed on a gold tailings storage (identified as storage #2) after a long period of dry windy weather. The windward slopes (where the wind amplification factor is high) had been grooved and

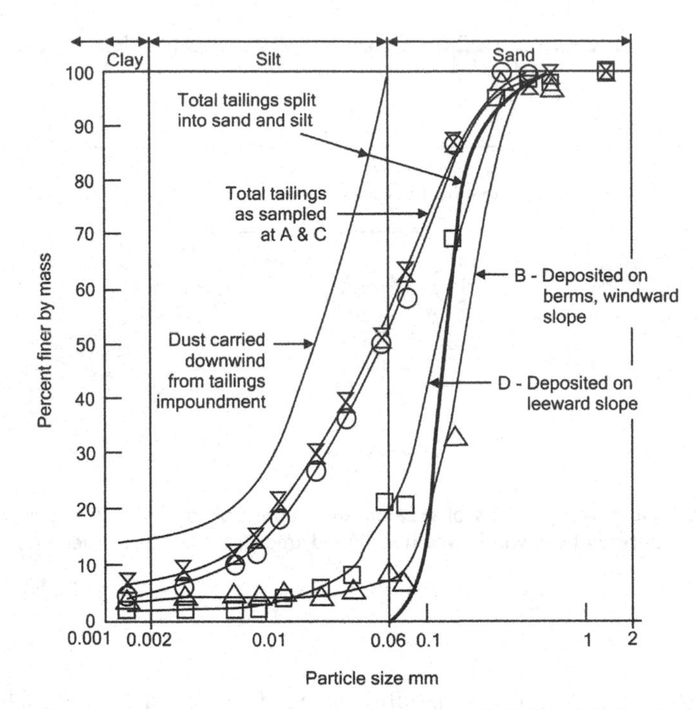

Figure 12.29 Particle size analyses for samples from points A, B, C and D (marked in Figure 12.28) on Storage #2.

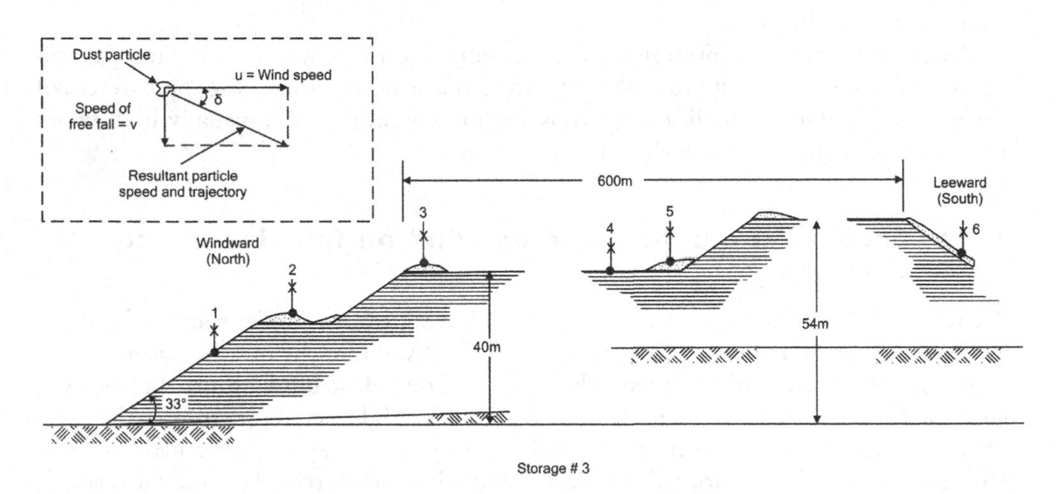

Figure 12.30 Patterns of erosion and deposition on gold tailings storage (#3) after period of dry windy weather. Inset diagram: Trajectory of falling dust particle blown by wind (see Section 12.2).

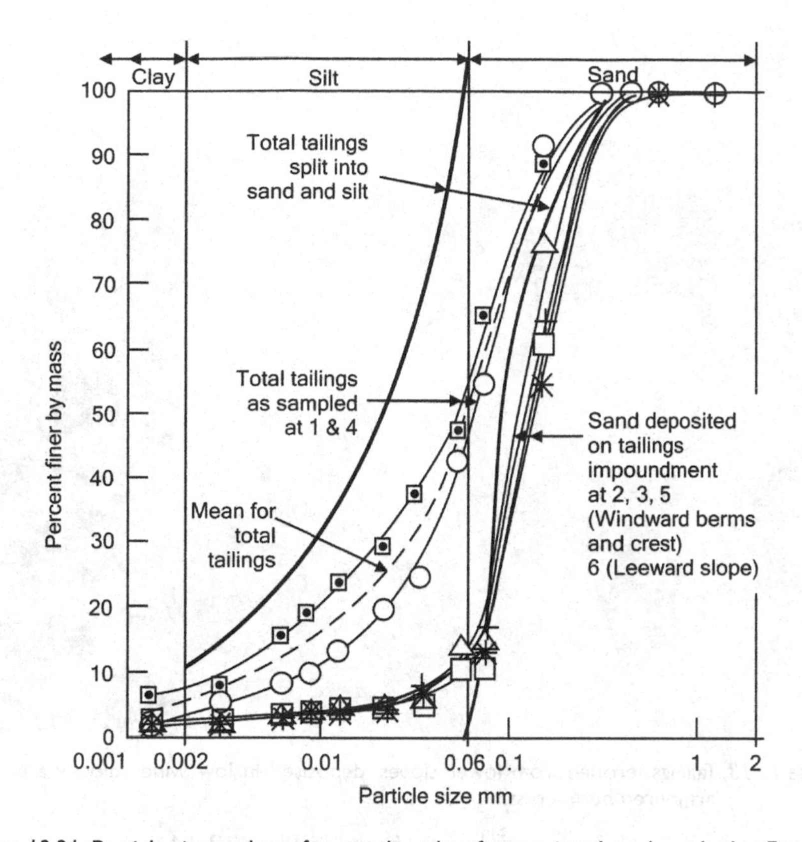

Figure 12.31 Particle size analyses for samples taken from points 1 to 6, marked in Figure 12.30.

swept clean by the wind. Although the process could not be witnessed by eye, erosion of these windward slopes must have provided the loose material deposited on the windward berms and just beyond the windward crest as there was no other source for this material. The leeward slope, from the crest down, was completely covered by loose material deposited as the wind passed off the top of the impoundment and the amplification factor fell to zero in its lee. The top surface of the impoundment was also swept clean, but showed no signs of grooving or surface erosion.

Plate 12.13 shows the windward slope shortly below the crest. The berm below the crest had been armoured to reduce wind erosion, and the loose tailings deposited on the berm were eroded from lower down on the slope and deposited in the low wind velocity area just past the crest of the berm (see Figure 12.25). The un-armoured crest of the dyke can be seen to the top left of the photograph. The "castellations" in the crest were formed because the un-armoured temporary crest wall had been formed by dozing tailings up to the crest from within the storage. The high sections of the castellations mark the presence of more densely compacted material that was formed by direct pushes of the dozer blade. The low portions, eroded away by the wind, were composed of the looser material in between.

Plate 12.14 shows the leeward or down-wind slope covered by loose tailings that have blown right across the top of the storage and been deposited in the calm air in the

Plate 12.13 Tailings, eroded from lower slopes, deposited in low wind velocity area to lee of armoured berm crest.

lee of the storage. The unarmoured crest shown in Plate 12.14 was formed in exactly the same way as that shown in Plate 12.13, but was not exposed to the scouring of the wind, and therefore remained intact. The unprotected upper slopes above the experimental sections of the Doornkop experiment, shown in Plates 12.4 and 12.5 were also lee slopes during the dry windy months and can also be seen to be blanketed with loose material. These two photographs were taken near the end of the wet season during which most of the loose material is eroded away by water.

Samples of the total tailings composing the original slope surfaces were taken at points A and C in Figure 12.28 and samples of the loose wind-deposited material were taken at points B and D. The particle size analyses for these samples are given in Figure 12.29. The analyses of the total tailings from points A and C were near-identical. The analyses for the loose re-deposited material from points B and D were considerably coarser than the total tailings, with material from B (deposited on a windward berm) being slightly coarser than that from D (deposited on the leeward slope). If the analysis for the total tailings is split into its sand and silt fractions (sand particles being those coarser than 0.06 mm), the grading of the sand is found to be very similar to that of the samples B and D. This indicates that the loose re-deposited material (B and D) consists of the tailings with the silt sized particles winnowed out of it by the wind.

The patterns of erosion and deposition shown in Figure 12.30 for a second gold tailings storage, identified as storage #3 (at a different mine about 20 km from storage #2) are very similar to those shown in Figure 12.28, with windward slopes grooved, top

Plate 12.14 Tailings deposited on leeward slope of tailings storage.

surfaces swept clean by the wind and loose material re-deposited on the windward berm and just beyond the windward crest. This pattern was repeated at the higher section of the storage to the south. Again, the leeward slope beyond the leeward crest was completely covered with loose material. Samples of total tailings were taken at points 1 and 4 in Figure 12.30, and of loose re-deposited material at points 2, 3, 5 and 6. The results of particle size analyses carried out on these samples are shown in Figure 12.31.

On storage #3, the surface of the area where sample 4 was taken showed signs of slight grooving and erosion (Figure 12.30). The area was not extensive and the erosion is thought to have been caused by eddies set up when the wind was blowing obliquely to the plane of the nearby slope. The analyses for the loose re-deposited material were all very similar in this case, and that for sample 4 (from the top of the impoundment) was slightly finer than sample 1 (from the windward slope). Splitting the particle size curves into silt and sand fractions (Figure 12.31) gave a very similar result to that shown in Figure 12.29. (The diagram inset on Figure 12.30 refers to section 12.12 below and analyses the likely travel distance of dust particles of various sizes once the wind-borne dust blows clear of the storage top surface.)

Although measurements of total erosion (Figure 12.5), wind speed amplification on full size storages (Figure 12.24) and wind tunnel tests (Figure 12.25) all show that there is little erosion from the near horizontal top surface of an impoundment, it is also clear that much of the material eroded from the windward slope is transported across the top of a storage. This is shown by the blanket of re-deposited winnowed sand found on the leeward slopes (Figures 12.28 and 12.30). Also, in periods of high wind speed, clouds of dust are clearly visible blowing across and off the top surfaces (e.g. Plate 12.12).

12.12 Analysis of particle movement in the wind

A question that inevitably arises is: How is this material transported without causing erosion to the surface over which it travels? The answer is not known for certain at present. It may be that both sand and fine dust are transported air-borne in the high wind speed amplification zone above the top surface (see Figure (12.27), or that the fine dust is transported in this zone, while the sand moves across the surface by saltation. Saltation (Bagnold, 1941) is a process whereby sand grains are picked up by the wind, are air-borne for some distance, bounce off the surface and are again air-borne for a distance, thus progressing by a "skipping" motion. The calculations given below show that even if the wind is at the threshold speed u_t, it is very likely that sand particles will saltate in "skips" of 50 to 100 m, while silt particles, once they become air-borne, will remain in the air.

The diagram inset on Figure 12.30 illustrates the dynamics of an isolated airborne particle of tailings blown by wind. The particle is assumed to be dragged and carried at a horizontal speed u, equal to that of the wind, and simultaneously to be falling at a vertical speed v, given by Stokes' law. According to Stokes' equation, the falling velocity would be

$$v = \gamma_s d^2/18\eta \qquad (12.4)$$

where γ_s is the solid unit weight of the particle, taken as $26.5\,\text{kN/m}^3$, d is the equivalent particle diameter in mm; and η is the viscosity of air, taken as 1.5×10^{-8} kPas $= 4.1 \times 10^{-12}$ kPah. Hence

$$v = 360d^2 \text{ km/h, with d in mm} \qquad (12.4a)$$

If d = 0.06 mm (the dividing size between silt and sand), v = 1.3 km/h, or 0.36 m/s.

Suppose that the wind speed at the surface is 50 km/h (13.9 m/s), and that a grain of diameter 0.06 mm is projected into the air by the wind, to a height of 1 m. It will then simultaneously be travelling at a horizontal speed of 13.9 m/s and falling at a speed of 0.36 m/s. It will take 2.75 s to fall to the surface, and in that time will travel a distance down wind of 38 m. If it continues in this way, it will impact the surface 26 times to cover a distance of 1 km (26 saltations). Once it reaches the leeward edge of the storage (in 16 saltations in the case of storage #2), it will fall into the calm air downwind of the edge and travel with constant terminal velocity but rapidly reducing horizontal velocity and be deposited onto the leeward slope.

For a particle of diameter 0.006 mm, the falling velocity will be 0.0036 m/s, and the distance of each 1 m high saltation will increase from 38 to 3800 m. Hence, once a particle as small as this has been projected into the air, even to a distance of only 1m above the surface, it will effectively remain air-borne indefinitely.

12.13 Summary of points to be remembered

12.13.1 Measurements on natural soil slopes in the water deficit climate of South Africa have shown that they erode at a rate of 5 to 50 T/ha/y. When attempting to reduce the rate of erosion of tailings slopes, this range of erosion rates should be considered the target to be aimed for. Erosion rates for unprotected slopes of gold tailings storages have been measured at up to 500T/ha/y. There is therefore a pressing need to implement measures to protect the surfaces of tailings storages against erosion.

12.13.2 Studies have shown that rates of erosion of unprotected slopes of gold tailings depend on the angle of the slope and its length. The maximum rate of erosion occurs at slope angles of 31 to 37°, and decreases as the slope angle decreases. Flat and horizontal slopes erode very little. The overall rate of erosion of slopes increases roughly linearly with slope length.

12.13.3 The total erosion of a tailings slope is made up of components due to water and wind. Erosion by water predominates in the wet season of a year, and wind erosion predominates in the dry season. Limited studies show that more erosion may be caused by wind than by water.

12.13.4 Considering the requirement that the storage capacity for tailings solids of a given site must be maximized, together with the requirements for overall shear stability and the requirement to limit rates of surface erosion, as well as the practical advantage of a slope that can be worked mechanically, an acceptable slope angle for tailings slopes appears to lie in the range of 15–20°. A round figure of 1 vertical on 3 horizontal or 18° appears to be a good compromise.

12.13.5 Berms, step-ins and ramps are necessary to provide access to the slopes of a tailings storage during its operational phase, as well as after decommissioning and closure. They should be carefully preplanned and be located, designed and constructed to maximise their functionality. They should be easily accessible to motor vehicles, and traversable in both directions between at least two entry and exit points.

12.13.6 A number of geotechnical means of preventing or reducing erosion have been devised, experimented with and applied on a full scale with reasonable success. These include gravel mulching of surfaces with flat (0–10°) inclinations, rock cladding for steeper slopes and rock armouring, in which the surface texture of a natural vegetated slope is simulated. Various forms of rock cladding have been shown experimentally (Doornkop experiment) to reduce erosion rates to 20% of the erosion rate for an untreated slope. Only limited measurements exist for gravel-mulched or rock armoured surfaces. Various forms of rock cladding have been shown by the FS S6 experiment to reduce cumulative surface runoff to 5% of cumulative rainfall, whereas cumulative runoff from the untreated control section was 25% of cumulative rainfall. The 80% reduction of runoff is the probable main reason for the effectiveness of rock cladding in reducing rates of erosion. A full scale trial of rock cladding at the Cooke L1 tailings storage showed that even on relatively steep (25°) slopes, rock cladding is effective in reducing erosion rates to 80T/ha/y.

12.13.7 Earlier (1980s) work on stabilizing the surfaces of tailings slopes with cement or lime showed promise and should be revisited at least to investigate current potential costs in situations where waste rock is either not available, or could more profitably be used as concrete aggregate. The problem experienced with attack by salt crystallization could be investigated further if current cost estimates prove promising. Recent experiments have shown that blocks or bricks of cement-stabilized tailings can be produced that are suitable to use in rock cladding or rock armouring.

12.13.8 The wind speed over flat terrain increases logarithmically with height above the ground surface. The ratio of the wind speed at a given height above the surface to the speed at a standard height is the wind speed amplification factor. The wind begins to erode soil particles from the surface once a threshold speed is exceeded. The threshold speed increases as the particle size of a surface soil or soil-like material increases.

12.13.9 In situ measurements of wind speed amplification profiles on a natural sand dune and a tailings storage show that the highest wind speed amplification factor occurs just upwind of the windward crest, and that features such as berms and flattening of the slope angle provide protection from the wind. The amplification factor over the horizontal top surface of an impoundment is relatively low.

12.13.10 It follows that upper windward slopes, and particularly the areas just below windward crests are most susceptible to wind erosion. The top surfaces are much less susceptible to significant wind erosion by wind.

12.13.11 Wind tunnel tests on model waste storages give further detail of wind amplification factors and are in good agreement with field observations and measurements. The wind tunnel tests demonstrate the advantages of using flatter slope angles to reduce wind erosion and of lining the crest of a tailings storage with a substantial crest wall that is resistant to wind erosion.

12.13.12 The wind tunnel tests also demonstrated the protective action of windrows placed across the top surface of a storage at extended intervals to protect against the erosive effects of particularly high wind speeds.

12.13.13 Observations of deposition patterns of tailings eroded by the wind support conclusions 12.13.8 and 1213.9. Particle size analyses of samples taken from the originally deposited total tailings and from loose eroded material re-deposited from the air in areas protected from the wind show that tailings eroded from the windward slopes and re-deposited, in sheltered areas and on the leeward slopes, consist of sand from which the silt fraction has been winnowed by the wind. It can be inferred that the fine silty material has been carried downwind, forming the clouds of dust that cause a dust nuisance downwind of the storage. This conclusion is supported by the simple calculations of particle travel distance given in Section 12.12.

12.13.14 It is usually not necessary to protect the lower slopes of a waste storage against wind erosion, but the upper slopes should be protected. At present, rock cladding (see Tables 12.1a and 12.1b) appears to provide the most cost-effective, long lasting and physically effective protection, not only against wind erosion, but also against total erosion. Apart from providing a substantial erosion-resistant crest wall, and a few widely spaced, zig-zagged, erosion-resistant windrows to guard against extreme wind conditions, it is not necessary to protect the near-horizontal top surfaces of waste storages against wind erosion.

12.13.15 The source of down-wind dust is erosion from the upper portions of upwind side slopes. To prevent this erosion , areas of slopes that are vulnerable to wind erosion should be armoured, as described earlier.

References

Amponsah-Da Costa, F.: *A Strategy for Reducing Erosion of the Slopes of Mine Waste Deposits.* Ph.D Thesis. Johannesburg: University of the Witwatersrand, 2000.

Bagnold, R.A.: *The Physics of Blown Sand and Desert Dunes.* London, U.K.: Methuen, 1941.

Blight, G.E.: Migration of Sub-Grade Salts Damages Thin Pavements. *Transport. Eng. J., ASCE 102(TE 4)* (1976), pp. 779–791. New York, USA: American Society of Civil Engineers.

Blight, G.E.: Erosion Losses from the Surfaces of Gold-Tailings Dams. *J. S. Afr. I. Min Metall. 89(1)* (1989), pp. 23–29.

Blight, G.E., & Caldwell, J.A.: The Abatement of Pollution from Abandoned Gold-Residue Dams. *J. S. Afr. I. Min. Metall. 84(1)* (1984), pp. 1–9.

Blight, G.E., & Amponsah-Da Costa, F.: Towards the 1000-year Erosion-Free Tailings Dam Slope – A Study in South Africa. In: D.H. Barker, A.J. Watson, S. Sombatpanit, B. Northcutt, A.R. Maglinao, (eds): *Ground and Water Bioengineering for Erosion Control and Slope Stabilization.* New Hampshire, U.S.A.: Science Publishers, 2004, pp. 365–376.

Blight, G.E., Rea, C.E., Caldwell, J.A., & Davidson, K.W.: Environmental Protection of Abandoned Tailings Dams. In: *10th International Conference on Soil Mechanics and Foundation Engineering*, Stockholm, Sweden, Vol. 1, 1981, pp. 303–308.

Chamber of Mines of South Africa: Guidelines for Environmental Protection, Vol. 1/79, The Engineering Design, Operation and Closure of Metalliferous, Diamond and Coal Residue Deposits, Vol. 2/79. The Protection of Residue Deposits Against Water and Wind Erosion. (Vol. 1 revised 1983 and 1996), 1979.

du Preez, J.: *Unpublished Collection of Measured Profiles of Conductivity, pH and Sulphur Beneath the Surfaces of Gold Tailings Storages in South Africa*, 1995.

Fourie, A.B., Blight, G.E., & Barnard, N.: Contribution of Chemical Hardening to Strength Gain of Fly Ash. In: *Geotechnics for Developing Africa.* Rotterdam, Netherlands: Balkema, ISBN 90 5809 0825, 1999, pp. 35–40.

Mrost, M.: The Rehabilitation of Gold Mine Tailings Deposits, In: S.A. *Inst. Civil Engrs, 5th Quinquennial Convention*, Johannesburg, South Africa, 1973, pp. 14–21.

Oke, T.R.: *Boundary Layer Climates.* London, UK: Methuen, 1978.

Smith, G.M., Waugh, W.J., & Kastens, M.K.: Analog of the Long Term Performance of Vegetated Rocky Slopes for Landfill Covers. In: *Tailings and Mine Waste '97.* Rotterdam, Netherlands: Balkema, 1997, pp. 291–300.

van Rensburg, L.D., Strydom, M.G., du Preez, C.C., Bennie, A.J.P., le Roux, P.A.L., & Pretorius, J.P.: Prediction of Salt Balances in Irrigated Soils Along the Lower Vaal River, South Africa. *Water SA. 34(1)* (2008), pp. 10–15.

Chapter 13

The use of mine waste for backfilling of mining voids and as a construction material

An obvious solution to the problem of storing mining waste is to use it to backfill the voids created by mining. There are a number of reasons, however, why this can only ever be a partial solution. The most obvious and insuperable difficulty is that of bulking. Rock with an in situ dry density of, say, 2500 kg/m^3 is mined, crushed and milled to allow of the extraction of its mineral content. After re-deposition of the waste it has a dry density of, say, 1500 kg/m^3. The original 1 m^3 volume has increased or bulked to $2500/1500 = 1.67$ m^3. Also, where the tailings are placed back in the underground workings, the voids will have decreased in size due to the fact that the surrounding rock will be subject to high compressive and shearing stresses and the excavations will have distorted, contracted or closed under the stress. A further problem will be that reprocessing the waste at a later date to extract the remaining mineral content will be uneconomical.

13.1 Applications of backfilling

Backfilling of mining voids, however, can be and is used for several purposes, even recognizing the above difficulties:

- Where extensive areas of mineral deposits at shallow depth are mined by open cast or strip-mining methods, the usual procedure is to create a void when stripping the overburden strata to first expose the mineral deposit. The overburden is cast back in the opposite direction to the direction of mining. The ore, for example coal, is removed and the next strip of overburden is cast back or backfilled into the initial void in order to expose the next strip of ore, and so on. The backfilled area may be very extensive and may be needed for future surface developments, e.g. agriculture, roads and housing. In this case, future compressive settlement of the mined out and backfilled area may become an issue.
- Backfilling can be used to stabilise near-surface strata that have been undermined by shallow underground mining, so that surface structures can safely be built over an under-mined area.
- Backfill may be used to support the mine workings as mining proceeds, especially in ultra-deep mines. In this case the strength of the backfilled tailings may be augmented by incorporating cement or by reinforcing the backfill with steel reinforcing mesh.

- Backfilling of voids may be used to provide access for progressive mining of adjacent ore, or to support blocks of ore prior to mining them.

This chapter will discuss most of these aspects.

13.1.1 *The settlement of coal strip-mine backfill*

Coal deposits often occur as horizontal to sub-horizontal seams at relatively shallow depths of 20 to 70 m. If this is the case, the current most economical method of mining is to strip off and stockpile the top-soil – to use for later rehabilitation of the surface – and then to strip the overburden strata and excavate and remove the coal. This process takes place over a working face that may extend laterally for several hundred metres. While coal is being excavated and removed for beneficiation and sale, dragline excavators are occupied in stripping overburden and exposing the coal stratum ahead of the working face. The stripped overburden – usually consisting of soil and soft to hard sandstone and shale fragments of random sizes – is cast back by the dragline into the void behind the working face. The mine backfill is deposited in the form of a series of rows of conical piles with no attempt at compaction. At the end of the mining operation, the cones of spoil are flattened to form a regular surface, the top-soil is spread and the area rehabilitated. Because of the bulking of the rock when excavated, the final surface is usually above the original ground level, unless the coal seams removed are particularly thick.

Because strip mining covers large areas, there has been an ongoing interest in the long term settlement of coal mine backfill and its effect on structures built over rehabilitated mine land. The results of several studies have been published (e.g. Charles, et al., 1977, Charles, et al., 1984, Reed and Hughes, 1990). These have established the following mechanisms for compression of the backfill, and hence settlement of the surface:

- Immediate compression of the backfill occurs under increasing overburden, during the backfilling process, as particles are forced closer together causing distortion and crushing of point contacts.
- Time-dependent sloughing and slaking of rock fragments, resulting in creep settlement, occur because of relief of the original in situ stresses in the intact strata. Oxidation of minerals such as pyrite, now exposed to air, accentuate the particle breakdown.
- The regional water table has usually been lowered, by pumping from de-watering holes, to below the level of the deepest coal seam to be mined. Once mining ceases, the water table will gradually recover and re-establish itself within the backfill. As a result, effective stresses in the rock fragments will be decreased and further "collapse" type settlement may occur as these crush and compress.

A survey of experience in the U.K. (Charles et al., 1977, Charles et al., 1984, Reed and Hughes, 1990, Hills and Denby, 1996) shows that immediate compression may amount to 1 to 2% of the backfill thickness, time-dependent or creep settlement may amount to an additional 1%, and collapse settlement may be up to 2% of backfill thickness. However, these figures will depend on the nature and state of weathering of the backfill. Also, in water-deficient climates, permanent water tables are deep – often deeper than 30 to 50 m. Thus the collapse component of mine backfill settlement can be expected to be either absent, or considerably less than in a water surplus climate, the U.K., for example, where permanent water tables are much shallower.

13.1.2 *Settlement of strip mine backfill in a water-deficient area*

As a comparison with British experience in a water surplus climate, the following study was made in a water deficit climate:

The Lethabo power station in South Africa is coal-fired, fed from an adjacent "captive" opencast colliery, the New Vaal Colliery. The power station has a dry ashing system that can handle 500–600 T/hour of ash (moistened to a water content of 15% to reduce dusting). To avoid sterilizing virgin ground, it was decided to stack the ash over the mined-out and backfilled area of the mine (van Wyk, 1998). The mechanical stacking system creates a front stack that is 35 to 38 m high above the level of the mine backfill, and then adds a back stack of between 8 and 12 m high above the top of the front stack. (See Plate 1.5). The ash is not compacted and has a stacked density of about 1000 kg/m^3. Thus the loading superimposed on the surface of the mine backfill is up to 500 kPa.

There was a major concern that, as each front stack advanced, differential settlement of the backfill might compromise the stability and verticality of the mechanical stacker, at least resulting in misalignment and mechanical problems, and at worst causing the stacker to topple. A small-scale test embankment built after mining had started, but before the power station was commissioned, indicated that the settlement should not be excessive (Day, 1992). Nevertheless, it was decided to carry out settlement measurements under the full loading of the stacked ash at the start of full-scale operations.

The layout of the settlement monitoring exercise is shown in Figure 13.1. A series of 3 vibrating wire settlement sensors was installed (as indicated in Figure 13.1a) in a row ahead of the advancing series of front stacks (as shown in Figure 13.1b). To protect the settlement sensors, each was mounted on a concrete block and covered by a metre of selected hand-placed fine backfill. The sensors were spaced 30 m apart, starting at the toe of the existing front stack and spaced so that the second sensor would be under the center of the next front stack and the third at its toe. A vibrating wire piezometer was installed at a depth of 40m below the backfill surface to locate and monitor the level of the groundwater in the backfill.

Figure 13.2 shows the settlement-time curves recorded as two successive front stacks of ash of 35 m average height were placed over the settlement cells. Cells 1 and 2 settled very rapidly as the first new front stack advanced to cover them in a period of about 10 days. Cell 3 was only partly covered and therefore did not settle by much.

The site was left undisturbed for a period of about 150 days during which a slight time-dependent settlement was recorded by cells 1 and 2. The second new front stack was then placed, completing the loading of all three settlement cells. During the 300 days after the second loading, slight time-dependent settlement was registered by all three cells, with cell 3 showing a second acceleration of settlement a year after loading had been completed. The piezometer showed no substantial change in water level during this time. The increased settlement was probably caused by some form of collapse or degradation of particles in the mine backfill.

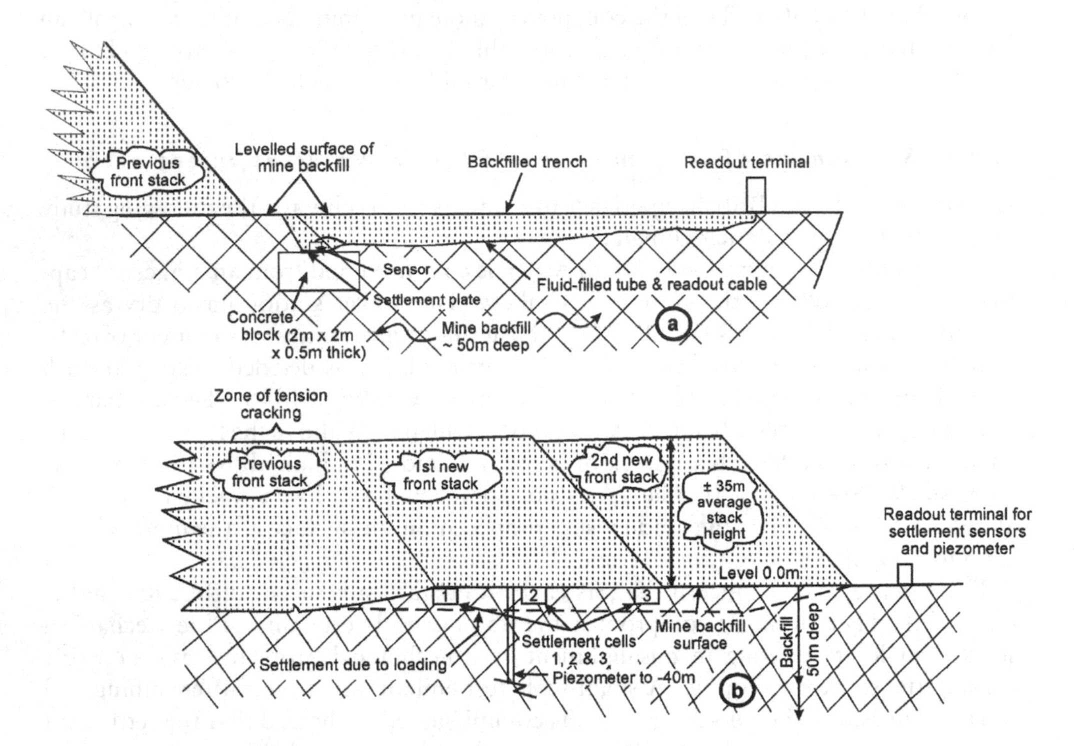

Figure 13.1 Layout of full-scale settlement monitoring exercise.

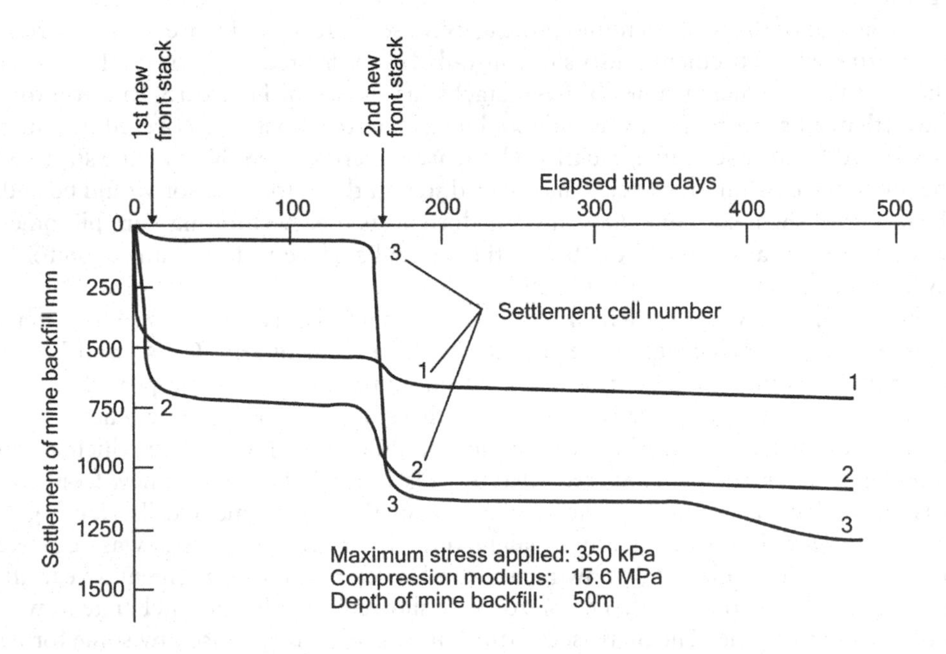

Figure 13.2 Settlement-time relationships observed in full-scale settlement monitoring exercise.

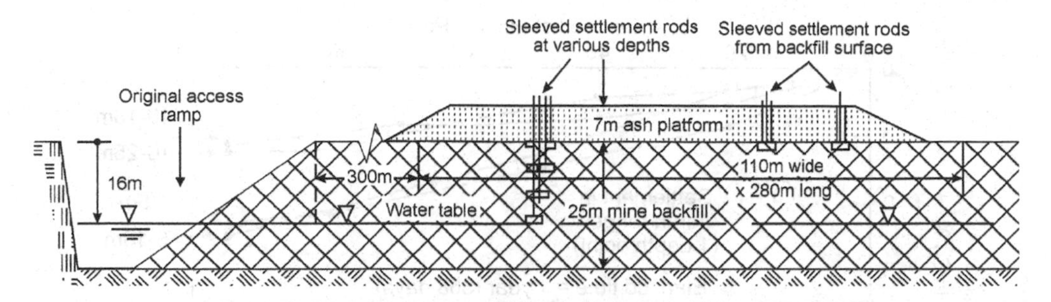

Figure 13.3 Layout of 7 m hydraulic fill ash platform placed on mine backfill.

Relatively small (up to 200 mm) differential movements occurred that had little effect on the operation or stability of the stacker. However, because the ash self-cements itself within a short time after placing, settlement near the toe of the previous front stack (see Figure 13.1) caused severe and deep tension cracks to open parallel to the crest. Similar cracks appeared in the first new front stack as the second new front stack was placed. These were closed by dozing, to keep out rainwater, as soon as they opened and had no further effect.

13.1.3 *Settlement of mine backfill under the load of hydraulically placed ash*

The Matla power station, an older station than Lethabo, has a wet ashing system, with the ash being disposed of in hydraulic fill impoundments like mine tailings storages. A new ash impoundment was needed and it was decided to place it over a mined-out and backfilled section of the station's captive colliery the adjacent Kriel Colliery. Because of uncertainty as to how the hydraulically placed ash would react to differential settlement of the mine backfill, a large-scale test wall of ash measuring 280 m long by 110 m wide was placed hydraulically on the mine backfill, as shown in Figure 13.3 (see van Wyk and Blight, 1996). The chosen site was adjacent to one of the original mine access ramps, and this void was used as a return-water reservoir for the ashing. The water table had re-established at a depth of 16 m in the backfill and remained constant during building the test wall.

The settlement of the mine backfill was measured by a set of 8 sleeved rods set in concrete blocks on the mine backfill surface, and a second set of sleeved rods drilled into the mine backfill and anchored at depths of 1, 5, 10 and 16 m below the backfill surface. This enabled the distribution of settlement with depth to be measured.

The test wall was built to a height of 7 m, which with an ash bulk density of 1300 kg/m^3, represents a maximum loading of 91 kPa. As the ash and the mine backfill are relatively pervious and the ash was placed as a 1:1 ash:water slurry, the settlement, measured over the construction period of a year (368 days), probably includes some creep and collapse settlement, as well as immediate compression of the mine backfill.

The settlement versus imposed stress relationships are summarized in Figure 13.4. Figure 13.4a shows the distribution of settlement strain with depth, as measured by means of the sleeved rods set at various depths. The larger strains occurred between

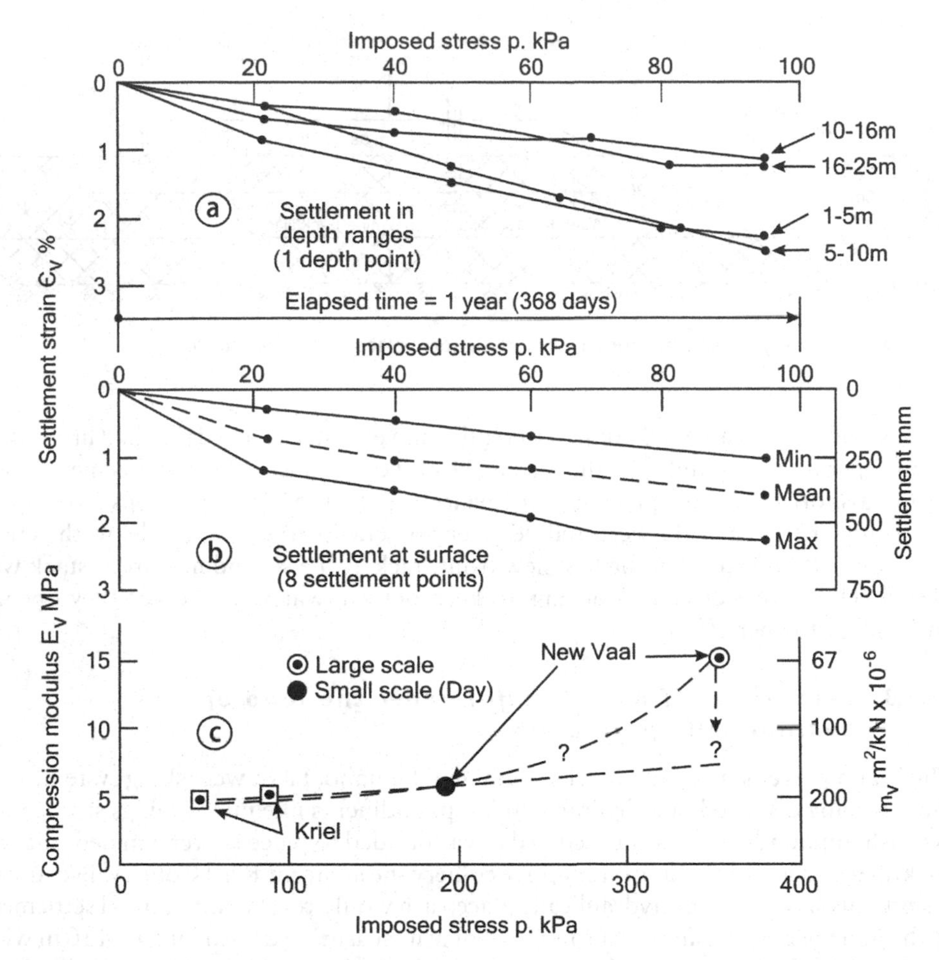

Figure 13.4 a: and b: Summary of settlement measurements on mine backfill loaded by hydraulically-placed ash wall. c: Comparison of vertical compression modulus (E_v) and vertical compressibility (m_v) of mine backfill measured in two field experiments.

depths of 1–5 m and 5–10 m. The deeper layers, 10–16 m and 16–25 m, preloaded with larger overburdens, strained approximately half as much as the shallower layers. In considering this result, it must be remembered that the backfill had been deposited in conical piles, with the lower material being dropped a considerable distance and then being covered by a considerable overburden, both of which would have produced more compaction than higher up. The tops of the backfill cones were then leveled off by dozing which must have resulted in a looser condition in the upper zone of backfill.

Figure 13.4b shows minimum, maximum and mean settlement strain-stress and settlement-stress relationships measured by means of the surface settlement points. It is notable that the range of settlement strains bracketed in Figure 13.4b is very close to the range covered in Figure 13.4a.

13.1.4 *Summary of settlement measurements*

The settlement measurements at New Vaal and Kriel collieries are summarized in Figure 13.4c as a relationship between average compression modulus, E_v (or its inverse, vertical compressibility, m_v) and the imposed stress p. The compression modulus is defined by

$$E_v = dp/d\varepsilon_v \text{ in MPa}$$

The vertical compressibility is the inverse

$$m_v = 1/E_v = d\varepsilon_v/dp \text{ or } de/dp(1 + e_o) \text{ in } m^2/kN$$

where the subscript v indicates the vertical direction, ε_v is the vertical strain, e the current void ratio, and e_o the original void ratio.

The data for Kriel are quite comparable with the result obtained by Day (1992) for a small scale test at New Vaal, but the large scale test at New Vaal indicated a much higher compression modulus (15 MPa as compared with 5 to 6 MPa measured by Day), i.e. a less compressible fill.

However, the large scale tests at New Vaal excluded the collapse settlement component, and the final value of compression modulus, as indicated in Figure 13.4c, may be less than that reached after 460 days (Figure 13.2) if the water table eventually rises above its present 40 m depth.

13.2 Backfilling of shallow underground mine workings to stabilize the surface

Figure 1.1 shows the very considerable settlement that can occur over long periods of time on land that has been undermined at depths of up to 250 m. In 1979 it became necessary to re-align a major road in the area referred to in Figure 1.1. Figure 13.5 shows a section and plan of the area where the re-aligned road was intended to cross the outcrop of the old workings that had been mined out and abandoned in the 1880's. The strata below the re-alignment, above the South Reef had been stabilized a few years previously, by drilling a series of holes from the surface to intersect the tabular South Reef void and injecting a slurry of sand cement down the holes to fill the void. The abandoned stope had been considered too dangerous for men to enter and the extent to which the backfilling had been effective was not known. The surface continued to settle, it was presumed because of closure of the deeper Main Reef Leader void. This void was entered, inspected and considered safe to work in. It was decided to fill the Main Reef Leader void by means of a sand tailings fill which would mostly be uncemented, but would be retained by a surrounding sand-cement bund or wall. The extent of the filling is shown in the plan in Figure 13.5.

13.2.1 *Properties of sand tailings fill*

The pH of the sand tailings that were available for use as fill averaged 4.2 and the sulfate content was 60 mg/kg (0.006%). The sand was medium to fine with $D_{90} = 0.25$ mm, $D_{50} = 0.1$ mm and $D_{10} = 0.06$ mm. The addition of only 1% of Portland Blast Furnace

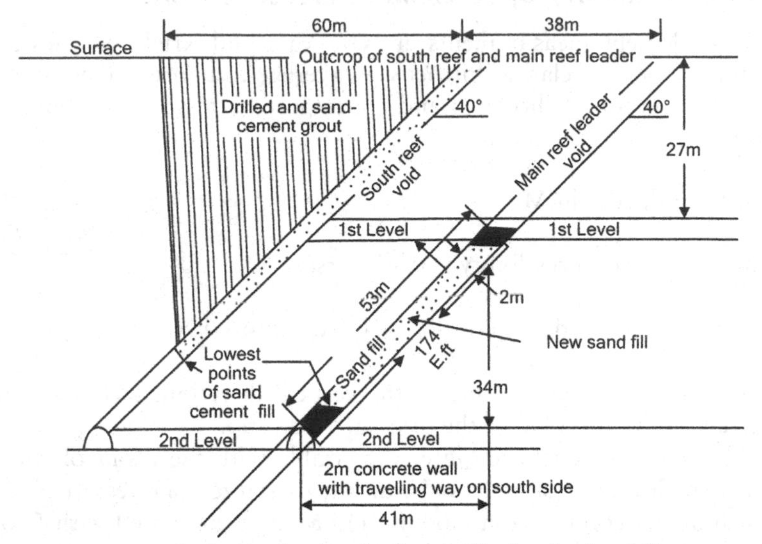

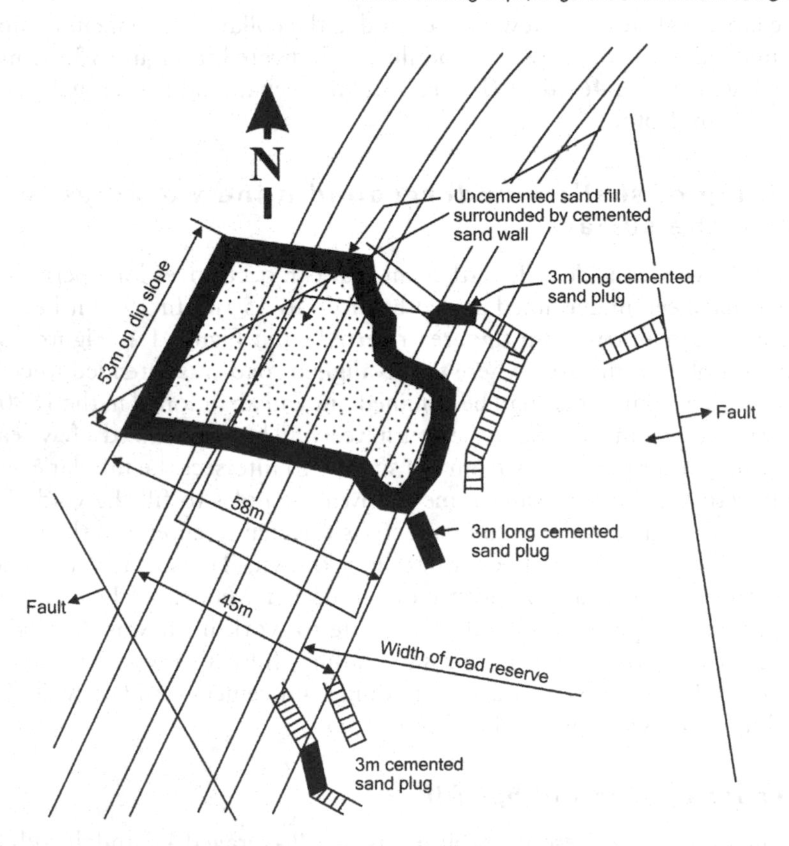

Figure 13.5 Section and plan showing extent of fill and location of cemented sand plugs in stabilized crossing of old mine workings.

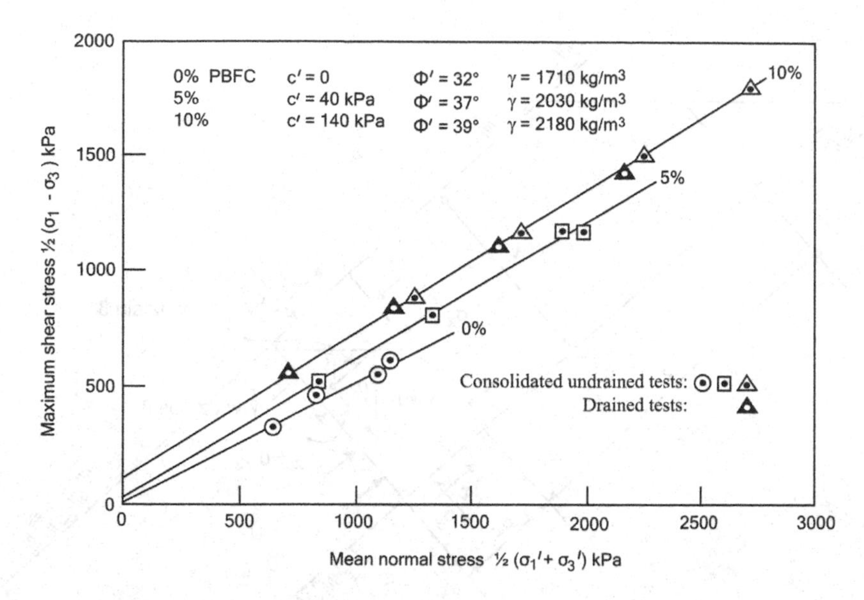

Figure 13.6 Shear strength properties of sand with various cement contents.

cement (PBFC) raised the pH to 12.7. According to available information (Neville, 1981) the sulfate content should not cause any problem with the PBFC, which was regarded as sulfate resistant. This was important because the stope was very steep (40° to horizontal), and the wall of sand cement was considered essential to retain the sand fill in place.

A series of tests was performed on the sand tailings to investigate the stability of the sand fill in the stope. Previous work on placing pumped sand tailings slurries had shown that the minimum water content to produce a pumpable slurry is that at which the groove in a standard Atterberg liquid limit test closes at 1 blow. This was used to determine the water contents (ranging from 31.5 to 33%) at which specimens for triaxial testing were prepared. Specimens were prepared at cement contents of 0%, 5% and 10% of PBFC and these were tested after 7 days of curing at constant water content and 25°C.

The results of the triaxial shear tests are shown in Figure 13.6. As φ^{I} in each of the tests was less than the 40° inclination of the stope, it appeared that 10% of PBFC would be required and that the cohesion of 140 kPa would be crucial. In all of the tests, the sand or sand cement dilated strongly. Hence there was no likelihood of liquefaction occurring if mining-induced seismic loading should occur.

13.2.2 *Stability analysis for sand fill*

The stability analysis was based on the forces shown in Figure 13.7.

For equilibrium of the element in the z direction

$$w\sigma_z + \gamma w \sin\beta dz = w(\sigma_z + d\sigma_z) + 2\tau dz$$

$$\text{i.e.}\quad d\sigma_z = (\gamma \sin\beta - 2\tau/w)dz \qquad (13.1)$$

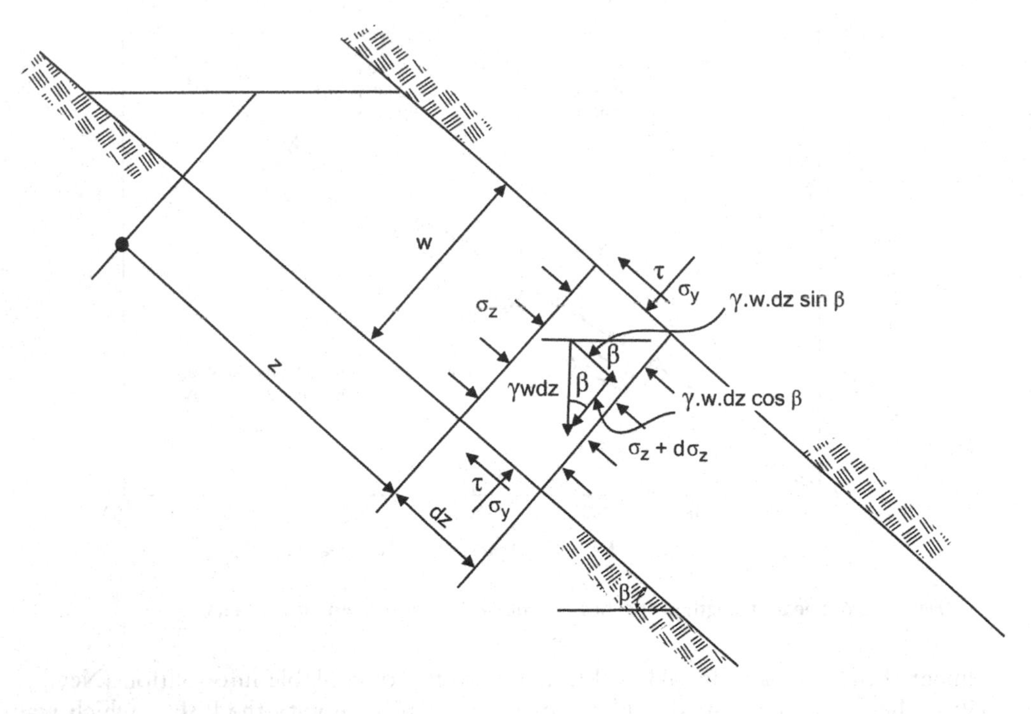

Figure 13.7 Forces acting on element of backfill in inclined stope.

Also $\tau = c^I + \sigma_y^I \tan \varphi^I$
If the fill is perfectly stable and at rest, σ_y^I would be related to σ_z^I by
$\sigma_y^I = K\sigma_z^I$ where K would approximate to K_o, the coefficient of earth pressure at rest.

$$\text{i.e.} \quad \tau = c^I + K\sigma_z^I \tan \varphi^I$$

$$\text{and} \quad d\sigma_z = [\gamma \sin \beta - 2/w(c^I + K\sigma_z^I \tan \varphi^I)]dz \qquad (13.2)$$

for a perfectly stable fill

$$d\sigma_z/dz = 0,$$

i.e. the frictional forces on any element must completely balance the disturbing forces.
i.e. $\gamma \sin \beta = 2/w(c^I + K\sigma_z^I \tan \varphi^I)$
or $(w\gamma/2\sin\beta - c^I)/K\tan \varphi^I = \sigma_z^I$
at the open lower end of the stope σ_z^I must be zero,
i.e. $c^I \geqq w\gamma \sin \beta/2$
For the case in question, with $w = 2$ m, $\gamma = 18.5$ kN/m^3 and $\beta = 40°$

$$c^I \geqq 2 \cdot 18.5 \cdot \sin 40°/2 = 12\,\text{kPa}$$

The test results shown in Figure 13.6 indicate that the required cohesion at the lower end of the fill is available if the average strength obtains. Capillary cohesion alone will eventually, once all free water has drained away, average a value of

$$c_{cap} = w/2\cos\beta \cdot \gamma_w \cdot \tan\varphi^I = u_{cap}\tan\varphi^I \quad (13.3)$$

Taking $\varphi^I = 40°$

$$c_{cap} = 11\,kPa$$

Hence very little additional contribution from cementation would be required.

To give a complete distribution of vertical and lateral stress in the column of fill, equation (13.2) must be integrated.

The integration depends on the distribution of u and whether u is dependent on z, on σ_z or on both u and σ_z. The integration can be made formally or numerically.

For example, while the fill is being placed and while it is draining subsequent to placing, u could be represented by the linear relationship $u = u_0 + B\sigma_z$. In this case

$$\sigma_z = 1/A[\gamma\sin\beta - 2/w(c^I - u_0K\tan\varphi^I)][1 - e^{-Az}] \quad (13.4a)$$

$$A = 2K\tan\varphi^I(1 - B)/w \quad (13.4b)$$

and $\sigma_h = K(\sigma_v - u) + u$

Under long term conditions equations (13.4)a and (13.4)b would apply with $B = 0$ and $u = u_{cap}$ given by equation (13.3).

As an example, take $w = 1.5\,m$, $\gamma = 18.5\,kN/m^3$, $\beta = 40°$, $B = 0.5$, $u_0 = 10\,kPa$, $K = 0.5$, $c^I = 12\,kPa$, $\varphi^I = 32°$

Then $\sigma_z = 0.30(1 - e^{-0.21z})$, from which it is obvious that the stress parallel to the stope will be negligible, e.g.

at $z = 0\,m$ $\quad \sigma_z = 0$

$z = 53\,m$ $\quad \sigma_z = 0.3\,kPa$

Thus once the sand slurry had drained, the barricade constructed to retain it would have been very adequate. Details of the retaining barricade are shown in Figure 13.8.

13.2.3 *Post-construction re-assessment of backfilling*

The backfilling was carried out and completed early in 1980. Six months later, one of the City Council engineers, in casual conversation with the contractor who had carried out the work, discovered that instead of using PBFC (Portland blast furnace cement), GGBS (ground granulated blast furnace slag) had been used. As no activator had been added and GGBS requires an activator to become cementitious, it appeared likely that the sand would not have been cemented.

A careful in situ inspection carried out by the author showed that the backfill appeared perfectly stable. A number of undisturbed block samples was cut from the lower portion of the fill and drained shear box tests were carried out. These gave minimum shear strength parameters of $c^I = 0$, $\varphi^I = 45°$ and mean parameters of $c^I = 15\,kPa$, $\varphi^I = 45°$ (see Figure 13.9) which were an improvement on the original

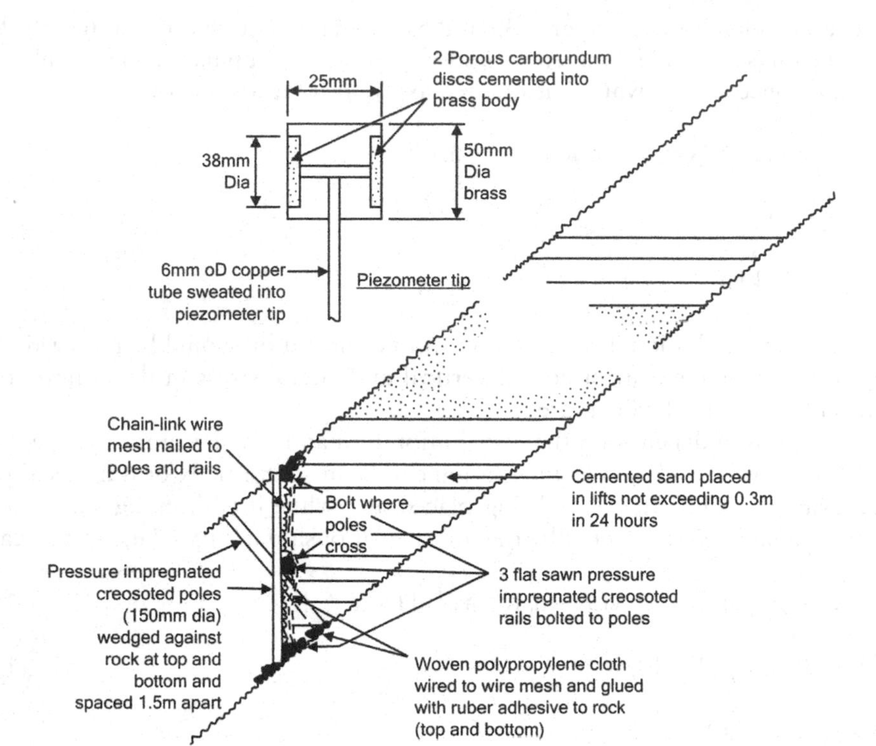

Figure 13.8 Method of retaining cemented sand, pumped into place, using wire mesh barrier backed by polyprophlene cloth. The piezometers were made, but contractors "forgot" to install them.

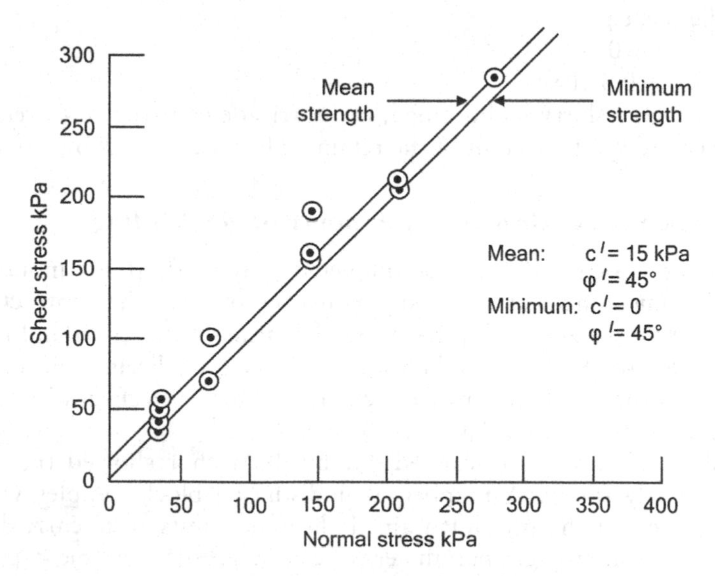

Figure 13.9 Results of shear box tests on undisturbed block specimens taken from fill in situ.

Plate 13.1 Confined space in typical deep gold mine stope. Gold-bearing reef is darker horizontal band in rock face behind miner.

laboratory tests (Figure 13.6). This result, together with the analysis by equations (13.4)a and b lead to the conclusion that the fill's stability was satisfactory.

The initial tests that showed the soluble sulfate content of the sand to be 0.006% had probably not accounted for the sulfate content in the form of gypsum, which is relatively insoluble. As gypsum acts as an activator for GGBS, it was concluded that the GGBS had been sufficiently activated to produce the strengths shown in Figure 13.9. 29 years later, the road shows no signs of settlement and the backfilling operation can be counted as a success.

13.3 The properties of mine waste as a structural underground support in narrow stopes

In the ultra-deep (2 to 4 km) gold mines of India, Ghana and South Africa, the gold usually occurs in thin, sheet-like semi-continuous reefs, often no more than 100 mm thick. The reef is mined by means of narrow tabular slot-like stopes. These stopes may be no more than 1 m high, just wide or high enough to provide access to the reef. Plate 13.1 shows the cramped conditions in a gold mining stope. The miner has his back to the working face of the reef. Once a stope has been mined out, its roof and floor, or hanging wall and footwall, converge and it gradually closes. However, if stresses become too high, over too large an area, an explosive closure or rock-burst may occur. In a rock-burst, large areas of mined-out stope may close instantaneously and catastrophically.

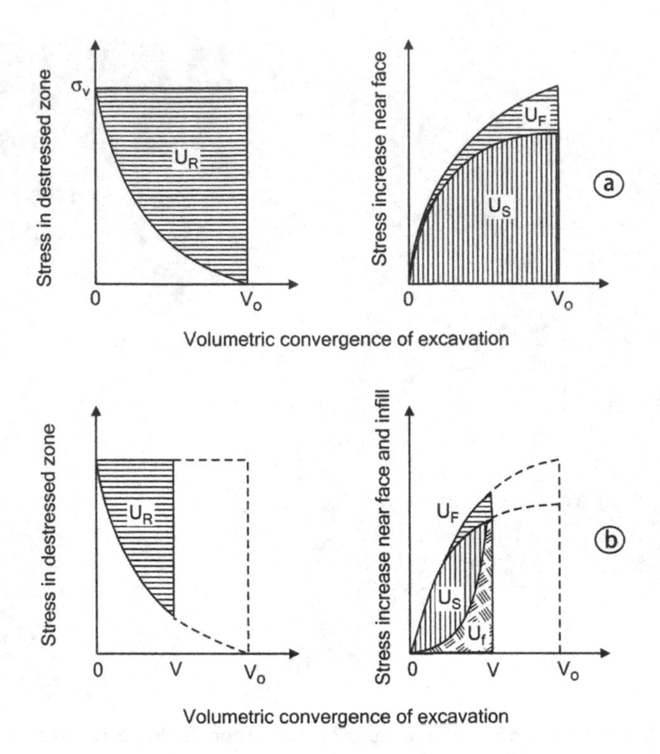

Figure 13.10 Energy changes in vicinity of: a: unfilled stope excavation, and b: filled stope.

To control closure and reduce the incidence of rock-bursts, the stopes are supported in a variety of ways, one of the most effective being to fill the stope with hydraulically-placed tailings backfill which is often cemented to increase its stiffness and strength.

To limit and control the possible collapse of large contiguous areas of stope, barrier pillars, i.e. continuous walls of rock are left at suitable intervals. These become subject to enormous vertical stress, as the overburden stress in the virgin unmined rock may be as high as 50 MPa at a depth of 2 km. If these pillars are surrounded by a stiff mine waste fill, lateral support will develop passively as the pillar starts to crush and dilate laterally. The lateral support limits the vertical deflexion of the pillar and hence assists its support and energy absorption functions.

13.3.1 *Energy releases during underground mining and their relation to the properties of backfill*

When an unsupported excavation is made in a mass of stressed rock, energy is released from the destressed rock surrounding the excavation and, if the rock does not fail, the released energy is redistributed and stored as strain energy in highly stressed zones at the perimeter of the excavation. If, however, partial failure of the rock occurs, part of the released energy continues to be stored as strain energy in the unfailed rock and part is consumed as energy of fracture. Figure 13.10a illustrates the energy balance diagrammatically (Blight et al., 1977). σ_v is the average virgin rock stress and V_o is the volumetric convergence of the excavation. U_R is the energy released by the excavation and U_s is the energy stored in the surrounding rock. If no fracture occurs around the excavation, $U_s = U_R$. If fracture does occur, the stored energy is reduced by U_F, the energy of fracture. In the case of a catastrophic failure of the excavation the entire stored energy is converted into fracture energy and $U_F = U_R$.

If the excavation is filled with a compressible fill, part of the released energy is absorbed in compressing the fill. The energy absorbed by the fill is represented by U_f in Figure 13.10b and for conservation of energy, $U_R = U_S + U_f + U_F$. Because filling reduces the volumetric convergence of the excavation from V_o to V, the released, stored and fracture energies are less for a filled excavation and the potential for catastrophic failure is reduced. It is obvious from Figure 13.10b that the less compressible or stiffer the fill, the smaller will be V and U_R and the more favourable will be the support conditions achieved.

The stiffness obtainable using mine waste as a backfill can cover a large range. Hydraulic fill tailings, placed as a slurry, have a relatively low stiffness (over a range of 10% compression) of from 50 to 200 MPa. Higher stiffnesses can be obtained by adding a binder or cement to the slurry. Relatively stiff fills can be prepared from a mixture of crushed aggregate and cemented tailings. These can have stiffnesses (over 2½% compression) ranging from 400 to 1400 MPa.

13.3.2 *Laboratory measurements of support properties of fills*

When the fill is in place in the excavation, it will be subjected to essentially one dimensional vertical compression. To simulate this condition in the laboratory, specimens of fill are cast in standard steel AASHTO compaction moulds to form specimens 150 mm in diameter and 100 mm high. (AASHTO = American Association of State Highway and Transportation Officials). The specimens are sealed against moisture loss by means of a 10 mm thick layer of paraffin wax and cured at 40°C (or an approximate underground temperature) for 7 days before testing. The cylindrical walls of the moulds are instrumented with vertical and horizontal electric resistance strain gauges spaced at 120° around the circumference of the mould. This enables both vertical and circumferential strains in the mould cylinder to be measured as the fill is compressed vertically. Vertical strains are used to correct the applied load for wall friction between the mould and the fill and circumferential strains to assess the lateral stress developed in the fill. As the vertical compression can only occur by expulsion of water from the pores of the fill, compression tests are carried out at a rate slow enough to enable the pore water to escape without building up any water pressure in the fill. (See Sections 4.6.2 and 4.6.4.)

Fills are often described as "soft" or "stiff". Soft fills usually consist of the coarse fraction of a tailings, separated from the fines by cycloning and stiffened by the addition of cement. Ordinary Portland cement (OPC) may be used or alternatively Portland Blast Furnace Cement (PBFC) or ground granulated blast furnace cement (GGBS). GGBS consists of slag that has been quenched by pouring the molten slag into water. This produces a chemically unstable glass or pozzolan that will react with alkaline substances, e.g. OPC, slaked lime ($Ca_2(OH)_2$) or caustic soda (NaOH), to produce a cementing action. Power station fly ash (PFA) is another frequently used pozzolan.

Stiff fills are stiffened by combining cycloned tailings with coarse aggregate and cement to produce what is, in effect, a low strength mass concrete. GGBS and PFA can also be activated with a sodium silicate (water glass) solution.

Figure 13.11 shows the results of confined compression tests on three stiff fills, together with representative values of the chord modulus E_{ch} and the coefficient of zero lateral strain K_o.

Figure 13.12a summarizes the results of compression tests on soft and stiff fills and shows how both materials progressively stiffen as they approach a limiting strain at which the fill becomes, under these conditions of constraint, to all intents and purposes, incompressible.

Figure 13.12b summarizes the relationship between vertical and lateral stresses in both stiff and soft fills as they are subjected to laterally confined compression. The ratio of lateral to vertical stress in stiff fills varies from about 0.2 to 0.42 whereas in soft fills the ratio varies from about 0.4 to 0.75. These lateral stress ratios illustrate the supportive effect that a fill can potentially have on a loaded rock pillar, although it must be remembered that appreciable vertical strains are necessary to develop the lateral stresses.

13.3.3 *The drainage of backfill after placing*

Backfill is usually placed as a viscous slurry having a relative density (RD) of 1.7 to 1.8 (unit weight of 17 to 18 kN/m^3), gravity being used as the transportation mechanism. Excess water has to drain from the slurry for it to become a frictional solid, with the solid particles in contact and interlocked. If a binder or cement has been used, then, as it hydrates and develops chemical bonds, the interlocked particles become cemented together and the cohesive strength can increase considerably. Some of the interstitial water is consumed by the hydration of the cement, causing a self-desiccation process (e.g. Helsinki, et al., 2007.)

The drainage process (shown diagrammatically in Figure 13.13) can be divided into two phases. In the first, excess water leaves the fill by draining from free surfaces and into fissures in the footwall, and the particles move closer together and into contact. However, the pore spaces in the fill remain fully saturated with water at this stage. The second phase starts once a threshold value of negative pore pressure has been reached, and air starts to enter the pores of the fill, displacing more water. This threshold negative pore pressure is known as the air entry pressure because it is numerically equal to the air pressure that would start to displace water from the pores if the pore-water pressure were atmospheric. (See Section 10.7.) Figure 10.9 illustrates the principle of air entry and drainage by means of a test on a specimen of total (i.e. uncycloned) tailings. The air entry pressure is the pressure at which the air permeability of the specimen ceases to be zero. Water will be expelled first from the coarsest pores at the free surface of a backfill. Thereafter, successively smaller pores will drain as the pore pressure becomes more negative.

Figure 13.14 shows saturated drainage-time curves measured in the laboratory for a specimen of typical cyclone-underflow backfill. On the horizontal axis (time scale) of the figure, the actual times are given in (minutes) for the specimen (38 mm in diameter by 76 mm long), as well as time factors in days per m^2 to allow for the dimensions of the specimen. The values of $\Delta\sigma$ shown in Figure 13.14 represent the increments

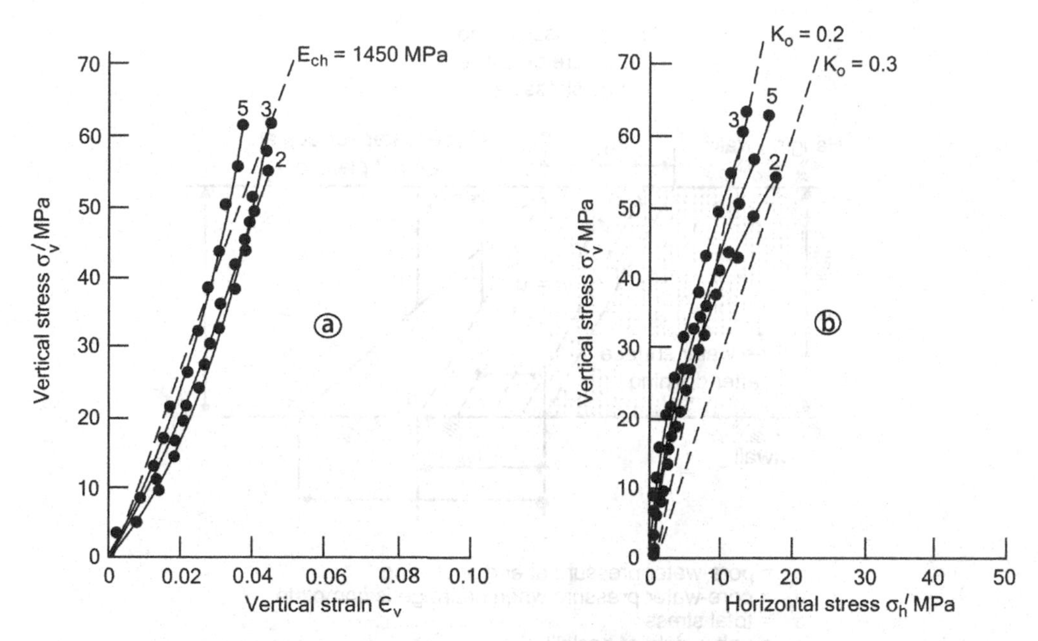

Figure 13.11 Confined compression characteristics of three typical stiff fills: a: vertical stress σ_v^I versus vertical strain, b: vertical stress σ_v^I versus horizontal stress σ_h^I. E_{ch} is typical chord modulus at $\sigma_v^I = 60$ MPa. K_0 are typical chord values of stress ratio σ_h^I/σ_v^I.

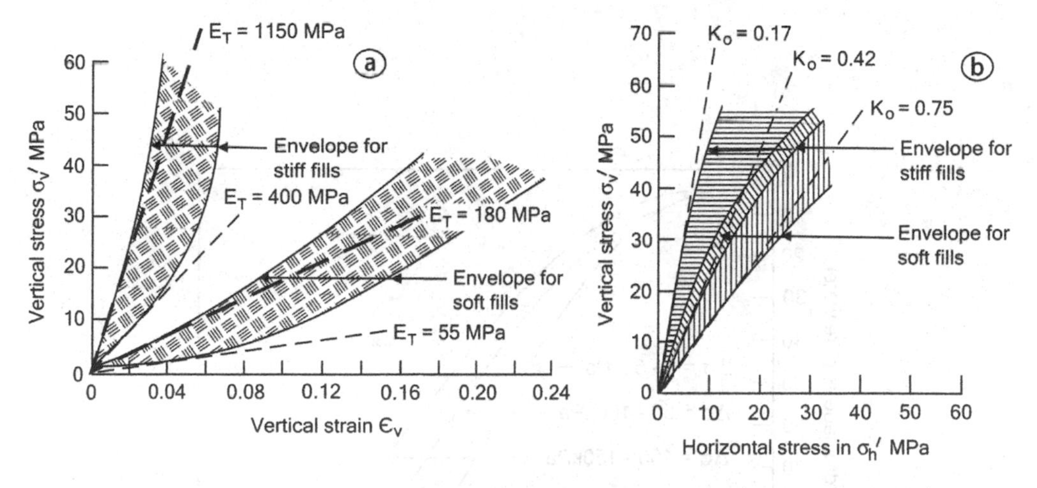

Figure 13.12 Ranges of compression characteristics for soft and stiff fills. Vertical stress σ_v^I versus: a: vertical strain ϵ_v and tangential modulus E_T, b: horizontal stress σ_h^I and tangential stress ratios for zero lateral strain.

of all-round stress, σ, applied to the specimen to cause the drainage to take place. It should be noted that drainage occurs more slowly and with a reduced volume as the applied stress is progressively increased and the solid particles move closer together, thus reducing the void space in the fill and increasing its stiffness.

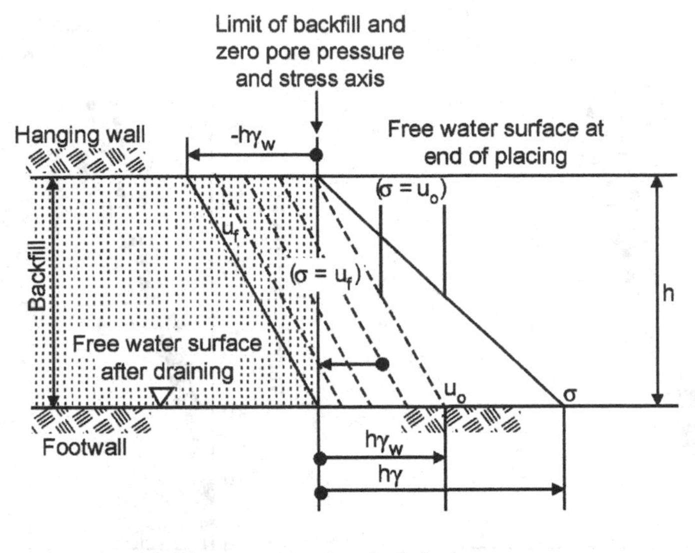

Figure 13.13 Drainage of water from backfill placed hydraulically in stope.

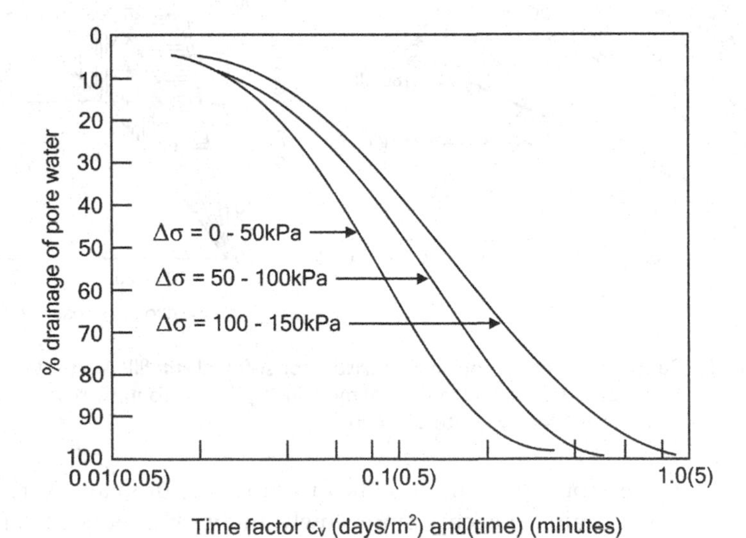

Figure 13.14 Typical drainage-time curves for a tailings fill.

On the basis of Figure 13.14 the drainage of water from a panel of backfill 2 m long (on strike) with a stope width or height of 1.5 m can be calculated to take about 1 day to occur. As the pore pressure moves into the negative range (Figure 13.13) and becomes numerically equal to the air entry pressure, air will start to enter the pores of the fill, and more water will drain out, thus prolonging the period of initial drainage.

Figure 13.13 is intended to represent narrow tabular stopes (1–2 m wide). The process of drainage in large high stopes (e.g. 20 m × 20 m in plan and 50 m in height) has recently been studied by Helsinki, et al. (2007).

13.3.4 *Early strength of cemented backfill*

At early times after placement, before chemical bonds from the cement have developed appreciably, the strength of backfill will arise entirely from the frictional effective stress term $(\sigma - u)\tan\varphi^I$ in the shear strength equation:

$$\tau = c^I + (\sigma - u)\tan\varphi^I \quad (4.14)$$

Because the total stress in the fill, σ, will be small until some compression has occurred as a result of closure, the negative pore pressure (−u) can be expected to provide a significant component of the early strength. (See Section 13.2.2.) For a fully drained fill (Figure 13.13) when h = 1.5 m, the average value of u would be −7.5 kPa and the average value of σ would be 9 kPa. Hence, $\sigma - u = 9 + 7.5 = 16.5$ kPa and the average shear strength would be $16.5\tan\varphi^I$. If $\varphi^I = 36°$, $\tau = 12$ kPa. Evaporation of water from the free surface of the fill can result in more highly negative values for the pore-water pressure than the 7.5 kPa quoted here.

Figure 13.15a illustrates the growth of vane shear strength with time for a specimen of uncemented cycloned backfill. For the strength measurements, each specimen of the backfill slurry was placed in a standard soil-compaction mould (AASHTO mould) of 150 mm diameter by 150 mm high. The moulds have perforated bases, which were lined with a disc of geofabric to allow clear water to drain from the backfill. Once the excess water had drained from the specimens, evaporation from their upper surfaces caused the pore-water pressure to slowly become more negative. The shear strengths were measured by means of a hand-held laboratory shear vane. As Figure 13.15a shows, little strength developed during the first 24 hours after the placement. Thereafter, the strength developed more rapidly, although two of the specimens did not reach the expected fully drained strength of 12 kPa (see earlier) within 96 hours of placement. Figure 13.15b shows the growth of vane shear strength for the same backfill cemented with various percentages of GGBS activated with sodium silicate. With the addition of a binder, the strength of 12kPa was reached within 5 hours of placement.

13.3.5 *Shear strength of cemented backfill*

Figure 13.16 shows the results of a series of shear strength measurements on a cycloned fill cemented with various proportions of GGBS activated with sodium silicate. The strengths were measured by means of a shear-box at 24 hours after placement. It will be noted that the angle of shearing resistance, φ^I, is constant for all binder contents, and that the cohesion increases with increasing binder content. However, it should

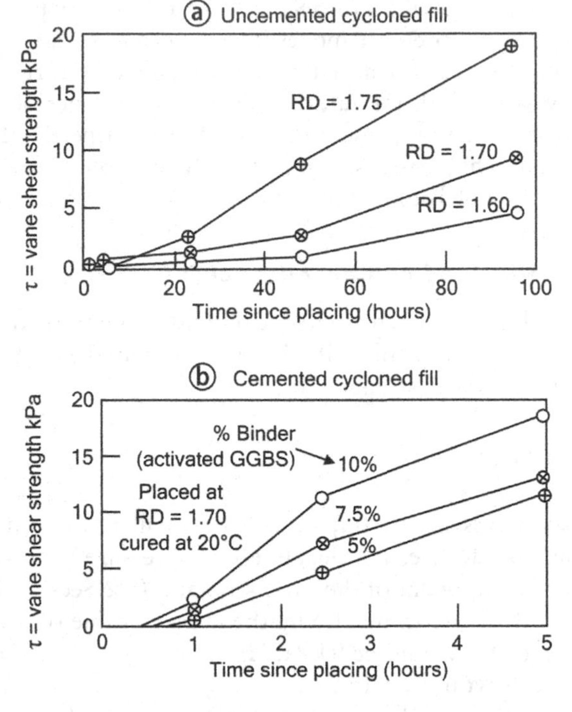

Figure 13.15 Growth of strength with time: a: uncemented tailings fill, b: cemented tailings fill.

also be noted that results reported previously in Figure 13.6 showed that, with certain materials, the angle of shearing resistance also increases with increasing binder content.

13.3.6 *Compressibility of cemented backfill*

The resistance to compression of backfill, which controls the closure of a backfilled stope arises largely from its shear strength. Compressibility is usually measured by means of the confined compression or oedometer test, in which the fill specimen, contained laterally in a metal ring, is subjected to vertical compression. (See Section 4.5.6). Figure 13.17a shows the results of a series of confined compression tests at low pressures on a cycloned fill cemented with various percentages of GGBS activated with sodium silicate. The relative density at placement was 1.70 which corresponds with a void ratio of 1.5. It will be noted that the compression curves all have the same slope (C_c) of 0.0055, for all binder contents, up to a stress of about 50 kPa. The increase in the initial void ratio as the binder content was increased resulted from the progressive increase of the sodium silicate content and its gelling effect on water in the voids. While the gelled water is retained in the fill under relatively low stresses, it is expelled as the stresses increase from the low stress (kPa) range to the high stress (MPa) range. The result is that the compressibility of fill containing gelled water is more than that of fill containing only free water after initial drainage. The effect of the delayed expulsion of water decreases as the RD at placement increases, as seen in Figure 13.17b. In this

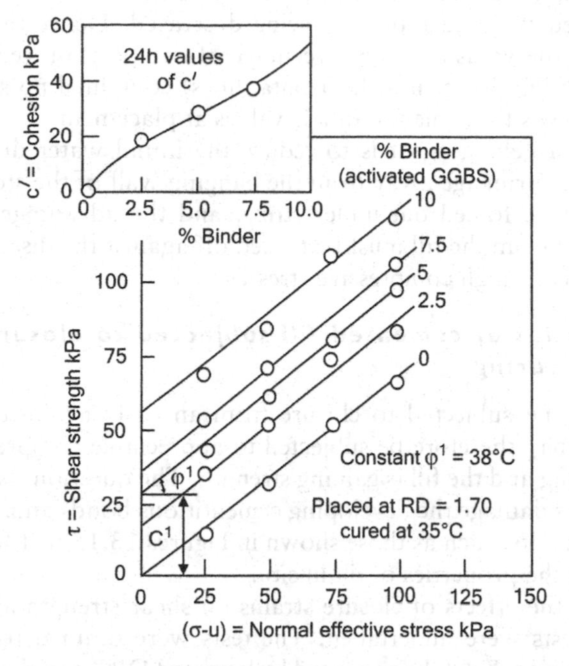

Figure 13.16 Effect of increasing binder content on 24 hour strength of cemented tailings fill (GGBS activated with sodium silicate.)

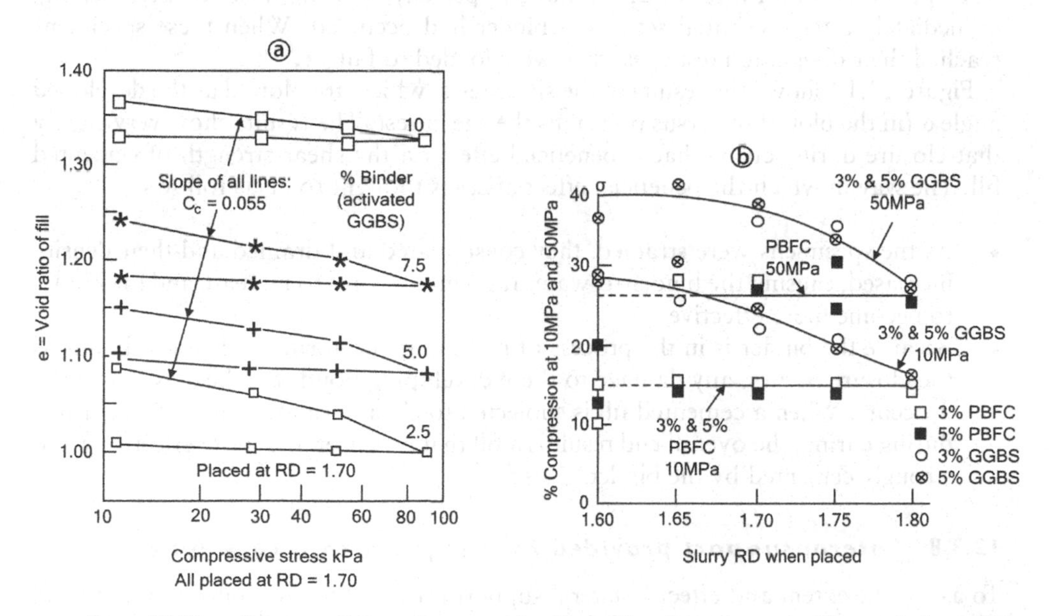

Figure 13.17 a: Compressibility of cemented tailings fill at low stress. b: Effects of slurry RD (when placed) on compressibility of cemented tailings at high stresses.

figure, the backfill cemented with GGBS also contained sodium silicate. As the slurry RD was increased the void ratio at placing decreased. Hence there was less gelled water present in the voids and the percentage of compression reduced. The backfill cemented with PBFC (lower near-horizontal lines) contained no sodium silicate and the compression was the same for all RD values at placement.

The object of a gelling agent is to reduce the initial water drainage from a fill, and hence reduce shrinkage away from the hanging wall of the stope. Ultimately the gelled water will be forced out under stress, and the advantage of reduced initial drainage of water from the fill must be traded off against the disadvantage of greater fill compressibility at high compressive stresses.

13.3.7 *Properties of cemented fill subjected to closure during curing*

Fills will usually be subjected to closure from an early age after their placement. A cemented fill may therefore be subjected to appreciable compressive strains while the binder is curing and the fill is gaining strength. The question that arises is whether the closure strains damage the developing cementitious bonds and, therefore, whether laboratory test results, such as those shown in Figures 13.15 to 13.17 can possibly be representative of the properties of fill in situ.

To investigate the effects of closure strains on shear strength and compressibility, three series of tests were undertaken. The tests were drained triaxial compression tests, the binder was 6% GGBS activated by lime and OPC and the fill was a cycloned tailings. In the first sub-series of tests the specimens were tested at ages of 1, 2, 5, 7 and 14 days. These specimens were regarded as having been subject to zero closure during curing. In the second and third sub-series of tests, the specimens were subjected to compressive strain rates of 2.5% and 5% per day, starting 4 hours after casting, immediately after the initial set of the binder had occurred. When these specimens reached their designated test ages, they were loaded to failure.

Figure 13.18 shows the results of the shear tests, which are plotted as the developed angle α (in the plot of q^I versus p^I) versus the age at test. The results show very clearly that closure during curing had a beneficial effect on the shear strength of cemented fill. The way in which the beneficial effect arises is thought to be as follows.

- As the specimens were strained, they consolidated and drained and their density increased, causing the binder-to-water ratio in the pores to increase and the binder to become more effective.
- Because the binder is in the process of hydrating and forming chemical bonds as the closure occurs, any damage to these developing bonds can heal as quickly as it occurs. When a cemented fill is subjected to closure rates of up to 5% per day during curing, the overall end result is a fill that is denser and more effectively and strongly cemented by the binder.

13.3.8 *Lateral support provided by a stiff fill to a rock pillar*

To assess the extent and effect of lateral support provided to rock pillars by a stiff fill encasement, a series of tests was performed in which quartzite drill cores (representing the rock pillars) were embedded in either soft or stiff fill contained in the strain-gauged

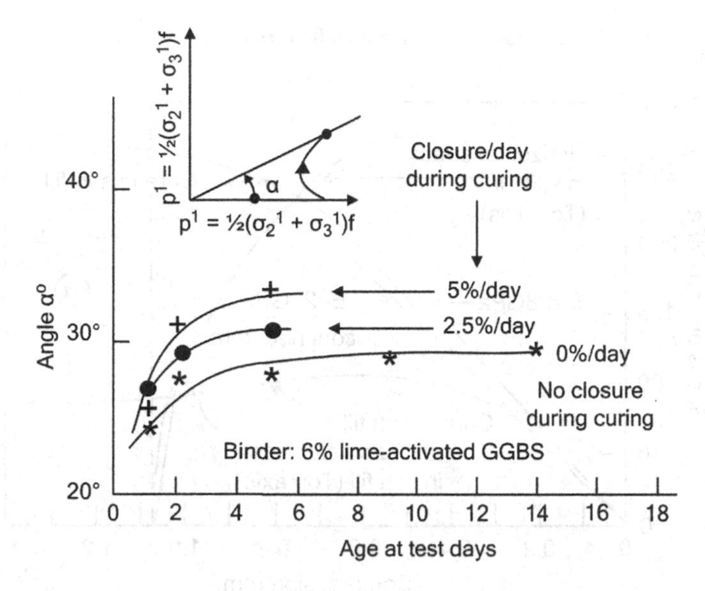

Figure 13.18 Effect of closure during curing on strength of cemented tailings backfill.

moulds used to assess the properties of the fill. The composite rock core-fill system was loaded over its full top area by means of a rigid steel piston. The stress-compression relationship for the cores and the relationship between compression of the core and lateral stress in the fill were recorded.

A typical set of these relationships is shown in Figure 13.19a. The figure shows that the soft fill provided very little lateral support to the core, as the strengths of the unconfined core and the core supported by soft fill were almost identical. Relatively little lateral stress was generated in the soft fill, but there was enough lateral support to maintain a post failure strength of about 85% of the peak strength.

A considerable lateral stress was generated in the stiff fill and Figure 13.19a shows that a peak strength was not reached by the core even at a compression of 1.4 mm or 1.75%. This shows that surrounding a highly stressed rock pillar with a soft fill will not materially improve the strength of the pillar, but will provide a high level of residual resistance. A stiff fill, on the other hand, can considerably increase the strength of a pillar, as well as providing a resistance that continues to relatively large strains. Figure 13.19a also shows that because the support mechanism is passive and the fill must be laterally compressed in order to provide lateral support, surrounding a rock pillar with fill has little or no effect on the pre-failure compression modulus of the pillar.

13.3.9 *Modeling the lateral support mechanism*

In the analysis set out below, it is assumed that closure takes place sufficiently slowly so that the pore pressure is always zero, i.e. effective stress equals total stress:

Figure 13.19b represents a cylindrical pillar of initial height S and diameter D which is compressed at approximately constant volume by an amount v. The lateral expansion

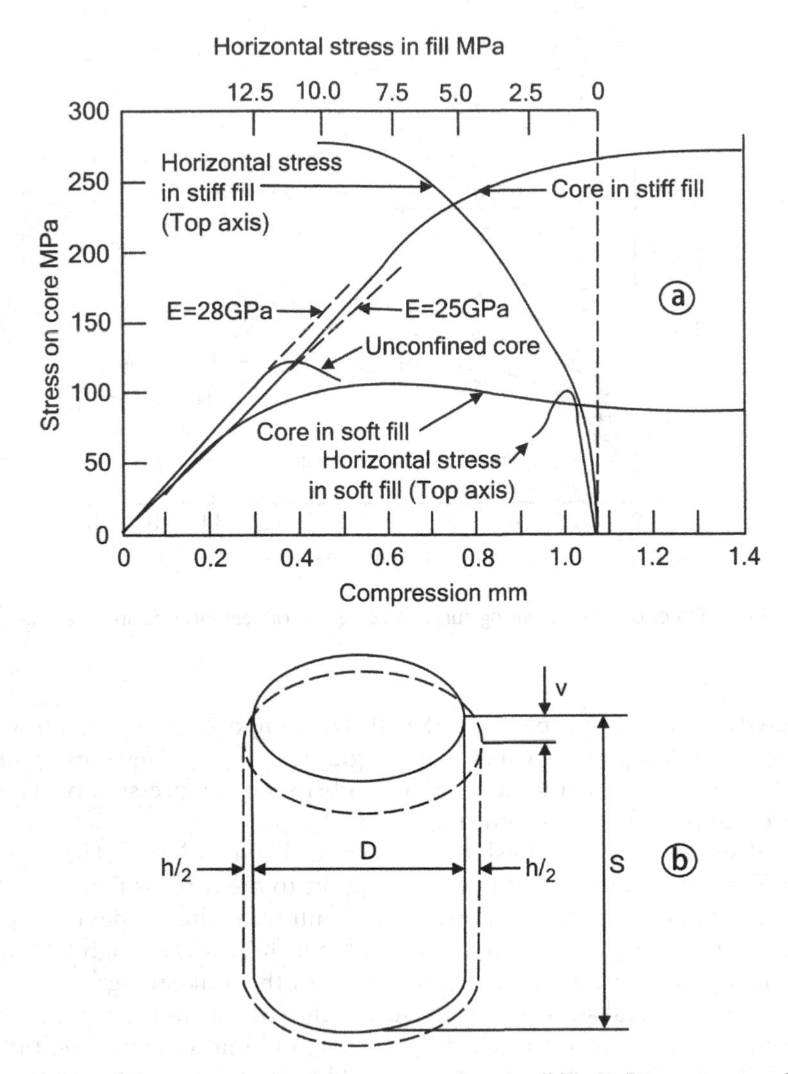

Figure 13.19 a: Effect on rock pillar behaviour in soft and stiff fills. b: Dimensions used in analysis of pillar/fill system.

h of the pillar is then given approximately by:

$$h/D = v/2S \tag{13.22}$$

The pressure required to expand a cylindrical pillar into a surrounding elastic fill is given by

$$\Delta\sigma_h = 2G \cdot h/D = G \cdot v/S \tag{13.23}$$

(The process of expansion can be likened to the expansion of a cylindrical cavity in an elastic medium. See, e.g. Baguelin et al., 1978.)

$\Delta\sigma_h$ is the expanding pressure, i.e. the supporting pressure on the pillar
and G is the shear modulus of the fill.
For an elastic material, the shear modulus is related to the elastic modulus E by

$$G = E/2(1+\nu) \tag{13.24}$$

where ν is Poisson's ratio,
ν and E for the fill can be evaluated from the results of one-dimensional compression tests such as those described by Figures 13.11 and 13.12.
It can be shown that if a material is subjected to vertical one-dimensional compression the ratio of horizontal to vertical stresses is

$$\sigma_h/\sigma_v = K_o = \nu/(1-\nu) \tag{13.25a}$$

$$\text{or} \quad \nu = K_o/(1+K_o) \tag{13.25b}$$

Also, for such a process, the vertical strain

$$\varepsilon_v = v/S = \sigma_v/E \cdot (1+2K_o)/(1+K_o)$$

$$\text{Hence} \quad G = \sigma_v S/2v \cdot (1-K_o)(1+K_o)/(1+2K_o) \tag{13.26}$$

or, from equation 13.23

$$\Delta\sigma_h = \sigma_v(1-K_o)(1+K_o)/2(1+2K_o) \tag{13.27}$$

In equation (13.27) σ_v represents the vertical stress in the fill surrounding the pillar, while $\Delta\sigma_h$ represents the additional horizontal stress generated in the fill when the pillar and the fill have both compressed by an amount v.
The additional vertical strength of the pillar at this stage will be given by

$$\Delta\sigma_v^P = \Delta\sigma_h(1+\sin\varphi^P)/(1-\sin\varphi^P) \tag{13.28}$$

where φ^P is the angle of shearing resistance of the cohesionless fractured rock of which the pillar now consists.
For a very stiff fill at a compression of 2% (see Figure 13.12) $K_o = 0.2$ and $\sigma_v = 20$ MPa.
Hence $\Delta\sigma_h = 7$ MPa
Reference to Figure 13.19a will show that in the model tests referred to earlier, $\Delta\sigma_v^P$ was about 100 MPa when $\Delta\sigma_h$ was 7 MPa. i.e. the ratio $(1+\sin\varphi^P)/(1-\sin\varphi^P)$ was 14 which corresponds to a value for φ^P of 60°.

This is a perfectly possible value for a confined fractured rock.

For a good soft fill at a compression of 2%, $K_o = 0.45$ and $\sigma_v = 5$ MPa. Here $\Delta\sigma_h = 1$ MPa which is again similar to the measured value in Figure 13.19a.

If the failure strain for the confined pillar is taken to be the same as that for the unconfined core, $\Delta\sigma_v^P$ was about 12 MPa which corresponds to a value for φ^P of 58°.

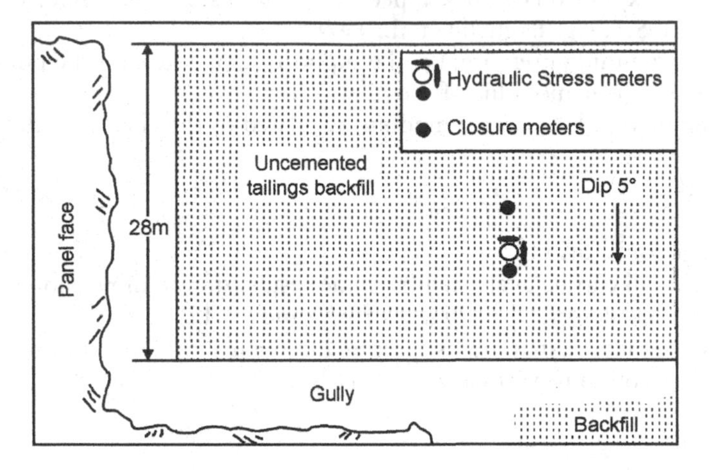

Figure 13.20 Layout of experiment to measure stresses in backfill at 1600 m depth.

Hence there is a reasonable correspondence between the results of the model tests and the predictions of equations 13.27 and 13.28 if suitable values are used for φ^P.

13.4 Measurements in situ of stresses and strains in fills at great depth

This section will describe one of a number of down-the-mine experiments that have been performed to assess the behaviour of structural fills when subjected to the large in situ stresses that are a feature of mining at great depths (Clark, 1988).

Figure 13.20 shows a plan of an experiment in a South African gold mine in which instruments were installed in a panel of uncemented cycloned tailings backfill placed in a 1 m wide (i.e. 1 m in height) stope at a depth of 1600 m below surface. The backfill, a fine silty sand in particle size range, was placed at a relative density of 1.65 (a water content of 60%). After settling and draining, measured void ratios ranged from 0.32 to 0.34. Previous tests in which piezometers had been installed in a similar fill had shown that placement pore pressures (Section 13.3.3) dissipated completely in less than 24 hours after placing. As closure rates were expected to be less than 0.2% per day, it was considered unnecessary to monitor pore pressures which could be considered to remain at zero throughout.

The stress meters consisted of oil-filled flat jacks placed normal to the three principal directions, i.e. normal to dip, parallel to dip and normal to strike. The closure meters, installed normal to dip between the foot and hanging walls of the stope consisted of a pair of telescoping steel tubes with their free ends anchored to the foot and hanging walls. Closure between the two tubes was measured by comparing the ends of two steel Bowden cables, one fastened to each end of one of the steel tubes. The cables were lead from the closure meters, out of the fill through a carefully radiused steel tube. Closure results are reported as the average of the two closure meters.

Figure 13.21 shows (a) the observed relationships between stress measurements by the three stress meters and time and also (b) between closure strain and time.

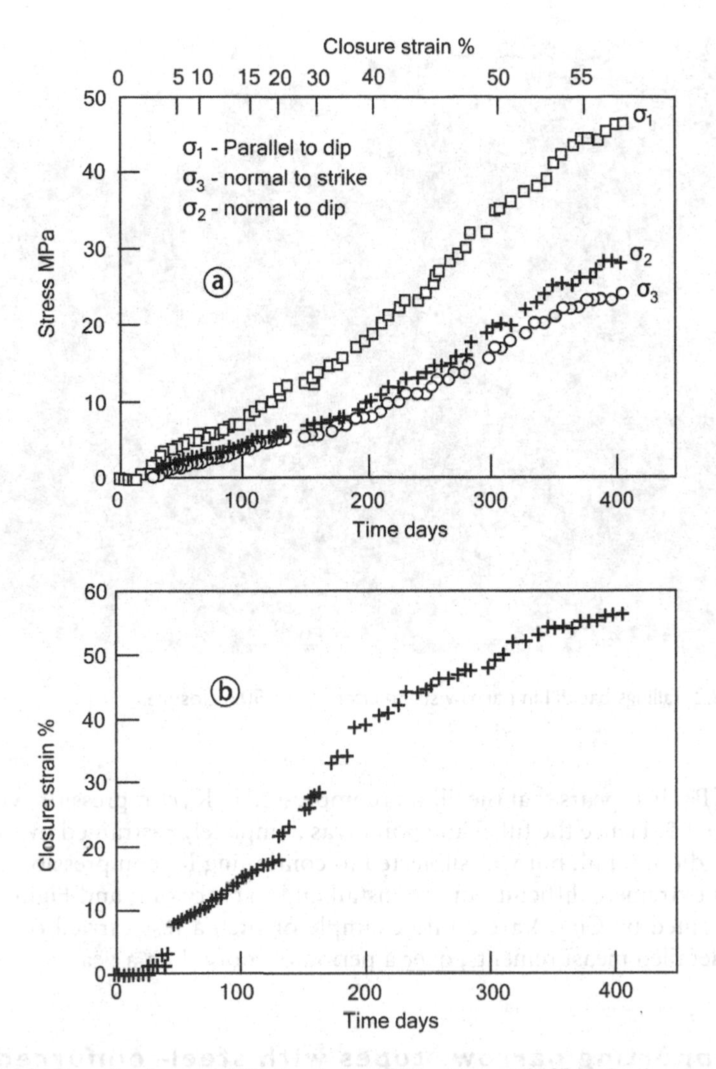

Figure 13.21 Growth of stress with time in backfill in stope at 1600 m depth.

Figure 13.21a also gives an approximate closure strain scale. Closure occurs as mining proceeds away from the point of measurement, and more and more load is transferred from the solid rock into the backfill. The stress in the backfill increases commensurately, and in this case, the closure approached 60%. Plate 13.2 shows the side of a tailings fill when the closure was about 50%. The fractured timber poles indicate the amount of compression that has occurred.

Figure 13.22 shows the major and minor principal stresses (σ_1^I and σ_3^I) displayed as a p-q stress path. The K_f line was established from triaxial shear tests on samples of the fill. Figure 13.22a shows detailed test results for values of ½($\sigma_1^I + \sigma_3^I$) of up to 11.5 MPa, and Figure 13.22b shows overall results over the whole range of ½($\sigma_1^I + \sigma_3^I$)

Plate 13.2 Tailings backfill in narrow stope after about 50% closure.

up to 38 MPa. It appears that the fill was compressed in K_o compression with a value of $K_o = 0.4$ to 0.5. Hence the fill at any point was completely restrained by the surrounding fill and did not fail, but was subjected to continuing K_o compression. This type of field test is extremely difficult both to install and to carry out, and Figures 13.21 and 13.22 (obtained by Clark) are a rare example of such a test carried out successfully and with detailed measurements, over a period of more than a year.

13.5 Supporting narrow stopes with steel-reinforced granular tailings backfill

The incorporation of horizontal steel reinforcing into a tailings stope fill has the effect of stiffening and strengthening the fill against the vertical loads generated by closure of the stope. The lateral strain generated by the closure (see Figure 13.19b) is resisted by the horizontal reinforcing, which, in turn, generates a horizontal stress in the fill. As shown by equation (13.27), for elastic conditions in the fill, if $K_o = 0.5$, $\Delta\sigma_h = 0.125\sigma_v$ or $\sigma_v = 8\sigma_h$. For failure conditions in the fill, $\sigma_v = K_p\sigma_h$, where K_p is the passive pressure coefficient. For a cohesionless ($c^I = o$, φ^I) granular material:

$$K_p = \tan^2(45° + \varphi^I/2) = (1 + \sin\varphi^I)/(1 - \sin\varphi^I) \qquad (13.29)$$

For $\varphi^I = 30°$, $K_p = 3.0$ and $\sigma_v^I = 3\sigma_h^I$

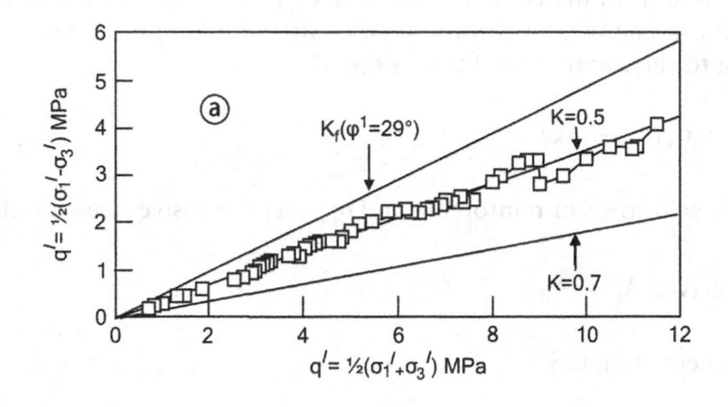

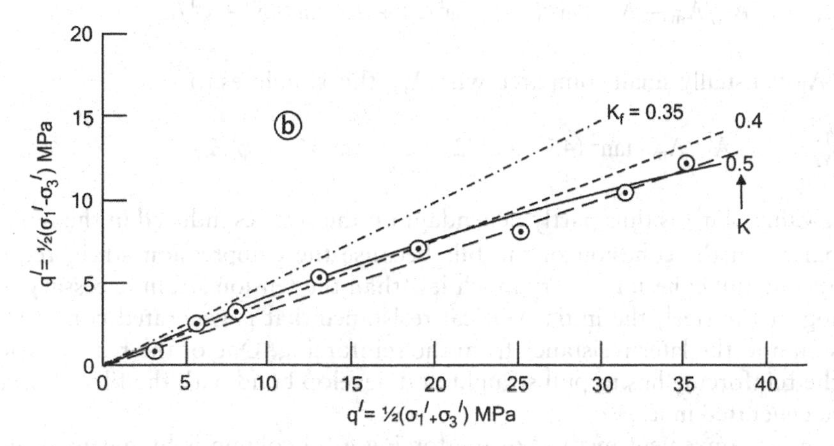

Figure 13.22 **Stress path for fill corresponding with stress measurements shown in Figure 13.21a. a: at relatively low stresses, and b: over whole stress range.**

For a cohesive (c^I) material, the passive pressure would be

$$\sigma_v^I = \sigma_h^I \tan^2(45^\circ + \varphi^I/2) + 2c^I \tan(45^\circ + \varphi^I/2) \qquad (13.30)$$

and for $\varphi^I = 30^\circ$

$$\sigma_v^I = 3.0\sigma_h^I + 3.46c^I$$

Supposing $\sigma_v^I = 1000$ kPa and $c^i = 0$, then $\sigma_h^I = 333$ kPa
If $\sigma_v^I = 1000$ kPa and $c^I = 100$ kPa $\sigma_h^I = 218$ kPa

These examples show that a vertical stress σ_v^I applied to a tailings fill can be resisted by generating a horizontal confining stress σ_h^I in the fill and that cohesion in the fill also assists in resisting vertical stress.

Suppose that a fill material is reinforced with cross-sectional area A_R of horizontal steel reinforcing per vertical area A_m of fill. Equating the tension at yield of the reinforcing to the compressive force in the fill:

$$\sigma_y A_R = \sigma^I_{hy}(A_m - A_R)$$

where σ_y = yield stress of reinforcing and σ^I_{hy} = compressive stress in fill, i.e.

$$\sigma^I_{hy} = \sigma_y A_R/(A_m - A_R) \quad (13.31)$$

Hence from equation (13.30)

$$\sigma^I_{vy} = \sigma_y A_R/A_m - A_R)\tan^2(45° + \varphi^I/2) + 2c^I \tan(45° + \varphi^I/2)$$

Since A_R is usually small compared with A_m, this simplifies to

$$\sigma^I_{vy} = \sigma_y \cdot A_R/A_m \cdot \tan^2(45° + \varphi^I/2) + 2c^I \tan(45° + \varphi^I/2) \quad (13.32)$$

The value of σ^I_{vy} is thus partly dependant on the stresses induced in the reinforcing, and partly on the cohesion of the fill. Because the compression strain required to develop the full cohesion is very much less than the tension strain necessary to cause yielding of the steel, the initial vertical resistance that is generated comes from the cohesion and the later resistance from the reinforcing. One of the reasons for this is that the reinforcing has to pull straight and develop bond with the fill before tension can be generated in it.

The most convenient method of reinforcing a fill column is by means of layers of steel mesh at regular vertical spacings. If the horizontal spacing of the wires in the mesh is h and the vertical spacing of the mats of mesh is v, it follows that

$$A_R/A_m = A_R/vh$$

where A_R is now the cross-sectional area of a single wire. Equation 13.32 can then be modified to

$$\sigma^I_{vy} = \sigma_y \cdot A_R/vh \cdot \tan^2(45° + \varphi^I/2) + 2c^I \tan(45° + \varphi^I/2) \quad (13.32a)$$

Figure 13.23a shows the results of a compression strain versus compressive stress test on a rectangular column or wall consisting of uncemented power station bottom ash reinforced with steel mesh, in which $A_R = 3.14\ mm^2$, v = 50 mm and h = 25 mm. It will be seen that the compressive resistance increased almost linearly with compression until the steel yielded when the compressive stress reached 4.86 MPa. When the wall was unloaded and reloaded at a stress of 2.39 MPa, the unloading and reloading moduli were almost the same, and very much higher (328 MPa) than the modulus for first loading (32MPa). This is to be expected as Figure 13.23a is simply a consolidation curve rotated anticlockwise through 90° (see Figure 4.11).

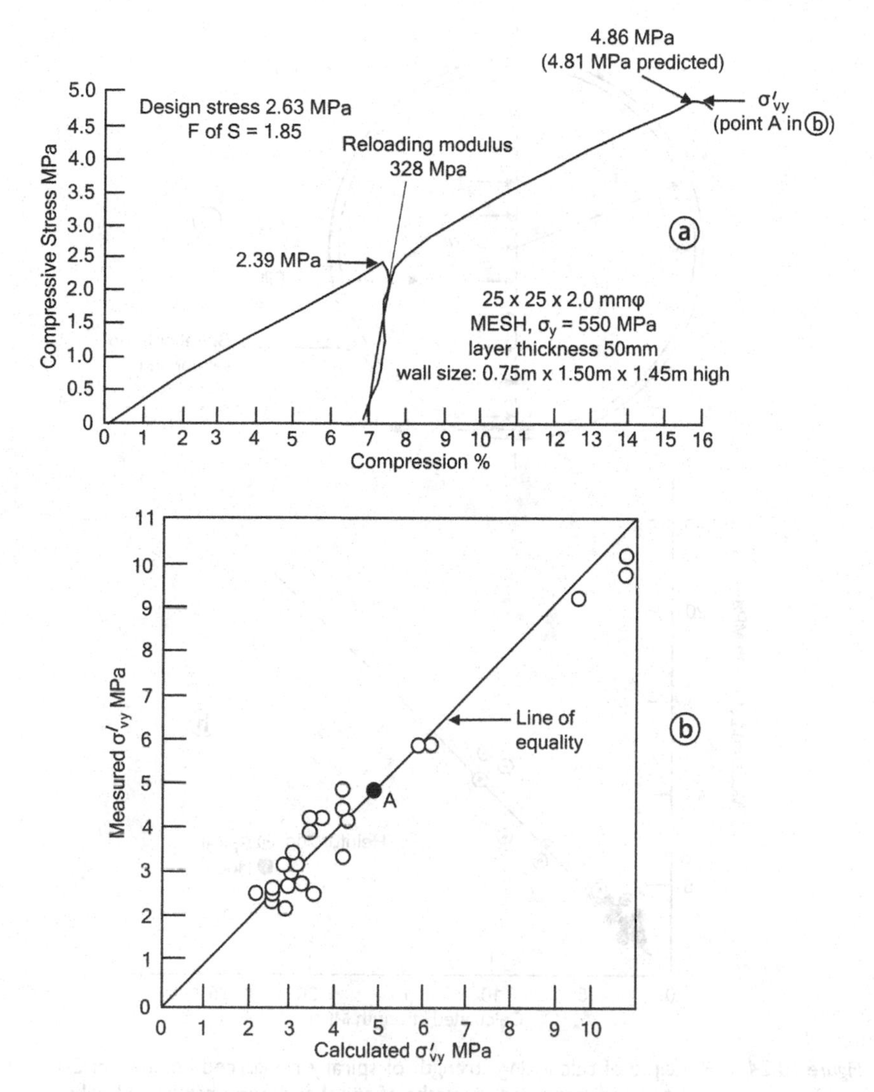

Figure 13.23 a: Stress-compression curve for wall of horizontally reinforced uncemented ash. b: Relationship between calculated and measured strengths of walls of horizontally reinforced granular material.

Figure 13.23b compares calculated failure stresses (σ_f) with corresponding measured values for 25 tests on columns and walls of reinforced, uncemented granular material (both quartzite sand and power station bottom ash), with a range of values of A_R, v and h. Point A represents the test shown in Figure 13.23a.

Figure 13.23a immediately suggests a way of increasing the stiffness of a support of reinforced granular material by building the support almost to roof level and then pre-loading it by jacking off the roof. The space created by the jacking can then be filled with timber wedged in place, or with a pressurized grout-filled bag.

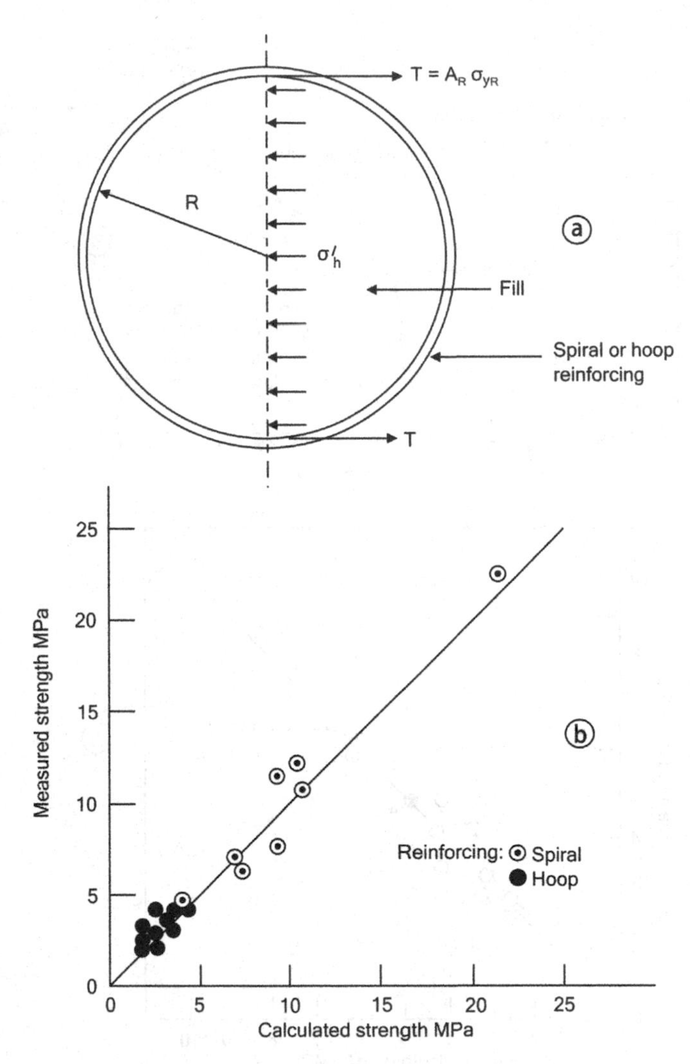

Figure 13.24 a: Principle of calculating strength of spirally reinforced column. b: Comparison of calculated and measured strengths of spiral and hoop reinforced columns.

13.5.1 *Spirally or hoop-reinforced supports*

As an alternative to embedding reinforcing mesh in the fill, a cylindrical column of fill can be reinforced by surrounding it with a steel spiral or a series of steel hoops. A radial confining stress is then developed as shown in Figure 13.24a.

For horizontal equilibrium of the column, at yield of the reinforcing,

$$2Rp\sigma^{I}_{hy} = 2T = 2A_R\sigma_y, \text{Or}\, \sigma^{I}_{hy} = A_R\sigma_y/Rp, \text{and}$$

$$\sigma^{I}_{vy} = K_p\sigma_{hy} = K_pA_R\sigma_y/Rp$$

For a cohesive material

$$\sigma^{I}_{vy} = \sigma_y A_R/Rp \cdot \tan^2(45^\circ + \varphi^I/2) + 2c^I \tan(45^\circ + \varphi^I/2)$$

in which p is the pitch or vertical spacing of the reinforcing coils or hoops.

Figure 13.24b compares predicted and measured values of σ^{I}_{vy} for hoop and spirally reinforced columns. Of the two systems, separate hoops are preferable to a continuous spiral, because the entire confining stress is lost if one break occurs in a spiral, whereas this does not happen with separate hoops. Note also that the hoop or spiral reinforcing can be replaced by a thin walled steel pipe and equation (13.32)a will apply with A_R/p replaced by the thickness of the pipe wall, t.

Bond between the reinforcement and the fill plays no part in a spirally reinforced column. However, to transfer the confining stress into the fill, it is necessary to have a retaining membrane between the fill and the reinforcing. In the tests carried out so far, a woven polypropylene geotextile has been successfully used for this purpose.

Comparing equations (13.32)a and (13.32)b, the same value of σ^{I}_{vy} can be reached if

$$A_R/vh(\text{wall}) = A_R/Rp(\text{column})$$

Taking some practical values, if $v = h = 50$ mm, $p = 25$ mm and $R = 500$ mm

$$A_R(\text{wall}) = A_R(\text{col}) \cdot 50 \times 50/(500 \times 25) = 0.2A_R(\text{col})$$

However, the mesh requires wires with area A_R in two directions, hence, actually

$$A_R(\text{wall}) = 0.4A_R(\text{col})$$

If the radius of the column is reduced to 200 mm, keeping $p = 25$ mm

$$A_R(\text{wall}) = A_R(\text{col})$$

Hence a mesh-reinforced wall or square pillar is equally economic in steel required regardless of plan dimension, but hoop-reinforced columns become more and more demanding of steel as they increase in radius.

13.5.2 *Development of bond between mesh and fill*

The tension developed in the reinforcing of a horizontally reinforced granular material depends on stress transfer by bond between the reinforcing and the reinforced material. The mechanism of bond development is illustrated in Figure 13.25a.

The average bond stress transmitted to a reinforcing wire over a length x by friction is

$$\sigma^{I}_{vy}(1 + 1/K_p) \tan \varphi^I \pi D x/2$$

where the wire diameter is D

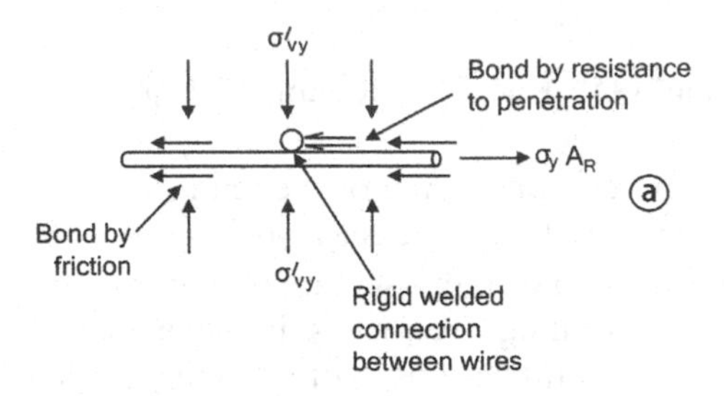

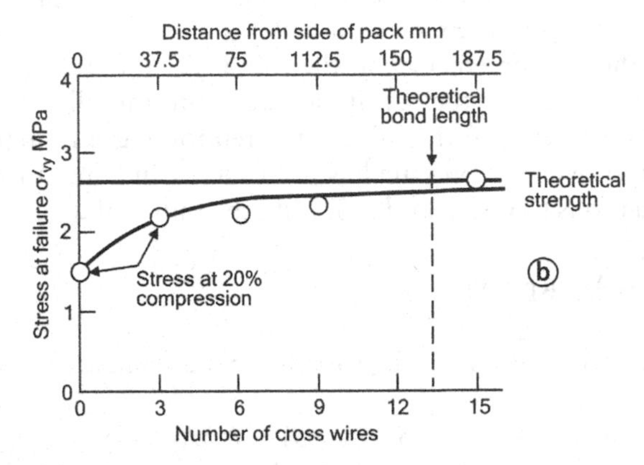

Figure 13.25 a: Mechanism of bond development in reinforced granular material. b: Investigation of bond length requirements for reinforced granular material.

The most convenient form of horizontal reinforcing for a wall consists of a system of orthogonal wires rigidly bonded to each other at points where they cross. Readily available systems consist of either square or rectangular welded mesh or twisted diamond mesh. The wires that run at an angle to the direction of tension under consideration contribute considerably to the bond because they must penetrate the reinforced material before the reinforcing can slip relative to the material. If the bond wires in an orthogonal system are spaced ℓ apart, the load transferred through these wires in length x will be a minimum of:

$$\sigma^{\mathrm{I}}_{\mathrm{vy}}\mathrm{h}x/\ell(\mathrm{d}/\mathrm{K_p} + 2\mathrm{d}\tan\varphi^{\mathrm{I}})$$

where d is the diameter of the orthogonal bond wires that are spaced h apart. (Usually, d = D)

Full bond resistance will be developed in a bond length x given by: $\pi D^2\sigma_y/4 = x\sigma^I_{vy}$, i.e.

$$x = \pi D^2/4 \cdot (\sigma_y/\sigma^{\mathrm{I}}_{\mathrm{vy}})/[D/2(1 + 1/K_p)\tan\varphi^{\mathrm{I}} + h/\ell(d/K_p + 2d\tan\varphi^{\mathrm{I}})] \quad (13.33)$$

Example: If $\sigma^{\mathrm{I}}_{\mathrm{vy}} = 5$ MPa, $\sigma_y = 600$ MPa, $h = \ell = 25$ mm
$D = d = 2.5$ mm, $K_p = 3.5$, $\tan\varphi^{\mathrm{I}} = 0.7$
$x = 127$ mm

Hence full bond could be achieved with 127/25 = (5.09) = 6 cross wires.

Figure 13.25b illustrates the results of a series of tests on reinforced walls that was designed to investigate the validity of equation (13.33). The figure shows the stress at failure in a series of tests on walls in which the number of bond wires at each side of the wall was increased progressively from zero. The spacing of the bond wires was 12.5mm and the theoretical bond length was 163mm, hence full bond could theoretically be achieved with 13 bond wires or a bond length of 163mm. The results in Figure 13.25b confirm this.

13.6 The behaviour of steel mesh-reinforced square columns of cemented cyclone tailings underflow (Grout packs)

This section will deal with the characteristics of steel mesh-reinforced square columns of cemented cyclone tailings, known as grout packs that, developed from the theory set out in Section 13.5, are in increasing use as replacements for timber grillage supports, referred to as timber packs. The requirement for support in narrow stopes at great depth has been stated by Ortlepp (1978) as: "There exists the somewhat paradoxical requirement that a stope support must be very stiff, and, at the same time, have an almost unrestricted ability to yield". In other words, a pack should be stiff in accepting load up to the yield stress in the steel, (which should equal the required strength), and then maintain load up to a compression of 30% or more. Packs of uncemented tailings are unable to provide this characteristic unless they have been pre-loaded to the minimum required load and chocked against the hanging wall (or roof) to maintain the pre-load. (See Figure 13.23a). The cohesion provided by the cementing action in grout packs can, however, provide the initial stiffness, and if annealed steel is used for reinforcing, their ductility can be considerably extended.

The packs are constructed of geofabric bags or pillows filled with grout and separated, one from another, by horizontal sheets of welded steel mesh. Plate 13.3 shows a grout pack installed in a stope, while Plate 13.1 shows a timber pack (to the left of the miner).

Figure 13.26a shows the effect of annealing on the stress-strain characteristics of reinforcing wire. (Complete sheets of welded mesh are heated in a furnace to temperatures of about 650°C which is above "red heat". The temperature is maintained for long enough to allow the locked-in strain-hardening in the steel caused by cold-drawing the wire to be relieved and the mesh is then slowly cooled in the furnace back to ambient temperature.) Annealing considerably reduces the yield stress of the steel, but extends the failure strain by a factor of 12 to 15.

The effect of using annealed mesh in grout packs is shown in Figure 13.26b which records the results of compression tests on cubical model grout packs (A and B) reinforced with un-annealed wire of two different diameters (2.5 mm and 3.15 mm), but the same value of A_R/vh equal to 0.0022 and annealed wire also with $A_R/vh = 0.0022$ (C). As the figure shows, both the maximum load and the ductility of the packs with annealed reinforcing was considerably improved.

Figure 13.27a shows the effect of grout cement content and grout strength, on the stress compression characteristics of cubical packs with constant steel reinforcing content. The grout strength is expressed as the mean test cube strength after 7 days of

Plate 13.3 Grout pack installed in stope.

curing at constant water content. It is clear from the curves that the initial stiff portion of the characteristic up to 1 to 2% strain arises from the strength of the grout and that the later linear characteristic is produced by continuing yielding of the steel.

Figure 13.27b shows the effect of changing the steel content at constant grout strength. The effect of increasing steel content is to increase the slope of the linear part of the characteristic, together with some extension to higher stresses of the initial stiff characteristic.

Finally, Figure 13.28 shows the effect of varying the width to height ratio, i.e. the slenderness of a pack at constant grout strength and reinforcing content. The stiffness and strength increase progressively as W/H increases, i.e. as slenderness H/W decreases. When W/H exceeds 2, the packs become virtually indestructible, even though they continue to compress under increasing load.

13.7 The use of geotextiles for temporary retention of backfill in narrow stopes during hydraulic placing

Tailings backfill usually consists of the coarser fraction of the total tailings that has been separated by cycloning (the cyclone underflow). This is gravitated underground

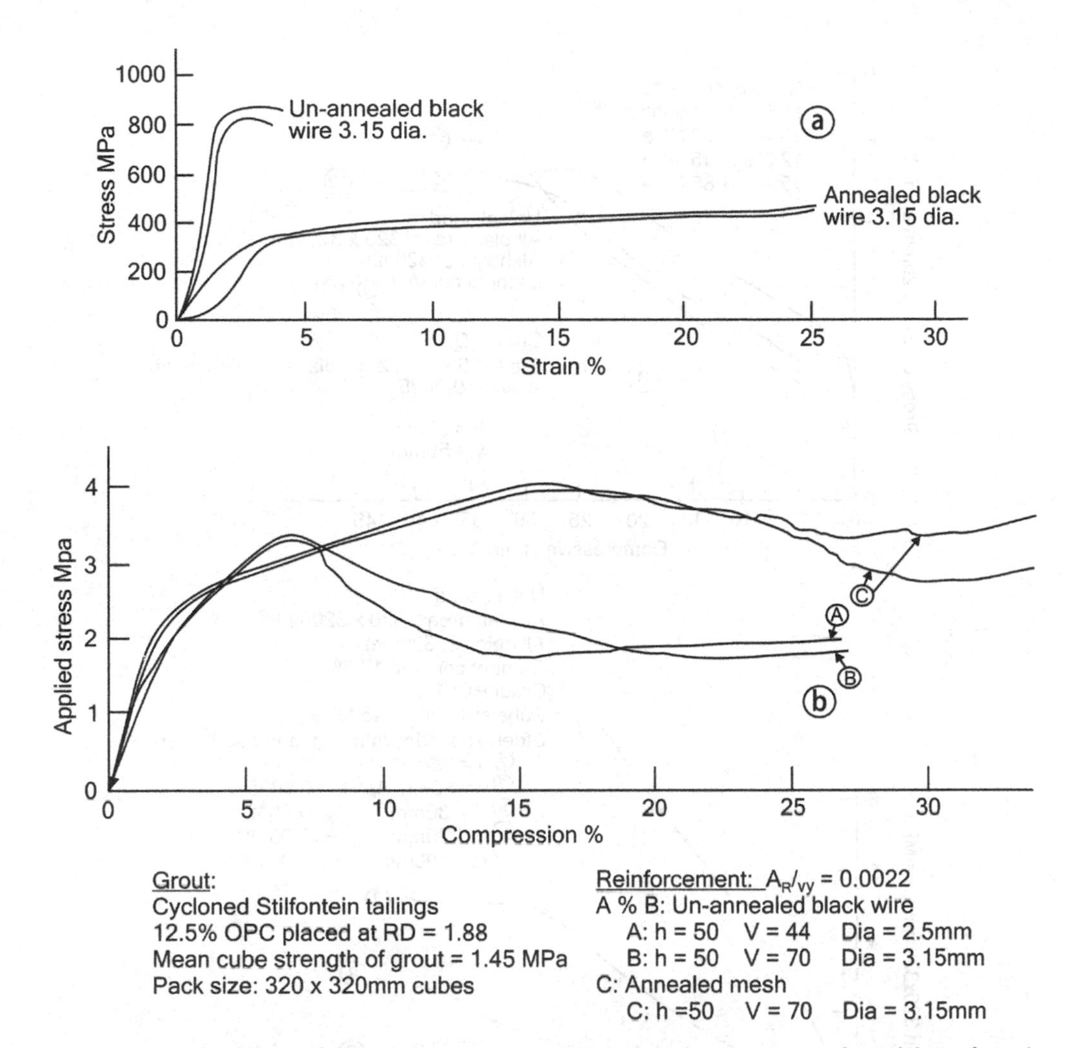

Figure 13.26 Effect of annealing on stress-compression characteristics of model reinforced cemented grout packs: a. Typical stress-strain characteristics of unannealed and annealed wire. b. Stress-compression characteristics of model grout packs reinforced with high tensile (A & B) and annealed mesh (C).

to the site of the backfilling. The process of hydraulic backfilling is illustrated by Figure 13.29. The stope to be filled is lined with a geotextile bag having a cross-section of up to 4 m wide and with a height to suit the stope. The bags have lengths of up to 30 m, and are placed (empty) in the stope to be filled. The backfill slurry is introduced from the high end of the stope through a flexible pipe or pipes and flucculant and/or cement in the correct dosage is introduced at the nozzle. The bag fills from the bottom up, as shown, and excess water separates from the solids and both runs off the top and through the sides of the geofabric, as well as percolating out of the fill at the lower end.

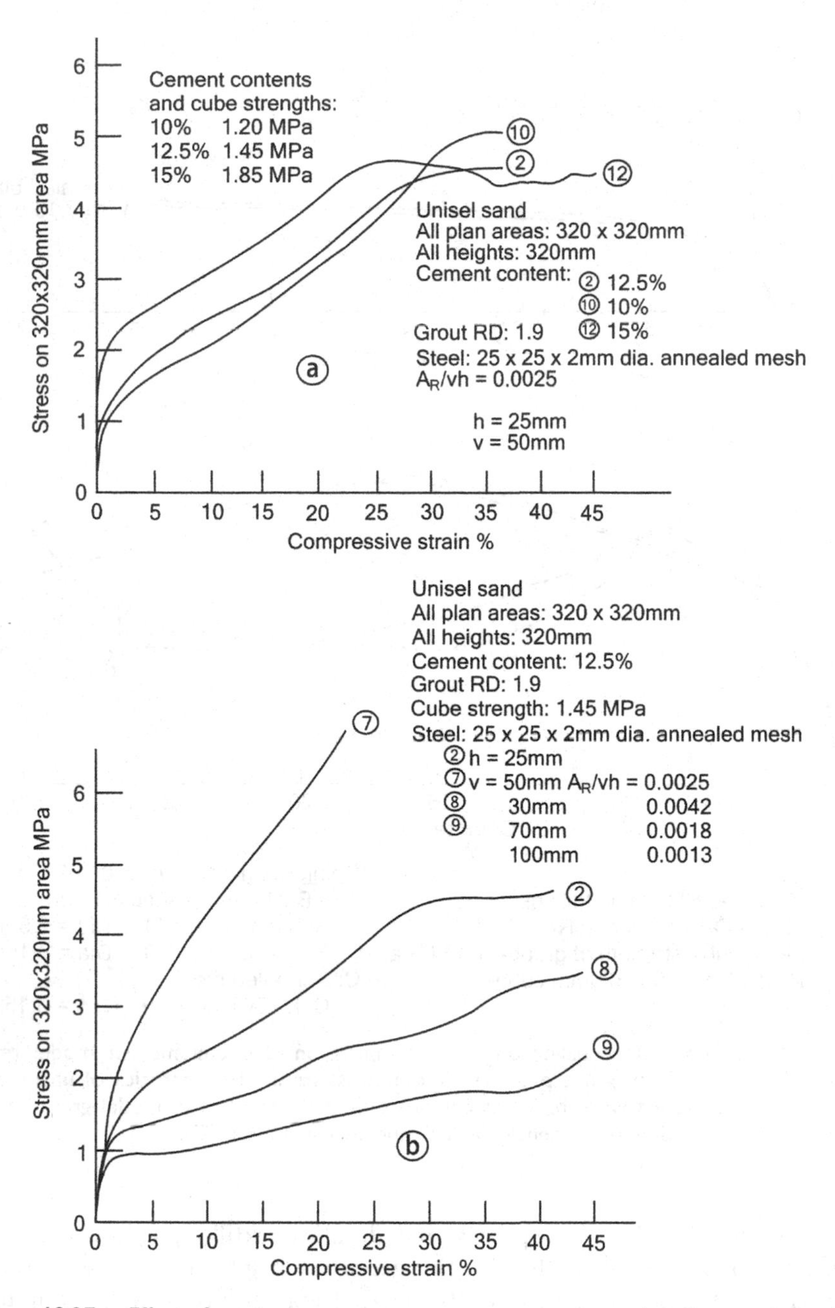

Figure 13.27 a: Effect of varying cement content on stress-compression characteristics. b: Effect of varying steel content on stress-compression characteristics.

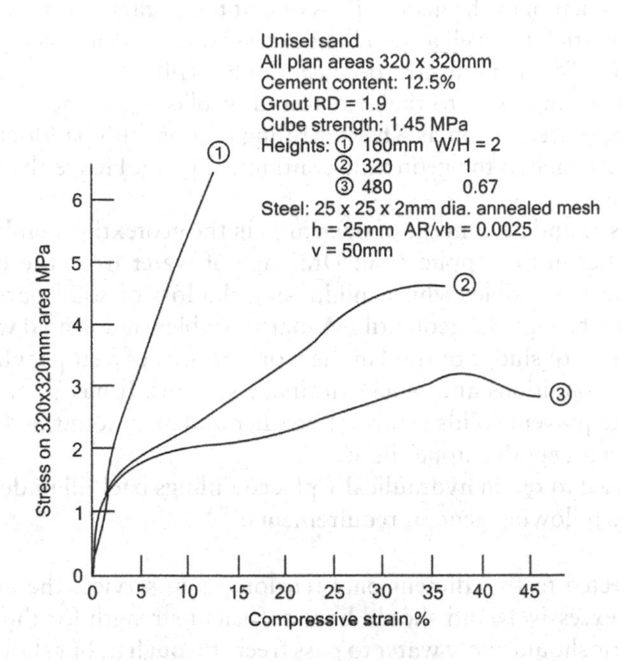

Figure 13.28 Effect of varying width/height (W/H) ratios on stress-compression characteristics.

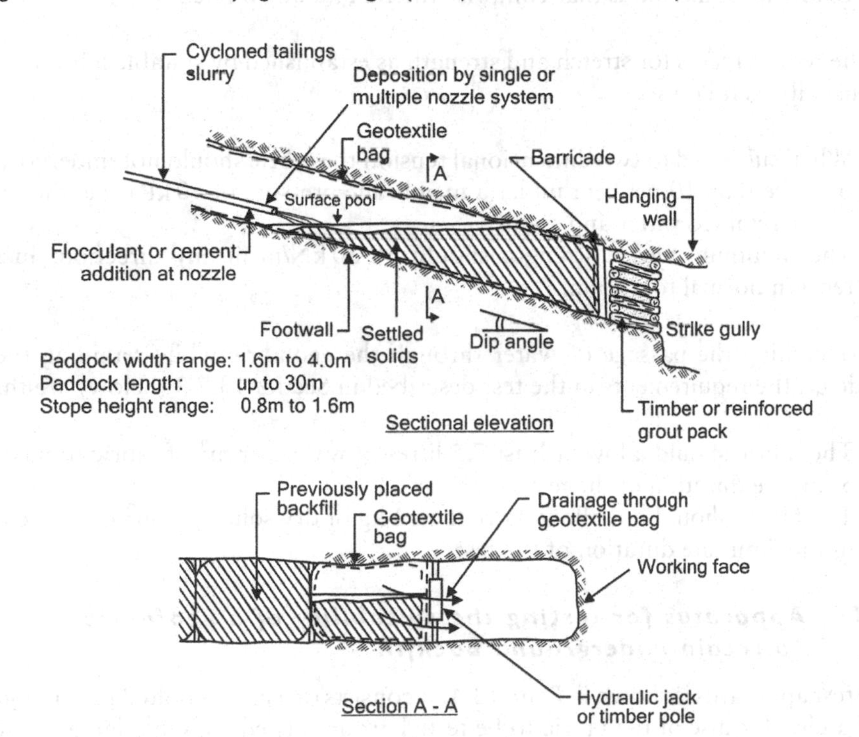

Figure 13.29 Geotextile bags used as temporary containment for backfill in narrow stopes.

The primary function of the geotextile is one of temporary containment, i.e. it must retain the solid particles as well as allowing excess water to drain as rapidly as possible from the backfill. This is necessary because the backfill must drain, consolidate and become self-supporting prior to the next round of blasting at the working face. This blasting can take place as soon as 8 h after filling of a backfill paddock and inevitably results in some damage to the geotextile containment bag. Hence the requirement for early self-support.

The opening size and percentage of open area in the geotextile membrane are important characteristics in this application. Drainage of water from the backfill must be impeded as little as possible, whilst minimising the loss of solid particles (including cement particles) through the geotextile. A major problem associated with solids losses is the accumulation of sludge or mud in the working area of a stope which creates hazardous working conditions and blocks drainage systems. It has been recognized that it is unrealistic to prevent solids escape altogether and an amount 3–4% loss of solids is regarded as an acceptable upper limit.

A geofabric used to retain hydraulically placed tailings backfill underground should comply with the following general requirements:

- When subjected to two dimensional tension, as in service, the geofabric should not stretch excessively and should have sufficient strength for the purpose.
- The geofabric should allow water to pass freely through it, but should substantially retain all of the solids that comprise the fill and any binder.

The requirements for stretch and strength as established by suitable laboratory tests are usually as follows:

- When subjected to two dimensional tension the fabric should not undergo a strain of more than 10 per cent under a membrane pressure of 50 kPa (i.e. the pressure of the retained water and solids.)
- The minimum tensile strength should be 30 kN/m in any direction, including tension normal to a seam.

Concerning the passage of water through the geofabric while retaining the solid particles, the requirements in the test described in Section13.7.3 (below) are that:

- The fabric should allow at least 7.5 litres of water per m^2 of fabric to pass in the 5 minute duration of the test.
- The fabric should not allow more than 1 kg of dry solids per m^2 of fabric to pass in the 5 minute duration of the test.

13.7.1 *Apparatus for testing the suitability of a geofabric to retain underground backfill*

The test apparatus is shown in Plate 13.4. It consists of a pair of bolted pipe flanges that grip a circular disc of the fabric to be tested around its edges. One flange is attached to a short open ended length of pipe, the other to a short closed ended length. The steel disc that forms the closure to the closed end of the pipe is equipped with a central

Plate 13.4 Apparatus for measuring two-dimensional strain in geofabric under pressure.

filling spout, and an inlet for compressed air. The filling spout can be sealed by means of a threaded cap.

The disc of geofabric under test is clamped between the flanges equipped with two rubber flange sealing rings. The pipe diameter is 300 mm, the flanges are 400 mm in diameter, and the length of pipe on either side of each flange is 100 mm.

13.7.2 *Test for stretch (strain) under two-dimensional tension*

A disc of geofabric is cut out to the correct size and holes punched to match those in the flanges. The geofabric is backed with a disc of thin plastic sheeting. The two flanges are assembled with the open-ended pipe uppermost and the plastic backing under the geofabric. A dial gauge or LVDT is mounted, as shown in Plate 13.4 with its spindle resting on a small disc of rigid material which is glued to the centre of the geofabric disc. A measured air pressure is applied to the underside of the geofabric and the relationship between the rise of the centre of the geofabric and pressure is recorded. Referring to Figure 13.30a, the following geometrical relationships apply:

$$\text{Length of arc } A = C + 5B^2/2C \tag{13.34}$$

$$\text{Strain in geofabric } \varepsilon = (A - C)/C$$

$$\varepsilon = 5B^2/2C^2 \tag{13.35}$$

$$\text{Radius of curvature } R = (C^2 + 4B^2)/8B \tag{13.36}$$

$$\text{Tension in geofabric } T = pR/2 \tag{13.37}$$

Where p = applied air pressure.

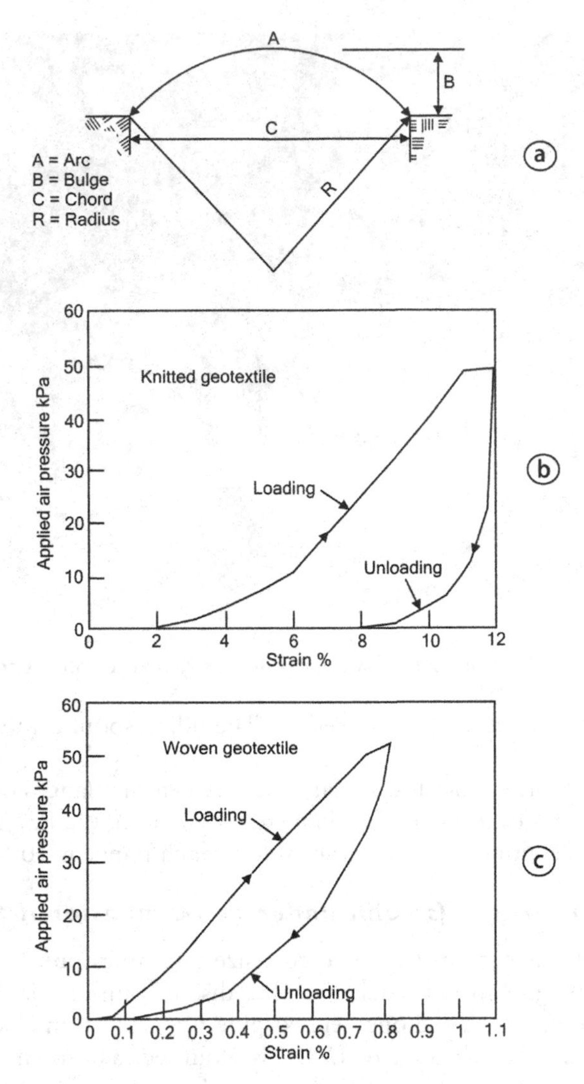

Figure 13.30 a: Geometric relationships for analysis of two-dimensional stretch under pressure. Variation of geotextile strain with applied air pressure measured in pressure apparatus for b: knitted geotextile, c: woven geotextile.

Hence the strain or stretch in the geofabric at a pressure p = 50 kPa can be determined. Figures 13.30b and c show typical pressure versus strain recordings from the tests described above. Note the differences in strain between knitted and woven geotextiles at the same pressure.

13.7.3 *Test for retention of solids*

A fresh disc of geofabric is mounted in the apparatus, without the plastic backing disc. The apparatus is inverted so that the closed end is uppermost, as shown in Plate 13.5.

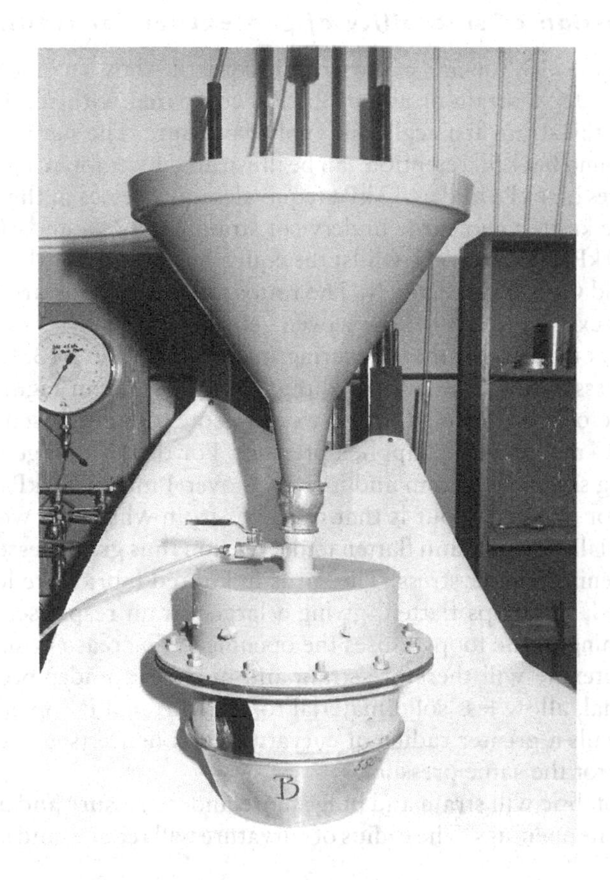

Plate 13.5 Apparatus of Plate 13.4 inverted to measure drainage rate of water and retention of solids.

The apparatus rests on, or in a metal basin (B in Plate 13.5) that has been weighed to the nearest gram.

10 kg of slurry is prepared at an appropriate slurry RD (e.g. 1.7). (A Marcy mud balance can be used for this purpose). The cap is removed from the filling spout and the slurry, agitated to ensure that its RD is uniform, is poured onto the geofabric. A large plastic funnel placed in the filling spout is used for the filling operation.

After 5 minutes the basin and contents (water and solids) are removed and weighed. After drying in a drying oven at 105°C for 24 hours, the basin and dried contents are reweighed. The results are expressed in terms of kg/m^2 for the mass of solids passing the geofabric and as litres/m^2 for the water passing. (The area of a 300 mm diameter circle is 0.071 m^2).

Typical results for a geofabric are as follows:

Solids + water passing : 564 g
Solids passing : 6.6 g = 0.093 *kg*/m^2
Water passing : 558 g = 7.86 ℓ/m^2

These results are for a geofabric for which the strain under an air pressure of 50 kPa was 6%.

13.7.4 *Discussion of suitability of geotextiles for retaining backfill*

Considering Figures 13.30b and c, the woven geotextile shows a much stiffer response, developing only 0.5% strain at about 30 kPa, compared with the knitted geotextile which develops this strain at a negligible applied pressure. The significance with regard to the underground backfill retention can be illustrated by comparing the strains developed at pressures of 4 kPa and at 40 kPa (equivalent to stresses in the field) for the two geotextiles. The knitted geotextile underwent strains of 2.6% and 10% for pressures of 4 kPa and 40 kPa, respectively, whilst the equivalent values for the woven geotextile were 0.08% and 0.68% respectively. The ratios of these strains are therefore 3.8 for the knitted geotextile and 8.5 for the woven geotextile.

Measurements made with the measuring microscope (Plate 13.4) during the tests using the air pressure apparatus showed that for the results in Figures 13.30b and c, the average size of the openings of the woven geotextile decreased from 0.5 mm at zero stress to 0.4 mm at 50 kPa applied pressure. For the knitted geotextile the initial average opening size was 0.5 mm and increased over 1 mm at 50 kPa.

The reason for this behaviour is that the tapes from which the woven geotextile is made are essentially straight and flatten under tension thus giving lesser strain response and smaller openings under stress. The yarns in knitted fabrics are looped and as the fabric is stressed, the loops flatten, giving a larger strain response, and at the same time, the flattening of the loops causes the openings to increase in size.

A woven geotextile will therefore strain and bulge less under pressure, and other things been equal, allow less solid material to pass through its openings. However, a lesser bulge entails a greater radius of curvature, and hence (see equation (13.37)) a greater tension for the same pressure.

A knitted geofabric will strain and bulge more under pressure and allow more solids to pass through its openings. The radius of curvature will reduce, and so will the tension in the fabric.

As examples, suppose two geofabrics, the woven and knitted ones that feature in Figures 13.30b and c, are considered for a backfill bag measuring 1.5 m high by 1.5 m wide.

From equations (13.34) and (13.35)

$$(A - C)/C = \varepsilon \quad \text{and} \quad A = C(1 + \varepsilon)$$

For the knitted fabric: $A = 1.5(1 + 0.12) = 1.68\,\text{m}$
For the woven fabric: $A = 1.5(1 + 0.008) = 1.51\,\text{m}$

$$B = \sqrt{[2C(A - C)/5]}$$

For the knitted: $B = 0.33\,\text{m} = \text{bulge}$
For the woven: $B = 0.08\,\text{m}$
Tension in geofabric for $p = 50\,\text{kPa} = 2.94\,\text{m}$ of slurry depth

$$R = (C^2 + 4B^2)/8B$$

For the knitted: $R = 1.02\,\text{m}$ and $T = 26\,\text{kN/m}$
For the woven: $R = 3.55\,\text{m}$ and $T = 89\,\text{kN/m}$

As the comparisons show, the bulge is moderate in both cases, but the tension in the woven geofabric is 3.4 times that in the knitted material. Hence, on the basis of tension in the free face of the bag, the knitted geofabric is preferable. However, the tensions T should also be compared with the tensile strengths of the two geofabrics.

13.8 The use of mine and industrial wastes in surface construction

Mine wastes, such as overburden from open cast operations, waste rock from mine development and tailings have been used extensively as construction materials in road construction, both as fill for embankments and as aggregate in road pavements. Waste rock has also been used as a concrete aggregate. These materials can and have performed very successfully. However, incautious use, without careful investigation of the physical, mineralogical and geochemical properties of the wastes has also resulted in some disastrous failures. This also applies to industrial wastes. Quenched blast furnace slag and power station fly ash have been used successfully as pozzolanic or cement replacement materials for many years, and phospho-gypsum waste from the manufacture of phosphoric acid for use as an agricultural fertilizer has been used to produce building panels as ceiling and dry-walling boards. Unquenched slags have been used successfully as concrete aggregates and furnace and boiler bottom ashes have been used successfully to produce building bricks and breeze blocks. However, some slags are expansive and have caused problems when used as fill under buildings.

This section will describe three not uncommon problems that have resulted from the use of mine wastes in construction, but there may be others, possibly not experienced as yet.

13.8.1 *The use of ancient alluvial sands as a structural backfill*

An alluvial diamond mine, on the coast, mines diamonds from the ancient sea floor which is presently a few metres above sea level and is covered by an overburden of ancient beach sands from a time when the sea level was much higher than at present. The mining operation consists of removing the sand overburden from above the rock that formed the sea floor and then recovering the diamondiferous gravels that lie immediately above the rock floor. The gravels are then put through washing and heavy medium plants that separate the diamonds from the gravel. The washing plant operates by passing the gravels through a series of sieves and riffled tables by gravity. To achieve this a high ramp is constructed of sand overburden, supported at its highest point by a retaining wall against which the plant is constructed. The gravels are dumped into a hopper at the top of the ramp and work their way through the plant from one stage to the next by gravity flow, from the top of the plant to the bottom.

In the case to be considered, a complex of reinforced earth walls, 41m high was constructed with its reinforcing strips embedded in compacted ancient beach sand. At the time of construction the Tweepad walls were the highest reinforced walls ever built (Blight and Dane, 1989). Plate 13.6 is a view of the wall complex with the tipping hopper at the highest point and the rest of the diamond recovery plant below this. Figure 13.31 is a section through the 27 m highest wall (K) and the wall below it, 15.75 m high Wall A.

Plate 13.6 Complex of reinforced earth walls at Tweepad.

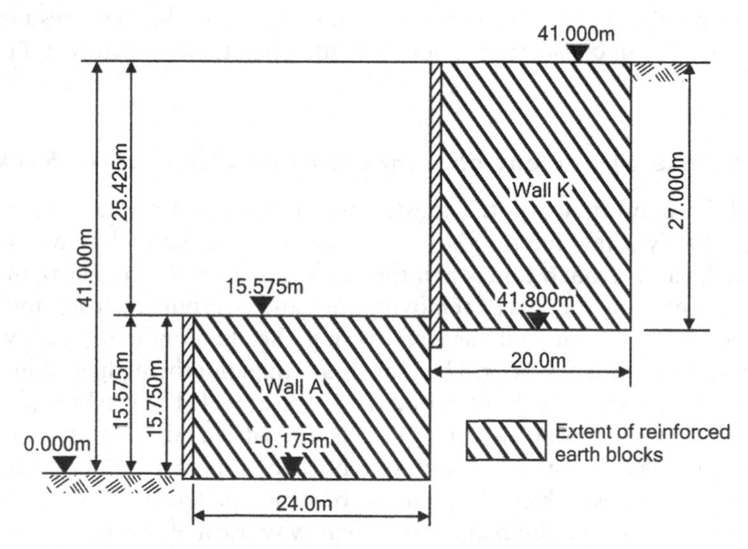

Figure 13.31 Section through reinforced earth walls at Tweepad that were constructed with sand overburden from mine.

The Tweepad area has a desert climate, with an average annual precipitation of 92 mm. Evaporation from a free water surface averages 1940 mm. The mean temperatures are a minimum of 18°C and a maximum of 34°C.

A failure of a reinforced earth wall at another mine on the coast resulted from corrosion of the reinforcing strips. As a result, it was decided to investigate the more

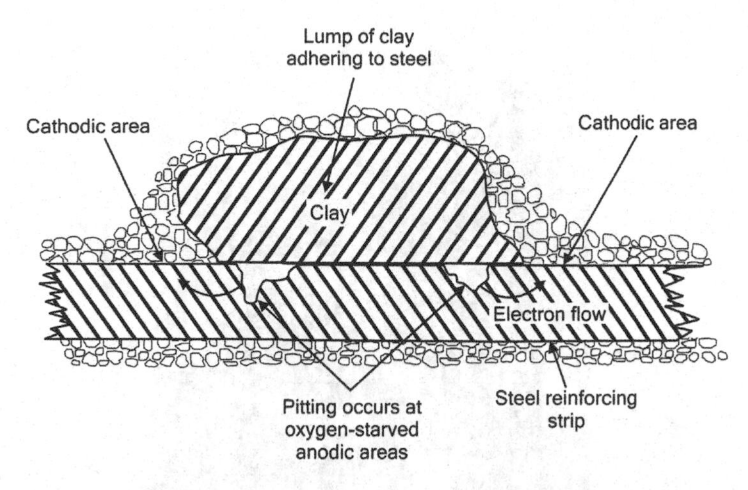

Figure 13.32 The mechanism of pitting corrosion of steel reinforcing strips.

recently constructed walls at Tweepad (3 years after construction) for signs of similar deterioration. It was found that the reinforcement of these walls was also deteriorating, as a result of severe pitting corrosion. Monitoring exercises were subsequently carried out at regular intervals to follow the process of deterioration and to assess the safety of the walls. The walls were finally demolished and rebuilt after 8 years of the 30-year design life.

Once monitoring of the progress of deterioration was started, it soon became apparent that the fine fraction of the fill occurred as discrete lumps of clay, which was not uniformly disseminated through the sand. Wherever a clay lump rested in contact with a reinforcing strip, severe pitting corrosion had occurred. It was obvious that corrosion was occurring mainly as a result of the formation of differential aeration cells (Allen & Lewis, 1979) as shown in Figure 13.32. The contact between the relatively pervious sand and the steel was well oxygenated, whereas that between the clay and the steel was oxygen-starved. This caused the steel near the perimeter of the clay lump to become anodic and corrode over a limited area, forming pits.

This conclusion was confirmed when it was found that the reinforcement of a similar reinforced earth wall complex (at the nearby Koingnaas mine) that had been backfilled with a uniform clean sand was quite free of corrosion after 6 years, even though the chloride content was much higher than at Tweepad.

Initially, monitoring of the progress of corrosion was accomplished by breaking out facing panels and retrieving strips from behind the wall face. Plate 13.7 shows some of the corroded strips retrieved in this way. Replacement strips were then welded into place. As the safety condition of the walls became more critical, it was decided to sink a shaft behind each wall, located approximately on the theoretical locus of maximum strip tension, i.e. one third of the length of the reinforcing strips back from the wall facing. As each set of strips was exposed in the shaft, usually two at each level, targets for a demountable Demec strain gauge were glued on. After taking an initial reading, the strip was cut at one end by means of an oxy-acetylene torch. The elastic shortening of the strip was then measured, and hence the in-situ tension deduced.

Plate 13.7 Corroded reinforcing strips taken from behind facing panels of walls.

Once the shaft had been completed, the shaft casing was progressively withdrawn and replacement strips were welded in place. The recovered strips were tested for strength, and elongation at failure was measured over a 100 mm gauge length straddling the break. The effect of the pitting corrosion was not only to reduce the load-bearing cross-section of the reinforcing strips, but also to embrittle them by introducing stress concentrators in the form of the irregularly-shaped corrosion pits.

The influence of the corrosion on both strength and ductility is illustrated in Figure 13.33, in which the strength values have been normalized with respect to the mean strength of new strips. The uncorroded strips taken from the Koingnaas wall show the usually expected relationship between ultimate tensile strength and ductility: an increase in ductility corresponds to a slight decrease in strength.

Strips sampled from Tweepad after 3 years already showed a marked reduction in strength. They also showed the effect of embrittlement by the presence of corrosion pitting, in that ductility decreased with decreasing strength. The results of tests on 7½-year-old strips showed this tendency. Figure 13.34 shows the results of strain-release measurements of tensions in the strips exposed in the shaft behind wall K which were similar to those found in shafts behind the other walls. In each case a considerable scatter occurred in measured tensions. It was quite common to find one of a pair of adjacent strips highly stressed while the other carried almost no load. This was evident not only from the spring-back measurements, but also from the audio-note given out by uncut individual strips when struck with a spade.

This appears to be almost an inherent characteristic of reinforced earth that results from the method of construction. If a strip is bent by the passage of a piece of

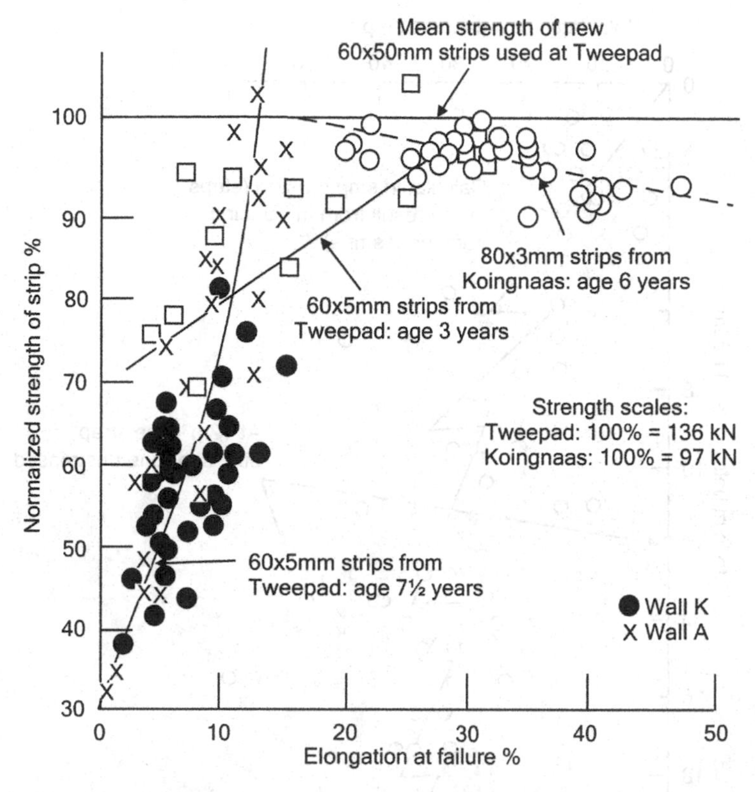

Figure 13.33 Effect of pitting corrosion on strength and ductility of reinforcing strips at Tweepad. (Walls A & K shown, other walls were similar.)

earth-moving machinery, while an adjacent strip is not, the bent strip, being effectively shortened, will carry more load. Plate 13.8 shows two strips exposed in the shaft behind wall K after they were cut at the end nearest the camera. The strip on the left was heavily loaded and sprang clear of the backfill when cut. The other strip carried almost no load and hardly moved when cut. The measured tensions shown in Figure 13.34 have been compared with tensions calculated by conventional analysis of reinforced earth walls (e.g. McKittrick, 1978). The steps in the calculated line correspond to changes in the number of strips at a particular level. The measurements roughly followed the predicted trends, but with very wide scatter and with considerable differences in tension between adjacent strips at the same level.

Figure 13.35 shows the profile of strip strengths, with depth, as measured on strips recovered from the shaft behind wall A. The shafts behind walls H and K gave similar results. The diagram illustrates the large scatter in measurements, which is also reflected in Figure 13.36. The shaft results showed that the deterioration of strength was by no means uniform with either depth or lateral extent.

In the case of Figure 13.35, relatively little deterioration is shown down to a depth of 8 m. Below 8 m, however, the deterioration became severe. The observed deterioration did not correlate with the salt content of the fill, or variations of moisture content. The salt content varied roughly linearly with depth from 0.5% by dry mass of fill

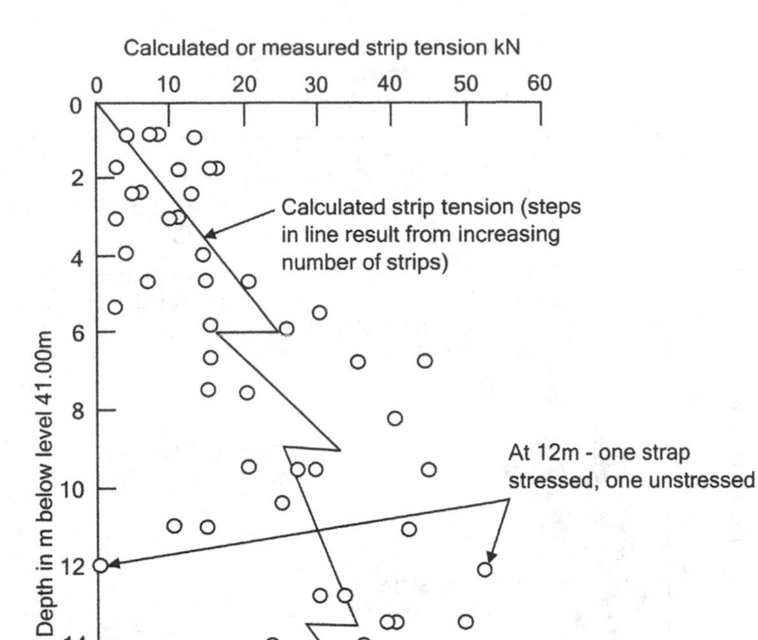

Figure 13.34 Comparison of calculated strip loads with loads deduced from strain release measurements made in shaft behind wall K.

at the surface to 0.3% at a depth of 12 m. The moisture content also varied roughly linearly with depth, from 11% at the surface to 6% at 12 m. Hence the more severe corrosion between depths of 8m and 12m is not obviously explained by correlation with fill characteristics. It was of particular concern that two of the measured strip strengths were less than the 95% upper limit to the measured strip tensions (at 11 m depth). Hence there was a distinct threat that Wall A could fail at any time.

Figure 13.36 shows the observed deterioration of strip strength with time for 60 mm × 5 mm reinforcing strips. As in Figure 13.33, the strength axis of the diagram has been normalized with respect to the mean strength of new strips. It is interesting to note that the 'most pessimistic estimated life of structure' was determined after a time of 4 years on a basis of linear extrapolation. This estimate did not require subsequent revision. Figure 13.36 relates to average conditions for all the walls of the complex. The walls were demolished and rebuilt after a life of 8 years. The mine closed 22 years later. The rebuilt walls, constructed with uniform, clean beach sand were then still in a serviceable state.

Plate 13.8 Reinforcing strips exposed in shaft excavated in backfill behind wall K.

13.8.2 *Damage to road pavements by upward migration of soluble salts*

Mine waste rock, ash, clinker, slag, and other rock-like industrial wastes frequently contain soluble salts. If waste containing salts is used to construct road or pavement layers, and especially in arid to semi-arid climates, evaporation will cause the salts to migrate to the surface in solution and may cause physical damage to road or pavement surfacings when they crystallize out at or just below the surface. (Blight 1976). Figure 13.37 shows profiles of soluble salt content produced in a compacted crusher-run pavement base of quartzite mine waste rock (a) by upward migration of soluble salts caused by an evaporation gradient towards the surface and (b) by downward leaching of the near-surface salt by infiltrating rain. The figure also shows (c) a salt profile within the thin asphalt surfacing of a road pavement damaged by salt crystallization and (d) a salt profile for an undamaged surfacing of thicker asphalt in which case the crystallized salt is confined to the base layer.

Typically, the first sign of damage to a road surface is the appearance of white streaks and patches on the surface after a light rain shower. The damage then progresses to the formation of salt-filled blisters in the surfacing, as shown in Plate 13.9. Depending on the size and intensity of the blistering (individual blisters may be as large as 150 mm in dia. and 50 mm high), traffic may break up the blistered surface and cause pot-holes to form.

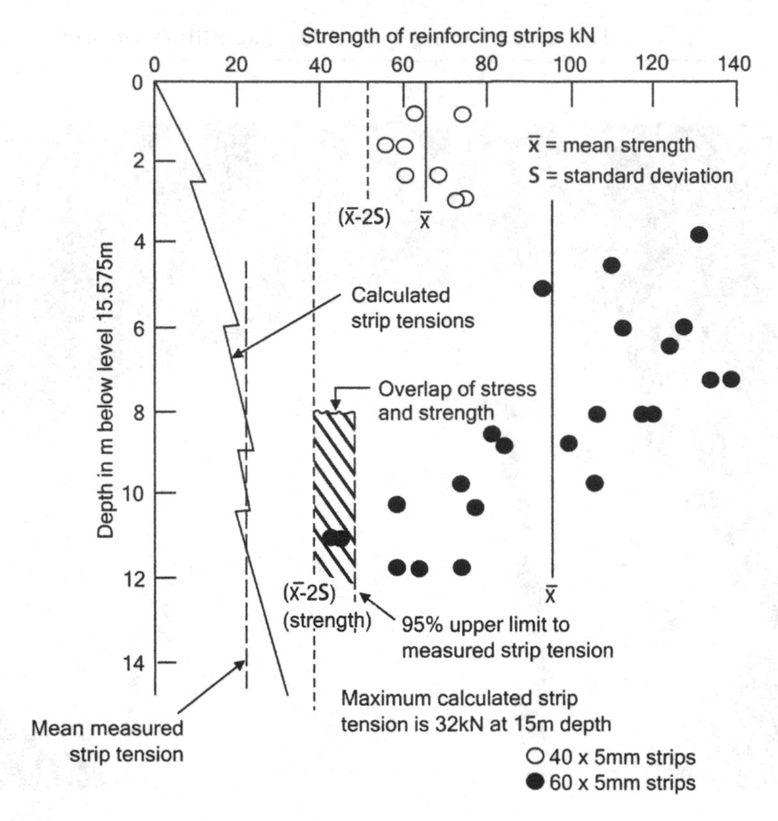

Figure 13.35 Results of strength tests on reinforcing strips taken from shaft behind wall A, compared with measured and calculated tensions.

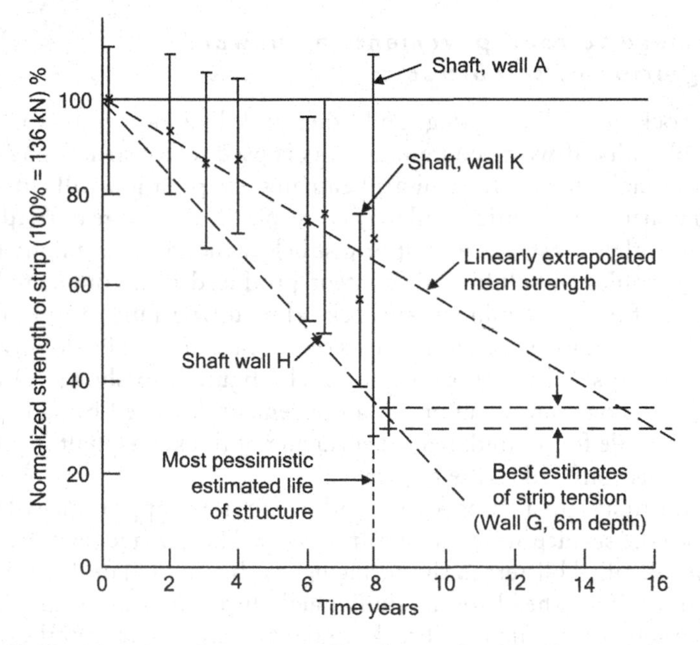

Figure 13.36 Deterioration of strip strength with time for 60 × 5 mm reinforcing strips (error bars show two standard deviations).

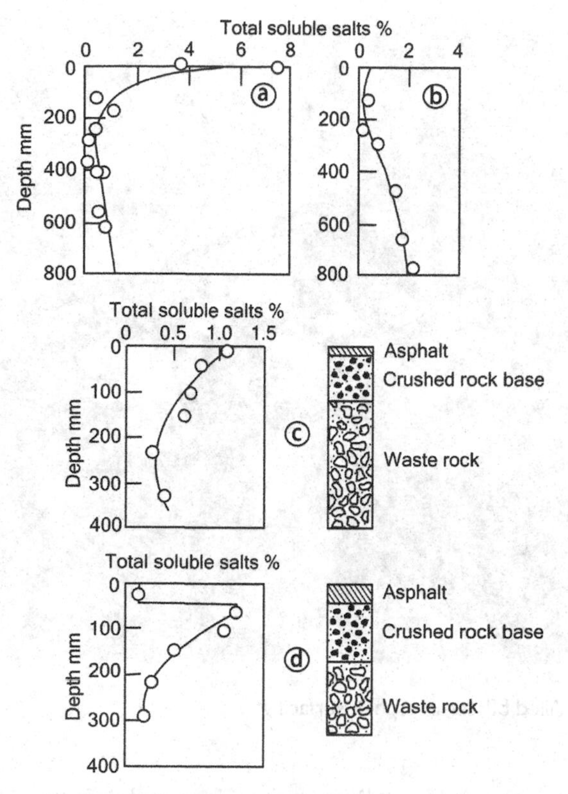

Figure 13.37 a: Salt raised to surface by evaporation. b: Salt leached downwards by rain. c: Damaged asphalt surfacing; pavement layers and salt profile. d: Undamaged asphalt surfacing; pavement layers and salt profile.

Figure 13.37c and d give two clues as to how to overcome the problem of salt blistering:

- For new construction, material containing more than a defined maximum of soluble salts should not be used.
- It will be noted from Figure 13.37c that the damaged surfacing was only about 10 mm thick, whereas (Figure 13.37d) the undamaged one was 40 mm thick. Moreover, the salt content of the damaged surfacing was higher than that of the material below it, showing that salt had concentrated within the pores of the asphalt. This indicates that the surfacing was relatively permeable and had allowed salt-laden pore water to move upwards and evaporate at the surface, depositing the salt. In contrast, the salt content of the undamaged surfacing was negligible, showing that it was sufficiently impervious, not to have allowed pore water to evaporate through it. The migration of salt shown in Figure 13.36d must have taken place before the surfacing was applied. Thus an effective preventive measure would be to use thicker surfacing, designed to be impervious.
- As a third possible prophylactic measure, the solubility of the salts can be substantially reduced by treating the aggregate with a high calcium slaked lime. This

Plate 13.9 Salt-filled blister in asphalt surfacing.

converts the more soluble sodium, magnesium and iron sulphates to calcium sulphate which has a much lower solubility of only 0.02 g/litre, and reduces the mobility of the salts.

Further investigation in several localities showed that salt blistering did not occur, provided that the overall soluble salt content of the road aggregate was less than 0.2% by dry mass. Also, if the ratio of the asphalt permeability in mm/s to the surfacing thickness in mm was less than 30×10^{-6}/s, no salt blistering was observed. Of the three palliative measures listed above, the first, selection of material containing a low salt content, is the most useful for new construction. Sources of crushed rock, or other granular material should be routinely tested for soluble salt content. If the salt content exceeds 0.2% by dry mass, it should be rejected for use as road or surfacing aggregate, or it may be treated with lime, to immobilize the salts.

For the repair of salt-damaged surfacings, an overlay of impervious asphalt is used in a thickness to give a maximum ratio of permeability to thickness of 30×10^{-6}/s where the permeability is expressed in mm/s and the thickness is in mm.

13.8.3 *Concrete aggregate made from mine waste rock that is susceptible to akali-aggregate reaction (AAR)*

Alkali-aggregate reaction (AAR) occurs when the alkali associated with the cement content of a concrete reacts with the concrete aggregate. The reaction is expansive and results in disruption of the concrete. Because of the expansive nature of AAR, the disruption of reinforced concrete is restrained by the reinforcing, and patterns of surface cracking reflect the effects of this restraint. It is quite possible that mine waste rock, especially quartzite containing strained quartz (quartz that has been under high stress for millions of years) or opaline quartz will be susceptible to AAR. The examples that follow were observed on concrete made with Witwatersrand quartzite waste rock from the ultra-deep South African gold mines. AAR is a slow expansive reaction that causes the concrete to crack, with damage usually only becoming evident 10 to 20 years after the end of construction.

Cracking caused by AAR may vary from continuous cracks aligned parallel to the direction of the major reinforcing in a compressive member to severe omni-directional block cracking of the surface of relatively unstressed areas. Cracks are usually not static, but progressively widen as time proceeds and they penetrate more deeply into the heart of the R.C. member. Figure 13.38a shows some crack width-time relationships that have been observed on the surface of a reinforced concrete (R.C.) member (Blight et al, 1989). The growth in width of the cracks is influenced by the weather. Close examination of Figure 13.38a will show that an acceleration of movement occurs in the second half of each year (i.e. each rainy season) and that the rate of expansion slows during May to August (the dry season). This structure was 18 years old at the start of year 1 in the figure. The numbers in the figure identify each crack. Plate 13.10 shows a pier and columns of a R.C. freeway structure that have deteriorated severely as a result of AAR.

Increasing crack widths are accompanied by swelling strains on the surface of the concrete. Figure 13.38b shows a set of surface strains measured on the same R.C. member (Alexander, et al., 1992) that is the subject of Figure 13.38a. At this time the concrete was 30 years old. The surface cracks had been sealed and repaired and the concrete coated with an impervious coating, but the expansion still continued. The measurements were carried but with a 400 mm gauge length Demec gauge and were corrected for temperature by means of thermocouples embedded in the concrete. As stated earlier, the months of May to October were the dry season in which no rain falls, and the rain started again in November, as the sudden increase in the strains clearly shows. Because the swelling of the concrete is restrained by the reinforcing, swelling pressures are generated by the restrained swelling.

Figure 13.39a shows the result of swell-under-load tests undertaken on cores from the R.C. member referred to in Figure 13.38. The tests were made to assess if the concrete still had an appreciable swelling potential after 30 years of exposure to the elements. The constant stress applied to each specimen is indicated next to each curve. It will be seen that free swells of over 450×10^{-6} were recorded and that strains of close to 250×10^{-6} occurred under a confining stress of 1 MPa although results were very variable. Extrapolation of the measurements to zero strain indicate that the swelling pressure in the fully constrained concrete could rise to as high as 2 MPa, even after 30 years of exposure to weather.

The swelling pressure generated by AAR, acting against the restraint of the reinforcing ensures that AAR has relatively little effect on the compressive strength of concrete. Even when the prestressing effect is removed by coring the concrete, the compressive strength remains surprisingly high, presumably because of mechanical interlock between the disrupted blocks of concrete. Figure 13.39b shows the results

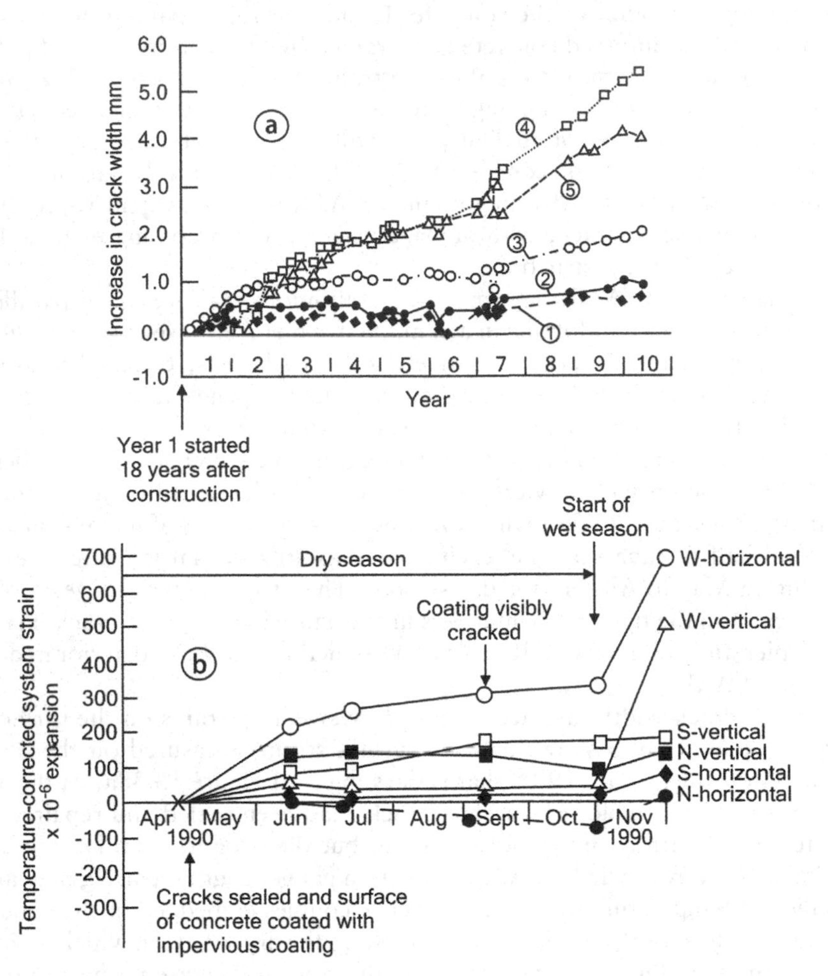

Figure 13.38 a: Relationships between crack width and time for surface cracks in 18 year-old R.C. member damaged by AAR. b: Measured expansion of surface of same member when 30 years-old.

of a series of compression and splitting tensile strength tests on sets of cores from two series of concrete structures having different concretes but the same design strength of 30 MPa. Although the cores were taken from structures that had apparently been badly affected by AAR, Figure 13.39b shows that the mean strength for concrete 1 still exceeded the design strength. The mean compressive strength for concrete 2 was still up to specification, although the splitting tensile strength had been reduced. The direct tensile strength was not measured, but was likely to have been close to zero. This does not really matter, as in design of RC, the direct tensile strength is conventionally assumed to be zero.

To summarize:

- Concrete subject to attack by AAR undergoes an expansion that results in unsightly and alarming surface cracking. The cracks usually appear up to two decades after

Plate 13.10 Effects of AAR attack on piers and columns of R.C. freeway structure.

construction and may continue to widen and develop for a further 30 years or more.

- The restraint provided by the reinforcing in reinforced concrete allows a swelling pressure to develop in the concrete. Swelling pressures of up to 4 MPa appear possible, and even 30 year-old concrete may develop swelling pressures as high as 2 MPa.
- The restrained swelling acts as a prestress on the concrete which, even though disrupted, retains most of its compressive and shear strength. Even its splitting tensile strength, although reduced, does not disappear, although the direct tensile strength probably does approach zero. The disruption caused by AAR also reduces the elastic and creep moduli of the concrete.

Apart from laboratory tests on cores taken from an affected structure, the safety and structural integrity of the structure can be evaluated by measurements made during service under actual service loads, or by means of full-scale load testing to the design load. Measurements made in this way have shown that AAR has relatively little overall effect on the performance of R.C. structures.

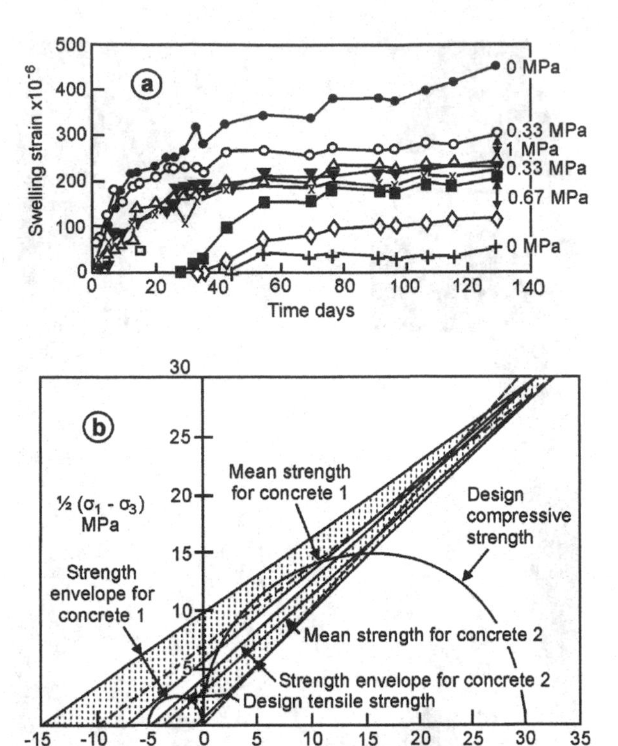

Figure 13.39 a: Swell-under-load versus time relationships for cores of 25 year-old concrete. b: Comparison of strength of concrete deteriorated by AAR with original specified strength.

References

Alexander, M.G., Blight, G.E., & Lampacher, B.J.: Pre-demolition tests on structural concrete damaged by AAR. In: *Proc., 9th Int. Conf. Alkali-Aggregate Reaction Concrete*. London, U.K., 1992, pp. 1–7.

Allen, M.D., & Lewis, D.A.: *Cathodic Protection in Civil Engineering*. London, U.K.: The Institution of Civil Engineers, 1979, pp. 79–94.

Baguelin, F., Jezequel, J.F., & Shields, D.H.: The Pressuremeter and Foundation Engineering, *Trans Tech, Zurich, Switzerland*, 1978.

Blight, G.E.: Migration of Subgrade Salts Damages Thin Pavements. *Transport. Eng. J., ASCE, 102(TE 4)*, (1976), pp. 779–791.

Blight, G.E., More O'Ferral, R.C., & Avalle, D.L.: Cemented Tailings Fill for Mining Excavations. In: 9th *Int. Conf. Soil Mech. Found. Eng.*, Tokyo, Vol. 1, 1977, pp. 47–54.

Blight, G.E., & Clarke, L.E.: Design and Properties of Stiff Fill for Lateral Support. In: *Symp. Mining Backfill*. Lulea, Sweden. Rotterdam, Netherlands: A.A. Balkema, 1983, pp. 306–310.

Blight, G.E., Alexander, M.G., Ralph, T.K., & Lewis, B.A.: Effect of Alkali-Aggregate Reaction on the Performance of a Reinforced Concrete Structure over a Six-Year Period. *Mag. Concr. Res. 41(147)* (1989), pp. 67–77.

Blight, G.E., & Dane, M.S.W.: Deterioration of a Wall Complex Constructed of Reinforced Earth. *Geotechnique 39(1)* (1989), pp. 47–53.
Blight, G.E., & Spearing, A.J.S.: The Properties of Cemented Silicate Backfill for Use in Narrow, Hard-Rock, Tabular Mines. *J. S. Afr. Inst. Mining Metall. Jan/Feb* (1996), pp. 17–28.
Charles, J.A., Naismith, W.A., & Burford, D.: Settlement of Backfill at Horsley Restored Open Cast Coal Mining Site. In: 1*st Conf. Large Ground Movements Their Effect Struct.*, Cardiff, U.K., 1977, pp. 320–330.
Charles, J.A., Hughes, D.B., & Burford, D.: The Effect of a Rise of Water Table on the Settlement of Backfill at Horsley Restored Coal Mining Site (1973–1983). In: 3rd *Conf. Ground Movements Struct.*, Cardiff, U.K., 1984, pp. 423–442.
Clark, I.H.: The Strength and Deformation Behaviour of Backfill in Tabular Deep-Level Mining Excavations. PhD Thesis, University of the Witwatersrand, 1988.
Day, P.: Determination of Parameters for Opencast Pit Backfill by Means of Large Scale Tests. In: *Symp. Constr. Mined Areas, S. Afr. Inst. Civil Eng.*, Pretoria, South Africa, 1992, pp. 73–78.
Dison, L. & Blight, G.E.: Reinforced Cemented Tailings Fill. *Backfill in South African Mines,* Johannesburg, South Africa: S.A. Institute of Mining and Metallurgy, 1988, pp. 91–108.
Fourie, A.B., & Blight, G.E.: Tests to Determine the Suitability of a Geotextile for Use in the Containment of Hydraulically Placed Backfill. *Geotext. Geomembr. 14* (1996), pp. 465–480.
Hahn, J.A., Dison, L., & Blight, G.E.: Supports of Reinforced Granular Fill. In: *Symp. Mining Backfill*, Lulea, Sweden. Rotterdam, Netherlands: A.A. Balkema, 1983, pp. 300–306.
Helsinki, M., Fahey, M., & Fourie, A.B.: Numerical Modeling of Cemented Mine Backfill Deposition, *J. Geotech. Geoenviron. Eng., ASCE, 133(10)* (2007), pp. 1308–1319.
Hill, C.W.W., & Denby, B.: The Prediction of Opencast Backfill Settlement. *Proc. Inst. Civil Engrs, U.K., Geotech. Eng., 134* (1996), pp. 160–176.
McKittrick, D.P.: Reinforced Earth: Application of Theory and Research to Practice. In: *Symp. Soil Reinforcing Stabilizing Tech. Eng. Practice.* University of New South Wales separate paper, Sydney Australia, 1978, p.44.
Neville, A.M.: *Properties of Concrete.* 3rd Edition. London, U.K.: Pitman International, 1981, pp. 208 and 214.
Ortlepp, W.D.: Basic Engineering Principles in the Design of Sandwich Pack Support. *J. S. Afr. Inst. Mining Metall., 78* (1978), pp. 275–287.
Reed, S.M., & Hughes, D.B.: Long Term Settlement of Opencast Mine Backfills – Case Studies from The North East of England. *Reclamation, Treatment and Utilization of Coal Mining Wastes.* Rotterdam, Netherlands: Balkema, 1990, pp. 141–155.

Subject Index

Printed in the United States
by Baker & Taylor Publisher Services

Printed in the United States
by Baker & Taylor Publisher Services